To Kitchener Publi

Gifted to you, the reader of the book
This is the science which I learned and I took
I hope this helps you understand this field
After reading it, I hope it's off the hook

Benyamin Ghojogh (author of book)

April 26, 2023

Elements of Dimensionality Reduction and Manifold Learning

Benyamin Ghojogh • Mark Crowley •
Fakhri Karray • Ali Ghodsi

Elements of Dimensionality Reduction and Manifold Learning

Springer

Benyamin Ghojogh
Department of Electrical and Computer Engineering
University of Waterloo
Waterloo, ON, Canada

Mark Crowley
Department of Electrical and Computer Engineering
University of Waterloo
Waterloo, ON, Canada

Fakhri Karray
Department of Electrical and Computer Engineering
University of Waterloo
Waterloo, ON, Canada

Ali Ghodsi
Department of Statistics and Actuarial Science & David R. Cheriton School of Computer Science
University of Waterloo
Waterloo, ON, Canada

ISBN 978-3-031-10601-9 ISBN 978-3-031-10602-6 (eBook)
https://doi.org/10.1007/978-3-031-10602-6

This Springer imprint is published by the registered company Springer Nature Switzerland AG
The registered company address is: Gewerbestrasse 11, 6330 Cham, Switzerland

Benyamin Ghojogh:

To my lovely parents, Shokuh Azam Zolfaghari and Yousef Ghojogh, who have put a lot of efforts for me and my brother, Aydin.

Mark Crowley:

To my wife Lily, who always helps me to see and explore new dimensions.

Fakhri Karray:

To the memory of my father and mother. To my wife, Neila, and my sons, Amir, Malek, and Bacem. To my family at large.

Ali Ghodsi:

To my mother and the memory of my father. To my sons, Soroush and Maseeh, with the love of a proud father.

Preface

How This Book Can Be Useful

Motivation

Dimensionality reduction, also known as manifold learning, is an area of machine learning used for extracting informative features from data for better representation of data or separation between classes. With the explosion of interest and advances in machine learning, there has been a corresponding increased need for educational and reference books to explain various aspects of machine learning. However, there has not been a comprehensive text tackling the various methods in dimensionality reduction, manifold learning, and feature extraction that integrate with modern machine learning theory and practice.

This book presents a cohesive review of linear and nonlinear dimensionality reduction and manifold learning. Three main aspects of dimensionality reduction are covered—spectral dimensionality reduction, probabilistic dimensionality reduction, and neural network-based dimensionality reduction—which have geometric, probabilistic, and information-theoretic points of view to dimensionality reduction, respectively. This book delves into basic concepts and recent developments in the field of dimensionality reduction and manifold learning, providing the reader with a comprehensive understanding. The necessary background and preliminaries on linear algebra, optimization, and kernels are also highlighted to ensure a comprehensive understanding of the algorithms. The tools introduced in this book can be applied to various applications involving feature extraction, image processing, computer vision, and signal processing.

Targeted Readers

This book provides the required understanding to extract, transform, and interpret the structure of data. It is intended for academics, students, and industry professionals:

- Academic researchers and students can use this book as a textbook for machine learning and dimensionality reduction.
- Data scientists, machine learning scientists, computer vision scientists, and computer scientists can use this book as a reference for both technical and applied concepts. It can also be helpful to statisticians in the field of statistical learning and applied mathematicians in the fields of manifolds and subspace analysis.
- Industry professionals, including applied engineers, data engineers, and engineers in various fields of science dealing with machine learning, can use this as a guidebook for feature extraction from their data, as the raw data in industry often require preprocessing.

This book is structured as a reference textbook so that it can be used for advanced courses, as an in-depth supplementary resource or for researchers or practitioners who want to learn about dimensionality reduction and manifold learning. The book is grounded in theory but provides thorough explanations and diverse examples to improve the reader's comprehension of the advanced topics. Advanced methods are explained in a step-by-step manner so that readers of all levels can follow the reasoning and come to a deep understanding of the concepts. This book does not assume an advanced theoretical background in machine learning and provides the necessary background, although an undergraduate-level background in linear algebra and calculus is recommended.

Corresponding Courses

The book can be a resource for instructors teaching advanced undergraduate or graduate level courses in engineering, computer science, mathematics, and science. There are various corresponding courses that can use this book as their textbook. Some of these courses are machine learning, data science, artificial intelligence, unsupervised machine learning, data clustering, dimensionality reduction, manifold learning, manifold embedding, feature extraction, feature embedding, feature engineering, data visualization, etc. This book can also be considered a reference book for dimensionality reduction and manifold learning. It can also be seen as a history book for dimensionality reduction as a particular field of machine learning, as it presents the development of these concepts from inception.

Organization of the Book

This book is divided into four main sections: preliminaries and background concepts, spectral or geographic methods, probabilistic methods, and neural network-based methods.

Preliminaries and Background

The preliminaries and background provide a comprehensive review of linear algebra, including the eigenvalue decomposition, the singular value decomposition, and linear projection. This is followed by introducing the reproducing kernel Hilbert space, kernels, the Nyström method, and the Hilbert Schmidt Independence Criterion (HSIC). Lastly, the essential optimization background for machine learning and dimensionality reduction is reviewed, where the first order methods (gradient methods), the second order methods (Newton's method, interior method, etc.), and distributed optimization such as the Alternating Direction Method of Multipliers (ADMM) are explained.

Spectral Dimensionality Reduction

The methods in spectral dimensionality reduction reduce to eigenvalue decomposition, have a geometric point of view, and unfold the manifold in a lower dimensional subspace. This section of the book is presented to introduce the geometrical approaches to dimensionality reduction. This section covers unsupervised and supervised principal component analysis, multidimensional scaling, Fisher discriminant analysis (also called linear discriminant analysis), Sammon mapping, Isomap, Laplacian-based methods (including spectral clustering, Laplacian eigenmap, locality preserving projection, graph embedding, and diffusion map), and locally linear embedding. It also explains that these methods can be unified as kernel PCA with different kernels, where the maximum variance unfolding method (semidefinite programming) can find the best low-rank kernel among possible kernels. Lastly, spectral metric learning is introduced.

Probabilistic Dimensionality Reduction

In probabilistic dimensionality reduction, it is assumed that there is a low-dimensional latent variable on which the high-dimensional variable is conditioned and that the latent variable should be inferred and discovered. This section of the book is presented to introduce the probabilistic and stochastic approaches to dimensionality reduction. The benefit of the probabilistic approach over the spectral methods is its ability to handle missing data. Within the probabilistic dimensionality reduction section, factor analysis, variational autoencoder, probabilistic PCA, probabilistic metric learning, linear and nonlinear random projections, kernel dimensionality reduction, Stochastic Neighbor Embedding (SNE), t-SNE, and Uniform Manifold Approximation and Projection (UMAP) are introduced.

Neural Network-Based Dimensionality Reduction

The neural network-based dimensionality reduction takes an information theoretic point of view, where the middle of the neural network or autoencoder is seen as a bottleneck of information that keeps only the useful and important information. This section of the book is presented to introduce the neural network-based and deep learning approaches to dimensionality reduction. In this section, the restricted Boltzmann machine and deep belief network are covered, which are fundamental dimensionality reduction methods in a network structure. The autoencoder is introduced, where the latent embedding space is encoded by a middle layer of a possibly deep autoencoder. Deep metric learning, which encodes the data in an embedding space trained by a deep neural network, is also explained. Variational inference and variational autoencoder are covered, where Bayesian analysis is implemented in an autoencoder structure. Lastly, adversarial learning, generative adversarial autoencoder, and adversarial autoencoder are discussed.

Other Related Books

There have been other books written on dimensionality reduction and manifold learning that the reader may find useful. Some of them are listed in the following:

- [2] follows a topological-based approach to dimensionality reduction and manifold learning. It introduces the concepts of topology and manifold through high-dimensional visualization. It does not cover the probabilistic and neural network-based approaches of dimensionality reduction, and moreover, many important spectral methods are omitted.
- [3] provides a review of topological manifolds and Riemannian manifolds, followed by the spectral methods, i.e., PCA, MDS, Isomap, Locally Linear Embedding (LLE), Laplacian eigenmap, diffusion map, Hessian eigenmap, and kernel PCA, as well as Laplacian eigenmap and density preserving maps. Sample complexity in manifold learning, manifold alignment, and the Nyström approximation are introduced, and pure-math concepts, including Ricci flow

and 3-manifold, are explored. Finally, it covers additional applications in object morphing, image manifolds, and motion analysis.

- [5] takes a machine learning approach to dimensionality reduction and manifold learning. It introduces the geometry of data and the Riemannian manifold, followed by linear methods in the spectral approach, i.e., PCA, MDS, and random projection. It reviews several nonlinear methods in the spectral approach, i.e., Isomap, MVU, LLE, Local Tangent Space Alignment (LTSA), Laplacian eigenmap, Hessian LLE, and diffusion maps. Lastly, it covers the Nyström approximation to introduce the fast versions of these methods.
- [4] introduces PCA, MDS, Isomap, MVU, diffusion maps, LLE, Laplacian eigenmap, LTSA, and kernel PCA. Then, it reviews topology, intrinsic dimensionality, out-of-sample extension, and the Nyström approximation.
- [6] covers PCA, LLE, Locality Preserving Projection (LPP), Fisher Discriminant Analysis (FDA), kernel-based methods, sparse and low-rank methods, and neural network-based methods.
- [1] covers spectral and probabilistic methods, specifically intrinsic dimensionality estimation, PCA, probabilistic PCA, kernel PCA, canonical correlation analysis, LDA, sufficient dimensionality reduction, the Nyström approximation, MDS, Isomap, LLE, and the graphical models.

Existing books on dimensionality reduction usually do not have the categorization used in this book. This book thoroughly covers the important algorithms in each of the spectral, probabilistic, and neural network-based categories.

References

1. Christopher JC Burges. *Dimension reduction: A guided tour* Now Publishers Inc, 2010.
2. John A Lee and Michel Verleysen. *Nonlinear dimensionality reduction* Springer Science & Business Media, 2007.
3. Yunqian Ma and Yun Fu. *Manifold learning theory and applications* Vol. 434. CRC Press Boca Raton, FL, 2012.
4. Harry Strange and Reyer Zwiggelaar. *Open Problems in Spectral Dimensionality Reduction* Springer, 2014.
5. Jianzhong Wang. *Geometric structure of high-dimensional data and dimensionality reduction* Vol. 5. Springer, 2012.
6. Haitao Zhao et al. *Feature Learning and Understanding: Algorithms and Applications* Springer Nature, 2020.

Waterloo, ON, Canada — Benyamin Ghojogh
Waterloo, ON, Canada — Mark Crowley
Waterloo, ON, Canada — Fakhri Karray
Waterloo, ON, Canada — Ali Ghodsi

Acknowledgment

The authors of this book thank Dr. Bethany Davidson-Eng and Springer's editors for copyediting the book. They also thank Stephanie Filsinger, at the Institutional Research of the University of Waterloo, for her support. The authors acknowledge the efforts of Paul Drougas, Shanthini Kamaraj, and Subhalakshmi Mounissamy, at Springer, during the publication process of the book. Last but not least, they thank all those who provided valuable feedback during the preparation of the book.

Contents

Chapter 1
Introduction

The world is in the Hilbert space,
And is vast and all-encompassing.
But it is so simple,
And falls on a low-dimensional submanifold.

1.1 Introduction

Machine learning is a field of science where a mathematical model learns to represent, classify, regress, or cluster data and/or makes appropriate decisions. This book introduces dimensionality reduction, also known as manifold learning, which is a field of machine learning. Dimensionality reduction transforms data to another lower-dimensional subspace for better representation of data. This chapter defines dimensionality reduction and enumerates its main categories as an introduction to the next chapters of the book.

1.1.1 Dataset

Consider the measurement of a quantity. This quantity could be:

- personal health data, including blood pressure, blood sugar, and blood fat,
- images from a specific scene but taken from different perspectives,
- images from several categories of animals, such as cat, dog, frog, etc.,
- medical images, such as digital pathology image patches, including both healthy and tumorous tissues, or

any other measured signal. The quantity can be multidimensional, i.e., a set of values, and therefore, every quantity can be considered a multidimensional data point in a Euclidean space. Let the dimensionality of this space be d, meaning that

B. Ghojogh et al., *Elements of Dimensionality Reduction and Manifold Learning*,
https://doi.org/10.1007/978-3-031-10602-6_1

every quantity is a d-dimensional vector, or data point, in $\mathbb{R}^d$. The set of d values for the quantity can be called *features* of the quantity. Multiple measurements of a quantity can exist, each of which is a d-dimensional data point. Therefore, there will be a set of d-dimensional data points, called a *dataset*. For example, the quantity can be an image, whose features are its pixels. The dataset can be a set of images from a specific scene but with different perspectives and angles.

1.1.2 Manifold Hypothesis

Each feature of a data point does not carry an equal amount of information. For example, some pixels of an image are background regions with limited information, while other pixels contain important objects that describe the scene in the image. This means that data points can be significantly compressed to preserve the most informative features while eliminating those with limited information. In other words, the d-dimensional data points of a dataset usually do not cover the entire d-dimensional Euclidean space, but they lie on a specific lower-dimensional structure in the space.

Consider the illustration in Fig. 1.1, where several three-dimensional points exist in $\mathbb{R}^3$. These points can represent any measurement, such as personal health measurements, including blood pressure, blood sugar, and blood fat. As demonstrated in Fig. 1.1, the points of the dataset have a structure in a two-dimensional space. The three-dimensional Euclidean space is called the *input space*, and the two-dimensional space, which has a lower dimensionality than the input space, is called the *subspace*, the *submanifold*, or the *embedding space*. The subspace can be either *linear* or *nonlinear*, depending on whether a linear (hyper)plane passes through the

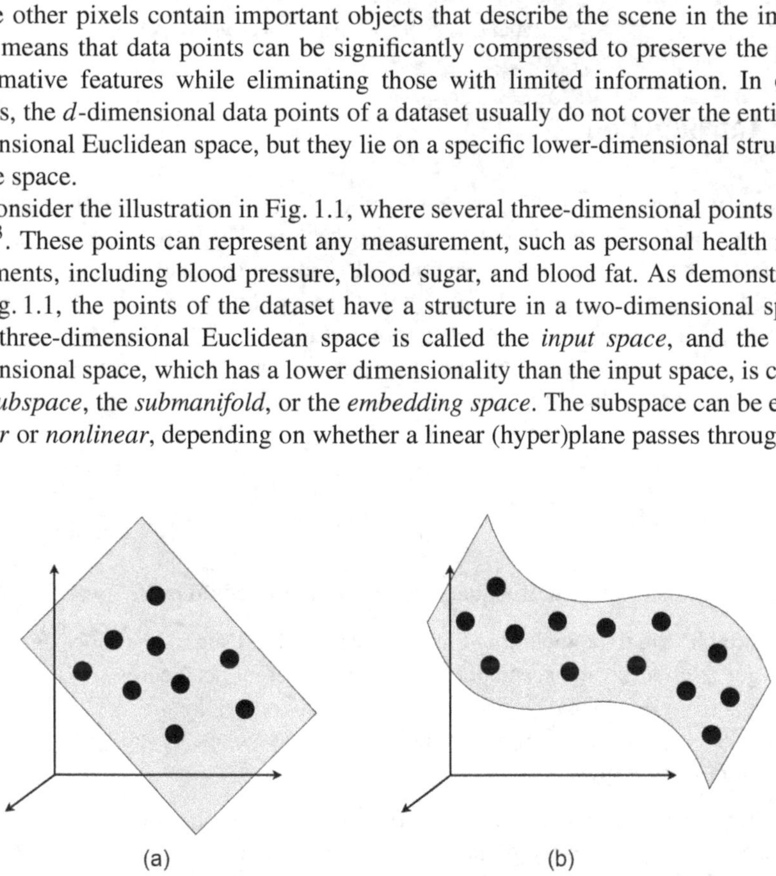

Fig. 1.1 Sets of three-dimensional data points in $\mathbb{R}^3$, lying on a two-dimensional (**a**) linear subspace and (**b**) a nonlinear submanifold

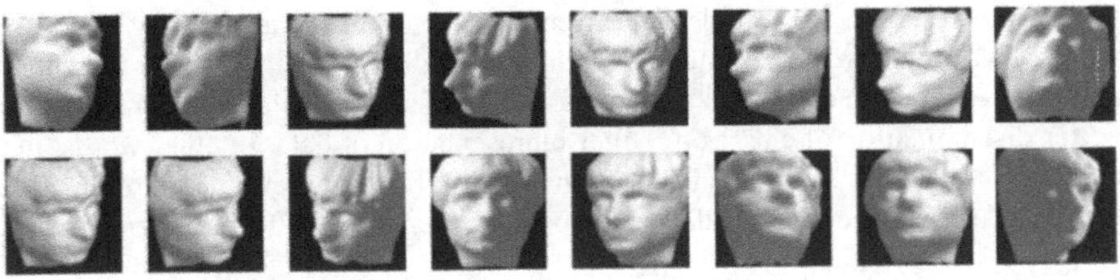

Fig. 1.2 Different perspectives of a sculpture (credit of the images is for [61])

points. Usually, subspace and submanifold are used for linear and nonlinear lower-dimensional spaces, respectively. Linear and nonlinear subspaces are depicted in Fig. 1.1a and b, respectively.

Whether the points of a dataset lie on a space is a hypothesis, but this hypothesis is usually true because the data points typically represent a natural signal, such as an image. When the data acquisition process is natural, the data will have a defined structure. For example, in the dataset where there are multiple images from different angles depicting the same scene, the objects of the scene remain the same, but the point of view changes (see Fig. 1.2). This hypothesis is called the manifold hypothesis [14]. Its formal definition is as follows. According to the manifold hypothesis, data points of a dataset lie on a submanifold or subspace with lower dimensionality. In other words, the dataset in $\mathbb{R}^d$ lies on an embedded submanifold [38] with local dimensionality less than d [14]. According to this hypothesis, the data points most often lie on a submanifold with high probability [64].

1.1.3 Feature Engineering

Due to the manifold hypothesis, a dataset can be compressed while preserving most of the important information. Therefore, engineering and processing can be applied to the features for the sake of compression [4]. Feature engineering can be seen as a preprocessing stage, where the dimensionality of the data is reduced. Assume d and p denote the dimensionality of the input space and the subspace, respectively, where $p \in (0, d]$. Feature engineering is a map from a d-dimensional Euclidean space to a p-dimensional Euclidean space, i.e., $\mathbb{R}^d \to \mathbb{R}^p$. The dimensionality of the subspace is usually much smaller than the dimensionality of the space, i.e. $p \ll d$, because most of the information usually exists in only a few features.

Feature engineering is divided into two broad approaches—feature selection and feature extraction [22]. In feature selection, the p most informative features of the d-dimensional data vector are selected so the features of the transformed data points are a subset of the original features. In feature extraction, however, the d-dimensional data vector is transformed to a p-dimensional data vector, where the p new features are completely different from the original features. In other words, data points are represented in another lower-dimensional space. Both feature selection and feature extraction are used for compression, which results in either the better discrimination of classes or better representation of data. In other words, the compressed data by feature engineering may have a better representation of the data or may separate the classes of data. This book concentrates on feature extraction.

1.2 Dimensionality Reduction and Manifold Learning

Feature extraction is also referred to as dimensionality reduction, manifold learning [12], subspace learning, submanifold learning, manifold unfolding, embedding, encoding, and representation learning [7, 70]. This book uses manifold learning and dimensionality reduction interchangeably for feature extraction. Manifold learning techniques can be used in a variety of ways, including:

- Data dimensionality reduction: Produce a compact (compressed) low-dimensional encoding of a given high-dimensional dataset.
- Data visualization: Provide an interpretation of a given dataset in terms of intrinsic degrees of freedom, usually as a byproduct of data dimensionality reduction.
- Preprocessing for supervised learning: Simplify, reduce, and clean the data for subsequent supervised training.

In dimensionality reduction, the data points are mapped to a lower-dimensional subspace either linearly or nonlinearly. Dimensionality reduction methods can be grouped into three categories—spectral dimensionality reduction, probabilistic dimensionality reduction, and (artificial) neural network-based dimensionality reduction [19] (see Fig. 1.3). These categories are introduced in the following subsections.

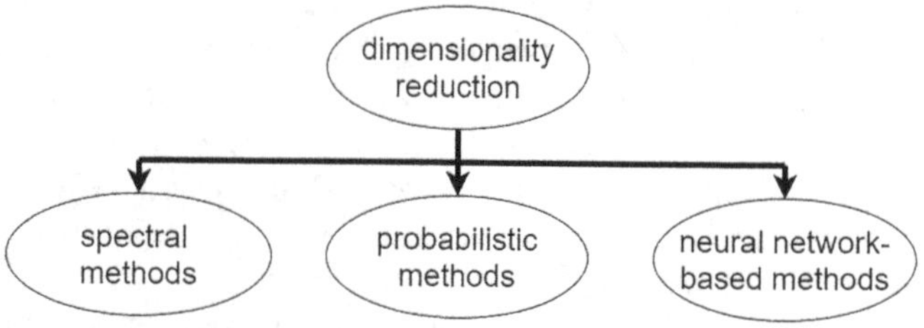

Fig. 1.3 Categories of dimensionality reduction

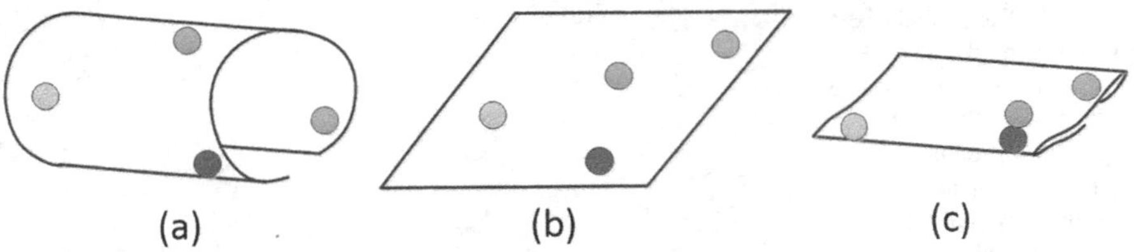

Fig. 1.4 (**a**) A 2D nonlinear submanifold where the data exist in the 3D original space, (**b**) the correct unfolded manifold, and (**c**) applying linear dimensionality reduction, which takes Euclidean distances into account

1.2.1 *Spectral Dimensionality Reduction*

Consider several points in a three-dimensional Euclidean space. Assume these points lie on a nonlinear submanifold with a local dimensionality of two, as illustrated in Fig. 1.4a. This means that there is no need to have three features to represent each of these points. Rather, two features can represent most of the data points' information if the two features demonstrate the 2D coordinates on the submanifold. A nonlinear dimensionality reduction method can unfold this manifold correctly, as depicted in Fig. 1.4b. However, a linear dimensionality reduction method cannot properly find a correct underlying 2D representation of the data points. Figure 1.4c demonstrates that a linear method ruins the relative structure of the data points. This is because a linear method uses the Euclidean distances between the points, while a nonlinear method considers the geodesic distances along the nonlinear submanifold. If the submanifold is linear, the linear method is able to obtain the lower-dimensional structure of the data. Spectral dimensionality reduction methods typically have a geometric perspective and attempt to find the linear or nonlinear submanifold of the data. These methods are often reduced to a generalized eigenvalue problem [20].

1.2.2 *Probabilistic Dimensionality Reduction*

The data can be considered a multidimensional random variable. It is possible to assume a lower-dimensional latent random variable, which the data random variable is conditioned on (see the probabilistic graphical mode in Fig. 1.5). This latent variable contains the most important information of the data. Consider Fig. 1.2 where several pictures of a sculpture are taken from a variety of angles. The point of view can be modeled by two left-to-right and bottom-to-up angles. In this example, the data points are high-dimensional images with the dimensionality of pixels. However, these images differ only in terms of the two perspectives. Therefore, the data points are conditioned on a two-dimensional latent variable. Probabilistic dimensionality reduction methods attempt to find the lower-dimensional latent variable from the higher-dimensional data.

Fig. 1.5 The probabilistic graphical model of a probabilistic dimensionality reduction method. The high-dimensional data point $\boldsymbol{x}_i$ is conditioned on a low-dimensional latent random variable z_i

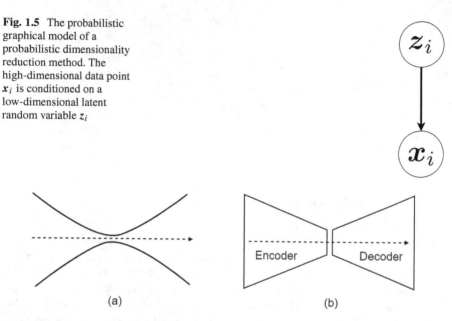

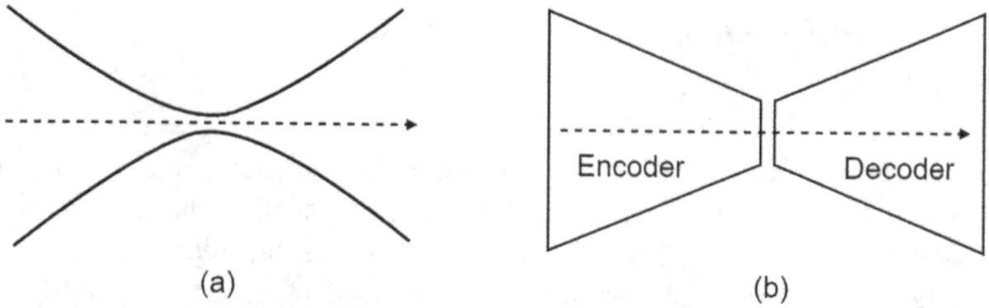

Fig. 1.6 (**a**) Passing the high-dimensional data through a lower-dimensional bottleneck for extracting the most important information, and (**b**) implementing the bottleneck using an autoencoder with an encoder and a decoder

1.2.3 Neural Network-Based Dimensionality Reduction

It is possible to pass high-dimensional data through a lower-dimensional bottleneck and then reconstruct it, as demonstrated in Fig. 1.6a. As the data should pass through the bottleneck and be reconstructed almost perfectly afterwards, the most informative features of the data should be preserved at the bottleneck. Therefore, the representation of the data at the bottleneck is a proper candidate for dimensionality reduction. This bottleneck can be implemented using an autoencoder neural network, where the data are compressed between the encoder and the decoder (see Fig. 1.6b).

Neural network-based dimensionality reduction methods take an information theoretic perspective [2, 58], where the network is used as an information bottleneck [57, 63] for extracting low-dimensional informative features. These methods use deep or shallow neural networks for feature extraction.

1.2.4 Other Types of Dimensionality Reduction

It is noteworthy that there are additional dimensionality reduction approaches to those mentioned above. Riemannian manifold learning [40] assumes that the data points belong to a Riemannian manifold, such as Grassmannian, Stiefel, and Symmetric Positive Definite (SPD) manifolds, which use the concepts of differential geometry. In this book, the theory of dimensionality reduction is introduced and reviewed through spectral, probabilistic, and neural network-based methods while providing insight into Riemannian methods within each of these categories.

1.3 A Brief History of Dimensionality Reduction

In the literature, various linear and nonlinear methods have been proposed for dimensionality reduction and manifold learning.

1.3.1 History of Spectral Dimensionality Reduction

Principal Component Analysis (PCA) [34] was first proposed by Pearson in 1901 [47]. It was the first spectral dimensionality reduction method and one of the first methods in linear subspace learning. It is unsupervised, meaning that it does not use any class labels. *Fisher Discriminant Analysis (FDA)* [16], proposed by Fisher in 1936 [15], was the first supervised spectral dimensionality reduction method. PCA and FDA are based on the scatter, i.e., variance of the data. A proper subspace preserves either the relative similarity or relative dissimilarity of the data points after transformation of data from the input space to the subspace. This was the goal of *Multidimensional Scaling (MDS)* [13], which preserves the relative similarities of data points in its subspace. In later MDS approaches, the cost function was changed to preserve the distances between points [37], which developed into *Sammon mapping* [52]. Sammon mapping is considered to be the first nonlinear dimensionality reduction method.

Figure 1.4 demonstrates that a linear algorithm cannot perform well on nonlinear data. For nonlinear data, two approaches can be used:

- a nonlinear algorithm should be designed to handle nonlinear data, or
- the nonlinear data should be modified to become linear. In this case, the data should be transformed to another space to become linearly separable in that space. Then, the transformed data, which now have a linear pattern, will be able to use a linear approach. This approach is called *kernelization* in machine learning.

The *kernel PCA* [54, 55] uses the PCA and the kernel trick [32] to transform data to a high-dimensional space so that it becomes roughly linear within that space. *Kernel FDA* [44, 45] was also proposed to manipulate nonlinear data in a supervised manner using representation theory [3]. Representation theory can be used for kernelization; it will be introduced in Chap. 3.

Isometric Mapping (Isomap) [61] and *locally linear embedding* [51] emerged as important dimensionality reduction methods in 2000, following the introduction of Sammon mapping. Unlike Kernel PCA/FDA, Isomap and Locally Linear Embedding (LLE) do not rely on transforming the data. Isomap uses the geodesic distances, rather than the Euclidean distances, in the kernel of MDS [28, 60]. LLE reconstructs every point by its k-Nearest Neighbours (kNN) to learn the submanifold and locally fit the data in the low-dimensional subspace [53].

Spectral clustering [46, 67], which considers the dataset as a graph and uses the Laplacian of the data graph for clustering, appeared at roughly the same time as Isomap and LLE. Researchers adopted the idea of spectral clustering with the goal of using it for dimensionality reduction. The result was the *Laplacian eigenmap* [6].

While the spectral dimensionality reduction methods were developing, it was determined that most of them are essentially a kernel PCA with different kernels. More specifically, they all are learning eigenfunctions of their kernels [8, 9]. This unification of spectral algorithms, as kernel PCA with different kernels, encouraged researchers to find the optimal kernel for every specific dataset. This was the main goal of *semidefinite embedding* [66], which was later renamed to *maximum variance unfolding* [65]. This method uses semidefinite programming, introduced in Chap. 4, to optimize the kernel matrix.

Metric learning, another dimensionality reduction approach, is a family of methods, including spectral, probabilistic, and neural network-based algorithms. It learns a suitable distance metric, rather than the Euclidean distance, for better discrimination between classes in the subspace [69]. Apart from metric learning, many of the spectral methods were developed after the proposal of the *Hilbert-Schmidt Independence Criterion (HSIC)* [26]. HSIC could estimate the dependence of two random variables with the linear correlation of their kernels. One of the spectral dimensionality reduction methods, which used the HSIC, was the *supervised PCA* [5]. Supervised PCA attempted to maximize the dependence of the class labels and the projected data to its subspace. It can be demonstrated that PCA is a special case of the supervised PCA, where the labels are not used. Another recent advancement of spectral methods, which also uses the HSIC, is the *Roweis Discriminant Analysis (RDA)* [21]. RDA combined PCA, the supervised PCA, and FDA into a generalized family of algorithms for linear subspace learning.

1.3.2 History of Probabilistic Dimensionality Reduction

The probabilistic variants of many of the spectral dimensionality reduction methods were gradually developed and proposed. For example, *probabilistic PCA* [50, 62] is the stochastic version of PCA, and it assumes that data are obtained by the addition of noise to the linear projection of a latent random variable. It can be demonstrated that PCA is a special case of probabilistic PCA, where the variance of noise tends to zero. The probabilistic PCA, itself, is a special case of *factor analysis* [17]. In the factor analysis, the different dimensions of noise can be correlated, while they are uncorrelated and isometric in the probabilistic PCA. The factor analysis and the probabilistic PCA use expectation maximization and variational inference on the data.

The linear spectral dimensionality reduction methods, such as PCA and FDA, learn a projection matrix from the data for better data representation or more separation between classes. In 1984, it was surprisingly determined that if a random matrix is used for the linear projection of the data, without being learned from the data, it represents data well! The correctness of this mystery of *random projection* was proven by the *Johnson-Lindenstrauss lemma* [33], which put a bound on the error of preservation of the distances in the subspace. Later, nonlinear variants of random projection were developed, including *random Fourier features* [48] and *random kitchen sinks* [49].

Sufficient Dimension Reduction (SDR) is another family of probabilistic methods, whose first method was proposed in 1991 [39]. It is used for finding a transformation of data to a lower-dimensional space, which does not change the conditional of labels given data. Therefore, the subspace is sufficient for predicting labels from projected data onto the subspace. SDR was mainly proposed for high-dimensional regression, where the regression labels are used. Later, *Kernel Dimensionality Reduction (KDR)* was proposed [18] as a method in the family of SDR for dimensionality reduction in machine learning.

Stochastic Neighbour Embedding (SNE) was proposed in 2003 [30] and took a probabilistic approach to dimensionality reduction. It attempted to preserve the probability of a point being a neighbour of others in the subspace. A problem with SNE was that it could not find an optimal subspace because it was not possible for it to preserve all the important information of high-dimensional data in the low-dimensional subspace. Therefore, t-SNE was proposed [41], which used another distribution with more capacity in the subspace. This allowed t-SNE to preserve a larger amount of information in the low-dimensional subspace. A recent successful probabilistic dimensionality reduction method is the *Uniform Manifold Approximation and Projection (UMAP)* [43], which is widely used for data visualization. Today, both t-SNE and UMAP are used for high-dimensional data visualization, especially in the visualization of extracted features in deep learning. They have also been widely used for visualizing high-dimensional genome data.

1.3.3 History of Neural Network-Based Dimensionality Reduction

Neural networks are machine learning models modeled after the neural structure of the human brain. Neural networks are currently powerful tools for representation learning and dimensionality reduction. In the 1990s, researchers' interest in neural networks decreased; this was called the winter of neural networks. This winter occurred mainly because networks could not become deep, as gradients vanished after many layers of network during optimization. The success of *kernel support vector machines* [10] also exaggerated this winter. In 2006, Hinton and Salakhutdinov demonstrated that a network's weights can be initialized using energy-based training, where the layers of the network are considered stacks of *Restricted Boltzmann Machines (RBM)* [1, 31]. RBM is a two-layer structure of neurons, whose weights between the two layers are trained using maximum likelihood estimation [68]. This initialization saved the neural network from the vanishing gradient problem and ended the neural networks' winter. A deep network using RBM training was named the *deep belief network* [29]. Although, later, the proposal of the *ReLU activation function* [23] and the *dropout technique* [59] made it possible to train deep neural networks with random initial weights [24].

In fundamental machine learning, people often extract features using traditional dimensionality reduction and then apply the classification, regression, or clustering task afterwards. However, modern deep learning extracts features and learns embedding spaces in the layers of the network; this process is called *end-to-end*. Therefore, deep learning can be seen as performing a form of dimensionality reduction as part of its model. One problem with end-to-end models is that they are harder to troubleshoot if the performance is not satisfactory on a part of the data. The insights and meaning of the data coming from representation learning are critical to fully understand a model's performance. Some of these insights can be useful for improving or understanding how deep neural networks operate. Researchers often visualize the extracted features of a neural network to interpret and analyze why deep learning is working properly on their data.

Deep metric learning [35] utilizes deep neural networks for extracting low-dimensional descriptive features from data at the last or one-to-last layer of the network. *Siamese networks* [11] are important network structures for deep metric learning. They contain several identical networks that share their weights, but have different inputs. *Contrastive loss* [27] and *triplet loss* [56] are two well-known loss functions that were proposed for training Siamese networks. *Deep reconstruction autoencoders* also make it possible to capture informative features at the bottleneck between the encoder and decoder.

As mentioned in Sect. 1.3.2, factor analysis [17] and probabilistic PCA [50, 62] demonstrated that variational inference and Bayesian learning can be used for dimensionality reduction. *Variational autoencoder*, proposed in 2014 [36], modeled variational inference and Bayesian learning in an autoencoder network structure. In this network, the parameters of the conditional distribution of a latent variable are trained at the bottleneck of an autoencoder, where the latent variable is sampled randomly from this distribution before being fed to the decoder of the autoencoder.

In 2014, *adversarial learning* was introduced in machine learning [25], where two networks compete with each other. One of the networks attempts to generate new data, while the other network tries to recognize that the generated data are not the original data. The networks improve gradually until the generator network can generate very realistic data, fooling the discriminative network. A year later, *adversarial autoencoder* was proposed [42], which uses adversarial learning in an autoencoder structure for dimensionality reduction at its bottleneck. The decoder of both the variational and adversarial autoencoder can be considered as generative models.

1.4 Chapter Summary

Dimensionality reduction, also referred to as manifold learning and feature extraction, is one of the important areas of machine learning and data science. This chapter started with introducing a dataset, explained the intuition of the manifold hypothesis, and then defined feature engineering. The main categories of dimensionality reduction, i.e. the spectral, probabilistic, and neural network-based approaches, were introduced. Finally, a history of dimensionality reduction in each of these categories was provided. This book will provide further historical information and cover the fundamental algorithms in the three categories of dimensionality reduction.

References

1. David H Ackley, Geoffrey E Hinton, and Terrence J Sejnowski. "A learning algorithm for Boltzmann machines". In: *Cognitive science* 9.1 (1985), pp. 147–169.
2. Zeyuan Allen-Zhu, Yuanzhi Li, and Yingyu Liang. "Learning and generalization in overparameterized neural networks, going beyond two layers". In: *Advances in neural information processing systems* 32 (2019).
3. Jonathan L Alperin. *Local representation theory: Modular representations as an introduction to the local representation theory of finite groups*. Vol. 11. Cambridge University Press, 1993.
4. Shaeela Ayesha, Muhammad Kashif Hanif, and Ramzan Talib. "Overview and comparative study of dimensionality reduction techniques for high dimensional data". In: *Information Fusion* 59 (2020), pp. 44–58.
5. Elnaz Barshan et al. "Supervised principal component analysis: Visualization, classification and regression on subspaces and submanifolds". In: *Pattern Recognition* 44.7 (2011), pp. 1357–1371.
6. Mikhail Belkin and Partha Niyogi. "Laplacian eigenmaps and spectral techniques for embedding and clustering". In: *Advances in neural information processing systems* 14 (2001), pp. 585–591.
7. Yoshua Bengio, Aaron Courville, and Pascal Vincent. "Representation learning: A review and new perspectives". In: *IEEE transactions on pattern analysis and machine intelligence* 35.8 (2013), pp. 1798–1828.
8. Yoshua Bengio et al. "Out-of-sample extensions for LLE, Isomap, MDS, eigenmaps, and spectral clustering". In: *Advances in neural information processing systems* 16 (2003), pp. 177–184.
9. Yoshua Bengio et al. *Spectral clustering and kernel PCA are learning eigenfunctions*. Tech. rep. Departement d'Informatique et Recherche Operationnelle, Technical Report 1239, 2003.
10. Bernhard E Boser, Isabelle M Guyon, and Vladimir N Vapnik. "A training algorithm for optimal margin classifiers". In: *Proceedings of the fifth annual workshop on Computational learning theory*. 1992, pp. 144–152.

11. Jane Bromley et al. "Signature verification using a "Siamese" time delay neural network". In: *International Journal of Pattern Recognition and Artificial Intelligence* 7.04 (1993), pp. 669–688.
12. Lawrence Cayton. *Algorithms for manifold learning*. Tech. rep. University of California at San Diego, 2005.
13. Trevor F Cox and Michael AA Cox. *Multidimensional scaling*. Chapman and hall/CRC, 2000.
14. Charles Fefferman, Sanjoy Mitter, and Hariharan Narayanan. "Testing the manifold hypothesis". In: *Journal of the American Mathematical Society* 29.4 (2016), pp. 983–1049.
15. Ronald A Fisher. "The use of multiple measurements in taxonomic problems". In: *Annals of eugenics* 7.2 (1936), pp. 179–188.
16. Jerome Friedman, Trevor Hastie, and Robert Tibshirani. *The elements of statistical learning*. Vol. 2. Springer series in statistics New York, NY, USA, 2009.
17. Benjamin Fruchter. *Introduction to factor analysis*. Van Nostrand, 1954.
18. Kenji Fukumizu, Francis R Bach, and Michael I Jordan. "Kernel dimensionality reduction for supervised learning". In: *Advances in neural information processing systems*. Vol. 16. 2003.
19. Benyamin Ghojogh. "Data Reduction Algorithms in Machine Learning and Data Science". PhD thesis. University of Waterloo, 2021.
20. Benyamin Ghojogh, Fakhri Karray, and Mark Crowley. "Eigenvalue and generalized eigenvalue problems: Tutorial". In: *arXiv preprint arXiv:1903.11240* (2019).
21. Benyamin Ghojogh, Fakhri Karray, and Mark Crowley. "Generalized subspace learning by Roweis discriminant analysis". In: *International Conference on Image Analysis and Recognition*. Springer. 2020, pp. 328–342.
22. Benyamin Ghojogh et al. "Feature selection and feature extraction in pattern analysis: A literature review". In: *arXiv preprint arXiv:1905.02845* (2019).
23. Xavier Glorot, Antoine Bordes, and Yoshua Bengio. "Deep sparse rectifier neural networks". In: *Proceedings of the fourteenth international conference on artificial intelligence and statistics*. JMLR Workshop and Conference Proceedings. 2011, pp. 315–323.
24. Ian Goodfellow, Yoshua Bengio, and Aaron Courville. *Deep learning*. MIT Press, 2016.
25. Ian Goodfellow et al. "Generative adversarial nets". In: *Advances in neural information processing systems*. Vol. 27. 2014.
26. Arthur Gretton et al. "Measuring statistical dependence with Hilbert-Schmidt norms". In: *International conference on algorithmic learning theory*. Springer. 2005, pp. 63–77.
27. Raia Hadsell, Sumit Chopra, and Yann LeCun. "Dimensionality reduction by learning an invariant mapping". In: *2006 IEEE Computer Society Conference on Computer Vision and Pattern Recognition (CVPR'06)* Vol. 2. IEEE. 2006, pp. 1735–1742.
28. Ji Hun Ham et al. "A kernel view of the dimensionality reduction of manifolds". In: *International Conference on Machine Learning*. 2004.
29. Geoffrey E Hinton, Simon Osindero, and Yee-Whye Teh. "A fast learning algorithm for deep belief nets". In: *Neural computation* 18.7 (2006), pp. 1527–1554.
30. Geoffrey E Hinton and Sam T Roweis. "Stochastic neighbor embedding". In: *Advances in neural information processing systems*. 2003, pp. 857–864.
31. Geoffrey E Hinton and Terrence J Sejnowski. "Optimal perceptual inference". In: *Proceedings of the IEEE conference on Computer Vision and Pattern Recognition* Vol. 448. IEEE, 1983.
32. Thomas Hofmann, Bernhard Schölkopf, and Alexander J Smola. "Kernel methods in machine learning". In: *The annals of statistics* (2008), pp. 1171–1220.
33. William B Johnson and Joram Lindenstrauss. "Extensions of Lipschitz mappings into a Hilbert space". In: *Contemporary mathematics* 26 (1984).
34. Ian Jolliffe. *Principal component analysis*. Springer, 2011.
35. Mahmut Kaya and Hasan S¸akir Bilge. "Deep metric learning: A survey". In: *Symmetry* 11.9 (2019), p. 1066.
36. Diederik P Kingma and Max Welling. "Auto-encoding variational Bayes". In: *International Conference on Learning Representations*. 2014.
37. John A Lee and Michel Verleysen. *Nonlinear dimensionality reduction* Springer Science & Business Media, 2007.

38. John M Lee. *Introduction to Smooth Manifolds*. Springer Science & Business Media, 2013, pp. 1–31.
39. Ker-Chau Li. “Sliced inverse regression for dimension reduction”. In: *Journal of the American Statistical Association* 86.414 (1991), pp. 316–327.
40. Tong Lin and Hongbin Zha. “Riemannian manifold learning”. In: *IEEE Transactions on Pattern Analysis and Machine Intelligence* 30.5 (2008), pp. 796–809.
41. Laurens van der Maaten and Geoffrey Hinton. “Visualizing data using t-SNE”. In: *Journal of machine learning research* 9.Nov (2008), pp. 2579–2605.
42. Alireza Makhzani et al. “Adversarial autoencoders”. In: *arXiv preprint arXiv:1511.05644* (2015).
43. Leland McInnes, John Healy, and James Melville. “UMAP: Uniform manifold approximation and projection for dimension reduction”. In: *arXiv preprint arXiv:1802.03426* (2018).
44. Sebastian Mika et al. “Fisher discriminant analysis with kernels”. In: *Proceedings of the 1999 IEEE signal processing society workshop on Neural networks for signal processing IX*. IEEE. 1999, pp. 41–48.
45. Sebastian Mika et al. “Invariant feature extraction and classification in kernel spaces”. In: *Advances in neural information processing systems*. 2000, pp. 526–532.
46. Andrew Ng, Michael Jordan, and Yair Weiss. “On spectral clustering: Analysis and an algorithm”. In: *Advances in neural information processing systems* 14 (2001), pp. 849–856.
47. Karl Pearson. “LIII. On lines and planes of closest fit to systems of points in space”. In: *The London, Edinburgh, and Dublin Philosophical Magazine and Journal of Science* 2.11 (1901), pp. 559–572.
48. Ali Rahimi and Benjamin Recht. “Random Features for Large-Scale Kernel Machines”. In: *Advances in neural information processing systems*. Vol. 20. 2007.
49. Ali Rahimi and Benjamin Recht. “Weighted sums of random kitchen sinks: replacing minimization with randomization in learning”. In: *Advances in neural information processing systems*. 2008, pp. 1313–1320.
50. Sam Roweis. “EM algorithms for PCA and SPCA”. In: *Advances in neural information processing systems* 10 (1997), pp. 626–632.
51. Sam T Roweis and Lawrence K Saul. “Nonlinear dimensionality reduction by locally linear embedding”. In: *Science* 290.5500 (2000), pp. 2323–2326.
52. John W Sammon. “A nonlinear mapping for data structure analysis”. In: *IEEE Transactions on computers* 100.5 (1969), pp. 401–409.
53. Lawrence K Saul and Sam T Roweis. “Think globally fit locally: unsupervised learning of low dimensional manifolds”. In: *Journal of machine learning research* 4.Jun (2003), pp. 119–155.
54. Bernhard Schölkopf, Alexander Smola, and Klaus-Robert Müller. “Kernel principal component analysis”. In: *International conference on artificial neural networks*. Springer. 1997, pp. 583–588.
55. Bernhard Schölkopf, Alexander Smola, and Klaus-Robert Müller. “Nonlinear component analysis as a kernel eigenvalue problem”. In: *Neural computation* 10.5 (1998), pp. 1299–1319.
56. Florian Schroff, Dmitry Kalenichenko, and James Philbin. “FaceNet: A unified embedding for face recognition and clustering”. In: *Proceedings of the IEEE conference on computer vision and pattern recognition*. 2015, pp. 815–823.
57. Ravid Shwartz-Ziv and Naftali Tishby. “Opening the black box of deep neural networks via information”. In: *arXiv preprint arXiv:1703.00810* (2017).
58. Mahdi Soltanolkotabi, Adel Javanmard, and Jason D Lee. “Theoretical insights into the optimization landscape of over-parameterized shallow neural networks”. In: *IEEE Transactions on Information Theory* 65.2 (2018), pp. 742–769.
59. Nitish Srivastava et al. “Dropout: a simple way to prevent neural networks from overfitting”. In: *The journal of machine learning research* 15.1 (2014), pp. 1929–1958.
60. Harry Strange and Reyer Zwiggelaar. *Open Problems in Spectral Dimensionality Reduction*. Springer, 2014.
61. Joshua B Tenenbaum, Vin De Silva, and John C Langford. “A global geometric framework for nonlinear dimensionality reduction”. In: *Science* 290.5500 (2000), pp. 2319–2323.

62. Michael E Tipping and Christopher M Bishop. "Probabilistic principal component analysis". In: *Journal of the Royal Statistical Society: Series B (Statistical Methodology)* 61.3 (1999), pp. 611–622.
63. Naftali Tishby, Fernando C Pereira, and William Bialek. "The information bottleneck method". In: *The 37th annual Allerton Conference on Communication, Control, and Computing* 1999, pp. 368–377.
64. Jianzhong Wang. *Geometric structure of high-dimensional data and dimensionality reduction*. Vol. 5. Springer, 2012.
65. Kilian Q Weinberger and Lawrence K Saul. "An introduction to nonlinear dimensionality reduction by maximum variance unfolding". In: *Proceedings of the AAAI Conference on Artificial Intelligence*. Vol. 6. 2006, pp. 1683–1686.
66. Kilian Q Weinberger, Fei Sha, and Lawrence K Saul. "Learning a kernel matrix for nonlinear dimensionality reduction". In: *Proceedings of the twenty-first international conference on Machine learning*. 2004, p. 106.
67. Yair Weiss. "Segmentation using eigenvectors: a unifying view". In: *Proceedings of the seventh IEEE international conference on computer vision*. Vol. 2. IEEE. 1999, pp. 975–982.
68. Max Welling, Michal Rosen-Zvi, and Geoffrey E Hinton. "Exponential Family Harmoniums with an Application to Information Retrieval." In: *Advances in neural information processing systems*. Vol. 4. 2004, pp. 1481–1488.
69. Liu Yang and Rong Jin. "Distance metric learning: A comprehensive survey". In: *Michigan State University* 2.2 (2006), p. 4.
70. Guoqiang Zhong et al. "An overview on data representation learning: From traditional feature learning to recent deep learning". In: *The Journal of Finance and Data Science* 2.4 (2016), pp. 265–278.

Part I
Preliminaries and Background

Chapter 2
Background on Linear Algebra

2.1 Introduction

Machine learning, especially dimensionality reduction, can be understood in terms of linear algebra, probability theory, and optimization. Therefore, linear algebra is an essential tool for machine learning in general and dimensionality reduction in particular. Linear algebra is a field of mathematics that addresses linear operations on matrices and vectors. Every data point in machine learning is represented by a vector; therefore, the dataset can be represented as a matrix, being dealt with by linear algebra. This chapter provides a comprehensive review of linear algebra, including the centering matrix, linear projection, Rayleigh-Ritz quotient, eigenvalue and generalized eigenvalue problems, and singular value decomposition.

2.2 The Centering Matrix

Consider a matrix $\boldsymbol{A} \in \mathbb{R}^{\alpha \times \beta}$, which is represented by its rows, $\boldsymbol{A} = [\boldsymbol{a}_1, \ldots, \boldsymbol{a}_\alpha]^\top$ or by its columns, $\boldsymbol{A} = [\boldsymbol{b}_1, \ldots, \boldsymbol{b}_\beta]$, where $\boldsymbol{a}_i$ and $\boldsymbol{b}_j$ denote the i-th row and j-th column of $\boldsymbol{A}$, respectively. Note that the vectors are column vectors.

The *left centering matrix* is defined as:

$$\mathbb{R}^{\alpha \times \alpha} \ni \boldsymbol{H} := \boldsymbol{I} - (1/\alpha)\boldsymbol{1}\boldsymbol{1}^\top, \tag{2.1}$$

where $\boldsymbol{1} = [1, \ldots, 1]^\top \in \mathbb{R}^\alpha$ and $\boldsymbol{I} \in \mathbb{R}^{\alpha \times \alpha}$ is the identity matrix. Left multiplying this matrix by $\boldsymbol{A}$, i.e., $\boldsymbol{H}\boldsymbol{A}$, removes the mean of rows of $\boldsymbol{A}$ from all of its rows:

$$\boldsymbol{H}\boldsymbol{A} \overset{(2.1)}{=} \boldsymbol{A} - (1/\alpha)\boldsymbol{1}\boldsymbol{1}^\top\boldsymbol{A} = (\boldsymbol{A}^\top - \boldsymbol{\mu}_{\text{rows}})^\top, \tag{2.2}$$

B. Ghojogh et al., *Elements of Dimensionality Reduction and Manifold Learning*,
https://doi.org/10.1007/978-3-031-10602-6_2

where the column vector $\boldsymbol{\mu}_{\text{rows}} \in \mathbb{R}^{\beta}$ is the mean of the rows of $\boldsymbol{A}$.

The *right centering matrix* is defined as:

$$\mathbb{R}^{\beta\times\beta} \ni \boldsymbol{H} := \boldsymbol{I} - (1/\beta)\mathbf{1}\mathbf{1}^\top, \tag{2.3}$$

where $\mathbf{1} = [1,\dots,1]^\top \in \mathbb{R}^{\beta}$ and $\boldsymbol{I} \in \mathbb{R}^{\beta\times\beta}$ is the identity matrix. Right multiplying this matrix to $\boldsymbol{A}$, i.e., $\boldsymbol{AH}$, removes the mean of the columns of $\boldsymbol{A}$ from all of its columns:

$$\boldsymbol{AH} \overset{(2.3)}{=} \boldsymbol{A} - (1/\beta)\boldsymbol{A}\mathbf{1}\mathbf{1}^\top = \boldsymbol{A} - \boldsymbol{\mu}_{\text{cols}}, \tag{2.4}$$

where the column vector $\boldsymbol{\mu}_{\text{cols}} \in \mathbb{R}^{\alpha}$ is the mean of the columns of $\boldsymbol{A}$.

Both left and right centering matrices can be used at the same time to have a double-centered matrix $\boldsymbol{A}$:

$$\begin{aligned}
\boldsymbol{HAH} &= (\boldsymbol{I}_\alpha - (1/\alpha)\mathbf{1}_\alpha\mathbf{1}_\alpha^\top)\boldsymbol{A}(\boldsymbol{I}_\beta - (1/\beta)\mathbf{1}_\beta\mathbf{1}_\beta^\top) \\
&= (\boldsymbol{A} - (1/\alpha)\mathbf{1}_\alpha\mathbf{1}_\alpha^\top\boldsymbol{A})(\boldsymbol{I}_\beta - (1/\beta)\mathbf{1}_\beta\mathbf{1}_\beta^\top) \\
&= \boldsymbol{A} - (1/\alpha)\mathbf{1}_\alpha\mathbf{1}_\alpha^\top\boldsymbol{A} - (1/\beta)\boldsymbol{A}\mathbf{1}_\beta\mathbf{1}_\beta^\top + (1/(\alpha\beta))\mathbf{1}_\alpha\mathbf{1}_\alpha^\top\boldsymbol{A}\mathbf{1}_\beta\mathbf{1}_\beta^\top.
\end{aligned} \tag{2.5}$$

This operation is commonly done for a kernel[1] (see [9, Appendix A]). The second term removes the mean of the rows of $\boldsymbol{A}$ according to Eq. (2.2) and the third term removes the mean of the columns of $\boldsymbol{A}$ according to Eq. (2.4). The last term, however, adds the overall mean of $\boldsymbol{A}$ back to it. Assume the matrix $\boldsymbol{\mu}_{\text{all}} \in \mathbb{R}^{\alpha\times\beta}$, whose elements are the overall mean of $\boldsymbol{A}$, is:

$$\boldsymbol{\mu}_{\text{all}} := (1/(\alpha\beta))\mathbf{1}_\alpha\mathbf{1}_\alpha^\top\boldsymbol{A}\mathbf{1}_\beta\mathbf{1}_\beta^\top, \tag{2.6}$$

$$\boldsymbol{\mu}_{\text{all}}(.,.) = \frac{1}{\alpha\beta}\sum_{i=1}^{\alpha}\sum_{j=1}^{\beta}\boldsymbol{A}(i,j), \tag{2.7}$$

where $\boldsymbol{A}(i,j)$ is the (i,j)-th element of $\boldsymbol{A}$ and $\boldsymbol{\mu}_{\text{all}}(.,.)$ is every element of $\boldsymbol{A}$. Therefore, *double-centering* for $\boldsymbol{A}$ is defined as:

$$\boldsymbol{HAH} = (\boldsymbol{A}^\top - \boldsymbol{\mu}_{\text{rows}})^\top - \boldsymbol{\mu}_{\text{cols}} + \boldsymbol{\mu}_{\text{all}}, \tag{2.8}$$

which removes both the row and column means of $\boldsymbol{A}$ but adds back the overall mean. Note that if the matrix $\boldsymbol{A}$ is a square matrix, the left and right centering matrices are equal with the same dimensionality as the matrix $\boldsymbol{A}$. In computer programming, the usage of centering matrices might cause precision errors. Therefore, in computer programming:

[1] Refer to Chap. 3 for the definition of a kernel.

$$\begin{aligned}
\boldsymbol{HA} &\approx (\boldsymbol{A}^\top - \boldsymbol{\mu}_{\text{rows}})^\top, \\
\boldsymbol{AH} &\approx \boldsymbol{A} - \boldsymbol{\mu}_{\text{cols}}, \\
\boldsymbol{HAH} &\approx (\boldsymbol{A}^\top - \boldsymbol{\mu}_{\text{rows}})^\top - \boldsymbol{\mu}_{\text{cols}} + \boldsymbol{\mu}_{\text{all}},
\end{aligned}$$

which are good enough approximations.

The centering matrix is symmetric because:

$$\begin{aligned}
\boldsymbol{H}^\top &= (\boldsymbol{I} - (1/\alpha)\mathbf{1}\mathbf{1}^\top)^\top = \boldsymbol{I}^\top - (1/\alpha)(\mathbf{1}\mathbf{1}^\top)^\top \\
&= \boldsymbol{I} - (1/\alpha)\mathbf{1}\mathbf{1}^\top \overset{(2.1)}{=} \boldsymbol{H}.
\end{aligned} \tag{2.9}$$

The centering matrix is also considered to be idempotent:

$$\boldsymbol{H}^k = \underbrace{\boldsymbol{H}\boldsymbol{H}\cdots\boldsymbol{H}}_{k \text{ times}} = \boldsymbol{H}, \tag{2.10}$$

where k is a positive integer. This is because of the following:

$$\begin{aligned}
\boldsymbol{HH} &= (\boldsymbol{I} - (1/\alpha)\mathbf{1}\mathbf{1}^\top)(\boldsymbol{I} - (1/\alpha)\mathbf{1}\mathbf{1}^\top) \\
&= \boldsymbol{I} - (1/\alpha)\mathbf{1}\mathbf{1}^\top - (1/\alpha)\mathbf{1}\mathbf{1}^\top + (1/\alpha^2)\mathbf{1}\underbrace{\mathbf{1}^\top\mathbf{1}}_{\alpha}\mathbf{1}^\top \\
&= \boldsymbol{I} - (1/\alpha)\mathbf{1}\mathbf{1}^\top - (1/\alpha)\mathbf{1}\mathbf{1}^\top + (1/\alpha)\mathbf{1}\mathbf{1}^\top = \boldsymbol{I} - (1/\alpha)\mathbf{1}\mathbf{1}^\top \overset{(2.1)}{=} \boldsymbol{H}.
\end{aligned}$$

Therefore:

$$\boldsymbol{H}^k = \underbrace{(\boldsymbol{H}\cdots\underbrace{(\boldsymbol{H}\underbrace{(\boldsymbol{H}\boldsymbol{H})}_{\boldsymbol{H}}))))))}_{\boldsymbol{H}}}_{\boldsymbol{H}} = \boldsymbol{H}.$$

□

The following example provides an illustration of this concept.

$$\boldsymbol{A} = \begin{bmatrix} 1 & 2 & 3 \\ 4 & 3 & 1 \end{bmatrix},$$

whose row mean, column mean, and overall mean matrices are:

$$\begin{aligned}
\boldsymbol{\mu}_{\text{rows}} &= [2.5, 2.5, 2]^\top, \\
\boldsymbol{\mu}_{\text{cols}} &= [2, 2.66]^\top,
\end{aligned}$$

$$\boldsymbol{\mu}_{\text{all}} = \begin{bmatrix} 2.33 & 2.33 & 2.33 \\ 2.33 & 2.33 & 2.33 \end{bmatrix},$$

respectively. The left, right, and double centering of $\boldsymbol{A}$ are, respectively, as follows:

$$\boldsymbol{HA} = \begin{bmatrix} -1.5 & -0.5 & 1 \\ 1.5 & 0.5 & -1 \end{bmatrix},$$

$$\boldsymbol{AH} = \begin{bmatrix} -1 & 0 & 1 \\ 1.34 & 0.34 & -1.66 \end{bmatrix},$$

$$\boldsymbol{HAH} = \boldsymbol{HA} - \boldsymbol{\mu}_{\text{cols}} + \boldsymbol{\mu}_{\text{all}} = \begin{bmatrix} -1.17 & -0.17 & 1.33 \\ 1.17 & 0.17 & -1.33 \end{bmatrix}.$$

2.3 Linear Projection

2.3.1 A Projection Point of View

Assume there is a data point $\boldsymbol{x} \in \mathbb{R}^d$. The aim is to project this data point onto the vector space spanned by p vectors $\{\boldsymbol{u}_1, \ldots, \boldsymbol{u}_p\}$, where each vector is d-dimensional and usually $p \ll d$. These vectors can be stacked columnwise in matrix $\boldsymbol{U} = [\boldsymbol{u}_1, \ldots, \boldsymbol{u}_p] \in \mathbb{R}^{d \times p}$. In other words, the goal is to project $\boldsymbol{x}$ onto the column space of $\boldsymbol{U}$, denoted by $\mathbb{C}\text{ol}(\boldsymbol{U})$.

The projection of $\boldsymbol{x} \in \mathbb{R}^d$ onto $\mathbb{C}\text{ol}(\boldsymbol{U}) \in \mathbb{R}^p$ and then its representation in $\mathbb{R}^d$ (its reconstruction) can be seen as a linear system of equations:

$$\mathbb{R}^d \ni \widehat{\boldsymbol{x}} := \boldsymbol{U}\boldsymbol{\beta}, \tag{2.11}$$

where the unknown coefficients $\boldsymbol{\beta} \in \mathbb{R}^p$ should be found.

If $\boldsymbol{x}$ lies in the $\mathbb{C}\text{ol}(\boldsymbol{U})$ or $\textbf{span}\{\boldsymbol{u}_1, \ldots, \boldsymbol{u}_p\}$, this linear system has an exact solution, $\widehat{\boldsymbol{x}} = \boldsymbol{x} = \boldsymbol{U}\boldsymbol{\beta}$. However, if $\boldsymbol{x}$ does not lie in this space, there is no solution $\boldsymbol{\beta}$ as $\boldsymbol{x} = \boldsymbol{U}\boldsymbol{\beta}$. In this case, the projection of $\boldsymbol{x}$ onto $\mathbb{C}\text{ol}(\boldsymbol{U})$ or $\textbf{span}\{\boldsymbol{u}_1, \ldots, \boldsymbol{u}_p\}$ and then its reconstruction should be solved. Specifically, the aim is to solve for Eq. (2.11). In this case, $\widehat{\boldsymbol{x}}$ and $\boldsymbol{x}$ are different and there is a residual between the original data $\boldsymbol{x}$ and its reconstruction:

$$\boldsymbol{r} = \boldsymbol{x} - \widehat{\boldsymbol{x}} = \boldsymbol{x} - \boldsymbol{U}\boldsymbol{\beta}, \tag{2.12}$$

This residual should be small because the reconstructed data should be as similar to the original data as possible. For an illustration of this concept, see Fig. 2.1, which demonstrates two sets of image reconstructions with good and weak qualities of reconstruction.

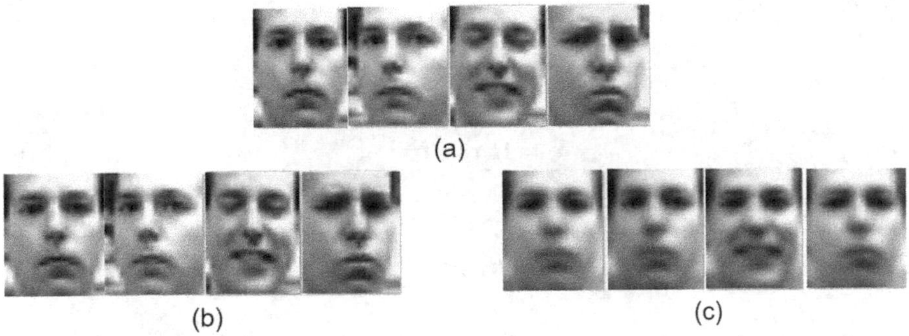

Fig. 2.1 An example of four images that are (**a**) the original images, (**b**) reconstructed images with a good approximation, and (**c**) reconstructed images with a weak approximation

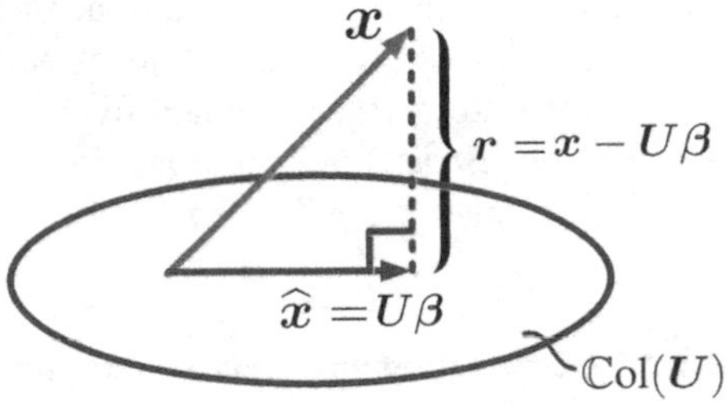

Fig. 2.2 The residual and projection onto the column space of $\boldsymbol{U}$

As shown in Fig. 2.2, the smallest residual vector is orthogonal to $\mathbb{C}\mathrm{ol}(\boldsymbol{U})$; therefore:

$$\boldsymbol{x} - \boldsymbol{U\beta} \perp \boldsymbol{U} \implies \boldsymbol{U}^\top(\boldsymbol{x} - \boldsymbol{U\beta}) = 0 \implies \boldsymbol{\beta} = (\boldsymbol{U}^\top \boldsymbol{U})^{-1}\boldsymbol{U}^\top \boldsymbol{x}. \tag{2.13}$$

It is noteworthy to mention that Eq. (2.13) is also the formula for coefficients in linear regression [3], where the input data are the rows of $\boldsymbol{U}$ and the labels are $\boldsymbol{x}$. However, our goal here is different. Inserting Eq. (2.13) into Eq. (2.11) results in the following expression:

$$\widehat{\boldsymbol{x}} = \boldsymbol{U}(\boldsymbol{U}^\top \boldsymbol{U})^{-1}\boldsymbol{U}^\top \boldsymbol{x}.$$

The following is defined:

$$\mathbb{R}^{d\times d} \ni \boldsymbol{\Pi} := \boldsymbol{U}(\boldsymbol{U}^\top \boldsymbol{U})^{-1}\boldsymbol{U}^\top, \tag{2.14}$$

as the *projection matrix* because it projects $\boldsymbol{x}$ onto $\mathbb{C}\mathrm{ol}(\boldsymbol{U})$ (and reconstructs back). Note that $\boldsymbol{\Pi}$ is also referred to as the *hat matrix* in the literature because it puts a 'hat' on top of $\boldsymbol{x}$.

If the vectors $\{\boldsymbol{u}_1, \ldots, \boldsymbol{u}_p\}$ are orthonormal (the matrix $\boldsymbol{U}$ is orthogonal), there is $\boldsymbol{U}^\top = \boldsymbol{U}^{-1}$ and thus $\boldsymbol{U}^\top \boldsymbol{U} = \boldsymbol{I}$. Therefore, Eq. (2.14) is simplified to:

$$\boldsymbol{\Pi} = \boldsymbol{U}\boldsymbol{U}^\top. \tag{2.15}$$

Therefore:

$$\widehat{\boldsymbol{x}} = \boldsymbol{\Pi}\,\boldsymbol{x} = \boldsymbol{U}\boldsymbol{U}^\top\boldsymbol{x}. \tag{2.16}$$

2.3.2 *Projection and Reconstruction*

Equation (2.16) is called a *projection onto the subspace* where $\boldsymbol{U}^\top\boldsymbol{x}$ projects $\boldsymbol{x}$ onto the row space of $\boldsymbol{U}$, i.e., $\mathbb{C}\text{ol}(\boldsymbol{U}^\top)$ (projection onto a space spanned by d vectors that are p-dimensional). It is a *subspace* because $p \leq d$, where p and d are the dimensionality of the subspace and the original $\boldsymbol{x}$, respectively. $\boldsymbol{U}(\boldsymbol{U}^\top\boldsymbol{x})$ then projects the projected data back onto the column space of $\boldsymbol{U}$, i.e., $\mathbb{C}\text{ol}(\boldsymbol{U})$ (projection onto a space spanned by p vectors that are d-dimensional). This step is called *reconstruction* from projection and the aim is to have the residual between $\boldsymbol{x}$ and its reconstruction $\widehat{\boldsymbol{x}}$ be small (see the example in Fig. 2.1).

2.3.2.1 Projection and Reconstruction for Noncentered Data

If there exist n training data points, i.e., $\{\boldsymbol{x}_i\}_{i=1}^n$, the projection of a training data point $\boldsymbol{x}$ is:

$$\mathbb{R}^p \ni \widetilde{\boldsymbol{x}} := \boldsymbol{U}^\top\boldsymbol{x}. \tag{2.17}$$

The reconstruction of a training data point $\boldsymbol{x}$ after projection onto the subspace is:

$$\mathbb{R}^d \ni \widehat{\boldsymbol{x}} := \boldsymbol{U}\boldsymbol{U}^\top\boldsymbol{x} = \boldsymbol{U}\widetilde{\boldsymbol{x}}. \tag{2.18}$$

If the n data points are stacked columnwise in a matrix then $\boldsymbol{X} = [\boldsymbol{x}_1, \ldots, \boldsymbol{x}_n] \in \mathbb{R}^{d\times n}$. The projection and reconstruction, Eqs. (2.17) and (2.18), for the whole training data are:

$$\mathbb{R}^{p\times n} \ni \widetilde{\boldsymbol{X}} := \boldsymbol{U}^\top\boldsymbol{X}, \tag{2.19}$$

$$\mathbb{R}^{d\times n} \ni \widehat{\boldsymbol{X}} := \boldsymbol{U}\boldsymbol{U}^\top\boldsymbol{X} = \boldsymbol{U}\widetilde{\boldsymbol{X}}, \tag{2.20}$$

where $\widetilde{\boldsymbol{X}} = [\widetilde{\boldsymbol{x}}_1, \ldots, \widetilde{\boldsymbol{x}}_n]$ and $\widehat{\boldsymbol{X}} = [\widehat{\boldsymbol{x}}_1, \ldots, \widehat{\boldsymbol{x}}_n]$ are the data projected onto the subspace and the reconstructed data, respectively.

It is possible to project a new data point onto the subspace for $\boldsymbol{X}$, where the new data point is not a column of $\boldsymbol{X}$. In other words, the new data point has not contributed to constructing the subspace. This new data point is referred to as a "test data point" or "out-of-sample data" in the literature. Equation (2.19) is the projection of the training data $\boldsymbol{X}$ onto the subspace. If $\boldsymbol{x}_t$ denotes an out-of-sample data point, its projection onto the subspace ($\widetilde{\boldsymbol{x}}_t$) and its reconstruction ($\widehat{\boldsymbol{x}}_t$) are:

$$\mathbb{R}^p \ni \widetilde{\boldsymbol{x}}_t = \boldsymbol{U}^\top \boldsymbol{x}_t, \tag{2.21}$$

$$\mathbb{R}^d \ni \widehat{\boldsymbol{x}}_t = \boldsymbol{U}\boldsymbol{U}^\top \boldsymbol{x}_t = \boldsymbol{U}\widetilde{\boldsymbol{x}}_t. \tag{2.22}$$

If n_t out-of-sample data points, $\mathbb{R}^{d\times n_t} \ni \boldsymbol{X}_t = [\boldsymbol{x}_{t,1}, \ldots, \boldsymbol{x}_{t,n_t}]$, are considered, the projection and reconstruction are:

$$\mathbb{R}^{p\times n_t} \ni \widetilde{\boldsymbol{X}}_t = \boldsymbol{U}^\top \boldsymbol{X}_t, \tag{2.23}$$

$$\mathbb{R}^{d\times n_t} \ni \widehat{\boldsymbol{X}}_t = \boldsymbol{U}\boldsymbol{U}^\top \boldsymbol{X}_t = \boldsymbol{U}\widetilde{\boldsymbol{X}}_t, \tag{2.24}$$

respectively.

2.3.2.2 Projection and Reconstruction for Centered Data

Let the mean of training data points be:

$$\mathbb{R}^d \ni \boldsymbol{\mu}_x := \frac{1}{n}\sum_{i=1}^{n} \boldsymbol{x}_i. \tag{2.25}$$

Some subspace learning methods require data points to be centered. In this case, the projection of the training data point $\boldsymbol{x}$ is:

$$\mathbb{R}^p \ni \widetilde{\boldsymbol{x}} := \boldsymbol{U}^\top \breve{\boldsymbol{x}}, \tag{2.26}$$

where $\breve{\boldsymbol{x}}$ is the centered data point:

$$\mathbb{R}^d \ni \breve{\boldsymbol{x}} := \boldsymbol{x} - \boldsymbol{\mu}_x. \tag{2.27}$$

The reconstruction of the training data point $\boldsymbol{x}$ after projection onto the subspace is:

$$\mathbb{R}^d \ni \widehat{\boldsymbol{x}} := \boldsymbol{U}\boldsymbol{U}^\top \breve{\boldsymbol{x}} + \boldsymbol{\mu}_x = \boldsymbol{U}\widetilde{\boldsymbol{x}} + \boldsymbol{\mu}_x, \tag{2.28}$$

where the mean is readded because it was removed prior to projection.

If the n data points are stacked columnwise in a matrix $\boldsymbol{X} = [\boldsymbol{x}_1, \ldots, \boldsymbol{x}_n] \in \mathbb{R}^{d\times n}$, they first need to be centered:

$$\mathbb{R}^{d\times n} \ni \breve{\boldsymbol{X}} := \boldsymbol{X}\boldsymbol{H} = \boldsymbol{X} - \boldsymbol{\mu}_x, \tag{2.29}$$

where $\breve{\boldsymbol{X}} = [\breve{\boldsymbol{x}}_1, \dots, \breve{\boldsymbol{x}}_n] = [\boldsymbol{x}_1 - \boldsymbol{\mu}_x, \dots, \boldsymbol{x}_n - \boldsymbol{\mu}_x]$ is the centered data and $\mathbb{R}^{n \times n} \ni \boldsymbol{H} := \boldsymbol{I} - (1/n)\boldsymbol{1}\boldsymbol{1}^\top$ is the centering matrix. The projection and reconstruction, Eqs. (2.26) and (2.28), for the whole training data are:

$$\mathbb{R}^{p \times n} \ni \widetilde{\boldsymbol{X}} := \boldsymbol{U}^\top \breve{\boldsymbol{X}}, \tag{2.30}$$

$$\mathbb{R}^{d \times n} \ni \widehat{\boldsymbol{X}} := \boldsymbol{U}\boldsymbol{U}^\top \breve{\boldsymbol{X}} + \boldsymbol{\mu}_x = \boldsymbol{U}\widetilde{\boldsymbol{X}} + \boldsymbol{\mu}_x, \tag{2.31}$$

where $\widetilde{\boldsymbol{X}} = [\widetilde{\boldsymbol{x}}_1, \dots, \widetilde{\boldsymbol{x}}_n]$ and $\widehat{\boldsymbol{X}} = [\widehat{\boldsymbol{x}}_1, \dots, \widehat{\boldsymbol{x}}_n]$ are the data projected onto the subspace and the reconstructed data, respectively.

The projection of a test data point $\boldsymbol{x}_t$ onto the subspace ($\widetilde{\boldsymbol{x}}_t$) and its reconstruction ($\widehat{\boldsymbol{x}}_t$) are:

$$\mathbb{R}^{p} \ni \widetilde{\boldsymbol{x}}_t = \boldsymbol{U}^\top \breve{\boldsymbol{x}}_t, \tag{2.32}$$

$$\mathbb{R}^{d} \ni \widehat{\boldsymbol{x}}_t = \boldsymbol{U}\boldsymbol{U}^\top \breve{\boldsymbol{x}}_t + \boldsymbol{\mu}_x = \boldsymbol{U}\widetilde{\boldsymbol{x}}_t + \boldsymbol{\mu}_x, \tag{2.33}$$

where:

$$\mathbb{R}^{d} \ni \breve{\boldsymbol{x}}_t := \boldsymbol{x}_t - \boldsymbol{\mu}_x, \tag{2.34}$$

is the centered out-of-sample data point, which is centered using the mean of training data. Note that for centering the out-of-sample data point(s), the mean of the training data should be used and not the out-of-sample data because each test point should be processed individually regardless of the other test points. The projection and reconstruction of the combined out-of-sample points are:

$$\mathbb{R}^{p \times n_t} \ni \widetilde{\boldsymbol{X}}_t = \boldsymbol{U}^\top \breve{\boldsymbol{X}}_t, \tag{2.35}$$

$$\mathbb{R}^{d \times n_t} \ni \widehat{\boldsymbol{X}}_t = \boldsymbol{U}\boldsymbol{U}^\top \breve{\boldsymbol{X}}_t + \boldsymbol{\mu}_x = \boldsymbol{U}\widetilde{\boldsymbol{X}}_t + \boldsymbol{\mu}_x, \tag{2.36}$$

respectively, where:

$$\mathbb{R}^{d \times n_t} \ni \breve{\boldsymbol{X}}_t := \boldsymbol{X}_t - \boldsymbol{\mu}_x. \tag{2.37}$$

2.4 Rayleigh-Ritz Quotient

The *Rayleigh-Ritz quotient* or *Rayleigh quotient* is defined as [2, 8]:

$$\mathbb{R} \ni R(\boldsymbol{A}, \boldsymbol{x}) := \frac{\boldsymbol{x}^\top \boldsymbol{A}\boldsymbol{x}}{\boldsymbol{x}^\top \boldsymbol{x}}, \tag{2.38}$$

where $\boldsymbol{A}$ is a symmetric matrix and $\boldsymbol{x}$ is a nonzero vector:

$$\boldsymbol{A} = \boldsymbol{A}^\top, \quad \boldsymbol{x} \neq \boldsymbol{0}. \tag{2.39}$$

One of the properties of the Rayleigh-Ritz quotient is:

$$R(\boldsymbol{A}, c\boldsymbol{x}) = R(\boldsymbol{A}, \boldsymbol{x}), \tag{2.40}$$

where c is a scalar. This is because of the following:

$$R(\boldsymbol{A}, c\boldsymbol{x}) = \frac{(c\boldsymbol{x})^\top \boldsymbol{A}\, c\boldsymbol{x}}{(c\boldsymbol{x})^\top c\boldsymbol{x}} \overset{(a)}{=} \frac{c\boldsymbol{x}^\top \boldsymbol{A}\, c\boldsymbol{x}}{c\boldsymbol{x}^\top c\boldsymbol{x}} \overset{(b)}{=} \frac{c^2}{c^2} \times \frac{\boldsymbol{x}^\top \boldsymbol{A}\, \boldsymbol{x}}{\boldsymbol{x}^\top \boldsymbol{x}} \overset{(2.38)}{=} R(\boldsymbol{A}, \boldsymbol{x}),$$

where (a) and (b) are because c is a scalar. Consider the optimization problem of the Rayleigh-Ritz quotient:

$$\underset{\boldsymbol{x}}{\text{minimize/maximize}} \quad R(\boldsymbol{A}, \boldsymbol{x}). \tag{2.41}$$

According to Eq. (2.40), this is equivalent to the following problem [2]:

$$\begin{aligned} &\underset{\boldsymbol{x}}{\text{minimize/maximize}} && R(\boldsymbol{A}, \boldsymbol{x}) \\ &\text{subject to} && ||\boldsymbol{x}||_2 = 1. \end{aligned} \tag{2.42}$$

This equivalence is explained more in the following. Let $\boldsymbol{y} := (1/||\boldsymbol{x}||_2)\,\boldsymbol{x}$. The Rayleigh-Ritz quotient is:

$$R(\boldsymbol{A}, \boldsymbol{y}) = \frac{\boldsymbol{y}^\top \boldsymbol{A}\, \boldsymbol{y}}{\boldsymbol{y}^\top \boldsymbol{y}} = \frac{1/||\boldsymbol{x}||_2^2}{1/||\boldsymbol{x}||_2^2} \times \frac{\boldsymbol{x}^\top \boldsymbol{A}\, \boldsymbol{x}}{\boldsymbol{x}^\top \boldsymbol{x}} = R(\boldsymbol{A}, \boldsymbol{x}).$$

Due to the following:

$$||\boldsymbol{y}||_2^2 = \frac{1}{||\boldsymbol{x}||_2^2} \times ||\boldsymbol{x}||_2^2 = 1 \implies ||\boldsymbol{y}||_2 = 1,$$

$R(\boldsymbol{A}, \boldsymbol{y})$ should be optimized subject to $||\boldsymbol{y}||_2 = 1$. Changing the dummy variable[2] $\boldsymbol{y}$ to $\boldsymbol{x}$ gives Eq. (2.42).

Another equivalent problem for Eq. (2.41) is:

$$\begin{aligned} &\underset{\boldsymbol{x}}{\text{minimize/maximize}} && \boldsymbol{x}^\top \boldsymbol{A}\, \boldsymbol{x} \\ &\text{subject to} && ||\boldsymbol{x}||_2 = 1, \end{aligned} \tag{2.43}$$

[2] A dummy variable is a variable that can be renamed to any variable without affecting the operation.

obtained by inserting the constraint $\|\boldsymbol{x}\|_2^2 = \boldsymbol{x}^\top\boldsymbol{x} = 1$ in Eqs. (2.38) and (2.42). Note that the constraint in Eqs. (2.42) and (2.43) can be equal to any constant which is proven similarly. Moreover, the value of the constant in the constraint is not important because it will be removed after taking the derivative of the Lagrangian during optimization [1].

2.4.1 Generalized Rayleigh-Ritz Quotient

The *generalized Rayleigh-Ritz quotient* or *generalized Rayleigh quotient* is defined as [8]:

$$\mathbb{R} \ni R(\boldsymbol{A}, \boldsymbol{B}; \boldsymbol{x}) := \frac{\boldsymbol{x}^\top \boldsymbol{A}\,\boldsymbol{x}}{\boldsymbol{x}^\top \boldsymbol{B}\,\boldsymbol{x}}, \tag{2.44}$$

where $\boldsymbol{A}$ and $\boldsymbol{B}$ are symmetric matrices and $\boldsymbol{x}$ is a nonzero vector:

$$\boldsymbol{A} = \boldsymbol{A}^\top, \quad \boldsymbol{B} = \boldsymbol{B}^\top, \quad \boldsymbol{x} \neq \boldsymbol{0}. \tag{2.45}$$

If the symmetric $\boldsymbol{B}$ is positive definite:

$$\boldsymbol{B} \succ 0, \tag{2.46}$$

it has a Cholesky decomposition:

$$\boldsymbol{B} = \boldsymbol{C}\boldsymbol{C}^\top, \tag{2.47}$$

where $\boldsymbol{C}$ is a lower triangular matrix. In case $\boldsymbol{B} \succ 0$, the generalized Rayleigh-Ritz quotient can be converted to a Rayleigh-Ritz quotient:

$$R(\boldsymbol{A}, \boldsymbol{B}; \boldsymbol{x}) = R(\boldsymbol{D}, \boldsymbol{C}^\top\boldsymbol{x}), \tag{2.48}$$

where:

$$\boldsymbol{D} := \boldsymbol{C}^{-1}\boldsymbol{A}\boldsymbol{C}^{-\top}. \tag{2.49}$$

This is because:

$$\text{RHS} = R(\boldsymbol{D}, \boldsymbol{C}^\top\boldsymbol{x}) \overset{(2.38)}{=} \frac{(\boldsymbol{C}^\top\boldsymbol{x})^\top \boldsymbol{D}\,(\boldsymbol{C}^\top\boldsymbol{x})}{(\boldsymbol{C}^\top\boldsymbol{x})^\top(\boldsymbol{C}^\top\boldsymbol{x})}$$

$$\overset{(2.49)}{=} \frac{\boldsymbol{x}^\top\boldsymbol{C}\boldsymbol{C}^{-1}\boldsymbol{A}(\boldsymbol{C}\boldsymbol{C}^{-1})^\top\boldsymbol{x}}{\boldsymbol{x}^\top(\boldsymbol{C}\boldsymbol{C}^\top)\boldsymbol{x}} \overset{(a)}{=} \frac{\boldsymbol{x}^\top\boldsymbol{A}\,\boldsymbol{x}}{\boldsymbol{x}^\top\boldsymbol{B}\,\boldsymbol{x}} \overset{(2.44)}{=} R(\boldsymbol{A}, \boldsymbol{B}; \boldsymbol{x}) = \text{LHS}$$

□

where RHS and LHS are short for right and left hand sides and (a) is because of Eq. (2.47) and $\boldsymbol{C}\boldsymbol{C}^{-1} = \boldsymbol{I}$ because $\boldsymbol{C}$ is a square matrix.

Similarly, one of the properties of the generalized Rayleigh-Ritz quotient is:

$$R(\boldsymbol{A}, \boldsymbol{B}; c\boldsymbol{x}) = R(\boldsymbol{A}, \boldsymbol{B}; \boldsymbol{x}), \tag{2.50}$$

where c is a scalar. The proof is that:

$$R(\boldsymbol{A}, \boldsymbol{B}; c\boldsymbol{x}) = \frac{(c\boldsymbol{x})^\top \boldsymbol{A}\, c\boldsymbol{x}}{(c\boldsymbol{x})^\top \boldsymbol{B}\, c\boldsymbol{x}} \overset{(a)}{=} \frac{c\boldsymbol{x}^\top \boldsymbol{A}\, c\boldsymbol{x}}{c\boldsymbol{x}^\top \boldsymbol{B}\, c\boldsymbol{x}} \overset{(b)}{=} \frac{c^2}{c^2} \times \frac{\boldsymbol{x}^\top \boldsymbol{A}\, \boldsymbol{x}}{\boldsymbol{x}^\top \boldsymbol{B}\, \boldsymbol{x}} \overset{(2.44)}{=} R(\boldsymbol{A}, \boldsymbol{B}; \boldsymbol{x}),$$

where (a) and (b) are because c is a scalar. Consider the optimization problem of the generalized Rayleigh-Ritz quotient:

$$\underset{\boldsymbol{x}}{\text{minimize/maximize}} \quad R(\boldsymbol{A}, \boldsymbol{B}; \boldsymbol{x}). \tag{2.51}$$

According to Eq. (2.50), it has an equivalent form:

$$\begin{aligned} &\underset{\boldsymbol{x}}{\text{minimize/maximize}} && \boldsymbol{x}^\top \boldsymbol{A}\, \boldsymbol{x} \\ &\text{subject to} && \boldsymbol{x}^\top \boldsymbol{B}\, \boldsymbol{x} = 1, \end{aligned} \tag{2.52}$$

for the same reason as the Rayleigh-Ritz quotient. The constraint can be equal to any constant because in the derivative of Lagrangian, the constant will be dropped.

2.5 Eigenvalue and Generalized Eigenvalue Problems

Eigenvalue and generalized eigenvalue problems play important roles in different fields of science, including machine learning, physics, statistics, and mathematics. In the eigenvalue problem, the eigenvectors of a matrix represent the most important and informative directions of that matrix. For example, if the matrix is a covariance matrix of data, the eigenvectors represent the directions of the spread or variance of data and the corresponding eigenvalues are the magnitude of the spread in these directions [7]. These directions are impacted by another matrix in the generalized eigenvalue problem. If the other matrix is the identity matrix, this impact is cancelled and the eigenvalue problem captures the directions of the maximum spread.

2.5.1 *Introducing Eigenvalue and Generalized Eigenvalue Problems*

This section introduces the eigenvalue problem and the generalized eigenvalue problem.

2.5.1.1 Eigenvalue Problem

The eigenvalue problem [6, 11] of a symmetric matrix $\boldsymbol{A} \in \mathbb{R}^{d\times d}$ is defined as:

$$\boldsymbol{A}\boldsymbol{\phi}_i = \lambda_i \boldsymbol{\phi}_i, \quad \forall i \in \{1, \dots, d\}, \tag{2.53}$$

and in matrix form, it is:

$$\boldsymbol{A}\boldsymbol{\Phi} = \boldsymbol{\Phi}\boldsymbol{\Lambda}, \tag{2.54}$$

where the columns of $\mathbb{R}^{d\times d} \ni \boldsymbol{\Phi} := [\boldsymbol{\phi}_1, \dots, \boldsymbol{\phi}_d]$ are the eigenvectors and diagonal elements of $\mathbb{R}^{d\times d} \ni \boldsymbol{\Lambda} := \textbf{diag}([\lambda_1, \dots, \lambda_d]^\top)$ are the eigenvalues. Note that $\boldsymbol{\phi}_i \in \mathbb{R}^d$ and $\lambda_i \in \mathbb{R}$.

For the eigenvalue problem, the matrix $\boldsymbol{A}$ can be nonsymmetric. If the matrix is symmetric, its eigenvectors are orthogonal/orthonormal and if it is nonsymmetric, its eigenvectors are not orthogonal/orthonormal. Equation (2.54) can be restated as:

$$\boldsymbol{A}\boldsymbol{\Phi} = \boldsymbol{\Phi}\boldsymbol{\Lambda} \implies \boldsymbol{A}\underbrace{\boldsymbol{\Phi}\boldsymbol{\Phi}^\top}_{\boldsymbol{I}} = \boldsymbol{\Phi}\boldsymbol{\Lambda}\boldsymbol{\Phi}^\top \implies \boldsymbol{A} = \boldsymbol{\Phi}\boldsymbol{\Lambda}\boldsymbol{\Phi}^\top = \boldsymbol{\Phi}\boldsymbol{\Lambda}\boldsymbol{\Phi}^{-1}, \tag{2.55}$$

where $\boldsymbol{\Phi}^\top = \boldsymbol{\Phi}^{-1}$ because $\boldsymbol{\Phi}$ is an orthogonal matrix. Moreover, note that there is always $\boldsymbol{\Phi}^\top\boldsymbol{\Phi} = \boldsymbol{I}$ for orthogonal $\boldsymbol{\Phi}$, but there is only $\boldsymbol{\Phi}\boldsymbol{\Phi}^\top = \boldsymbol{I}$ if "all" columns of the orthogonal $\boldsymbol{\Phi}$ exist (it is not truncated, i.e., it is a square matrix). Equation (2.55) is referred to as "eigenvalue decomposition", "eigen-decomposition", or "spectral decomposition".

2.5.1.2 Generalized Eigenvalue Problem

The generalized eigenvalue problem [6, 8] of two symmetric matrices $\boldsymbol{A} \in \mathbb{R}^{d\times d}$ and $\boldsymbol{B} \in \mathbb{R}^{d\times d}$ is defined as:

$$\boldsymbol{A}\boldsymbol{\phi}_i = \lambda_i \boldsymbol{B}\boldsymbol{\phi}_i, \quad \forall i \in \{1, \dots, d\}, \tag{2.56}$$

and in matrix form, it is:

$$\boldsymbol{A}\boldsymbol{\Phi} = \boldsymbol{B}\boldsymbol{\Phi}\boldsymbol{\Lambda}, \tag{2.57}$$

where the columns of $\mathbb{R}^{d\times d} \ni \boldsymbol{\Phi} := [\boldsymbol{\phi}_1, \dots, \boldsymbol{\phi}_d]$ are the eigenvectors and diagonal elements of $\mathbb{R}^{d\times d} \ni \boldsymbol{\Lambda} := \mathbf{diag}([\lambda_1, \dots, \lambda_d]^\top)$ are the eigenvalues. Note that $\boldsymbol{\phi}_i \in \mathbb{R}^d$ and $\lambda_i \in \mathbb{R}$.

The generalized eigenvalue problem of Eq. (2.56) or (2.57) is denoted by $(\boldsymbol{A}, \boldsymbol{B})$. The $(\boldsymbol{A}, \boldsymbol{B})$ is called a "pair" or "pencil" [8], and the order in the pair matters, according to Eq. (2.57). The $\boldsymbol{\Phi}$ and $\boldsymbol{\Lambda}$ are called the generalized eigenvectors and eigenvalues of $(\boldsymbol{A}, \boldsymbol{B})$. The $(\boldsymbol{\Phi}, \boldsymbol{\Lambda})$ or $(\boldsymbol{\phi}_i, \lambda_i)$ is called the "eigenpair" of the pair $(\boldsymbol{A}, \boldsymbol{B})$ in the literature [8]. Comparing Eqs. (2.53) and (2.56) or Eqs. (2.54) and (2.57) demonstrates that the eigenvalue problem is a special case of the generalized eigenvalue problem where $\boldsymbol{B} = \boldsymbol{I}$.

2.5.2 Generalized Eigenvalue Optimization

This section introduces the optimization problems that result in a (generalized) eigenvalue problem. If $\boldsymbol{B} = \boldsymbol{I}$, these optimization problems reduce to eigenvalue problems.

Optimization Form 1 Consider the following optimization problem with the variable $\boldsymbol{\phi} \in \mathbb{R}^d$:

$$\begin{aligned} \underset{\boldsymbol{\phi}}{\text{maximize}} \quad & \boldsymbol{\phi}^\top \boldsymbol{A}\, \boldsymbol{\phi}, \\ \text{subject to} \quad & \boldsymbol{\phi}^\top \boldsymbol{B}\, \boldsymbol{\phi} = 1, \end{aligned} \tag{2.58}$$

where $\boldsymbol{A} \in \mathbb{R}^{d\times d}$ and $\boldsymbol{B} \in \mathbb{R}^{d\times d}$. The Lagrangian [1] for Eq. (2.58) is:

$$\mathcal{L} = \boldsymbol{\phi}^\top \boldsymbol{A}\, \boldsymbol{\phi} - \lambda\, (\boldsymbol{\phi}^\top \boldsymbol{B}\, \boldsymbol{\phi} - 1),$$

where $\lambda \in \mathbb{R}$ is the Lagrange multiplier. Equating the derivative of Lagrangian to zero gives us:

$$\mathbb{R}^d \ni \frac{\partial \mathcal{L}}{\partial \boldsymbol{\phi}} = 2\boldsymbol{A}\boldsymbol{\phi} - 2\lambda \boldsymbol{B}\boldsymbol{\phi} \overset{\text{set}}{=} 0 \implies \boldsymbol{A}\boldsymbol{\phi} = \lambda \boldsymbol{B}\boldsymbol{\phi},$$

which is a generalized eigenvalue problem $(\boldsymbol{A}, \boldsymbol{B})$ according to Eq. (2.56), where $\boldsymbol{\phi}$ is the eigenvector and λ is the eigenvalue. As Eq. (2.58) is a *maximization* problem, the eigenvector is the one having the largest eigenvalue. If Eq. (2.58) is a *minimization* problem, the eigenvector is the one having the smallest eigenvalue.[3]

[3] Refer to Chap. 4 for more information on maximization and minimization problems.

Optimization Form 2 Consider the following optimization problem with the variable $\boldsymbol{\Phi} \in \mathbb{R}^{d \times d}$:

$$\begin{aligned} \underset{\boldsymbol{\Phi}}{\text{maximize}} \quad & \textbf{tr}(\boldsymbol{\Phi}^\top \boldsymbol{A}\, \boldsymbol{\Phi}), \\ \text{subject to} \quad & \boldsymbol{\Phi}^\top \boldsymbol{B}\, \boldsymbol{\Phi} = \boldsymbol{I}, \end{aligned} \tag{2.59}$$

where $\boldsymbol{A} \in \mathbb{R}^{d \times d}$ and $\boldsymbol{B} \in \mathbb{R}^{d \times d}$ and $\textbf{tr}(.)$ denotes the trace of matrix. Note that according to the properties of the trace, the objective function can be any of these: $\textbf{tr}(\boldsymbol{\Phi}^\top \boldsymbol{A}\, \boldsymbol{\Phi}) = \textbf{tr}(\boldsymbol{\Phi}\boldsymbol{\Phi}^\top \boldsymbol{A}) = \textbf{tr}(\boldsymbol{A}\boldsymbol{\Phi}\boldsymbol{\Phi}^\top)$. The Lagrangian [1] for Eq. (2.59) is:

$$\mathcal{L} = \textbf{tr}(\boldsymbol{\Phi}^\top \boldsymbol{A}\, \boldsymbol{\Phi}) - \textbf{tr}\big(\boldsymbol{\Lambda}^\top (\boldsymbol{\Phi}^\top \boldsymbol{B}\, \boldsymbol{\Phi} - \boldsymbol{I})\big),$$

where $\boldsymbol{\Lambda} \in \mathbb{R}^{d \times d}$ is a diagonal matrix (see Appendix 2.1) whose entries are the Lagrange multipliers. Equating the derivative of $\mathcal{L}$ to zero gives us:

$$\mathbb{R}^{d \times d} \ni \frac{\partial \mathcal{L}}{\partial \boldsymbol{\Phi}} = 2\,\boldsymbol{A}\boldsymbol{\Phi} - 2\,\boldsymbol{B}\boldsymbol{\Phi}\boldsymbol{\Lambda} \overset{\text{set}}{=} \boldsymbol{0} \implies \boldsymbol{A}\boldsymbol{\Phi} = \boldsymbol{B}\boldsymbol{\Phi}\boldsymbol{\Lambda},$$

which is an eigenvalue problem $(\boldsymbol{A}, \boldsymbol{B})$ according to Eq. (2.57). The columns of $\boldsymbol{\Phi}$ are the eigenvectors of $\boldsymbol{A}$ and the diagonal elements of $\boldsymbol{\Lambda}$ are the eigenvalues. As Eq. (2.59) is a *maximization* problem, the eigenvalues and eigenvectors in $\boldsymbol{\Lambda}$ and $\boldsymbol{\Phi}$ are sorted from the largest to smallest eigenvalues. If Eq. (2.59) is a *minimization* problem, the eigenvalues and eigenvectors in $\boldsymbol{\Lambda}$ and $\boldsymbol{\Phi}$ are sorted from the smallest to largest eigenvalues (see Appendix 2.2).

Optimization Form 3 Consider the following optimization problem with the variable $\boldsymbol{\phi} \in \mathbb{R}^{d}$:

$$\begin{aligned} \underset{\boldsymbol{\phi}}{\text{minimize}} \quad & ||\boldsymbol{X} - \boldsymbol{\phi}\boldsymbol{\phi}^\top \boldsymbol{X}||_F^2, \\ \text{subject to} \quad & \boldsymbol{\phi}^\top \boldsymbol{\phi} = 1, \end{aligned} \tag{2.60}$$

where $\boldsymbol{X} \in \mathbb{R}^{d \times n}$ and $\|.\|_F$ denotes the Frobenius norm. The objective function in Eq. (2.60) is simplified as:

$$\begin{aligned} ||\boldsymbol{X} - \boldsymbol{\phi}\boldsymbol{\phi}^\top \boldsymbol{X}||_F^2 &= \textbf{tr}\big((\boldsymbol{X} - \boldsymbol{\phi}\boldsymbol{\phi}^\top \boldsymbol{X})^\top (\boldsymbol{X} - \boldsymbol{\phi}\boldsymbol{\phi}^\top \boldsymbol{X})\big) \\ &= \textbf{tr}\big((\boldsymbol{X}^\top - \boldsymbol{X}^\top \boldsymbol{\phi}\boldsymbol{\phi}^\top)(\boldsymbol{X} - \boldsymbol{\phi}\boldsymbol{\phi}^\top \boldsymbol{X})\big) \\ &= \textbf{tr}\big(\boldsymbol{X}^\top \boldsymbol{X} - 2\boldsymbol{X}^\top \boldsymbol{\phi}\boldsymbol{\phi}^\top \boldsymbol{X} + \boldsymbol{X}^\top \boldsymbol{\phi}\, \underbrace{\boldsymbol{\phi}^\top \boldsymbol{\phi}}_{1} \boldsymbol{\phi}^\top \boldsymbol{X}\big) \\ &= \textbf{tr}(\boldsymbol{X}^\top \boldsymbol{X} - \boldsymbol{X}^\top \boldsymbol{\phi}\boldsymbol{\phi}^\top \boldsymbol{X}) \\ &= \textbf{tr}(\boldsymbol{X}^\top \boldsymbol{X}) - \textbf{tr}(\boldsymbol{X}^\top \boldsymbol{\phi}\boldsymbol{\phi}^\top \boldsymbol{X}) \end{aligned}$$

$$= \mathbf{tr}(X^\top X) - \mathbf{tr}(XX^\top \phi\phi^\top)$$
$$= \mathbf{tr}(X^\top X - XX^\top \phi\phi^\top).$$

The Lagrangian [1] is:

$$\mathcal{L} = \mathbf{tr}(X^\top X) - \mathbf{tr}(XX^\top \phi\phi^\top) + \lambda(\phi^\top \phi - 1),$$

where λ is the Lagrange multiplier. Equating the derivative of $\mathcal{L}$ to zero gives:

$$\mathbb{R}^d \ni \frac{\partial \mathcal{L}}{\partial \phi} = -2\,XX^\top \phi + 2\,\lambda\,\phi \overset{\text{set}}{=} \mathbf{0} \implies XX^\top \phi = \lambda\,\phi \implies A\,\phi = \lambda\,\phi,$$

which is an eigenvalue problem for $A := XX^\top$ according to Eq. (2.53), where ϕ is the eigenvector and λ is the eigenvalue.

Optimization Form 4 Consider the following optimization problem with the variable $\Phi \in \mathbb{R}^{d\times d}$:

$$\begin{aligned} \underset{\Phi}{\text{minimize}} \quad & ||X - \Phi\,\Phi^\top X||_F^2, \\ \text{subject to} \quad & \Phi^\top \Phi = I, \end{aligned} \tag{2.61}$$

where $X \in \mathbb{R}^{d\times n}$. Similar to Eq. (2.60), the objective function in Eq. (2.61) is simplified as:

$$||X - \Phi\Phi^\top X||_F^2 = \mathbf{tr}(X^\top X - XX^\top \Phi\Phi^\top)$$

The Lagrangian [1] is:

$$\mathcal{L} = \mathbf{tr}(X^\top X) - \mathbf{tr}(XX^\top \Phi\Phi^\top) + \mathbf{tr}\big(\Lambda^\top(\Phi^\top \Phi - I)\big),$$

where $\Lambda \in \mathbb{R}^{d\times d}$ is a diagonal matrix (see Appendix 2.1) including Lagrange multipliers. Equating the derivative of $\mathcal{L}$ to zero gives:

$$\mathbb{R}^{d\times d} \ni \frac{\partial \mathcal{L}}{\partial \Phi} = -2\,XX^\top \Phi + 2\,\Phi\Lambda \overset{\text{set}}{=} \mathbf{0} \implies XX^\top \Phi = \Phi\Lambda \implies A\Phi = \Phi\Lambda,$$

which is an eigenvalue problem for A according to Eq. (2.54). The columns of Φ are the eigenvectors of A and the diagonal elements of Λ are the eigenvalues.

Optimization Form 5 Consider the following optimization problem [8] with the variable $\phi \in \mathbb{R}^d$:

$$\underset{\boldsymbol{\phi}}{\text{maximize}} \quad \frac{\boldsymbol{\phi}^\top \boldsymbol{A}\,\boldsymbol{\phi}}{\boldsymbol{\phi}^\top \boldsymbol{B}\,\boldsymbol{\phi}}. \tag{2.62}$$

According to the generalized Rayleigh-Ritz quotient method [2], this optimization problem can be restated as (see Sect. 2.4):

$$\begin{aligned} &\underset{\boldsymbol{\phi}}{\text{maximize}} \quad \boldsymbol{\phi}^\top \boldsymbol{A}\,\boldsymbol{\phi}, \\ &\text{subject to} \quad \boldsymbol{\phi}^\top \boldsymbol{B}\,\boldsymbol{\phi} = 1, \end{aligned} \tag{2.63}$$

The Lagrangian [1] is:

$$\mathcal{L} = \boldsymbol{\phi}^\top \boldsymbol{A}\,\boldsymbol{\phi} - \lambda(\boldsymbol{\phi}^\top \boldsymbol{B}\,\boldsymbol{\phi} - 1),$$

where λ is the Lagrange multiplier. Equating the derivative of $\mathcal{L}$ to zero results in:

$$\frac{\partial \mathcal{L}}{\partial \boldsymbol{w}} = 2\,\boldsymbol{A}\,\boldsymbol{\phi} - 2\,\lambda\,\boldsymbol{B}\,\boldsymbol{\phi} \overset{\text{set}}{=} \boldsymbol{0} \implies 2\,\boldsymbol{A}\,\boldsymbol{\phi} = 2\,\lambda\,\boldsymbol{B}\,\boldsymbol{\phi} \implies \boldsymbol{A}\,\boldsymbol{\phi} = \lambda\,\boldsymbol{B}\,\boldsymbol{\phi},$$

which is a generalized eigenvalue problem $(\boldsymbol{A}, \boldsymbol{B})$ according to Eq. (2.56), where $\boldsymbol{\phi}$ is the eigenvector and λ is the eigenvalue. As Eq. (2.62) is a *maximization* problem, the eigenvector has the largest eigenvalue. If Eq. (2.62) is a *minimization* problem, the eigenvector has the smallest eigenvalue.

2.5.3 Solution to the Eigenvalue problem

This section introduces the solution to the eigenvalue problem. Consider Eq. (2.53):

$$\boldsymbol{A}\boldsymbol{\phi}_i = \lambda_i \boldsymbol{\phi}_i \implies (\boldsymbol{A} - \lambda_i \boldsymbol{I})\,\boldsymbol{\phi}_i = \boldsymbol{0}, \tag{2.64}$$

which is a linear system of equations. According to the Cramer's rule in linear algebra, a linear system of equations has nontrivial solutions if and only if the determinant vanishes. Therefore:

$$\mathbf{det}(\boldsymbol{A} - \lambda_i \boldsymbol{I}) = 0, \tag{2.65}$$

where **det**(.) denotes the determinant of the matrix. Equation (2.65) results in a d-degree polynomial equation that has d roots (answers). Note that if $\boldsymbol{A}$ is not full rank (if it is a singular matrix), some of the roots will be zero. Moreover, if $\boldsymbol{A}$ is positive semidefinite, i.e., $\boldsymbol{A} \succeq 0$, all the roots are nonnegative.

The roots (answers) from Eq. (2.65) are the eigenvalues of $\boldsymbol{A}$. After finding the roots, Eq. (2.64) is used to find the corresponding eigenvector $\boldsymbol{\phi}_i \in \mathbb{R}^d$ for every eigenvalue. Note that putting the root in Eq. (2.64) results in an eigenvector that can be normalized because its direction, and not its magnitude, carries information. The information of magnitude exists in its corresponding eigenvalue.

2.5.4 *Solution to Generalized Eigenvalue Problem*

This section introduces the solution to the generalized eigenvalue problem. Recall Eq. (2.62). Let ρ be a fraction of the Rayleigh-Ritz quotient (see Sect. 2.4):

$$\rho(\boldsymbol{u}; \boldsymbol{A}, \boldsymbol{B}) := \frac{\boldsymbol{u}^\top \boldsymbol{A}\, \boldsymbol{u}}{\boldsymbol{u}^\top \boldsymbol{B}\, \boldsymbol{u}}, \quad \forall \boldsymbol{u} \neq \boldsymbol{0}. \tag{2.66}$$

The ρ is stationary at $\boldsymbol{\phi} \neq \boldsymbol{0}$ if and only if:

$$(\boldsymbol{A} - \lambda\, \boldsymbol{B})\, \boldsymbol{\phi} = \boldsymbol{0}, \tag{2.67}$$

for some scalar λ [8]. Equation (2.67) is a linear system of equations, which can also be obtained from Eq. (2.56):

$$\boldsymbol{A}\boldsymbol{\phi}_i = \lambda_i \boldsymbol{B}\boldsymbol{\phi}_i \implies (\boldsymbol{A} - \lambda_i \boldsymbol{B})\, \boldsymbol{\phi}_i = \boldsymbol{0}. \tag{2.68}$$

As mentioned earlier, the eigenvalue problem is a special case of the generalized eigenvalue problem (where $\boldsymbol{B} = \boldsymbol{I}$), as seen when comparing Eqs. (2.64) and (2.68). According to Cramer's rule, a linear system of equations has nontrivial solutions if and only if the determinant vanishes. Therefore:

$$\mathbf{det}(\boldsymbol{A} - \lambda_i\, \boldsymbol{B}) = 0. \tag{2.69}$$

Similar to the explanations for Eq. (2.65), the roots of Eq. (2.69) can be solved. However, note that Eq. (2.69) is obtained from Eq. (2.56) or (2.62), where only one eigenvector $\boldsymbol{\phi}$ is considered. To solve Eq. (2.57) in a general case, there exist two solutions for the generalized eigenvalue problem—a simplified solution and a rigorous method.

2.5.4.1 The Simplified Solution

Consider Eq. (2.57). If $\boldsymbol{B}$ is not singular (is invertible), it is possible to left-multiply the expressions by $\boldsymbol{B}^{-1}$:

$$\boldsymbol{B}^{-1}\boldsymbol{A}\boldsymbol{\Phi} = \boldsymbol{\Phi}\boldsymbol{\Lambda} \overset{(a)}{\implies} \boldsymbol{C}\boldsymbol{\Phi} = \boldsymbol{\Phi}\boldsymbol{\Lambda}, \tag{2.70}$$

where (a) is because $\boldsymbol{C} = \boldsymbol{B}^{-1}\boldsymbol{A}$. Equation (2.70) is the eigenvalue problem for $\boldsymbol{C}$ according to Eq. (2.54) and can be solved using the approach of Eq. (2.65). Note that even if $\boldsymbol{B}$ is singular, it is possible to use a numeric trick and slightly strengthen its main diagonal to make it have a full rank:

$$(\boldsymbol{B} + \varepsilon \boldsymbol{I})^{-1}\boldsymbol{A}\boldsymbol{\Phi} = \boldsymbol{\Phi}\boldsymbol{\Lambda} \implies \boldsymbol{C}\boldsymbol{\Phi} = \boldsymbol{\Phi}\boldsymbol{\Lambda}, \tag{2.71}$$

where ε is a very small positive number, e.g., $\varepsilon = 10^{-5}$, which is large enough to make $\boldsymbol{B}$ full rank. Note that although this numeric trick helps in numerical stability, it reduces the accuracy of the solution.

2.5.4.2 The Rigorous Solution

Although the simplified solution is faster to process, its accuracy may not be sufficient in some cases. There exists a rigorous method to solve the generalized eigenvalue problem in a more accurate manner [4]. Consider the eigenvalue problem for $\boldsymbol{B}$:

$$\boldsymbol{B}\boldsymbol{\Phi}_B = \boldsymbol{\Phi}_B\boldsymbol{\Lambda}_B, \tag{2.72}$$

where $\boldsymbol{\Phi}_B$ and $\boldsymbol{\Lambda}_B$ are the eigenvectors and eigenvalues of $\boldsymbol{B}$, respectively. Then:

$$\boldsymbol{B}\boldsymbol{\Phi}_B = \boldsymbol{\Phi}_B\boldsymbol{\Lambda}_B \implies \boldsymbol{\Phi}_B^{-1}\boldsymbol{B}\boldsymbol{\Phi}_B = \underbrace{\boldsymbol{\Phi}_B^{-1}\boldsymbol{\Phi}_B}_{\boldsymbol{I}}\boldsymbol{\Lambda}_B = \boldsymbol{\Lambda}_B \overset{(a)}{\implies} \boldsymbol{\Phi}_B^\top\boldsymbol{B}\boldsymbol{\Phi}_B = \boldsymbol{\Lambda}_B, \tag{2.73}$$

where (a) is because $\boldsymbol{\Phi}_B$ is an orthogonal matrix (its columns are orthonormal) and thus $\boldsymbol{\Phi}_B^{-1} = \boldsymbol{\Phi}_B^\top$. Then $\boldsymbol{\Lambda}_B^{-1/2}$ is multiplied by Eq. (2.73) from the left and right hand sides:

$$\boldsymbol{\Lambda}_B^{-1/2}\boldsymbol{\Phi}_B^\top\boldsymbol{B}\boldsymbol{\Phi}_B\boldsymbol{\Lambda}_B^{-1/2} = \boldsymbol{\Lambda}_B^{-1/2}\boldsymbol{\Lambda}_B\boldsymbol{\Lambda}_B^{-1/2} = \boldsymbol{I} \implies \breve{\boldsymbol{\Phi}}_B^\top\boldsymbol{B}\breve{\boldsymbol{\Phi}}_B = \boldsymbol{I},$$

where:

$$\breve{\boldsymbol{\Phi}}_B := \boldsymbol{\Phi}_B\boldsymbol{\Lambda}_B^{-1/2}. \tag{2.74}$$

$\breve{\boldsymbol{A}}$ is defined as:

$$\breve{\boldsymbol{A}} := \breve{\boldsymbol{\Phi}}_B^\top\boldsymbol{A}\breve{\boldsymbol{\Phi}}_B. \tag{2.75}$$

The $\breve{\boldsymbol{A}}$ is symmetric because:

$$\breve{\boldsymbol{A}}^\top = (\breve{\boldsymbol{\Phi}}_B^\top \boldsymbol{A} \breve{\boldsymbol{\Phi}}_B)^\top \overset{(a)}{=} \breve{\boldsymbol{\Phi}}_B^\top \boldsymbol{A} \breve{\boldsymbol{\Phi}}_B = \breve{\boldsymbol{A}},$$

where (a) notices that $\boldsymbol{A}$ is symmetric. The eigenvalue problem for $\breve{\boldsymbol{A}}$ is:

$$\breve{\boldsymbol{A}} \boldsymbol{\Phi}_A = \boldsymbol{\Phi}_A \boldsymbol{\Lambda}_A, \tag{2.76}$$

where $\boldsymbol{\Phi}_A$ and $\boldsymbol{\Lambda}_A$ are the eigenvector and eigenvalue matrices of $\breve{\boldsymbol{A}}$. Left-multiplying $\boldsymbol{\Phi}_A^{-1}$ to Eq. (2.76) results in:

$$\boldsymbol{\Phi}_A^{-1} \breve{\boldsymbol{A}} \boldsymbol{\Phi}_A = \underbrace{\boldsymbol{\Phi}_A^{-1} \boldsymbol{\Phi}_A}_{\boldsymbol{I}} \boldsymbol{\Lambda}_A \overset{(a)}{\Longrightarrow} \boldsymbol{\Phi}_A^\top \breve{\boldsymbol{A}} \boldsymbol{\Phi}_A = \boldsymbol{\Lambda}_A, \tag{2.77}$$

where (a) is because $\boldsymbol{\Phi}_A$ is an orthogonal matrix (its columns are orthonormal), so that $\boldsymbol{\Phi}_A^{-1} = \boldsymbol{\Phi}_A^\top$. Note that $\boldsymbol{\Phi}_A$ is an orthogonal matrix because $\breve{\boldsymbol{A}}$ is symmetric (if the matrix is symmetric, its eigenvectors are orthogonal/orthonormal). Equation (2.77) is diagonalizing matrix $\breve{\boldsymbol{A}}$, meaning that it makes this matrix diagonal. Inserting Eq. (2.75) into Eq. (2.77) results in:

$$\boldsymbol{\Phi}_A^\top \breve{\boldsymbol{\Phi}}_B^\top \boldsymbol{A} \breve{\boldsymbol{\Phi}}_B \boldsymbol{\Phi}_A = \boldsymbol{\Lambda}_A \overset{(2.74)}{\Longrightarrow} \boldsymbol{\Phi}_A^\top \boldsymbol{\Lambda}_B^{-1/2} \boldsymbol{\Phi}_B^\top \boldsymbol{A} \boldsymbol{\Phi}_B \boldsymbol{\Lambda}_B^{-1/2} \boldsymbol{\Phi}_A = \boldsymbol{\Lambda}_A \Longrightarrow \boldsymbol{\Phi}^\top \boldsymbol{A} \boldsymbol{\Phi} = \boldsymbol{\Lambda}_A, \tag{2.78}$$

where:

$$\boldsymbol{\Phi} := \breve{\boldsymbol{\Phi}}_B \boldsymbol{\Phi}_A = \boldsymbol{\Phi}_B \boldsymbol{\Lambda}_B^{-1/2} \boldsymbol{\Phi}_A. \tag{2.79}$$

Since $\boldsymbol{I}$ is a diagonal matrix, $\boldsymbol{\Phi}$ also diagonalizes $\boldsymbol{B}$:

$$\begin{aligned}\boldsymbol{\Phi}^\top \boldsymbol{B} \boldsymbol{\Phi} &\overset{(2.79)}{=} (\boldsymbol{\Phi}_B \boldsymbol{\Lambda}_B^{-1/2} \boldsymbol{\Phi}_A)^\top \boldsymbol{B} (\boldsymbol{\Phi}_B \boldsymbol{\Lambda}_B^{-1/2} \boldsymbol{\Phi}_A) = \boldsymbol{\Phi}_A^\top \boldsymbol{\Lambda}_B^{-1/2} (\boldsymbol{\Phi}_B^\top \boldsymbol{B} \boldsymbol{\Phi}_B) \boldsymbol{\Lambda}_B^{-1/2} \boldsymbol{\Phi}_A \\ &\overset{(2.73)}{=} \boldsymbol{\Phi}_A^\top \underbrace{\boldsymbol{\Lambda}_B^{-1/2} \boldsymbol{\Lambda}_B \boldsymbol{\Lambda}_B^{-1/2}}_{\boldsymbol{I}} \boldsymbol{\Phi}_A = \boldsymbol{\Phi}_A^\top \boldsymbol{\Phi}_A \overset{(a)}{=} \boldsymbol{\Phi}_A^{-1} \boldsymbol{\Phi}_A = \boldsymbol{I},\end{aligned} \tag{2.80}$$

where (a) is because $\boldsymbol{\Phi}_A$ is an orthogonal matrix. From Eq. (2.80), the following is obtained:

$$\boldsymbol{\Phi}^\top \boldsymbol{B} \boldsymbol{\Phi} = \boldsymbol{I} \Longrightarrow \boldsymbol{\Phi}^\top \boldsymbol{B} \boldsymbol{\Phi} \boldsymbol{\Lambda}_A = \boldsymbol{\Lambda}_A \overset{(2.78)}{\Longrightarrow} \boldsymbol{\Phi}^\top \boldsymbol{B} \boldsymbol{\Phi} \boldsymbol{\Lambda}_A = \boldsymbol{\Phi}^\top \boldsymbol{A} \boldsymbol{\Phi} \overset{(a)}{\Longrightarrow} \boldsymbol{B} \boldsymbol{\Phi} \boldsymbol{\Lambda}_A = \boldsymbol{A} \boldsymbol{\Phi}, \tag{2.81}$$

where (a) is because $\boldsymbol{\Phi} \neq \boldsymbol{0}$.

Comparing equations (2.57) and (2.81) demonstrates that:

$$\boldsymbol{\Lambda}_A = \boldsymbol{\Lambda}. \tag{2.82}$$

$\boldsymbol{\Phi}_B, \boldsymbol{\Lambda}_B \leftarrow \boldsymbol{B}\boldsymbol{\Phi}_B = \boldsymbol{\Phi}_B\boldsymbol{\Lambda}_B$
$\breve{\boldsymbol{\Phi}}_B \leftarrow \breve{\boldsymbol{\Phi}}_B = \boldsymbol{\Phi}_B\boldsymbol{\Lambda}_B^{-1/2} \approx \boldsymbol{\Phi}_B(\boldsymbol{\Lambda}_B^{1/2} + \varepsilon\boldsymbol{I})^{-1}$
$\breve{\boldsymbol{A}} \leftarrow \breve{\boldsymbol{A}} = \breve{\boldsymbol{\Phi}}_B^\top \boldsymbol{A}\breve{\boldsymbol{\Phi}}_B$
$\boldsymbol{\Phi}_A, \boldsymbol{\Lambda}_A \leftarrow \breve{\boldsymbol{A}}\boldsymbol{\Phi}_A = \boldsymbol{\Phi}_A\boldsymbol{\Lambda}_A$
$\boldsymbol{\Lambda} \leftarrow \boldsymbol{\Lambda} = \boldsymbol{\Lambda}_A$
$\boldsymbol{\Phi} \leftarrow \boldsymbol{\Phi} = \breve{\boldsymbol{\Phi}}_B\boldsymbol{\Phi}_A$
return $\boldsymbol{\Phi}$ and $\boldsymbol{\Lambda}$

Algorithm 2.1: Solution to the generalized eigenvalue problem $\boldsymbol{A}\boldsymbol{\Phi} = \boldsymbol{B}\boldsymbol{\Phi}\boldsymbol{\Lambda}$

To summarize, the following steps are required to find $\boldsymbol{\Phi}$ and $\boldsymbol{\Lambda}$ in Eq. (2.57) (note that $\boldsymbol{A}$ and $\boldsymbol{B}$ are given):

1. Find $\boldsymbol{\Phi}_B$ and $\boldsymbol{\Lambda}_B$ using Eq. (2.72).
2. Find $\breve{\boldsymbol{\Phi}}_B$ using Eq. (2.74). In the case that $\boldsymbol{\Lambda}_B^{1/2}$ is singular in Eq. (2.74), use the numeric hack $\breve{\boldsymbol{\Phi}}_B \approx \boldsymbol{\Phi}_B(\boldsymbol{\Lambda}_B^{1/2} + \varepsilon\boldsymbol{I})^{-1}$ where ε is a very small positive number, e.g., $\varepsilon = 10^{-5}$, large enough to make $\boldsymbol{\Lambda}_B^{1/2}$ full rank.
3. Find $\breve{\boldsymbol{A}}$ using Eq. (2.75).
4. Find $\boldsymbol{\Phi}_A$ and $\boldsymbol{\Lambda}_A$ using Eq. (2.76). Note that $\boldsymbol{\Lambda}$ can be found using Eq. (2.82), $\boldsymbol{\Lambda}$.
5. Find $\boldsymbol{\Phi}$ using Eq. (2.79).

The above instructions are given as an algorithm in Algorithm 2.1.

2.6 Singular Value Decomposition

Singular Value Decomposition (SVD) [10] is one of the most well-known and effective matrix decomposition methods. There are different methods for obtaining this decomposition, one of which is Jordan's algorithm [10]. This section does not explain how to obtain SVD, but instead introduces different forms of SVD and their properties. SVD has two different forms, i.e., complete and incomplete, which are explained in the following.

Consider a matrix $\boldsymbol{A} \in \mathbb{R}^{\alpha\times\beta}$. The *complete SVD* decomposes the matrix as:

$$\mathbb{R}^{\alpha\times\beta} \ni \boldsymbol{A} = \boldsymbol{U}\boldsymbol{\Sigma}\boldsymbol{V}^\top, \tag{2.83}$$
$$\boldsymbol{U} \in \mathbb{R}^{\alpha\times\alpha}, \quad \boldsymbol{V} \in \mathbb{R}^{\beta\times\beta}, \quad \boldsymbol{\Sigma} \in \mathbb{R}^{\alpha\times\beta},$$

where the columns of $\boldsymbol{U}$ and the columns of $\boldsymbol{V}$ are called *left singular vectors* and *right singular vectors*, respectively. In complete SVD, $\boldsymbol{\Sigma}$ is a *rectangular* diagonal matrix whose main diagonal includes the *singular values*. In the cases with $\alpha > \beta$ and $\alpha < \beta$, this matrix is in the following forms:

$$\boldsymbol{\Sigma} = \begin{bmatrix} \sigma_1 & 0 & 0 \\ \vdots & \ddots & \vdots \\ 0 & 0 & \sigma_\beta \\ 0 & 0 & 0 \\ \vdots & \vdots & \vdots \\ 0 & 0 & 0 \end{bmatrix} \text{ and } \begin{bmatrix} \sigma_1 & 0 & 0 & 0 \cdots 0 \\ \vdots & \ddots & \vdots & 0 \cdots 0 \\ 0 & 0 & \sigma_\alpha & 0 \cdots 0 \end{bmatrix},$$

respectively. In other words, the number of singular values is $\min(\alpha, \beta)$.

The *incomplete SVD* decomposes the matrix as:

$$\mathbb{R}^{\alpha\times\beta} \ni \boldsymbol{A} = \boldsymbol{U}\boldsymbol{\Sigma}\boldsymbol{V}^\top, \tag{2.84}$$
$$\boldsymbol{U} \in \mathbb{R}^{\alpha\times k}, \quad \boldsymbol{V} \in \mathbb{R}^{\beta\times k}, \quad \boldsymbol{\Sigma} \in \mathbb{R}^{k\times k},$$

where [5]:

$$k := \min(\alpha, \beta), \tag{2.85}$$

and the columns of $\boldsymbol{U}$ and the columns of $\boldsymbol{V}$ are called *left singular vectors* and *right singular vectors*, respectively. In incomplete SVD, $\boldsymbol{\Sigma}$ is a *square* diagonal matrix whose main diagonal includes the *singular values*. The matrix $\boldsymbol{\Sigma}$ is in the form:

$$\boldsymbol{\Sigma} = \begin{bmatrix} \sigma_1 & 0 & 0 \\ \vdots & \ddots & \vdots \\ 0 & 0 & \sigma_k \end{bmatrix}.$$

Note that in both complete and incomplete SVD, the left singular vectors are orthonormal and the right singular vectors are also orthonormal; therefore, $\boldsymbol{U}$ and $\boldsymbol{V}$ are both orthogonal matrices so:

$$\boldsymbol{U}^\top\boldsymbol{U} = \boldsymbol{I}, \tag{2.86}$$
$$\boldsymbol{V}^\top\boldsymbol{V} = \boldsymbol{I}. \tag{2.87}$$

If these orthogonal matrices are not truncated and thus are square matrices, e.g. for complete SVD, there are also:

$$\boldsymbol{U}\boldsymbol{U}^\top = \boldsymbol{I}, \tag{2.88}$$
$$\boldsymbol{V}\boldsymbol{V}^\top = \boldsymbol{I}. \tag{2.89}$$

Proposition 2.1 *In both complete and incomplete SVD of matrix* **A**, *the left and right singular vectors are the eigenvectors of* $\boldsymbol{A}\boldsymbol{A}^\top$ *and* $\boldsymbol{A}^\top\boldsymbol{A}$, *respectively, and the singular values are the square root of eigenvalues of either* $\boldsymbol{A}\boldsymbol{A}^\top$ *or* $\boldsymbol{A}^\top\boldsymbol{A}$.

Proof There is:

$$\boldsymbol{A}\boldsymbol{A}^\top = (\boldsymbol{U}\boldsymbol{\Sigma}\boldsymbol{V}^\top)(\boldsymbol{U}\boldsymbol{\Sigma}\boldsymbol{V}^\top)^\top = \boldsymbol{U}\boldsymbol{\Sigma}\underbrace{\boldsymbol{V}^\top\boldsymbol{V}}_{\boldsymbol{I}}\boldsymbol{\Sigma}\boldsymbol{U}^\top = \boldsymbol{U}\boldsymbol{\Sigma}\boldsymbol{\Sigma}\boldsymbol{U}^\top = \boldsymbol{U}\boldsymbol{\Sigma}^2\boldsymbol{U}^\top,$$

which is the eigen-decomposition [4] of $\boldsymbol{A}\boldsymbol{A}^\top$ where the columns of $\boldsymbol{U}$ are the eigenvectors and the diagonal of $\boldsymbol{\Sigma}^2$ are the eigenvalues so the diagonal of $\boldsymbol{\Sigma}$ are the square root of eigenvalues. There is also:

$$\boldsymbol{A}^\top\boldsymbol{A} = (\boldsymbol{U}\boldsymbol{\Sigma}\boldsymbol{V}^\top)^\top(\boldsymbol{U}\boldsymbol{\Sigma}\boldsymbol{V}^\top) = \boldsymbol{V}\boldsymbol{\Sigma}\underbrace{\boldsymbol{U}^\top\boldsymbol{U}}_{\boldsymbol{I}}\boldsymbol{\Sigma}\boldsymbol{V}^\top = \boldsymbol{V}\boldsymbol{\Sigma}\boldsymbol{\Sigma}\boldsymbol{V}^\top = \boldsymbol{V}\boldsymbol{\Sigma}^2\boldsymbol{V}^\top,$$

which is the eigenvalue decomposition of $\boldsymbol{A}^\top\boldsymbol{A}$ where the columns of $\boldsymbol{V}$ are the eigenvectors and the diagonal of $\boldsymbol{\Sigma}^2$ are the eigenvalues, so the diagonal of $\boldsymbol{\Sigma}$ are the square root of eigenvalues. □

2.7 Chapter Summary

Linear algebra is one of the most important tools required in machine learning and dimensionality reduction. This chapter reviewed the linear algebra concepts that are necessary in machine learning and dimensionality reduction. It covered the centering matrix, linear projection, the Rayleigh-Ritz quotient, eigenvalue and generalized eigenvalue problems, and the singular value decomposition. These concepts and tools will form the foundation for future chapters in this book. For example, linear projection is used in all linear methods for dimensionality reduction and most of the spectral methods reduce to a generalized eigenvalue problem.

Appendix 2.1: Proof of Why the Lagrange Multiplier is Diagonal in the Eigenvalue Problem

In both eigenvalue and generalized eigenvalue problems, the Lagrange multiplier matrix is diagonal. For example, consider the Lagrangian of Eq. (2.59) with $\boldsymbol{B} = \boldsymbol{I}$, i.e.,

$$\mathcal{L} = \textbf{tr}(\boldsymbol{\Phi}^\top\boldsymbol{A}\,\boldsymbol{\Phi}) - \textbf{tr}\big(\boldsymbol{\Lambda}^\top(\boldsymbol{\Phi}^\top\boldsymbol{\Phi} - \boldsymbol{I})\big), \tag{2.90}$$

in which $\boldsymbol{\Lambda}$ is a diagonal matrix. This proves why this Lagrange multiplier matrix is diagonal. A similar proof can be provided for the Lagrange multiplier in the generalized eigenvalue problem.

Discussion on the Number of Constraints

First note that the orthogonality constraint $\boldsymbol{\Phi}^\top \boldsymbol{\Phi} = \boldsymbol{I}$ embeds $d \times d$ constraints but, as some of them are symmetric, there are only $d(d-1)/2 + d$ constraints. In other words, d constraints are for the d columns of $\boldsymbol{\Phi}$ to have unit length and $d(d-1)/2$ constraints are for every two columns of $\boldsymbol{\Phi}$ to be orthogonal. Let $\{\lambda_1, \dots, \lambda_d\}$ be the Lagrange multipliers for columns of $\boldsymbol{\Phi}$ to have unit length and let $\{\lambda^1, \dots, \lambda^{d(d-1)/2}\}$ be the Lagrange multipliers for the columns of $\boldsymbol{\Phi}$ to be orthogonal. Then, the Lagrangian is:

$$
\begin{aligned}
\mathcal{L} = &\,\mathbf{tr}(\boldsymbol{\Phi}^\top \boldsymbol{A}\,\boldsymbol{\Phi}) - \lambda_1(\boldsymbol{\phi}_1^\top \boldsymbol{\phi}_1 - 1) - \dots - \lambda_d(\boldsymbol{\phi}_d^\top \boldsymbol{\phi}_d - 1) \\
&- \lambda^1(\boldsymbol{\phi}_1^\top \boldsymbol{\phi}_2 - 0) - \lambda^1(\boldsymbol{\phi}_2^\top \boldsymbol{\phi}_1 - 0) - \dots \\
&- \lambda^{d(d-1)/2}(\boldsymbol{\phi}_{d-1}^\top \boldsymbol{\phi}_d - 0) - \lambda^{d(d-1)/2}(\boldsymbol{\phi}_d^\top \boldsymbol{\phi}_{d-1} - 0),
\end{aligned}
$$

where each λ^j is repeated twice because of the symmetry of the inner product between the two columns. The dual variables can be gathered together using trace to obtain Eq. (2.90). Setting the derivative of this equation to zero results in $\boldsymbol{A\Phi} = \boldsymbol{\Phi\Lambda}$, which is in the form of an eigenvalue problem (see Eq. (2.54)). Therefore, one of the possible solutions is the eigenvalue problem of $\boldsymbol{\Phi}$, in which the eigenvalue matrix $\boldsymbol{\Lambda}$ is diagonal. Therefore, in one of the solutions, the Lagrange multiplier matrix must be diagonal. In fact, this problem can have many solutions. The following demonstrates that if $\boldsymbol{\Phi}$ is the optimal solution for Eq. (2.59) with $\boldsymbol{B} = \boldsymbol{I}$, any matrix $\boldsymbol{M} := \boldsymbol{\Phi V}$ is also a solution if $\boldsymbol{V}$ is an orthogonal matrix.

Proof of Being Diagonal

Consider the Lagrangian, i.e., Eq. (2.90). The following proves why the Lagrange multiplier in this equation is a diagonal matrix. Assume that $\boldsymbol{\Lambda}$ is not necessarily diagonal, but is known to be symmetric because of the discussions in Sect. 2.7. As the matrix $\boldsymbol{\Lambda}$ is symmetric, it can have its own eigenvalue decomposition as (see Eq. (2.55)):

$$
\boldsymbol{\Lambda} = \boldsymbol{V\Omega V}^\top, \tag{2.91}
$$

where $\boldsymbol{V}$ and $\boldsymbol{\Omega}$ are the eigenvectors and eigenvalues of $\boldsymbol{\Lambda}$. The eigenvectors are orthonormal so the matrix $\boldsymbol{V}$ is a nontruncated orthogonal matrix; therefore:

$$\boldsymbol{V}^\top \boldsymbol{V} = \boldsymbol{V}\boldsymbol{V}^\top = \boldsymbol{I}. \tag{2.92}$$

The Lagrangian, Eq. (2.90), can be restated as:

$$\begin{aligned}\mathcal{L} &= \textbf{tr}(\boldsymbol{\Phi}^\top \boldsymbol{A}\, \boldsymbol{\Phi}) - \textbf{tr}\big(\boldsymbol{\Lambda}^\top (\boldsymbol{\Phi}^\top \boldsymbol{\Phi} - \boldsymbol{I})\big) \overset{(2.91)}{=} \textbf{tr}(\boldsymbol{\Phi}^\top \boldsymbol{A}\, \boldsymbol{\Phi}) - \textbf{tr}\big(\boldsymbol{V}\boldsymbol{\Omega}\boldsymbol{V}^\top (\boldsymbol{\Phi}^\top \boldsymbol{\Phi} - \boldsymbol{I})\big) \\ &\overset{(a)}{=} \textbf{tr}(\boldsymbol{\Phi}^\top \boldsymbol{A}\, \boldsymbol{\Phi}) - \textbf{tr}\big(\boldsymbol{\Omega}\boldsymbol{V}^\top (\boldsymbol{\Phi}^\top \boldsymbol{\Phi} - \boldsymbol{I})\boldsymbol{V}\big) = \textbf{tr}(\boldsymbol{\Phi}^\top \boldsymbol{A}\, \boldsymbol{\Phi}) - \textbf{tr}\big(\boldsymbol{\Omega}(\boldsymbol{V}^\top \boldsymbol{\Phi}^\top \boldsymbol{\Phi}\boldsymbol{V} - \boldsymbol{V}^\top \boldsymbol{V})\big) \\ &\overset{(b)}{=} \textbf{tr}(\boldsymbol{\Phi}^\top \boldsymbol{A}\, \boldsymbol{\Phi}) - \textbf{tr}\big(\boldsymbol{\Omega}(\boldsymbol{M}^\top \boldsymbol{M} - \boldsymbol{I})\big) \overset{(c)}{=} \textbf{tr}(\boldsymbol{\Phi}^\top \boldsymbol{A}\, \boldsymbol{\Phi}) - \textbf{tr}\big(\boldsymbol{\Omega}^\top (\boldsymbol{M}^\top \boldsymbol{M} - \boldsymbol{I})\big),\end{aligned} \tag{2.93}$$

where (a) is because of the cyclic property of trace, (b) is because of Eq. (2.92) and our definition $\boldsymbol{M} := \boldsymbol{\Phi}\boldsymbol{V}$, and (c) is because the matrix $\boldsymbol{\Omega}$ is diagonal (see Eq. (2.91)) so it is equal to its transpose. This is stated as follows:

$$\boldsymbol{M} = \boldsymbol{\Phi}\boldsymbol{V} \implies \boldsymbol{M}\boldsymbol{V}^\top = \boldsymbol{\Phi}\boldsymbol{V}\boldsymbol{V}^\top \overset{(2.92)}{=} \boldsymbol{\Phi} \implies \boldsymbol{\Phi}^\top = \boldsymbol{V}\boldsymbol{M}^\top.$$

Therefore, the first term in Eq. (2.93) can be restated as:

$$\textbf{tr}(\boldsymbol{\Phi}^\top \boldsymbol{A}\boldsymbol{\Phi}) = \textbf{tr}(\boldsymbol{V}\boldsymbol{M}^\top \boldsymbol{A}\boldsymbol{M}\boldsymbol{V}^\top) \overset{(a)}{=} \textbf{tr}(\boldsymbol{V}^\top \boldsymbol{V}\boldsymbol{M}^\top \boldsymbol{A}\boldsymbol{M}) \overset{(2.92)}{=} \textbf{tr}(\boldsymbol{M}^\top \boldsymbol{A}\boldsymbol{M}).$$

Consequently, Eq. (2.93), which is the Lagrangian, can be stated as:

$$\mathcal{L} = \textbf{tr}(\boldsymbol{M}^\top \boldsymbol{A}\boldsymbol{M}) - \textbf{tr}\big(\boldsymbol{\Omega}^\top (\boldsymbol{M}^\top \boldsymbol{M} - \boldsymbol{I})\big), \tag{2.94}$$

Comparing Eqs. (2.90) and (2.94) demonstrates that a change of variable $\boldsymbol{M} = \boldsymbol{\Phi}\boldsymbol{V}$ may occur and state the Lagrange multiplier as a diagonal matrix (because $\boldsymbol{\Omega}$ is diagonal in Eq. (2.94)). As both $\boldsymbol{\Lambda}$ and $\boldsymbol{\Omega}$ are dummy variables and can have any name, it can initially be assumed that $\boldsymbol{\Lambda}$ is diagonal.

Appendix 2.2: Discussion of the Sorting of Eigenvectors and Eigenvalues

The second derivative demonstrates the curvature direction of a function. Consider Eq. (2.59) with $\boldsymbol{B} = \boldsymbol{I}$, whose Lagrangian is Eq. (2.90). The second derivative of the Lagrangian is:

$$\mathbb{R}^{d\times d} \ni \frac{\partial^2 \mathcal{L}}{\partial \boldsymbol{\Phi}^2} = 2\,\boldsymbol{A}^\top - 2\,\boldsymbol{\Lambda}^\top = 2(\boldsymbol{A} - \boldsymbol{\Lambda})^\top \overset{\text{set}}{=} \boldsymbol{0} \implies \boldsymbol{A} = \boldsymbol{\Lambda}.$$

Accordingly, for the second derivative, matrix $\boldsymbol{A}$ is equal to eigenvalues $\boldsymbol{\Lambda}$. Matrix $\boldsymbol{A}$ is in the objective function of the optimization problem. Therefore, if the optimization problem of Eq. (2.59) is maximization, the eigenvalues, and their corresponding eigenvectors, should be sorted from the largest to the smallest. Conversely, for minimization, the sort is from the smallest to the largest.

Likewise, consider Eq. (2.59), with a matrix $\boldsymbol{B}$, for the generalized eigenvalue problem. The second derivative of the Lagrangian is:

$$\mathbb{R}^{d\times d} \ni \frac{\partial^2 \mathcal{L}}{\partial \boldsymbol{\Phi}^2} = 2\boldsymbol{A}^\top - 2\boldsymbol{B}^\top \boldsymbol{\Lambda}^\top = 2(\boldsymbol{A} - \boldsymbol{\Lambda}\boldsymbol{B})^\top \overset{\text{set}}{=} \boldsymbol{0} \implies \boldsymbol{A} = \boldsymbol{\Lambda}\boldsymbol{B}.$$

Therefore, for the second derivative, matrix $\boldsymbol{A}$ is related to eigenvalues $\boldsymbol{\Lambda}$, and a similar analysis holds. A similar analysis can also be performed for other forms of optimization for eigenvalues and generalized eigenvalue problems.

References

1. Stephen Boyd and Lieven Vandenberghe. *Convex optimization*. Cambridge University Press, 2004.
2. Ernie Croot. *The Rayleigh Principle for Finding Eigenvalues*. Tech. rep. Online, Accessed: March 2019. Georgia Institute of Technology School of Mathematics, 2005.
3. Jerome Friedman, Trevor Hastie, and Robert Tibshirani. *The elements of statistical learning*. Vol. 2. Springer series in statistics New York, NY, USA, 2009.
4. Benyamin Ghojogh, Fakhri Karray, and Mark Crowley. "Eigenvalue and generalized eigenvalue problems: Tutorial". In: *arXiv preprint arXiv:1903.11240* (2019).
5. Gene H Golub and Christian Reinsch. "Singular value decomposition and least squares solutions". In: *Numerische mathematik* 14.5 (1970), pp. 403–420.
6. Gene H. Golub and Charles F. Van Loan. *Matrix computations*. Vol. 3. The Johns Hopkins University Press, 2012.
7. Ian Jolliffe. *Principal component analysis*. Springer, 2011.
8. Beresford N Parlett. "The Symmetric Eigenvalue Problem". In: *Classics in Applied Mathematics* 20 (1998).
9. Bernhard Schölkopf, Alexander Smola, and Klaus-Robert Müller. "Nonlinear component analysis as a kernel eigenvalue problem". In: *Neural computation* 10.5 (1998), pp. 1299–1319.
10. Gilbert W Stewart. "On the early history of the singular value decomposition". In: *SIAM review* 35.4 (1993), pp. 551–566.
11. James Hardy Wilkinson. *The algebraic eigenvalue problem* Vol. 662. Oxford Clarendon, 1965.

Chapter 3
Background on Kernels

3.1 Introduction

In functional analysis—a field of mathematics—there are various spaces of either data points or functions. For example, the Euclidean space is a subset of the Hilbert space, while the Hilbert space itself is a subset of the Banach space. The Hilbert space is a space of functions and its dimensionality is often considered to be high. The Reproducing Kernel Hilbert Space (RKHS), which was first proposed in [3], is a special case of the Hilbert space with several properties. To be more specific, it is a Hilbert space of functions with reproducing kernels [11]. After Aronszajn's initial work on RKHS [3], Aizerman et al. expanded this work to further develop RKHS concepts [1]. RKHS and eigenfunctions are used in theoretical physics, in such fields as quantum mechanics. It is also very useful for working on kernels in machine learning, because a kernel is a measure of similarity, meaning it can capture the similarity between data points. This chapter introduces the concepts related to RKHS and kernels and, later, demonstrates how to use kernels in machine learning and dimensionality reduction. More specifically, this chapter will review the required background on kernels, including the reproducing kernel Hilbert space, Mercer kernel, feature map, well-known kernels, kernel construction from distance matrix, kernel centering and normalization, eigenfunctions, kernelization techniques, difference measures using kernels, kernel factorization, and the Nyström method. For more information about kernels, refer to [13].

3.2 Reproducing Kernel Hilbert Space

The RKHS has three main properties, (1) reproducing; (2) containing Kernels; and (3) being a Hilbert Space. Section 3.2.1 introduces the second property, i.e., containing kernels. The third property is provided in Sect. 3.2.2 to explain

B. Ghojogh et al., *Elements of Dimensionality Reduction and Manifold Learning*,
https://doi.org/10.1007/978-3-031-10602-6_3

why RKHS is a Hilbert space. The reproducing property is then demonstrated in Sect. 3.2.3. Lastly, the effectiveness of RKHS is demonstrated by the Mercer's theorem in Sect. 3.2.4.

3.2.1 Mercer Kernel and Gram Matrix

Definition 3.1 (Mercer Kernel [23]) The function $k : \mathcal{X}^2 \rightarrow \mathbb{R}$ is a Mercer kernel function (also known as kernel function) where:

1. it is symmetric: $k(\boldsymbol{x}, \boldsymbol{y}) = k(\boldsymbol{y}, \boldsymbol{x})$, and
2. its corresponding kernel matrix $\boldsymbol{K}(i, j) = k(\boldsymbol{x}_i, \boldsymbol{x}_j), \forall i, j \in \{1, \dots, n\}$ is positive semidefinite: $\boldsymbol{K} \succeq 0$.

The corresponding kernel matrix of a Mercer kernel is a Mercer kernel matrix.

For the proof of these two properties of a Mercer kernel, see [13]. By convention, unless otherwise stated, the term kernel refers to a Mercer kernel.

Definition 3.2 (Gram Matrix or Kernel Matrix) The matrix $\boldsymbol{K} \in \mathbb{R}^{n \times n}$ is a Gram matrix, also known as a Gramian matrix or a kernel matrix, whose (i, j)-th element is:

$$\boldsymbol{K}(i, j) := k(\boldsymbol{x}_i, \boldsymbol{x}_j), \quad \forall i, j \in \{1, \dots, n\}. \tag{3.1}$$

Here, the square kernel matrix applied on a set of n data instances was defined, which means the kernel is an $n \times n$ matrix. Although, a kernel matrix may be found between two sets of data instances (see Sect. 3.7) and it can be computed using the inner product between pulled data to the feature space (this will be explained in Sect. 3.3).

3.2.2 RKHS as a Hilbert Space

Definition 3.3 (RKHS [3, 11]) A Reproducing Kernel Hilbert Space (RKHS) is a Hilbert space $\mathcal{H}$ of functions $f : \mathcal{X} \rightarrow \mathbb{R}$ with a reproducing kernel $k : \mathcal{X}^2 \rightarrow \mathbb{R}$ where $k(\boldsymbol{x}, .) \in \mathcal{H}$ and $f(\boldsymbol{x}) = \langle k(\boldsymbol{x}, .), f \rangle$.

Consider the kernel function $k(\boldsymbol{x}, \boldsymbol{y})$ which is a function of two variables. Suppose, for n points, one of the variables is fixed to have $k(\boldsymbol{x}_1, \boldsymbol{y}), k(\boldsymbol{x}_2, \boldsymbol{y}), \dots, k(\boldsymbol{x}_n, \boldsymbol{y})$. These are all functions of the variable $\boldsymbol{y}$. RKHS is a function space that is the set of all possible linear combinations of these functions [20]:

$$\mathcal{H} := \Big\{ f(.) = \sum_{i=1}^{n} \alpha_i \, k(\boldsymbol{x}_i, .) \Big\} \overset{(a)}{=} \Big\{ f(.) = \sum_{i=1}^{n} \alpha_i \, k_{\boldsymbol{x}_i}(.) \Big\}, \tag{3.2}$$

where (a) is because of the notation $k_{\boldsymbol{x}}(.) := k(\boldsymbol{x}, .)$. This equation demonstrates that the bases of the RKHS are kernels. The proof of this equation is obtained by considering both Eqs. (3.9) and (3.13) together (note that for better organization, it is better that these equations be presented later).

According to Eq. (3.2), every function in the RKHS can be written as a linear combination. Consider two functions in this space represented as $f = \sum_{i=1}^{n} \alpha_i \, k(\boldsymbol{x}_i, \boldsymbol{y})$ and $g = \sum_{j=1}^{n} \beta_j \, k(\boldsymbol{x}, \boldsymbol{y}_j)$; therefore, the inner product in an RKHS is calculated as:

$$\begin{aligned}
\langle f, g \rangle_k &\overset{(3.2)}{=} \Big\langle \sum_{i=1}^{n} \alpha_i \, k(\boldsymbol{x}_i, .), \sum_{j=1}^{n} \beta_j \, k(\boldsymbol{y}_j, .) \Big\rangle_k \overset{(a)}{=} \Big\langle \sum_{i=1}^{n} \alpha_i \, k(\boldsymbol{x}_i, .), \sum_{j=1}^{n} \beta_j \, k(., \boldsymbol{y}_j) \Big\rangle_k \\
&= \sum_{i=1}^{n} \sum_{j=1}^{n} \alpha_i \, \beta_j \, k(\boldsymbol{x}_i, \boldsymbol{y}_j),
\end{aligned} \tag{3.3}$$

where (a) is because a kernel is symmetric.

3.2.3 Reproducing Property

In Eq. (3.3), only one component for g is needed to have $g(\boldsymbol{x}) = \sum_{j=1}^{n} \beta_j \, k(\boldsymbol{x}_i, \boldsymbol{x}) = \beta k(\boldsymbol{x}, \boldsymbol{x})$, where $\beta = 1$ is taken to have $g(\boldsymbol{x}) = k(\boldsymbol{x}, \boldsymbol{x}) = k_{\boldsymbol{x}}(.)$. In other words, assume the function $g(\boldsymbol{x})$ is a kernel in the RKHS space. The inner product of the function $f(\boldsymbol{x}) = \sum_{i=1}^{n} \alpha_i \, k(\boldsymbol{x}_i, \boldsymbol{x})$ in the space, according to Eq. (3.3), is:

$$\langle f(\boldsymbol{x}), g(\boldsymbol{x}) \rangle_k = \langle f, k_{\boldsymbol{x}}(.) \rangle_k = \Big\langle \sum_{i=1}^{n} \alpha_i \, k(\boldsymbol{x}_i, \boldsymbol{x}), k(\boldsymbol{x}, \boldsymbol{x}) \Big\rangle_k = \sum_{i=1}^{n} \alpha_i \, k(\boldsymbol{x}_i, \boldsymbol{x}) \overset{(a)}{=} f(\boldsymbol{x}), \tag{3.4}$$

where (a) is because the considered function was $f(\boldsymbol{x}) = \sum_{i=1}^{n} \alpha_i \, k(\boldsymbol{x}_i, \boldsymbol{x})$. As Eq. (3.4) demonstrates, the function f is reproduced from the inner product of that function with one of the kernels in the space, demonstrating the reproducing property of the RKHS space. A special case of Eq. (3.4) is to use $f(\boldsymbol{x}) = g(\boldsymbol{x}) = k_{\boldsymbol{x}}$ to have $\langle k_{\boldsymbol{x}}, k_{\boldsymbol{x}} \rangle_k = k(\boldsymbol{x}, \boldsymbol{x})$.

3.2.4 Mercer's Theorem

Mercer's theorem was first proposed by Mercer [23]. Kernels, introduced in Sect. 3.2.1, were gradually found to be very useful in mathematics when this theorem was proposed. It demonstrated the importance and functionality of kernel functions; the effectiveness of the Mercer kernel is justified by the Mercer's theorem. Much later, kernels were also found to be useful in machine learning with the introduction of the kernel Support Vector Machine (SVM) [39]. Theorem 3.1 introduces this concept.

Theorem 3.1 (Mercer's Theorem [23]) *Suppose* $k : [a, b] \times [a, b] \to \mathbb{R}$ *is a continuous symmetric positive semidefinite kernel which is bounded:*

$$\sup_{\boldsymbol{x},\boldsymbol{y}} k(\boldsymbol{x}, \boldsymbol{y}) < \infty. \tag{3.5}$$

The operator T_k *takes a function* $f(\boldsymbol{x})$ *as its argument and outputs the function:*

$$T_k f(\boldsymbol{x}) := \int_a^b k(\boldsymbol{x}, \boldsymbol{y}) f(\boldsymbol{y})\, d\boldsymbol{y}, \tag{3.6}$$

which is a Fredholm integral equation [32]. The operator T_k *is called the Hilbert–Schmidt integral operator [30, Chapter 8]. This output function is positive semidefinite:*

$$\iint k(\boldsymbol{x}, \boldsymbol{y}) f(\boldsymbol{y})\, d\boldsymbol{x}\, d\boldsymbol{y} \geq 0. \tag{3.7}$$

Lastly, there is a set of orthonormal bases $\{\psi_i(.)\}_{i=1}^{\infty}$ *of* $L_2(a, b)$ *consisting of eigenfunctions of* T_K*, such that the corresponding sequence of eigenvalues* $\{\lambda_i\}_{i=1}^{\infty}$ *are nonnegative:*

$$\int k(\boldsymbol{x}, \boldsymbol{y})\, \psi_i(\boldsymbol{y})\, d\boldsymbol{y} = \lambda_i\, \psi_i(\boldsymbol{x}). \tag{3.8}$$

The eigenfunctions corresponding to the nonzero eigenvalues are continuous on $[a, b]$ *and* k *can be represented as [1]:*

$$k(\boldsymbol{x}, \boldsymbol{y}) = \sum_{i=1}^{\infty} \lambda_i\, \psi_i(\boldsymbol{x})\, \psi_i(\boldsymbol{y}), \tag{3.9}$$

where the convergence is absolute and uniform. For proof of this theorem, refer to [13].

3.3 Feature Map and Pulling Function

Thus far, the kernel and RKHS have been reviewed, setting the foundation to introduce how to construct a kernel matrix from available data points. Let $\mathcal{X} := \{\boldsymbol{x}_i\}_{i=1}^n$ be the set of data in the input space (note that the input space is the original space of data). The t-dimensional (perhaps infinite dimensional) feature space (or Hilbert space) is denoted by $\mathcal{H}$.

Definition 3.4 (Feature Map or Pulling Function) The following mapping is defined as:

$$\boldsymbol{\phi} : \mathcal{X} \to \mathcal{H}, \tag{3.10}$$

to transform data from the input space to the feature space, i.e., Hilbert space. In other words, this mapping pulls data to the feature space:

$$\boldsymbol{x} \mapsto \boldsymbol{\phi}(\boldsymbol{x}). \tag{3.11}$$

The function $\boldsymbol{\phi}(\boldsymbol{x})$ is called the feature map or pulling function. The feature map is a (possibly infinite-dimensional) vector whose elements are [25]:

$$\boldsymbol{\phi}(\boldsymbol{x}) = [\phi_1(\boldsymbol{x}), \phi_2(\boldsymbol{x}), \dots]^\top := [\sqrt{\lambda_1}\,\boldsymbol{\psi}_1(\boldsymbol{x}), \sqrt{\lambda_2}\,\boldsymbol{\psi}_2(\boldsymbol{x}), \dots]^\top, \tag{3.12}$$

where $\{\boldsymbol{\psi}_i\}$ and $\{\lambda_i\}$ are eigenfunctions and eigenvalues of the kernel operator (see Eq. (3.8)). Note that eigenfunctions will be explained more in Sect. 3.6.

Let t denote the dimensionality of $\boldsymbol{\phi}(\boldsymbol{x})$. The feature map may be infinite or finite dimensional, i.e. t can be infinity; it is usually a very large number. Considering both Eqs. (3.9) and (3.12) demonstrates that:

$$k(\boldsymbol{x}, \boldsymbol{y}) = \big\langle \boldsymbol{\phi}(\boldsymbol{x}), \boldsymbol{\phi}(\boldsymbol{y}) \big\rangle_k = \boldsymbol{\phi}(\boldsymbol{x})^\top \boldsymbol{\phi}(\boldsymbol{y}). \tag{3.13}$$

Therefore, the kernel between two points is the inner product of pulled data points to the feature space. Suppose the feature maps of all points $\boldsymbol{X} \in \mathbb{R}^{d \times n}$ are stacked columnwise in:

$$\boldsymbol{\Phi}(\boldsymbol{X}) := [\boldsymbol{\phi}(\boldsymbol{x}_1), \boldsymbol{\phi}(\boldsymbol{x}_2), \dots, \boldsymbol{\phi}(\boldsymbol{x}_n)], \tag{3.14}$$

which is $t \times n$ dimensional and t may be infinity or a large number. The kernel matrix defined in Definition 3.2 can be calculated as:

$$\mathbb{R}^{n \times n} \ni \boldsymbol{K} = \big\langle \boldsymbol{\Phi}(\boldsymbol{X}), \boldsymbol{\Phi}(\boldsymbol{X}) \big\rangle_k = \boldsymbol{\Phi}(\boldsymbol{X})^\top \boldsymbol{\Phi}(\boldsymbol{X}). \tag{3.15}$$

Equations (3.13) and (3.15) illustrate that there is no need to compute a kernel using eigenfunctions, but a simple inner product suffices for kernel computation. This is the beauty of kernel methods which are simple to compute.

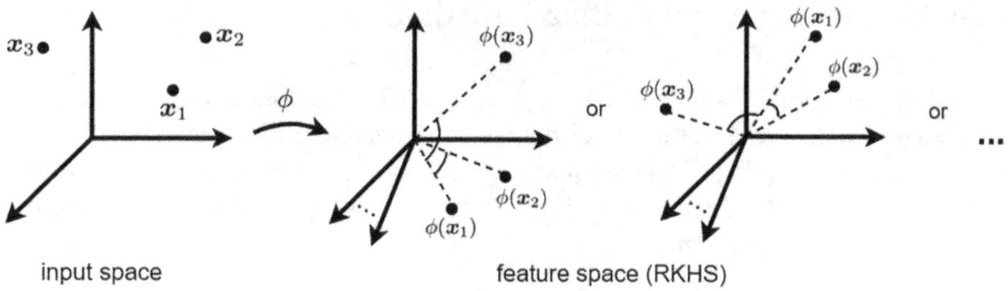

Fig. 3.1 Pulling data from the input space to the feature space (RKHS). The explicit locations of pulled points are not necessarily known, but the relative similarity (inner product) of pulled data points is known in the feature space

Definition 3.5 (Input Space and Feature Space [36]) The space in which data X exist is called the input space, also known as the original space. This space is denoted by $\mathcal{X}$ and is usually an $\mathbb{R}^d$ Euclidean space. The RKHS to which the data have been pulled is called the feature space. Data can be pulled from the input to feature space using kernels.

Remark 3.1 (Kernel Is a Measure of Similarity) The Inner product is a measure of similarity in terms of angles of vectors or in terms of location of points with respect to origin. According to Eq. (3.13), a kernel can be seen as an inner product between feature maps of points. This means that a kernel is a measure of similarity between points and this similarity is computed in the feature space rather than the input space.

Pulling data to the feature space is performed using kernels, which is the inner product of points in a RKHS according to Eq. (3.13). Therefore, the relative similarity (inner product) of pulled data points is known by the kernel. However, in most kernels, it is not possible to find an explicit expression for the pulled data points. Therefore, the exact location of pulled data points to RKHS is not necessarily known, but the relative similarity of pulled points, which is the kernel, is known. An exceptional kernel is the linear kernel in which there is $\phi(x) = x$. This concept is illustrated in Fig. 3.1.

3.4 Well-Known Kernel Functions

3.4.1 Frequently Used Kernels

Many different kernel functions exist that are widely used in machine learning [31]. Some of the most well-known kernels are introduced below.

Linear Kernel The linear kernel is the simplest kernel, which is the inner product of points:

$$k(\boldsymbol{x}, \boldsymbol{y}) := \boldsymbol{x}^\top \boldsymbol{y}. \tag{3.16}$$

Comparing this with Eq. (3.13) demonstrates that in the linear kernel there is $\boldsymbol{\phi}(\boldsymbol{x}) = \boldsymbol{x}$, meaning that in this kernel, the feature map is explicitly known. Note that $\boldsymbol{\phi}(\boldsymbol{x}) = \boldsymbol{x}$ demonstrates that data are not pulled to any other space in the linear kernel, but in the input space the inner products of points are calculated to obtain the feature space. Remark 3.4 of Sect. 3.7 will discuss that a linear kernel may or may not be useful depending on the type of kernelization.

Radial Basis Function (RBF) or Gaussian Kernel RBF kernel has a scaled Gaussian (or normal) distribution, where the normalization factor of distribution is typically ignored. Therefore, it is also called the Gaussian kernel. The RBF kernel is formulated as:

$$k(\boldsymbol{x}, \boldsymbol{y}) := \exp(-\gamma \, \|\boldsymbol{x} - \boldsymbol{y}\|_2^2) = \exp(-\frac{\|\boldsymbol{x} - \boldsymbol{y}\|_2^2}{\sigma^2}), \tag{3.17}$$

where $\gamma := 1/\sigma^2$ and σ^2 is the variance of the kernel. A proper value for this parameter is $\gamma = 1/d$, where d is the dimensionality of the data. This kernel function is the most well-known kernel used in machine learning. For example, it is often used in kernel Support Vector Machine (SVM) [39]. The RBF kernel has also been widely used in RBF networks, where the data are represented as linear combinations of several RBF basis functions. This kernel function has also been used in kernel density estimation, where a probability density function is estimated and fitted from existing data points.

Laplacian Kernel The Laplacian kernel, also called the Laplace kernel, is similar to the RBF kernel, but with the ℓ_1 norm rather than the squared ℓ_2 norm. The Laplacian kernel is described as:

$$k(\boldsymbol{x}, \boldsymbol{y}) := \exp(-\gamma \, \|\boldsymbol{x} - \boldsymbol{y}\|_1) = \exp(-\frac{\|\boldsymbol{x} - \boldsymbol{y}\|_1}{\sigma^2}), \tag{3.18}$$

where $\|\boldsymbol{x} - \boldsymbol{y}\|_1$ is also called the Manhattan distance because the ℓ_1 distance between two points in a city like Manhattan is similar to summing the length of streets between the two points. A proper value for this parameter is $\gamma = 1/d$, where d is the dimensionality of data. In some cases, the Laplacian kernel may perform better than the Gaussian kernel because of betting on the sparsity principle [18], as the ℓ_1 norm makes algorithm sparse. According to this principle, if there exist both dense and sparse versions of an algorithm, the sparse version will likely perform better. Note that the ℓ_2 norm in the RBF kernel is also more sensitive to noise; however, the computation and derivative of the ℓ_1 norm is more difficult than that of the ℓ_2 norm.

Sigmoid Kernel The sigmoid kernel is a hyperbolic tangent function applied on the inner product of points. It is described as follows:

$$k(\boldsymbol{x}, \boldsymbol{y}) := \tanh(\gamma \boldsymbol{x}^\top \boldsymbol{y} + c), \tag{3.19}$$

where $\gamma > 0$ is the slope and c is the intercept. Some proper values for these parameters are $\gamma = 1/d$ and $c = 1$, where d is the dimensionality of the data. Note that the hyperbolic tangent function is used widely for activation functions in neural networks.

Polynomial Kernel Polynomial kernel applies a polynomial function with degree δ (a positive integer) on inner product of points:

$$k(\boldsymbol{x}, \boldsymbol{y}) := (\gamma \boldsymbol{x}^\top \boldsymbol{y} + c)^d, \tag{3.20}$$

where $\gamma > 0$ is the slope and c is the intercept. Some proper values for these parameters are $\gamma = 1/d$ and $c = 1$, where d is the dimensionality of the data.

Cosine Kernel According to Remark 3.1, a kernel is a measure of similarity and computes the inner product between points in the feature space. Cosine kernel computes the similarity between points. It is obtained from the formula of cosine and the inner product:

$$k(\boldsymbol{x}, \boldsymbol{y}) := \cos(\boldsymbol{x}, \boldsymbol{y}) = \frac{\boldsymbol{x}^\top \boldsymbol{y}}{\|\boldsymbol{x}\|_2 \, \|\boldsymbol{y}\|_2}. \tag{3.21}$$

The normalization in the denominator projects the points onto a unit hypersphere so that the inner product measures the similarity of their angles regardless of their lengths. Note that angle-based measures such as cosine are found to work better for face recognition compared to Euclidean distances [27].

3.4.2 *Kernel Construction from Distance Metric*

It is also possible to construct a kernel matrix using a distance metric applied to the data points. The squared Euclidean distance between $\boldsymbol{x}_i$ and $\boldsymbol{x}_j$ is $d_{ij}^2 = ||\boldsymbol{x}_i - \boldsymbol{x}_j||_2^2$. It can be simplified as follows:

$$\begin{aligned} d_{ij}^2 &= ||\boldsymbol{x}_i - \boldsymbol{x}_j||_2^2 = (\boldsymbol{x}_i - \boldsymbol{x}_j)^\top (\boldsymbol{x}_i - \boldsymbol{x}_j) \\ &= \boldsymbol{x}_i^\top \boldsymbol{x}_i - \boldsymbol{x}_i^\top \boldsymbol{x}_j - \boldsymbol{x}_j^\top \boldsymbol{x}_i + \boldsymbol{x}_j^\top \boldsymbol{x}_j = \boldsymbol{x}_i^\top \boldsymbol{x}_i - 2\boldsymbol{x}_i^\top \boldsymbol{x}_j + \boldsymbol{x}_j^\top \boldsymbol{x}_j = \boldsymbol{G}_{ii} - 2\boldsymbol{G}_{ij} + \boldsymbol{G}_{jj}, \end{aligned}$$

where $\mathbb{R}^{n\times n} \ni \boldsymbol{G} := \boldsymbol{X}^\top \boldsymbol{X}$ is the linear Gram matrix (see Definition 3.2). If $\mathbb{R}^n \ni \boldsymbol{g} := [g_1, \dots, g_n] = [\boldsymbol{G}_{11}, \dots, \boldsymbol{G}_{nn}] = \textbf{diag}(\boldsymbol{G})$, then:

$$\begin{aligned}
d_{ij}^2 &= g_i - 2\boldsymbol{G}_{ij} + g_j, \\
\boldsymbol{D} &= \boldsymbol{g}\boldsymbol{1}^\top - 2\boldsymbol{G} + \boldsymbol{1}\boldsymbol{g}^\top = \boldsymbol{1}\boldsymbol{g}^\top - 2\boldsymbol{G} + \boldsymbol{g}\boldsymbol{1}^\top,
\end{aligned}$$

where $\boldsymbol{1}$ is the vector of ones and $\boldsymbol{D}$ is the distance matrix with squared Euclidean distance (d_{ij}^2 as its elements). $\boldsymbol{H}$ denotes the centering matrix, $\mathbb{R}^{n\times n} \ni \boldsymbol{H} := \boldsymbol{I} - \frac{1}{n}\boldsymbol{1}_n\boldsymbol{1}_n^\top$, and $\boldsymbol{I}$ is the identity matrix (see Chap. 2 for centering matrix). The matrix $\boldsymbol{D}$ is double-centered as follows:

$$\begin{aligned}
\boldsymbol{H}\boldsymbol{D}\boldsymbol{H} &= (\boldsymbol{I} - \frac{1}{n}\boldsymbol{1}\boldsymbol{1}^\top)\boldsymbol{D}(\boldsymbol{I} - \frac{1}{n}\boldsymbol{1}\boldsymbol{1}^\top) \\
&= (\boldsymbol{I} - \frac{1}{n}\boldsymbol{1}\boldsymbol{1}^\top)(\boldsymbol{1}\boldsymbol{g}^\top - 2\boldsymbol{G} + \boldsymbol{g}\boldsymbol{1}^\top)(\boldsymbol{I} - \frac{1}{n}\boldsymbol{1}\boldsymbol{1}^\top) \\
&= \big[\underbrace{(\boldsymbol{I} - \frac{1}{n}\boldsymbol{1}\boldsymbol{1}^\top)\boldsymbol{1}}_{=\boldsymbol{0}}\boldsymbol{g}^\top - 2(\boldsymbol{I} - \frac{1}{n}\boldsymbol{1}\boldsymbol{1}^\top)\boldsymbol{G} + (\boldsymbol{I} - \frac{1}{n}\boldsymbol{1}\boldsymbol{1}^\top)\boldsymbol{g}\boldsymbol{1}^\top\big](\boldsymbol{I} - \frac{1}{n}\boldsymbol{1}\boldsymbol{1}^\top) \\
&= -2(\boldsymbol{I} - \frac{1}{n}\boldsymbol{1}\boldsymbol{1}^\top)\boldsymbol{G}(\boldsymbol{I} - \frac{1}{n}\boldsymbol{1}\boldsymbol{1}^\top) + (\boldsymbol{I} - \frac{1}{n}\boldsymbol{1}\boldsymbol{1}^\top)\boldsymbol{g}\underbrace{\boldsymbol{1}^\top(\boldsymbol{I} - \frac{1}{n}\boldsymbol{1}\boldsymbol{1}^\top)}_{=\boldsymbol{0}} \\
&= -2(\boldsymbol{I} - \frac{1}{n}\boldsymbol{1}\boldsymbol{1}^\top)\boldsymbol{G}(\boldsymbol{I} - \frac{1}{n}\boldsymbol{1}\boldsymbol{1}^\top) = -2\,\boldsymbol{H}\boldsymbol{G}\boldsymbol{H}.
\end{aligned}$$

$$\therefore \quad \boldsymbol{H}\boldsymbol{G}\boldsymbol{H} = \boldsymbol{H}\boldsymbol{X}^\top\boldsymbol{X}\boldsymbol{H} = -\frac{1}{2}\boldsymbol{H}\boldsymbol{D}\boldsymbol{H}. \tag{3.22}$$

Note that $(\boldsymbol{I} - \frac{1}{n}\boldsymbol{1}\boldsymbol{1}^\top)\boldsymbol{1} = \boldsymbol{0}$ and $\boldsymbol{1}^\top(\boldsymbol{I} - \frac{1}{n}\boldsymbol{1}\boldsymbol{1}^\top) = \boldsymbol{0}$ because removing the row mean of $\boldsymbol{1}$ and column mean of $\boldsymbol{1}^\top$ results in the zero vectors. If data $\boldsymbol{X}$ are already centered, i.e., the mean has been removed ($\boldsymbol{X} \leftarrow \boldsymbol{X}\boldsymbol{H}$), Eq. (3.22) becomes:

$$\boldsymbol{X}^\top\boldsymbol{X} = -\frac{1}{2}\boldsymbol{H}\boldsymbol{D}\boldsymbol{H}. \tag{3.23}$$

According to the kernel trick, Eq. (3.56), a general kernel matrix can be written rather than the linear Gram matrix in Eq. (3.23), resulting in [12]:

$$\mathbb{R}^{n\times n} \ni \boldsymbol{K} = \boldsymbol{\Phi}(\boldsymbol{X})^\top\boldsymbol{\Phi}(\boldsymbol{X}) = -\frac{1}{2}\boldsymbol{H}\boldsymbol{D}\boldsymbol{H}. \tag{3.24}$$

This kernel is double-centered because of $\boldsymbol{HDH}$. It is important to note that Eq. (3.24) can be used for unifying the spectral dimensionality reduction methods as special cases of kernel principal component analysis with different kernels (see [7, 17], [38, Table 2.1]). Further details can be found in Chaps. 7 and 10.

Lemma 3.1 (Distance-Based Kernel Is a Mercer Kernel) *The kernel constructed from a valid distance metric, i.e. Eq. (3.24), is a Mercer kernel.*

Proof The kernel is symmetric because:

$$\boldsymbol{K}^\top = -\frac{1}{2}\boldsymbol{H}^\top \boldsymbol{D}^\top \boldsymbol{H}^\top \overset{(a)}{=} -\frac{1}{2}\boldsymbol{H}\boldsymbol{D}\boldsymbol{H} = \boldsymbol{K},$$

where (a) is because $\boldsymbol{H}$ and $\boldsymbol{D}$ are symmetric matrices. Moreover, the kernel is positive semidefinite because:

$$\boldsymbol{K} = -\frac{1}{2}\boldsymbol{H}\boldsymbol{D}\boldsymbol{H} = \boldsymbol{\Phi}(\boldsymbol{X})^\top \boldsymbol{\Phi}(\boldsymbol{X}) \implies \boldsymbol{v}^\top \boldsymbol{K} \boldsymbol{v} = \boldsymbol{v}^\top \boldsymbol{\Phi}(\boldsymbol{X})^\top \boldsymbol{\Phi}(\boldsymbol{X}) \boldsymbol{v} = \|\boldsymbol{\Phi}(\boldsymbol{X})\boldsymbol{v}\|_2^2 \geq 0, \quad \forall \boldsymbol{v} \in \mathbb{R}^n.$$

Therefore, according to Definition 3.1, this kernel is a Mercer kernel. □

Remark 3.2 (Kernel Construction from Metric) One can use any valid distance metric satisfying the following properties:

1. nonnegativity: $\boldsymbol{D}(\boldsymbol{x}, \boldsymbol{y}) \geq 0$,
2. equal points: $\boldsymbol{D}(\boldsymbol{x}, \boldsymbol{y}) = 0 \iff \boldsymbol{x} = \boldsymbol{y}$,
3. symmetry: $\boldsymbol{D}(\boldsymbol{x}, \boldsymbol{y}) = \boldsymbol{D}(\boldsymbol{y}, \boldsymbol{x})$,
4. triangular inequality: $\boldsymbol{D}(\boldsymbol{x}, \boldsymbol{y}) \leq \boldsymbol{D}(\boldsymbol{x}, \boldsymbol{z}) + \boldsymbol{D}(\boldsymbol{z}, \boldsymbol{y})$,

to calculate elements of distance matrix $\boldsymbol{D}$ in Eq. (3.23). It is important that the used distance matrix should be a valid distance matrix. Using various distance metrics in Eq. (3.23) results in various useful kernels.

3.5 Kernel Centering and Normalization

3.5.1 Kernel Centering

In some cases, there is a need to center the pulled data in the feature space. For this, the kernel matrix should be centered in such a way that the mean of the pulled dataset becomes zero. For kernel centering, the theory, introduced in Sects. 3.5.1.1 and 3.5.1.2, should be adhered to, This theory is based on [34] and [35, Appendix A]. An example of use of kernel centering in machine learning is kernel principal component analysis which will be introduced in Chap. 5.

3.5.1.1 Centering the Kernel of Training Data

Assume there is a training dataset $\boldsymbol{X} = [\boldsymbol{x}_1, \dots, \boldsymbol{x}_n] \in \mathbb{R}^{d \times n}$ and some out-of-sample data $\boldsymbol{X}_t = [\boldsymbol{x}_{t,1}, \dots, \boldsymbol{x}_{t,n_t}] \in \mathbb{R}^{d \times n_t}$. Consider the kernel matrix for the training data $\mathbb{R}^{n \times n} \ni \boldsymbol{K} := \boldsymbol{\Phi}(\boldsymbol{X})^\top \boldsymbol{\Phi}(\boldsymbol{X})$, whose (i, j)-th element is $\mathbb{R} \ni \boldsymbol{K}(i, j) = \boldsymbol{\phi}(\boldsymbol{x}_i)^\top \boldsymbol{\phi}(\boldsymbol{x}_j)$. The goal is to first center the pulled training data in the feature space:

$$\breve{\boldsymbol{\phi}}(\boldsymbol{x}_i) := \boldsymbol{\phi}(\boldsymbol{x}_i) - \frac{1}{n} \sum_{k=1}^{n} \boldsymbol{\phi}(\boldsymbol{x}_k). \tag{3.25}$$

If the pulled training dataset is centered, the (i, j)-th element of the kernel matrix becomes:

$$\begin{aligned} \breve{\boldsymbol{K}}(i, j) &:= \breve{\boldsymbol{\phi}}(\boldsymbol{x}_i)^\top \breve{\boldsymbol{\phi}}(\boldsymbol{x}_j) \\ &\overset{(3.25)}{=} \Big(\boldsymbol{\phi}(\boldsymbol{x}_i) - \frac{1}{n} \sum_{k_1=1}^{n} \boldsymbol{\phi}(\boldsymbol{x}_{k_1})\Big)^\top \Big(\boldsymbol{\phi}(\boldsymbol{x}_j) - \frac{1}{n} \sum_{k_2=1}^{n} \boldsymbol{\phi}(\boldsymbol{x}_{k_2})\Big) \\ &= \boldsymbol{\phi}(\boldsymbol{x}_i)^\top \boldsymbol{\phi}(\boldsymbol{x}_j) - \frac{1}{n} \sum_{k_1=1}^{n} \boldsymbol{\phi}(\boldsymbol{x}_{k_1})^\top \boldsymbol{\phi}(\boldsymbol{x}_j) - \frac{1}{n} \sum_{k_2=1}^{n} \boldsymbol{\phi}(\boldsymbol{x}_i)^\top \boldsymbol{\phi}(\boldsymbol{x}_{k_2}) + \frac{1}{n^2} \sum_{k_1=1}^{n} \sum_{k_2=1}^{n} \boldsymbol{\phi}(\boldsymbol{x}_{k_1})^\top \boldsymbol{\phi}(\boldsymbol{x}_{k_2}). \end{aligned} \tag{3.26}$$

In matrix form, this is written as:

$$\mathbb{R}^{n \times n} \ni \breve{\boldsymbol{K}} = \boldsymbol{K} - \frac{1}{n} \boldsymbol{1}_{n \times n} \boldsymbol{K} - \frac{1}{n} \boldsymbol{K} \boldsymbol{1}_{n \times n} + \frac{1}{n^2} \boldsymbol{1}_{n \times n} \boldsymbol{K} \boldsymbol{1}_{n \times n} = \boldsymbol{H} \boldsymbol{K} \boldsymbol{H}, \tag{3.27}$$

where $\boldsymbol{H}$ is the centering matrix (see Chap. 2). Another proof for double-centering a kernel is as follows. If the centered pulled data $\boldsymbol{\Phi}(\boldsymbol{X})$ are demonstrated as:

$$\breve{\boldsymbol{\Phi}}(\boldsymbol{X}) := \boldsymbol{\Phi}(\boldsymbol{X}) \boldsymbol{H}, \tag{3.28}$$

then, the centered kernel is:

$$\breve{\boldsymbol{K}} = \breve{\boldsymbol{\Phi}}(\boldsymbol{X})^\top \breve{\boldsymbol{\Phi}}(\boldsymbol{X}) = \big(\boldsymbol{\Phi}(\boldsymbol{X}) \boldsymbol{H}\big)^\top \big(\boldsymbol{\Phi}(\boldsymbol{X}) \boldsymbol{H}\big) \overset{(a)}{=} \boldsymbol{H} \boldsymbol{\Phi}(\boldsymbol{X})^\top \boldsymbol{\Phi}(\boldsymbol{X}) \boldsymbol{H} \overset{(b)}{=} \boldsymbol{H} \boldsymbol{K} \boldsymbol{H},$$

where (a) and (b) are the results of the centering matrix being symmetric and idempotent, respectively (see Chap. 2).

Equation (3.27), known as the *double-centered kernel*, is the kernel matrix when the pulled training data in the feature space are centered. Additionally, the double-centered kernel does not have a rowwise and columnwise mean (meaning their summations are zero). Therefore, after kernel centering, the equations are as follows:

$$\frac{1}{n}\sum_{i=1}^{n}\breve{\boldsymbol{\phi}}(\boldsymbol{x}_i) = \boldsymbol{0}, \tag{3.29}$$

$$\sum_{i=1}^{n}\sum_{j=1}^{n}\breve{\boldsymbol{K}}(i,j) = 0. \tag{3.30}$$

3.5.1.2 Centering the Kernel Between Training and Out-of-sample Data

Now, consider the kernel matrix between the training data and the out-of-sample data $\mathbb{R}^{n\times n_t} \ni \boldsymbol{K}_t := \boldsymbol{\Phi}(\boldsymbol{X})^\top \boldsymbol{\Phi}(\boldsymbol{X}_t)$. whose (i,j)-th element is $\mathbb{R} \ni \boldsymbol{K}_t(i,j) = \boldsymbol{\phi}(\boldsymbol{x}_i)^\top \boldsymbol{\phi}(\boldsymbol{x}_{t,j})$. The aim is to center the pulled training data in the feature space, i.e., Eq. (3.25). Moreover, the out-of-sample data should be centered using the mean of the training (and not out-of-sample) data:

$$\breve{\boldsymbol{\phi}}(\boldsymbol{x}_{t,i}) := \boldsymbol{\phi}(\boldsymbol{x}_{t,i}) - \frac{1}{n}\sum_{k=1}^{n}\boldsymbol{\phi}(\boldsymbol{x}_k). \tag{3.31}$$

If the pulled training and out-of-sample data are centered, the (i,j)-th element of the kernel matrix becomes:

$$\begin{aligned}
\breve{\boldsymbol{K}}_t(i,j) &:= \breve{\boldsymbol{\phi}}(\boldsymbol{x}_i)^\top \breve{\boldsymbol{\phi}}(\boldsymbol{x}_{t,j}) \\
&\overset{(a)}{=} \Big(\boldsymbol{\phi}(\boldsymbol{x}_i) - \frac{1}{n}\sum_{k_1=1}^{n}\boldsymbol{\phi}(\boldsymbol{x}_{k_1})\Big)^\top \Big(\boldsymbol{\phi}(\boldsymbol{x}_{t,j}) - \frac{1}{n}\sum_{k_2=1}^{n}\boldsymbol{\phi}(\boldsymbol{x}_{k_2})\Big) \\
&= \boldsymbol{\phi}(\boldsymbol{x}_i)^\top \boldsymbol{\phi}(\boldsymbol{x}_{t,j}) - \frac{1}{n}\sum_{k_1=1}^{n}\boldsymbol{\phi}(\boldsymbol{x}_{k_1})^\top \boldsymbol{\phi}(\boldsymbol{x}_{t,j}) - \frac{1}{n}\sum_{k_2=1}^{n}\boldsymbol{\phi}(\boldsymbol{x}_i)^\top \boldsymbol{\phi}(\boldsymbol{x}_{k_2}) \\
&\quad + \frac{1}{n^2}\sum_{k_1=1}^{n}\sum_{k_2=1}^{n}\boldsymbol{\phi}(\boldsymbol{x}_{k_1})^\top \boldsymbol{\phi}(\boldsymbol{x}_{k_2}),
\end{aligned}$$

where (a) is because of Eqs. (3.25) and (3.31). Therefore, the double-centered kernel matrix over training and out-of-sample data is [34], [35, Appendix A]:

$$\mathbb{R}^{n\times n_t} \ni \breve{\boldsymbol{K}}_t = \boldsymbol{K}_t - \frac{1}{n}\boldsymbol{1}_{n\times n}\boldsymbol{K}_t - \frac{1}{n}\boldsymbol{K}\boldsymbol{1}_{n\times n_t} + \frac{1}{n^2}\boldsymbol{1}_{n\times n}\boldsymbol{K}\boldsymbol{1}_{n\times n_t}, \tag{3.32}$$

where $\mathbb{R}^{n\times n_t} \ni \boldsymbol{1}_{n\times n_t} := \boldsymbol{1}_n \boldsymbol{1}_{n_t}^\top$ and $\mathbb{R}^{n_t} \ni \boldsymbol{1}_{n_t} := [1, \dots, 1]^\top$. Equation (3.32) is the kernel matrix when the pulled training data in the feature space are centered and the pulled out-of-sample data are centered using the mean of pulled training data.

If there is one out-of-sample $\boldsymbol{x}_t$, Eq. (3.32) becomes:

$$\mathbb{R}^n \ni \breve{\boldsymbol{k}}_t = \boldsymbol{k}_t - \frac{1}{n}\boldsymbol{1}_{n\times n}\boldsymbol{k}_t - \frac{1}{n}\boldsymbol{K}\boldsymbol{1}_n + \frac{1}{n^2}\boldsymbol{1}_{n\times n}\boldsymbol{K}\boldsymbol{1}_n, \tag{3.33}$$

where:

$$\mathbb{R}^n \ni \boldsymbol{k}_t = \boldsymbol{k}_t(\boldsymbol{X}, \boldsymbol{x}_t) := \boldsymbol{\Phi}(\boldsymbol{X})^\top \boldsymbol{\phi}(\boldsymbol{x}_t) = [\boldsymbol{\phi}(\boldsymbol{x}_1)^\top \boldsymbol{\phi}(\boldsymbol{x}_t), \dots, \boldsymbol{\phi}(\boldsymbol{x}_n)^\top \boldsymbol{\phi}(\boldsymbol{x}_t)]^\top, \tag{3.34}$$

$$\mathbb{R}^n \ni \breve{\boldsymbol{k}}_t = \breve{\boldsymbol{k}}_t(\boldsymbol{X}, \boldsymbol{x}_t) := \breve{\boldsymbol{\Phi}}(\boldsymbol{X})^\top \breve{\boldsymbol{\phi}}(\boldsymbol{x}_t) = [\breve{\boldsymbol{\phi}}(\boldsymbol{x}_1)^\top \breve{\boldsymbol{\phi}}(\boldsymbol{x}_t), \dots, \breve{\boldsymbol{\phi}}(\boldsymbol{x}_n)^\top \breve{\boldsymbol{\phi}}(\boldsymbol{x}_t)]^\top, \tag{3.35}$$

where $\breve{\boldsymbol{\Phi}}(\boldsymbol{X})$ and $\breve{\boldsymbol{\phi}}(\boldsymbol{x}_t)$ are according to Eqs. (3.25) and (3.31), respectively. Note that Eq. (3.27) or (3.32) can be restated as the following lemma.

Lemma 3.2 (Kernel Centering [10]) *The pulled data to the feature space can be centered by kernel centering. The kernel matrix* $\boldsymbol{K}(\boldsymbol{x}, \boldsymbol{y})$ *is centered as:*

$$\begin{aligned}\breve{\boldsymbol{K}}(\boldsymbol{x}, \boldsymbol{y}) &= \big(\boldsymbol{\phi}(\boldsymbol{x}) - \mathbb{E}_x[\boldsymbol{\phi}(\boldsymbol{x})]\big)^\top \big(\boldsymbol{\phi}(\boldsymbol{x}) - \mathbb{E}_x[\boldsymbol{\phi}(\boldsymbol{x})]\big) \\ &= \boldsymbol{K}(\boldsymbol{x}, \boldsymbol{y}) - \mathbb{E}_x[\boldsymbol{K}(\boldsymbol{x}, \boldsymbol{y})] - \mathbb{E}_y[\boldsymbol{K}(\boldsymbol{x}, \boldsymbol{y})] + \mathbb{E}_x[\mathbb{E}_y[\boldsymbol{K}(\boldsymbol{x}, \boldsymbol{y})]].\end{aligned} \tag{3.36}$$

Note that $\mathbb{E}_x[\boldsymbol{K}(\boldsymbol{x}, \boldsymbol{y})]$, $\mathbb{E}_y[\boldsymbol{K}(\boldsymbol{x}, \boldsymbol{y})]$, *and* $\mathbb{E}_x[\mathbb{E}_y[\boldsymbol{K}(\boldsymbol{x}, \boldsymbol{y})]]$ *are the average of rows, average of columns, and total average of rows and columns of the kernel matrix, respectively.*

Proof The explained derivations for Eqs. (3.25) and (3.31) and the definition of expectation complete the proof. □

3.5.2 Kernel Normalization

According to Eq. (3.13), a kernel value can be large if the pulled vectors to the feature map have a large length. Therefore, in practical computations and optimization, it is sometimes required to normalize the kernel matrix. This may occur especially when the kernel-based algorithm has an iterative optimization. In such cases, the solution can gradually explode if normalization is not performed after every optimization iteration.

Lemma 3.3 (Cosine Normalization of Kernel [28]) *The kernel matrix* $\boldsymbol{K} \in \mathbb{R}^{n\times n}$ *can be normalized as:*

$$\boldsymbol{K}(i, j) \leftarrow \frac{\boldsymbol{K}(i, j)}{\sqrt{\boldsymbol{K}(i, i)\boldsymbol{K}(j, j)}}, \quad \forall i, j \in \{1, \dots, n\}. \tag{3.37}$$

Proof Cosine normalizes points onto a unit hypersphere and then computes the similarity of points using an inner product. Cosine is computed by Eq. (3.21) and according to the relation of norm and inner product, it is:

$$\cos(\boldsymbol{x}_i, \boldsymbol{x}_j) = \frac{\boldsymbol{x}_i^\top \boldsymbol{x}_j}{\|\boldsymbol{x}_i\|_2 \, \|\boldsymbol{x}_j\|_2} = \frac{\boldsymbol{x}_i^\top \boldsymbol{x}_j}{\sqrt{\|\boldsymbol{x}_i\|_2^2 \, \|\boldsymbol{x}_j\|_2^2}} = \frac{\boldsymbol{x}_i^\top \boldsymbol{x}_j}{\sqrt{\boldsymbol{x}_i^\top \boldsymbol{x}_i \, \boldsymbol{x}_j^\top \boldsymbol{x}_j}}.$$

According to Remark 3.1, kernel is also a measure of similarity. Using the kernel trick, Eq. (3.55), the cosine similarity (which is already normalized) is kernelized as:

$$\boldsymbol{K}(i, j) = \frac{\boldsymbol{\phi}(\boldsymbol{x}_i)^\top \boldsymbol{\phi}(\boldsymbol{x}_j)}{\sqrt{\boldsymbol{\phi}(\boldsymbol{x}_i)^\top \boldsymbol{\phi}(\boldsymbol{x}_i) \, \boldsymbol{\phi}(\boldsymbol{x}_j)^\top \boldsymbol{\phi}(\boldsymbol{x}_j)}} \overset{(3.55)}{=} \frac{\boldsymbol{K}(i, j)}{\sqrt{\boldsymbol{K}(i, i)\boldsymbol{K}(j, j)}}.$$

□

3.6 Eigenfunctions

3.6.1 Eigenfunctions

Recall the eigenvalue problem for a matrix $\boldsymbol{A}$ (see Chap. 2):

$$\boldsymbol{A}\,\boldsymbol{\phi}_i = \lambda_i\,\boldsymbol{\phi}_i, \quad \forall i \in \{1, \dots, d\}, \tag{3.38}$$

where $\boldsymbol{\phi}_i$ and λ_i are the i-th eigenvector and eigenvalue of $\boldsymbol{A}$, respectively. The following introduces the eigenfunction problem, which has a similar form but for an operator rather than a matrix.

Definition 3.6 (Eigenfunction) Consider a linear operator O, which can be applied on a function f. If applying this operator to the function results in a multiplication of the function to a constant:

$$Of = \lambda f, \tag{3.39}$$

then the function f is an eigenfunction for the operator O and the constant λ is the corresponding eigenvalue. Note that the form of the eigenfunction problem is:

$$\text{Operator (function } f) = \text{constant } \times \text{function } f. \tag{3.40}$$

Derivatives and kernel functions are both examples of operators in this equation. $e^{\lambda x}$ is an eigenfunction of derivative because $\frac{d}{dx}e^{\lambda x} = \lambda e^{\lambda x}$, for example. Note that eigenfunctions have applications in many fields of science, including machine learning [10] and theoretical physics (quantum mechanics). Machine learning uses eigenfunctions for the out-of-sample extension of algorithms to be able to perform on test data. This will be explained in Chap. 10.

Recall that in the eigenvalue problem, the eigenvectors illustrate the most important or informative directions of the matrix and the corresponding eigenvalue demonstrates the amount of importance. Likewise, in the eigenfunction problem of an operator, the eigenfunction is the most important function of the operator and the corresponding eigenvalue demonstrates the amount of this importance. This connection between the eigenfunction and the eigenvalue problems can be proven through the following theorem:

Theorem 3.2 (Connection of Eigenfunction and Eigenvalue Problems) *If it is assumed that the operator and the function are a matrix and a vector, respectively, the eigenfunction problem is converted to an eigenvalue problem, where the vector is the eigenvector of the matrix. See [13] for proof.*

3.6.2 Use of Eigenfunctions for Spectral Embedding

Within a Hilbert space $\mathcal{H}$ of functions, the data in the input space are $\mathcal{X} = \{\boldsymbol{x}_i \in \mathbb{R}^d\}_{i=1}^n$. In this space, it is possible to consider an operator for the kernel function K_p as [9, Section 3]:

$$(K_p f)(\boldsymbol{x}) := \int k(\boldsymbol{x}, \boldsymbol{y})\, f(\boldsymbol{y})\, p(\boldsymbol{y})\, d\boldsymbol{y}, \tag{3.41}$$

where $f \in \mathcal{H}$ and the density function $p(\boldsymbol{y})$ can be approximated empirically. A discrete approximation of this operator is [40]:

$$(K_{p,n} f)(\boldsymbol{x}) := \frac{1}{n} \sum_{i=1}^{n} k(\boldsymbol{x}, \boldsymbol{x}_i)\, f(\boldsymbol{x}_i), \tag{3.42}$$

which converges to Eq. (3.41), if $n \to \infty$. Note that this equation is also mentioned in [10, Section 2], [7, Section 4].

Lemma 3.4 (Relation of Eigenvalues of Eigenvalue Problem and Eigenfunction Problem for Kernel [9, Proposition 1], [10, Theorem 1], [7, Section 4]) *Assume λ_k denotes the k-th eigenvalue for eigenfunction decomposition of the operator K_p and δ_k denotes the k-th eigenvalue for the eigenvalue problem of the matrix $\boldsymbol{K} \in \mathbb{R}^{n \times n}$. Therefore:*

$$\delta_k = n\, \lambda_k. \tag{3.43}$$

Proof Building upon [8, proof of Proposition 3], according to Eq. (3.39), the eigenfunction problems for the operators K_p and $K_{p,n}$ (discrete version) are:

$$\begin{aligned} (K_p f_k)(\boldsymbol{x}) &= \lambda_k f_k(\boldsymbol{x}), \quad \forall k \in \{1, \dots, n\}, \\ (K_{p,n} f_k)(\boldsymbol{x}) &= \lambda_k f_k(\boldsymbol{x}), \quad \forall k \in \{1, \dots, n\}, \end{aligned} \tag{3.44}$$

where $f_k(.)$ is the k-th eigenfunction and λ_k is the corresponding eigenvalue. Consider the kernel matrix defined by Definition 3.2. The eigenvalue problem for the kernel matrix is (see Chap. 2):

$$\boldsymbol{K}\boldsymbol{v}_k = \delta_k \boldsymbol{v}_k, \quad \forall k \in \{1, \dots, n\}, \tag{3.45}$$

where $\boldsymbol{v}_k$ is the k-th eigenvector and δ_k is the corresponding eigenvalue. According to Eqs. (3.42) and (3.44):

$$\frac{1}{n}\sum_{i=1}^{n} k(\boldsymbol{x}, \boldsymbol{x}_i)\, f(\boldsymbol{x}_i) = \lambda_k f_k(\boldsymbol{x}), \quad \forall k \in \{1, \dots, n\}.$$

Evaluating the equation at $\boldsymbol{x}_i \in \mathcal{X}$ [7, Section 4], results in:

$$\frac{1}{n}\boldsymbol{K} f_k = \lambda_k f_k \implies \boldsymbol{K} f_k = n\lambda_k f_k, \quad \forall k \in \{1, \dots, n\}.$$

According to Theorem 3.2, eigenfunctions can be seen as eigenvectors, which results in:

$$\boldsymbol{K} f_k = n\lambda_k f_k \implies \boldsymbol{K}\boldsymbol{v}_k = n\lambda_k \boldsymbol{v}_k, \tag{3.46}$$

Comparing Eqs. (3.45) and (3.46) results in Eq. (3.43). □

Lemma 3.5 (Relation of Eigenvalues of Kernel and Covariance in the Feature Space [35]) *Consider the covariance of pulled data to the feature space:*

$$\boldsymbol{C}_H := \frac{1}{n}\sum_{i=1}^{n} \breve{\phi}(\boldsymbol{x}_i)\breve{\phi}(\boldsymbol{x}_i)^\top, \tag{3.47}$$

where $\breve{\phi}(\boldsymbol{x}_i)$ is the centered pulled data defined by Eq. (3.31). The $\boldsymbol{C}_H$ is $t \times t$ dimensional where t can be infinite. Assume η_k denotes the k-th eigenvalue $\boldsymbol{C}_H$ and δ_k denotes the k-th eigenvalue of centered kernel $\breve{\boldsymbol{K}}$. There is:

$$\delta_k = n\,\eta_k. \tag{3.48}$$

Proof Based on [35, Section 2], the eigenvalue problem for this covariance matrix is:

$$\eta_k \, \boldsymbol{u}_k = \boldsymbol{C}_H \, \boldsymbol{u}_k, \quad \forall k \in \{1, \dots, n\},$$

where $\boldsymbol{u}_k$ is the k-th eigenvector and η_k is its corresponding eigenvalue (see Chap. 2). Left multiplying this equation with $\breve{\phi}(\boldsymbol{x}_j)^\top$ results in:

$$\eta_k \, \breve{\phi}(\boldsymbol{x}_j)^\top \boldsymbol{u}_k = \breve{\phi}(\boldsymbol{x}_j)^\top \boldsymbol{C}_H \, \boldsymbol{u}_k, \quad \forall k \in \{1, \dots, n\}. \tag{3.49}$$

As $\boldsymbol{u}_k$ is the eigenvector of the covariance matrix in the feature space, it lies in the feature space; meaning that according to Lemma 3.7 (presented later in Sect. 3.7.2), it is represented as:

$$\boldsymbol{u}_k = \frac{1}{\sqrt{\delta_k}} \sum_{\ell=1}^{n} v_\ell \, \breve{\phi}(\boldsymbol{x}_\ell), \tag{3.50}$$

where pulled data to the feature space are assumed to be centered, v_ℓ's are the coefficients in representation, and the normalization by $1/\sqrt{\delta_k}$ is the result of a normalization used in [10, Section 4]. Substituting Eqs. (3.50) and (3.47) in Eq. (3.49) results in:

$$\eta_k \, \breve{\phi}(\boldsymbol{x}_j)^\top \sum_{\ell=1}^{n} v_\ell \, \breve{\phi}(\boldsymbol{x}_\ell) = \breve{\phi}(\boldsymbol{x}_j)^\top \frac{1}{n} \sum_{i=1}^{n} \breve{\phi}(\boldsymbol{x}_i) \breve{\phi}(\boldsymbol{x}_i)^\top \sum_{\ell=1}^{n} v_\ell \, \breve{\phi}(\boldsymbol{x}_\ell),$$

where normalization factors are simplified from the sides. The right-hand side, as the summations are finite, can be rearranged. Rearranging the terms in this equation translates to:

$$\eta_k \sum_{\ell=1}^{n} v_\ell \, \breve{\phi}(\boldsymbol{x}_j)^\top \breve{\phi}(\boldsymbol{x}_\ell) = \frac{1}{n} \sum_{\ell=1}^{n} v_\ell \Big(\breve{\phi}(\boldsymbol{x}_j)^\top \sum_{i=1}^{n} \breve{\phi}(\boldsymbol{x}_i) \Big) \Big(\breve{\phi}(\boldsymbol{x}_i)^\top \, \breve{\phi}(\boldsymbol{x}_\ell) \Big).$$

Considering Eqs. (3.15) and (3.26), this equation can be written in matrix form $\eta_k \breve{\boldsymbol{K}} \boldsymbol{v}_k = \frac{1}{n} \breve{\boldsymbol{K}}^2 \boldsymbol{v}_k$, where $\boldsymbol{v}_k := [v_1, \dots, v_n]^\top$. As $\breve{\boldsymbol{K}}$ is positive semidefinite, it is often nonsingular. For nonzero eigenvalues, it is possible to left multiply this equation to $\breve{\boldsymbol{K}}^{-1}$ to have:

$$n \, \eta_k \, \boldsymbol{v}_k = \breve{\boldsymbol{K}} \boldsymbol{v}_k,$$

which is the eigenvalue problem for $\breve{\boldsymbol{K}}$, where $\boldsymbol{v}$ is the eigenvector and $\delta_k = n \, \eta_k$ is the eigenvalue (cf. Eq. (3.45)). □

Lemma 3.6 (Relation of Eigenfunctions and Eigenvectors for Kernel [9, Proposition 1], [10, Theorem 1]) *Consider a training dataset* $\{\boldsymbol{x}_i \in \mathbb{R}^d\}_{i=1}^n$ *and the eigenvalue problem (3.45), where* $\boldsymbol{v}_k \in \mathbb{R}^n$ *and* δ_k *are the* k*-th eigenvector and eigenvalue of matrix* $\boldsymbol{K} \in \mathbb{R}^{n \times n}$*. If* v_{ki} *is the* i*-th element of vector* $\boldsymbol{v}_k$*, the eigenfunction for the point* $\boldsymbol{x}$ *and the* i*-th training point* $\boldsymbol{x}_i$ *are:*

$$f_k(\boldsymbol{x}) = \frac{\sqrt{n}}{\delta_k} \sum_{i=1}^{n} v_{ki}\, \breve{k}(\boldsymbol{x}_i, \boldsymbol{x}), \tag{3.51}$$

$$f_k(\boldsymbol{x}_i) = \sqrt{n}\, v_{ki}, \tag{3.52}$$

respectively, where $\breve{k}(\boldsymbol{x}_i, \boldsymbol{x})$ *is the centered kernel. If* $\boldsymbol{x}$ *is a training point,* $\breve{k}(\boldsymbol{x}_i, \boldsymbol{x})$ *is the centered kernel over training data and if* $\boldsymbol{x}$ *is an out-of-sample point, then* $\breve{k}(\boldsymbol{x}_i, \boldsymbol{x}) = \breve{k}_t(\boldsymbol{x}_i, \boldsymbol{x})$ *is between the training set and the out-of-sample point (n.b. kernel centering is explained in Sect. 3.5.1).*

Proof The proof for Eq. (3.51) can be found in [10, proof of Theorem 1]. Equation (3.52) is claimed in [10, Proposition 1], and the proof can be found in [10, proof of Theorem 1, Eq. 7]. □

It is noteworthy that Eq. (3.51) is similar and related to the Nyström approximation of eigenfunctions of kernel operator which will be explained in Lemma 3.10.

Theorem 3.3 (Embedding from Eigenfunctions of Kernel Operator [9, Proposition 1], [10, Section 4]) *Consider a dimensionality reduction algorithm that embeds data into a low-dimensional embedding space. Let the embedding of the point* $\boldsymbol{x}$ *be* $\mathbb{R}^p \ni \boldsymbol{y}(\boldsymbol{x}) = [y_1(\boldsymbol{x}), \dots, y_p(\boldsymbol{x})]^\top$*, where* $p \leq n$*. The* k*-th dimension of this embedding is:*

$$y_k(\boldsymbol{x}) = \sqrt{\delta_k}\, \frac{f_k(\boldsymbol{x})}{\sqrt{n}} = \frac{1}{\sqrt{\delta_k}} \sum_{i=1}^{n} v_{ki}\, \breve{k}(\boldsymbol{x}_i, \boldsymbol{x}), \tag{3.53}$$

where $\breve{k}(\boldsymbol{x}_i, \boldsymbol{x})$ *is the centered training or out-of-sample kernel depending on whether* $\boldsymbol{x}$ *is a training or an out-of-sample point (n.b. kernel centering will be explained in Sect. 3.5.1). For proof, see [13].*

Theorem 3.3 has been widely used for out-of-sample (test data) embedding in many spectral dimensionality reduction algorithms [9]. Most spectral dimensionality reduction algorithms, such as multidimensional scaling, Isomap, and Laplacian eigenmap, can be extended to out-of-sample embedding using this theorem. This will be explained more in Chap. 10.

Corollary 3.1 (Embedding from Eigenvectors of Kernel Matrix) *Consider the eigenvalue problem for the kernel matrix, i.e. Eq. (3.45), where* $\boldsymbol{v}_k = [v_{k1}, \dots, v_{kn}]^\top$ *and* δ_k *are the* k*-th eigenvector and eigenvalue of kernel, respectively. According to Eqs. (3.52) and (3.53), it is possible to compute the* *embedding of point* $\boldsymbol{x}$*, denoted by* $\boldsymbol{y}(\boldsymbol{x}) = [y_1(\boldsymbol{x}), \ldots, y_p(\boldsymbol{x})]^\top$ *(where* $p \leq n$*) using the eigenvector of kernel as:*

$$y_k(\boldsymbol{x}) = \sqrt{\delta_k}\, \frac{1}{\sqrt{n}} (\sqrt{n}) v_{ki} = \sqrt{\delta_k}\, v_{ki}. \tag{3.54}$$

Equation (3.54) is used in several dimensionality reduction methods, such as maximum variance unfolding. The details of this will be explained in Chap. 10.

3.7 Kernelization Techniques

Linear algorithms cannot properly handle nonlinear patterns of data. When dealing with nonlinear data, if the algorithm is linear, two solutions exist to maximize performance:

1. Either the linear method should be modified to become nonlinear or a completely new nonlinear algorithm should be proposed to be able to handle nonlinear data. Some examples of this category are nonlinear dimensionality methods, such as locally linear embedding and Isomap; or
2. The nonlinear data should be modified to become more linear in pattern. That is, a transformation should be applied on data so that the pattern of data becomes roughly linear or easier to process by the linear algorithm. Some examples of this category are kernel versions of linear methods, such as kernel Principal Component Analysis (PCA), kernel Fisher Discriminant Analysis (FDA), and kernel Support Vector Machine (SVM).

The second approach is called kernelization in machine learning, and is defined as:

Definition 3.7 (Kernelization) In machine learning and data science, kernelization means a slight change in algorithm formulation (without any modification in the idea of algorithm) so that the pulled data to the RKHS, rather than the raw data, are used as the input of the algorithm.

Figure 3.2 demonstrates how kernelization for transforming data can help separate classes for better classification. These classes may not be linearly separable using a linear hyperplane. A pulling function to RKHS may be able to transform the data points in a way that they become linearly separable, or at least better separated, in the new space. This concept has been widely used for kernel SVM [39].

Kernelization can be useful for enabling linear algorithms to handle nonlinear data better. Nevertheless, it should be noted that nonlinear algorithms can also be kernelized to handle nonlinear data perhaps better by transforming data. Generally, there exist two main approaches for kernelization in machine learning, kernel trick and kernelization using representation theory. These two approaches are related in theory, but have two ways for kernelization.

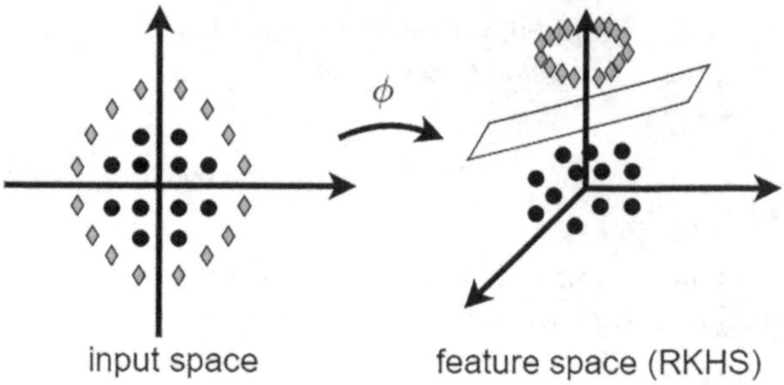

Fig. 3.2 Transforming data to RKHS using kernels to make the nonlinear pattern of data more linear. For example, here the classes have become linearly separable (by a linear hyperplane) after kernelization

3.7.1 Kernelization by Kernel Trick

Recall Eqs. (3.13) and (3.15) where a kernel can be computed by the inner product between pulled data instances to the RKHS. One technique to kernelize an algorithm is called a kernel trick. This technique attempts to formulate the algorithm formulas or optimization in a way that data always appear as an inner product of data instances and not as a data instance alone. In other words, the formulation of the algorithm should only have $\boldsymbol{x}^\top \boldsymbol{x}$, $\boldsymbol{x}^\top \boldsymbol{X}$, $\boldsymbol{X}^\top \boldsymbol{x}$, or $\boldsymbol{X}^\top \boldsymbol{X}$ and not a single $\boldsymbol{x}$ or $\boldsymbol{X}$. In this way, the kernel trick replaces $\boldsymbol{x}^\top \boldsymbol{x}$ with $\boldsymbol{\phi}(\boldsymbol{x})^\top \boldsymbol{\phi}(\boldsymbol{x})$ and uses Eq. (3.13) or (3.15). To better explain this, the kernel trick applies the following mapping:

$$\boldsymbol{x}^\top \boldsymbol{x} \mapsto \boldsymbol{\phi}(\boldsymbol{x})^\top \boldsymbol{\phi}(\boldsymbol{x}) \overset{(3.13)}{=} k(\boldsymbol{x}, \boldsymbol{x}). \tag{3.55}$$

Therefore, the inner products of points are all replaced with the kernel between points. The matrix form of the kernel trick is:

$$\boldsymbol{X}^\top \boldsymbol{X} \mapsto \boldsymbol{\Phi}(\boldsymbol{X})^\top \boldsymbol{\Phi}(\boldsymbol{X}) \overset{(3.15)}{=} \boldsymbol{K}(\boldsymbol{X}, \boldsymbol{X}) \in \mathbb{R}^{n \times n}. \tag{3.56}$$

Most often, the kernel matrix is computed over one dataset, meaning its dimensionality is $n \times n$. However, in some cases, the kernel matrix is computed between two sets of data instances with sample sizes n_1 and n_2 for example, i.e., datasets $\boldsymbol{X}_1 := [\boldsymbol{x}_{1,1}, \ldots, \boldsymbol{x}_{1,n_1}]$ and $\boldsymbol{X}_2 := [\boldsymbol{x}_{2,1}, \ldots, \boldsymbol{x}_{2,n_2}]$. In this case, the kernel matrix has size $n_1 \times n_2$ and the kernel trick is:

$$\boldsymbol{x}_{1,i}^\top \boldsymbol{x}_{1,j} \mapsto \boldsymbol{\phi}(\boldsymbol{x}_{1,i})^\top \boldsymbol{\phi}(\boldsymbol{x}_{1,j}) \overset{(3.13)}{=} k(\boldsymbol{x}_{1,i}, \boldsymbol{x}_{1,j}), \tag{3.57}$$

$$\boldsymbol{X}_1^\top \boldsymbol{X}_2 \mapsto \boldsymbol{\Phi}(\boldsymbol{X}_1)^\top \boldsymbol{\Phi}(\boldsymbol{X}_2) \overset{(3.15)}{=} \boldsymbol{K}(\boldsymbol{X}_1, \boldsymbol{X}_2) \in \mathbb{R}^{n_1 \times n_2}. \tag{3.58}$$

The kernel trick can be used in machine learning. In the cases of kernel PCA [34, 35] and kernel SVM [39], a "dual" method for the algorithm is proposed, which only uses the inner product of data. Then, the dual algorithm is kernelized using the kernel trick.

3.7.2 Kernelization by Representation Theory

Some algorithms cannot be formulated to have data only in an inner product form, nor does their dual have this form. The Fisher Discriminant Analysis (FDA) [24], demonstrates this concept and must use a different technique for kernelization.

Lemma 3.7 (Representation of Function Using Bases [24]) *Consider an RKHS denoted by $\mathcal{H}$. Any function $f \in \mathcal{H}$ lies in the span of all points in the RKHS, i.e.,*

$$f = \sum_{i=1}^{n} \alpha_i \, \phi(\boldsymbol{x}_i). \tag{3.59}$$

Remark 3.3 (Justification by Representation Theory) According to representation theory [2], any function in the space can be represented as a linear combination of bases of the space, since the function is in the space and the space is spanned by the bases. Now, assume the space is RKHS. This would mean that any function should lie in the RKHS spanned by the pulled data points to the feature space. This justifies Lemma 3.7 using representation theory.

3.7.2.1 Kernelization for Vector Solution

To kernelize an algorithm whose optimization variable or solution is the vector/direction $\boldsymbol{u} \in \mathbb{R}^d$ in the input space, the solution must be pulled to RKHS by Eq. (3.11) to obtain $\phi(\boldsymbol{u})$. According to Lemma 3.7, this pulled solution must lie in the span of all pulled training points $\{\phi(\boldsymbol{x}_i)\}_{i=1}^n$ as:

$$\phi(\boldsymbol{u}) = \sum_{i=1}^{n} \alpha_i \, \phi(\boldsymbol{x}_i) = \boldsymbol{\Phi}(\boldsymbol{X})\, \boldsymbol{\alpha}, \tag{3.60}$$

which is t dimensional, and t may be infinite. Note that $\boldsymbol{\Phi}(\boldsymbol{X})$ is defined by Eq. (3.14) and is $t \times n$ dimensional. The vector $\boldsymbol{\alpha} := [\alpha_1, \ldots, \alpha_n]^\top \in \mathbb{R}^n$ contains the coefficients. According to Eq. (3.11), $\boldsymbol{u}$ can be replaced with $\phi(\boldsymbol{u})$ in the algorithm. If, by this replacement, the terms $\phi(\boldsymbol{x}_i)^\top \phi(\boldsymbol{x}_i)$ or $\boldsymbol{\Phi}(\boldsymbol{X})^\top \boldsymbol{\Phi}(\boldsymbol{X})$ appear, the Eq. (3.13) can be used and $\phi(\boldsymbol{x}_i)^\top \phi(\boldsymbol{x}_i)$ can be replaced with $k(\boldsymbol{x}_i, \boldsymbol{x}_i)$ or Eq. (3.15) can be used to replace $\boldsymbol{\Phi}(\boldsymbol{X})^\top \boldsymbol{\Phi}(\boldsymbol{X})$ with $\boldsymbol{K}(\boldsymbol{X}, \boldsymbol{X})$—this kernelizes the method. The steps of kernelization by representation theory can be summarized in the following process:

- Step 1: $\boldsymbol{u} \rightarrow \phi(\boldsymbol{u})$
- Step 2: Replace $\phi(\boldsymbol{u})$ with Eq. (3.60) in the algorithm formulation
- Step 3: Some $\phi(\boldsymbol{x}_i)^\top \phi(\boldsymbol{x}_i)$ or $\boldsymbol{\Phi}(\boldsymbol{X})^\top \boldsymbol{\Phi}(\boldsymbol{X})$ terms appear in the formulation
- Step 4: Use Eq. (3.13) or (3.15)

- Step 5: Solve (optimize) the algorithm where the variable to find is $\boldsymbol{\alpha}$, rather than $\boldsymbol{u}$

Usually, the goal of the algorithm results in the kernel. For example, if $\boldsymbol{u}$ is a projection direction, the desired projected data are obtained as follows:

$$\boldsymbol{u}^\top \boldsymbol{x}_i \overset{(3.11)}{\mapsto} \phi(\boldsymbol{u})^\top \phi(\boldsymbol{x}_i) \overset{(3.60)}{=} \boldsymbol{\alpha}^\top \boldsymbol{\Phi}(\boldsymbol{X})^\top \phi(\boldsymbol{x}_i) \overset{(3.15)}{=} \boldsymbol{\alpha}^\top \boldsymbol{k}(\boldsymbol{X}, \boldsymbol{x}_i), \tag{3.61}$$

where $\boldsymbol{k}(\boldsymbol{X}, \boldsymbol{x}_i) \in \mathbb{R}^n$ is the kernel between all n training points and point $\boldsymbol{x}_i$. This equation demonstrates that the desired goal is based on the kernel.

3.7.2.2 Kernelization for Matrix Solution

Usually, the algorithm has multiple directions/vectors as its solution. In other words, its solution is a matrix $\boldsymbol{U} = [\boldsymbol{u}_1, \ldots, \boldsymbol{u}_p] \in \mathbb{R}^{d \times p}$. In this case, Eq. (3.60) is used for all p vectors and, in a matrix form, there is:

$$\boldsymbol{\Phi}(\boldsymbol{U}) = \boldsymbol{\Phi}(\boldsymbol{X})\, \boldsymbol{A}, \tag{3.62}$$

where $\mathbb{R}^{n \times p} \ni \boldsymbol{A} := [\boldsymbol{\alpha}_1, \ldots, \boldsymbol{\alpha}_p]^\top$, and $\boldsymbol{\Phi}(\boldsymbol{U})$ is $t \times p$ dimensional where t may be infinite. Similarly, the following steps should be performed to kernelize the algorithm:

- Step 1: $\boldsymbol{U} \to \phi(\boldsymbol{U})$
- Step 2: Replace $\phi(\boldsymbol{U})$ with Eq. (3.62) in the algorithm formulation
- Step 3: Some $\boldsymbol{\Phi}(\boldsymbol{X})^\top \boldsymbol{\Phi}(\boldsymbol{X})$ terms appear in the formulation
- Step 4: Use Eq. (3.15)
- Step 5: Solve (optimize) the algorithm where the variable to find is $\boldsymbol{A}$ rather than $\boldsymbol{U}$

The goal of the algorithm usually results in kernel. For example, if $\boldsymbol{U}$ is a projection matrix onto its column space, then:

$$\boldsymbol{U}^\top \boldsymbol{x}_i \overset{(3.11)}{\mapsto} \phi(\boldsymbol{U})^\top \phi(\boldsymbol{x}_i) \overset{(3.62)}{=} \boldsymbol{A}^\top \boldsymbol{\Phi}(\boldsymbol{X})^\top \phi(\boldsymbol{x}_i) \overset{(3.15)}{=} \boldsymbol{A}^\top \boldsymbol{k}(\boldsymbol{X}, \boldsymbol{x}_i), \tag{3.63}$$

where $\boldsymbol{k}(\boldsymbol{X}, \boldsymbol{x}_i) \in \mathbb{R}^n$ is the kernel between all n training points and point $\boldsymbol{x}_i$.

In many machine learning algorithms, the solution $\boldsymbol{U} \in \mathbb{R}^{d \times p}$ is a projection matrix for projecting d-dimensional data onto a p-dimensional subspace. Some example methods, that have used kernelization by representation theory, are kernel Fisher discriminant analysis (FDA) [24] and kernel supervised principal component analysis [5].

Remark 3.4 (Linear Kernel in Kernelization) Using the kernel trick, the kernelized algorithm with a linear kernel is equivalent to the nonkernelized algorithm. This is because in a linear kernel, there is $\phi(\boldsymbol{x}) = \boldsymbol{x}$ and $k(\boldsymbol{x}, \boldsymbol{y}) = \boldsymbol{x}^\top \boldsymbol{y}$ according to

Eq. (3.16). The kernel trick, Eq. (3.57), maps data as $\boldsymbol{x}^\top \boldsymbol{y} \mapsto \phi(\boldsymbol{x})^\top \phi(\boldsymbol{y}) = \boldsymbol{x}^\top \boldsymbol{y}$ for a linear kernel. Therefore, a linear kernel does not have any effect when using the kernel trick. Examples for this are kernel PCA [34, 35] and kernel SVM [39], which are equivalent to PCA and SVM, respectively, if the linear kernel is used.

However, the kernel trick does have an impact when using kernelization by representation theory. This is because it first pulls the solution and represents it as a span of bases and then uses the kernel trick. This is in contrast to kernelization by the kernel trick, which applies the kernel trick directly to the data points in the input space. Therefore, a kernelized algorithm using representation theory with a linear kernel is not equivalent to a nonkernelized algorithm. Examples of this are kernel FDA [24] and kernel supervised PCA [5] which are different from FDA and supervised PCA, respectively, even if a linear kernel is used.

3.8 Difference Measures Using Kernels

Kernels can be used for measuring the difference between data points or random variables. For example, they can be used for measuring the distance of points in the RKHS (Sect. 3.8.1), calculating dependence between two random variables (Sect. 3.8.2), and computing the difference of moments between two random distributions (Sect. 3.8.3).

3.8.1 Distance in RKHS

Lemma 3.8 (Distance in RKHS [33]) *The squared Euclidean distance between points in the feature space is:*

$$\|\phi(\boldsymbol{x}_i) - \phi(\boldsymbol{x}_j)\|_k^2 = k(\boldsymbol{x}_i, \boldsymbol{x}_i) + k(\boldsymbol{x}_j, \boldsymbol{x}_j) - 2k(\boldsymbol{x}_i, \boldsymbol{x}_j). \tag{3.64}$$

Proof

$$\begin{aligned}
&\|\phi(\boldsymbol{x}_i) - \phi(\boldsymbol{x}_j)\|_k^2 = \big(\phi(\boldsymbol{x}_i) - \phi(\boldsymbol{x}_j)\big)^\top \big(\phi(\boldsymbol{x}_i) - \phi(\boldsymbol{x}_j)\big) \\
&= \phi(\boldsymbol{x}_i)^\top \phi(\boldsymbol{x}_i) + \phi(\boldsymbol{x}_j)^\top \phi(\boldsymbol{x}_j) - \phi(\boldsymbol{x}_i)^\top \phi(\boldsymbol{x}_j) - \phi(\boldsymbol{x}_j)^\top \phi(\boldsymbol{x}_i) \\
&= \phi(\boldsymbol{x}_i)^\top \phi(\boldsymbol{x}_i) + \phi(\boldsymbol{x}_j)^\top \phi(\boldsymbol{x}_j) - 2\phi(\boldsymbol{x}_i)^\top \phi(\boldsymbol{x}_j) \overset{(3.13)}{=} k(\boldsymbol{x}_i, \boldsymbol{x}_i) + k(\boldsymbol{x}_j, \boldsymbol{x}_j) - 2k(\boldsymbol{x}_i, \boldsymbol{x}_j).
\end{aligned}$$

□

3.8.2 Hilbert-Schmidt Independence Criterion (HSIC)

Measuring the correlation between two random variables is easier than measuring the dependence between them because correlation is just "linear" dependence. According to Hein and Bousquet [19], two random variables X and Y are independent if, and only if, any of their bounded continuous functions are uncorrelated. Therefore, if the samples of two random variables $\{\boldsymbol{x}\}_{i=1}^n$ and $\{\boldsymbol{y}\}_{i=1}^n$ are mapped to two different ("separable") RKHSs and have $\phi(\boldsymbol{x})$ and $\phi(\boldsymbol{y})$, it is possible to measure the correlation of $\phi(\boldsymbol{x})$ and $\phi(\boldsymbol{y})$ in Hilbert space, which provides an estimation of the dependence of $\boldsymbol{x}$ and $\boldsymbol{y}$ in the input space.

The correlation of $\phi(\boldsymbol{x})$ and $\phi(\boldsymbol{y})$ can be computed by the Hilbert-Schmidt norm of the cross-covariance of them [16]. Note that the squared Hilbert-Schmidt norm of a matrix $\boldsymbol{A}$ is [6]:

$$||\boldsymbol{A}||_{HS}^2 := \mathbf{tr}(\boldsymbol{A}^\top \boldsymbol{A}), \tag{3.65}$$

and the cross-covariance matrix of two vectors $\boldsymbol{x}$ and $\boldsymbol{y}$ is [16]:

$$\mathbb{C}\mathrm{ov}(\boldsymbol{x}, \boldsymbol{y}) := \mathbb{E}\Big[\big(\boldsymbol{x} - \mathbb{E}(\boldsymbol{x})\big)\big(\boldsymbol{y} - \mathbb{E}(\boldsymbol{y})\big)\Big]. \tag{3.66}$$

Using the explained intuition, an empirical estimation of the Hilbert-Schmidt Independence Criterion (HSIC) is introduced [16]:

$$\mathrm{HSIC}(X, Y) := \frac{1}{(n-1)^2}\,\mathbf{tr}(\boldsymbol{K}_x \boldsymbol{H} \boldsymbol{K}_y \boldsymbol{H}), \tag{3.67}$$

where $\boldsymbol{K}_x := \boldsymbol{\phi}(\boldsymbol{x})^\top \boldsymbol{\phi}(\boldsymbol{x})$ and $\boldsymbol{K}_y := \boldsymbol{\phi}(\boldsymbol{y})^\top \boldsymbol{\phi}(\boldsymbol{y})$ are the kernels over $\boldsymbol{x}$ and $\boldsymbol{y}$, respectively. The term $1/(n-1)^2$ is used for normalization. The matrix $\boldsymbol{H}$ is the centering matrix (see Chap. 2). Note that HSIC double-centers one of the kernels and then computes the Hilbert-Schmidt norm between kernels. HSIC measures the dependence of two random variable vectors $\boldsymbol{x}$ and $\boldsymbol{y}$. Note that HSIC $= 0$ and HSIC > 0 mean that $\boldsymbol{x}$ and $\boldsymbol{y}$ are independent and dependent, respectively. The greater the HSIC is, the greater the dependence they have.

Lemma 3.9 (Independence of Random Variables Using Cross-Covariance [14, Theorem 5]) *Two random variables X and Y are independent if and only if $\mathbb{C}ov(f(\boldsymbol{x}), f(\boldsymbol{y})) = 0$ for any pair of bounded continuous functions (f, g). Because of the relation of HSIC with the cross-covariance of variables, two random variables are independent if and only if HSIC$(X, Y) = 0$.*

3.8.3 Maximum Mean Discrepancy (MMD)

Maximum Mean Discrepancy (MMD), also known as the kernel two-sample test and proposed in [15], is a measure for difference of distributions. For comparison of two distributions, the difference of all moments of the two distributions can be found. However, as the number of moments is infinite, it is intractable to calculate the difference of all moments. MMD proposed that, to do this tractably, both distributions need to be pulled to the feature space and then the distance of all pulled data points from distributions in RKHS need to be computed. This difference will be a suitable estimate for the difference of all moments in the input space.

MMD is a semimetric and uses distance in the RKHS [33] (see Lemma 3.8). Consider probability density functions $\mathbb{P}$ and $\mathbb{Q}$ and samples $\{\boldsymbol{x}_i\}_{i=1}^n \sim \mathbb{P}$ and $\{\boldsymbol{y}_i\}_{i=1}^n \sim \mathbb{Q}$. The squared MMD between them is:

$$
\begin{aligned}
\text{MMD}^2(\mathbb{P}, \mathbb{Q}) &:= \Big\| \frac{1}{n}\sum_{i=1}^n \boldsymbol{\phi}(\boldsymbol{x}_i) - \frac{1}{n}\sum_{i=1}^n \boldsymbol{\phi}(\boldsymbol{y}_i) \Big\|_k^2 \\
&\overset{(3.64)}{=} \frac{1}{n^2}\sum_{i=1}^n\sum_{j=1}^n k(\boldsymbol{x}_i, \boldsymbol{x}_j) + \frac{1}{n^2}\sum_{i=1}^n\sum_{j=1}^n k(\boldsymbol{y}_i, \boldsymbol{y}_j) - \frac{2}{n^2}\sum_{i=1}^n\sum_{j=1}^n k(\boldsymbol{x}_i, \boldsymbol{y}_j) \\
&= \mathbb{E}_x[\boldsymbol{K}(\boldsymbol{x}, \boldsymbol{y})] + \mathbb{E}_y[\boldsymbol{K}(\boldsymbol{x}, \boldsymbol{y})] - 2\mathbb{E}_x[\mathbb{E}_y[\boldsymbol{K}(\boldsymbol{x}, \boldsymbol{y})]],
\end{aligned} \tag{3.68}
$$

where $\mathbb{E}_x[\boldsymbol{K}(\boldsymbol{x}, \boldsymbol{y})]$, $\mathbb{E}_y[\boldsymbol{K}(\boldsymbol{x}, \boldsymbol{y})]$, and $\mathbb{E}_x[\mathbb{E}_y[\boldsymbol{K}(\boldsymbol{x}, \boldsymbol{y})]]$ are the average of rows, average of columns, and total average of rows and columns of the kernel matrix, respectively. Note that $\text{MMD} \geq 0$ where $\text{MMD} = 0$ means the two distributions are equivalent if the kernel used is characteristic. MMD has been widely used in machine learning, such as generative moment matching networks [22]. In these networks, MMD is used for matching all moments between the distributions of a dataset and the generated data so that the generated data become similar to the real data.

Remark 3.5 (Equivalence of HSIC and MMD [37]) After the development of HSIC and MMD measures, it was found out that they are equivalent.

3.9 Factorization of Kernel and the Nyström Method

3.9.1 Singular Value and Eigenvalue Decompositions of Kernel

The Singular Value Decomposition (SVD) of the pulled data to the feature space is:

$$
\mathbb{R}^{t \times n} \ni \boldsymbol{\Phi}(\boldsymbol{X}) = \boldsymbol{U}\boldsymbol{\Sigma}\boldsymbol{V}^\top, \tag{3.69}
$$

where $\boldsymbol{U} \in \mathbb{R}^{t \times n}$ and $\boldsymbol{V} \in \mathbb{R}^{n \times n}$ are orthogonal matrices and contain the left and right singular vectors, respectively, and $\boldsymbol{\Sigma} \in \mathbb{R}^{n \times n}$ is a diagonal matrix with singular values. Note that here, notations such as $\mathbb{R}^{t \times n}$ are used to demonstrate the dimensionality of the matrices and this notation does not imply a Euclidean space.

As mentioned previously, pulled data might not be available. This means that Eq. (3.69) cannot be performed; however the kernel is available. Using the SVD of pulled data in the formulation of the kernel results in:

$$\boldsymbol{K} \overset{(3.15)}{=} \boldsymbol{\Phi}(\boldsymbol{X})^\top \boldsymbol{\Phi}(\boldsymbol{X}) \overset{(3.69)}{=} (\boldsymbol{U\Sigma V}^\top)^\top (\boldsymbol{U\Sigma V}^\top) = \boldsymbol{V\Sigma} \underbrace{\boldsymbol{U}^\top \boldsymbol{U}}_{=\boldsymbol{I}} \boldsymbol{\Sigma V}^\top = \boldsymbol{V\Sigma\Sigma V}^\top = \boldsymbol{V\Sigma}^2 \boldsymbol{V}^\top,$$

$$\implies \boldsymbol{KV} = \boldsymbol{V\Sigma}^2 \underbrace{\boldsymbol{V}^\top \boldsymbol{V}}_{=\boldsymbol{I}} \implies \boldsymbol{KV} = \boldsymbol{V\Sigma}^2. \tag{3.70}$$

If $\boldsymbol{\Delta} = \mathbf{diag}([\delta_1, \ldots, \delta_n]^\top) := \boldsymbol{\Sigma}^2$ is used, the equation becomes:

$$\boldsymbol{KV} = \boldsymbol{V\Delta}, \tag{3.71}$$

which is the matrix of Eq. (3.45), i.e. the eigenvalue problem (see Chap. 2) for the kernel matrix with $\boldsymbol{V}$ and $\boldsymbol{\Delta}$ as eigenvectors and eigenvalues, respectively. Therefore, for SVD on the pulled dataset, the Eigenvalue Decomposition (EVD) can be applied on the kernel, where the eigenvectors of the kernel are equal to right singular vectors of the pulled dataset and the eigenvalues of the kernel are the squared singular values of the pulled dataset. This technique has been used in kernel principal component analysis.

3.9.2 Nyström Method for Approximation of Eigenfunctions

Nyström method was first proposed in [26] and was initially used for approximating the eigenfunctions of an operator (or of a matrix corresponding to an operator). The following lemma provides the Nyström approximation for the eigenfunctions of the kernel operator defined by Eq. (3.41).

Lemma 3.10 (Nyström Approximation of Eigenfunction [4, Chapter 3]) *Consider a training dataset $\{\boldsymbol{x}_i \in \mathbb{R}^d\}_{i=1}^n$ and the eigenfunction problem (3.39) where $f_k \in \mathcal{H}$ and λ_k are the k-th eigenfunction and eigenvalue of kernel operator defined by Eq. (3.41) or (3.42). The eigenfunction can be approximated by Nyström method:*

$$f_k(\boldsymbol{x}) \approx \frac{1}{n\lambda_k} \sum_{i=1}^{n} k(\boldsymbol{x}_i, \boldsymbol{x})\, f_k(\boldsymbol{x}_i), \tag{3.72}$$

where $k(\boldsymbol{x}_i, \boldsymbol{x})$ is the kernel (or centered kernel) corresponding to the kernel operator. See [13] for proof.

3.9.3 Nyström Method for Kernel Completion and Approximation

The kernel matrix can usually be approximated due to its low rank. This is important because in big data when $n \gg 1$, constructing the kernel matrix is both time-consuming and intractable to store on a computer, i.e., its computation will run indefinitely and will eventually raise a memory error. Therefore, it is desired to compute the kernel function between a subset of data points (called landmarks) and then approximate the rest of the kernel matrix using this subset of the kernel matrix. Nyström approximation can be used to accomplish this goal. It is a technique used to approximate a positive semidefinite matrix using merely a subset of its columns (or rows) [41, Section 1.2].

A positive semidefinite matrix $\mathbb{R}^{n\times n} \ni \boldsymbol{K} \succeq 0$ can be described as:

$$\mathbb{R}^{n\times n} \ni \boldsymbol{K} = \left[\begin{array}{c|c} \boldsymbol{A} & \boldsymbol{B} \\ \hline \boldsymbol{B}^\top & \boldsymbol{C} \end{array}\right], \tag{3.73}$$

where $\boldsymbol{A} \in \mathbb{R}^{m\times m}$, $\boldsymbol{B} \in \mathbb{R}^{m\times(n-m)}$, and $\boldsymbol{C} \in \mathbb{R}^{(n-m)\times(n-m)}$, and in which $m \ll n$. This positive semidefinite matrix can be a kernel (or Gram) matrix.

The Nyström approximation says if there are small parts of this matrix, i.e. $\boldsymbol{A}$ and $\boldsymbol{B}$, it is possible to approximate $\boldsymbol{C}$, and thus the whole matrix $\boldsymbol{K}$. Assume $m = 2$ (containing two points, a and b) and $n = 5$ (containing three other points, c, d, and e). If the similarity (or distance) between points a and b is known, resulting in matrix $\boldsymbol{A}$, as well as the similarity (or distance) of points c, d, and e from points a and b, resulting in matrix $\boldsymbol{B}$, there are few possibilities for the locations of c, d, and e, which is matrix $\boldsymbol{C}$. This is because of the positive semidefiniteness of matrix $\boldsymbol{K}$. The points selected in submatrix $\boldsymbol{A}$ are named *landmarks*. Note that the landmarks can be selected randomly from the columns/rows of matrix $\boldsymbol{K}$ and, without loss of generality, they can be merged to form a submatrix at the top-left corner of the matrix. For the Nyström approximation, some methods have been proposed for sampling more important columns/rows of the matrix specifically rather than randomly [21]. This field of research, such as [21], is named "column subset selection" or "landmark selection" for Nyström approximation.

As the matrix $\boldsymbol{K}$ is positive semidefinite, by definition, it can be written as $\boldsymbol{K} = \boldsymbol{O}^\top \boldsymbol{O}$. If $\boldsymbol{O} = [\boldsymbol{R}, \boldsymbol{S}]$ is taken, where $\boldsymbol{R}$ contains the selected columns (landmarks) of $\boldsymbol{O}$ and $\boldsymbol{S}$ are the other columns of $\boldsymbol{O}$, this results in:

$$\boldsymbol{K} = \boldsymbol{O}^\top \boldsymbol{O} = \begin{bmatrix} \boldsymbol{R}^\top \\ \boldsymbol{S}^\top \end{bmatrix} [\boldsymbol{R}, \boldsymbol{S}] = \begin{bmatrix} \boldsymbol{R}^\top \boldsymbol{R} & \boldsymbol{R}^\top \boldsymbol{S} \\ \boldsymbol{S}^\top \boldsymbol{R} & \boldsymbol{S}^\top \boldsymbol{S} \end{bmatrix} \overset{(3.73)}{=} \begin{bmatrix} \boldsymbol{A} & \boldsymbol{B} \\ \boldsymbol{B}^\top & \boldsymbol{C} \end{bmatrix}. \tag{3.74}$$

Therefore, $\boldsymbol{A} = \boldsymbol{R}^\top \boldsymbol{R}$. The eigenvalue decomposition (see Chap. 2) of $\boldsymbol{A}$ results in:

$$\boldsymbol{A} = \boldsymbol{U}\boldsymbol{\Sigma}\boldsymbol{U}^\top \tag{3.75}$$

$$\implies \boldsymbol{R}^\top \boldsymbol{R} = \boldsymbol{U}\boldsymbol{\Sigma}\boldsymbol{U}^\top \implies \boldsymbol{R} = \boldsymbol{\Sigma}^{(1/2)}\boldsymbol{U}^\top. \tag{3.76}$$

Moreover, $\boldsymbol{B} = \boldsymbol{R}^\top \boldsymbol{S}$ which results in:

$$\boldsymbol{B} = (\boldsymbol{\Sigma}^{(1/2)}\boldsymbol{U}^\top)^\top \boldsymbol{S} = \boldsymbol{U}\boldsymbol{\Sigma}^{(1/2)}\boldsymbol{S} \overset{(a)}{\implies} \boldsymbol{U}^\top \boldsymbol{B} = \boldsymbol{\Sigma}^{(1/2)}\boldsymbol{S} \implies \boldsymbol{S} = \boldsymbol{\Sigma}^{(-1/2)}\boldsymbol{U}^\top \boldsymbol{B}, \tag{3.77}$$

where (a) is because $\boldsymbol{U}$ is orthogonal (in the eigenvalue decomposition). Lastly:

$$\boldsymbol{C} = \boldsymbol{S}^\top \boldsymbol{S} = \boldsymbol{B}^\top \boldsymbol{U}\boldsymbol{\Sigma}^{(-1/2)}\boldsymbol{\Sigma}^{(-1/2)}\boldsymbol{U}^\top \boldsymbol{B} = \boldsymbol{B}^\top \boldsymbol{U}\boldsymbol{\Sigma}^{-1}\boldsymbol{U}^\top \boldsymbol{B} \overset{(3.75)}{=} \boldsymbol{B}^\top \boldsymbol{A}^{-1}\boldsymbol{B}. \tag{3.78}$$

Therefore, Eq. (3.73) transforms into:

$$\boldsymbol{K} \approx \left[\begin{array}{c|c} \boldsymbol{A} & \boldsymbol{B} \\ \hline \boldsymbol{B}^\top & \boldsymbol{B}^\top \boldsymbol{A}^{-1}\boldsymbol{B} \end{array}\right]. \tag{3.79}$$

Lemma 3.11 (Impact of Size of Submatrix A on Nyström approximation) *By increasing m, the approximation of Eq. (3.79) becomes more accurate. If the rank of $\boldsymbol{K}$ is at most m, this approximation is exact.*

Proof Equation (3.78) provides the inverse of $\boldsymbol{A}$. In order to have this inverse, the matrix $\boldsymbol{A}$ must not be singular. For having a full-rank $\boldsymbol{A} \in \mathbb{R}^{m \times m}$, the rank of $\boldsymbol{A}$ should be m. This results in m being an upper bound on the rank of $\boldsymbol{K}$ and a lower bound on the number of landmarks. In practice, it is recommended to use a higher number of landmarks to create a more accurate approximation; however, there is a trade-off with the speed of calculations. □

Corollary 3.2 *Since $m \ll n$, the Nyström approximation works well, especially for the low-rank matrices, because a small $\boldsymbol{A}$ and therefore a small number of landmarks is needed for approximation. Usually, due to the manifold hypothesis, data points fall on a submanifold; signifying that the kernel (similarity) matrix or the distance matrix usually has a low rank. Therefore, the Nyström approximation works well for many kernel-based or distance-based dimensionality reduction methods. For example, it is used in most of the spectral dimensionality reduction methods for embedding big data. Details will be further explained in Chaps. 7, 8, and 10.*

3.9.4 Use of Nyström Approximation for Landmark Spectral Embedding

The spectral dimensionality reduction methods usually reduce to an eigenvalue problem. Therefore, they cannot handle big data, where $n \gg 1$. To address this issue, there exist landmark methods that approximate the embedding of all points using the embedding of landmarks. Big data, i.e. $n \gg 1$, results in large kernel matrices. Utilizing landmarks can reduce computations, a technique that is named the Nyström approximation, which is used for kernel approximation and completion. The Nyström approximation can be used to make the spectral methods, such as locally linear embedding and Multidimensional Scaling (MDS) scalable, suitable for big data embedding. It is demonstrated in [29] that all landmark MDS methods are Nyström approximations.

3.10 Chapter Summary

This chapter introduced the necessary background on kernels and RKHS, including the Mercer's theorem to demonstrate why kernels work, the well-known kernel functions in machine learning, and eigenfunctions and their use in spectral dimensionality reduction. Two main kernelization techniques, i.e. kernel trick and representation theory, were explained in detail. The reader was also introduced to several difference measures using kernels, low-rank kernel factorization and the Nyström method for kernel approximation.

All concepts covered in this chapter will be used as a foundation for the following chapters. Many of the dimensionality reduction methods have their own kernel versions, in which kernel functions are used. Eigenfunctions will be used for the unified spectral framework introduced in Chap. 10. They are also going to be used for out-of-sample embedding in spectral methods. The Nyström method will be used for extension of many spectral methods for embedding big data. The factorization of the kernel matrix will also be used in several spectral methods, such as in kernel principal component analysis in Chap. 5.

References

1. Mark A Aizerman, E. M. Braverman, and L. I. Rozonoer. "Theoretical foundations of the potential function method in pattern recognition learning". In: *Automation and remote control* 25 (1964), pp. 821–837.
2. Jonathan L Alperin. *Local representation theory: Modular representations as an introduction to the local representation theory of finite groups*. Cambridge University Press, 1993.
3. Nachman Aronszajn. "Theory of reproducing kernels". In: *Transactions of the American mathematical society* 68.3 (1950), pp. 337–404.
4. Christopher TH Baker. *The numerical treatment of integral equations*. Clarendon Press, 1978.
5. Elnaz Barshan et al. "Supervised principal component analysis: Visualization, classification and regression on subspaces and submanifolds". In: *Pattern Recognition* 44.7 (2011), pp. 1357–1371.
6. Jordan Bell. "Trace class operators and Hilbert-Schmidt operators". In: *Department of Mathematics, University of Toronto, Technical Report* (2016).
7. Yoshua Bengio et al. "Learning eigenfunctions links spectral embedding and kernel PCA". In: *Neural computation* 16.10 (2004), pp. 2197–2219.

8. Yoshua Bengio et al. *Learning eigenfunctions of similarity: linking spectral clustering and kernel PCA*. Tech. rep. Departement d'Informatique et Recherche Operationnelle, 2003.
9. Yoshua Bengio et al. "Out-of-sample extensions for LLE, Isomap, MDS, eigenmaps, and spectral clustering". In: *Advances in neural information processing systems* 16 (2003), pp. 177–184.
10. Yoshua Bengio et al. *Spectral clustering and kernel PCA are learning eigenfunctions*. Tech. rep. Departement d'Informatique et Recherche Operationnelle, Technical Report 1239, 2003.
11. Alain Berlinet and Christine Thomas-Agnan. *Reproducing kernel Hilbert spaces in probability and statistics*. Springer Science & Business Media, 2011.
12. Michael AA Cox and Trevor F Cox. "Multidimensional scaling". In: *Handbook of data visualization*. Springer, 2008, pp. 315–347.
13. Benyamin Ghojogh et al. "Reproducing Kernel Hilbert Space, Mercer's Theorem, Eigenfunctions, Nyström Method, and Use of Kernels in Machine Learning: Tutorial and Survey". In: *arXiv preprint arXiv:2106.08443* (2021).
14. Arthur Gretton and László Györfi. "Consistent nonparametric tests of independence". In: *The Journal of Machine Learning Research* 11 (2010), pp. 1391–1423.
15. Arthur Gretton et al. "A kernel method for the two-sample-problem". In: *Advances in neural information processing systems* 19 (2006), pp. 513–520.
16. Arthur Gretton et al. "Measuring statistical dependence with Hilbert-Schmidt norms". In: *International conference on algorithmic learning theory* Springer. 2005, pp. 63–77.
17. Jihun Ham et al. "A kernel view of the dimensionality reduction of manifolds". In: *Proceedings of the twenty-first international conference on Machine learning*. 2004, p. 47.
18. Trevor Hastie, Robert Tibshirani, and Jerome Friedman. *The elements of statistical learning: data mining, inference, and prediction*. Springer Science & Business Media, 2009.
19. Matthias Hein and Olivier Bousquet. "Kernels, associated structures and generalizations". In: *Max-Planck-Institut fuer biologische Kybernetik, Technical Report* (2004).
20. George Kimeldorf and Grace Wahba. "Some results on Tchebycheffian spline functions". In: *Journal of mathematical analysis and applications* 33.1 (1971), pp. 82–95.
21. Sanjiv Kumar, Mehryar Mohri, and Ameet Talwalkar. "Sampling techniques for the Nyström method". In: *Artificial Intelligence and Statistics*. PMLR. 2009, pp. 304–311.
22. Yujia Li, Kevin Swersky, and Rich Zemel "Generative moment matching networks". In: *International Conference on Machine Learning*. PMLR. 2015, pp. 1718–1727.
23. J Mercer. "Functions of positive and negative type and their connection with the theory of integral equations". In: *Philosophical Transactions of the Royal Society* A.209 (1909), pp. 415–446.
24. Sebastian Mika et al. "Fisher discriminant analysis with kernels". In: *Neural networks for signal processing IX: Proceedings of the 1999 IEEE signal processing society workshop*. IEEE. 1999, pp. 41–48.
25. Ha Quang Minh, Partha Niyogi, and Yuan Yao. "Mercer's theorem, feature maps, and smoothing". In: *International Conference on Computational Learning Theory*. Springer. 2006, pp. 154–168.
26. Evert J Nyström. "Über die praktische Auflösung von Integralgleichungen mit Anwendungen auf Randwertaufgaben". In: *Acta Mathematica* 54.1 (1930), pp. 185–204.
27. Vytautas Perlibakas. "Distance measures for PCA-based face recognition". In: *Pattern recognition letters* 25.6 (2004), pp. 711–724.
28. Julien Ah-Pine. "Normalized kernels as similarity indices". In: *Pacific-Asia Conference on Knowledge Discovery and Data Mining*. Springer. 2010, pp. 362–373.
29. John Platt. "FastMap, MetricMap, and Landmark MDS are all Nystrom Algorithms". In: *AISTATS*. 2005.
30. Michael Renardy and Robert C Rogers. *An introduction to partial differential equations*. Vol. 13. Springer Science & Business Media, 2006.
31. José Luis Rojo-Álvarez et al. *Digital signal processing with Kernel methods*. Wiley Online Library, 2018.

32. Erhard Schmidt. "Über die Auflösung linearer Gleichungen mit unendlich vielen Unbekannten". In: *Rendiconti del Circolo Matematico di Palermo (1884–1940)* 25.1 (1908), pp. 53–77.
33. Bernhard Schölkopf. "The kernel trick for distances". In: *Advances in neural information processing systems* (2001), pp. 301–307.
34. Bernhard Schölkopf, Alexander Smola, and Klaus-Robert Müller. "Kernel principal component analysis". In: *International conference on artificial neural networks* Springer. 1997, pp. 583–588.
35. Bernhard Schölkopf, Alexander Smola, and Klaus-Robert Müller. "Nonlinear component analysis as a kernel eigenvalue problem". In: *Neural computation* 10.5 (1998), pp. 1299–1319.
36. Bernhard Schölkopf et al. "Input space versus feature space in kernel-based methods". In: *IEEE transactions on neural networks* 10.5 (1999), pp. 1000–1017.
37. Dino Sejdinovic et al. "Equivalence of distance-based and RKHS-based statistics in hypothesis testing". In: *The Annals of Statistics* (2013), pp. 2263–2291.
38. Harry Strange and Reyer Zwiggelaar. *Open Problems in Spectral Dimensionality Reduction*. Springer, 2014.
39. Vladimir Vapnik. *The nature of statistical learning theory* Springer Science & Business Media, 1995.
40. Christopher Williams and Matthias Seeger. "The effect of the input density distribution on kernel-based classifiers". In: *Proceedings of the 17th international conference on machine learning*. 2000.
41. Christopher Williams and Matthias Seeger. "Using the Nyström method to speed up kernel machines". In: *Proceedings of the 14th annual conference on neural information processing systems*. CONF. 2001, pp. 682–688.

Chapter 4
Background on Optimization

4.1 Introduction

It is not wrong to say that almost all machine learning algorithms, including the dimensionality reduction methods, reduce to optimization. Many of the optimization methods can be explained in terms of the Karush-Kuhn-Tucker (KKT) conditions [30], proposed in [29, 32]. Therefore, the KKT conditions play an important role in optimization. Most well-known optimization algorithms can be categorized into the first-order optimization and second-order optimization. The first-order methods are based on the gradient (i.e., the first-order derivative), while the second-order methods make use of Hessian (i.e., the second-order derivative) or approximation of Hessian as well as the gradient. The second-order methods are usually faster than the first-order methods because they use the Hessian information. However, computation of Hessian or approximation of Hessian in second-order methods is time-consuming and difficult. This might be the reason why most machinc learning algorithms, such as backpropagation for neural networks [46], use first-order methods.

Recently, distributed optimization has also gained more attention. Distributed optimization has two benefits. First, it makes the problem able to run in parallel on several servers. Second, it can be used to solve problems with multiple optimization variables. Especially, for the second reason, it has been widely used in machine learning and signal processing. The two most well-known distributed optimization approaches are alternating optimization [28, 34] and Alternating Direction Method of Multipliers (ADMM) [6, 17, 22]. Alternating optimization alternates between optimizing over variables one-by-one iteratively. ADMM is based on dual decomposition [2, 10, 13] and augmented Lagrangian [25, 44].

B. Ghojogh et al., *Elements of Dimensionality Reduction and Manifold Learning*,
https://doi.org/10.1007/978-3-031-10602-6_4

All the abovementioned concepts will be introduced in this chapter. This chapter reviews the required background on optimization, including the optimization concepts, the KKT conditions, first-order and second-order optimization, and distributed optimization. Proofs of some claims in this chapter are in [19]. The reader can refer to [19] for more information on optimization methods.

4.2 Notations and Preliminaries

4.2.1 Preliminaries on Sets and Norms

Definition 4.1 (Minimum, Maximum, Infimum, and Supremum) A minimum and maximum of a function $f : \mathbb{R}^d \rightarrow \mathbb{R}$, $f : \boldsymbol{x} \mapsto f(\boldsymbol{x})$, with a domain $\mathcal{D}$, are defined as:

$$\min_{\boldsymbol{x}} f(\boldsymbol{x}) \leq f(\boldsymbol{y}), \quad \forall \boldsymbol{y} \in \mathcal{D},$$

$$\max_{\boldsymbol{x}} f(\boldsymbol{x}) \geq f(\boldsymbol{y}), \quad \forall \boldsymbol{y} \in \mathcal{D},$$

respectively. The minimum and maximum of a function belong to the range of the function and the infimum and supremum are the lower bound and upper bound of that function, respectively:

$$\inf_{\boldsymbol{x}} f(\boldsymbol{x}) := \max\{z \in \mathbb{R} \,|\, z \leq f(\boldsymbol{x}), \forall \boldsymbol{x} \in \mathcal{D}\},$$

$$\sup_{\boldsymbol{x}} f(\boldsymbol{x}) := \min\{z \in \mathbb{R} \,|\, z \geq f(\boldsymbol{x}), \forall \boldsymbol{x} \in \mathcal{D}\}.$$

Depending on the function, the infimum and supremum of the function may or may not belong to the range of the function. Figure 4.1 provides examples of a minimum, maximum, infimum, and supremum of functions. The minimum and maximum of a function are also the infimum and supremum of function, respectively, but the converse is not necessarily true. If the minimum and maximum of a function are the minimum and maximum in the entire domain of the function, they are the global minimum and global maximum, respectively. See Fig. 4.2 for examples of global minimum and maximum.

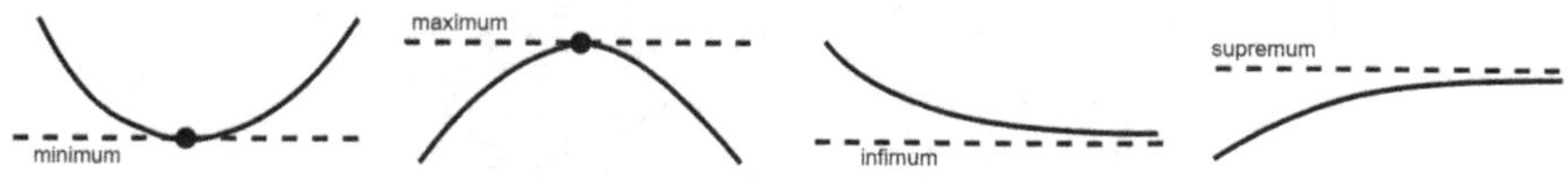

Fig. 4.1 Minimum, maximum, infimum, and supremum of example functions

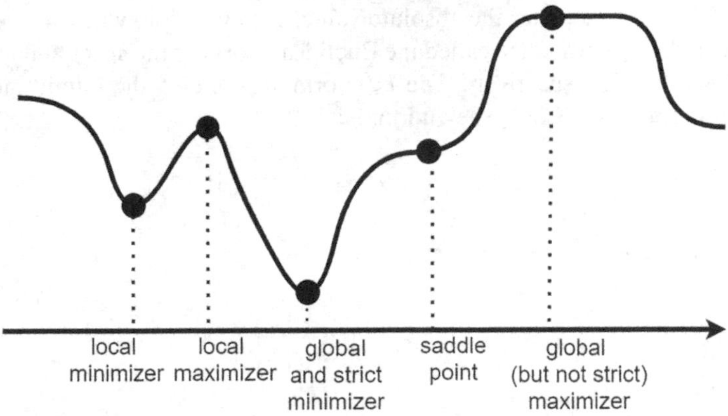

Fig. 4.2 Examples of stationary points such as local and global extreme points, strict and nonstrict extreme points, and saddle point

Lemma 4.1 (Inner Product) *Consider two vectors* $\boldsymbol{x} = [x_1, \ldots, x_d]^\top \in \mathbb{R}^d$ *and* $\boldsymbol{y} = [y_1, \ldots, y_d]^\top \in \mathbb{R}^d$. *Their inner product, also called a dot product, is:*

$$\langle \boldsymbol{x}, \boldsymbol{y} \rangle = \boldsymbol{x}^\top \boldsymbol{y} = \sum_{i=1}^{d} x_i \, y_i.$$

If $\boldsymbol{X}_{ij}$ *denotes the* (i, j)*-th element of matrix* $\boldsymbol{X}$*, the inner product of matrices* $\boldsymbol{X}$ *and* $\boldsymbol{Y}$ *is defined to be:*

$$\langle \boldsymbol{X}, \boldsymbol{Y} \rangle = \mathbf{tr}(\boldsymbol{X}^\top \boldsymbol{Y}) = \sum_{i=1}^{d_1} \sum_{j=1}^{d_2} \boldsymbol{X}_{i,j} \, \boldsymbol{Y}_{i,j},$$

where $\mathbf{tr}(.)$ *denotes the trace of the matrix.*

Definition 4.2 (Norm) A function $\|\cdot\| : \mathbb{R}^d \to \mathbb{R}, \|\cdot\| : \boldsymbol{x} \mapsto \|\boldsymbol{x}\|$ is a norm if it satisfies the following conditions:

1. $\|\boldsymbol{x}\| \geq 0, \forall \boldsymbol{x}$
2. $\|a\boldsymbol{x}\| = |a| \, \|\boldsymbol{x}\|, \forall \boldsymbol{x}$ and all scalars a
3. $\|\boldsymbol{x}\| = 0$ if and only if $\boldsymbol{x} = 0$
4. Triangle inequality: $\|\boldsymbol{x} + \boldsymbol{y}\| \leq \|\boldsymbol{x}\| + \|\boldsymbol{y}\|$.

Definition 4.3 (Important Norms) Some important norms for a vector $\boldsymbol{x} = [x_1, \ldots, x_d]^\top$ are as follows. The ℓ_p norm is:

$$\|\boldsymbol{x}\|_p := \big(|x_1|^p + \cdots + |x_d|^p\big)^{1/p},$$

where $p \geq 1$ and $|.|$ denotes the absolute value. Two well-known ℓ_p norms are the ℓ_1 norm and the ℓ_2 norm (also called the Euclidean norm or the spectral norm), with $p = 1$ and $p = 2$, respectively. The ℓ_∞ norm, also called the infinity norm, the maximum norm, or the Chebyshev norm, is:

$$\|\boldsymbol{x}\|_\infty := \max\{|x_1|, \ldots, |x_d|\}.$$

For the matrix $\boldsymbol{X} \in \mathbb{R}^{d_1 \times d_2}$, the ℓ_p norm is:

$$\|\boldsymbol{X}\|_p := \sup_{y \neq 0} \frac{\|\boldsymbol{X}\boldsymbol{y}\|_p}{\|\boldsymbol{y}\|_p}.$$

A special case for this is the ℓ_2 norm ($p = 2$), where the spectral norm is related to the largest singular value of the matrix:

$$\|\boldsymbol{X}\|_2 = \sup_{y \neq 0} \frac{\|\boldsymbol{X}\boldsymbol{y}\|_2}{\|\boldsymbol{y}\|_2} = \sqrt{\lambda_{\max}(\boldsymbol{X}^\top \boldsymbol{X})} = \sigma_{\max}(\boldsymbol{X}),$$

where $\lambda_{\max}(\boldsymbol{X}^\top \boldsymbol{X})$ and $\sigma_{\max}(\boldsymbol{X})$ denote the largest eigenvalue of $\boldsymbol{X}^\top \boldsymbol{X}$ and the largest singular value of $\boldsymbol{X}$, respectively. Other special cases are the maximum-absolute-column-sum norm ($p = 1$) and the maximum-absolute-row-sum norm ($p = \infty$):

$$\|\boldsymbol{X}\|_1 = \sup_{y \neq 0} \frac{\|\boldsymbol{X}\boldsymbol{y}\|_1}{\|\boldsymbol{y}\|_1} = \max_{1 \leq j \leq d_2} \sum_{i=1}^{d_1} |\boldsymbol{X}_{i,j}|,$$

$$\|\boldsymbol{X}\|_\infty = \sup_{y \neq 0} \frac{\|\boldsymbol{X}\boldsymbol{y}\|_\infty}{\|\boldsymbol{y}\|_\infty} = \max_{1 \leq i \leq d_1} \sum_{j=1}^{d_2} |\boldsymbol{X}_{i,j}|.$$

The formulation of the Frobenius norm for a matrix is similar to the formulation of the ℓ_2 norm for a vector:

$$\|\boldsymbol{X}\|_F := \sqrt{\sum_{i=1}^{d_1} \sum_{j=1}^{d_2} \boldsymbol{X}_{i,j}^2},$$

where $\boldsymbol{X}_{ij}$ denotes the (i, j)-th element of $\boldsymbol{X}$.

The $\ell_{2,1}$ norm of matrix $\boldsymbol{X}$ is:

$$\|\boldsymbol{X}\|_{2,1} := \sum_{i=1}^{d_1} \sqrt{\sum_{j=1}^{d_2} \boldsymbol{X}_{i,j}^2}.$$

The Schatten ℓ_p norm of matrix $\boldsymbol{X}$ is:

$$\|\boldsymbol{X}\|_p := \Big(\sum_{i=1}^{\min(d_1,d_2)} \big(\sigma_i(\boldsymbol{X})\big)^p\Big)^{1/p},$$

where $\sigma_i(\boldsymbol{X})$ denotes the i-th singular value of $\boldsymbol{X}$. A special case of the Schatten norm, with $p = 1$, is called the nuclear norm or the trace norm [14]:

$$\|\boldsymbol{X}\|_* := \sum_{i=1}^{\min(d_1,d_2)} \sigma_i(\boldsymbol{X}) = \mathbf{tr}\Big(\sqrt{\boldsymbol{X}^\top \boldsymbol{X}}\Big),$$

which is the summation of the singular values of the matrix. Note that similar to how the ℓ_1 norm of the vector is used for sparsity, the nuclear norm is used to impose sparsity on the matrix.

Lemma 4.2 *The following statements hold:*

$$\|\boldsymbol{x}\|_2^2 = \boldsymbol{x}^\top \boldsymbol{x} = \langle \boldsymbol{x}, \boldsymbol{x}\rangle,$$
$$\|\boldsymbol{X}\|_F^2 = \mathbf{tr}(\boldsymbol{X}^\top \boldsymbol{X}) = \langle \boldsymbol{X}, \boldsymbol{X}\rangle,$$

which are convex and in quadratic forms.

Definition 4.4 (Cone and Dual Cone) A set $\mathcal{K} \subseteq \mathbb{R}^d$ is a cone if the following conditions are met:

1. it contains the origin, i.e., $\mathbf{0} \in \mathcal{K}$,
2. $\mathcal{K}$ is a convex set,
3. for each $\boldsymbol{x} \in \mathcal{K}$ and $\lambda \geq 0$, there is a $\lambda \boldsymbol{x} \in \mathcal{K}$.

The dual cone of a cone $\mathcal{K}$ is:

$$\mathcal{K}^* := \{\boldsymbol{y} \mid \boldsymbol{y}^\top \boldsymbol{x} \geq 0, \forall \boldsymbol{x} \in \mathcal{K}\}.$$

Definition 4.5 (Proper Cone [7]) A convex cone $\mathcal{K} \subseteq \mathbb{R}^d$ is a proper cone if the following conditions are met:

1. $\mathcal{K}$ is closed, i.e., it contains its boundary,
2. $\mathcal{K}$ is solid, i.e., its interior is nonempty,
3. $\mathcal{K}$ is pointed, i.e., it contains no line. In other words, it is not a two-sided cone around the origin.

Definition 4.6 (Generalized Inequality [7]) A generalized inequality, defined by a proper cone $\mathcal{K}$, is:

$$\boldsymbol{x} \succeq_{\mathcal{K}} \boldsymbol{y} \iff \boldsymbol{x} - \boldsymbol{y} \in \mathcal{K}.$$

This means $\boldsymbol{x} \succeq_{\mathcal{K}} \boldsymbol{y} \iff \boldsymbol{x} - \boldsymbol{y} \in \textbf{int}(\mathcal{K})$. Note that $\boldsymbol{x} \succeq_{\mathcal{K}} \boldsymbol{y}$ can also be stated as $\boldsymbol{x} - \boldsymbol{y} \succeq_{\mathcal{K}} \boldsymbol{0}$.

Definition 4.7 (Important Examples for Generalized Inequality) The generalized inequality defined by the nonnegative orthant, $\mathcal{K} = \mathbb{R}^d_+$, is the default inequality for vectors $\boldsymbol{x} = [x_1, \ldots, x_d]^\top$, $\boldsymbol{y} = [y_1, \ldots, y_d]^\top$:

$$\boldsymbol{x} \succeq \boldsymbol{y} \iff \boldsymbol{x} \succeq_{\mathbb{R}^d_+} \boldsymbol{y}.$$

This means componentwise inequality:

$$\boldsymbol{x} \succeq \boldsymbol{y} \iff x_i \geq y_i, \quad \forall i \in \{1, \ldots, d\}.$$

The generalized inequality defined by the positive definite cone, $\mathcal{K} = \mathbb{S}^d_+$, is the default inequality for symmetric matrices $\boldsymbol{X}, \boldsymbol{Y} \in \mathbb{S}^d$:

$$\boldsymbol{X} \succeq \boldsymbol{Y} \iff \boldsymbol{X} \succeq_{\mathbb{S}^d_+} \boldsymbol{Y}.$$

This means that $(\boldsymbol{X} - \boldsymbol{Y})$ is positive semidefinite. Note that if the inequality is strict, i.e., $\boldsymbol{X} \succ \boldsymbol{Y}$, it indicates that $(\boldsymbol{X} - \boldsymbol{Y})$ is positive definite. In conclusion, $\boldsymbol{x} \succeq \boldsymbol{0}$ means all elements of vector $\boldsymbol{x}$ are nonnegative and $\boldsymbol{X} \succeq \boldsymbol{0}$ indicates that the matrix $\boldsymbol{X}$ is considered to be positive semidefinite.

4.2.2 *Preliminaries on Optimization*

Definition 4.8 (Lipschitz Smoothness) A function $f(.)$ with a domain of $\mathcal{D}$ is considered to be Lipschitz smooth (or Lipschitz continuous) if it satisfies the following:

$$|f(\boldsymbol{x}) - f(\boldsymbol{y})| \leq L \, \|\boldsymbol{x} - \boldsymbol{y}\|_2, \quad \forall \boldsymbol{x}, \boldsymbol{y} \in \mathcal{D}. \tag{4.1}$$

A function with Lipschitz smoothness (with a Lipschitz constant L) is called L-smooth.

Lemma 4.3 *Consider a differentiable function $f(.)$, with domain of $\mathcal{D}$, and whose gradient is L-smooth, i.e., $|\nabla f(\boldsymbol{x}) - \nabla f(\boldsymbol{y})| \leq L \, \|\boldsymbol{x} - \boldsymbol{y}\|_2, \forall \boldsymbol{x}, \boldsymbol{y} \in \mathcal{D}$; then for any $\boldsymbol{x}, \boldsymbol{y} \in \mathcal{D}$ the following is held true:*

$$f(\boldsymbol{y}) \leq f(\boldsymbol{x}) + \nabla f(\boldsymbol{x})^\top (\boldsymbol{y} - \boldsymbol{x}) + \frac{L}{2} \|\boldsymbol{y} - \boldsymbol{x}\|_2^2. \tag{4.2}$$

Definition 4.9 (Convex Function) A function $f(.)$ with a domain $\mathcal{D}$ is considered convex if:

$$f\big(\alpha \boldsymbol{x} + (1-\alpha)\boldsymbol{y}\big) \leq \alpha f(\boldsymbol{x}) + (1-\alpha) f(\boldsymbol{y}), \tag{4.3}$$

$\forall \boldsymbol{x}, \boldsymbol{y} \in \mathcal{D}$, where $\alpha \in [0, 1]$. Moreover, if the function $f(.)$ is differentiable, it is considered convex if:

$$f(\boldsymbol{x}) \geq f(\boldsymbol{y}) + \nabla f(\boldsymbol{y})^\top (\boldsymbol{x} - \boldsymbol{y}), \tag{4.4}$$

$\forall \boldsymbol{x}, \boldsymbol{y} \in \mathcal{D}$. Moreover, if the function $f(.)$ is twice differentiable, it is considered convex if its second-order derivative is positive semidefinite:

$$\nabla^2 f(\boldsymbol{x}) \succeq \boldsymbol{0}, \quad \forall \boldsymbol{x} \in \mathcal{D}. \tag{4.5}$$

Definition 4.10 (Local and Global Minimizers) A point $\boldsymbol{x} \in \mathcal{D}$ is a local minimizer of function $f(.)$ if and only if:

$$\exists \epsilon > 0 : \forall \boldsymbol{y} \in \mathcal{D}, \; \|\boldsymbol{y} - \boldsymbol{x}\|_2 \leq \epsilon \implies f(\boldsymbol{x}) \leq f(\boldsymbol{y}), \tag{4.6}$$

meaning that in an ϵ-neighborhood of $\boldsymbol{x}$, the value of the function is minimum at $\boldsymbol{x}$. A point $\boldsymbol{x} \in \mathcal{D}$ is a global minimizer of function $f(.)$ if and only if:

$$f(\boldsymbol{x}) \leq f(\boldsymbol{y}), \quad \forall \boldsymbol{y} \in \mathcal{D}. \tag{4.7}$$

Figure 4.2 provides examples of local minimizer and maximizer.

Definition 4.11 (Strict Minimizers) In Eqs. (4.6) and (4.7), $f(\boldsymbol{x}) < f(\boldsymbol{y})$ indicates that the minimizer is a strict local minimizer, where $f(\boldsymbol{x}) \leq f(\boldsymbol{y})$ indicates a global minimizer. Figure 4.2 provides examples of strict/nonstrict minimizer and maximizer.

Lemma 4.4 (Minimizer in Convex Function) *In a convex function, any local minimizer is a global minimizer.*

Corollary 4.1 *In a convex function, there exists only one minimum value, which is the global minimum value. In such function, there may be one or several local minimizers but all of them have the same minimum function value.*

Lemma 4.5 (Gradient of a Convex Function at the Minimizer Point) *When the function $f(.)$ is convex and differentiable, a point $\boldsymbol{x}^*$ is a minimizer if and only if $\nabla f(\boldsymbol{x}^*) = \boldsymbol{0}$.*

Definition 4.12 (Stationary, Extremum, and Saddle Points) In the general (not-necessarily-convex) function $f(.)$, the point $\boldsymbol{x}^*$ is considered to be stationary if and only if $\nabla f(\boldsymbol{x}^*) = \boldsymbol{0}$. A saddle point is a point that passes through it flips the sign of the second derivative to the opposite sign. Minimizer and maximizer points (locally or globally) minimize and maximize the function, respectively. A saddle point is neither a minimizer nor maximizer, although the gradient at a saddle point is zero. Both the minimizer and maximizer are also called the extremum points. As Fig. 4.2 demonstrates, stationary points can be either a minimizer, a maximizer, or a saddle point of function.

Lemma 4.6 (First-Order Optimality Condition [37, Theorem 1.2.1]) *If $\boldsymbol{x}^*$ is a local minimizer for a differentiable function $f(.)$, then:*

$$\nabla f(\boldsymbol{x}^*) = \boldsymbol{0}. \tag{4.8}$$

Note that if $f(.)$ is convex, this equation is a necessary and sufficient condition for a minimizer.

Note that if setting the derivative to zero, i.e. Eq. (4.8), the optimization is completed once a closed-form solution for $\boldsymbol{x}^*$ is provided. Otherwise, a random initialized solution is required while an iterative update is used on the gradient. First-order or second-order methods can be used for iterative optimization (see Sects. 4.5 and 4.6).

Definition 4.13 (Arguments of Minimization and Maximization) In the domain of a function, the point that minimizes (resp. maximizes) the function $f(.)$ is the argument for the minimization (resp. maximization) of the function. The minimizer and maximizer of the function are denoted by $\arg\min_{\boldsymbol{x}} f(\boldsymbol{x})$ and $\arg\max_{\boldsymbol{x}} f(\boldsymbol{x})$, respectively.

Remark 4.1 It is possible to convert maximization to minimization and vice versa by:

$$\begin{aligned}
&\underset{\boldsymbol{x}}{\text{maximize}} \;\; f(\boldsymbol{x}) = -\underset{\boldsymbol{x}}{\text{minimize}} \;\; \big(-f(\boldsymbol{x})\big), \\
&\underset{\boldsymbol{x}}{\text{minimize}} \;\; f(\boldsymbol{x}) = -\underset{\boldsymbol{x}}{\text{maximize}} \;\; \big(-f(\boldsymbol{x})\big).
\end{aligned} \tag{4.9}$$

Similar conversions are possible for the arguments of maximization and minimization. However, as the sign of the optimal value of the function is not important in the argument, there should not be a negative sign before maximization and minimization:

$$\begin{aligned}
&\arg\max_{\boldsymbol{x}} f(\boldsymbol{x}) = \arg\min_{\boldsymbol{x}} \big(-f(\boldsymbol{x})\big), \\
&\arg\min_{\boldsymbol{x}} f(\boldsymbol{x}) = \arg\max_{\boldsymbol{x}} \big(-f(\boldsymbol{x})\big).
\end{aligned} \tag{4.10}$$

Definition 4.14 (Proximal Mapping/Operator [41]) The proximal mapping or proximal operator of a convex function $g(.)$ is:

$$\textbf{prox}_{\lambda g}(\boldsymbol{x}) := \arg\min_{\boldsymbol{u}} \Big(g(\boldsymbol{u}) + \frac{1}{2\lambda}\|\boldsymbol{u} - \boldsymbol{x}\|_2^2\Big). \tag{4.11}$$

Lemma 4.7 (Projection onto Set) *An indicator function $\mathbb{I}(.)$ is considered to be zero if its condition is satisfied and if not, it is considered to be infinite. The proximal mapping of the indicator function to a convex set $\mathcal{S}$, i.e. $\mathbb{I}(\boldsymbol{x} \in \mathcal{S})$, is the projection of point $\boldsymbol{x}$ onto set $\mathcal{S}$. Therefore, the projection of $\boldsymbol{x}$ onto set $\mathcal{S}$, denoted by $\Pi_{\mathcal{S}}(\boldsymbol{x})$, is defined as:*

$$\Pi_{\mathcal{S}}(\boldsymbol{x}) := \textbf{prox}_{\mathbb{I}(.\in\mathcal{S})}(\boldsymbol{x}) = \arg\min_{\boldsymbol{u}\in\mathcal{S}} \Big(\frac{1}{2}\|\boldsymbol{u} - \boldsymbol{x}\|_2^2\Big). \tag{4.12}$$

This projection simply means projecting point $\boldsymbol{x}$ onto the closest point of the set from point $\boldsymbol{x}$; hence, the vector connecting points $\boldsymbol{x}$ and $\Pi_{\mathcal{S}}(\boldsymbol{x})$ is orthogonal to set $\mathcal{S}$.

Proof

$$\textbf{prox}_{\mathbb{I}(.\in\mathcal{S})}(\boldsymbol{x}) = \arg\min_{\boldsymbol{u}} \Big(\mathbb{I}(\boldsymbol{x} \in \mathcal{S}) + \frac{1}{2}\|\boldsymbol{u} - \boldsymbol{x}\|_2^2\Big) \overset{(a)}{=} \arg\min_{\boldsymbol{u}\in\mathcal{S}} \Big(\frac{1}{2}\|\boldsymbol{u} - \boldsymbol{x}\|_2^2\Big),$$

where (a) is because $\mathbb{I}(\boldsymbol{x} \in \mathcal{S})$ becomes infinity if $\boldsymbol{x} \notin \mathcal{S}$. □

4.2.3 Preliminaries on Derivative

Remark 4.2 (Dimensionality of Derivative) Consider a function $f : \mathbb{R}^{d_1} \to \mathbb{R}^{d_2}$, $f : \boldsymbol{x} \mapsto f(\boldsymbol{x})$. Derivative of function $f(\boldsymbol{x}) \in \mathbb{R}^{d_2}$ with respect to $\boldsymbol{x} \in \mathbb{R}^{d_1}$ has dimensionality $(d_1 \times d_2)$. This is because altering every element of $\boldsymbol{x} \in \mathbb{R}^{d_1}$ can change every element of $f(\boldsymbol{x}) \in \mathbb{R}^{d_2}$. The (i, j)-th element of the $(d_1 \times d_2)$-dimensional derivative states that the amount of change in the j-th element of $f(\boldsymbol{x})$ results from changing the i-th element of $\boldsymbol{x}$.

Note that it is possible to use a transpose of the derivative as the derivative. This is possible as long as the dimensionality of other terms in the equations for optimization coincide (i.e., they are all transposed). In that case, the dimensionality of the derivative is $(d_2 \times d_1)$, where the (i, j)-th element of the derivative states the amount of change in the i-th element of $f(\boldsymbol{x})$ is the result of changing the j-th element of $\boldsymbol{x}$.

The following are examples of derivatives:

- If the function is $f : \mathbb{R} \to \mathbb{R}$, $f : x \mapsto f(x)$, the derivative $(\partial f(x)/\partial x) \in \mathbb{R}$ is a scalar because changing the scalar x can change the scalar $f(x)$.

- If the function is $f : \mathbb{R}^d \to \mathbb{R}, f : \boldsymbol{x} \mapsto f(\boldsymbol{x})$, the derivative $(\partial f(\boldsymbol{x})/\partial \boldsymbol{x}) \in \mathbb{R}^d$ is a vector because changing every element of the vector $\boldsymbol{x}$ can change the scalar $f(\boldsymbol{x})$.
- If the function is $f : \mathbb{R}^{d_1 \times d_2} \to \mathbb{R}, f : \boldsymbol{X} \mapsto f(\boldsymbol{X})$, the derivative $(\partial f(\boldsymbol{X})/\partial \boldsymbol{X}) \in \mathbb{R}^{d_1 \times d_2}$ is a matrix because changing every element of the matrix $\boldsymbol{X}$ can change the scalar $f(\boldsymbol{X})$.
- If the function is $f : \mathbb{R}^{d_1} \to \mathbb{R}^{d_2}, f : \boldsymbol{x} \mapsto f(\boldsymbol{x})$, the derivative $(\partial f(\boldsymbol{x})/\partial \boldsymbol{x}) \in \mathbb{R}^{d_1 \times d_2}$ is a matrix because changing every element of the vector $\boldsymbol{x}$ can change every element of the vector $f(\boldsymbol{x})$.
- If the function is $f : \mathbb{R}^{d_1 \times d_2} \to \mathbb{R}^{d_3}, f : \boldsymbol{X} \mapsto f(\boldsymbol{X})$, the derivative $(\partial f(\boldsymbol{X})/\partial \boldsymbol{X})$ is a $(d_1 \times d_2 \times d_3)$-dimensional tensor because changing every element of the matrix $\boldsymbol{X}$ can change every element of the vector $f(\boldsymbol{X})$.
- If the function is $f : \mathbb{R}^{d_1 \times d_2} \to \mathbb{R}^{d_3 \times d_4}, f : \boldsymbol{X} \mapsto f(\boldsymbol{X})$, the derivative $(\partial f(\boldsymbol{X})/\partial \boldsymbol{X})$ is a $(d_1 \times d_2 \times d_3 \times d_4)$-dimensional tensor because changing every element of the matrix $\boldsymbol{X}$ can change every element of the matrix $f(\boldsymbol{X})$.

In other words, the derivative of a scalar with respect to a scalar is a scalar. The derivative of a scalar with respect to a vector is a vector. The derivative of a scalar with respect to a matrix is a matrix. The derivative of a vector with respect to a vector is a matrix. The derivative of a vector with respect to a matrix is a rank-3 tensor. The derivative of a matrix with respect to a matrix is a rank-4 tensor.

Definition 4.15 (Gradient, Jacobian, and Hessian) Consider the function $f : \mathbb{R}^d \to \mathbb{R}, f : \boldsymbol{x} \mapsto f(\boldsymbol{x})$. In optimizing function f, the derivative of the function with respect to its variable $\boldsymbol{x}$ is called the gradient, denoted by:

$$\nabla f(\boldsymbol{x}) := \frac{\partial f(\boldsymbol{x})}{\partial \boldsymbol{x}} \in \mathbb{R}^d.$$

The second derivative of the function with respect to its derivative is called the Hessian matrix, denoted by

$$\boldsymbol{B} = \nabla^2 f(\boldsymbol{x}) := \frac{\partial^2 f(\boldsymbol{x})}{\partial \boldsymbol{x}^2} \in \mathbb{R}^{d \times d}.$$

The Hessian matrix is symmetric. If the function is convex, its Hessian matrix is positive semidefinite.

If the function is multidimensional, i.e., $f : \mathbb{R}^{d_1} \to \mathbb{R}^{d_2}, f : \boldsymbol{x} \mapsto f(\boldsymbol{x})$, the gradient becomes a Jacobian matrix, stated as follows:

$$\boldsymbol{J} := \Big[\frac{\partial f}{\partial x_1}, \dots, \frac{\partial f}{\partial x_{d_1}}\Big]^\top = \begin{bmatrix} \frac{\partial f_1}{\partial x_1} & \dots & \frac{\partial f_{d_2}}{\partial x_{d_1}} \\ \vdots & \ddots & \vdots \\ \frac{\partial f_1}{\partial x_{d_1}} & \dots & \frac{\partial f_{d_2}}{\partial x_{d_1}} \end{bmatrix} \in \mathbb{R}^{d_1 \times d_2},$$

where $\boldsymbol{x} = [x_1, \ldots, x_{d_1}]^\top$ and $f(\boldsymbol{x}) = [f_1, \ldots, f_{d_2}]^\top$.

Corollary 4.2 (Technique for Calculating Derivative) *According to the size of the derivative, it is possible to calculate the derivatives. To find the correct derivative for multiplications of matrices (or vectors), temporarily assume some dimensionality for every matrix and find the correct dimensionality of matrices in the derivative. As an example, for a matrix* $\boldsymbol{X} \in \mathbb{R}^{a\times b}$*, the following derivative exists:*

$$\mathbb{R}^{a\times b} \ni \frac{\partial}{\partial \boldsymbol{X}}\big(\mathbf{tr}(\boldsymbol{A}\boldsymbol{X}\boldsymbol{B})\big) = \boldsymbol{A}^\top \boldsymbol{B}^\top = (\boldsymbol{B}\boldsymbol{A})^\top. \tag{4.13}$$

This is calculated by assuming that $\boldsymbol{A} \in \mathbb{R}^{c\times a}$ *and* $\boldsymbol{B} \in \mathbb{R}^{b\times c}$*. Matrix multiplication is determined by* $\boldsymbol{AXB}$ *and its size is calculated as* $\boldsymbol{AXB} \in \mathbb{R}^{c\times c}$ *because the argument of the trace should be a square matrix. The derivative* $\partial(\mathbf{tr}(\boldsymbol{AXB}))/\partial \boldsymbol{X}$ *has a size* $\mathbb{R}^{a\times b}$ *because* $\mathbf{tr}(\boldsymbol{AXB})$ *is a scalar and* $\boldsymbol{X}$ *is* $(a \times b)$*-dimensional. The derivative should be the multiplication of* $\boldsymbol{A}$ *and* $\boldsymbol{B}$ *because* $\mathbf{tr}(\boldsymbol{AXB})$ *is linear with respect to* $\boldsymbol{X}$*. Based on the assumed sizes of* $\boldsymbol{A}$ *and* $\boldsymbol{B}$*,* $\boldsymbol{A}^\top\boldsymbol{B}^\top$ *is considered to be the desired size, and these matrices can be multiplied. This means that this is the correct derivative.*

Lemma 4.8 (Derivative of Matrix with Respect to Matrix) *As explained in Remark 4.2, the derivative of a matrix with respect to another matrix is a tensor. Working with tensors is difficult; however, a Kronecker product can be used for representing a tensor as a matrix. This is the Magnus-Neudecker convention [36] in which all matrices are vectorized. For example, if* $\boldsymbol{X} \in \mathbb{R}^{a\times b}$*,* $\boldsymbol{A} \in \mathbb{R}^{c\times a}$*, and* $\boldsymbol{B} \in \mathbb{R}^{b\times d}$*, then:*

$$\mathbb{R}^{(cd)\times(ab)} \ni \frac{\partial}{\partial \boldsymbol{X}}(\boldsymbol{AXB}) = \boldsymbol{B}^\top \otimes \boldsymbol{A}, \tag{4.14}$$

where $\otimes$ *denotes the Kronecker product.*[1]

Remark 4.3 (Chain Rule in Matrix Derivatives) When having composite functions (i.e., function of function), a chain rule can be used for a derivative. For the derivative of a matrix with respect to a matrix, this chain rule can become difficult but can be done by checking the compatibility of dimensions in matrix multiplications. For this, Lemma 4.8 and a vectorization technique in which the matrix is vectorized should be used. The vectorization technique is explained in the following.

[1] The Kronecker product of two matrices is a block matrix, where every element of the first matrix is multiplied by every element of the second matrix.

Let $\mathbf{vec}(.)$ denote the vectorization of an $\mathbb{R}^{a\times b}$ matrix to an $\mathbb{R}^{ab}$ vector. Additionally, let $\mathbf{vec}^{-1}_{a\times b}(.)$ be the devectorization of a $\mathbb{R}^{ab}$ vector to a $\mathbb{R}^{a\times b}$ matrix. The following provides an example calculation of derivative by chain rule:

$$f(\boldsymbol{S}) = \mathbf{tr}(\boldsymbol{ASB}), \;\; \boldsymbol{S} = \boldsymbol{C}\widehat{\boldsymbol{M}}\boldsymbol{D}, \;\; \widehat{\boldsymbol{M}} = \frac{\boldsymbol{M}}{\|\boldsymbol{M}\|_F^2},$$

where $\boldsymbol{A} \in \mathbb{R}^{c\times a}$, $\boldsymbol{S} \in \mathbb{R}^{a\times b}$, $\boldsymbol{B} \in \mathbb{R}^{b\times c}$, $\boldsymbol{C} \in \mathbb{R}^{a\times d}$, $\widehat{\boldsymbol{M}} \in \mathbb{R}^{d\times d}$, $\boldsymbol{D} \in \mathbb{R}^{d\times b}$, and $\boldsymbol{M} \in \mathbb{R}^{d\times d}$. Some of the derivatives are:

$$\begin{aligned}
\mathbb{R}^{a\times b} &\ni \frac{\partial f(\boldsymbol{S})}{\partial \boldsymbol{S}} \overset{(4.13)}{=} (\boldsymbol{BA})^\top. \\
\mathbb{R}^{ab\times d^2} &\ni \frac{\partial \boldsymbol{S}}{\partial \widehat{\boldsymbol{M}}} \overset{(4.14)}{=} \boldsymbol{D}^\top \otimes \boldsymbol{C}, \\
\mathbb{R}^{d^2\times d^2} &\ni \frac{\partial \widehat{\boldsymbol{M}}}{\partial \boldsymbol{M}} \overset{(a)}{=} \frac{1}{\|\boldsymbol{M}\|_F^4}\big(\|\boldsymbol{M}\|_F^2 \boldsymbol{I}_{d^2} - 2\boldsymbol{M}\otimes\boldsymbol{M}\big) \\
&= \frac{1}{\|\boldsymbol{M}\|_F^2}\big(\boldsymbol{I}_{d^2} - \frac{2}{\|\boldsymbol{M}\|_F^2}\boldsymbol{M}\otimes\boldsymbol{M}\big),
\end{aligned}$$

where (a) is because of the formula for the derivative of fraction and $\boldsymbol{I}_{d^2}$ is a $(d^2 \times d^2)$-dimensional identity matrix. By the chain rule, the final derivative is:

$$\mathbb{R}^{d\times d} \ni \frac{\partial f}{\boldsymbol{M}} = \mathbf{vec}^{-1}_{d\times d}\Big(\big(\frac{\partial \widehat{\boldsymbol{M}}}{\partial \boldsymbol{M}}\big)^\top \big(\frac{\partial \boldsymbol{S}}{\partial \widehat{\boldsymbol{M}}}\big)^\top \mathbf{vec}\big(\frac{\partial f(\boldsymbol{S})}{\partial \boldsymbol{S}}\big)\Big).$$

Note that the chain rule in matrix derivatives usually is stated right to left in matrix multiplications, while transpose is used for matrices in multiplication.

More formulas for matrix derivatives can be found in the matrix cookbook [42] and similar resources. This section focused on derivatives in real spaces. When working with complex data (containing an imaginary part), a complex derivative is needed. The reader can refer to [26] and [8, Chapter 7, Complex Derivatives] for techniques in complex derivatives.

4.3 Optimization Problem

This section reviews the standard forms for convex optimization and explains why these forms are important. Note that the term "programming" refers to solving optimization problems.

General Optimization Problem Consider the function $f : \mathbb{R}^d \to \mathbb{R}$, $f : \boldsymbol{x} \mapsto f(\boldsymbol{x})$. Let the domain of function be $\mathcal{D}$ where $\boldsymbol{x} \in \mathcal{D}$, $\boldsymbol{x} \in \mathbb{R}^d$. The unconstrained minimization of a cost function $f(.)$ is:

$$\underset{x}{\text{minimize}} \quad f(\boldsymbol{x}), \tag{4.15}$$

where $\boldsymbol{x}$ is called the *optimization variable* and the function $f(.)$ is called the *objective function* or the *cost function*. This is an unconstrained problem where the optimization variable $\boldsymbol{x}$ needs only be in the domain of the function, i.e., $\boldsymbol{x} \in \mathcal{D}$, while minimizing the function $f(.)$.

The optimization problem can be constrained, where the optimization variable $\boldsymbol{x}$ should satisfy some equality and/or inequality constraints, in addition to being in the domain of the function, while minimizing the function $f(.)$. In a constrained optimization problem, it is possible to minimize the function $f(\boldsymbol{x})$, while satisfying m_1 inequality constraints and m_2 equality constraints, by:

$$\begin{aligned} \underset{x}{\text{minimize}} \quad & f(\boldsymbol{x}) \\ \text{subject to} \quad & y_i(\boldsymbol{x}) \leq 0, \ i \in \{1, \ldots, m_1\}, \\ & h_i(\boldsymbol{x}) = 0, \ i \in \{1, \ldots, m_2\}, \end{aligned} \tag{4.16}$$

where $f(\boldsymbol{x})$ is the objective function, every $y_i(\boldsymbol{x}) \leq 0$ is an inequality constraint, and every $h_i(\boldsymbol{x}) = 0$ is an equality constraint. Note that if some of the inequality constraints are not in the form $y_i(\boldsymbol{x}) \leq 0$, they can be restated as:

$$\begin{aligned} y_i(\boldsymbol{x}) \geq 0 &\implies -y_i(\boldsymbol{x}) \leq 0, \\ y_i(\boldsymbol{x}) \leq c &\implies y_i(\boldsymbol{x}) - c \leq 0. \end{aligned}$$

Therefore, all inequality constraints can be written in the form $y_i(\boldsymbol{x}) \leq 0$. Furthermore, according to Eq. (4.9), if optimization problem (4.16) is a maximization problem, rather than minimization, it can be converted to a maximization by multiplying its objective function to -1:

$$\begin{aligned} & \underset{x}{\text{maximize}} \quad f(\boldsymbol{x}) \\ & \text{subject to constraints} \end{aligned} \equiv \begin{aligned} & \underset{x}{\text{minimize}} \quad -f(\boldsymbol{x}) \\ & \text{subject to constraints} \end{aligned} \tag{4.17}$$

Definition 4.16 (Feasible Point) Point $\boldsymbol{x}$ for optimization problem (4.16) is feasible if:

$$\begin{aligned} & \boldsymbol{x} \in \mathcal{D}, \text{ and} \\ & y_i(\boldsymbol{x}) \leq 0, \quad \forall i \in \{1, \ldots, m_1\}, \text{ and} \\ & h_i(\boldsymbol{x}) = 0, \quad \forall i \in \{1, \ldots, m_2\}. \end{aligned} \tag{4.18}$$

The constrained optimization problem can also be stated as:

$$\begin{aligned} \underset{\boldsymbol{x}}{\text{minimize}} \quad & f(\boldsymbol{x}) \\ \text{subject to} \quad & \boldsymbol{x} \in \mathcal{S}, \end{aligned} \tag{4.19}$$

where $\mathcal{S}$ is the feasible set of constraints.

Convex Optimization Problem A convex optimization problem is of the form:

$$\begin{aligned} \underset{\boldsymbol{x}}{\text{minimize}} \quad & f(\boldsymbol{x}) \\ \text{subject to} \quad & y_i(\boldsymbol{x}) \leq 0, \; i \in \{1, \dots, m_1\}, \\ & \boldsymbol{A}\boldsymbol{x} = \boldsymbol{b}, \end{aligned} \tag{4.20}$$

where the functions $f(.)$ and $y_i(.)$, for all i, are convex functions, and the equality constraints are affine functions. The feasible set of a convex problem is a convex set.

Semidefinite Programming A semidefinite programming (SDP) problem is:

$$\begin{aligned} \underset{\boldsymbol{X}}{\text{minimize}} \quad & \mathbf{tr}(\boldsymbol{C}\boldsymbol{X}) \\ \text{subject to} \quad & \boldsymbol{X} \succeq \boldsymbol{0}, \\ & \mathbf{tr}(\boldsymbol{D}_i\boldsymbol{X}) \leq \boldsymbol{e}_i, \quad i \in \{1, \dots, m_1\}, \\ & \mathbf{tr}(\boldsymbol{A}_i\boldsymbol{X}) = \boldsymbol{b}_i, \quad i \in \{1, \dots, m_2\}, \end{aligned} \tag{4.21}$$

where the optimization variable $\boldsymbol{X}$ belongs to the positive semidefinite cone $\mathbb{S}^d_+$, $\mathbf{tr}(.)$ denotes the trace of matrix, $\boldsymbol{C}, \boldsymbol{D}_i, \boldsymbol{A}_i \in \mathbb{S}^d, \forall i$, and $\mathbb{S}^d$ denotes the cone of $(d \times d)$ symmetric matrices. The trace terms may be written in summation forms. Note that $\mathbf{tr}(\boldsymbol{C}^\top \boldsymbol{X})$ is the inner product of two matrices, $\boldsymbol{C}$ and $\boldsymbol{X}$, and if matrix $\boldsymbol{C}$ is symmetric, this inner product is equal to $\mathbf{tr}(\boldsymbol{C}\boldsymbol{X})$. Another form for SDP is:

$$\begin{aligned} \underset{\boldsymbol{x}}{\text{minimize}} \quad & \boldsymbol{c}^\top \boldsymbol{x} \\ \text{subject to} \quad & \Big(\sum_{i=1}^{d} x_i \boldsymbol{F}_i\Big) + \boldsymbol{G} \preceq \boldsymbol{0}, \\ & \boldsymbol{A}\boldsymbol{x} = \boldsymbol{b}, \end{aligned} \tag{4.22}$$

where $\boldsymbol{x} = [x_1, \dots, x_d]^\top$, $\boldsymbol{G}, \boldsymbol{F}_i \in \mathbb{S}^d, \forall i$, and $\boldsymbol{A}$, $\boldsymbol{b}$, and $\boldsymbol{c}$ are constant matrices/vectors.

Optimization Toolboxes All the standard optimization forms can be restated as SDP because their constraints can be written as belonging to some cones (see

Definitions 4.6 and 4.7); making them special SDP cases. The interior-point method, or the barrier method, introduced in Sect. 4.6.4, can be used for solving various optimization problems, including SDP [7, 38]. Optimization toolboxes, such as CVX [23], often use the interior-point method (see Sect. 4.6.4) for solving optimization problems, such as SDP. Note that the interior-point method is iterative, and solving SDP is usually time consuming, especially for large matrices. If the optimization problem is a convex optimization problem (e.g. SDP is a convex problem), it has only one local optimum, which is the global optimum (see Corollary 4.1).

Epigraph Form It is possible to convert the optimization problem (4.16) to its *epigraph form*:

$$
\begin{aligned}
\underset{\boldsymbol{x},t}{\text{minimize}} \quad & t \\
\text{subject to} \quad & f(\boldsymbol{x}) - t \leq 0, \\
& y_i(\boldsymbol{x}) \leq 0, \; i \in \{1, \ldots, m_1\}, \\
& h_i(\boldsymbol{x}) = 0, \; i \in \{1, \ldots, m_2\},
\end{aligned}
\tag{4.23}
$$

because it is possible to minimize an upper bound t on the objective function, rather than minimize the objective function. Likewise, for a maximization problem, the lower bound of the objective function can be maximized rather than maximizing the objective function. The upper/lower bound does not necessarily need to be t; it can be any upper/lower bound function for the objective function. This technique is valuable as sometimes optimizing an upper/lower bound function is simpler than the objective function itself. An example of this in machine learning is maximizing the evidence lower bound, rather than the likelihood, in maximum likelihood estimation.

4.4 Karush-Kuhn-Tucker Conditions

Many of the optimization algorithms are reduced to, and can be explained by, the Karush-Kuhn-Tucker (KKT) conditions. Therefore, KKT conditions are fundamental requirements for optimization. In this section, the KTT conditions are explained.

4.4.1 The Lagrangian Function

4.4.1.1 Lagrangian and Dual Variables

Definition 4.17 (Lagrangian and Dual Variables) The Lagrangian function for optimization problem (4.16) is $\mathcal{L} : \mathbb{R}^d \times \mathbb{R}^{m_1} \times \mathbb{R}^{m_2} \to \mathbb{R}$, with a domain $\mathcal{D} \times$

$\mathbb{R}^{m_1} \times \mathbb{R}^{m_2}$, which is defined as follows:

$$\mathcal{L}(\boldsymbol{x}, \boldsymbol{\lambda}, \boldsymbol{\nu}) := f(\boldsymbol{x}) + \sum_{i=1}^{m_1} \lambda_i y_i(\boldsymbol{x}) + \sum_{i=1}^{m_2} \nu_i h_i(\boldsymbol{x}) = f(\boldsymbol{x}) + \boldsymbol{\lambda}^\top \boldsymbol{y}(\boldsymbol{x}) + \boldsymbol{\nu}^\top \boldsymbol{h}(\boldsymbol{x}), \tag{4.24}$$

where $\{\lambda_i\}_{i=1}^{m_1}$ and $\{\nu_i\}_{i=1}^{m_2}$ are the Lagrange multipliers, also called the dual variables, corresponding to inequality and equality constraints, respectively. Note that $\boldsymbol{\lambda} := [\lambda_1, \dots, \lambda_{m_1}]^\top \in \mathbb{R}^{m_1}$, $\boldsymbol{\nu} := [\nu_1, \dots, \nu_{m_2}]^\top \in \mathbb{R}^{m_2}$, $\boldsymbol{y}(\boldsymbol{x}) := [y_1(\boldsymbol{x}), \dots, y_{m_1}(\boldsymbol{x})]^\top \in \mathbb{R}^{m_1}$, and $\boldsymbol{h}(\boldsymbol{x}) := [h_1(\boldsymbol{x}), \dots, h_{m_2}(\boldsymbol{x})]^\top \in \mathbb{R}^{m_2}$. Equation (4.24) is also called the Lagrange relaxation of the optimization problem (4.16).

4.4.1.2 Sign of Terms in Lagrangian

In some papers, the plus sign behind $\sum_{i=1}^{m_2} \nu_i h_i(\boldsymbol{x})$ is replaced with the negative sign. As $h_i(\boldsymbol{x})$ is for the equality constraint, its sign is not important in the Lagrangian function. However, the sign of the term $\sum_{i=1}^{m_1} \lambda_i y_i(\boldsymbol{x})$ is important because the sign of the inequality constraint is important. The sign of $\{\lambda_i\}_{i=1}^{m_1}$ will be discussed in Corollary 4.3. Moreover, according to Eq. (4.17), if problem (4.16) is a maximization problem rather than minimization, the Lagrangian function is $\mathcal{L}(\boldsymbol{x}, \boldsymbol{\lambda}, \boldsymbol{\nu}) = -f(\boldsymbol{x}) + \sum_{i=1}^{m_1} \lambda_i y_i(\boldsymbol{x}) + \sum_{i=1}^{m_2} \nu_i h_i(\boldsymbol{x})$ instead of Eq. (4.24).

4.4.1.3 Interpretation of Lagrangian

Lagrangian can be interpreted using penalty. As Eq. (4.16) states, the goal is to minimize the objective function $f(\boldsymbol{x})$. A cost function is created consisting of the objective function. It is important that the optimization problem constraints are satisfied while minimizing the objective function. Therefore, the cost function is penalized if the constraints are not satisfied. The constraints can be added to the objective function as the regularization (or penalty) terms, thereby minimizing the regularized cost. The dual variables $\boldsymbol{\lambda}$ and $\boldsymbol{\nu}$ can be seen as the regularization parameters, which weight the penalties compared to the objective function $f(\boldsymbol{x})$. This regularized cost function is the Lagrangian function or the Lagrangian relaxation of problem (4.16). Minimization of the regularized cost function minimizes the function $f(\boldsymbol{x})$, while trying to satisfy the constraints.

4.4.1.4 Lagrange Dual Function

Definition 4.18 (Lagrange Dual Function) The Lagrange dual function (also called the dual function) $g : \mathbb{R}^{m_1} \times \mathbb{R}^{m_2} \to \mathbb{R}$ is defined as:

$$g(\boldsymbol{\lambda}, \boldsymbol{\nu}) := \inf_{\boldsymbol{x}\in\mathcal{D}} \mathcal{L}(\boldsymbol{x}, \boldsymbol{\lambda}, \boldsymbol{\nu}) = \inf_{\boldsymbol{x}\in\mathcal{D}} \Big(f(\boldsymbol{x}) + \sum_{i=1}^{m_1} \lambda_i y_i(\boldsymbol{x}) + \sum_{i=1}^{m_2} \nu_i h_i(\boldsymbol{x})\Big). \tag{4.25}$$

Note that the dual function g is a concave function. Section 4.4.2 will describe that maximization of this concave function is achieved in a so-called dual problem.

Lemma 4.9 (Dual Function as a Lower Bound) *If $\boldsymbol{\lambda} \succeq \mathbf{0}$, then the dual function is a lower bound for f^*, i.e., $g(\boldsymbol{\lambda}, \boldsymbol{\nu}) \leq f^*$.*

The optimal point $\boldsymbol{x}^*$ minimizes the Lagrangian function because a Lagrangian is the relaxation of the optimization problem to an unconstrained problem (see Sect. 4.4.1.3). On the other hand, according to Eq. (4.25), the dual function is the minimum of Lagrangian with respect to $\boldsymbol{x}$. Therefore, the dual function can be written as:

$$g(\boldsymbol{\lambda}, \boldsymbol{\nu}) \overset{(4.25)}{=} \inf_{\boldsymbol{x}\in\mathcal{D}} \mathcal{L}(\boldsymbol{x}, \boldsymbol{\lambda}, \boldsymbol{\nu}) = \mathcal{L}(\boldsymbol{x}^*, \boldsymbol{\lambda}, \boldsymbol{\nu}). \tag{4.26}$$

Corollary 4.3 (Nonnegativity of Dual Variables for Inequality Constraints) *From Lemma 4.9, it is concluded that for having the dual function as a lower bound for the optimum function, the dual variable $\{\lambda_i\}_{i=1}^{m_1}$ for inequality constraints (less than or equal to zero) should be nonnegative, i.e.:*

$$\boldsymbol{\lambda} \succeq \mathbf{0} \quad or \quad \lambda_i \geq 0, \ \forall i \in \{1, \dots, m_1\}. \tag{4.27}$$

Note that if the inequality constraints are greater than or equal to zero, then $\lambda_i \leq 0, \ \forall i$ because $y_i(\boldsymbol{x}) \geq 0 \implies -y_i(\boldsymbol{x}) \leq 0$. It is assumed that the inequality constraints are less than or equal to zero. If some of the inequality constraints are greater than or equal to zero, they are converted to less than or equal to zero by multiplying them to -1.

4.4.2 *The Dual Problem, Weak and Strong Duality, and Slater's Condition*

According to Eq. (4.9), the dual function is a lower bound for the optimum function, i.e., $g(\boldsymbol{\lambda}, \boldsymbol{\nu}) \leq f^*$. To maximize $g(\boldsymbol{\lambda}, \boldsymbol{\nu})$, it is important to find the best lower bound with respect to the dual variables $\boldsymbol{\lambda}$ and $\boldsymbol{\nu}$. Moreover, Eq. (4.27) states that the dual variables for inequalities must be nonnegative. Therefore, the following optimization is considered:

$$\begin{aligned} & \underset{\boldsymbol{\lambda}, \boldsymbol{\nu}}{\text{maximize}} && g(\boldsymbol{\lambda}, \boldsymbol{\nu}) \\ & \text{subject to} && \boldsymbol{\lambda} \succeq \mathbf{0}. \end{aligned} \tag{4.28}$$

This equation is called the *Lagrange dual optimization problem* for Eq. (4.16). Equation (4.16) is referred to as the *primal optimization problem*. The variable of problem (4.16), i.e. $\boldsymbol{x}$, is called the *primal variable* while the variables of problem (4.28), i.e. $\boldsymbol{\lambda}$ and $\boldsymbol{\nu}$, are called the *dual variables*. Assuming that the solutions of the dual problem are denoted by $\boldsymbol{\lambda}^*$ and $\boldsymbol{\nu}^*$, the notation $g^* := g(\boldsymbol{\lambda}^*, \boldsymbol{\nu}^*) = \sup_{\boldsymbol{\lambda}, \boldsymbol{\nu}} g$ can be used.

Definition 4.19 (Weak and Strong Duality) For all convex and nonconvex problems, the optimum dual problem is a lower bound for the optimum function:

$$g^* \leq f^* \quad \text{i.e.,} \quad g(\boldsymbol{\lambda}^*, \boldsymbol{\nu}^*) \leq f(\boldsymbol{x}^*). \tag{4.29}$$

This is called the weak duality. Some optimization problems have a strong duality, which is when the optimum dual problem is equal to the optimum function:

$$g^* = f^* \quad \text{i.e.,} \quad g(\boldsymbol{\lambda}^*, \boldsymbol{\nu}^*) = f(\boldsymbol{x}^*). \tag{4.30}$$

The strong duality usually holds for convex optimization problems.

Corollary 4.4 *Equations (4.29) and (4.30) demonstrate that the optimum dual function, g^*, always provides a lower bound for the optimum primal function, f^*.*

The primal optimization problem, i.e. Eq. (4.16), is a minimization so its cost function is similar to a bowl as illustrated in Fig. 4.3. The dual optimization problem, i.e. Eq. (4.28), is maximization so its cost function is like a reversed bowl, as demonstrated in Fig. 4.3. The domains for primal and dual problems are the domain of primal variable $\boldsymbol{x}$ and the domain of dual variables $\boldsymbol{\lambda}$ and $\boldsymbol{\nu}$, respectively. As this figure demonstrates, the optimal $\boldsymbol{x}^*$ corresponds to the optimal $\boldsymbol{\lambda}^*$ and $\boldsymbol{\nu}^*$. There is a possible nonnegative gap between the two bowls. In the best case, this gap is zero. If the gap is zero, there is strong duality; otherwise, a weak duality exists.

If optimization is iterative, the solution is updated iteratively until convergence. First-order and second-order numerical optimization, which will be introduced later, are iterative. In optimization, the series of primal optimal and dual optimal solutions converge to the optimal solution and the dual optimal, respectively. The function values converge to the local minimum and the dual function values converge to the optimal (maximum) dual function. Let the superscript (k) denote the value of the variable at iteration k:

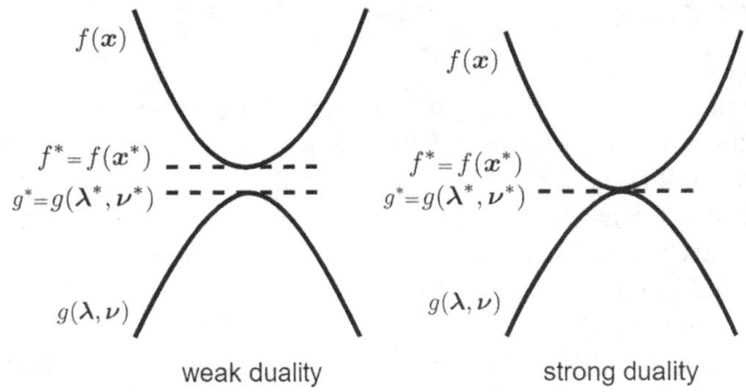

Fig. 4.3 Illustration of weak duality and strong duality

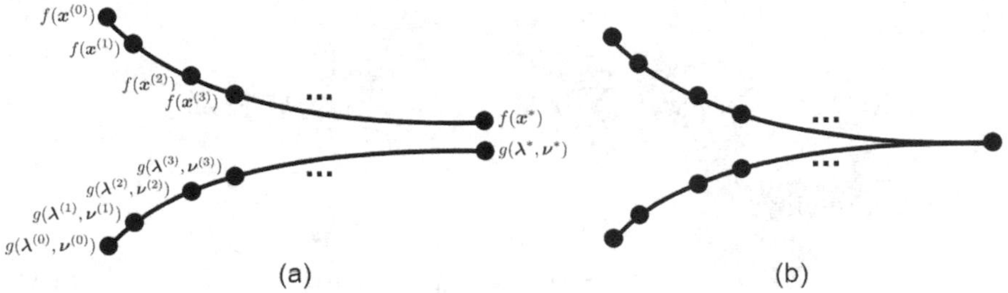

Fig. 4.4 Progress of iterative optimization: (**a**) gradual minimization of the primal function and maximization of dual function and (**b**) the primal optimal and dual optimal reach each other and become equal if strong duality holds

$$
\begin{aligned}
&\{x^{(0)}, x^{(1)}, x^{(2)}, \dots\} \to x^*, \\
&\{\nu^{(0)}, \nu^{(1)}, \nu^{(2)}, \dots\} \to \nu^*, \\
&\{\lambda^{(0)}, \lambda^{(1)}, \lambda^{(2)}, \dots\} \to \lambda^*, \\
&f(x^{(0)}) \geq f(x^{(1)}) \geq f(x^{(2)}) \geq \cdots \geq f(x^*), \\
&g(\lambda^{(0)}, \nu^{(0)}) \leq g(\lambda^{(1)}, \nu^{(1)}) \leq \cdots \leq g(\lambda^*, \nu^*).
\end{aligned} \tag{4.31}
$$

Therefore, the value of the function decreases, but the value of the dual function increases. As Fig. 4.4 depicts, they reach each other if strong duality holds; otherwise, there will be a gap between them after convergence. Note that if the optimization problem is a convex problem, the solution will be the global solution; otherwise, the solution is local.

Corollary 4.5 *As every iteration of a numerical optimization must satisfy either the weak or strong duality, the optimum dual function at every iteration always provides a lower bound for the optimum primal function at that iteration:*

$$
g(\lambda^{(k)}, \nu^{(k)}) \leq f(x^{(k)}), \quad \forall k. \tag{4.32}
$$

4.4.3 KKT Conditions

In 1952, Kuhn and Tucker published an important paper proposing several conditions in optimization [32]. However, later, it was found that there is a master's thesis by Karush in 1939 at the University of Chicago, Illinois [29]. That thesis had also proposed the conditions; however, researchers, including Kuhn and Tucker, were not aware of that thesis. Therefore, these conditions were named after all three of them as the *Karush-Kuhn-Tucker (KKT) conditions* [30]. In the following, these conditions are introduced.

The primal optimal variable $\boldsymbol{x}^*$ and the dual optimal variables $\boldsymbol{\lambda}^* = [\lambda_1^*, \ldots, \lambda_{m_1}^*]^\top$, $\boldsymbol{\nu}^* = [\nu_1^*, \ldots, \nu_{m_2}^*]^\top$ must satisfy the KKT conditions [29, 32]. The KKT conditions are summarized as follows:

1. Stationarity condition:

$$\nabla_x \mathcal{L}(\boldsymbol{x}, \boldsymbol{\lambda}, \boldsymbol{\nu}) = \nabla_x f(\boldsymbol{x}) + \sum_{i=1}^{m_1} \lambda_i \nabla_x y_i(\boldsymbol{x}) + \sum_{i=1}^{m_2} \nu_i \nabla_x h_i(\boldsymbol{x}) = 0. \tag{4.33}$$

2. Primal feasibility:

$$y_i(\boldsymbol{x}^*) \leq 0, \quad \forall i \in \{1, \ldots, m_1\}, \tag{4.34}$$

$$h_i(\boldsymbol{x}^*) = 0, \quad \forall i \in \{1, \ldots, m_2\}. \tag{4.35}$$

3. Dual feasibility:

$$\boldsymbol{\lambda} \succeq \boldsymbol{0} \quad \text{or} \quad \lambda_i \geq 0,\ \forall i \in \{1, \ldots, m_1\}. \tag{4.36}$$

4. Complementary slackness:

$$\lambda_i^* \, y_i(\boldsymbol{x}^*) = 0, \quad \forall i \in \{1, \ldots, m_1\}. \tag{4.37}$$

As listed above, KKT conditions impose constraints on the optimal dual variables of inequality constraints because the sign of inequalities is important. Refer to [19] for derivation of these conditions.

The constraint in the dual problem (4.28) is already satisfied by the dual feasibility in the KKT conditions. Therefore, the constraint of the dual problem can be ignored, as it is automatically satisfied by dual feasibility:

$$\underset{\boldsymbol{\lambda}, \boldsymbol{\nu}}{\text{maximize}} \quad g(\boldsymbol{\lambda}, \boldsymbol{\nu}), \tag{4.38}$$

which should provide $\boldsymbol{\lambda}^*$, $\boldsymbol{\nu}^*$, and $g^* = g(\boldsymbol{\lambda}^*, \boldsymbol{\nu}^*)$. This is an unconstrained optimization problem and for solving it, the derivative of $g(\boldsymbol{\lambda}, \boldsymbol{\nu})$ with respect to

$\boldsymbol{\lambda}$ and $\boldsymbol{\nu}$ should be set to zero:

$$\nabla_{\boldsymbol{\lambda}} g(\boldsymbol{\lambda}, \boldsymbol{\nu}) = 0 \overset{(4.26)}{\Longrightarrow} \nabla_{\boldsymbol{\lambda}} \mathcal{L}(\boldsymbol{x}^*, \boldsymbol{\lambda}, \boldsymbol{\nu}) = 0. \tag{4.39}$$

$$\nabla_{\boldsymbol{\nu}} g(\boldsymbol{\lambda}, \boldsymbol{\nu}) = 0 \overset{(4.26)}{\Longrightarrow} \nabla_{\boldsymbol{\nu}} \mathcal{L}(\boldsymbol{x}^*, \boldsymbol{\lambda}, \boldsymbol{\nu}) = 0. \tag{4.40}$$

Note that setting the derivatives of Lagrangian with respect to dual variables always gives back the corresponding constraints in the primal optimization problem. Equations (4.33), (4.39), and (4.40) state that the primal and dual residuals must be zero.

Finally, Eqs. (4.25) and (4.38) can be summarized into the following max-min optimization problem:

$$\sup_{\boldsymbol{\lambda}, \boldsymbol{\nu}} g(\boldsymbol{\lambda}, \boldsymbol{\nu}) \overset{(4.25)}{=} \sup_{\boldsymbol{\lambda}, \boldsymbol{\nu}} \inf_{\boldsymbol{x}} \mathcal{L}(\boldsymbol{x}, \boldsymbol{\lambda}, \boldsymbol{\nu}) = \mathcal{L}(\boldsymbol{x}^*, \boldsymbol{\lambda}^*, \boldsymbol{\nu}^*). \tag{4.41}$$

4.5 First-Order Optimization: Gradient Methods

4.5.1 *Gradient Descent*

Gradient descent is one of the fundamental first-order methods. It was first suggested by Cauchy in 1874 [33] and Hadamard in 1908 [24], and its convergence was later analyzed in [9].

4.5.1.1 Step of Update

The unconstrained optimization problem (4.15) is described as $\boldsymbol{x}^* := \arg\min_{\boldsymbol{x}} f(\boldsymbol{x})$ and $f^* := \min_{\boldsymbol{x}} f(\boldsymbol{x}) = f(\boldsymbol{x}^*)$. A random feasible initial point is the beginning of numerical optimization for unconstrained optimization and it is iteratively updated by step $\Delta\boldsymbol{x}$:

$$\boldsymbol{x}^{(k+1)} := \boldsymbol{x}^{(k)} + \Delta\boldsymbol{x}, \tag{4.42}$$

until it converges to (or gets sufficiently close to) the desired optimal point $\boldsymbol{x}^*$. Note that the step $\Delta\boldsymbol{x}$ is also denoted by $\boldsymbol{p}$ in the optimization literature, i.e., $\boldsymbol{p} := \Delta\boldsymbol{x}$. Let the function $f(.)$ be differentiable and its gradient be L-smooth. Setting $\boldsymbol{x} = \boldsymbol{x}^{(k)}$ and $\boldsymbol{y} = \boldsymbol{x}^{(k+1)} = \boldsymbol{x}^{(k)} + \Delta\boldsymbol{x}$ in Eq. (4.2), results in:

$$f(\boldsymbol{x}^{(k)} + \Delta\boldsymbol{x}) \leq f(\boldsymbol{x}^{(k)}) + \nabla f(\boldsymbol{x}^{(k)})^\top \Delta\boldsymbol{x} + \frac{L}{2}\|\Delta\boldsymbol{x}\|_2^2$$

$$\implies f(\boldsymbol{x}^{(k)} + \Delta\boldsymbol{x}) - f(\boldsymbol{x}^{(k)}) \leq \nabla f(\boldsymbol{x}^{(k)})^\top \Delta\boldsymbol{x} + \frac{L}{2}\|\Delta\boldsymbol{x}\|_2^2. \tag{4.43}$$

Until reaching the minimum, the cost function $f(.)$ needs to be reduced in every iteration; therefore:

$$f(\boldsymbol{x}^{(k)} + \Delta\boldsymbol{x}) - f(\boldsymbol{x}^{(k)}) < 0. \tag{4.44}$$

According to Eq. (4.43), one way to achieve Eq. (4.44) is:

$$\nabla f(\boldsymbol{x}^{(k)})^\top \Delta\boldsymbol{x} + \frac{L}{2}\|\Delta\boldsymbol{x}\|_2^2 < 0.$$

Therefore, $\nabla f(\boldsymbol{x}^{(k)})^\top \Delta\boldsymbol{x} + \frac{L}{2}\|\Delta\boldsymbol{x}\|_2^2$ should be minimized with respect to $\Delta\boldsymbol{x}$:

$$\underset{\Delta\boldsymbol{x}}{\text{minimize}} \;\; \nabla f(\boldsymbol{x}^{(k)})^\top \Delta\boldsymbol{x} + \frac{L}{2}\|\Delta\boldsymbol{x}\|_2^2. \tag{4.45}$$

This function is convex with respect to $\Delta\boldsymbol{x}$ and can be optimized by setting its derivative to zero:

$$\frac{\partial}{\partial \Delta\boldsymbol{x}}\Big(\nabla f(\boldsymbol{x}^{(k)})^\top \Delta\boldsymbol{x} + \frac{L}{2}\|\Delta\boldsymbol{x}\|_2^2\Big) = \nabla f(\boldsymbol{x}^{(k)}) + L\Delta\boldsymbol{x} \overset{\text{set}}{=} \boldsymbol{0}$$

$$\implies \Delta\boldsymbol{x} = -\frac{1}{L}\nabla f(\boldsymbol{x}^{(k)}). \tag{4.46}$$

Using Eq. (4.46) in Eq. (4.43) results in:

$$f(\boldsymbol{x}^{(k)} + \Delta\boldsymbol{x}) - f(\boldsymbol{x}^{(k)}) \leq -\frac{1}{2L}\|\nabla f(\boldsymbol{x}^{(k)})\|_2^2 \leq 0,$$

which satisfies Eq. (4.44). Equation (4.46) means that it is better to move toward a scale of minus gradient for updating the solution. This inspires the name of the *gradient descent* algorithm.

The problem is that the Lipschitz constant L is not often known or is hard to compute. Therefore, instead of Eq. (4.46), the following can be used:

$$\Delta\boldsymbol{x} = -\eta\nabla f(\boldsymbol{x}^{(k)}), \text{ i.e., } \boldsymbol{x}^{(k+1)} := \boldsymbol{x}^{(k)} - \eta\nabla f(\boldsymbol{x}^{(k)}), \tag{4.47}$$

where $\eta > 0$ is the step size, also called the learning rate in the data science literature. Note that if the optimization problem is maximization rather than minimization, the step should be $\Delta\boldsymbol{x} = \eta\nabla f(\boldsymbol{x}^{(k)})$, rather than Eq. (4.47). In that case, the name of the method is *gradient ascent*.

Using Eq. (4.47) into Eq. (4.43) results in:

$$f(\boldsymbol{x}^{(k)} + \Delta\boldsymbol{x}) - f(\boldsymbol{x}^{(k)}) \leq -\eta\|\nabla f(\boldsymbol{x}^{(k)})\|_2^2 + \frac{L}{2}\eta^2\|\nabla f(\boldsymbol{x}^{(k)})\|_2^2 \tag{4.48}$$
$$= \eta(\frac{L}{2}\eta - 1)\|\nabla f(\boldsymbol{x}^{(k)})\|_2^2$$

If $\boldsymbol{x}^{(k)}$ is not a stationary point, then $\|\nabla f(\boldsymbol{x}^{(k)})\|_2^2 > 0$. Noticing $\eta > 0$, to satisfy Eq. (4.44), there should be:

$$\frac{L}{2}\eta - 1 < 0 \implies \eta < \frac{2}{L}. \tag{4.49}$$

On the other hand, minimizing Eq. (4.48) by setting its derivative with respect to η to zero results in:

$$\frac{\partial}{\partial\eta}(-\eta\|\nabla f(\boldsymbol{x}^{(k)})\|_2^2 + \frac{L}{2}\eta^2\|\nabla f(\boldsymbol{x}^{(k)})\|_2^2) = -\|\nabla f(\boldsymbol{x}^{(k)})\|_2^2 + L\eta\|\nabla f(\boldsymbol{x}^{(k)})\|_2^2$$
$$= (-1 + L\eta)\|\nabla f(\boldsymbol{x}^{(k)})\|_2^2 \overset{\text{set}}{=} 0 \implies \eta = \frac{1}{L}.$$

If:

$$\eta < \frac{1}{L}, \tag{4.50}$$

then Eq. (4.48) becomes:

$$f(\boldsymbol{x}^{(k)} + \Delta\boldsymbol{x}) - f(\boldsymbol{x}^{(k)}) \leq -\frac{1}{L}\|\nabla f(\boldsymbol{x}^{(k)})\|_2^2 + \frac{1}{2L}\|\nabla f(\boldsymbol{x}^{(k)})\|_2^2$$
$$= -\frac{1}{2L}\|\nabla f(\boldsymbol{x}^{(k)})\|_2^2 < 0 \implies f(\boldsymbol{x}^{(k+1)}) \leq f(\boldsymbol{x}^{(k)}) - \frac{1}{2L}\|\nabla f(\boldsymbol{x}^{(k)})\|_2^2. \tag{4.51}$$

Equation (4.50) means that there should be an upper bound, dependent on the Lipschitz constant, on the step size; indicating that L is still needed. Equation (4.51) demonstrates that every iteration of a gradient descent decreases the cost function:

$$f(\boldsymbol{x}^{(k+1)}) \leq f(\boldsymbol{x}^{(k)}), \tag{4.52}$$

and the amount of this decrease depends on the norm of the gradient at that iteration. In conclusion, the series of solutions converges to the optimal solution, while the function value decreases iteratively until the local minimum:

$$\{\boldsymbol{x}^{(0)}, \boldsymbol{x}^{(1)}, \boldsymbol{x}^{(2)}, \dots\} \to \boldsymbol{x}^*,$$
$$f(\boldsymbol{x}^{(0)}) \geq f(\boldsymbol{x}^{(1)}) \geq f(\boldsymbol{x}^{(2)}) \geq \cdots \geq f(\boldsymbol{x}^*).$$

4.5.1.2 Line-Search

As was demonstrated in Sect. 4.5.1.1, the step size of gradient descent requires knowledge of the Lipschitz constant for the smoothness of the gradient. Meaning a suitable step size η, by a search named the *line-search*, can be found. The line-search of every optimization iteration begins with $\eta = 1$ and halves it, $\eta \leftarrow \eta/2$, if it does not satisfy Eq. (4.44) with step $\Delta \boldsymbol{x} = -\eta \nabla f(\boldsymbol{x}^{(k)})$:

$$f(\boldsymbol{x}^{(k)} - \eta \nabla f(\boldsymbol{x}^{(k)})) < f(\boldsymbol{x}^{(k)}). \tag{4.53}$$

This halving step size is repeated until this equation is satisfied, i.e., until a decrease in the objective function. Note that this decrease will occur when the step size becomes small enough to satisfy Eq. (4.50).

4.5.1.3 Backtracking Line-Search

A more sophisticated line-search method is the *Armijo line-search* or the *Armijo-Goldstein condition* [1], also called the *backtracking line-search*. Rather than Eq. (4.53), it checks if the cost function is *sufficiently* decreased:

$$f(\boldsymbol{x}^{(k)} + \boldsymbol{p}) \leq f(\boldsymbol{x}^{(k)}) + c\, \boldsymbol{p}^\top f(\boldsymbol{x}^{(k)}), \tag{4.54}$$

where $c \in (0.0.5]$ is the parameter of the Armijo line-search and $\boldsymbol{p} = \Delta \boldsymbol{x}$ is the search direction for updating. The value of c should be small, e.g., $c = 10^{-4}$ [39].

In gradient descent, the search direction is $\boldsymbol{p} = \Delta \boldsymbol{x} = -\eta \nabla f(\boldsymbol{x}^{(k)})$ according to Eq. (4.47). Therefore, for gradient descent:

$$f(\boldsymbol{x}^{(k)} - \eta \nabla f(\boldsymbol{x}^{(k)})) \leq f(\boldsymbol{x}^{(k)}) - \eta\, \gamma \|\nabla f(\boldsymbol{x}^{(k)})\|_2^2. \tag{4.55}$$

Note that a more sophisticated line-search is possible with *Wolfe conditions* [49].

4.5.1.4 Convergence Criterion

For all numerical optimization methods, including gradient descent, several methods exist for the convergence criterion to stop updating the solution and terminate optimization. Some of them are:

- Small norm of gradient: $\|\nabla f(\boldsymbol{x}^{(k+1)})\|_2 \leq \epsilon$, where ϵ is a small positive number. The reason for this criterion is the first-order optimality condition (see Lemma 4.6).
- Small change of cost function: $|f(\boldsymbol{x}^{(k+1)}) - f(\boldsymbol{x}^{(k)})| \leq \epsilon$.
- Small change of gradient of function: $|\nabla f(\boldsymbol{x}^{(k+1)}) - \nabla f(\boldsymbol{x}^{(k)})| \leq \epsilon$.

- Reaching the maximum desired number of iterations, denoted by $\max_k$: $k + 1 < \max_k$.

4.5.2 *Stochastic Gradient Methods*

4.5.2.1 Stochastic Gradient Descent

The idea of stochastic approximation was first proposed in [45] and was first used for machine learning in [4]. Consider a dataset that contains n data points, $\{\boldsymbol{a}_i \in \mathbb{R}^d\}_{i=1}^n$ and their labels $\{l_i \in \mathbb{R}\}_{i=1}^n$. Let the cost function $f(.)$ be decomposed into summation of n terms $\{f_i(\boldsymbol{x})\}_{i=1}^n$. Some well-known examples for the cost function terms are:

- Least squares error: $f_i(\boldsymbol{x}) = 0.5(\boldsymbol{a}_i^\top \boldsymbol{x} - l_i)^2$,
- Absolute error: $f_i(\boldsymbol{x}) = \boldsymbol{a}_i^\top \boldsymbol{x} - l_i$,
- Hinge loss (for $l_i \in \{-1, 1\}$): $f_i(\boldsymbol{x}) = \max(0, 1 - l_i \boldsymbol{a}_i^\top \boldsymbol{x})$.
- Logistic loss (for $l_i \in \{-1, 1\}$): $\log(\frac{1}{1+\exp(-l_i \boldsymbol{a}_i^\top \boldsymbol{x})})$.

The optimization problem (4.15) becomes:

$$\underset{\boldsymbol{x}}{\text{minimize}} \quad \frac{1}{n} \sum_{i=1}^{n} f_i(\boldsymbol{x}). \tag{4.56}$$

In this case, the full gradient is the average gradient, i.e:

$$\nabla f(\boldsymbol{x}) = \frac{1}{n} \sum_{i=1}^{n} \nabla f_i(\boldsymbol{x}), \tag{4.57}$$

so Eq. (4.46) becomes $\Delta \boldsymbol{x} = -(1/(Ln)) \sum_{i=1}^n \nabla f_i(\boldsymbol{x}^{(k)})$. This is what gradient descent uses in Eq. (4.42) for updating the solution at every iteration. However, calculation of this full gradient is time-consuming and inefficient for large values of n, especially as it needs to be recalculated at every iteration. *Stochastic Gradient Descent (SGD)*, also called the *stochastic gradient method*, approximates gradient descent stochastically and samples (i.e., bootstraps) one of the points at every iteration for updating the solution. Therefore, it uses:

$$\boldsymbol{x}^{(k+1)} := \boldsymbol{x}^{(k)} - \eta^{(k)} \nabla f_i(\boldsymbol{x}^{(k)}), \tag{4.58}$$

rather than Eq. (4.47), where $\eta^{(k)}$ is the step size at iteration k.

4.5.2.2 Minibatch Stochastic Gradient Descent

Gradient descent uses the entire n data points, where SGD uses one randomly sampled point at every iteration. For large datasets, gradient descent is very slow and intractable in every iteration, while SGD will need a significant number of iterations to roughly cover all data and has a low accuracy in convergence to the optimal point. There does, however, exist a middle case where a batch of b randomly sampled points at every iteration can be used, making it the best suited batchwise approach for large datasets [5]. This method is called the *minibatch SGD* or the *hybrid deterministic-stochastic gradient* method.

Usually, before the start of optimization, the n data points are randomly divided into $\lfloor n/b \rfloor$ batches of size b. This is equivalent to simple random sampling for sampling points into batches without replacement. The dataset is represented by $\mathcal{D}$ (where $|\mathcal{D}| = n$) and the i-th batch is represented by $\mathcal{B}_i$ (where $|\mathcal{B}_i| = b$). The batches are disjointed:

$$\bigcup_{i=1}^{\lfloor n/b \rfloor} \mathcal{B}_i = \mathcal{D}, \tag{4.59}$$

$$\mathcal{B}_i \cap \mathcal{B}_j = \varnothing, \quad \forall i, j \in \{1, \dots, \lfloor n/b \rfloor\},\ i \neq j. \tag{4.60}$$

Another less-used approach for creating batches is to sample points for a batch during optimization. This is equivalent to bootstrapping for sampling points into batches with replacement. In this case, the batches are no longer disjointed, and Eqs. (4.59) and (4.60) do not hold.

Definition 4.20 (Epoch) In minibatch SGD, when all $\lfloor n/b \rfloor$ batches of data are used for optimization once, an *epoch* is completed. After completion of an epoch, the next epoch is started and epochs are repeated until convergence of optimization.

In minibatch SGD, if the k-th iteration of optimization uses the k'-th batch, the update of solution is completed as follows:

$$\boldsymbol{x}^{(k+1)} := \boldsymbol{x}^{(k)} - \eta^{(k)} \frac{1}{b} \sum_{i \in \mathcal{B}_{k'}} \nabla f_i(\boldsymbol{x}^{(k)}). \tag{4.61}$$

The scale factor $1/b$ can be dropped for simplicity. Minibatch SGD is used significantly in machine learning, especially in neural networks. Because of dividing data into batches, minibatch SGD can be solved on parallel servers as a distributed optimization method.

4.5.3 Projected Gradient Method

The *projected gradient method* [27], also called the *gradient projection method* and the *projected gradient descent*, considers the constrained optimization problem (4.19), which can be restated to:

$$\underset{\boldsymbol{x}}{\text{minimize}} \quad f(\boldsymbol{x}) + \mathbb{I}(\boldsymbol{x} \in \mathcal{S}), \tag{4.62}$$

because the indicator function becomes infinity if its condition is not satisfied. The solution is updated as follows:

$$\begin{aligned} \boldsymbol{x}^{(k+1)} &:= \mathbf{prox}_{\eta^{(k)}\mathbb{I}(.\in\mathcal{S})}\big(\boldsymbol{x}^{(k)} - \eta^{(k)}\nabla f(\boldsymbol{x}^{(k)})\big) \\ &\overset{(4.12)}{=} \Pi_{\mathcal{S}}\big(\boldsymbol{x}^{(k)} - \eta^{(k)}\nabla f(\boldsymbol{x}^{(k)})\big). \end{aligned} \tag{4.63}$$

In other words, the projected gradient method performs a step of gradient descent and then projects the solution onto the set of constraints. This procedure is repeated until convergence of the solution.

Lemma 4.10 (Projection onto the Cone of Orthogonal Matrices [41, Section 6.7.2]) *A function* $g : \mathbb{R}^{d_1 \times d_2} \rightarrow \mathbb{R}$ *is orthogonally invariant if* $g(\boldsymbol{U}\boldsymbol{X}\boldsymbol{V}^\top) = g(\boldsymbol{X})$, *for all* $\boldsymbol{U} \in \mathbb{R}^{d_1 \times d_1}$, $\boldsymbol{X} \in \mathbb{R}^{d_1 \times d_2}$, *and* $\boldsymbol{V} \in \mathbb{R}^{d_2 \times d_2}$ *where* $\boldsymbol{U}$ *and* $\boldsymbol{V}$ *are orthogonal matrices. Let g be a convex and orthogonally invariant function, and it works on the singular values of a matrix variable* $\boldsymbol{X} \in \mathbb{R}^{d_1 \times d_2}$, *i.e.,* $g = \widehat{g} \circ \sigma$ *where the function* $\sigma(\boldsymbol{X})$ *provides the vector of singular values of* $\boldsymbol{X}$. *In this case, there is:*

$$\mathbf{prox}_{\lambda,g}(\boldsymbol{X}) = \boldsymbol{U}\,\mathbf{diag}\Big(\mathbf{prox}_{\lambda,g}\big(\sigma(\boldsymbol{X})\big)\Big)\,\boldsymbol{V}^\top, \tag{4.64}$$

where $\mathbf{diag}(.)$ *makes a diagonal matrix with its input as the diagonal, and* $\boldsymbol{U} \in \mathbb{R}^{d_1 \times d_1}$ *and* $\boldsymbol{V} \in \mathbb{R}^{d_2 \times d_2}$ *are the matrices of the left and right singular vectors of* $\boldsymbol{X}$, *respectively.*

In the constraint for projection onto the cone of orthogonal matrices, i.e., $\boldsymbol{X}^\top\boldsymbol{X} = \boldsymbol{I}$, *the function g deals with the singular values of* $\boldsymbol{X}$ *because from the Singular Value Decomposition (SVD) of* $\boldsymbol{X}$, *the following is obtained:* $\boldsymbol{X} \overset{SVD}{=} \boldsymbol{U}\boldsymbol{\Sigma}\boldsymbol{V}^\top \implies \boldsymbol{X}^\top\boldsymbol{X} = \boldsymbol{U}\boldsymbol{\Sigma}\boldsymbol{V}^\top\boldsymbol{V}\boldsymbol{\Sigma}\boldsymbol{U}^\top \overset{(a)}{=} \boldsymbol{U}\boldsymbol{\Sigma}^2\boldsymbol{U}^\top \overset{set}{=} \boldsymbol{I} \implies \boldsymbol{U}\boldsymbol{\Sigma}^2\boldsymbol{U}^\top\boldsymbol{U} = \boldsymbol{U} \overset{(b)}{\implies} \boldsymbol{U}\boldsymbol{\Sigma}^2 = \boldsymbol{U} \implies \boldsymbol{\Sigma} = \boldsymbol{I}$, *where (a) and (b) are because* $\boldsymbol{U}$ *and* $\boldsymbol{V}$ *are orthogonal matrices. Therefore, the constraint* $\boldsymbol{X}^\top\boldsymbol{X} = \boldsymbol{I}$ *(i.e., projecting onto the cone of orthogonal matrices) can be modelled by Eq. (4.64), which is simplified to setting all singular values of* $\boldsymbol{X}$ *to one:*

$$\mathbf{prox}_{\lambda,g}(\boldsymbol{X}) = \Pi_{\mathcal{O}} = \boldsymbol{U}\boldsymbol{I}\boldsymbol{V}^\top, \tag{4.65}$$

where $\boldsymbol{I} \in \mathbb{R}^{d_1 \times d_2}$ is a rectangular identity matrix and $\mathcal{O}$ denotes the cone of orthogonal matrices. If the constraint is scaled orthogonality, i.e. $\boldsymbol{X}^\top \boldsymbol{X} = \lambda \boldsymbol{I}$ with λ as the scale, the projection sets all singular values to λ by $\boldsymbol{U}(\lambda \boldsymbol{I})\boldsymbol{V}^\top = \lambda \boldsymbol{U}\boldsymbol{I}\boldsymbol{V}^\top$.

Although most often the projected gradient method is used for Eq. (4.63), there are a few other variants of the projected gradient methods, such as [12]:

$$\boldsymbol{y}^{(k)} := \Pi_{\mathcal{S}}\big(\boldsymbol{x}^{(k)} - \eta^{(k)} \nabla f(\boldsymbol{x}^{(k)})\big), \tag{4.66}$$

$$\boldsymbol{x}^{(k+1)} := \boldsymbol{x}^{(k)} + \gamma^{(k)}(\boldsymbol{y}^{(k)} - \boldsymbol{x}^{(k)}), \tag{4.67}$$

where $\eta^{(k)}$ and $\gamma^{(k)}$ are positive step sizes at iteration k. In this alternating approach, an additional variable $\boldsymbol{y}$ is found by gradient descent, followed by projection. Then, $\boldsymbol{x}$ is updated to get close to the found $\boldsymbol{y}$ while staying close to the previous solution by line-search.

4.6 Second-Order Optimization: Newton's Method

4.6.1 Newton's Method from the Newton-Raphson Root Finding Method

The root of a function $f : \boldsymbol{x} \mapsto f(\boldsymbol{x})$ can be found by solving the equation $f(\boldsymbol{x}) \overset{\text{set}}{=} 0$. The closer to the root over iterations, the more likely the root of the function can be found iteratively. One of the iterative root-finding methods is the *Newton-Raphson method* [47]. In every iteration, it finds the next solution as:

$$\boldsymbol{x}^{(k+1)} := \boldsymbol{x}^{(k)} - \frac{f(\boldsymbol{x}^{(k)})}{\nabla f(\boldsymbol{x}^{(k)})}, \tag{4.68}$$

where $\nabla f(\boldsymbol{x}^{(k)})$ is the derivative of the function with respect to $\boldsymbol{x}$. According to Eq. (4.8), in unconstrained optimization, the extremum (minimum or maximum) of the function can be found by setting its derivative to zero, i.e., $\nabla f(\boldsymbol{x}) \overset{\text{set}}{=} 0$. Recall that Eq. (4.68) was used for solving $f(\boldsymbol{x}) \overset{\text{set}}{=} 0$. Therefore, for solving Eq. (4.8), $f(\boldsymbol{x})$ should be replaced with $\nabla f(\boldsymbol{x})$ in Eq. (4.68):

$$\boldsymbol{x}^{(k+1)} := \boldsymbol{x}^{(k)} - \eta^{(k)} \frac{\nabla f(\boldsymbol{x}^{(k)})}{\nabla^2 f(\boldsymbol{x}^{(k)})}, \tag{4.69}$$

where $\nabla^2 f(\boldsymbol{x}^{(k)})$ is the second derivative of function with respect to $\boldsymbol{x}$ and which includes a step size at iteration k denoted by $\eta^{(k)} > 0$. This step size can be either fixed or adaptive. If $\boldsymbol{x}$ is multivariate, i.e., $\boldsymbol{x} \in \mathbb{R}^d$, then Eq. (4.69) is written as:

$$\boldsymbol{x}^{(k+1)} := \boldsymbol{x}^{(k)} - \eta^{(k)}\big(\nabla^2 f(\boldsymbol{x}^{(k)})\big)^{-1}\nabla f(\boldsymbol{x}^{(k)}), \tag{4.70}$$

where $\nabla f(\boldsymbol{x}^{(k)}) \in \mathbb{R}^d$ is the gradient of the function with respect to $\boldsymbol{x}$ and $\nabla^2 f(\boldsymbol{x}^{(k)}) \in \mathbb{R}^{d\times d}$ is the Hessian matrix with respect to $\boldsymbol{x}$. This optimization method, or *Newton's method*, is a second-order method because of the second derivative or the Hessian.

4.6.2 Newton's Method for Unconstrained Optimization

Consider the following optimization problem:

$$\underset{\boldsymbol{x}}{\text{minimize}} \quad f(\boldsymbol{x}). \tag{4.71}$$

where $f(.)$ is a convex function. Iterative optimization can be first-order or second-order, and updates the solution iteratively as in Eq. (4.42). The update continues until $\Delta\boldsymbol{x}$ becomes very small, which is the convergence of optimization. In first-order optimization, the updating step is $\Delta\boldsymbol{x} := -\nabla f(\boldsymbol{x})$. Near the optimal point $\boldsymbol{x}^*$, the gradient is very small, so the second-order Taylor series expansion of the function becomes:

$$\begin{aligned} f(\boldsymbol{x}) &\approx f(\boldsymbol{x}^*) + \underbrace{\nabla f(\boldsymbol{x}^*)^\top}_{\approx 0}(\boldsymbol{x}-\boldsymbol{x}^*) + \frac{1}{2}(\boldsymbol{x}-\boldsymbol{x}^*)^\top \nabla^2 f(\boldsymbol{x}^*)(\boldsymbol{x}-\boldsymbol{x}^*) \\ &\approx f(\boldsymbol{x}^*) + \frac{1}{2}(\boldsymbol{x}-\boldsymbol{x}^*)^\top \nabla^2 f(\boldsymbol{x}^*)(\boldsymbol{x}-\boldsymbol{x}^*). \end{aligned} \tag{4.72}$$

This demonstrates that the function is almost quadratic near the optimal point. Following this intuition, Newton's method uses Hessian $\nabla^2 f(\boldsymbol{x})$ in its updating step:

$$\Delta\boldsymbol{x} := -\nabla^2 f(\boldsymbol{x})^{-1}\nabla f(\boldsymbol{x}). \tag{4.73}$$

In the optimization literature, this equation is sometimes restated to:

$$\nabla^2 f(\boldsymbol{x})\,\Delta\boldsymbol{x} := -\nabla f(\boldsymbol{x}). \tag{4.74}$$

4.6.3 Newton's Method for Equality-Constrained Optimization

The optimization problem may have equality constraints:

$$\begin{aligned} \underset{x}{\text{minimize}} \quad & f(\boldsymbol{x}) \\ \text{subject to} \quad & \boldsymbol{A}\boldsymbol{x} = \boldsymbol{b}. \end{aligned} \tag{4.75}$$

After an updating step by $\boldsymbol{p} = \Delta \boldsymbol{x}$, this optimization becomes:

$$\begin{aligned} \underset{x}{\text{minimize}} \quad & f(\boldsymbol{x} + \boldsymbol{p}) \\ \text{subject to} \quad & \boldsymbol{A}(\boldsymbol{x} + \boldsymbol{p}) = \boldsymbol{b}. \end{aligned} \tag{4.76}$$

The Lagrangian of this optimization problem is:

$$\mathcal{L} = f(\boldsymbol{x} + \boldsymbol{p}) + \boldsymbol{\nu}^\top (\boldsymbol{A}(\boldsymbol{x} + \boldsymbol{p}) - \boldsymbol{b}),$$

where $\boldsymbol{\nu}$ is the dual variable. The second-order Taylor series expansion of the function $f(\boldsymbol{x} + \boldsymbol{p})$ is:

$$f(\boldsymbol{x} + \boldsymbol{p}) \approx f(\boldsymbol{x}) + \nabla f(\boldsymbol{x})^\top \boldsymbol{p} + \frac{1}{2} \boldsymbol{p}^\top \nabla^2 f(\boldsymbol{x}^*)\, \boldsymbol{p}. \tag{4.77}$$

Substituting this into the Lagrangian results in:

$$\mathcal{L} = f(\boldsymbol{x}) + \nabla f(\boldsymbol{x})^\top \boldsymbol{p} + \frac{1}{2} \boldsymbol{p}^\top \nabla^2 f(\boldsymbol{x}^*)\, \boldsymbol{p} + \boldsymbol{\nu}^\top (\boldsymbol{A}(\boldsymbol{x} + \boldsymbol{p}) - \boldsymbol{b}).$$

According to Eqs. (4.33) and (4.40) in KKT conditions, the primal and dual residuals must be zero:

$$\begin{aligned} & \nabla_x \mathcal{L} = \nabla f(\boldsymbol{x}) + \nabla^2 f(\boldsymbol{x})^\top \boldsymbol{p} + \boldsymbol{p}^\top \underbrace{\nabla^3 f(\boldsymbol{x}^*)}_{\approx \boldsymbol{0}}\, \boldsymbol{p} + \boldsymbol{A}^\top \boldsymbol{\nu} \overset{\text{set}}{=} \boldsymbol{0} \\ & \implies \nabla^2 f(\boldsymbol{x})^\top \boldsymbol{p} + \boldsymbol{A}^\top \boldsymbol{\nu} = -\nabla f(\boldsymbol{x}), \qquad (4.78) \\ & \nabla_\nu \mathcal{L} = \boldsymbol{A}(\boldsymbol{x} + \boldsymbol{p}) - \boldsymbol{b} \overset{(a)}{=} \boldsymbol{A}\boldsymbol{p} \overset{\text{set}}{=} \boldsymbol{0}, \qquad (4.79) \end{aligned}$$

where $\nabla^3 f(\boldsymbol{x}^*) \approx 0$ is because the gradient of function at the optimal point vanishes according to Eq. (4.8) and (a) is the result of the constraint $\boldsymbol{A}\boldsymbol{x} - \boldsymbol{b} = \boldsymbol{0}$ in problem (4.75). Equations (4.78) and (4.79) can be written as a system of equations:

$$\begin{bmatrix} \nabla^2 f(\boldsymbol{x})^\top & \boldsymbol{A}^\top \\ \boldsymbol{A} & \boldsymbol{0} \end{bmatrix} \begin{bmatrix} \boldsymbol{p} \\ \boldsymbol{\nu} \end{bmatrix} = \begin{bmatrix} -\nabla f(\boldsymbol{x}) \\ \boldsymbol{0} \end{bmatrix}. \tag{4.80}$$

Solving this system of equations provides the desired step $\boldsymbol{p}$ (i.e., $\Delta \boldsymbol{x}$) for updating the solution at the iteration.

Starting with Nonfeasible Initial Point Newton's method can even start with a nonfeasible point, which does not satisfy all of the constraints. If the initial point for optimization is not a feasible point, i.e., $\boldsymbol{Ax} - \boldsymbol{b} \neq \boldsymbol{0}$, Eq. (4.79) becomes:

$$\nabla_{\boldsymbol{\nu}} \mathcal{L} = \boldsymbol{A}(\boldsymbol{x} + \boldsymbol{p}) - \boldsymbol{b} \overset{\text{set}}{=} \boldsymbol{0} \implies \boldsymbol{Ap} = -(\boldsymbol{Ax} - \boldsymbol{b}). \tag{4.81}$$

Therefore, for the first iteration, the following system is solved, rather than Eq. (4.80):

$$\begin{bmatrix} \nabla^2 f(\boldsymbol{x})^\top & \boldsymbol{A}^\top \\ \boldsymbol{A} & \boldsymbol{0} \end{bmatrix} \begin{bmatrix} \boldsymbol{p} \\ \boldsymbol{\nu} \end{bmatrix} = - \begin{bmatrix} \nabla f(\boldsymbol{x}) \\ \boldsymbol{Ax} - \boldsymbol{b} \end{bmatrix}, \tag{4.82}$$

and Eq. (4.82) is used for the rest of the iterations because the next points will be in the feasibility set (since the solutions are forced to satisfy $\boldsymbol{Ax} = \boldsymbol{b}$).

4.6.4 *Interior-Point and Barrier Methods: Newton's Method for Inequality-Constrained Optimization*

The optimization problem may possess inequality constraints:

$$\begin{aligned} & \underset{\boldsymbol{x}}{\text{minimize}} && f(\boldsymbol{x}) \\ & \text{subject to} && y_i(\boldsymbol{x}) \leq 0, \quad i \in \{1, \dots, m_1\}, \\ & && \boldsymbol{Ax} = \boldsymbol{b}. \end{aligned} \tag{4.83}$$

Constrained optimization problems can be solved using *Barrier methods*, also known as *interior-point methods* [7, 38, 43, 50], *Unconstrained Minimization Technique (UMT)* or *Sequential UMT (SUMT)* [16], because they convert the problem to an unconstrained problem and solve it iteratively. Interior-point methods were first proposed by Dikin [11].

The interior-point methods, convert inequality-constrained problems to equality-constrained or unconstrained problems. Ideally, this conversion can be done using the indicator function $\mathbb{I}(.)$, which is zero if its input condition is satisfied and is infinity otherwise.[2] The problem is converted to:

[2] The indicator function in the optimization literature is not like the indicator in data science which is one if its input condition is satisfied and zero otherwise.

$$\begin{aligned} \underset{\boldsymbol{x}}{\text{minimize}} \quad & f(\boldsymbol{x}) + \sum_{i=1}^{m_1} \mathbb{I}(y_i(\boldsymbol{x}) \leq 0) \\ \text{subject to} \quad & \boldsymbol{A}\boldsymbol{x} = \boldsymbol{b}. \end{aligned} \tag{4.84}$$

The indicator function is not differentiable because it is not smooth:

$$\mathbb{I}(y_i(\boldsymbol{x}) \leq 0) := \begin{cases} 0 & \text{if } y_i(\boldsymbol{x}) \leq 0 \\ \infty & \text{if } y_i(\boldsymbol{x}) > 0. \end{cases} \tag{4.85}$$

Therefore, it can be approximated with differentiable functions called the *barrier functions* [7, 37]. A barrier function is logarithm, named the *logarithmic barrier* or *log barrier*. It approximates the indicator function by:

$$\mathbb{I}(y_i(\boldsymbol{x}) \leq 0) \approx -\frac{1}{t} \log(-y_i(\boldsymbol{x})), \tag{4.86}$$

where $t > 0$ (usually a large number such as $t = 10^6$) and the approximation becomes more accurate by $t \to \infty$. It changes the problem to:

$$\begin{aligned} \underset{\boldsymbol{x}}{\text{minimize}} \quad & f(\boldsymbol{x}) - \frac{1}{t} \sum_{i=1}^{m_1} \log(-y_i(\boldsymbol{x})) \\ \text{subject to} \quad & \boldsymbol{A}\boldsymbol{x} = \boldsymbol{b}. \end{aligned} \tag{4.87}$$

This optimization problem is an equality-constrained optimization problem, which was covered in Sect. 4.6.3. Note that there exist many approximations for the barrier; the most commonly used one is the logarithmic barrier.

The iterative solutions of the interior-point method satisfy Eq. (4.31) and follow Fig. 4.4. If the optimization problem is a convex problem, the solution of the interior-point method is the global solution; otherwise, the solution is local. The interior-point and barrier methods are used in many optimization toolboxes such as CVX [23].

4.7 Distributed Optimization

4.7.1 Alternating Optimization

When there are several optimization variables, it is possible to alternate between optimizing each of these variables. This technique is called *alternating optimization* [34] (also see [28, Chapter 4]). Consider the following multivariate optimization problem:

$$\underset{\{\boldsymbol{x}_i\}_{i=1}^m}{\text{minimize}} \quad f(\boldsymbol{x}_1, \ldots, \boldsymbol{x}_m), \tag{4.88}$$

where the objective function depends on m variables. *Alternating optimization* alternates between updating every variable, while assuming other variables are set to their latest updated values as constants. After random feasible initialization, it updates solutions as [34]:

$$\begin{aligned}
\boldsymbol{x}_1^{(k+1)} &:= \arg\min_{\boldsymbol{x}_1} f(\boldsymbol{x}_1, \boldsymbol{x}_2^{(k)}, \ldots, \boldsymbol{x}_{m-1}^{(k)}, \boldsymbol{x}_m^{(k)}), \\
\boldsymbol{x}_2^{(k+1)} &:= \arg\min_{\boldsymbol{x}_2} f(\boldsymbol{x}_1^{(k+1)}, \boldsymbol{x}_2, \ldots, \boldsymbol{x}_{m-1}^{(k)}, \boldsymbol{x}_m^{(k)}), \\
&\vdots \\
\boldsymbol{x}_m^{(k+1)} &:= \arg\min_{\boldsymbol{x}_m} f(\boldsymbol{x}_1^{(k+1)}, \boldsymbol{x}_2^{(k+1)}, \ldots, \boldsymbol{x}_{m-1}^{(k+1)}, \boldsymbol{x}_m),
\end{aligned}$$

until convergence. Any optimization methods, including first-order and second-order methods, can be used for each of the optimization lines above. In most cases, alternating optimization is robust to changing the order of updates of variables.

Remark 4.4 If the function $f(\boldsymbol{x}_1, \ldots, \boldsymbol{x}_m)$ is decomposable in terms of variables, i.e., if $f(\boldsymbol{x}_1, \ldots, \boldsymbol{x}_m) = \sum_{i=1}^m f_i(\boldsymbol{x}_i)$, then the alternating optimization can be simplified to:

$$\begin{aligned}
\boldsymbol{x}_1^{(k+1)} &:= \arg\min_{\boldsymbol{x}_1} f_1(\boldsymbol{x}_1), \\
\boldsymbol{x}_2^{(k+1)} &:= \arg\min_{\boldsymbol{x}_2} f_2(\boldsymbol{x}_2), \\
&\vdots \\
\boldsymbol{x}_m^{(k+1)} &:= \arg\min_{\boldsymbol{x}_m} f_m(\boldsymbol{x}_m),
\end{aligned}$$

because other terms become constant in optimization. The above updates indicate that if the function is completely decomposable in terms of variables, the variable updates are independent and can be done independently. Therefore, in that case, alternating optimization is reduced to m independent optimization problems, each of which can be solved by any optimization method.

The alternating optimization methods can also be used for constrained problems:

$$\begin{aligned}
&\underset{\{\boldsymbol{x}_i\}_{i=1}^m}{\text{minimize}} && f(\boldsymbol{x}_1, \ldots, \boldsymbol{x}_m) \\
&\text{subject to} && \boldsymbol{x}_i \in \mathcal{S}_i, \quad \forall i \in \{1, \ldots, m\}.
\end{aligned} \tag{4.89}$$

In this case, every line of the optimization is a constrained problem:

$$\begin{aligned}
\boldsymbol{x}_1^{(k+1)} &:= \arg\min_{\boldsymbol{x}_1} \Big(f(\boldsymbol{x}_1, \boldsymbol{x}_2^{(k)}, \dots, \boldsymbol{x}_{m-1}^{(k)}, \boldsymbol{x}_m^{(k)}), \text{ s.t. } \boldsymbol{x}_1 \in \mathcal{S}_1\Big), \\
\boldsymbol{x}_2^{(k+1)} &:= \arg\min_{\boldsymbol{x}_2} \Big(f(\boldsymbol{x}_1^{(k+1)}, \boldsymbol{x}_2, \dots, \boldsymbol{x}_{m-1}^{(k)}, \boldsymbol{x}_m^{(k)}), \text{ s.t. } \boldsymbol{x}_2 \in \mathcal{S}_2\Big), \\
&\vdots \\
\boldsymbol{x}_m^{(k+1)} &:= \arg\min_{\boldsymbol{x}_m} \Big(f(\boldsymbol{x}_1^{(k+1)}, \boldsymbol{x}_2^{(k+1)}, \dots, \boldsymbol{x}_{m-1}^{(k+1)}, \boldsymbol{x}_m), \text{ s.t. } \boldsymbol{x}_m \in \mathcal{S}_m\Big).
\end{aligned}$$

Any constrained optimization methods, such as projected gradient method, proximal methods, and interior-point methods, can be used for each of the optimization lines above.

Finally, it is noteworthy that practical experiments have demonstrated that there is usually no need to use a complete optimization until convergence for every step in the alternating optimization, either unconstrained or constrained. Often, a single step of updating, such as a step of gradient descent or the projected gradient method, is enough for the algorithm to work.

4.7.2 Dual Ascent and Dual Decomposition Methods

Consider the following problem:

$$\begin{aligned}
&\underset{\boldsymbol{x}}{\text{minimize}} \quad f(\boldsymbol{x}) \\
&\text{subject to} \quad \boldsymbol{A}\boldsymbol{x} = \boldsymbol{b}.
\end{aligned} \tag{4.90}$$

The Lagrangian is:

$$\mathcal{L}(\boldsymbol{x}, \boldsymbol{\nu}) = f(\boldsymbol{x}) + \boldsymbol{\nu}^\top (\boldsymbol{A}\boldsymbol{x} - \boldsymbol{b}).$$

The dual function is:

$$g(\boldsymbol{\nu}) = \inf_{\boldsymbol{x}} \mathcal{L}(\boldsymbol{x}, \boldsymbol{\nu}). \tag{4.91}$$

The dual problem maximizes $g(\boldsymbol{\nu})$:

$$\boldsymbol{\nu}^* = \arg\max_{\boldsymbol{\nu}} g(\boldsymbol{\nu}). \tag{4.92}$$

Therefore, the optimal primal variable is:

$$x^* = \arg\min_{x} \mathcal{L}(x, \nu^*). \tag{4.93}$$

To solve Eq. (4.92), the derivative of the dual function with respect to the dual variable results in:

$$\nabla_\nu g(\nu) \overset{(4.91)}{=} \nabla_\nu(\inf_{x} \mathcal{L}(x, \nu)) \overset{(4.93)}{=} \nabla_\nu(f(x^*) + \nu^\top(Ax^* - b)) = Ax^* - b.$$

The dual problem is a maximization problem, meaning the gradient ascent (see Sect. 4.5.1) can be used iteratively to update the dual variable with this gradient. It is possible to alternate between updating the optimal primal and dual variables:

$$x^{(k+1)} := \arg\min_{x} \mathcal{L}(x, \nu^{(k)}), \tag{4.94}$$

$$\nu^{(k+1)} := \nu^{(k)} + \eta^{(k)}(Ax^{(k+1)} - b), \tag{4.95}$$

where k is the iteration index and $\eta^{(k)}$ is the step size (also called the learning rate) at iteration k. Equation (4.94) can be performed by any optimization method. The gradient of $\mathcal{L}(x, \nu^{(k)})$ with respect to x is computed; if setting this gradient to zero does not result in a closed-form x, gradient descent (see Sect. 4.5.1) can be used to perform Eq. (4.94). Some papers approximate Eq. (4.94) by one step or a few steps of gradient descent, rather than a complete gradient descent until convergence. If one step is used, Eq. (4.94) is written as:

$$x^{(k+1)} := x^{(k)} - \gamma \nabla_x \mathcal{L}(x, \nu^{(k)}), \tag{4.96}$$

where $\gamma > 0$ is the step size. It has been empirically demonstrated that even one step of gradient descent for Eq. (4.94) works properly for the whole alternating algorithm.

Iterations are continued until the primal and dual variables converge to stable values. When nearing convergence, $(Ax^{k+1} - b) \rightarrow 0$ happens so that updates to the dual variable are not needed, according to Eq. (4.95). This means that the result of convergence is $(Ax^{k+1} - b) \approx 0$, so that the constraint in Eq. (4.90) is satisfied. In other words, the update of the dual variable in Eq. (4.95) satisfies the constraint. This method is known as the *dual ascent* method because it uses gradient ascent for updating the dual variable.

If the objective function can be distributed and decomposed on b blocks $\{x_i\}_{i=1}^b$, namely:

$$f(x) = f_1(x_1) + \cdots + f_1(x_b),$$

then the total Lagrangian is the summation of b Lagrangian functions:

$$\mathcal{L}_i(x_i, \nu) = f(x_i) + \nu^\top(Ax_i - b),$$

$$\mathcal{L}(\boldsymbol{x}_i, \boldsymbol{\nu}) = \sum_{i=1}^{b} \big(f(\boldsymbol{x}_i) + \boldsymbol{\nu}^\top (\boldsymbol{A}\boldsymbol{x}_i - \boldsymbol{b})\big).$$

Equation (4.94) can be divided into b updates, each for one of the blocks.

$$\boldsymbol{x}_i^{(k+1)} := \arg\min_{\boldsymbol{x}_i} \mathcal{L}(\boldsymbol{x}, \boldsymbol{\nu}^{(k)}), \quad \forall i \in \{1, \dots, b\}, \tag{4.97}$$

$$\boldsymbol{\nu}^{(k+1)} := \boldsymbol{\nu}^{(k)} + \eta^{(k)}(\boldsymbol{A}\boldsymbol{x}^{(k+1)} - \boldsymbol{b}). \tag{4.98}$$

This is called *dual decomposition* which is developed by decomposition techniques, such as the Dantzig-Wolfe decomposition [10], Bender's decomposition [2], and Lagrangian decomposition [13]. The dual decomposition methods can divide a problem into subproblems and solve them in parallel. Therefore, it can be used for big data but they are usually slow to converge.

4.7.3 Augmented Lagrangian Method (Method of Multipliers)

Assume the objective function in Eq. (4.90) is regularized by a penalty on not satisfying the constraint:

$$\begin{aligned} \underset{\boldsymbol{x}}{\text{minimize}} \quad & f(\boldsymbol{x}) + \frac{\rho}{2}\|\boldsymbol{A}\boldsymbol{x} - \boldsymbol{b}\|_2^2 \\ \text{subject to} \quad & \boldsymbol{A}\boldsymbol{x} = \boldsymbol{b}, \end{aligned} \tag{4.99}$$

where $\rho > 0$ is the regularization parameter.

Definition 4.21 (Augmented Lagrangian [25, 44]) The Lagrangian for problem (4.99) is:

$$\mathcal{L}_\rho(\boldsymbol{x}, \boldsymbol{\nu}) := f(\boldsymbol{x}) + \boldsymbol{\nu}^\top (\boldsymbol{A}\boldsymbol{x} - \boldsymbol{b}) + \frac{\rho}{2}\|\boldsymbol{A}\boldsymbol{x} - \boldsymbol{b}\|_2^2. \tag{4.100}$$

This Lagrangian is called the *augmented Lagrangian* for problem (4.90).

This augmented Lagrangian can be used in Eqs. (4.94) and (4.95):

$$\boldsymbol{x}^{(k+1)} := \arg\min_{\boldsymbol{x}} \mathcal{L}_\rho(\boldsymbol{x}, \boldsymbol{\nu}^{(k)}), \tag{4.101}$$

$$\boldsymbol{\nu}^{(k+1)} := \boldsymbol{\nu}^{(k)} + \rho(\boldsymbol{A}\boldsymbol{x}^{(k+1)} - \boldsymbol{b}), \tag{4.102}$$

where ρ is used for the step size of updating the dual variable. This method is called the *augmented Lagrangian method* or the *method of multipliers* [3, 25, 44].

4.7.4 *Alternating Direction Method of Multipliers*

Alternating Direction Method of Multipliers (ADMM) [6, 17, 22] has been used in many recent machine learning and signal processing papers. There are two main reasons to use ADMM (and other distributed methods) in machine learning and signal processing: (1) it makes the problem distributed and parallelizable on several servers; and (2) it makes it possible to solve an optimization problem with multiple variables. An example use of ADMM in machine learning is [18], which learns locally linear embedding over multiple blocks of an image. Another example is [15], which uses ADMM for low-rank feature extraction.

4.7.4.1 ADMM Algorithm

Consider the following problem:

$$\begin{aligned} \underset{\boldsymbol{x}_1, \boldsymbol{x}_2}{\text{minimize}} \quad & f_1(\boldsymbol{x}_1) + f_2(\boldsymbol{x}_2) \\ \text{subject to} \quad & \boldsymbol{A}\boldsymbol{x}_1 + \boldsymbol{B}\boldsymbol{x}_2 = \boldsymbol{c}, \end{aligned} \tag{4.103}$$

which is an optimization over two variables $\boldsymbol{x}_1$ and $\boldsymbol{x}_2$. The augmented Lagrangian for this problem is:

$$\mathcal{L}_\rho(\boldsymbol{x}_1, \boldsymbol{x}_2, \boldsymbol{\nu}) = f_1(\boldsymbol{x}_1) + f_2(\boldsymbol{x}_2) + \boldsymbol{\nu}^\top(\boldsymbol{A}\boldsymbol{x}_1 + \boldsymbol{B}\boldsymbol{x}_2 - \boldsymbol{c}) + \frac{\rho}{2}\|\boldsymbol{A}\boldsymbol{x}_1 + \boldsymbol{B}\boldsymbol{x}_2 - \boldsymbol{c}\|_2^2. \tag{4.104}$$

It is possible to alternate between updating the primal variables $\boldsymbol{x}_1$ and $\boldsymbol{x}_2$ and the dual variable $\boldsymbol{\nu}$ until convergence of these variables:

$$\boldsymbol{x}_1^{(k+1)} := \arg\min_{\boldsymbol{x}_1} \mathcal{L}_\rho(\boldsymbol{x}_1, \boldsymbol{x}_2^{(k)}, \boldsymbol{\nu}^{(k)}), \tag{4.105}$$

$$\boldsymbol{x}_2^{(k+1)} := \arg\min_{\boldsymbol{x}_2} \mathcal{L}_\rho(\boldsymbol{x}_1^{(k+1)}, \boldsymbol{x}_2, \boldsymbol{\nu}^{(k)}), \tag{4.106}$$

$$\boldsymbol{\nu}^{(k+1)} := \boldsymbol{\nu}^{(k)} + \rho(\boldsymbol{A}\boldsymbol{x}_1^{(k+1)} + \boldsymbol{B}\boldsymbol{x}_2^{(k+1)} - \boldsymbol{c}). \tag{4.107}$$

It is important to note that the order in which the primal and dual variables are updated is important, and the dual variable should be updated after the primal variables. However, the order in which you update the primal variables is not important. This method is called the Alternating Direction Method of Multipliers (ADMM) [17, 22]. A good tutorial and survey paper on ADMM is [6] where the details of the ADMM method are explained.

As previously discussed, Eqs. (4.105) and (4.106) can be performed by any optimization method, such as calculating the gradient of an augmented Lagrangian with respect to $\boldsymbol{x}_1$ and $\boldsymbol{x}_2$, and using a few (or even one) iterations of the gradient descent for each of these equations.

4.7.4.2 Simplifying Equations in ADMM

The last term in the augmented Lagrangian, Eq. (4.104), can be restated as:

$$\begin{aligned}
&\boldsymbol{\nu}^\top(\boldsymbol{A}\boldsymbol{x}_1 + \boldsymbol{B}\boldsymbol{x}_2 - \boldsymbol{c}) + \frac{\rho}{2}\|\boldsymbol{A}\boldsymbol{x}_1 + \boldsymbol{B}\boldsymbol{x}_2 - \boldsymbol{c}\|_2^2 \\
&= \boldsymbol{\nu}^\top(\boldsymbol{A}\boldsymbol{x}_1 + \boldsymbol{B}\boldsymbol{x}_2 - \boldsymbol{c}) + \frac{\rho}{2}\|\boldsymbol{A}\boldsymbol{x}_1 + \boldsymbol{B}\boldsymbol{x}_2 - \boldsymbol{c}\|_2^2 + \frac{1}{2\rho}\|\boldsymbol{\nu}\|_2^2 - \frac{1}{2\rho}\|\boldsymbol{\nu}\|_2^2 \\
&= \frac{\rho}{2}\Big(\|\boldsymbol{A}\boldsymbol{x}_1 + \boldsymbol{B}\boldsymbol{x}_2 - \boldsymbol{c}\|_2^2 + \frac{1}{\rho^2}\|\boldsymbol{\nu}\|_2^2 + \frac{2}{\rho}\boldsymbol{\nu}^\top(\boldsymbol{A}\boldsymbol{x}_1 + \boldsymbol{B}\boldsymbol{x}_2 - \boldsymbol{c})\Big) - \frac{1}{2\rho}\|\boldsymbol{\nu}\|_2^2 \\
&\overset{(a)}{=} \frac{\rho}{2}\Big\|\boldsymbol{A}\boldsymbol{x}_1 + \boldsymbol{B}\boldsymbol{x}_2 - \boldsymbol{c} + \frac{1}{\rho}\boldsymbol{\nu}\Big\|_2^2 - \frac{1}{2\rho}\|\boldsymbol{\nu}\|_2^2 \overset{(b)}{=} \frac{\rho}{2}\Big\|\boldsymbol{A}\boldsymbol{x}_1 + \boldsymbol{B}\boldsymbol{x}_2 - \boldsymbol{c} + \boldsymbol{u}\Big\|_2^2 - \frac{1}{2\rho}\|\boldsymbol{\nu}\|_2^2.
\end{aligned}$$

where (a) is because of the square of summation of two terms, and (b) is because $\boldsymbol{u} := (1/\rho)\boldsymbol{\nu}$. The last term $-(1/(2\rho))\|\boldsymbol{\nu}\|_2^2$ is constant with respect to the primal variables $\boldsymbol{x}_1$ and $\boldsymbol{x}_2$, meaning it is possible to drop that term from the Lagrangian when updating the primal variables. Therefore, the Lagrangian can be restated as:

$$\mathcal{L}_\rho(\boldsymbol{x}_1, \boldsymbol{x}_2, \boldsymbol{u}) = f_1(\boldsymbol{x}_1) + f_2(\boldsymbol{x}_2) + \frac{\rho}{2}\Big\|\boldsymbol{A}\boldsymbol{x}_1 + \boldsymbol{B}\boldsymbol{x}_2 - \boldsymbol{c} + \boldsymbol{u}\Big\|_2^2 + \text{constant}. \tag{4.108}$$

For updating $\boldsymbol{x}_1$ and $\boldsymbol{x}_2$, the terms $f_2(\boldsymbol{x}_2)$ and $f(\boldsymbol{x}_1)$ are constant, respectively, and can be dropped (because arg min is important here and not the minimum value).

Therefore, Eqs. (4.105), (4.106), and (4.107) can be restated as:

$$\boldsymbol{x}_1^{(k+1)} := \arg\min_{\boldsymbol{x}_1}\Big(f_1(\boldsymbol{x}_1) + \frac{\rho}{2}\Big\|\boldsymbol{A}\boldsymbol{x}_1 + \boldsymbol{B}\boldsymbol{x}_2^{(k)} - \boldsymbol{c} + \boldsymbol{u}^{(k)}\Big\|_2^2\Big), \tag{4.109}$$

$$\boldsymbol{x}_2^{(k+1)} := \arg\min_{\boldsymbol{x}_2}\Big(f_2(\boldsymbol{x}_2) + \frac{\rho}{2}\Big\|\boldsymbol{A}\boldsymbol{x}_1^{(k+1)} + \boldsymbol{B}\boldsymbol{x}_2 - \boldsymbol{c} + \boldsymbol{u}^{(k)}\Big\|_2^2\Big), \tag{4.110}$$

$$\boldsymbol{u}^{(k+1)} := \boldsymbol{u}^{(k)} + \rho(\boldsymbol{A}\boldsymbol{x}_1^{(k+1)} + \boldsymbol{B}\boldsymbol{x}_2^{(k+1)} - \boldsymbol{c}). \tag{4.111}$$

Again, Eqs. (4.109) and (4.110) can be performed by one or a few steps of gradient descent or any other optimization methods. The convergence of ADMM for nonconvex and nonsmooth functions is analyzed and proven in [48].

4.7.5 ADMM Algorithm for General Optimization Problems and Any Number of Variables

4.7.5.1 Distributed Optimization

ADMM can be extended to several equality and inequality constraints for several optimization variables [20, 21]. The following optimization problem with m optimization variables and an equality and inequality constraint for every variable is described as:

$$
\begin{aligned}
\underset{\{\boldsymbol{x}_i\}_{i=1}^m}{\text{minimize}} \quad & \sum_{i=1}^{m} f_i(\boldsymbol{x}_i) \\
\text{subject to} \quad & y_i(\boldsymbol{x}_i) \leq 0, \; i \in \{1, \dots, m\}, \\
& h_i(\boldsymbol{x}_i) = 0, \; i \in \{1, \dots, m\}.
\end{aligned} \tag{4.112}
$$

Every inequality constraint can be converted to an equality constraint by the following technique [20, 21]:

$$
y_i(\boldsymbol{x}_i) \leq 0 \quad \equiv \quad y_i'(\boldsymbol{x}_i) := \big(\max(0, y_i(\boldsymbol{x}_i))\big)^2 = 0.
$$

Therefore, the problem becomes:

$$
\begin{aligned}
\underset{\{\boldsymbol{x}_i\}_{i=1}^m}{\text{minimize}} \quad & \sum_{i=1}^{m} f_i(\boldsymbol{x}_i) \\
\text{subject to} \quad & y_i'(\boldsymbol{x}_i) = 0, \; i \in \{1, \dots, m\}, \\
& h_i(\boldsymbol{x}_i) = 0, \; i \in \{1, \dots, m\}.
\end{aligned}
$$

Having dual variables $\boldsymbol{\lambda} = [\lambda_1, \dots, \lambda_m]^\top$ and $\boldsymbol{\nu} = [\nu_1, \dots, \nu_m]^\top$ and the regularization parameter $\rho > 0$, the augmented Lagrangian for this problem is:

$$
\begin{aligned}
\mathcal{L}_\rho(\{\boldsymbol{x}_i\}_{i=1}^m, \boldsymbol{\nu}', \boldsymbol{\nu}) &= \sum_{i=1}^{m} f_i(\boldsymbol{x}_i) + \sum_{i=1}^{m} \lambda_i y_i'(\boldsymbol{x}_i) + \sum_{i=1}^{m} \nu_i h_i(\boldsymbol{x}_i) \\
&\quad + \frac{\eta}{2} \sum_{i=1}^{m} (y_i'(\boldsymbol{x}_i))^2 + \frac{\rho}{2} \sum_{i=1}^{m} (h_i(\boldsymbol{x}_i))^2 \\
&= \sum_{i=1}^{m} f_i(\boldsymbol{x}_i) + \boldsymbol{\lambda}^\top \boldsymbol{y}'(\boldsymbol{x}) + \boldsymbol{\nu}^\top \boldsymbol{h}(\boldsymbol{x}) + \frac{\rho}{2} \|\boldsymbol{y}'(\boldsymbol{x})\|_2^2 + \frac{\rho}{2} \|\boldsymbol{h}(\boldsymbol{x})\|_2^2,
\end{aligned} \tag{4.113}
$$

where $\mathbb{R}^m \ni \boldsymbol{y}'(\boldsymbol{x}) := [y_1'(\boldsymbol{x}_1), \dots, y_m'(\boldsymbol{x}_m)]^\top$ and $\mathbb{R}^m \ni \boldsymbol{h}(\boldsymbol{x}) := [h_1(\boldsymbol{x}_1), \dots, h_m(\boldsymbol{x}_m)]^\top$. Updating the primal and dual variables is performed as [20, 21]:

$$
\begin{aligned}
&\boldsymbol{x}_i^{(k+1)} := \arg\min_{\boldsymbol{x}_i} \mathcal{L}_\rho(\boldsymbol{x}_i, \lambda_i^{(k)}, \nu_i^{(k)}), \; \forall i \in \{1, \dots, m\}, \\
&\boldsymbol{\lambda}^{(k+1)} := \boldsymbol{\lambda}^{(k)} + \rho \, \boldsymbol{y}'(\boldsymbol{x}^{(k+1)}), \\
&\boldsymbol{\nu}^{(k+1)} := \boldsymbol{\nu}^{(k)} + \rho \, \boldsymbol{h}(\boldsymbol{x}^{(k+1)}).
\end{aligned}
$$

Note that as the Lagrangian is completely decomposable by the i indices, the optimization for every i-th primal or dual variable does not depend on other indices; in other words, the terms of other indices become constant for every index. The last terms in the augmented Lagrangian, Eq. (4.113), can be restated as:

$$\begin{aligned}
&\boldsymbol{\lambda}^\top \boldsymbol{y}'(\boldsymbol{x}) + \boldsymbol{\nu}^\top \boldsymbol{h}(\boldsymbol{x}) + \frac{\rho}{2}\|\boldsymbol{y}'(\boldsymbol{x})\|_2^2 + \frac{\rho}{2}\|\boldsymbol{h}(\boldsymbol{x})\|_2^2 \\
&= \boldsymbol{\lambda}^\top \boldsymbol{y}'(\boldsymbol{x}) + \frac{\rho}{2}\|\boldsymbol{y}'(\boldsymbol{x})\|_2^2 + \frac{1}{2\rho}\|\boldsymbol{\lambda}\|_2^2 - \frac{1}{2\rho}\|\boldsymbol{\lambda}\|_2^2 \\
&+ \boldsymbol{\nu}^\top \boldsymbol{h}(\boldsymbol{x}) + \frac{\rho}{2}\|\boldsymbol{h}(\boldsymbol{x})\|_2^2 + \frac{1}{2\rho}\|\boldsymbol{\nu}\|_2^2 - \frac{1}{2\rho}\|\boldsymbol{\nu}\|_2^2 \\
&= \frac{\rho}{2}\Big(\|\boldsymbol{y}'(\boldsymbol{x})\|_2^2 + \frac{1}{\rho^2}\|\boldsymbol{\lambda}\|_2^2 + \frac{2}{\rho}\boldsymbol{\lambda}^\top \boldsymbol{y}'(\boldsymbol{x})\Big) - \frac{1}{2\rho}\|\boldsymbol{\lambda}\|_2^2 \\
&+ \frac{\rho}{2}\Big(\|\boldsymbol{h}(\boldsymbol{x})\|_2^2 + \frac{1}{\rho^2}\|\boldsymbol{\nu}\|_2^2 + \frac{2}{\rho}\boldsymbol{\nu}^\top \boldsymbol{h}(\boldsymbol{x})\Big) - \frac{1}{2\rho}\|\boldsymbol{\nu}\|_2^2 \\
&= \frac{\rho}{2}\Big\|\boldsymbol{y}'(\boldsymbol{x}) + \frac{1}{\rho}\boldsymbol{\lambda}\Big\|_2^2 - \frac{1}{2\rho}\|\boldsymbol{\lambda}\|_2^2 + \frac{\rho}{2}\Big\|\boldsymbol{h}(\boldsymbol{x}) + \frac{1}{\rho}\boldsymbol{\nu}\Big\|_2^2 - \frac{1}{2\rho}\|\boldsymbol{\nu}\|_2^2 \\
&\overset{(a)}{=} \frac{\rho}{2}\big\|\boldsymbol{y}'(\boldsymbol{x}) + \boldsymbol{u}_\lambda\big\|_2^2 + \frac{\rho}{2}\big\|\boldsymbol{h}(\boldsymbol{x}) + \boldsymbol{u}_\nu\big\|_2^2 - \text{constant},
\end{aligned}$$

where (a) is because $\boldsymbol{u}_\lambda := (1/\rho)\boldsymbol{\lambda}$ and $\boldsymbol{u}_\nu := (1/\rho)\boldsymbol{\nu}$ are defined. Therefore, the Lagrangian can be restated as:

$$\begin{aligned}
\mathcal{L}_\rho(\{\boldsymbol{x}_i\}_{i=1}^m, \boldsymbol{u}_\lambda, \boldsymbol{u}_\nu) &= \sum_{i=1}^m f_i(\boldsymbol{x}_i) + \frac{\rho}{2}\big\|\boldsymbol{y}'(\boldsymbol{x}) + \boldsymbol{u}_\lambda\big\|_2^2 + \frac{\rho}{2}\big\|\boldsymbol{h}(\boldsymbol{x}) + \boldsymbol{u}_\nu\big\|_2^2 + \text{constant} \\
&= \sum_{i=1}^m f_i(\boldsymbol{x}_i) + \frac{\rho}{2}\sum_{i=1}^m \big[(y_i'(\boldsymbol{x}_i) + u_{\lambda,i})^2 + (h_i(\boldsymbol{x}_i) + u_{\nu,i})^2\big] + \text{constant},
\end{aligned}$$

where $u_{\lambda,i} = (1/\rho)\lambda_i$ and $u_{\nu,i} = (1/\rho)\nu_i$ are the i-th elements of $\boldsymbol{u}_\lambda$ and $\boldsymbol{u}_\nu$, respectively. As such, updating variables can be restated as:

$$\begin{aligned}
\boldsymbol{x}_i^{(k+1)} &:= \arg\min_{\boldsymbol{x}_i} \Big(f_i(\boldsymbol{x}_i) + \frac{\rho}{2}\big[(y_i'(\boldsymbol{x}_i) + u_{\lambda,i}^{(k)})^2 \\
&\quad + (h_i(\boldsymbol{x}_i) + u_{\nu,i}^{(k)})^2\big]\Big), \forall i \in \{1, \dots, m\}, \quad (4.114)\\
u_{\lambda,i}^{(k+1)} &:= u_{\lambda,i}^{(k)} + \rho\, y_i'(\boldsymbol{x}_i^{(k+1)}), \ \forall i \in \{1, \dots, m\} \quad (4.115)\\
u_{\nu,i}^{(k+1)} &:= u_{\nu,i}^{(k)} + \rho\, h_i(\boldsymbol{x}_i^{(k+1)}), \ \forall i \in \{1, \dots, m\}. \quad (4.116)
\end{aligned}$$

Use of ADMM for Distributed Optimization ADMM is one of the most well-known and utilized algorithms for distributed optimization. If the problem can be divided into several disjoint blocks (i.e., several primal variables), it is possible to solve the optimization for each primal variable on a separate core or server (see Eq. (4.114) for every i). Therefore, in every iteration of ADMM, the update of the primal variables can be performed in parallel by distributed servers. At the end of each iteration, the updated primal variables are gathered in a central server so that the update of the dual variable(s) is performed (see Eqs. (4.115) and (4.116)). The updated dual variable(s) are sent to the distributed servers to update their primal variables. This procedure is repeated until convergence of primal and dual variables. In this sense, ADMM is performed with a similar approach to *federated learning* [31, 35] which is a machine learning technique for learning over decentralized servers.

4.7.5.2 Making Optimization Problem Distributed

Using ADMM, it is possible to convert a nondistributed optimization problem to a distributed optimization problem to solve it more easily. Many recent machine learning and signal processing papers are using this technique. An example is [15] which uses ADMM to solve a multivariate optimization problem for feature extraction.

Univariate Optimization Problem Consider a regular nondistributed problem with one optimization variable $\boldsymbol{x}$:

$$
\begin{aligned}
\underset{\boldsymbol{x}}{\text{minimize}} \quad & \sum_{i=1}^{m} f_i(\boldsymbol{x}) \\
\text{subject to} \quad & y_i(\boldsymbol{x}) \leq 0, \;\; i \in \{1, \dots, m\}, \\
& h_i(\boldsymbol{x}) = 0, \;\; i \in \{1, \dots, m\}.
\end{aligned}
\tag{4.117}
$$

This problem can be stated as follows:

$$
\begin{aligned}
\underset{\{\boldsymbol{x}_i\}_{i=1}^m}{\text{minimize}} \quad & \sum_{i=1}^{m} f_i(\boldsymbol{x}_i) \\
\text{subject to} \quad & y_i(\boldsymbol{x}_i) \leq 0, \;\; i \in \{1, \dots, m\}, \\
& h_i(\boldsymbol{x}_i) = 0, \;\; i \in \{1, \dots, m\}, \\
& \boldsymbol{x}_i = \boldsymbol{z}, \;\; i \in \{1, \dots, m\},
\end{aligned}
\tag{4.118}
$$

where m variables are introduced $\{\boldsymbol{x}_i\}_{i=1}^m$ and the trick $\boldsymbol{x}_i = \boldsymbol{z}, \forall i$ is used to make them equal to one variable. Equation (4.118) is similar to Eq. (4.112), except that it has $2m$ equality constraints, rather than m equality constraints. Therefore, the ADMM updates similarly to Eqs. (4.114), (4.115), and (4.116) can be used, but with a slight change, introduced in the following, because of the additional m constraints. m new dual variables are used for constraints $\boldsymbol{x}_i = \boldsymbol{z}, \forall i$ and are updated in addition to other variables. The augmented Lagrangian also has some additional terms for the additional constraints. The Lagrangian and ADMM updates are similarly obtained as explained in Sect. 4.7.5.1. This is a good technique to make a problem distributed, use ADMM to solve it, and solve it in parallel servers.

Multivariate Optimization Problem A regular nondistributed problem with multiple optimization variables $\{\boldsymbol{x}_i\}_{i=1}^m$ is described as:

$$\begin{aligned} \underset{\{\boldsymbol{x}\}_{i=1}^m}{\text{minimize}} \quad & \sum_{i=1}^{m} f_i(\boldsymbol{x}_i) \\ \text{subject to} \quad & \boldsymbol{x}_i \in \mathcal{S}_i, \; i \in \{1, \ldots, m\}, \end{aligned} \tag{4.119}$$

where $\boldsymbol{x}_i \in \mathcal{S}_i$ can be any constraint, such as belonging to a set $\mathcal{S}_i$, an equality constraint, or an inequality constraint. The constraint can be embedded in the objective function using the indicator function:

$$\underset{\{\boldsymbol{x}\}_{i=1}^m}{\text{minimize}} \quad \sum_{i=1}^{m} \big(f_i(\boldsymbol{x}_i) + \phi_i(\boldsymbol{x}_i)\big),$$

where $\phi_i(\boldsymbol{x}_i) := \mathbb{I}(\boldsymbol{x}_i \in \mathcal{S}_i)$ is zero if $\boldsymbol{x}_i \in \mathcal{S}_i$ and is infinity otherwise. This problem can be stated as follows:

$$\begin{aligned} \underset{\{\boldsymbol{x}_i\}_{i=1}^m}{\text{minimize}} \quad & \sum_{i=1}^{m} \big(f_i(\boldsymbol{x}_i) + \phi_i(\boldsymbol{z}_i)\big) \\ \text{subject to} \quad & \boldsymbol{x}_i = \boldsymbol{z}_i, \; i \in \{1, \ldots, m\}, \end{aligned} \tag{4.120}$$

where the variable $\boldsymbol{z}_i$ is introduced for every $\boldsymbol{x}_i$, the introduced variable is used for the second term in the objective function, and then they are equated in the constraint.

As the constraints $\boldsymbol{x}_i - \boldsymbol{z}_i = \boldsymbol{0}, \forall i$ are equality constraints, Eqs. (4.109), (4.110), and (4.111) can be used as ADMM updates for this problem:

$$\boldsymbol{x}_i^{(k+1)} := \arg\min_{\boldsymbol{x}_i} \Big(f_i(\boldsymbol{x}_i) + \frac{\rho}{2}\|\boldsymbol{x}_i - \boldsymbol{z}_i^{(k)} + \boldsymbol{u}_i^{(k)}\|_2^2\Big), \forall i \in \{1, \ldots, m\}, \tag{4.121}$$

$$\boldsymbol{z}_i^{(k+1)} := \arg\min_{\boldsymbol{z}_i} \Big(\phi_i(\boldsymbol{z}_i) + \frac{\rho}{2}\|\boldsymbol{x}_i^{(k+1)} - \boldsymbol{z}_i + \boldsymbol{u}_i^{(k)}\|_2^2\Big), \forall i \in \{1, \ldots, m\}, \tag{4.122}$$

$$\boldsymbol{u}_i^{(k+1)} := \boldsymbol{u}_i^{(k)} + \rho(\boldsymbol{x}_i^{(k+1)} + \boldsymbol{z}_i^{(k+1)}), \; \forall i \in \{1, \ldots, m\}.$$

Comparing Eqs. (4.121) and (4.122) with Eq. (4.11) demonstrates that these ADMM updates can be written as proximal mappings:

$$\begin{aligned}
\boldsymbol{x}_i^{(k+1)} &:= \mathbf{prox}_{\frac{1}{\rho} f_i}(\boldsymbol{z}_i^{(k)} - \boldsymbol{u}_i^{(k)}), \; \forall i \in \{1, \dots, m\}, \\
\boldsymbol{z}_i^{(k+1)} &:= \mathbf{prox}_{\frac{1}{\rho} \phi_i}(\boldsymbol{x}_i^{(k+1)} + \boldsymbol{u}_i^{(k)}), \; \forall i \in \{1, \dots, m\}, \\
\boldsymbol{u}_i^{(k+1)} &:= \boldsymbol{u}_i^{(k)} + \rho(\boldsymbol{x}_i^{(k+1)} + \boldsymbol{z}_i^{(k+1)}), \; \forall i \in \{1, \dots, m\},
\end{aligned} \tag{4.123}$$

if $\|\boldsymbol{x}_i^{(k+1)} - \boldsymbol{z}_i + \boldsymbol{u}_i^{(k)}\|_2^2 = \|\boldsymbol{z}_i - \boldsymbol{x}_i^{(k+1)} - \boldsymbol{u}_i^{(k)}\|_2^2$ is noticed. Note that multiple researchers, such as in [40], only use $m = 1$. In that case, only two primal variables $\boldsymbol{x}$ and $\boldsymbol{z}$ are available.

According to Lemma 4.7, as the function $\phi_i(.)$ is an indicator function, Eq. (4.123) can be implemented by projection onto the set $\mathcal{S}_i$:

$$\boldsymbol{z}_i^{(k+1)} := \Pi_{\mathcal{S}_i}(\boldsymbol{x}_i^{(k+1)} + \boldsymbol{u}_i^{(k)}), \; \forall i \in \{1, \dots, m\}.$$

As an example, assume the variables are all matrices, $\boldsymbol{X}_i$, $\boldsymbol{Z}_i$, and $\boldsymbol{U}_i$. If the set $\mathcal{S}_i$ is the cone of orthogonal matrices, the constraint $\boldsymbol{X}_i \in \mathcal{S}_i$ would be $\boldsymbol{X}_i^\top \boldsymbol{X}_i = \boldsymbol{I}$. In this case, the update of matrix variable $\boldsymbol{Z}_i$ would be done by setting the singular values of $(\boldsymbol{x}_i^{(k+1)} + \boldsymbol{u}_i^{(k)})$ to one (see Lemma 4.10).

4.8 Chapter Summary

This chapter reviewed the necessary background on optimization. It began by introducing sets, norms, functions, derivatives, and standard optimization problems. This base understanding built the foundation to discuss the Lagrange function, the Lagrange dual function, the dual problem, and weak and strong duality. Next, KKT conditions, including the stationary condition, primal feasibility, dual feasibility, and complementary slackness were introduced. First-order optimization methods, including gradient descent, stochastic gradient descent, and the projected gradient method, were discussed in detail, and the second-order optimization methods, including Newton's method for unconstrained problems, Newton's method for constrained problems, and the interior-point method, were fully elaborated on. Lastly, the distributed optimization approaches, including alternating optimization, dual decomposition methods, augmented Lagrangian method, and the ADMM algorithm, were covered. These optimization algorithms will be used in later chapters of this book to solve the optimization problems in spectral, probabilistic, and neural network-based dimensionality reduction methods.

References

1. Larry Armijo. "Minimization of functions having Lipschitz continuous first partial derivatives". In: *Pacific Journal of mathematics* 16.1 (1966), pp. 1–3.
2. Jacques F Benders. "Partitioning procedures for solving mixed-variables programming problems". In: *Numerische mathematik* 4.1 (1962), pp. 238–252.
3. Dimitri P Bertsekas. "The method of multipliers for equality constrained problems". In: *Constrained optimization and Lagrange multiplier methods* (1982), pp. 96–157.
4. Léon Bottou et al. "Online learning and stochastic approximations". In: *On-line learning in neural networks* 17.9 (1998), p. 142.
5. Léon Bottou, Frank E Curtis, and Jorge Nocedal. "Optimization methods for large-scale machine learning". In: *SIAM Review* 60.2 (2018), pp. 223–311.
6. Stephen Boyd, Neal Parikh, and Eric Chu. *Distributed optimization and statistical learning via the alternating direction method of multipliers*. Now Publishers Inc, 2011.
7. Stephen Boyd and Lieven Vandenberghe. *Convex optimization*. Cambridge University Press, 2004.
8. Yidong D. Chong. *Complex Methods for the Sciences*. Tech. rep. Nanyang Technological University, 2021.
9. Haskell B Curry. "The method of steepest descent for non-linear minimization problems". In: *Quarterly of Applied Mathematics* 2.3 (1944), pp. 258–261.
10. George B Dantzig and Philip Wolfe. "Decomposition principle for linear programs". In: *Operations research* 8.1 (1960), pp. 101–111.
11. I.I. Dikin. "Iterative solution of problems of linear and quadratic programming". In: *Doklady Akademii Nauk*. Vol. 174. 4. Russian Academy of Sciences. 1967, pp. 747–748.
12. LM Grana Drummond and Alfredo N Iusem. "A projected gradient method for vector optimization problems". In: *Computational Optimization and applications* 28.1 (2004), pp. 5–29.
13. Hugh Everett III. "Generalized Lagrange multiplier method for solving problems of optimum allocation of resources". In: *Operations research* 11.3 (1963), pp. 399–417.
14. Ky Fan. "Maximum properties and inequalities for the eigenvalues of completely continuous operators". In: *Proceedings of the National Academy of Sciences of the United States of America* 37.11 (1951), p. 760.
15. Xiaozhao Fang et al. "Approximate low-rank projection learning for feature extraction". In: *IEEE transactions on neural networks and learning systems* 29.11 (2018), pp. 5228–5241.
16. Anthony V Fiacco and Garth P McCormick. "The sequential unconstrained minimization technique (SUMT) without parameters". In: *Operations Research* 15.5 (1967), pp. 820–827.
17. Daniel Gabay and Bertrand Mercier. "A dual algorithm for the solution of nonlinear variational problems via finite element approximation". In: *Computers & mathematics with applications* 2.1 (1976), pp. 17–40.
18. Benyamin Ghojogh, Fakhri Karray, and Mark Crowley. "Locally linear image structural embedding for image structure manifold learning". In: *International Conference on Image Analysis and Recognition* Springer. 2019, pp. 126–138.
19. Benyamin Ghojogh et al. "KKT Conditions, First-Order and Second-Order Optimization, and Distributed Optimization: Tutorial and Survey". In: *arXiv preprint arXiv:2110.01858* (2021).
20. Joachim Giesen and Sören Laue. "Combining ADMM and the Augmented Lagrangian Method for Efficiently Handling Many Constraints". In: *International Joint Conference on Artificial Intelligence* 2019, pp. 4525–4531.
21. Joachim Giesen and Sören Laue. "Distributed convex optimization with many convex constraints". In: *arXiv preprint arXiv:1610.02967* (2016).
22. R Glowinski and A Marrocco. "Finite element approximation and iterative methods of solution for 2-D nonlinear magnetostatic problems". In: *Proceeding of International Conference on the Computation of Electromagnetic Fields (COMPUMAG)*. 1976.

23. Michael Grant, Stephen Boyd, and Yinyu Ye. *CVX: Matlab software for disciplined convex programming*. 2009.
24. Jacques Hadamard. *Mémoire sur le problème d'analyse relatif à l'équilibre des plaques élastiques encastrées*. Vol. 33. Imprimerie nationale, 1908.
25. Magnus R Hestenes. "Multiplier and gradient methods". In: *Journal of optimization theory and applications* 4.5 (1969), pp. 303–320.
26. Are Hjorungnes and David Gesbert. "Complex-valued matrix differentiation: Techniques and key results". In: *IEEE Transactions on Signal Processing* 55.6 (2007), pp. 2740–2746.
27. Alfredo N Iusem. "On the convergence properties of the projected gradient method for convex optimization". In: *Computational & Applied Mathematics* 22 (2003), pp. 37–52.
28. Prateek Jain and Purushottam Kar. "Non-convex optimization for machine learning". In: *arXiv preprint arXiv:1712.07897* (2017).
29. William Karush. "Minima of functions of several variables with inequalities as side constraints". MA thesis. Department of Mathematics, University of Chicago, Chicago, Illinois, 1939.
30. Tinne Hoff Kjeldsen. "A contextualized historical analysis of the Kuhn–Tucker theorem in nonlinear programming: the impact of World War II". In: *Historia mathematica* 27.4 (2000), pp. 331–361.
31. Jakub Konečn, Brendan McMahan, and Daniel Ramage. "Federated optimization: Distributed optimization beyond the datacenter". In: *arXiv preprint arXiv:1511.03575* (2015).
32. Harold W Kuhn and Albert W Tucker. "Nonlinear programming". In: *Berkeley Symposium on Mathematical Statistics and Probability*. Berkeley: University of California Press. 1951, pp. 481–492.
33. Claude Lemaréchal. "Cauchy and the gradient method". In: *Doc Math Extra* 251.254 (2012), p. 10.
34. Qiuwei Li, Zhihui Zhu, and Gongguo Tang. "Alternating minimizations converge to second-order optimal solutions". In: *International Conference on Machine Learning*. 2019, pp. 3935–3943.
35. Tian Li et al. "Federated learning: Challenges, methods, and future directions". In: *IEEE Signal Processing Magazine* 37.3 (2020), pp. 50–60.
36. Jan R Magnus and Heinz Neudecker. "Matrix differential calculus with applications to simple, Hadamard, and Kronecker products". In: *Journal of Mathematical Psychology* 29.4 (1985), pp. 474–492.
37. Yurii Nesterov. *Lectures on convex optimization*. Vol. 137. Springer, 2018.
38. Yurii Nesterov and Arkadii Nemirovskii. *Interior-point polynomial algorithms in convex programming* SIAM, 1994.
39. Jorge Nocedal and Stephen Wright. *Numerical optimization* 2nd ed. Springer Science & Business Media, 2006.
40. Daniel Otero et al. "Alternate direction method of multipliers for unconstrained structural similarity-based optimization". In: *International Conference Image Analysis and Recognition*. Springer. 2018, pp. 20–29.
41. Neal Parikh and Stephen Boyd. "Proximal algorithms". In: *Foundations and Trends in optimization* 1.3 (2014), pp. 127–239.
42. Kaare Brandt Petersen and Michael Syskind Pedersen. "The matrix cookbook". In: *Technical University of Denmark* 15 (2012).
43. Florian A Potra and Stephen J Wright. "Interior-point methods". In: *Journal of computational and applied mathematics* 124.1–2 (2000), pp. 281–302.
44. Michael JD Powell. "A method for nonlinear constraints in minimization problems". In: *Optimization* (1969), pp. 283–298.
45. Herbert Robbins and Sutton Monro. "A stochastic approximation method". In: *The annals of mathematical statistics* (1951), pp. 400–407.
46. David E Rumelhart, Geoffrey E Hinton, and Ronald J Williams. "Learning representations by back-propagating errors". In: *nature* 323.6088 (1986), pp. 533–536.

47. Josef Stoer and Roland Bulirsch. *Introduction to numerical analysis*. Vol. 12. Springer Science & Business Media, 2013.
48. Yu Wang, Wotao Yin, and Jinshan Zeng. "Global convergence of ADMM in nonconvex nonsmooth optimization". In: *Journal of Scientific Computing* 78.1 (2019), pp. 29–63.
49. Philip Wolfe. "Convergence conditions for ascent methods". In: *SIAM review* 11.2 (1969), pp. 226–235.
50. Margaret Wright. "The interior-point revolution in optimization: history, recent developments, and lasting consequences". In: *Bulletin of the American mathematical society* 42.1 (2005), pp. 39–56.

Part II
Spectral Dimensionality Reduction

Chapter 5
Principal Component Analysis

5.1 Introduction

Principal Component Analysis (PCA) [17] is a very well-known and fundamental linear method for subspace learning and dimensionality reduction [11]. This method, which is also used for feature extraction, was first proposed by Pearson in 1901 [21]. To learn a nonlinear submanifold, kernel PCA was proposed by Schölkopf et al. [24, 25] in 1997. The kernel PCA maps the data to a high dimensional feature space hoping to fall on a linear manifold in that space. PCA and kernel PCA are unsupervised methods for subspace learning. To use the class labels in PCA, Supervised PCA (SPCA) was proposed [3], which scores the features of the $\boldsymbol{X}$ and reduces the features before applying PCA. This type of SPCA was mostly used in bioinformatics [18]. Afterwards, another type of SPCA [4] was proposed, which has a very solid theory and generalizes PCA for using labels. This SPCA also has a dual and kernel SPCA. PCA and SPCA have had many applications, such as eigenfaces [30, 31] and kernel eigenfaces [33], which are used for facial recognition. It has also been used for detecting the orientation of an image [20].

5.2 Principal Component Analysis

Consider a dataset of *instances* or *data points* $\{(\boldsymbol{x}_i, \boldsymbol{y}_i)\}_{i=1}^n$ with a sample size n and dimensionality $\boldsymbol{x}_i \in \mathbb{R}^d$ and $\boldsymbol{y}_i \in \mathbb{R}^\ell$. $\{\boldsymbol{x}_i\}_{i=1}^n$ are the input data to the model, and $\{\boldsymbol{y}_i\}_{i=1}^n$ are the observations or labels. In the training set, $\mathbb{R}^{d\times n} \ni \boldsymbol{X} := [\boldsymbol{x}_1, \ldots, \boldsymbol{x}_n]$ and $\mathbb{R}^{\ell\times n} \ni \boldsymbol{Y} := [\boldsymbol{y}_1, \ldots, \boldsymbol{y}_n]$ are defined. It is also possible to have an out-of-sample data point, $\boldsymbol{x}_t \in \mathbb{R}^d$, which is not in the training set. If there are n_t out-of-sample data points, $\{\boldsymbol{x}_{t,i}\}_1^{n_t}$, then $\mathbb{R}^{d\times n_t} \ni \boldsymbol{X}_t := [\boldsymbol{x}_{t,1}, \ldots, \boldsymbol{x}_{t,n_t}]$ is defined.

B. Ghojogh et al., *Elements of Dimensionality Reduction and Manifold Learning*,
https://doi.org/10.1007/978-3-031-10602-6_5

5.2.1 Projection and Reconstruction in PCA

Recall linear projection, which was introduced in Chap. 2. According to Chap. 2, the projection of a point $\boldsymbol{x}$ onto the column space of a projection matrix $\boldsymbol{U}$ is:

$$\mathbb{R}^p \ni \widetilde{\boldsymbol{x}} := \boldsymbol{U}^\top \breve{\boldsymbol{x}}, \tag{5.1}$$

where:

$$\mathbb{R}^d \ni \breve{\boldsymbol{x}} := \boldsymbol{x} - \boldsymbol{\mu}_x, \tag{5.2}$$

$$\mathbb{R}^d \ni \boldsymbol{\mu}_x := \frac{1}{n}\sum_{i=1}^{n} \boldsymbol{x}_i. \tag{5.3}$$

Reconstruction of the projected data is:

$$\mathbb{R}^d \ni \widehat{\boldsymbol{x}} := \boldsymbol{U}\boldsymbol{U}^\top \breve{\boldsymbol{x}} + \boldsymbol{\mu}_x = \boldsymbol{U}\widetilde{\boldsymbol{x}} + \boldsymbol{\mu}_x, \tag{5.4}$$

where the mean is added back because it was removed before projection. Note that in PCA, all data points should be centered, i.e., the mean should be removed first. Figure 5.1 demonstrates that data should be centered for correct calculation of the principal directions.

In some applications, centering the data does not make sense. For example, in natural language processing, centering the data can create negative measures, which does not translate well when the data are text-based. Therefore, data are occasionally not centered and PCA is applied to the noncentered data. This method is called Latent Semantic Indexing (LSI) or Latent Semantic Analysis (LSA) [9].

If n data points are stacked columnwise in a matrix $\boldsymbol{X} = [\boldsymbol{x}_1, \dots, \boldsymbol{x}_n] \in \mathbb{R}^{d\times n}$, the projection and reconstruction are:

$$\mathbb{R}^{d\times n} \ni \breve{\boldsymbol{X}} := \boldsymbol{X}\boldsymbol{H} = \boldsymbol{X} - \boldsymbol{\mu}_x, \tag{5.5}$$

$$\mathbb{R}^{p\times n} \ni \widetilde{\boldsymbol{X}} := \boldsymbol{U}^\top \breve{\boldsymbol{X}}, \tag{5.6}$$

$$\mathbb{R}^{d\times n} \ni \widehat{\boldsymbol{X}} := \boldsymbol{U}\boldsymbol{U}^\top \breve{\boldsymbol{X}} + \boldsymbol{\mu}_x = \boldsymbol{U}\widetilde{\boldsymbol{X}} + \boldsymbol{\mu}_x, \tag{5.7}$$

where $\mathbb{R}^{n\times n} \ni \boldsymbol{H} := \boldsymbol{I} - (1/n)\boldsymbol{1}\boldsymbol{1}^\top$ is the centering matrix (see Chap. 2), $\breve{\boldsymbol{X}} = [\breve{\boldsymbol{x}}_1, \dots, \breve{\boldsymbol{x}}_n] = [\boldsymbol{x}_1 - \boldsymbol{\mu}_x, \dots, \boldsymbol{x}_n - \boldsymbol{\mu}_x]$ is the centered data, $\widetilde{\boldsymbol{X}} = [\widetilde{\boldsymbol{x}}_1, \dots, \widetilde{\boldsymbol{x}}_n]$, and $\widehat{\boldsymbol{X}} = [\widehat{\boldsymbol{x}}_1, \dots, \widehat{\boldsymbol{x}}_n]$. The residual error between data and reconstructed data is:

$$\boldsymbol{r} = \boldsymbol{x} - \widehat{\boldsymbol{x}} = \boldsymbol{x} - \boldsymbol{U}\boldsymbol{\beta}, \tag{5.8}$$

assuming that data points are centered.

Consider out-of-sample (or test) data point $\boldsymbol{x}_t$. It is possible to have n_t out-of-sample data points, $\mathbb{R}^{d\times n_t} \ni \boldsymbol{X}_t = [\boldsymbol{x}_{t,1}, \dots, \boldsymbol{x}_{t,n_t}]$ all together. Projection and reconstruction of the out-of-sample data are as follows:

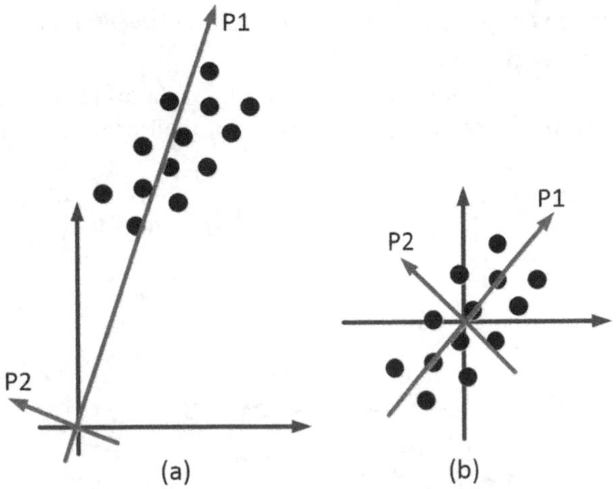

Fig. 5.1 The principal directions P1 and P2 for (**a**) noncentered and (**b**) centered data. As seen, the data should be centered for PCA. Centering the data makes it possible for PCA to find the correct most variant directions of the data

$$\mathbb{R}^p \ni \widetilde{\boldsymbol{x}}_t = \boldsymbol{U}^\top \breve{\boldsymbol{x}}_t, \tag{5.9}$$

$$\mathbb{R}^d \ni \widehat{\boldsymbol{x}}_t = \boldsymbol{U}\boldsymbol{U}^\top \breve{\boldsymbol{x}}_t + \boldsymbol{\mu}_x = \boldsymbol{U}\widetilde{\boldsymbol{x}}_t + \boldsymbol{\mu}_x, \tag{5.10}$$

$$\mathbb{R}^{d\times n_t} \ni \breve{\boldsymbol{X}}_t := \boldsymbol{X}_t - \boldsymbol{\mu}_x, \tag{5.11}$$

$$\mathbb{R}^{p\times n_t} \ni \widetilde{\boldsymbol{X}}_t = \boldsymbol{U}^\top \breve{\boldsymbol{X}}_t, \tag{5.12}$$

$$\mathbb{R}^{d\times n_t} \ni \widehat{\boldsymbol{X}}_t = \boldsymbol{U}\boldsymbol{U}^\top \breve{\boldsymbol{X}}_t + \boldsymbol{\mu}_x = \boldsymbol{U}\widetilde{\boldsymbol{X}}_t + \boldsymbol{\mu}_x. \tag{5.13}$$

Note that for centering the out-of-sample data point(s), the mean of the training data and not the out-of-sample data should be used.

5.2.2 PCA Using Eigen-Decomposition

5.2.2.1 Projection onto One Direction

In Eq. (5.4), if $p = 1$, $\boldsymbol{x}$ is projected onto only one vector $\boldsymbol{u}$ and is then reconstructed. If the mean is not added back, then:

$$\widehat{\boldsymbol{x}} = \boldsymbol{u}\boldsymbol{u}^\top \breve{\boldsymbol{x}}.$$

The squared length (squared ℓ_2-norm) of this reconstructed vector is:

$$\begin{aligned}||\widehat{\boldsymbol{x}}||_2^2 &= ||\boldsymbol{u}\boldsymbol{u}^\top \breve{\boldsymbol{x}}||_2^2 = (\boldsymbol{u}\boldsymbol{u}^\top \breve{\boldsymbol{x}})^\top (\boldsymbol{u}\boldsymbol{u}^\top \breve{\boldsymbol{x}}) \\ &= \breve{\boldsymbol{x}}^\top \boldsymbol{u}\underbrace{\boldsymbol{u}^\top \boldsymbol{u}}_{1}\boldsymbol{u}^\top \breve{\boldsymbol{x}} \overset{(a)}{=} \breve{\boldsymbol{x}}^\top \boldsymbol{u}\,\boldsymbol{u}^\top \breve{\boldsymbol{x}} \overset{(b)}{=} \boldsymbol{u}^\top \breve{\boldsymbol{x}}\,\breve{\boldsymbol{x}}^\top \boldsymbol{u},\end{aligned} \tag{5.14}$$

where (a) is because $\boldsymbol{u}$ is a unit (normal) vector, i.e., $\boldsymbol{u}^\top \boldsymbol{u} = ||\boldsymbol{u}||_2^2 = 1$, and (b) is because $\breve{\boldsymbol{x}}^\top \boldsymbol{u} = \boldsymbol{u}^\top \breve{\boldsymbol{x}} \in \mathbb{R}$.

Consider n data points $\{\boldsymbol{x}_i\}_{i=1}^n$ where $\{\breve{\boldsymbol{x}}_i\}_{i=1}^n$ are the centered data. The summation of the squared lengths of their projections $\{\widehat{\boldsymbol{x}}_i\}_{i=1}^n$ is:

$$\sum_{i=1}^n ||\widehat{\boldsymbol{x}}_i||_2^2 \overset{(5.14)}{=} \sum_{i=1}^n \boldsymbol{u}^\top \breve{\boldsymbol{x}}_i \breve{\boldsymbol{x}}_i^\top \boldsymbol{u} = \boldsymbol{u}^\top \Big(\sum_{i=1}^n \breve{\boldsymbol{x}}_i \breve{\boldsymbol{x}}_i^\top \Big) \boldsymbol{u}. \tag{5.15}$$

If $\breve{\boldsymbol{X}} = [\breve{\boldsymbol{x}}_1, \dots, \breve{\boldsymbol{x}}_n] \in \mathbb{R}^{d \times n}$, then:

$$\mathbb{R}^{d\times d} \ni \boldsymbol{S} := \sum_{i=1}^n \breve{\boldsymbol{x}}_i \breve{\boldsymbol{x}}_i^\top = \breve{\boldsymbol{X}} \breve{\boldsymbol{X}}^\top \overset{(5.5)}{=} \boldsymbol{X}\boldsymbol{H}\boldsymbol{H}^\top \boldsymbol{X}^\top \overset{(a)}{=} \boldsymbol{X}\boldsymbol{H}\boldsymbol{H}\boldsymbol{X}^\top \overset{(b)}{=} \boldsymbol{X}\boldsymbol{H}\boldsymbol{X}^\top, \tag{5.16}$$

where $\boldsymbol{S}$ is called the "covariance matrix" and (a) and (b) are because the centering matrix is symmetric and idempotent (see Chap. 2). If the data were already centered, then $\boldsymbol{S} = \boldsymbol{X}\boldsymbol{X}^\top$.

Plugging Eq. (5.16) into Eq. (5.15) results in:

$$\sum_{i=1}^n ||\widehat{\boldsymbol{x}}_i||_2^2 = \boldsymbol{u}^\top \boldsymbol{S} \boldsymbol{u}. \tag{5.17}$$

Note that $\boldsymbol{u}^\top \boldsymbol{S}\boldsymbol{u}$ is the variance of the projected data onto a PCA subspace. Thus, $\boldsymbol{u}^\top \boldsymbol{S}\boldsymbol{u} = \mathbb{V}\text{ar}(\boldsymbol{u}^\top \breve{\boldsymbol{X}})$. This is possible because when some nonrandom variable (here $\boldsymbol{u}$) is multiplied by the random data (here $\breve{\boldsymbol{X}}$), it will have a squared (quadratic) effect on the variance, and $\boldsymbol{u}^\top \boldsymbol{S}\boldsymbol{u}$ is quadratic in $\boldsymbol{u}$. Therefore, $\boldsymbol{u}^\top \boldsymbol{S}\boldsymbol{u}$ can be interpreted in two ways: (I) the squared length of reconstruction and (II) the variance of projection.

The following optimization finds a projection direction $\boldsymbol{u}$ that maximizes the squared length of reconstruction (or variance of projection):

$$\begin{aligned} \underset{\boldsymbol{u}}{\text{maximize}} \quad & \boldsymbol{u}^\top \boldsymbol{S}\boldsymbol{u}, \\ \text{subject to} \quad & \boldsymbol{u}^\top \boldsymbol{u} = 1, \end{aligned} \tag{5.18}$$

where the constraint ensures that the $\boldsymbol{u}$ is a unit (normal) vector. Consider the Lagrangian:

$$\mathcal{L} = \boldsymbol{u}^\top \boldsymbol{S}\boldsymbol{u} - \lambda(\boldsymbol{u}^\top \boldsymbol{u} - 1),$$

Taking derivative of the Lagrangian and setting it to zero results in:

$$\mathbb{R}^p \ni \frac{\partial \mathcal{L}}{\partial \boldsymbol{u}} = 2\boldsymbol{S}\boldsymbol{u} - 2\lambda \boldsymbol{u} \overset{\text{set}}{=} 0 \implies \boldsymbol{S}\boldsymbol{u} = \lambda \boldsymbol{u}. \tag{5.19}$$

Equation (5.19) is the eigen-decomposition of $\boldsymbol{S}$, where $\boldsymbol{u}$ and λ are the leading eigenvector and eigenvalue of $\boldsymbol{S}$, respectively (see Chap. 2). Note that the leading eigenvalue is the largest. This is because the optimization problem is a maximization problem (see Chap. 2). Therefore, if projected onto one PCA direction, the PCA direction $\boldsymbol{u}$ is the leading eigenvector of the covariance matrix. Note that the "PCA direction" is also called the "principal direction" or the "principal axis". The dimensions (features) of the data projected onto a PCA subspace are called "principal components".

5.2.2.2 Projection onto Span of Several Directions

In Eq. (5.4) or (5.7), if $p > 1$, $\breve{\boldsymbol{x}}$ or $\breve{\boldsymbol{X}}$ is projected onto a PCA subspace with dimensionality greater than one and then reconstruct back. If the mean is not added back into the equation, then:

$$\widehat{\boldsymbol{X}} = \boldsymbol{U}\boldsymbol{U}^\top \breve{\boldsymbol{X}}.$$

This means that every column of $\breve{\boldsymbol{X}}$, i.e. $\breve{\boldsymbol{x}}$, is projected onto a space spanned by the p vectors $\{\boldsymbol{u}_1, \ldots, \boldsymbol{u}_p\}$ each of which is d-dimensional. Therefore, the projected data are p-dimensional and the reconstructed data are d-dimensional.

The squared length (squared Frobenius Norm) of this reconstructed matrix is:

$$\begin{aligned}||\widehat{\boldsymbol{X}}||_F^2 &= ||\boldsymbol{U}\boldsymbol{U}^\top \breve{\boldsymbol{X}}||_F^2 = \mathbf{tr}\big((\boldsymbol{U}\boldsymbol{U}^\top \breve{\boldsymbol{X}})^\top (\boldsymbol{U}\boldsymbol{U}^\top \breve{\boldsymbol{X}})\big) \\ &= \mathbf{tr}(\breve{\boldsymbol{X}}^\top \boldsymbol{U} \underbrace{\boldsymbol{U}^\top \boldsymbol{U}}_{\boldsymbol{I}} \boldsymbol{U}^\top \breve{\boldsymbol{X}}) \overset{(a)}{=} \mathbf{tr}(\breve{\boldsymbol{X}}^\top \boldsymbol{U}\boldsymbol{U}^\top \breve{\boldsymbol{X}}) \overset{(b)}{=} \mathbf{tr}(\boldsymbol{U}^\top \breve{\boldsymbol{X}}\breve{\boldsymbol{X}}^\top \boldsymbol{U}),\end{aligned}$$

where $\mathbf{tr}(.)$ denotes the trace of the matrix, (a) is because $\boldsymbol{U}$ is an orthogonal matrix (its columns are orthonormal), and (b) is because $\mathbf{tr}(\breve{\boldsymbol{X}}^\top \boldsymbol{U}\boldsymbol{U}^\top \breve{\boldsymbol{X}}) = \mathbf{tr}(\breve{\boldsymbol{X}}\breve{\boldsymbol{X}}^\top \boldsymbol{U}\boldsymbol{U}^\top) = \mathbf{tr}(\boldsymbol{U}^\top \breve{\boldsymbol{X}}\breve{\boldsymbol{X}}^\top \boldsymbol{U})$. According to Eq. (5.16), the $\boldsymbol{S} = \breve{\boldsymbol{X}}\breve{\boldsymbol{X}}^\top$ is the covariance matrix; therefore:

$$||\widehat{\boldsymbol{X}}||_F^2 = \mathbf{tr}(\boldsymbol{U}^\top \boldsymbol{S}\,\boldsymbol{U}). \tag{5.20}$$

The following optimization finds several projection directions $\{\boldsymbol{u}_1, \ldots, \boldsymbol{u}_p\}$, as columns of $\boldsymbol{U} \in \mathbb{R}^{d \times p}$, which maximize the squared length of reconstruction (or variance of projection):

$$\begin{aligned}&\underset{\boldsymbol{U}}{\text{maximize}} \quad \mathbf{tr}(\boldsymbol{U}^\top \boldsymbol{S}\,\boldsymbol{U}), \\ &\text{subject to} \quad \boldsymbol{U}^\top \boldsymbol{U} = \boldsymbol{I},\end{aligned} \tag{5.21}$$

where the constraint ensures that $\boldsymbol{U}$ is an orthogonal matrix, as previously assumed. The Lagrangian is:

$$\mathcal{L} = \mathbf{tr}(\boldsymbol{U}^\top \boldsymbol{S}\,\boldsymbol{U}) - \mathbf{tr}\big(\boldsymbol{\Lambda}^\top(\boldsymbol{U}^\top \boldsymbol{U} - \boldsymbol{I})\big),$$

where $\boldsymbol{\Lambda} \in \mathbb{R}^{p\times p}$ is a diagonal matrix $\mathbf{diag}([\lambda_1, \ldots, \lambda_p]^\top)$ including the Lagrange multipliers.

$$\mathbb{R}^{d\times p} \ni \frac{\partial \mathcal{L}}{\partial \boldsymbol{U}} = 2\boldsymbol{S}\boldsymbol{U} - 2\boldsymbol{U}\boldsymbol{\Lambda} \overset{\text{set}}{=} 0 \implies \boldsymbol{S}\boldsymbol{U} = \boldsymbol{U}\boldsymbol{\Lambda}. \tag{5.22}$$

Equation (5.22) is the eigen-decomposition of $\boldsymbol{S}$, where the columns of $\boldsymbol{U}$ and the diagonal of $\boldsymbol{\Lambda}$ are the eigenvectors and eigenvalues of $\boldsymbol{S}$, respectively (see Chap. 2). The eigenvectors and eigenvalues are sorted from the leading (largest eigenvalue) to the trailing (smallest eigenvalue) because the optimization problem is a maximization problem. In conclusion, if projecting onto the PCA subspace or $\mathbf{span}\{\boldsymbol{u}_1, \ldots, \boldsymbol{u}_p\}$, the PCA directions $\{\boldsymbol{u}_1, \ldots, \boldsymbol{u}_p\}$ are the sorted eigenvectors of the covariance matrix of $\boldsymbol{X}$.

5.2.3 Properties of U

5.2.3.1 Rank of the Covariance Matrix

Two cases for $\breve{\boldsymbol{X}} \in \mathbb{R}^{d\times n}$ are considered:

1. If the original dimensionality of the data is greater than the number of data points, i.e., $d \geq n$, then $\mathbf{rank}(\breve{\boldsymbol{X}}) = \mathbf{rank}(\breve{\boldsymbol{X}}^\top) \leq n$. Therefore, $\mathbf{rank}(\boldsymbol{S}) = \mathbf{rank}(\breve{\boldsymbol{X}}\breve{\boldsymbol{X}}^\top) \leq \min\big(\mathbf{rank}(\breve{\boldsymbol{X}}), \mathbf{rank}(\breve{\boldsymbol{X}}^\top)\big) - 1 = n - 1$. Note that -1 is because the data are centered. For example, if there is only one data point, it becomes zero after centering and the rank should be zero.
2. If the original dimensionality of data is less than the number of data points, i.e., $d \leq n - 1$ (the -1 again is because of centering the data), then $\mathbf{rank}(\breve{\boldsymbol{X}}) = \mathbf{rank}(\breve{\boldsymbol{X}}^\top) \leq d$. Therefore, $\mathbf{rank}(\boldsymbol{S}) = \mathbf{rank}(\breve{\boldsymbol{X}}\breve{\boldsymbol{X}}^\top) \leq \min\big(\mathbf{rank}(\breve{\boldsymbol{X}}), \mathbf{rank}(\breve{\boldsymbol{X}}^\top)\big) = d$.

This results in either $\mathbf{rank}(\boldsymbol{S}) \leq n - 1$ or $\mathbf{rank}(\boldsymbol{S}) \leq d$.

5.2.3.2 Truncating U

Consider the following cases:

1. If $\mathbf{rank}(\boldsymbol{S}) = d$, then $p = d$ (which has d nonzero eigenvalues of $\boldsymbol{S}$), so that $\boldsymbol{U} \in \mathbb{R}^{d\times d}$. This means that the dimensionality of the PCA subspace is

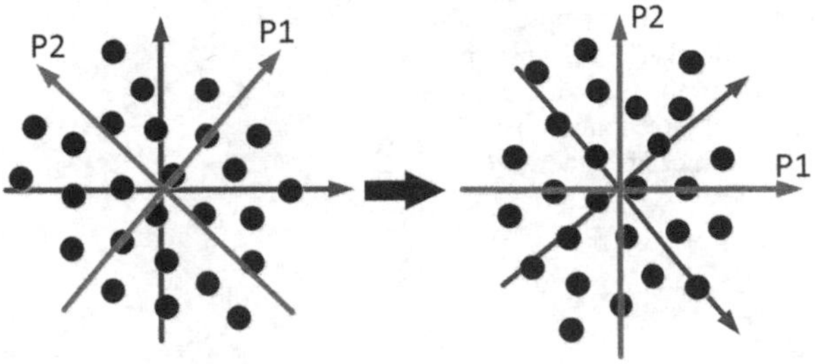

Fig. 5.2 Rotation of coordinates because of PCA. If the principal directions are considered as new coordinate axes, then it is similar to rotating data in space

d, which is equal to the dimensionality of the original space. Why does this happen? This happens because $\mathbf{rank}(\boldsymbol{S}) = d$ means that the data are spread wide enough in all dimensions of the original space, up to a possible rotation (see Fig. 5.2). Therefore, the dimensionality of the PCA subspace is equal to the original dimensionality. However, the PCA might merely rotate the coordinate axes. In this case, $\boldsymbol{U} \in \mathbb{R}^{d\times d}$ is a square orthogonal matrix so that $\mathbb{R}^{d\times d} \ni \boldsymbol{U}\boldsymbol{U}^\top = \boldsymbol{U}\boldsymbol{U}^{-1} = \boldsymbol{I}$ and $\mathbb{R}^{d\times d} \ni \boldsymbol{U}^\top\boldsymbol{U} = \boldsymbol{U}^{-1}\boldsymbol{U} = \boldsymbol{I}$ because $\mathbf{rank}(\boldsymbol{U}) = d$, $\mathbf{rank}(\boldsymbol{U}\boldsymbol{U}^\top) = d$, and $\mathbf{rank}(\boldsymbol{U}^\top\boldsymbol{U}) = d$. That is why in the literature, PCA is also referred to as coordinate rotation.

2. If $\mathbf{rank}(\boldsymbol{S}) < d$ and $n > d$, there are enough data points, but the data points exist on a subspace and do not fill the original space sufficiently wide in every direction. In this case, $\boldsymbol{U} \in \mathbb{R}^{d\times p}$ is not square and $\mathbf{rank}(\boldsymbol{U}) = p < d$ (there are p nonzero eigenvalues of $\boldsymbol{S}$). Therefore, $\mathbb{R}^{d\times d} \ni \boldsymbol{U}\boldsymbol{U}^\top \neq \boldsymbol{I}$ and $\mathbb{R}^{p\times p} \ni \boldsymbol{U}^\top\boldsymbol{U} = \boldsymbol{I}$ because $\mathbf{rank}(\boldsymbol{U}) = p$, $\mathbf{rank}(\boldsymbol{U}\boldsymbol{U}^\top) = p < d$, and $\mathbf{rank}(\boldsymbol{U}^\top\boldsymbol{U}) = p$.
3. If $\mathbf{rank}(\boldsymbol{S}) \leq n - 1 < d$, there are not enough data points to properly represent the original space, and the points have an "intrinsic dimensionality". An example is two three-dimensional points that are on a two-dimensional line (subspace). Similar to the previous case, the data points exist on a subspace and do not fill the original space sufficiently wide in every direction. The discussions about $\boldsymbol{U}$, $\boldsymbol{U}\boldsymbol{U}^\top$, and $\boldsymbol{U}^\top\boldsymbol{U}$ are similar to the previous case.

Note that with $\mathbf{rank}(\boldsymbol{S}) = d$ and thus $\boldsymbol{U} \in \mathbb{R}^{d\times d}$, there might be a need to "truncate" the matrix $\boldsymbol{U}$ to have $\boldsymbol{U} \in \mathbb{R}^{d\times p}$. Truncating $\boldsymbol{U}$ means that a subset of best (leading) eigenvectors is taken, rather than the whole d eigenvectors with nonzero eigenvalues. In this case, $\boldsymbol{U}\boldsymbol{U}^\top \neq \boldsymbol{I}$ and $\boldsymbol{U}^\top\boldsymbol{U} = \boldsymbol{I}$ hold. Truncating is used because the variance of the data in some directions might be noticeably smaller than other directions; in this case, it is possible to only keep the $p < d$ top eigenvectors (PCA directions) and "ignore" the PCA directions with smaller eigenvalues to have $\boldsymbol{U} \in \mathbb{R}^{d\times p}$. Figure 5.3 illustrates this case for a 2D example. Note that truncating can also be done, when $\boldsymbol{U} \in \mathbb{R}^{d\times p}$, to have $\boldsymbol{U} \in \mathbb{R}^{d\times q}$, where p is the number of nonzero eigenvalues of $\boldsymbol{S}$ and $q < p$. Therefore, as long as the columns of the matrix $\boldsymbol{U} \in \mathbb{R}^{d\times p}$ are orthonormal, there will always be $\boldsymbol{U}^\top\boldsymbol{U} = \boldsymbol{I}$, regardless of the value p. If the orthogonal matrix $\boldsymbol{U}$ is not truncated and thus is a square matrix, then $\boldsymbol{U}\boldsymbol{U}^\top = \boldsymbol{I}$ is also true.

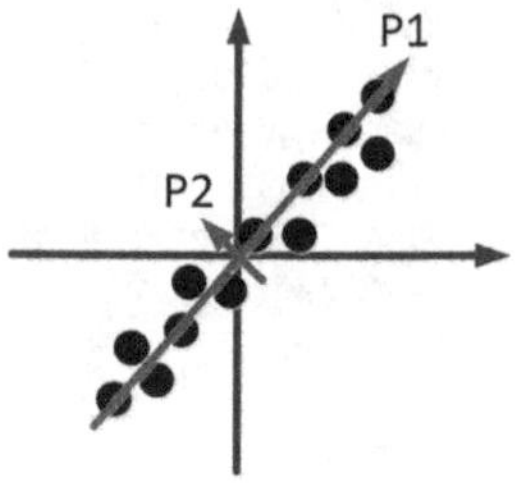

Fig. 5.3 A 2D example where the data are almost on a line and the second principal direction is very small and can be ignored

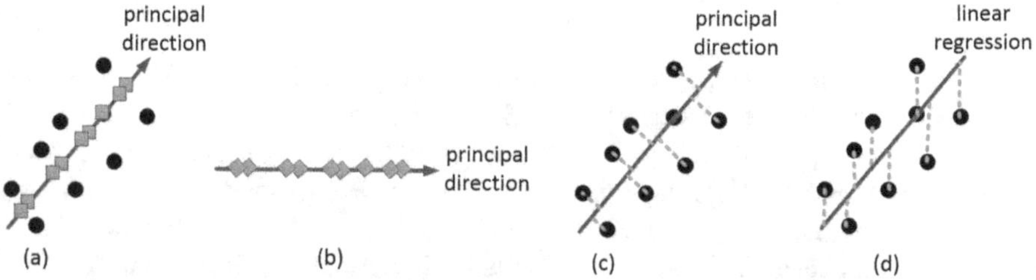

Fig. 5.4 (**a**) Projection of the black circle data points onto the principal direction, where the green square data points are the projected data. (**b**) The reconstruction coordinate of the data points. (**c**) The reconstruction error in PCA. (**d**) The least square error in linear regression

5.2.4 *Reconstruction Error in PCA*

5.2.4.1 Reconstruction in Linear Projection

When centering the data, Eq. (5.8) becomes $\boldsymbol{r} = \check{\boldsymbol{x}} - \widehat{\boldsymbol{x}}$ because the reconstructed data will also be centered according to Eq. (5.4). According to Eqs. (5.8), (5.2), and (5.4):

$$\boldsymbol{r} = \boldsymbol{x} - \widehat{\boldsymbol{x}} = \check{\boldsymbol{x}} + \boldsymbol{\mu}_x - \boldsymbol{U}\boldsymbol{U}^\top \check{\boldsymbol{x}} - \boldsymbol{\mu}_x = \check{\boldsymbol{x}} - \boldsymbol{U}\boldsymbol{U}^\top \check{\boldsymbol{x}}. \tag{5.23}$$

Figure 5.4 demonstrates the projection of a two-dimensional point (after the data have been centered) onto the first principal direction, its reconstruction, and its reconstruction error. As seen in this figure, the reconstruction error is different from the least square error in linear regression.

For n data points:

$$\boldsymbol{R} := \boldsymbol{X} - \widehat{\boldsymbol{X}} = \check{\boldsymbol{X}} + \boldsymbol{\mu}_x - \boldsymbol{U}\boldsymbol{U}^\top \check{\boldsymbol{X}} - \boldsymbol{\mu}_x = \check{\boldsymbol{X}} - \boldsymbol{U}\boldsymbol{U}^\top \check{\boldsymbol{X}}, \tag{5.24}$$

where $\mathbb{R}^{d\times n} \ni \boldsymbol{R} = [\boldsymbol{r}_1, \ldots, \boldsymbol{r}_n]$ is the matrix of residuals.

If the goal is to minimize the reconstruction error subject to the orthogonality of the projection matrix $\boldsymbol{U}$, then:

$$
\begin{aligned}
&\underset{U}{\text{minimize}} \quad ||\breve{X} - UU^\top \breve{X}||_F^2, \\
&\text{subject to} \quad U^\top U = I.
\end{aligned}
\tag{5.25}
$$

The objective function can be simplified:

$$
\begin{aligned}
||\breve{X} - UU^\top \breve{X}||_F^2 &= \mathbf{tr}\big((\breve{X} - UU^\top \breve{X})^\top (\breve{X} - UU^\top \breve{X})\big) \\
&= \mathbf{tr}\big((\breve{X}^\top - \breve{X}^\top UU^\top)(\breve{X} - UU^\top \breve{X})\big) \\
&= \mathbf{tr}(\breve{X}^\top \breve{X} - 2\breve{X}^\top UU^\top \breve{X} + \breve{X}^\top U \underbrace{U^\top U}_{I} U^\top \breve{X}) \\
&= \mathbf{tr}(\breve{X}^\top \breve{X} - \breve{X}^\top UU^\top \breve{X}) = \mathbf{tr}(\breve{X}^\top \breve{X}) - \mathbf{tr}(\breve{X}^\top UU^\top \breve{X}) \\
&= \mathbf{tr}(\breve{X}^\top \breve{X}) - \mathbf{tr}(\breve{X}\breve{X}^\top UU^\top).
\end{aligned}
$$

The Lagrangian is:

$$
\mathcal{L} = \mathbf{tr}(\breve{X}^\top \breve{X}) - \mathbf{tr}(\breve{X}\breve{X}^\top UU^\top) + \mathbf{tr}\big(\mathbf{\Lambda}^\top (U^\top U - I)\big),
$$

where $\mathbf{\Lambda} \in \mathbb{R}^{p \times p}$ is a diagonal matrix $\mathbf{diag}([\lambda_1, \ldots, \lambda_p]^\top)$ containing the Lagrange multipliers. Equating the derivative of the Lagrangian to zero results in:

$$
\begin{aligned}
\mathbb{R}^{d \times p} \ni \frac{\partial \mathcal{L}}{\partial U} &= -2\breve{X}\breve{X}^\top U + 2U\mathbf{\Lambda} \overset{\text{set}}{=} 0 \\
\implies \breve{X}\breve{X}^\top U &= U\mathbf{\Lambda} \overset{(5.16)}{\implies} SU = U\mathbf{\Lambda},
\end{aligned}
\tag{5.26}
$$

which is the eigenvalue problem (see Chap. 2) for the covariance matrix S. The same eigenvalue problem is found in the PCA. Therefore, the PCA subspace is the best linear projection in terms of reconstruction error. In other words, PCA has the least squared error in reconstruction.

5.2.4.2 Reconstruction in Autoencoder

It was demonstrated above that PCA is the best in reconstruction error for *linear* projection. If there are $m > 1$ successive linear projections, the reconstruction is:

$$
\widehat{X} = \underbrace{U_1 \cdots U_m}_{\text{reconstruct}} \underbrace{U_m^\top \cdots U_1^\top}_{\text{project}} \breve{X} + \boldsymbol{\mu}_x,
\tag{5.27}
$$

which can be seen as an undercomplete *autoencoder* [13] with $2m$ layers, without an activation function (or with identity activation functions $f(\boldsymbol{x}) = \boldsymbol{x}$). $\boldsymbol{\mu}_x$ is modelled

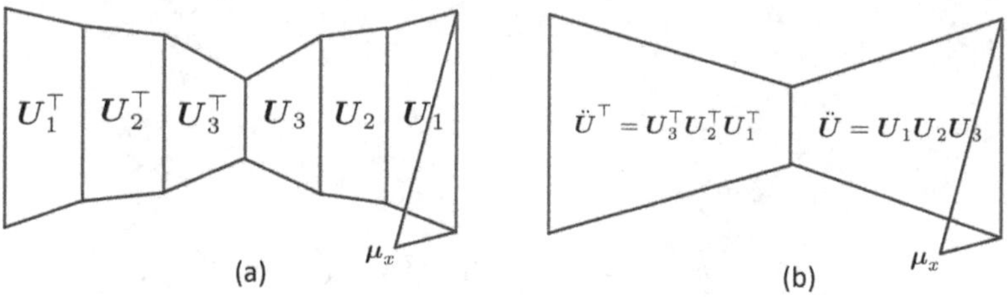

Fig. 5.5 (**a**) An example of an autoencoder with five hidden layers and linear activation functions, and (**b**) its reduction to an autoencoder with one hidden layer

by the intercepts, which are included as inputs to the neurons of the autoencoder layers. Figure 5.5 demonstrates this autoencoder. Since there is no nonlinearity between the projections, this can be defined as:

$$\ddot{\boldsymbol{U}} := \boldsymbol{U}_1 \cdots \boldsymbol{U}_m \implies \ddot{\boldsymbol{U}}^\top = \boldsymbol{U}_m^\top \cdots \boldsymbol{U}_1^\top, \quad \therefore \quad \widehat{\boldsymbol{X}} = \ddot{\boldsymbol{U}}\ddot{\boldsymbol{U}}^\top \boldsymbol{X} + \boldsymbol{\mu}_x. \tag{5.28}$$

Equation (5.28) demonstrates that the whole autoencoder can be reduced to an undercomplete autoencoder with one hidden layer, where the weight matrix is $\ddot{\boldsymbol{U}}$ (see Fig. 5.5). In other words, in an autoencoder neural network, every layer excluding the activation function behaves as a linear projection.

Comparing Eqs. (5.7) and (5.28) demonstrate that the whole autoencoder is reduced to PCA. Therefore, *PCA is equivalent to an undercomplete autoencoder with one hidden layer without an activation function*. Therefore, if the weights of such an autoencoder are trained by backpropagation [23], then they are roughly equal to the PCA directions. Moreover, as the PCA is the best linear projection in terms of reconstruction error, *if there is an undercomplete autoencoder with "one" hidden layer, it is best not to use an activation function*. Unfortunately, this is not considered by some papers in the literature.

An autoencoder with $2m$ hidden layers without activation functions reduces to linear PCA. This explains why in autoencoders with more than one layer, it is possible to use the nonlinear activation function $f(.)$ as:

$$\widehat{\boldsymbol{X}} = f^{-1}(\boldsymbol{U}_1 \ldots f^{-1}(\boldsymbol{U}_m f(\boldsymbol{U}_m^\top \ldots f(\boldsymbol{U}_1^\top \boldsymbol{X}) \ldots)) \ldots) + \boldsymbol{\mu}_x. \tag{5.29}$$

5.2.5 PCA Using Singular Value Decomposition

The PCA can be performed using the Singular Value Decomposition (SVD) of $\breve{\boldsymbol{X}}$, rather than the eigen-decomposition of $\boldsymbol{S}$. Consider the complete SVD of $\breve{\boldsymbol{X}}$ (see Chap. 2):

$$\mathbb{R}^{d\times n} \ni \breve{\boldsymbol{X}} = \boldsymbol{U\Sigma V}^\top, \tag{5.30}$$

where the columns of $\boldsymbol{U} \in \mathbb{R}^{d\times d}$ (called the left singular vectors) are the eigenvectors of $\breve{\boldsymbol{X}}\breve{\boldsymbol{X}}^\top$, the columns of $\boldsymbol{V} \in \mathbb{R}^{n\times n}$ (called the right singular vectors) are the eigenvectors of $\breve{\boldsymbol{X}}^\top\breve{\boldsymbol{X}}$, and $\boldsymbol{\Sigma} \in \mathbb{R}^{d\times n}$ is a rectangular diagonal matrix whose diagonal entries (called the singular values) are the square root of eigenvalues of $\breve{\boldsymbol{X}}\breve{\boldsymbol{X}}^\top$ and/or $\breve{\boldsymbol{X}}^\top\breve{\boldsymbol{X}}$. See Proposition 2.1 in Chap. 2 for proof of this claim.

According to Eq. (5.16), $\breve{\boldsymbol{X}}\breve{\boldsymbol{X}}^\top$ is the covariance matrix $\boldsymbol{S}$. In Eq. (5.22), the eigenvectors of $\boldsymbol{S}$ are the principal directions. On the other hand, here the columns of $\boldsymbol{U}$ are the eigenvectors of $\breve{\boldsymbol{X}}\breve{\boldsymbol{X}}^\top$. Therefore, SVD can be applied to $\breve{\boldsymbol{X}}$ and the left singular vectors (columns of $\boldsymbol{U}$) can be taken as the principal directions.

Interestingly, in the SVD of $\breve{\boldsymbol{X}}$, the columns of $\boldsymbol{U}$ are automatically sorted from largest to smallest singular values (eigenvalues) and there is no need to sort as was done when using eigenvalue decomposition for the covariance matrix.

5.2.6 Determining the Number of Principal Directions

Usually in PCA, the components with the smallest eigenvalues are cut off to reduce the data. There are different methods for estimating the best number of components to keep (denoted by p), such as using the Bayesian model selection [19], scree plot [6], and comparing the ratio $\lambda_j / \sum_{k=1}^{d} \lambda_k$ with a threshold [1], where λ_i denotes the eigenvalue related to the j-th principal component.

The scree plot [6] is a plot of the eigenvalues versus sorted components from the leading (having largest) to trailing (having smallest) eigenvalues. A threshold for the vertical (eigenvalue) axis chooses the components with the sufficiently large eigenvalues and removes the remaining components. A good threshold is considered to be where the eigenvalue drops significantly. In most datasets, a significant drop in eigenvalues occurs.

Another way to choose the most appropriate components is the ratio [1]:

$$\frac{\lambda_j}{\sum_{k=1}^{d} \lambda_k}, \tag{5.31}$$

for the j-th component. The features are then sorted from the largest to smallest ratio and the p best components or up to the component with a significant drop in the ratio is selected.

5.3 Dual Principal Component Analysis

Assume the case where the dimensionality of data is high and much greater than the sample size, i.e., $d \gg n$. In this case, the incomplete SVD of $\breve{\boldsymbol{X}}$ (see Chap. 2) is:

$$\breve{\boldsymbol{X}} = \boldsymbol{U}\boldsymbol{\Sigma}\boldsymbol{V}^\top, \tag{5.32}$$

where $\boldsymbol{U} \in \mathbb{R}^{d\times p}$ and $\boldsymbol{V} \in \mathbb{R}^{n\times p}$ contain the p leading left and right singular vectors of $\breve{\boldsymbol{X}}$, respectively, where p is the number of "nonzero" singular values of $\breve{\boldsymbol{X}}$ and usually $p \ll d$. Here, $\boldsymbol{\Sigma} \in \mathbb{R}^{p\times p}$ is a square matrix having the p largest non-zero singular values of $\breve{\boldsymbol{X}}$. As $\boldsymbol{\Sigma}$ is a square diagonal matrix and its diagonal includes nonzero entries (i.e., it is full-rank), it is invertible [12]. Therefore, $\boldsymbol{\Sigma}^{-1} = \textbf{diag}([\frac{1}{\sigma_1}, \dots, \frac{1}{\sigma_p}]^\top)$ if $\boldsymbol{\Sigma} = \textbf{diag}([\sigma_1, \dots, \sigma_p]^\top)$.

5.3.1 Projection

Recall Eq. (5.6) for projecting onto the PCA subspace: $\widetilde{\boldsymbol{X}} = \boldsymbol{U}^\top\breve{\boldsymbol{X}}$. On the other hand, according to Eq. (5.32):

$$\breve{\boldsymbol{X}} = \boldsymbol{U}\boldsymbol{\Sigma}\boldsymbol{V}^\top \implies \boldsymbol{U}^\top\breve{\boldsymbol{X}} = \underbrace{\boldsymbol{U}^\top\boldsymbol{U}}_{\boldsymbol{I}}\boldsymbol{\Sigma}\boldsymbol{V}^\top = \boldsymbol{\Sigma}\boldsymbol{V}^\top. \tag{5.33}$$

Whereas, according to Eqs. (5.6) and (5.33):

$$\therefore \quad \widetilde{\boldsymbol{X}} = \boldsymbol{\Sigma}\boldsymbol{V}^\top \tag{5.34}$$

Equation (5.34) can be used for projecting data onto a PCA subspace instead of Eq. (5.6). This is the projection of training data in dual PCA.

5.3.2 Reconstruction

According to Eq. (5.32):

$$\breve{\boldsymbol{X}} = \boldsymbol{U}\boldsymbol{\Sigma}\boldsymbol{V}^\top \implies \breve{\boldsymbol{X}}\boldsymbol{V} = \boldsymbol{U}\boldsymbol{\Sigma}\underbrace{\boldsymbol{V}^\top\boldsymbol{V}}_{\boldsymbol{I}} = \boldsymbol{U}\boldsymbol{\Sigma} \implies \boldsymbol{U} = \breve{\boldsymbol{X}}\boldsymbol{V}\boldsymbol{\Sigma}^{-1}. \tag{5.35}$$

Plugging Eq. (5.35) into Eq. (5.7) results in:

$$\widehat{\boldsymbol{X}} = \boldsymbol{U}\widetilde{\boldsymbol{X}} + \boldsymbol{\mu}_x \overset{(5.35)}{=} \breve{\boldsymbol{X}}\boldsymbol{V}\boldsymbol{\Sigma}^{-1}\widetilde{\boldsymbol{X}} + \boldsymbol{\mu}_x \overset{(5.34)}{=} \breve{\boldsymbol{X}}\boldsymbol{V}\underbrace{\boldsymbol{\Sigma}^{-1}\boldsymbol{\Sigma}}_{\boldsymbol{I}}\boldsymbol{V}^\top + \boldsymbol{\mu}_x$$

$$\implies \widehat{\boldsymbol{X}} = \breve{\boldsymbol{X}}\boldsymbol{V}\boldsymbol{V}^\top + \boldsymbol{\mu}_x. \tag{5.36}$$

Equation (5.36) can be used for the reconstruction of data instead of Eq. (5.7). This is the reconstruction of training data in dual PCA.

5.3.3 Out-of-Sample Projection

Recall Eq. (5.9) for the projection of an out-of-sample point $\boldsymbol{x}_t$ onto a PCA subspace. According to Eq. (5.35):

$$\boldsymbol{U}^\top \overset{(5.35)}{=} \boldsymbol{\Sigma}^{-\top} \boldsymbol{V}^\top \breve{\boldsymbol{X}}^\top \overset{(a)}{=} \boldsymbol{\Sigma}^{-1} \boldsymbol{V}^\top \breve{\boldsymbol{X}}^\top \tag{5.37}$$

$$\overset{(5.9)}{\Longrightarrow} \widetilde{\boldsymbol{x}}_t = \boldsymbol{\Sigma}^{-1} \boldsymbol{V}^\top \breve{\boldsymbol{X}}^\top \breve{\boldsymbol{x}}_t, \tag{5.38}$$

where (a) is because $\boldsymbol{\Sigma}^{-1}$ is diagonal and thus symmetric. Equation (5.38) can be used for projecting an out-of-sample data point onto a PCA subspace, instead of Eq. (5.9). This is out-of-sample projection in dual PCA. Considering all the n_t out-of-sample data points, the projection is:

$$\widetilde{\boldsymbol{X}}_t = \boldsymbol{\Sigma}^{-1} \boldsymbol{V}^\top \breve{\boldsymbol{X}}^\top \breve{\boldsymbol{X}}_t. \tag{5.39}$$

5.3.4 Out-of-Sample Reconstruction

Recall Eq. (5.10) for the reconstruction of an out-of-sample point $\boldsymbol{x}_t$. According to Eqs. (5.35) and (5.37):

$$\boldsymbol{U}\boldsymbol{U}^\top = \breve{\boldsymbol{X}} \boldsymbol{V} \boldsymbol{\Sigma}^{-1} \boldsymbol{\Sigma}^{-1} \boldsymbol{V}^\top \breve{\boldsymbol{X}}^\top \overset{(5.10)}{\Longrightarrow} \widehat{\boldsymbol{x}}_t = \breve{\boldsymbol{X}} \boldsymbol{V} \boldsymbol{\Sigma}^{-2} \boldsymbol{V}^\top \breve{\boldsymbol{X}}^\top \breve{\boldsymbol{x}}_t + \boldsymbol{\mu}_x. \tag{5.40}$$

Equation (5.40) can be used for the reconstruction of an out-of-sample data point instead of Eq. (5.10). This is out-of-sample reconstruction in dual PCA. Considering all the n_t out-of-sample data points, the reconstruction is:

$$\widehat{\boldsymbol{X}}_t = \breve{\boldsymbol{X}} \boldsymbol{V} \boldsymbol{\Sigma}^{-2} \boldsymbol{V}^\top \breve{\boldsymbol{X}}^\top \breve{\boldsymbol{X}}_t + \boldsymbol{\mu}_x. \tag{5.41}$$

5.3.5 Why Is Dual PCA Useful?

The dual PCA can be useful for two reasons:

1. As seen in Eqs. (5.34), (5.36), (5.38), and (5.40), the formulas for the dual PCA only include $\boldsymbol{V}$ and not $\boldsymbol{U}$. The columns of $\boldsymbol{V}$ are the eigenvectors of $\breve{\boldsymbol{X}}^\top \breve{\boldsymbol{X}} \in \mathbb{R}^{n \times n}$ and the columns of $\boldsymbol{U}$ are the eigenvectors of $\breve{\boldsymbol{X}} \breve{\boldsymbol{X}}^\top \in \mathbb{R}^{d \times d}$. In case the

dimensionality of the data is much high and greater than the sample size, i.e., $n \ll d$, the computation of eigenvectors of $\breve{X}^\top \breve{X}$ is easier and faster than $\breve{X}\breve{X}^\top$ and also requires less storage. Therefore, dual PCA is more efficient than direct PCA in this case, in terms of both speed and storage. Note that the results of PCA and dual PCA are exactly the same.

2. Some inner product forms, such as $\breve{X}^\top \breve{x}_t$, have appeared in the formulas of the dual PCA. This provides an opportunity for kernelizing the PCA to have the kernel PCA using the kernel trick.

5.4 Kernel Principal Component Analysis

PCA is a linear method because the projection is linear. In the event that the data points exist on a nonlinear submanifold, linear subspace learning might not be completely effective. For example, Fig. 5.6 demonstrates a 2D nonlinear manifold where the data points exist in the 3D original space. As the manifold is nonlinear, the geodesic distances of the points on the manifold are different from their Euclidean distances. If the manifold is unfolded correctly, the geodesic distances of the points on the manifold are preserved. However, PCA is linear using the Euclidean distance. When the linear PCA is applied to nonlinear data, the found subspace ruins the manifold. As the figure illustrates, the red and green points are no longer at a distance but are rather right next to each other.

There are two options to address this limitation of PCA because of its linearity. (1) change PCA to become a nonlinear method or (2) leave the PCA to be linear, but change the data so that it can fall on a linear or close to linear manifold. To change the data, its dimensionality is increased by mapping it to a feature space with higher dimensionality so that in the feature space it falls on a linear manifold. This is referred to as "blessing of dimensionality" in the literature [8], which is pursued using kernels [16]. A PCA method that uses a kernel of data is called a "kernel PCA" [24].

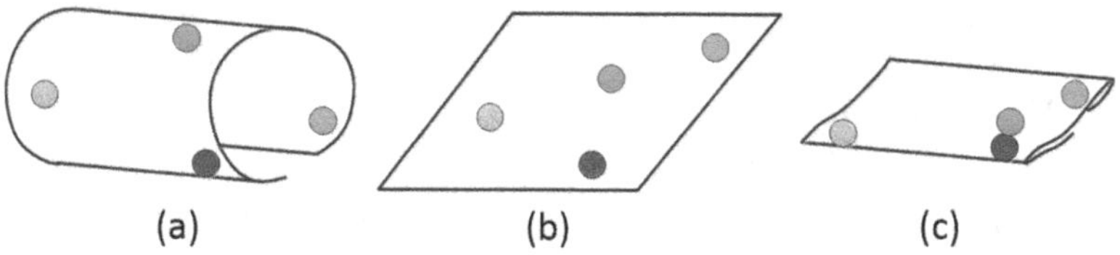

Fig. 5.6 (**a**) A 2D nonlinear manifold where the data exist in the 3D original space, (**b**) the correct unfolded manifold, and (**c**) applying the linear PCA, which takes Euclidean distances into account

5.4.1 Kernels and Hilbert Space

Refer to Chap. 3 for additional background on kernels. The kernel of two matrices $\boldsymbol{X}_1 \in \mathbb{R}^{d\times n_1}$ and $\boldsymbol{X}_2 \in \mathbb{R}^{d\times n_2}$ can be computed and have a "kernel matrix" (also called the "Gram matrix"):

$$\mathbb{R}^{n_1\times n_2} \ni \boldsymbol{K}(\boldsymbol{X}_1, \boldsymbol{X}_2) := \boldsymbol{\Phi}(\boldsymbol{X}_1)^\top \boldsymbol{\Phi}(\boldsymbol{X}_2), \tag{5.42}$$

where $\boldsymbol{\Phi}(\boldsymbol{X}_1) := [\boldsymbol{\phi}(\boldsymbol{x}_1), \dots, \boldsymbol{\phi}(\boldsymbol{x}_n)] \in \mathbb{R}^{t\times n_1}$ is the matrix of mapped $\boldsymbol{X}_1$ to the feature space. $\boldsymbol{\Phi}(\boldsymbol{X}_2) \in \mathbb{R}^{t\times n_2}$ is defined similarly. The kernel matrix of dataset $\boldsymbol{X} \in \mathbb{R}^{d\times n}$ can be computed over itself so that:

$$\mathbb{R}^{n\times n} \ni \boldsymbol{K}_x := \boldsymbol{K}(\boldsymbol{X}, \boldsymbol{X}) = \boldsymbol{\Phi}(\boldsymbol{X})^\top \boldsymbol{\Phi}(\boldsymbol{X}), \tag{5.43}$$

where $\boldsymbol{\Phi}(\boldsymbol{X}) := [\boldsymbol{\phi}(\boldsymbol{x}_1), \dots, \boldsymbol{\phi}(\boldsymbol{x}_n)] \in \mathbb{R}^{t\times n}$ is the pulled (mapped) data.

Note that in kernel methods, the pulled data $\boldsymbol{\Phi}(\boldsymbol{X})$ are usually not available and only the kernel matrix $\boldsymbol{K}(\boldsymbol{X}, \boldsymbol{X})$, which is the inner product of the pulled data with itself, is available. There exist different types of kernels (see Chap. 3). As discussed in Chap. 3, the centered pulled data to the feature space, $\breve{\boldsymbol{\Phi}}(\boldsymbol{X}) := \boldsymbol{\Phi}(\boldsymbol{X})\boldsymbol{H}$, yields a double-centered kernel:

$$\breve{\boldsymbol{K}}_x := \boldsymbol{H}\boldsymbol{K}_x\boldsymbol{H} = \breve{\boldsymbol{\Phi}}(\boldsymbol{X})^\top \breve{\boldsymbol{\Phi}}(\boldsymbol{X}), \tag{5.44}$$

where $\breve{\boldsymbol{K}}_x$ denotes the double-centered kernel matrix.

5.4.2 Projection

Applying an incomplete SVD on the centered pulled (mapped) data $\breve{\boldsymbol{\Phi}}(\boldsymbol{X})$ (see Chap. 2) results in:

$$\mathbb{R}^{t\times n} \ni \breve{\boldsymbol{\Phi}}(\boldsymbol{X}) = \boldsymbol{U}\boldsymbol{\Sigma}\boldsymbol{V}^\top, \tag{5.45}$$

where $\boldsymbol{U} \in \mathbb{R}^{t\times p}$ and $\boldsymbol{V} \in \mathbb{R}^{n\times p}$ contain the p leading left and right singular vectors of $\breve{\boldsymbol{\Phi}}(\boldsymbol{X})$, respectively, where p is the number of "nonzero" singular values of $\breve{\boldsymbol{\Phi}}(\boldsymbol{X})$ and usually $p \ll t$. Here, $\boldsymbol{\Sigma} \in \mathbb{R}^{p\times p}$ is a square matrix having the p largest nonzero singular values of $\breve{\boldsymbol{\Phi}}(\boldsymbol{X})$.

However, as previously mentioned, the pulled data are not necessarily available, meaning Eq. (5.45) cannot be done. The kernel, however, is available. Therefore, the eigen-decomposition (see Chap. 2) can be applied to the double-centered kernel:

$$\breve{\boldsymbol{K}}_x\boldsymbol{V} = \boldsymbol{V}\boldsymbol{\Lambda}, \tag{5.46}$$

where the columns of $\boldsymbol{V}$ and the diagonal of $\boldsymbol{\Lambda}$ are the eigenvectors and eigenvalues of $\breve{\boldsymbol{K}}_x$, respectively. The columns of $\boldsymbol{V}$ in Eq. (5.45) are the right singular vectors of $\breve{\boldsymbol{\Phi}}(\boldsymbol{X})$, which are equivalent to the eigenvectors of $\breve{\boldsymbol{\Phi}}(\boldsymbol{X})^\top \breve{\boldsymbol{\Phi}}(\boldsymbol{X}) = \breve{\boldsymbol{K}}_x$ (see Proposition 2.1 in Chap. 2). According to that Proposition 2.11 , the diagonal of $\boldsymbol{\Sigma}$ in Eq. (5.45) is equivalent to the square root of the eigenvalues of $\breve{\boldsymbol{K}}_x$.

Therefore, in practice where the pulling function is not necessarily available, Eq. (5.46) is used to find the $\boldsymbol{V}$ and $\boldsymbol{\Sigma}$ in Eq. (5.45). Equation (5.46) can be restated as:

$$\breve{\boldsymbol{K}}_x \boldsymbol{V} = \boldsymbol{V}\boldsymbol{\Sigma}^2, \tag{5.47}$$

to be compatible with Eq. (5.45). It is important to note that because of using Eq. (5.47) instead of Eq. (5.45), *the projection directions* $\boldsymbol{U}$ *are not available in kernel PCA to be observed or plotted.*

Similar to Eq. (5.34):

$$\breve{\boldsymbol{\Phi}}(\boldsymbol{X}) = \boldsymbol{U}\boldsymbol{\Sigma}\boldsymbol{V}^\top \implies \boldsymbol{U}^\top \breve{\boldsymbol{\Phi}}(\boldsymbol{X}) = \underbrace{\boldsymbol{U}^\top\boldsymbol{U}}_{\boldsymbol{I}}\boldsymbol{\Sigma}\boldsymbol{V}^\top = \boldsymbol{\Sigma}\boldsymbol{V}^\top$$

$$\therefore \quad \boldsymbol{\Phi}(\widetilde{\boldsymbol{X}}) = \boldsymbol{U}^\top \breve{\boldsymbol{\Phi}}(\boldsymbol{X}) = \boldsymbol{\Sigma}\boldsymbol{V}^\top, \tag{5.48}$$

where $\boldsymbol{\Sigma}$ and $\boldsymbol{V}$ are obtained from Eq. (5.47). Equation (5.48) is the projection of the training data in kernel PCA.

5.4.3 Reconstruction

Similar to Eq. (5.36):

$$\breve{\boldsymbol{\Phi}}(\boldsymbol{X}) = \boldsymbol{U}\boldsymbol{\Sigma}\boldsymbol{V}^\top \implies \breve{\boldsymbol{\Phi}}(\boldsymbol{X})\boldsymbol{V} = \boldsymbol{U}\boldsymbol{\Sigma}\underbrace{\boldsymbol{V}^\top\boldsymbol{V}}_{\boldsymbol{I}} = \boldsymbol{U}\boldsymbol{\Sigma} \implies \boldsymbol{U} = \breve{\boldsymbol{\Phi}}(\boldsymbol{X})\boldsymbol{V}\boldsymbol{\Sigma}^{-1}. \tag{5.49}$$

Therefore, the reconstruction is:

$$\boldsymbol{\Phi}(\widehat{\boldsymbol{X}}) = \boldsymbol{U}\boldsymbol{\Phi}(\widetilde{\boldsymbol{X}}) + \boldsymbol{\mu}_x \overset{(5.49)}{=} \breve{\boldsymbol{\Phi}}(\boldsymbol{X})\boldsymbol{V}\boldsymbol{\Sigma}^{-1}\boldsymbol{\Phi}(\widetilde{\boldsymbol{X}}) + \boldsymbol{\mu}_x \overset{(5.48)}{=} \breve{\boldsymbol{\Phi}}(\boldsymbol{X})\boldsymbol{V}\underbrace{\boldsymbol{\Sigma}^{-1}\boldsymbol{\Sigma}}_{\boldsymbol{I}}\boldsymbol{V}^\top + \boldsymbol{\mu}_x$$

$$\implies \boldsymbol{\Phi}(\widehat{\boldsymbol{X}}) = \breve{\boldsymbol{\Phi}}(\boldsymbol{X})\boldsymbol{V}\boldsymbol{V}^\top + \boldsymbol{\mu}_x. \tag{5.50}$$

However, $\breve{\boldsymbol{\Phi}}(\boldsymbol{X})$ is not necessarily available, which means that it is not possible to reconstruct the training data in kernel PCA.

5.4.4 Out-of-Sample Projection

Similar to Eq. (5.38):

$$U^\top \overset{(5.49)}{=} \Sigma^{-\top} V^\top \breve{\Phi}(X)^\top \overset{(a)}{=} \Sigma^{-1} V^\top \breve{\Phi}(X)^\top \implies \phi(\widetilde{x}_t) = U^\top \breve{\phi}(x_t) = \Sigma^{-1} V^\top \breve{\Phi}(X)^\top \breve{\phi}(x_t),$$
$$\implies \phi(\widetilde{x}_t) = \Sigma^{-1} V^\top \breve{k}_t, \tag{5.51}$$

where (a) is because Σ^{-1} is diagonal and thus symmetric, and $\breve{k}_t := \breve{\Phi}(X)^\top \breve{\phi}(x_t) \in \mathbb{R}^n$ is calculated as follows (see Chap. 3 for proof):

$$\mathbb{R}^n \ni \breve{k}_t = k_t - \frac{1}{n} \mathbf{1}_{n \times n} k_t - \frac{1}{n} K \mathbf{1}_n + \frac{1}{n^2} \mathbf{1}_{n \times n} K \mathbf{1}_n. \tag{5.52}$$

Equation (5.51) is the projection of out-of-sample data in kernel PCA. Considering all the n_t out-of-sample data points, X_t, the projection is:

$$\phi(\widetilde{X}_t) = \Sigma^{-1} V^\top \breve{K}_t, \tag{5.53}$$

where $\breve{K}_t$ is calculated by (see Chap. 3 for proof):

$$\mathbb{R}^{n \times n_t} \ni \breve{K}_t = K_t - \frac{1}{n} \mathbf{1}_{n \times n} K_t - \frac{1}{n} K \mathbf{1}_{n \times n_t} + \frac{1}{n^2} \mathbf{1}_{n \times n} K \mathbf{1}_{n \times n_t}, \tag{5.54}$$

5.4.5 Out-of-Sample Reconstruction

Similar to Eq. (5.40):

$$\implies UU^\top \overset{(5.49)}{=} \breve{\Phi}(X) V \Sigma^{-1} \Sigma^{-1} V^\top \breve{\Phi}(X)^\top \implies \phi(\widehat{x}_t) = \breve{\Phi}(X) V \Sigma^{-2} V^\top \breve{\Phi}(X)^\top \breve{\phi}(x_t) + \mu_x$$
$$\implies \phi(\widehat{x}_t) = \breve{\Phi}(X) V \Sigma^{-2} V^\top \breve{k}_t + \mu_x, \tag{5.55}$$

where $\breve{k}_t \in \mathbb{R}^n$ is calculated by Eq. (5.52). Considering all the n_t out-of-sample data points, X_t, the reconstruction is:

$$\Phi(\widehat{X}_t) = \breve{\Phi}(X) V \Sigma^{-2} V^\top \breve{K}_t + \mu_x, \tag{5.56}$$

where $\breve{K}_t$ is calculated by Eq. (5.54). In Eq. (5.55), $\breve{\Phi}(X)$ appears to the left of the expression and is not necessarily available; therefore, it is not possible to reconstruct an out-of-sample point in the kernel PCA. According to Eqs. (5.50) and (5.55), the kernel PCA is not able to reconstruct any data, whether training or out-of-sample.

5.4.6 Why Is Kernel PCA Useful?

The choice of the best kernel might be difficult, so the kernel PCA is not "always" effective in practice [12]. However, it provides useful theoretical insights for explaining the PCA, Multidimensional Scaling (MDS) [7], Isomap [28], Locally Linear Embedding (LLE) [22], and Laplacian Eigenmap (LE) [5]. These methods can be seen as special cases of kernel PCA with their own kernels (see [15] and chapter 2 in [27]). Further details will be explained in Chap. 10.

5.5 Supervised Principal Component Analysis Using Scoring

PCA is an unsupervised method. Algorithms have been proposed to convert PCA to a supervised method. The older version of Supervised Principal Component Analysis (SPCA) used scoring [3], which is explained in the following. In this version of SPCA, PCA is not a special case of SPCA. In SPCA using scoring, the similarity of every feature of data with the class labels is computed, the features are sorted, and then the features having the least similarity with the labels are removed. The larger the similarity of a feature with the labels, the better that feature is for discrimination in the embedded subspace.

Consider the training dataset $\mathbb{R}^{d\times n} \ni \boldsymbol{X} = [\boldsymbol{x}_1, \dots, \boldsymbol{x}_n] = [\boldsymbol{x}^1, \dots, \boldsymbol{x}^d]^\top$ where $\boldsymbol{x}_i \in \mathbb{R}^d$ and $\boldsymbol{x}^j \in \mathbb{R}^n$ are the i-th data point and the j-th feature, respectively. This type of SPCA is only for the classification task, meaning the dimensionality of the labels is one, $\ell = 1$. Thus, $\boldsymbol{Y} \in \mathbb{R}^{1\times n}$ and it can be defined to have $\mathbb{R}^n \ni \boldsymbol{y} := \boldsymbol{Y}^\top$.

The score of the j-th feature, $\boldsymbol{x}^j$, is:

$$\mathbb{R} \ni s_j := \frac{(\boldsymbol{x}^j)^\top \boldsymbol{y}}{||(\boldsymbol{x}^j)^\top \boldsymbol{x}^j||_2} = \frac{(\boldsymbol{x}^j)^\top \boldsymbol{y}}{\sqrt{(\boldsymbol{x}^j)^\top \boldsymbol{x}^j}}, \tag{5.57}$$

After computing the scores of all the features, they are sorted from largest to smallest score. Let $\boldsymbol{X}' \in \mathbb{R}^{d\times n}$ denote the training dataset whose features are sorted. Only the $q \leq d$ features with the largest scores are used and the other features are removed. Let:

$$\mathbb{R}^{q\times n} \ni \boldsymbol{X}'' := \boldsymbol{X}'(1 : q, :), \tag{5.58}$$

be the training dataset with q best features. Then, PCA is applied to $\boldsymbol{X}'' \in \mathbb{R}^{q\times n}$, rather than $\boldsymbol{X} \in \mathbb{R}^{d\times n}$. Applying PCA and kernel PCA on $\boldsymbol{X}''$ results in SPCA and kernel PCA, respectively. This type of SPCA was most often used and made popular in the bioinformatics field, for use with genome data analysis [18].

5.6 Supervised Principal Component Analysis Using HSIC

5.6.1 *Supervised PCA*

Supervised PCA (SPCA) [4] uses the Hilbert-Schmidt Independence Criterion (HSIC) [14], introduced in Chap. 3. The data $\boldsymbol{X} = [\boldsymbol{x}_1, \dots, \boldsymbol{x}_n] \in \mathbb{R}^{d \times n}$ and the labels $\boldsymbol{Y} = [\boldsymbol{y}_1, \dots, \boldsymbol{y}_n] \in \mathbb{R}^{\ell \times n}$, where ℓ is the dimensionality of the labels and typically $\ell = 1$. However, in the case the labels are encoded (e.g., one-hot-encoded) or SPCA is used for regression, then $\ell > 1$. SPCA attempts to maximize the dependence of the projected data points $\boldsymbol{U}^\top \boldsymbol{X}$ and the labels $\boldsymbol{Y}$. It uses a linear kernel for the projected data points:

$$\ddot{\boldsymbol{K}}_x = (\boldsymbol{U}^\top \boldsymbol{X})^\top (\boldsymbol{U}^\top \boldsymbol{X}) = \boldsymbol{X}^\top \boldsymbol{U}\boldsymbol{U}^\top \boldsymbol{X}, \tag{5.59}$$

and an arbitrary kernel $\boldsymbol{K}_y$ over $\boldsymbol{Y}$. For the classification task, one of the best choices for the $\boldsymbol{K}_y$ is the delta kernel [4], where the (i, j)-th element of the kernel is:

$$\boldsymbol{K}_y = \delta_{\boldsymbol{y}_i, \boldsymbol{y}_j} := \begin{cases} 1 \text{ if } \boldsymbol{y}_i = \boldsymbol{y}_j, \\ 0 \text{ if } \boldsymbol{y}_i \neq \boldsymbol{y}_j, \end{cases} \tag{5.60}$$

where $\delta_{\boldsymbol{y}_i, \boldsymbol{y}_j}$ is the Kronecker delta, which is one if $\boldsymbol{x}_i$ and $\boldsymbol{x}_j$ belong to the same class.

Another good choice for the kernel in the classification task is an arbitrary kernel (e.g., linear kernel $\boldsymbol{K}_y = \boldsymbol{Y}^\top \boldsymbol{Y}$) over $\boldsymbol{Y}$, where the columns of $\boldsymbol{Y}$ are one-hot encoded.[1] This is a good choice because the distances of classes will be equal; otherwise, some classes will fall closer than the others meaning there is no longer fairness between the classes.

One of the main advantages of the SPCA is that it can also be used for regression. In that case, a good choice for $\boldsymbol{K}_y$ is an arbitrary kernel (e.g., linear kernel $\boldsymbol{K}_y = \boldsymbol{Y}^\top \boldsymbol{Y}$) over $\boldsymbol{Y}$, where the columns of the $\boldsymbol{Y}$, i.e., labels, are the observations in the regression. Here, the distances of observations have meaning and should not be manipulated.

The HSIC in SPCA, for dependence of the projected data and the labels, is:

$$\text{HSIC} = \frac{1}{(n-1)^2} \textbf{tr}(\boldsymbol{X}^\top \boldsymbol{U}\boldsymbol{U}^\top \boldsymbol{X}\boldsymbol{H}\boldsymbol{K}_y\boldsymbol{H}), \tag{5.61}$$

where $\boldsymbol{U} \in \mathbb{R}^{d \times p}$ is the unknown projection matrix for projection onto the SPCA subspace. The desired dimensionality of the subspace is p and usually $p \ll d$.

---

[1] If there are c classes, one-hot encoding of the j-th class is a binary vector with all zeros except the j-th element.

HSIC should be maximized to maximize the dependence of $\boldsymbol{U}^\top \boldsymbol{X}$ and $\boldsymbol{Y}$. Therefore:

$$
\begin{aligned}
& \underset{\boldsymbol{U}}{\text{maximize}} \quad \mathbf{tr}(\boldsymbol{X}^\top \boldsymbol{U}\boldsymbol{U}^\top \boldsymbol{X}\boldsymbol{H}\boldsymbol{K}_y\boldsymbol{H}), \\
& \text{subject to} \quad \boldsymbol{U}^\top \boldsymbol{U} = \boldsymbol{I},
\end{aligned}
\tag{5.62}
$$

where the constraint ensures that $\boldsymbol{U}$ is an orthogonal matrix, i.e., the SPCA directions are orthonormal. The Lagrangian is:

$$
\begin{aligned}
\mathcal{L} &= \mathbf{tr}(\boldsymbol{X}^\top \boldsymbol{U}\boldsymbol{U}^\top \boldsymbol{X}\boldsymbol{H}\boldsymbol{K}_y\boldsymbol{H}) - \mathbf{tr}\big(\boldsymbol{\Lambda}^\top (\boldsymbol{U}^\top \boldsymbol{U} - \boldsymbol{I})\big) \\
&\overset{(a)}{=} \mathbf{tr}(\boldsymbol{U}\boldsymbol{U}^\top \boldsymbol{X}\boldsymbol{H}\boldsymbol{K}_y\boldsymbol{H}\boldsymbol{X}^\top) - \mathbf{tr}\big(\boldsymbol{\Lambda}^\top (\boldsymbol{U}^\top \boldsymbol{U} - \boldsymbol{I})\big),
\end{aligned}
$$

where (a) is because of the cyclic property of the trace and $\boldsymbol{\Lambda} \in \mathbb{R}^{p\times p}$ is a diagonal matrix $\mathbf{diag}([\lambda_1, \ldots, \lambda_p]^\top)$, including the Lagrange multipliers. Setting the derivative of the Lagrangian to zero results in:

$$
\mathbb{R}^{d\times p} \ni \frac{\partial \mathcal{L}}{\partial \boldsymbol{U}} = 2\boldsymbol{X}\boldsymbol{H}\boldsymbol{K}_y\boldsymbol{H}\boldsymbol{X}^\top \boldsymbol{U} - 2\boldsymbol{U}\boldsymbol{\Lambda} \overset{\text{set}}{=} \boldsymbol{0} \implies \boldsymbol{X}\boldsymbol{H}\boldsymbol{K}_y\boldsymbol{H}\boldsymbol{X}^\top \boldsymbol{U} = \boldsymbol{U}\boldsymbol{\Lambda},
\tag{5.63}
$$

which is the eigen-decomposition of $\boldsymbol{X}\boldsymbol{H}\boldsymbol{K}_y\boldsymbol{H}\boldsymbol{X}^\top$, where the columns of $\boldsymbol{U}$ and the diagonal of $\boldsymbol{\Lambda}$ are the eigenvectors and eigenvalues of $\boldsymbol{X}\boldsymbol{H}\boldsymbol{K}_y\boldsymbol{H}\boldsymbol{X}^\top$, respectively (see Chap. 2). The eigenvectors and eigenvalues are sorted from the leading (largest eigenvalue) to the trailing (smallest eigenvalue) because optimization is a maximization problem. Therefore, if projecting onto the SPCA subspace or $\mathbf{span}\{\boldsymbol{u}_1, \ldots, \boldsymbol{u}_p\}$, the SPCA directions $\{\boldsymbol{u}_1, \ldots, \boldsymbol{u}_p\}$ are the sorted eigenvectors of $\boldsymbol{X}\boldsymbol{H}\boldsymbol{K}_y\boldsymbol{H}\boldsymbol{X}^\top$. In other words, the columns of the projection matrix $\boldsymbol{U}$ in SPCA are the p leading eigenvectors of $\boldsymbol{X}\boldsymbol{H}\boldsymbol{K}_y\boldsymbol{H}\boldsymbol{X}^\top$.

Similar to PCA, the projection, projection of out-of-sample, reconstruction, and reconstruction of out-of-sample in SPCA are:

$$
\widetilde{\boldsymbol{X}} = \boldsymbol{U}^\top \boldsymbol{X}, \tag{5.64}
$$

$$
\widetilde{\boldsymbol{x}}_t = \boldsymbol{U}^\top \boldsymbol{x}_t, \tag{5.65}
$$

$$
\widehat{\boldsymbol{X}} = \boldsymbol{U}\boldsymbol{U}^\top \boldsymbol{X} = \boldsymbol{U}\widetilde{\boldsymbol{X}}, \tag{5.66}
$$

$$
\widehat{\boldsymbol{x}}_t = \boldsymbol{U}\boldsymbol{U}^\top \boldsymbol{x}_t = \boldsymbol{U}\widetilde{\boldsymbol{x}}_t, \tag{5.67}
$$

respectively. In SPCA, there is no need to center the data as the centering is already completed by $\boldsymbol{H}$ in HSIC. This becomes clearer in Sect. 5.6.2, which discusses that the PCA is a special case of the SPCA. Note that in the equations of SPCA, although not necessary,it is possible to center the data and in that case, the mean of embedding in the subspace will be zero.

Considering the n_t out-of-sample data points, the projection and reconstruction are as follows, respectively:

$$\widetilde{\boldsymbol{X}}_t = \boldsymbol{U}^\top \boldsymbol{X}_t, \tag{5.68}$$

$$\widehat{\boldsymbol{X}}_t = \boldsymbol{U}\boldsymbol{U}^\top \boldsymbol{X}_t = \boldsymbol{U}\widetilde{\boldsymbol{X}}_t. \tag{5.69}$$

5.6.2 PCA Is a Special Case of SPCA!

PCA is an unsupervised method. For unsupervised cases, the labels are not used; therefore, the similarities of the labels are not considered. If the similarities of labels are not considered, then the kernel over the labels becomes the identity matrix, $\boldsymbol{K}_y = \boldsymbol{I}$. According to Eq. (5.63), SPCA is the eigen-decomposition of $\boldsymbol{X}\boldsymbol{H}\boldsymbol{K}_y\boldsymbol{H}\boldsymbol{X}^\top$. In this case, this matrix becomes:

$$\begin{aligned}\boldsymbol{X}\boldsymbol{H}\boldsymbol{K}_y\boldsymbol{H}\boldsymbol{X}^\top &= \boldsymbol{X}\boldsymbol{H}\boldsymbol{I}\boldsymbol{H}\boldsymbol{X}^\top \overset{(a)}{=} \boldsymbol{X}\boldsymbol{H}\boldsymbol{I}\boldsymbol{H}^\top\boldsymbol{X}^\top \\ &= \boldsymbol{X}\boldsymbol{H}\boldsymbol{H}^\top\boldsymbol{X}^\top = (\boldsymbol{X}\boldsymbol{H})(\boldsymbol{X}\boldsymbol{H})^\top \overset{(5.5)}{=} \breve{\boldsymbol{X}}\breve{\boldsymbol{X}}^\top \overset{(5.16)}{=} \boldsymbol{S},\end{aligned}$$

where (a) is because of the symmetry of the centering matrix. The obtained $\boldsymbol{S}$ is the covariance matrix, whose eigenvectors are the PCA directions. Thus, if the similarities of labels are not considered, i.e.,the action is unsupervised, SPCA reduces to PCA.

5.6.3 Dual Supervised PCA

The SPCA can be formulated in dual form [4]. It was demonstrated that in SPCA, the columns of $\boldsymbol{U}$ are the eigenvectors of $\boldsymbol{X}\boldsymbol{H}\boldsymbol{K}_y\boldsymbol{H}\boldsymbol{X}^\top$. Consider the SVD of $\boldsymbol{K}_y$ (see Chap. 2):

$$\mathbb{R}^{n\times n} \ni \boldsymbol{K}_y = \boldsymbol{Q}\boldsymbol{\Omega}\boldsymbol{Q}^\top,$$

where $\boldsymbol{Q} \in \mathbb{R}^{n\times n}$ includes the left or right singular vectors and $\boldsymbol{\Omega} \in \mathbb{R}^{n\times n}$ contains the singular values of $\boldsymbol{K}_y$. Note that the left and right singular vectors are equal because $\boldsymbol{K}_y$ is symmetric and, thus, $\boldsymbol{K}_y\boldsymbol{K}_y^\top$ and $\boldsymbol{K}_y^\top\boldsymbol{K}_y$ are equal. As $\boldsymbol{\Omega}$ is a diagonal matrix with nonnegative entries, it can be decomposed to $\boldsymbol{\Omega} = \boldsymbol{\Omega}^{1/2}\boldsymbol{\Omega}^{1/2} = \boldsymbol{\Omega}^{1/2}(\boldsymbol{\Omega}^{1/2})^\top$, where the diagonal entries of $\boldsymbol{\Omega}^{1/2} \in \mathbb{R}^{n\times n}$ are the square root of the diagonal entries of $\boldsymbol{\Omega}$. Therefore, $\boldsymbol{K}_y$ can be decomposed into:

$$\boldsymbol{K}_y = \boldsymbol{Q}\boldsymbol{\Omega}^{1/2}(\boldsymbol{\Omega}^{1/2})^\top\boldsymbol{Q}^\top = (\boldsymbol{Q}\boldsymbol{\Omega}^{1/2})(\boldsymbol{Q}\boldsymbol{\Omega}^{1/2})^\top = \boldsymbol{\Delta}\boldsymbol{\Delta}^\top, \tag{5.70}$$

where:

$$\mathbb{R}^{n\times n} \ni \boldsymbol{\Delta} := \boldsymbol{Q}\boldsymbol{\Omega}^{1/2}. \tag{5.71}$$

Therefore:

$$\therefore \quad \boldsymbol{X}\boldsymbol{H}\boldsymbol{K}_y\boldsymbol{H}\boldsymbol{X}^\top \overset{(5.70)}{=} \boldsymbol{X}\boldsymbol{H}\boldsymbol{\Delta}\boldsymbol{\Delta}^\top\boldsymbol{H}\boldsymbol{X}^\top \overset{(a)}{=} (\boldsymbol{X}\boldsymbol{H}\boldsymbol{\Delta})(\boldsymbol{X}\boldsymbol{H}\boldsymbol{\Delta})^\top = \boldsymbol{\Psi}\boldsymbol{\Psi}^\top,$$

where (a) is because of the symmetry of the centering matrix and:

$$\mathbb{R}^{d\times n} \ni \boldsymbol{\Psi} := \boldsymbol{X}\boldsymbol{H}\boldsymbol{\Delta}. \tag{5.72}$$

Consider the incomplete SVD of $\boldsymbol{\Psi}$ (see Chap. 2):

$$\mathbb{R}^{d\times n} \ni \boldsymbol{\Psi} = \boldsymbol{U}\boldsymbol{\Sigma}\boldsymbol{V}^\top, \tag{5.73}$$

where $\boldsymbol{U} \in \mathbb{R}^{d\times p}$ and $\boldsymbol{V} \in \mathbb{R}^{d\times p}$ include the p leading left or right singular vectors of $\boldsymbol{\Psi}$, respectively, and $\boldsymbol{\Sigma} \in \mathbb{R}^{p\times p}$ contains the p largest singular values of $\boldsymbol{\Psi}$. $\boldsymbol{U}$ can be computed as:

$$\boldsymbol{\Psi} = \boldsymbol{U}\boldsymbol{\Sigma}\boldsymbol{V}^\top \implies \boldsymbol{\Psi}\boldsymbol{V} = \boldsymbol{U}\boldsymbol{\Sigma}\underbrace{\boldsymbol{V}^\top\boldsymbol{V}}_{\boldsymbol{I}} = \boldsymbol{U}\boldsymbol{\Sigma} \implies \boldsymbol{U} = \boldsymbol{\Psi}\boldsymbol{V}\boldsymbol{\Sigma}^{-1} \tag{5.74}$$

The projection of data $\boldsymbol{X}$ in dual SPCA is:

$$\widetilde{\boldsymbol{X}} \overset{(5.64)}{=} \boldsymbol{U}^\top\boldsymbol{X} \overset{(5.74)}{=} (\boldsymbol{\Psi}\boldsymbol{V}\boldsymbol{\Sigma}^{-1})^\top\boldsymbol{X} = \boldsymbol{\Sigma}^{-\top}\boldsymbol{V}^\top\boldsymbol{\Psi}^\top\boldsymbol{X} \overset{(5.72)}{=} \boldsymbol{\Sigma}^{-1}\boldsymbol{V}^\top\boldsymbol{\Delta}^\top\boldsymbol{H}\boldsymbol{X}^\top\boldsymbol{X}. \tag{5.75}$$

Note that $\boldsymbol{\Sigma}$ and $\boldsymbol{H}$ are symmetric.

Similarly, the out-of-sample projection in dual SPCA is:

$$\widetilde{\boldsymbol{x}}_t = \boldsymbol{\Sigma}^{-1}\boldsymbol{V}^\top\boldsymbol{\Delta}^\top\boldsymbol{H}\boldsymbol{X}^\top\boldsymbol{x}_t. \tag{5.76}$$

Considering all the n_t out-of-sample data points, the projection is:

$$\widetilde{\boldsymbol{X}}_t = \boldsymbol{\Sigma}^{-1}\boldsymbol{V}^\top\boldsymbol{\Delta}^\top\boldsymbol{H}\boldsymbol{X}^\top\boldsymbol{X}_t. \tag{5.77}$$

Reconstruction of $\boldsymbol{X}$ after the projection onto the SPCA subspace is:

$$\begin{aligned}\widehat{\boldsymbol{X}} &\overset{(5.66)}{=} \boldsymbol{U}\boldsymbol{U}^\top\boldsymbol{X} = \boldsymbol{U}\widetilde{\boldsymbol{X}} \overset{(a)}{=} \boldsymbol{\Psi}\boldsymbol{V}\boldsymbol{\Sigma}^{-1}\boldsymbol{\Sigma}^{-1}\boldsymbol{V}^\top\boldsymbol{\Delta}^\top\boldsymbol{H}\boldsymbol{X}^\top\boldsymbol{X} \\ &= \boldsymbol{\Psi}\boldsymbol{V}\boldsymbol{\Sigma}^{-2}\boldsymbol{V}^\top\boldsymbol{\Delta}^\top\boldsymbol{H}\boldsymbol{X}^\top\boldsymbol{X} \overset{(5.72)}{=} \boldsymbol{X}\boldsymbol{H}\boldsymbol{\Delta}\boldsymbol{V}\boldsymbol{\Sigma}^{-2}\boldsymbol{V}^\top\boldsymbol{\Delta}^\top\boldsymbol{H}\boldsymbol{X}^\top\boldsymbol{X}\end{aligned} \tag{5.78}$$

where (a) is because of Eqs. (5.74) and (5.75).

Similarly, the reconstruction of an out-of-sample data point in dual SPCA is:

$$\widehat{\boldsymbol{x}}_t = \boldsymbol{X}\boldsymbol{H}\boldsymbol{\Delta}\boldsymbol{V}\boldsymbol{\Sigma}^{-2}\boldsymbol{V}^\top\boldsymbol{\Delta}^\top\boldsymbol{H}\boldsymbol{X}^\top\boldsymbol{x}_t. \tag{5.79}$$

Considering all the n_t out-of-sample data points, the reconstruction is:

$$\widehat{\boldsymbol{X}}_t = \boldsymbol{X}\boldsymbol{H}\boldsymbol{\Delta}\boldsymbol{V}\boldsymbol{\Sigma}^{-2}\boldsymbol{V}^\top\boldsymbol{\Delta}^\top\boldsymbol{H}\boldsymbol{X}^\top\boldsymbol{X}_t. \tag{5.80}$$

Note that dual PCA was important because it provided the opportunity to kernelize the PCA. However, as explained in the next section, the kernel SPCA can be obtained directly from SPCA. Therefore, the dual SPCA might not be important for the sake of kernel SPCA. The dual SPCA has another benefit similar to the benefit of dual PCA. In Eqs. (5.75), (5.76), (5.78), and (5.79), $\boldsymbol{U}$ is not used, but $\boldsymbol{V}$ exists. In Eq. (5.73), the columns of $\boldsymbol{V}$ are the eigenvectors of $\boldsymbol{\Psi}^\top\boldsymbol{\Psi} \in \mathbb{R}^{n\times n}$, according to Proposition 2.1 in Chap. 2. On the other hand, the direct SPCA has the eigen-decomposition of $\boldsymbol{X}\boldsymbol{H}\boldsymbol{K}_y\boldsymbol{H}\boldsymbol{X}^\top \in \mathbb{R}^{d\times d}$ in Eq. (5.63), which is then used in Eqs. (5.64), (5.65), (5.66), and (5.67). In the event of large dimensionality, $d \gg n$, decomposition of the $n \times n$ matrix is faster and requires less storage, meaning that the dual SPCA would be more efficient.

5.6.4 *Kernel Supervised PCA*

The SPCA can be kernelized using either direct SPCA or dual SPCA [4]. These two approaches are explained in the following.

5.6.4.1 Kernel SPCA Using Direct SPCA

According to representation theory [2], any solution (direction) $\boldsymbol{u} \in \mathcal{H}$ must lie in the span of "all" the training vectors mapped to $\mathcal{H}$, i.e., $\boldsymbol{\Phi}(\boldsymbol{X}) = [\boldsymbol{\phi}(\boldsymbol{x}_1), \dots, \boldsymbol{\phi}(\boldsymbol{x}_n)] \in \mathbb{R}^{t\times n}$ (usually $t \gg d$). Note that $\mathcal{H}$ denotes the Hilbert space (feature space). Therefore:

$$\boldsymbol{u} = \sum_{i=1}^{n} \theta_i\, \boldsymbol{\phi}(\boldsymbol{x}_i) = \boldsymbol{\Phi}(\boldsymbol{X})\,\boldsymbol{\theta},$$

where $\boldsymbol{\theta} \in \mathbb{R}^n$ is the unknown vector of coefficients, and $\boldsymbol{u} \in \mathbb{R}^t$ is the kernel SPCA direction in the Hilbert space. The directions can be put together in $\mathbb{R}^{t\times p} \ni \boldsymbol{U} := [\boldsymbol{u}_1, \dots, \boldsymbol{u}_p]$:

$$\boldsymbol{U} = \boldsymbol{\Phi}(\boldsymbol{X})\,\boldsymbol{\Theta}, \tag{5.81}$$

where $\boldsymbol{\Theta} := [\boldsymbol{\theta}_1, \dots, \boldsymbol{\theta}_p] \in \mathbb{R}^{n\times p}$.

Equation (5.61) in the feature space becomes:

$$\text{HSIC} = \frac{1}{(n-1)^2}\,\mathbf{tr}(\boldsymbol{\Phi}(\boldsymbol{X})^\top \boldsymbol{U}\boldsymbol{U}^\top \boldsymbol{\Phi}(\boldsymbol{X})\boldsymbol{H}\boldsymbol{K}_y\boldsymbol{H}).$$

The $\mathbf{tr}(\boldsymbol{\Phi}(\boldsymbol{X})^\top \boldsymbol{U}\boldsymbol{U}^\top \boldsymbol{\Phi}(\boldsymbol{X})\boldsymbol{H}\boldsymbol{K}_y\boldsymbol{H})$ can be simplified to:

$$\begin{aligned}\mathbf{tr}(\boldsymbol{\Phi}(\boldsymbol{X})^\top \boldsymbol{U}\boldsymbol{U}^\top \boldsymbol{\Phi}(\boldsymbol{X})\boldsymbol{H}\boldsymbol{K}_y\boldsymbol{H}) &= \mathbf{tr}(\boldsymbol{U}\boldsymbol{U}^\top \boldsymbol{\Phi}(\boldsymbol{X})\boldsymbol{H}\boldsymbol{K}_y\boldsymbol{H}\boldsymbol{\Phi}(\boldsymbol{X})^\top)\\ &= \mathbf{tr}(\boldsymbol{U}^\top \boldsymbol{\Phi}(\boldsymbol{X})\boldsymbol{H}\boldsymbol{K}_y\boldsymbol{H}\boldsymbol{\Phi}(\boldsymbol{X})^\top \boldsymbol{U}) \end{aligned} \tag{5.82}$$

Plugging Eq. (5.81) into Eq. (5.82) results in:

$$\mathbf{tr}\big(\boldsymbol{\Theta}^\top \boldsymbol{\Phi}(\boldsymbol{X})^\top \boldsymbol{\Phi}(\boldsymbol{X})\boldsymbol{H}\boldsymbol{K}_y\boldsymbol{H}\boldsymbol{\Phi}(\boldsymbol{X})^\top \boldsymbol{\Phi}(\boldsymbol{X})\,\boldsymbol{\Theta}\big) = \mathbf{tr}(\boldsymbol{\Theta}^\top \boldsymbol{K}_x\boldsymbol{H}\boldsymbol{K}_y\boldsymbol{H}\boldsymbol{K}_x\boldsymbol{\Theta}), \tag{5.83}$$

where:

$$\mathbb{R}^{n\times n} \ni \boldsymbol{K}_x := \boldsymbol{\Phi}(\boldsymbol{X})^\top \boldsymbol{\Phi}(\boldsymbol{X}). \tag{5.84}$$

Note that the Eqs. (5.84) and (5.59) are different and should not be confused.

Moreover, the constraint of orthogonality of the projection matrix, i.e., $\boldsymbol{U}^\top\boldsymbol{U} = \boldsymbol{I}$, in the feature space becomes:

$$\boldsymbol{U}^\top\boldsymbol{U} = \big(\boldsymbol{\Phi}(\boldsymbol{X})\,\boldsymbol{\Theta}\big)^\top\big(\boldsymbol{\Phi}(\boldsymbol{X})\,\boldsymbol{\Theta}\big) = \boldsymbol{\Theta}^\top\boldsymbol{\Phi}(\boldsymbol{X})^\top\boldsymbol{\Phi}(\boldsymbol{X})\boldsymbol{\Theta} = \boldsymbol{\Theta}^\top\boldsymbol{K}_x\,\boldsymbol{\Theta}. \tag{5.85}$$

Therefore, the optimization problem is:

$$\begin{aligned} &\underset{\boldsymbol{\Theta}}{\text{maximize}} \quad \mathbf{tr}(\boldsymbol{\Theta}^\top \boldsymbol{K}_x\boldsymbol{H}\boldsymbol{K}_y\boldsymbol{H}\boldsymbol{K}_x\boldsymbol{\Theta}),\\ &\text{subject to} \quad \boldsymbol{\Theta}^\top\boldsymbol{K}_x\,\boldsymbol{\Theta} = \boldsymbol{I}, \end{aligned} \tag{5.86}$$

where the objective variable is the unknown $\boldsymbol{\Theta}$. The Lagrangian is:

$$\begin{aligned}\mathcal{L} &= \mathbf{tr}(\boldsymbol{\Theta}^\top \boldsymbol{K}_x\boldsymbol{H}\boldsymbol{K}_y\boldsymbol{H}\boldsymbol{K}_x\boldsymbol{\Theta}) - \mathbf{tr}\big(\boldsymbol{\Lambda}^\top(\boldsymbol{\Theta}^\top\boldsymbol{K}_x\,\boldsymbol{\Theta} - \boldsymbol{I})\big)\\ &= \mathbf{tr}(\boldsymbol{\Theta}\boldsymbol{\Theta}^\top \boldsymbol{K}_x\boldsymbol{H}\boldsymbol{K}_y\boldsymbol{H}\boldsymbol{K}_x) - \mathbf{tr}\big(\boldsymbol{\Lambda}^\top(\boldsymbol{\Theta}^\top\boldsymbol{K}_x\,\boldsymbol{\Theta} - \boldsymbol{I})\big),\end{aligned}$$

where $\boldsymbol{\Lambda} \in \mathbb{R}^{p\times p}$ is a diagonal matrix $\mathbf{diag}([\lambda_1, \dots, \lambda_p]^\top)$.

$$\begin{aligned}&\mathbb{R}^{n\times p} \ni \frac{\partial\mathcal{L}}{\partial\boldsymbol{\Theta}} = 2\boldsymbol{K}_x\boldsymbol{H}\boldsymbol{K}_y\boldsymbol{H}\boldsymbol{K}_x\boldsymbol{\Theta} - 2\boldsymbol{K}_x\boldsymbol{\Theta}\boldsymbol{\Lambda} \overset{\text{set}}{=} 0\\ &\implies \boldsymbol{K}_x\boldsymbol{H}\boldsymbol{K}_y\boldsymbol{H}\boldsymbol{K}_x\boldsymbol{\Theta} = \boldsymbol{K}_x\boldsymbol{\Theta}\boldsymbol{\Lambda},\end{aligned} \tag{5.87}$$

which is the generalized eigenvalue problem $(\boldsymbol{K}_x\boldsymbol{H}\boldsymbol{K}_y\boldsymbol{H}\boldsymbol{K}_x, \boldsymbol{K}_x)$ (see Chap. 2). $\boldsymbol{\Theta}$ and $\boldsymbol{\Lambda}$ are the eigenvector and eigenvalue matrices, respectively.

Note that in practice, Eq. (5.87) can be solved by left multiplying $\boldsymbol{K}_x^{-1}$ (hoping that it is positive definite and thus not singular):

$$\underbrace{\boldsymbol{K}_x^{-1}\boldsymbol{K}_x}_{\boldsymbol{I}}\boldsymbol{H}\boldsymbol{K}_y\boldsymbol{H}\boldsymbol{K}_x\boldsymbol{\Theta} = \boldsymbol{\Theta}\boldsymbol{\Lambda} \implies \boldsymbol{H}\boldsymbol{K}_y\boldsymbol{H}\boldsymbol{K}_x\boldsymbol{\Theta} = \boldsymbol{\Theta}\boldsymbol{\Lambda}, \tag{5.88}$$

which is the eigenvalue problem (see Chap. 2) for $\boldsymbol{H}\boldsymbol{K}_y\boldsymbol{H}\boldsymbol{K}_x$, where columns of $\boldsymbol{\Theta}$ are the eigenvectors of it and $\boldsymbol{\Lambda}$ includes its eigenvalues on its diagonal.

If the p leading eigenvectors are taken to have $\boldsymbol{\Theta} \in \mathbb{R}^{n\times p}$, the projection of $\boldsymbol{\Phi}(\boldsymbol{X}) \in \mathbb{R}^{t\times n}$ is:

$$\mathbb{R}^{p\times n} \ni \boldsymbol{\Phi}(\widetilde{\boldsymbol{X}}) = \boldsymbol{U}^\top\boldsymbol{\Phi}(\boldsymbol{X}) \overset{(5.81)}{=} \boldsymbol{\Theta}^\top\boldsymbol{\Phi}(\boldsymbol{X})^\top\boldsymbol{\Phi}(\boldsymbol{X}) = \boldsymbol{\Theta}^\top\boldsymbol{K}_x, \tag{5.89}$$

where $\mathbb{R}^{n\times n} \ni \boldsymbol{K}_x := \boldsymbol{\Phi}(\boldsymbol{X})^\top\boldsymbol{\Phi}(\boldsymbol{X})$. Similarly, the projection of the out-of-sample data point $\boldsymbol{\phi}(\boldsymbol{x}_t) \in \mathbb{R}^t$ is:

$$\mathbb{R}^p \ni \boldsymbol{\phi}(\widetilde{\boldsymbol{x}}_t) = \boldsymbol{U}^\top\boldsymbol{\phi}(\boldsymbol{x}_t) \overset{(5.81)}{=} \boldsymbol{\Theta}^\top\boldsymbol{\Phi}(\boldsymbol{X})^\top\boldsymbol{\phi}(\boldsymbol{x}_t) = \boldsymbol{\Theta}^\top\boldsymbol{k}_t, \tag{5.90}$$

where $\mathbb{R}^n \ni \boldsymbol{k}_t = \boldsymbol{k}_t(\boldsymbol{X}, \boldsymbol{x}_t) := \boldsymbol{\Phi}(\boldsymbol{X})^\top\boldsymbol{\phi}(\boldsymbol{x}_t)$. Considering all the n_t out-of-sample data points, $\boldsymbol{X}_t$, the projection is:

$$\mathbb{R}^{p\times n_t} \ni \boldsymbol{\phi}(\widetilde{\boldsymbol{X}}_t) = \boldsymbol{\Theta}^\top\boldsymbol{K}_t, \tag{5.91}$$

where $\mathbb{R}^{n\times n_t} \ni \boldsymbol{K}_t := \boldsymbol{\Phi}(\boldsymbol{X})^\top\boldsymbol{\Phi}(\boldsymbol{X}_t)$. In kernel SPCA, as in kernel PCA, it is not possible to reconstruct data, whether training or out-of-sample.

5.6.4.2 Kernel SPCA Using Dual SPCA

Another approach for kernelizing SPCA is by using the dual SPCA. Equation (5.72) in the t-dimensional feature space becomes:

$$\mathbb{R}^{t\times n} \ni \boldsymbol{\Psi} = \boldsymbol{\Phi}(\boldsymbol{X})\boldsymbol{H}\boldsymbol{\Delta}, \tag{5.92}$$

where $\boldsymbol{\Phi}(\boldsymbol{X}) = [\boldsymbol{\phi}(\boldsymbol{x}_1), \dots, \boldsymbol{\phi}(\boldsymbol{x}_n)] \in \mathbb{R}^{t\times n}$. Applying SVD (see Chap. 2) on the $\boldsymbol{\Psi}$ of Eq. (5.92) is similar to the form of Eq. (5.73). Similar to Eqs. (5.45) and (5.47), it is not necessary to have $\boldsymbol{\Phi}(\boldsymbol{X})$ in Eq. (5.92), as it is possible to obtain $\boldsymbol{V}$ and $\boldsymbol{\Sigma}$ as:

$$\big(\boldsymbol{\Delta}^\top\breve{\boldsymbol{K}}_x\boldsymbol{\Delta}\big)\boldsymbol{V} = \boldsymbol{V}\boldsymbol{\Sigma}^2, \tag{5.93}$$

where $\breve{\boldsymbol{K}}_x := \boldsymbol{H}\boldsymbol{K}_x\boldsymbol{H}\boldsymbol{\Delta}$ and the columns of $\boldsymbol{V}$ are the eigenvectors of (see Proposition 2.1 in Chap. 2):

$$\boldsymbol{\Psi}^\top\boldsymbol{\Psi} \overset{(a)}{=} \boldsymbol{\Delta}^\top\boldsymbol{H}\boldsymbol{\Phi}(\boldsymbol{X})^\top\boldsymbol{\Phi}(\boldsymbol{X})\boldsymbol{H}\boldsymbol{\Delta} \overset{(5.84)}{=} \boldsymbol{\Delta}^\top\boldsymbol{H}\boldsymbol{K}_x\boldsymbol{H}\boldsymbol{\Delta} = \boldsymbol{\Delta}^\top\breve{\boldsymbol{K}}_x\boldsymbol{\Delta},$$

where (a) is because of Eqs. (5.92) and symmetry of the centering matrix.

It is important to note that because of using Eq. (5.93) instead of Eq. (5.92), *the projection directions* $\boldsymbol{U}$ *are not available in kernel SPCA to be observed or plotted.*

Similar to Eqs. (5.73) and (5.74):

$$\boldsymbol{\Psi} = \boldsymbol{U}\boldsymbol{\Sigma}\boldsymbol{V}^\top \implies \boldsymbol{\Psi}\boldsymbol{V} = \boldsymbol{U}\boldsymbol{\Sigma}\underbrace{\boldsymbol{V}^\top\boldsymbol{V}}_{\boldsymbol{I}} = \boldsymbol{U}\boldsymbol{\Sigma} \implies \boldsymbol{U} = \boldsymbol{\Psi}\boldsymbol{V}\boldsymbol{\Sigma}^{-1}, \tag{5.94}$$

where $\boldsymbol{V}$ and $\boldsymbol{\Sigma}$ are obtained from Eq. (5.93).

The projection of data $\boldsymbol{\Phi}(\boldsymbol{X})$ is:

$$\begin{aligned}\boldsymbol{\Phi}(\widetilde{\boldsymbol{X}}) &= \boldsymbol{U}^\top\boldsymbol{\Phi}(\boldsymbol{X}) = (\boldsymbol{\Psi}\boldsymbol{V}\boldsymbol{\Sigma}^{-1})^\top\boldsymbol{\Phi}(\boldsymbol{X}) = \boldsymbol{\Sigma}^{-\top}\boldsymbol{V}^\top\boldsymbol{\Psi}^\top\boldsymbol{\Phi}(\boldsymbol{X}) \\ &\overset{(5.92)}{=} \boldsymbol{\Sigma}^{-1}\boldsymbol{V}^\top\boldsymbol{\Delta}^\top\boldsymbol{H}\boldsymbol{\Phi}(\boldsymbol{X})^\top\boldsymbol{\Phi}(\boldsymbol{X}) \overset{(5.84)}{=} \boldsymbol{\Sigma}^{-1}\boldsymbol{V}^\top\boldsymbol{\Delta}^\top\boldsymbol{H}\boldsymbol{K}_x.\end{aligned} \tag{5.95}$$

In the above calculation, the symmetry of $\boldsymbol{\Sigma}$ and $\boldsymbol{H}$ is noticed. Similarly, the out-of-sample projection in kernel SPCA is:

$$\boldsymbol{\phi}(\widetilde{\boldsymbol{x}}_t) = \boldsymbol{\Sigma}^{-1}\boldsymbol{V}^\top\boldsymbol{\Delta}^\top\boldsymbol{H}\boldsymbol{\Phi}(\boldsymbol{X})^\top\boldsymbol{\phi}(\boldsymbol{x}_t) = \boldsymbol{\Sigma}^{-1}\boldsymbol{V}^\top\boldsymbol{\Delta}^\top\boldsymbol{H}\,\boldsymbol{k}_t, \tag{5.96}$$

where $\mathbb{R}^n \ni \boldsymbol{k}_t := \boldsymbol{\Phi}(\boldsymbol{X})^\top\boldsymbol{\phi}(\boldsymbol{x}_t)$. Considering all the n_t out-of-sample data points, $\boldsymbol{X}_t$, the projection is:

$$\boldsymbol{\phi}(\widetilde{\boldsymbol{X}}_t) = \boldsymbol{\Sigma}^{-1}\boldsymbol{V}^\top\boldsymbol{\Delta}^\top\boldsymbol{H}\,\boldsymbol{K}_t. \tag{5.97}$$

where $\mathbb{R}^{n\times n_t} \ni \boldsymbol{K}_t := \boldsymbol{\Phi}(\boldsymbol{X})^\top\boldsymbol{\Phi}(\boldsymbol{X}_t)$.

Reconstruction of $\boldsymbol{\Phi}(\boldsymbol{X})$ after projection onto the SPCA subspace is:

$$\begin{aligned}\boldsymbol{\Phi}(\widehat{\boldsymbol{X}}) &= \boldsymbol{U}\boldsymbol{U}^\top\boldsymbol{\Phi}(\boldsymbol{X}) = \boldsymbol{U}\boldsymbol{\Phi}(\widetilde{\boldsymbol{X}}) \overset{(a)}{=} \boldsymbol{\Psi}\boldsymbol{V}\boldsymbol{\Sigma}^{-1}\boldsymbol{\Sigma}^{-1}\boldsymbol{V}^\top\boldsymbol{\Delta}^\top\boldsymbol{H}\boldsymbol{K}_x \\ &= \boldsymbol{\Psi}\boldsymbol{V}\boldsymbol{\Sigma}^{-2}\boldsymbol{V}^\top\boldsymbol{\Delta}^\top\boldsymbol{H}\boldsymbol{K}_x \overset{(5.92)}{=} \boldsymbol{\Phi}(\boldsymbol{X})\boldsymbol{H}\boldsymbol{\Delta}\boldsymbol{V}\boldsymbol{\Sigma}^{-2}\boldsymbol{V}^\top\boldsymbol{\Delta}^\top\boldsymbol{H}\boldsymbol{K}_x\end{aligned} \tag{5.98}$$

where (a) is because of Eqs. (5.94) and (5.95). Similarly, reconstruction of an out-of-sample data point in dual SPCA is:

$$\widehat{\boldsymbol{x}}_t = \boldsymbol{\Phi}(\boldsymbol{X})\boldsymbol{H}\boldsymbol{\Delta}\boldsymbol{V}\boldsymbol{\Sigma}^{-2}\boldsymbol{V}^\top\boldsymbol{\Delta}^\top\boldsymbol{H}\boldsymbol{\Phi}(\boldsymbol{X})^\top\boldsymbol{\phi}(\boldsymbol{x}_t) = \boldsymbol{\Phi}(\boldsymbol{X})\boldsymbol{H}\boldsymbol{\Delta}\boldsymbol{V}\boldsymbol{\Sigma}^{-2}\boldsymbol{V}^\top\boldsymbol{\Delta}^\top\boldsymbol{H}\,\boldsymbol{k}_t, \tag{5.99}$$

where $\mathbb{R}^n \ni \boldsymbol{k}_t := \boldsymbol{\Phi}(\boldsymbol{X})^\top \boldsymbol{\phi}(\boldsymbol{x}_t)$. However, in Eqs. (5.98) and (5.99), $\boldsymbol{\Phi}(\boldsymbol{X})$ is not necessarily available; therefore, in kernel SPCA, as in kernel PCA, it is not possible to reconstruct the data, whether training or out-of-sample.

5.7 Eigenfaces

This section introduces one of the most fundamental applications of PCA and its variants—facial recognition.

5.7.1 Projection Directions of Facial Images

PCA and kernel PCA can be trained using images of diverse faces, to learn the most important facial features, which account for the variation between faces. Here, two facial datasets, i.e. the Frey dataset [10] and the AT&T (ORL) face dataset [29], are used to illustrate this concept. The AT&T dataset has been used twice, i.e., (1) with human subjects as its classes and (2) with having and not having eyeglasses as its classes. Figure 5.7 demonstrates the top ten PCA directions for the PCA trained on these datasets. As demonstrated, the projection directions of a facial dataset are

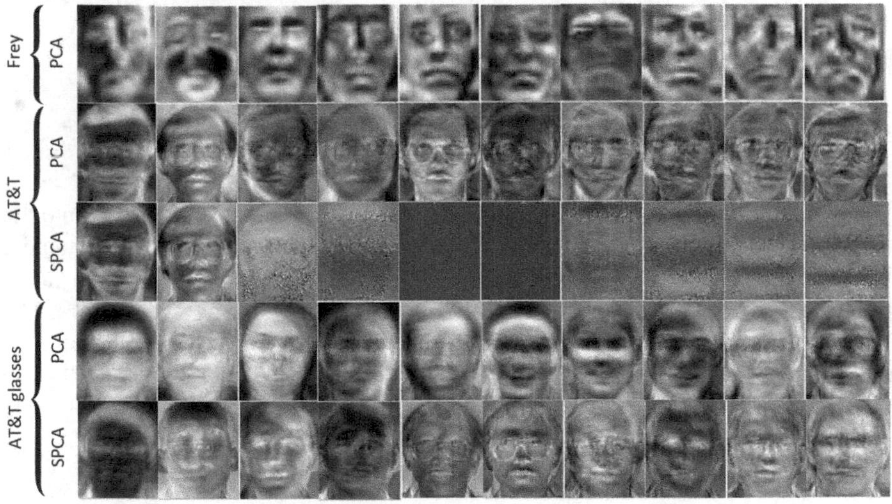

Fig. 5.7 The ghost faces: the leading eigenvectors of PCA and SPCA for Frey, AT&T, and AT&T glasses datasets

some facial features that are similar to ghost faces in terms of appearance. That is why the facial projection directions are also referred to as "ghost faces". The ghost faces in PCA are also referred to as "eigenfaces" [30, 31] because PCA uses eigenvalue decomposition of the covariance matrix.

In Fig. 5.7, the projection directions have captured different facial features that discriminate the data with respect to the maximum variance. The captured features are eyes, nose, cheeks, chin, lips, and eyebrows, which are the most important facial features. This figure does not include the projection directions of the kernel PCA because the projection directions are not available in kernel PCA. Note that facial recognition using kernel PCA is referred to as "kernel eigenfaces" [33]. The ghost faces (facial projection directions) of SPCA can be referred to as the "supervised eigenfaces". Facial recognition using the kernel SPCA can also be referred to as "kernel supervised eigenfaces". Figure 5.7 does not include the projection directions of the kernel SPCA because the projection directions are not available in kernel SPCA.

A comparison of the PCA and SPCA directions demonstrates that both PCA and SPCA capture eyeglasses as important discriminators. However, some Haar wavelet[2] like features [26] are captured as the projection directions in SPCA. Haar wavelets are important in face recognition and detection; for example, they have been used in the Viola-Jones face detector [32]. As demonstrated in Fig. 5.7, both PCA and SPCA have captured eyes as discriminators; however, SPCA has also focused on the frame of eyeglasses because of the usage of class labels. PCA has also captured other distracting facial features, such as forehead, cheeks, hair, mustache, etc, because it is not aware that the two classes are different, in terms of glasses, and sees the dataset as a whole.

5.7.2 Projection of Facial Images

Using the obtained projection directions, the facial images can be projected onto the PCA subspace. Similarly, projected images using kernel PCA can also be obtained. Figure 5.8 demonstrates the projection of both training and out-of-sample facial images of the Frey dataset onto the PCA, dual PCA, and kernel PCA subspaces. The kernels used were linear, RBF, and cosine. As seen in the figure, the out-of-sample data, although were not seen in the training phase, are projected very well. The model, somewhat, has extrapolated the projections so it has learned generalizable subspaces.

[2] Haar wavelet is a family of square-shaped filters which form wavelet bases.

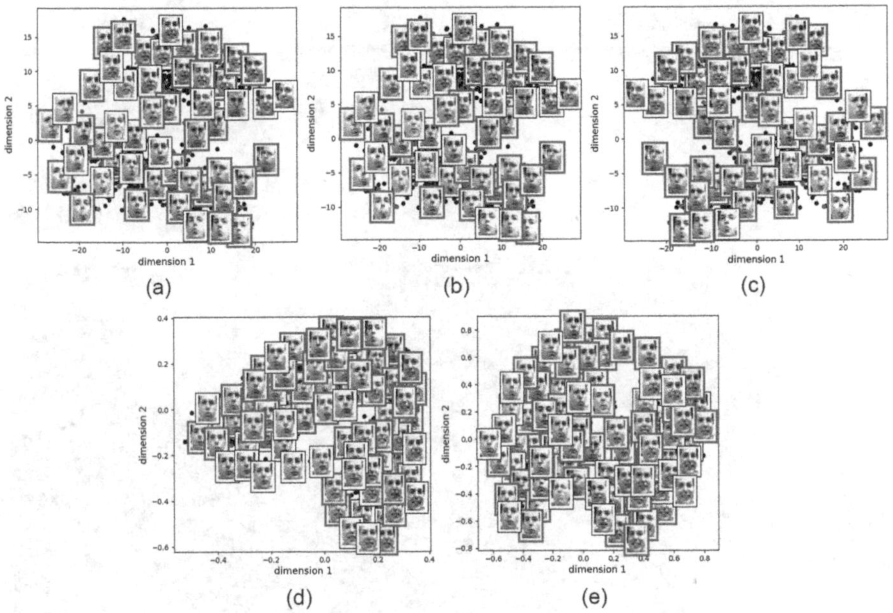

Fig. 5.8 Projection of the training and out-of-sample sets of the Frey dataset onto the subspaces of (**a**) PCA, (**b**) dual PCA, (**c**) kernel PCA (linear kernel), (**d**) kernel PCA (RBF kernel), and (**e**) kernel PCA (cosine kernel). The images with red frames are the out-of-sample images

5.7.3 Reconstruction of Facial Images

The facial images can be reconstructed after projection onto PCA and SPCA subspaces. The reconstruction of training and test images in the Frey and AT&T datasets are depicted in Fig. 5.9. Reconstructions have occurred using all and two top projection directions. As expected, the reconstructions using all projection directions are very similar to the original images. However, reconstruction using two leading projection directions is not prefect. Although, most important facial features are reconstructed because the leading projection directions carry most of the information.

5.8 Chapter Summary

PCA is the first proposed spectral dimensionality reduction method and is fundamental. This chapter introduced the unsupervised and supervised PCA methods and their dual variants. It was demonstrated that PCA can be obtained using both eigenvalue decomposition and singular value decomposition, which were introduced in Chap. 2. Kernel PCA was also introduced for both unsupervised and supervised PCA. It will be demonstrated in Chap. 10 that most of the spectral dimensionality

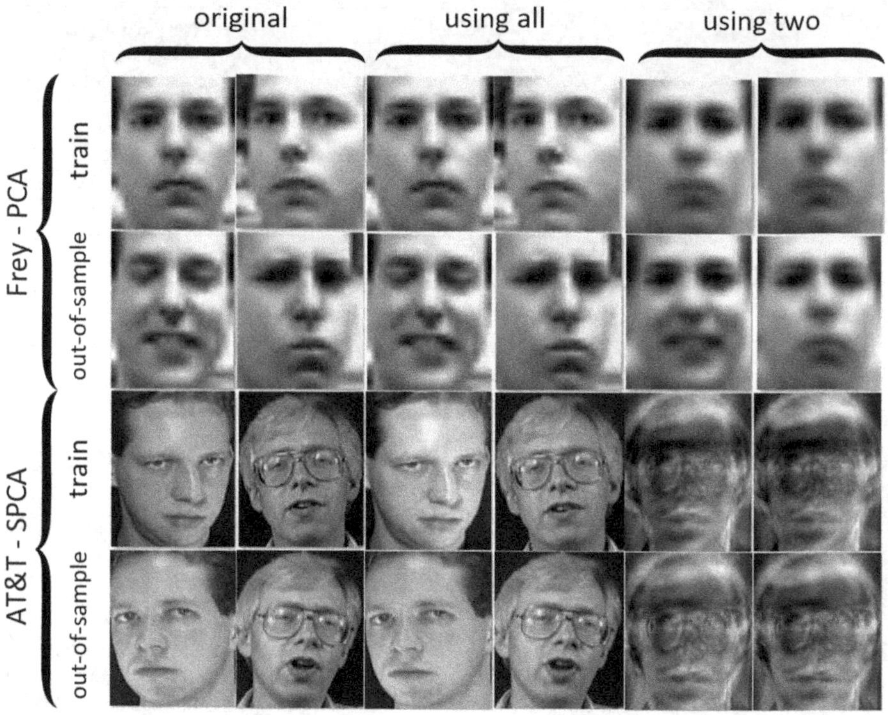

Fig. 5.9 The reconstructed faces using all and two of the leading eigenvectors of PCA and SPCA for Frey and AT&T datasets

reduction methods can be reduced to kernel PCA with different kernel functions. Therefore, kernel PCA is important for understanding the unification of spectral approaches. HSIC, which was introduced in Chap. 3 and used in the supervised PCA, is an important tool. It will also be used in various algorithms, such as guided LLE (Chap. 8) and supervised KDR (Chap. 15). The eigenface method introduced in this chapter is one of the fundamental applications of PCA in computer vision.

References

1. Hervé Abdi and Lynne J Williams. "Principal component analysis". In: *Wiley interdisciplinary reviews: computational statistics* 2.4 (2010), pp. 433–459.
2. Jonathan L Alperin. *Local representation theory: Modular representations as an introduction to the local representation theory of finite groups*. Vol. 11. Cambridge University Press, 1993.
3. Eric Bair et al. "Prediction by supervised principal components". In: *Journal of the American Statistical Association* 101.473 (2006), pp. 119–137.
4. Elnaz Barshan et al. "Supervised principal component analysis: Visualization, classification and regression on subspaces and submanifolds". In: *Pattern Recognition* 44.7 (2011), pp. 1357–1371.

5. Mikhail Belkin and Partha Niyogi. "Laplacian eigenmaps for dimensionality reduction and data representation". In: *Neural computation* 15.6 (2003), pp. 1373–1396.
6. Raymond B Cattell. "The scree test for the number of factors". In: *Multivariate behavioral research* 1.2 (1966), pp. 245–276.
7. Michael AA Cox and Trevor F Cox. "Multidimensional scaling". In: *Handbook of data visualization*. Springer, 2008, pp. 315–347.
8. David L Donoho. "High-dimensional data analysis: The curses and blessings of dimensionality". In: *AMS math challenges lecture* (2000).
9. Susan T Dumais. "Latent semantic analysis". In: *Annual review of information science and technology* 38.1 (2004), pp. 188–230.
10. Brendan Frey. *Frey face dataset*. https://cs.nyu.edu/~roweis/data.html. Accessed: June 2017.
11. Jerome Friedman, Trevor Hastie, and Robert Tibshirani. *The elements of statistical learning*. Vol. 2. Springer series in statistics New York, NY, USA: 2009.
12. Ali Ghodsi. "Dimensionality reduction: a short tutorial". In: *Department of Statistics and Actuarial Science, Univ. of Waterloo, Ontario, Canada* 37 (2006).
13. Ian Goodfellow, Yoshua Bengio, and Aaron Courville. *Deep learning*. MIT Press, 2016.
14. Arthur Gretton et al. "Measuring statistical dependence with Hilbert-Schmidt norms". In: *International conference on algorithmic learning theory*. Springer. 2005, pp. 63–77.
15. Ji Hun Ham et al. "A kernel view of the dimensionality reduction of manifolds". In: *International Conference on Machine Learning*. 2004.
16. Thomas Hofmann, Bernhard Schölkopf, and Alexander J Smola. "Kernel methods in machine learning". In: *The annals of statistics* (2008), pp. 1171–1220.
17. Ian Jolliffe. *Principal component analysis*. Springer, 2011.
18. Shuangge Ma and Ying Dai. "Principal component analysis based methods in bioinformatics studies". In: *Briefings in bioinformatics* 12.6 (2011), pp. 714–722.
19. Thomas P Minka. "Automatic choice of dimensionality for PCA". In: *Advances in neural information processing systems*. 2001, pp. 598–604.
20. Hoda Mohammadzade et al. "Critical object recognition in millimeter-wave images with robustness to rotation and scale". In: *JOSA A* 34.6 (2017), pp. 846–855.
21. Karl Pearson. "LIII. On lines and planes of closest fit to systems of points in space". In: *The London, Edinburgh, and Dublin Philosophical Magazine and Journal of Science* 2.11 (1901), pp. 559–572.
22. Sam T Roweis and Lawrence K Saul. "Nonlinear dimensionality reduction by locally linear embedding". In: *science* 290.5500 (2000), pp. 2323–2326.
23. David E Rumelhart, Geoffrey E Hinton, and Ronald J Williams. "Learning representations by back-propagating errors". In: *Nature* 323.6088 (1986), pp. 533–536.
24. Bernhard Schölkopf, Alexander Smola, and Klaus-Robert Müller. "Kernel principal component analysis". In: *International conference on artificial neural networks*. Springer. 1997, pp. 583–588.
25. Bernhard Schölkopf, Alexander Smola, and Klaus-Robert Müller. "Nonlinear component analysis as a kernel eigenvalue problem". In: *Neural computation* 10.5 (1998), pp. 1299–1319.
26. Radomir S Stanković and Bogdan J Falkowski. "The Haar wavelet transform: its status and achievements". In: *Computers & Electrical Engineering* 29.1 (2003), pp. 25–44.
27. Harry Strange and Reyer Zwiggelaar. *Open Problems in Spectral Dimensionality Reduction*. Springer, 2014.
28. Joshua B Tenenbaum, Vin De Silva, and John C Langford. "A global geometric framework for nonlinear dimensionality reduction". In: *science* 290.5500 (2000), pp. 2319–2323.
29. The Digital Technology Group. *AT&T Laboratories Cambridge*. https://www.cl.cam.ac.uk/research/dtg/attarchive/facedatabase.html. Accessed: June 2017. 1994.
30. Matthew Turk and Alex Pentland. "Eigenfaces for recognition". In: *Journal of cognitive neuroscience* 3.1 (1991), pp. 71–86.
31. Matthew A Turk and Alex P Pentland. "Face recognition using eigenfaces". In: *Computer Vision and Pattern Recognition, 1991. Proceedings CVPR'91., IEEE Computer Society Conference on* IEEE. 1991, pp. 586–591.

32. Yi-Qing Wang. “An analysis of the Viola-Jones face detection algorithm”. In: *Image Processing On Line* 4 (2014), pp. 128–148.
33. M-H Yang, Narendra Ahuja, and David Kriegman. “Face recognition using kernel eigenfaces”. In: *Image processing, 2000. proceedings. 2000 international conference on*. Vol. 1. IEEE. 2000, pp. 37–40.

Chapter 6
Fisher Discriminant Analysis

6.1 Introduction

Fisher Discriminant Analysis (FDA) [9] attempts to find a subspace that separates the classes as much as possible, while the data also become as spread as possible. It was first proposed in [6] by Sir. Ronald Aylmer Fisher (1890–1962), who was a genius in statistics. Fisher's work mostly concentrated on the statistics of genetics, and his paper [6], which proposed FDA, was the first to introduce the well-known Iris flower dataset. He proposed many important concepts in modern statistics, such as variance [5], FDA [6], Fisher information [8], and the Analysis of Variance (ANOVA) [7]. Much of his work was devoted to studying variance, making it no surprise that FDA focuses on variance and scatters.

The kernel FDA [15, 16] performs the goal of FDA in the feature space. The FDA has had many different applications, such as facial recognition (Fisherfaces) [2, 4, 27], action recognition (Fisherposes) [11], and gesture recognition [17]. Whereas, the kernel FDA has been used for facial recognition (kernel Fisherfaces) [14, 26] and palmprint recognition [22]. In the literature, FDA can be referred to as Linear Discriminant Analysis (LDA) or Fisher LDA (FLDA). This is because FDA and LDA [10] are equivalent, although LDA is a classification method and not a subspace learning algorithm. This chapter will prove why FDA and LDA are equivalent.

6.2 Projection and Reconstruction

6.2.1 Projection Formulation

Consider a dataset of *instances* or *data points* $\{(\boldsymbol{x}_i, \boldsymbol{y}_i)\}_{i=1}^n$ with sample size n and dimensionality $\boldsymbol{x}_i \in \mathbb{R}^d$ and $\boldsymbol{y}_i \in \mathbb{R}^\ell$. $\{\boldsymbol{x}_i\}_{i=1}^n$ are the input data to the model,

B. Ghojogh et al., *Elements of Dimensionality Reduction and Manifold Learning*,
https://doi.org/10.1007/978-3-031-10602-6_6

and $\{y_i\}_{i=1}^n$ are the observations (labels). The following are defined as $\mathbb{R}^{d\times n} \ni X := [x_1, \dots, x_n]$ and $\mathbb{R}^{\ell\times n} \ni Y := [y_1, \dots, y_n]$. It is also possible to have an out-of-sample data point, $x_t \in \mathbb{R}^d$, which is not in the training set. If there are n_t out-of-sample data points, $\{x_{t,i}\}_1^{n_t}$, they are defined as $\mathbb{R}^{d\times n_t} \ni X_t := [x_{t,1}, \dots, x_{t,n_t}]$. In a case where the observations $\{y_i\}_{i=1}^n$ come from a discrete set then the task is *classification*. Assume the dataset consists of c classes, $\{x_i^{(1)}\}_{i=1}^{n_1}, \dots, \{x_i^{(c)}\}_{i=1}^{n_c}$, where n_j denotes the sample size (cardinality) of the j-th class. Recall linear projection which was introduced in Chap. 2. In subspace learning, the projection of a vector $x \in \mathbb{R}^d$ onto the column space of $U \in \mathbb{R}^{d\times p}$ (a p-dimensional subspace spanned by $\{u_j\}_{j=1}^p$, where $u_j \in \mathbb{R}^d$) is defined as:

$$\mathbb{R}^p \ni \widetilde{x} := U^\top x, \tag{6.1}$$

$$\mathbb{R}^d \ni \widehat{x} := UU^\top x = U\widetilde{x}, \tag{6.2}$$

where $\widetilde{x}$ and $\widehat{x}$ denote the projection and reconstruction of x, respectively. If there are n data points, $\{x_i\}_{i=1}^n$, which can be stored column-wise in a matrix $X \in \mathbb{R}^{d\times n}$, the projection and reconstruction of X are defined as:

$$\mathbb{R}^{p\times n} \ni \widetilde{X} := U^\top X, \tag{6.3}$$

$$\mathbb{R}^{d\times n} \ni \widehat{X} := UU^\top X = U\widetilde{X}, \tag{6.4}$$

respectively.

If there is an out-of-sample data point x_t, which was not used in the calculation of U, the projection and reconstruction of it are defined as:

$$\mathbb{R}^p \ni \widetilde{x}_t := U^\top x_t, \tag{6.5}$$

$$\mathbb{R}^d \ni \widehat{x}_t := UU^\top x_t = U\widetilde{x}_t, \tag{6.6}$$

respectively. In case there are n_t out-of-sample data points, $\{x_{t,i}\}_{i=1}^{n_t}$, which can be stored columnwise in a matrix $X_t \in \mathbb{R}^{d\times n_t}$, the projection and reconstruction of X_t are defined as:

$$\mathbb{R}^{p\times n_t} \ni \widetilde{X}_t := U^\top X_t, \tag{6.7}$$

$$\mathbb{R}^{d\times n_t} \ni \widehat{X}_t := UU^\top X_t = U\widetilde{X}_t, \tag{6.8}$$

respectively.

Chapter 5 discussed that the covariance of the projected data is equal to the squared length of reconstructed data. If there is only one projection direction u, then:

$$||\widehat{X}||_F^2 = \sum_{i=1}^{n} ||\widehat{x}_i||_2^2 = u^\top S u. \tag{6.9}$$

If there are multiple projection directions in the projection matrix U, then:

$$||\widehat{X}||_F^2 = \textbf{tr}(U^\top S\, U). \tag{6.10}$$

6.3 Fisher Discriminant Analysis

6.3.1 One-Dimensional Subspace

6.3.1.1 Scatters in Two-Class Case

Assume there are two classes, $\{x_i^{(1)}\}_{i=1}^{n_1}$ and $\{x_i^{(2)}\}_{i=1}^{n_2}$, where n_1 and n_2 denote the sample size of the first class and second class, respectively, and $x_i^{(j)}$ denotes the i-th instance of the j-th class. If the data instances of the j-th class are projected onto a one-dimensional subspace (vector u) by $u^\top x_i^{(j)}$, the mean and the variance of the projected data are $u^\top \mu_j$ and $u^\top S_j u$, respectively, where μ_j and S_j are the mean and covariance matrix (scatter) of the j-th class. The mean of the j-th class is:

$$\mathbb{R}^d \ni \mu_j := \frac{1}{n_j} \sum_{i=1}^{n_j} x_i^{(j)}. \tag{6.11}$$

Similar to metric learning (see Chap. 11), after projection onto the one-dimensional subspace, the distance between the means of classes is:

$$\begin{aligned}
\mathbb{R} \ni d_B &:= (u^\top \mu_1 - u^\top \mu_2)^\top (u^\top \mu_1 - u^\top \mu_2) = (\mu_1 - \mu_2)^\top u u^\top (\mu_1 - \mu_2) \\
&\overset{(a)}{=} \textbf{tr}\big((\mu_1 - \mu_2)^\top u u^\top (\mu_1 - \mu_2)\big) \overset{(b)}{=} \textbf{tr}\big(u^\top (\mu_1 - \mu_2)(\mu_1 - \mu_2)^\top u\big) \\
&\overset{(c)}{=} u^\top (\mu_1 - \mu_2)(\mu_1 - \mu_2)^\top u \overset{(d)}{=} u^\top S_B\, u,
\end{aligned} \tag{6.12}$$

where (a) is because $(\mu_1 - \mu_2)^\top u u^\top (\mu_1 - \mu_2)$ is a scalar, (b) is because of the cyclic property of the trace, (c) is because $u^\top (\mu_1 - \mu_2)(\mu_1 - \mu_2)^\top u$ is a scalar, and (d) is because:

$$\mathbb{R}^{d \times d} \ni S_B := (\mu_1 - \mu_2)(\mu_1 - \mu_2)^\top, \tag{6.13}$$

as the *between-scatter* of classes (or the *interclass scatter/covariance*). Equation (6.12) can also be interpreted according to Eq. (6.9): the d_B is the variance of projection of the class means or the squared length of reconstruction of the class means.

The variance of projection is $\boldsymbol{u}^\top \boldsymbol{S}_j \boldsymbol{u}$ for the j-th class. If the variances of projections of the two classes are added together, the result is:

$$\mathbb{R} \ni d_W := \boldsymbol{u}^\top \boldsymbol{S}_1 \boldsymbol{u} + \boldsymbol{u}^\top \boldsymbol{S}_2 \boldsymbol{u} = \boldsymbol{u}^\top (\boldsymbol{S}_1 + \boldsymbol{S}_2)\, \boldsymbol{u} \overset{(a)}{=} \boldsymbol{u}^\top \boldsymbol{S}_W\, \boldsymbol{u}, \tag{6.14}$$

where:

$$\mathbb{R}^{d\times d} \ni \boldsymbol{S}_W := \boldsymbol{S}_1 + \boldsymbol{S}_2, \tag{6.15}$$

is the *within-scatter* of classes (or the *intraclass scatter/covariance*). According to Eq. (6.9), d_W is the summation of the projection variance of the class instances or the summation of the reconstruction length of the class instances.

6.3.1.2 Scatters in Multiclass Case: Variant 1

Assume $\{\boldsymbol{x}_i^{(j)}\}_{i=1}^{n_j}$ are the instances of the j-th class, where there are multiple classes. In this case, the *between-scatter* is defined as:

$$\mathbb{R}^{d\times d} \ni \boldsymbol{S}_B := \sum_{j=1}^{c} (\boldsymbol{\mu}_j - \boldsymbol{\mu})(\boldsymbol{\mu}_j - \boldsymbol{\mu})^\top, \tag{6.16}$$

where c is the number of classes and:

$$\mathbb{R}^{d} \ni \boldsymbol{\mu} := \frac{1}{\sum_{k=1}^{c} n_k} \sum_{j=1}^{c} n_j\, \boldsymbol{\mu}_j = \frac{1}{n} \sum_{i=1}^{n} \boldsymbol{x}_i, \tag{6.17}$$

is the weighted mean of means of classes or the total mean of the data. It is noteworthy that some researchers define the between-scatter in a weighted way:

$$\mathbb{R}^{d\times d} \ni \boldsymbol{S}_B := \sum_{j=1}^{c} n_j (\boldsymbol{\mu}_j - \boldsymbol{\mu})(\boldsymbol{\mu}_j - \boldsymbol{\mu})^\top. \tag{6.18}$$

If Eq. (6.15) is extended to c classes, the *within-scatter* is defined as:

$$\mathbb{R}^{d\times d} \ni \boldsymbol{S}_W := \sum_{j=1}^{c} \boldsymbol{S}_j \overset{(a)}{=} \sum_{j=1}^{c} \sum_{i=1}^{n_j} (\boldsymbol{x}_i^{(j)} - \boldsymbol{\mu}_j)(\boldsymbol{x}_i^{(j)} - \boldsymbol{\mu}_j)^\top, \tag{6.19}$$

where n_j is the sample size of the j-th class and (a) is because the scatter (covariance) of the j-th class is defined as:

$$\mathbb{R}^{d\times d} \ni \boldsymbol{S}_j = \sum_{i=1}^{n_j}(\boldsymbol{x}_i^{(j)} - \boldsymbol{\mu}_j)(\boldsymbol{x}_i^{(j)} - \boldsymbol{\mu}_j)^\top. \tag{6.20}$$

In this case, d_B and d_W are:

$$\mathbb{R} \ni d_B := \boldsymbol{u}^\top \boldsymbol{S}_B\, \boldsymbol{u}, \tag{6.21}$$

$$\mathbb{R} \ni d_W := \boldsymbol{u}^\top \boldsymbol{S}_W\, \boldsymbol{u}, \tag{6.22}$$

where $\boldsymbol{S}_B$ and $\boldsymbol{S}_W$ are Eqs. (6.16) and (6.19).

6.3.1.3 Scatters in Multiclass Case: Variant 2

There is another variant for the multiclass case in FDA, where the within-scatter is the same as Eq. (6.19). The between-scatter is, however, different. The *total scatter* is defined as the covariance matrix of the whole data, regardless of classes [23]:

$$\mathbb{R}^{d\times d} \ni \boldsymbol{S}_T := \frac{1}{n}\sum_{i=1}^{n}(\boldsymbol{x}_i - \boldsymbol{\mu})(\boldsymbol{x}_i - \boldsymbol{\mu})^\top, \tag{6.23}$$

where the total mean $\boldsymbol{\mu}$ is Eq. (6.17). The scaled total scatter can also be used by dropping the $1/n$ factor. On the other hand, the total scatter is equal to the summation of the within- and between-scatters:

$$\boldsymbol{S}_T = \boldsymbol{S}_W + \boldsymbol{S}_B. \tag{6.24}$$

Therefore, the between-scatter, in this variant, is obtained as:

$$\boldsymbol{S}_B := \boldsymbol{S}_T - \boldsymbol{S}_W. \tag{6.25}$$

6.3.1.4 Fisher Subspace: Variant 1

In FDA, the goal is to maximize the projection variance (scatter) of the means of classes and minimize the projection variance (scatter) of the class instances. In other words, the aim is to maximize d_B and minimize d_W. The reason is that after projection, it is important that the within scatter of every class is small and the between scatter of the classes is large; therefore, the instances of every class are closer together while the classes are farther apart. The two mentioned optimization problems are:

$$\underset{\boldsymbol{u}}{\text{maximize}} \quad d_B(\boldsymbol{u}), \tag{6.26}$$

$$\underset{\boldsymbol{u}}{\text{minimize}} \quad d_W(\boldsymbol{u}). \tag{6.27}$$

It is possible to merge these two optimization problems into a regularized optimization problem:

$$\underset{\boldsymbol{u}}{\text{maximize}} \quad d_B(\boldsymbol{u}) - \alpha\, d_W(\boldsymbol{u}), \tag{6.28}$$

where $\alpha > 0$ is the regularization parameter. Another way of merging Eqs. (6.26) and (6.27) is:

$$\underset{\boldsymbol{u}}{\text{maximize}} \quad f(\boldsymbol{u}) := \frac{d_B(\boldsymbol{u})}{d_W(\boldsymbol{u})} = \frac{\boldsymbol{u}^\top \boldsymbol{S}_B\, \boldsymbol{u}}{\boldsymbol{u}^\top \boldsymbol{S}_W\, \boldsymbol{u}}, \tag{6.29}$$

where $f(\boldsymbol{u}) \in \mathbb{R}$ is referred to as the *Fisher criterion* [24]. The Fisher criterion is a generalized Rayleigh-Ritz quotient (see Chap. 2). As was proven in Chap. 2, the optimization in Eq. (6.29) is equivalent to:

$$\begin{aligned} &\underset{\boldsymbol{u}}{\text{maximize}} \quad && \boldsymbol{u}^\top \boldsymbol{S}_B\, \boldsymbol{u} \\ &\text{subject to} && \boldsymbol{u}^\top \boldsymbol{S}_W\, \boldsymbol{u} = 1. \end{aligned} \tag{6.30}$$

The Lagrangian is:

$$\mathcal{L} = \boldsymbol{w}^\top \boldsymbol{S}_B\, \boldsymbol{w} - \lambda(\boldsymbol{w}^\top \boldsymbol{S}_W\, \boldsymbol{w} - 1),$$

where λ is the Lagrange multiplier. Equating the derivative of $\mathcal{L}$ to zero results in:

$$\begin{aligned} &\mathbb{R}^d \ni \frac{\partial \mathcal{L}}{\partial \boldsymbol{u}} = 2\,\boldsymbol{S}_B\,\boldsymbol{u} - 2\,\lambda\,\boldsymbol{S}_W\,\boldsymbol{u} \overset{\text{set}}{=} \boldsymbol{0} \\ &\implies 2\,\boldsymbol{S}_B\,\boldsymbol{u} = 2\,\lambda\,\boldsymbol{S}_W\,\boldsymbol{u} \implies \boldsymbol{S}_B\,\boldsymbol{u} = \lambda\,\boldsymbol{S}_W\,\boldsymbol{u}, \end{aligned} \tag{6.31}$$

which is a generalized eigenvalue problem $(\boldsymbol{S}_B, \boldsymbol{S}_W)$ (see Chap. 2), where $\boldsymbol{u}$ is the eigenvector with the largest eigenvalue (because the optimization is a maximization problem), and λ is the corresponding eigenvalue. $\boldsymbol{u}$ is referred to as the *Fisher direction* or *Fisher axis*. The projection and reconstruction are completed according to Eqs. (6.3) and (6.4), respectively, where $\boldsymbol{u} \in \mathbb{R}^d$ is used instead of $\boldsymbol{U} \in \mathbb{R}^{d \times p}$. The out-of-sample projection and reconstruction are performed according to Eqs. (6.7) and (6.8), respectively, with $\boldsymbol{u}$ rather than $\boldsymbol{U}$.

A possible solution to the generalized eigenvalue problem $(\boldsymbol{S}_B, \boldsymbol{S}_W)$ is (see Chap. 2):

$$\boldsymbol{S}_B\,\boldsymbol{u} = \lambda\,\boldsymbol{S}_W\,\boldsymbol{u} \implies \boldsymbol{S}_W^{-1}\boldsymbol{S}_B\,\boldsymbol{u} = \lambda\,\boldsymbol{u} \implies \boldsymbol{u} = \mathbf{eig}(\boldsymbol{S}_W^{-1}\boldsymbol{S}_B), \tag{6.32}$$

where **eig**(.) denotes the eigenvector of the matrix with the largest eigenvalue. The solution in Eq. (6.32) is very common for FDA, although it is not rigorous because $\boldsymbol{S}_w$ might be singular and not invertible. The diagonal of $\boldsymbol{S}_W$ can be strengthened to make it full rank and invertible (see Chap. 2):

$$\boldsymbol{u} = \mathbf{eig}((\boldsymbol{S}_W + \varepsilon \boldsymbol{I})^{-1} \boldsymbol{S}_B), \tag{6.33}$$

where ε is a very small positive number, large enough to make $\boldsymbol{S}_W$ full rank. In Sect. 6.5 of this chapter, robust FDA which tackles this problem will be discussed. On the other hand, the generalized eigenvalue problem $(\boldsymbol{S}_B, \boldsymbol{S}_W)$ has a rigorous solution, which does not require nonsingularity of $\boldsymbol{S}_W$ (see Chap. 2).

Another way to solve the optimization in Eq. (6.29) is obtained by taking the derivative from the Fisher criterion:

$$\mathbb{R}^d \ni \frac{\partial f(\boldsymbol{u})}{\partial \boldsymbol{u}} = \frac{1}{(\boldsymbol{u}^\top \boldsymbol{S}_W \boldsymbol{u})^2} \times \Big[(\boldsymbol{u}^\top \boldsymbol{S}_W \boldsymbol{u})(2\boldsymbol{S}_B \boldsymbol{u}) - (\boldsymbol{u}^\top \boldsymbol{S}_B \boldsymbol{u})(2\boldsymbol{S}_W \boldsymbol{u})\Big] \overset{\text{set}}{=} \boldsymbol{0}$$

$$\overset{(a)}{\Longrightarrow} \boldsymbol{S}_B \boldsymbol{u} = \frac{\boldsymbol{u}^\top \boldsymbol{S}_B \boldsymbol{u}}{\boldsymbol{u}^\top \boldsymbol{S}_W \boldsymbol{u}} \boldsymbol{S}_W \boldsymbol{u}, \tag{6.34}$$

where (a) is because $\boldsymbol{u}^\top \boldsymbol{S}_W \boldsymbol{u}$ is a scalar. Equation (6.34) is a generalized eigenvalue problem $(\boldsymbol{S}_B, \boldsymbol{S}_W)$ (see Chap. 2) with $\boldsymbol{u}$ and $(\boldsymbol{u}^\top \boldsymbol{S}_B \boldsymbol{u})/(\boldsymbol{u}^\top \boldsymbol{S}_W \boldsymbol{u})$ as the eigenvector with the largest eigenvalue (because the optimization is a maximization problem) and the corresponding eigenvalue, respectively. Therefore, *the Fisher criterion is the eigenvalue of the Fisher direction.*

6.3.1.5 Fisher Subspace: Variant 2

An additional way to find the FDA direction is to consider another version of the Fisher criterion. According to Eq. (6.25) for $\boldsymbol{S}_B$, the Fisher criterion becomes [23]:

$$f(\boldsymbol{u}) = \frac{\boldsymbol{u}^\top \boldsymbol{S}_B \boldsymbol{u}}{\boldsymbol{u}^\top \boldsymbol{S}_W \boldsymbol{u}} \overset{(6.25)}{=} \frac{\boldsymbol{u}^\top (\boldsymbol{S}_T - \boldsymbol{S}_W) \boldsymbol{u}}{\boldsymbol{u}^\top \boldsymbol{S}_W \boldsymbol{u}}$$

$$= \frac{\boldsymbol{u}^\top \boldsymbol{S}_T \boldsymbol{u} - \boldsymbol{u}^\top \boldsymbol{S}_W \boldsymbol{u}}{\boldsymbol{u}^\top \boldsymbol{S}_W \boldsymbol{u}} = \frac{\boldsymbol{u}^\top \boldsymbol{S}_T \boldsymbol{u}}{\boldsymbol{u}^\top \boldsymbol{S}_W \boldsymbol{u}} - 1. \tag{6.35}$$

The -1 is a constant and is dropped in the optimization; therefore:

$$\begin{aligned} \underset{\boldsymbol{u}}{\text{maximize}} \quad & \boldsymbol{u}^\top \boldsymbol{S}_T \boldsymbol{u} \\ \text{subject to} \quad & \boldsymbol{u}^\top \boldsymbol{S}_W \boldsymbol{u} = 1, \end{aligned} \tag{6.36}$$

whose solution is similarly obtained as:

$$\boldsymbol{S}_T\, \boldsymbol{u} = \lambda\, \boldsymbol{S}_W\, \boldsymbol{u}, \tag{6.37}$$

which is a generalized eigenvalue problem $(\boldsymbol{S}_T, \boldsymbol{S}_W)$ (see Chap. 2).

6.3.2 *Multidimensional Subspace*

The Fisher subspace is the span of several Fisher directions, $\{\boldsymbol{u}_j\}_{j=1}^p$ where $\boldsymbol{u}_j \in \mathbb{R}^d$, the d_B and d_W are defined as:

$$\mathbb{R} \ni d_B := \textbf{tr}(\boldsymbol{U}^\top \boldsymbol{S}_B\, \boldsymbol{U}), \tag{6.38}$$

$$\mathbb{R} \ni d_W := \textbf{tr}(\boldsymbol{U}^\top \boldsymbol{S}_W\, \boldsymbol{U}), \tag{6.39}$$

where $\mathbb{R}^{d\times p} \ni \boldsymbol{U} = [\boldsymbol{u}_1, \dots, \boldsymbol{u}_p]$. In this case, maximizing the *Fisher criterion* is:

$$\underset{\boldsymbol{U}}{\text{maximize}} \quad f(\boldsymbol{U}) := \frac{d_B(\boldsymbol{U})}{d_W(\boldsymbol{U})} = \frac{\textbf{tr}(\boldsymbol{U}^\top \boldsymbol{S}_B\, \boldsymbol{U})}{\textbf{tr}(\boldsymbol{U}^\top \boldsymbol{S}_W\, \boldsymbol{U})}. \tag{6.40}$$

The Fisher criterion $f(\boldsymbol{U})$ is a generalized Rayleigh-Ritz quotient (see Chap. 2). According to Chap. 2, the optimization in Eq. (6.40) is *approximately* equivalent to:

$$\begin{aligned} &\underset{\boldsymbol{U}}{\text{maximize}} \quad \textbf{tr}(\boldsymbol{U}^\top \boldsymbol{S}_B\, \boldsymbol{U}) \\ &\text{subject to} \quad \boldsymbol{U}^\top \boldsymbol{S}_W\, \boldsymbol{U} = \boldsymbol{I}. \end{aligned} \tag{6.41}$$

Note that this equivalence is exactly true for one projection vector $\boldsymbol{u}$, but it approximately holds for the projection matrix $\boldsymbol{U}$ having multiple projection directions. The Lagrangian is:

$$\mathcal{L} = \textbf{tr}(\boldsymbol{U}^\top \boldsymbol{S}_B\, \boldsymbol{U}) - \textbf{tr}\big(\boldsymbol{\Lambda}^\top (\boldsymbol{U}^\top \boldsymbol{S}_W\, \boldsymbol{U} - \boldsymbol{I})\big),$$

where $\boldsymbol{\Lambda} \in \mathbb{R}^{d\times d}$ is a diagonal matrix whose diagonal entries are the Lagrange multipliers. Equating the derivative of $\mathcal{L}$ to zero results in:

$$\begin{aligned} &\mathbb{R}^{d\times p} \ni \frac{\partial \mathcal{L}}{\partial \boldsymbol{U}} = 2\, \boldsymbol{S}_B\, \boldsymbol{U} - 2\, \boldsymbol{S}_W\, \boldsymbol{U} \boldsymbol{\Lambda} \overset{\text{set}}{=} \boldsymbol{0} \\ &\implies 2\, \boldsymbol{S}_B\, \boldsymbol{U} = 2\, \boldsymbol{S}_W\, \boldsymbol{U} \boldsymbol{\Lambda} \implies \boldsymbol{S}_B\, \boldsymbol{U} = \boldsymbol{S}_W\, \boldsymbol{U} \boldsymbol{\Lambda}, \end{aligned} \tag{6.42}$$

which is a generalized eigenvalue problem $(\boldsymbol{S}_B, \boldsymbol{S}_W)$. The columns of $\boldsymbol{U}$ are the eigenvectors sorted by the largest to smallest eigenvalues (because the optimization is a maximization problem) and the diagonal entries of $\boldsymbol{\Lambda}$ are the corresponding eigenvalues. The columns of $\boldsymbol{U}$ are referred to as the *Fisher directions* or *Fisher axes*. The projection and reconstruction are performed according to Eqs. (6.3) and (6.4), respectively. The out-of-sample projection and reconstruction are performed according to Eqs. (6.7) and (6.8), respectively.

One possible solution to the generalized eigenvalue problem $(\boldsymbol{S}_B, \boldsymbol{S}_W)$ is (see Chap. 2):

$$\boldsymbol{S}_B\,\boldsymbol{U} = \boldsymbol{S}_W\,\boldsymbol{U}\boldsymbol{\Lambda} \implies \boldsymbol{S}_W^{-1}\boldsymbol{S}_B\,\boldsymbol{U} = \boldsymbol{U}\boldsymbol{\Lambda} \implies \boldsymbol{U} = \mathbf{eig}(\boldsymbol{S}_W^{-1}\boldsymbol{S}_B), \tag{6.43}$$

where $\mathbf{eig}(.)$ denotes the eigenvectors of the matrix stacked columnwise. Again, it is possible to strengthen the diagonal:

$$\boldsymbol{U} = \mathbf{eig}((\boldsymbol{S}_W + \varepsilon\boldsymbol{I})^{-1}\boldsymbol{S}_B). \tag{6.44}$$

An additional way to solve the optimization in Eq. (6.40) is obtained by taking the derivative from the Fisher criterion:

$$\begin{aligned}
\mathbb{R}^{d\times p} \ni \frac{\partial f(\boldsymbol{U})}{\partial \boldsymbol{U}} &= \frac{1}{\big(\mathbf{tr}(\boldsymbol{U}^\top\boldsymbol{S}_W\,\boldsymbol{U})\big)^2}\times \\
&\Big[\mathbf{tr}(\boldsymbol{U}^\top\boldsymbol{S}_W\,\boldsymbol{U})(2\boldsymbol{S}_B\boldsymbol{U}) - \mathbf{tr}(\boldsymbol{U}^\top\boldsymbol{S}_B\,\boldsymbol{U})(2\boldsymbol{S}_W\boldsymbol{U})\Big] \overset{\text{set}}{=} \boldsymbol{0} \\
\overset{(a)}{\implies} \boldsymbol{S}_B\,\boldsymbol{U} &= \frac{\mathbf{tr}(\boldsymbol{U}^\top\boldsymbol{S}_B\,\boldsymbol{U})}{\mathbf{tr}(\boldsymbol{U}^\top\boldsymbol{S}_W\,\boldsymbol{U})}\,\boldsymbol{S}_W\,\boldsymbol{U},
\end{aligned} \tag{6.45}$$

where (a) is because $\mathbf{tr}(\boldsymbol{U}^\top\boldsymbol{S}_W\,\boldsymbol{U})$ is a scalar. Equation (6.45) is a generalized eigenvalue problem $(\boldsymbol{S}_B, \boldsymbol{S}_W)$ (see Chap. 2) with columns of $\boldsymbol{U}$ as the eigenvectors and $(\boldsymbol{u}_j^\top\boldsymbol{S}_B\,\boldsymbol{u}_j)/(\boldsymbol{u}_j^\top\boldsymbol{S}_W\,\boldsymbol{u}_j)$ as the j-th largest eigenvalue (because the optimization is a maximization problem).

Another way to find the FDA directions is to consider another version of the Fisher criterion. According to Eq. (6.25) for $\boldsymbol{S}_B$, the Fisher criterion becomes [23]:

$$f(\boldsymbol{U}) = \frac{\mathbf{tr}\big(\boldsymbol{U}^\top(\boldsymbol{S}_T - \boldsymbol{S}_W)\,\boldsymbol{U}\big)}{\mathbf{tr}(\boldsymbol{U}^\top\boldsymbol{S}_W\,\boldsymbol{U})} = \frac{\mathbf{tr}(\boldsymbol{U}^\top\boldsymbol{S}_T\,\boldsymbol{U})}{\mathbf{tr}(\boldsymbol{U}^\top\boldsymbol{S}_W\,\boldsymbol{U})} - 1. \tag{6.46}$$

The -1 is a constant and is dropped in the optimization; therefore:

$$\begin{aligned}
&\underset{\boldsymbol{U}}{\text{maximize}} && \mathbf{tr}(\boldsymbol{U}^\top\boldsymbol{S}_T\,\boldsymbol{U}) \\
&\text{subject to} && \boldsymbol{U}^\top\boldsymbol{S}_W\,\boldsymbol{U} = \boldsymbol{I},
\end{aligned} \tag{6.47}$$

whose solution is similarly obtained as:

$$\boldsymbol{S}_T\,\boldsymbol{U} = \boldsymbol{S}_W\,\boldsymbol{U}\boldsymbol{\Lambda}, \tag{6.48}$$

which is a generalized eigenvalue problem $(\boldsymbol{S}_T, \boldsymbol{S}_W)$ (see Chap. 2).

6.3.3 Discussion of the Dimensionality of the Fisher Subspace

In general, the rank of a covariance (scatter) matrix over the d-dimensional data with sample size n is at most $\min(d, n-1)$. The d is because the covariance matrix is a $d \times d$ matrix and the n is because it is iterated over n data instances for calculating the covariance matrix. The -1 is because of subtracting the mean in the calculation of the covariance matrix. For clarification, assume there is only one instance that becomes zero after removing the mean. This makes the covariance matrix a zero matrix.

According to Eq. (6.19), the rank of the $\boldsymbol{S}_W$ is at most $\min(d, n-1)$ because all the instances of all the classes are considered. Therefore, the rank of $\boldsymbol{S}_W$ is also at most $\min(d, n-1)$. According to Eq. (6.16), the rank of the $\boldsymbol{S}_B$ is at most $\min(d, c-1)$ because c iterations are in its calculation. In Eq. (6.43), the rank of $\boldsymbol{S}_W^{-1}\boldsymbol{S}_B$ is:

$$\begin{aligned}
&\mathbf{rank}(\boldsymbol{S}_W^{-1}\boldsymbol{S}_B) \leq \min\big(\mathbf{rank}(\boldsymbol{S}_W^{-1}), \mathbf{rank}(\boldsymbol{S}_B)\big) \\
&\leq \min\big(\min(d, n-1), \min(d, c-1)\big) = \min(d, n-1, c-1) \overset{(a)}{=} c-1,
\end{aligned} \tag{6.49}$$

where because of $c < d$ and $c < n$, there is (a). Therefore, the rank of $\boldsymbol{S}_W^{-1}\boldsymbol{S}_B$ is limited because of the rank of $\boldsymbol{S}_B$, which is at most $c-1$.

According to Eq. (6.43), the $c-1$ leading eigenvalues will be valid and the rest will be zero or very small. Therefore, p, which is the dimensionality of the Fisher subspace, is at most $c-1$. The $c-1$ leading eigenvectors are considered the Fisher directions and the rest of the eigenvectors are invalid and ignored.

6.4 Interpretation of FDA: The Example of a Man with Weak Eyes

Assume there is a man who has weak eyes and is colour blind. The man wants to separate (discriminate) several balls according to colour (classes), red and blue. However, he needs help because of his eye problems. First, consider his colour blindness. To help him, you separate the balls into two sets, one red and one blue. In other words, increase the distance between the balls based on colour to give him a clue about which balls belong to the same class. This is increasing the between-scatter of the two classes to help him. Second, consider his very weak eyes. Although the balls are almost separated according to colour, every ball looks blue to him. To make things easier for him, you place all the red balls in a group and all of the blue balls in a group. In other words, decrease the within-scatter of every class. In this way, the man sees every class as almost one blurry ball so that he can discriminate the classes better. Recall Eq. (6.43), which includes $\boldsymbol{S}_W^{-1}\boldsymbol{S}_B$. $\boldsymbol{S}_B$ implies an increase in the between-scatter, as was done to help the man with his colour blindness. $\boldsymbol{S}_W^{-1}$ implies a decrease in the within-scatter, as was done to help the man with his weak eyes. In conclusion, FDA increases the between-scatter and decreases the within-scatter (collapses each class [12]), at the same time, to better discriminate the classes.

6.5 Robust Fisher Discriminant Analysis

As discussed in Sect. 6.3.2, the matrix $\boldsymbol{S}_W$ may be singular or close to singular. One way to resolve this issue is to use the solution to the generalized eigenvalue problem (see Chap. 2) or to use Eq. (6.33) to make $\boldsymbol{S}_W$ a full-rank matrix. Another way to tackle the problem of singularity of $\boldsymbol{S}_W$ is the Robust FDA (RFDA) [3, 13]. In RFDA, the $\boldsymbol{S}_W$ is decomposed using eigenvalue decomposition (see Chap. 2):

$$\boldsymbol{S}_W = \boldsymbol{\Phi}^\top \boldsymbol{\Lambda} \boldsymbol{\Phi}, \tag{6.50}$$

where $\boldsymbol{\Phi}$ and $\boldsymbol{\Lambda} = \textbf{diag}([\lambda_1, \dots, \lambda_d]^\top)$ include the eigenvectors and eigenvalues of $\boldsymbol{S}_W$, respectively. The eigenvalues are sorted as $\lambda_1 \geq \dots \geq \lambda_d$ and the eigenvectors (columns of $\boldsymbol{\Phi}$) are sorted accordingly. If $\boldsymbol{S}_W$ is close to singularity, the first d' eigenvalues are valid and the remaining $(d - d')$ eigenvalues are either very small or zero. The appropriate d' is obtained as:

$$d' := \arg\min_{m} \left(\frac{\sum_{j=1}^{m} \lambda_j}{\sum_{k=1}^{d} \lambda_k} \geq 0.98 \right). \tag{6.51}$$

In RFDA, the $(d - d')$ invalid eigenvalues are replaced with λ_*:

$$\mathbb{R}^{d \times d} \ni \boldsymbol{\Lambda}' := \textbf{diag}([\lambda_1, \dots, \lambda_{d'}, \lambda_*, \dots, \lambda_*]^\top), \tag{6.52}$$

where [3]:

$$\lambda_* := \frac{1}{d - d'} \sum_{j=d'+1}^{d} \lambda_j. \tag{6.53}$$

Therefore, the S_W is replaced with S'_W:

$$\mathbb{R}^{d \times d} \ni S'_W := \Phi^\top \Lambda' \Phi, \tag{6.54}$$

and the robust Fisher directions are the eigenvectors of the generalized eigenvalue problem (S_B, S'_W) (see Chap. 2).

6.6 Comparison of FDA and PCA Directions

The FDA directions capture the directions where the instances of different classes fall apart and the instances in one class fall close together. On the other hand, the Principal Component Analysis (PCA) directions capture the directions where the data have maximum variance (spread), regardless of the classes (see Chap. 5). In some datasets the FDA and PCA are orthogonal, while in others, they are parallel. Other cases between these two extremes can occur, but they depend on the spread of classes in the dataset. Figure 6.1 demonstrates these cases for some two-dimensional datasets.

Moreover, considering Eq. (6.25) for S_B, the Fisher criterion becomes Eqs. (6.35) and (6.46) for one-dimensional and multidimensional Fisher subspaces, respectively. In these equations, -1 is a constant and is dropped in the optimization. This means that *the Fisher direction maximizes the total variance (spread) of the data, as also done in PCA, while at the same time, it minimizes the within-scatters of the classes (by making use of the class labels).* In other words, the optimization of FDA is equivalent to (Eq. (6.47) is repeated here):

$$\begin{aligned} & \underset{U}{\text{maximize}} && \mathbf{tr}(U^\top S_T\, U) \\ & \text{subject to} && U^\top S_W\, U = I, \end{aligned} \tag{6.55}$$

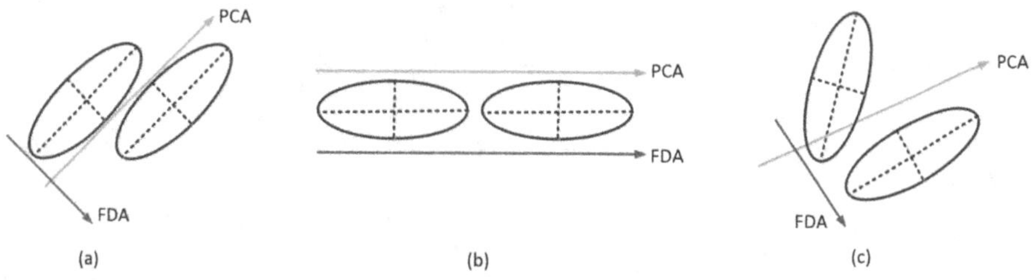

Fig. 6.1 Comparison of FDA and PCA directions for two-dimensional data with two classes: (**a**) a case where FDA and PCA are orthogonal, (**b**) a case where FDA and PCA are equivalent (parallel), and (**c**) a case between the two extreme cases of (**a**) and (**b**)

while the optimization of the PCA is (see Chap. 5):

$$\begin{aligned} &\underset{\boldsymbol{U}}{\text{maximize}} \quad \mathbf{tr}(\boldsymbol{U}^\top \boldsymbol{S}_T\, \boldsymbol{U}) \\ &\text{subject to} \quad \boldsymbol{U}^\top \boldsymbol{U} = \boldsymbol{I}. \end{aligned} \tag{6.56}$$

The solutions to Eqs. (6.55) and (6.56) are the generalized eigenvalue problem $(\boldsymbol{S}_T, \boldsymbol{S}_W)$ and the eigenvalue problem for $\boldsymbol{S}_T$, respectively (see Chap. 2).

6.7 On Equivalency of FDA and LDA

FDA is also referred to as Linear Discriminant Analysis (LDA) and Fisher LDA (FLDA) in the machine learning literature. Note that FDA is a manifold (subspace) learning method, and LDA [10] is a classification method. However, LDA can be seen as a metric learning method [10], and as metric learning is a subspace learning method (see Chap. 11), there is a connection between FDA and LDA.

FDA is a projection-based subspace learning method. Consider the projection vector $\boldsymbol{u}$. According to Eq. (6.1), the projection of data $\boldsymbol{x}$ is:

$$\boldsymbol{x} \mapsto \boldsymbol{u}^\top \boldsymbol{x}, \tag{6.57}$$

which can be done for all the data instances of every class. Thus, the mean and the covariance matrix of the class are transformed as:

$$\boldsymbol{\mu} \mapsto \boldsymbol{u}^\top \boldsymbol{\mu}, \tag{6.58}$$

$$\boldsymbol{\Sigma} \mapsto \boldsymbol{u}^\top \boldsymbol{\Sigma}\, \boldsymbol{u}, \tag{6.59}$$

respectively, because of the characteristics of the mean and variance.

According to Eq. (6.29), the Fisher criterion is the ratio of the between-class variance, σ_b^2, and within-class variance, σ_w^2:

$$f := \frac{\sigma_b^2}{\sigma_w^2} = \frac{(\boldsymbol{u}^\top \boldsymbol{\mu}_2 - \boldsymbol{u}^\top \boldsymbol{\mu}_1)^2}{\boldsymbol{u}^\top \boldsymbol{\Sigma}_2\, \boldsymbol{u} + \boldsymbol{u}^\top \boldsymbol{\Sigma}_1\, \boldsymbol{u}} = \frac{\big(\boldsymbol{u}^\top (\boldsymbol{\mu}_2 - \boldsymbol{\mu}_1)\big)^2}{\boldsymbol{u}^\top (\boldsymbol{\Sigma}_2 + \boldsymbol{\Sigma}_1)\, \boldsymbol{u}}, \tag{6.60}$$

where $\boldsymbol{\mu}_1$ and $\boldsymbol{\mu}_2$ are the means of the two classes and $\boldsymbol{\Sigma}_1$ and $\boldsymbol{\Sigma}_2$ are the covariances of the two classes. FDA maximizes the Fisher criterion:

$$\underset{\boldsymbol{u}}{\text{maximize}} \quad \frac{\big(\boldsymbol{u}^\top (\boldsymbol{\mu}_2 - \boldsymbol{\mu}_1)\big)^2}{\boldsymbol{u}^\top (\boldsymbol{\Sigma}_2 + \boldsymbol{\Sigma}_1)\, \boldsymbol{u}}, \tag{6.61}$$

which can be restated as:

$$\begin{aligned} &\underset{\boldsymbol{u}}{\text{maximize}} \quad \big(\boldsymbol{u}^\top(\boldsymbol{\mu}_2 - \boldsymbol{\mu}_1)\big)^2, \\ &\text{subject to} \quad \boldsymbol{u}^\top(\boldsymbol{\Sigma}_2 + \boldsymbol{\Sigma}_1)\,\boldsymbol{u} = 1, \end{aligned} \tag{6.62}$$

according to the Rayleigh-Ritz quotient method (see Chap. 2). The Lagrangian is:

$$\mathcal{L} = \big(\boldsymbol{u}^\top(\boldsymbol{\mu}_2 - \boldsymbol{\mu}_1)\big)^2 - \lambda\big(\boldsymbol{u}^\top(\boldsymbol{\Sigma}_2 + \boldsymbol{\Sigma}_1)\,\boldsymbol{u} - 1\big),$$

where λ is the Lagrange multiplier. Equating the derivative of $\mathcal{L}$ to zero results in:

$$\begin{aligned} &\frac{\partial \mathcal{L}}{\partial \boldsymbol{u}} = 2\,(\boldsymbol{\mu}_2 - \boldsymbol{\mu}_1)(\boldsymbol{\mu}_2 - \boldsymbol{\mu}_1)^\top \boldsymbol{u} - 2\,\lambda\,(\boldsymbol{\Sigma}_2 + \boldsymbol{\Sigma}_1)\,\boldsymbol{u} \overset{\text{set}}{=} \boldsymbol{0} \\ &\implies (\boldsymbol{\mu}_2 - \boldsymbol{\mu}_1)(\boldsymbol{\mu}_2 - \boldsymbol{\mu}_1)^\top \boldsymbol{u} = \lambda\,(\boldsymbol{\Sigma}_2 + \boldsymbol{\Sigma}_1)\,\boldsymbol{u}, \end{aligned}$$

which is a generalized eigenvalue problem $\big((\boldsymbol{\mu}_2 - \boldsymbol{\mu}_1)(\boldsymbol{\mu}_2 - \boldsymbol{\mu}_1)^\top, (\boldsymbol{\Sigma}_2 + \boldsymbol{\Sigma}_1)\big)$. The projection vector is the eigenvector of $(\boldsymbol{\Sigma}_2 + \boldsymbol{\Sigma}_1)^{-1}(\boldsymbol{\mu}_2 - \boldsymbol{\mu}_1)(\boldsymbol{\mu}_2 - \boldsymbol{\mu}_1)^\top$; therefore:

$$\boldsymbol{u} \propto (\boldsymbol{\Sigma}_2 + \boldsymbol{\Sigma}_1)^{-1}(\boldsymbol{\mu}_2 - \boldsymbol{\mu}_1)(\boldsymbol{\mu}_2 - \boldsymbol{\mu}_1)^\top. \tag{6.63}$$

On the other hand, in LDA, the decision function is [10]:

$$2\,\big(\boldsymbol{\Sigma}^{-1}(\boldsymbol{\mu}_2 - \boldsymbol{\mu}_1)\big)^\top \boldsymbol{x} + \boldsymbol{\mu}_1^\top \boldsymbol{\Sigma}^{-1} \boldsymbol{\mu}_1 - \boldsymbol{\mu}_2^\top \boldsymbol{\Sigma}^{-1} \boldsymbol{\mu}_2 + 2\ln(\frac{\pi_1}{\pi_2}) = 0, \tag{6.64}$$

where π_1 and π_2 are the prior distributions of the two classes. Moreover, in LDA, the covariance matrices are assumed to be equal [10]: $\boldsymbol{\Sigma}_1 = \boldsymbol{\Sigma}_2 = \boldsymbol{\Sigma}$. Therefore, in LDA, the Eq. (6.63) becomes [10]:

$$\boldsymbol{u} \propto (2\,\boldsymbol{\Sigma})^{-1}(\boldsymbol{\mu}_2 - \boldsymbol{\mu}_1)(\boldsymbol{\mu}_2 - \boldsymbol{\mu}_1)^\top \propto \boldsymbol{\Sigma}^{-1}(\boldsymbol{\mu}_2 - \boldsymbol{\mu}_1)(\boldsymbol{\mu}_2 - \boldsymbol{\mu}_1)^\top. \tag{6.65}$$

According to Eq. (6.57):

$$\boldsymbol{u}^\top \boldsymbol{x} \propto \big(\boldsymbol{\Sigma}^{-1}(\boldsymbol{\mu}_2 - \boldsymbol{\mu}_1)(\boldsymbol{\mu}_2 - \boldsymbol{\mu}_1)^\top\big)^\top \boldsymbol{x} \propto \big(\boldsymbol{\Sigma}^{-1}(\boldsymbol{\mu}_2 - \boldsymbol{\mu}_1)\big)^\top \boldsymbol{x}. \tag{6.66}$$

Comparing Eq. (6.66) with Eq. (6.64) shows that LDA and FDA are equivalent up to a scaling factor $\boldsymbol{\mu}_1^\top \boldsymbol{\Sigma}^{-1} \boldsymbol{\mu}_1 - \boldsymbol{\mu}_2^\top \boldsymbol{\Sigma}^{-1} \boldsymbol{\mu}_2 + 2\pi_1/\pi_2$. Note that this term is multiplied as an exponential factor before taking the logarithm to obtain Eq. (6.64), so this term is a scaling factor (see [10] for more details). It should be noted that in manifold (subspace) learning, the scale does not matter, as all the distances can scale similarly in the subspace without impacting the relative distances of points. Therefore, LDA and FDA are equivalent:

$$\text{LDA} \equiv \text{FDA}, \tag{6.67}$$

and *the FDA and LDA subspaces are the same subspace.* In other words, FDA followed by the use of Euclidean distance for classification in the subspace is equivalent to LDA. This sheds light on why LDA and FDA are used interchangeably in the literature. Note that LDA assumes *one* (and not several) Gaussian for every class [10], which is why FDA faces problems with multimodal data [18].

6.8 Kernel Fisher Discriminant Analysis

FDA is a linear dimensionality reduction method; therefore, it cannot be perfect for nonlinear data. It can be kernelized to be able to handle nonlinear patterns of data. This section introduces the kernel FDA.

6.8.1 *Kernels and Hilbert Space*

Recall Chap. 3, where it was noted that it is possible to compute the kernel of two matrices $\boldsymbol{X}_1 \in \mathbb{R}^{d \times n_1}$ and $\boldsymbol{X}_2 \in \mathbb{R}^{d \times n_2}$ and have a *kernel matrix* (also called the *Gram matrix*):

$$\mathbb{R}^{n_1 \times n_2} \ni \boldsymbol{K}(\boldsymbol{X}_1, \boldsymbol{X}_2) := \boldsymbol{\Phi}(\boldsymbol{X}_1)^\top \boldsymbol{\Phi}(\boldsymbol{X}_2), \tag{6.68}$$

where $\boldsymbol{\Phi}(\boldsymbol{X}_1) := [\boldsymbol{\phi}(\boldsymbol{x}_1), \ldots, \boldsymbol{\phi}(\boldsymbol{x}_n)] \in \mathbb{R}^{t \times n_1}$ is the matrix of mapped $\boldsymbol{X}_1$ to the feature space. The $\boldsymbol{\Phi}(\boldsymbol{X}_2) \in \mathbb{R}^{t \times n_2}$ is defined similarly. There can be a kernel matrix between a dataset $\boldsymbol{X} \in \mathbb{R}^{d \times n}$ and itself:

$$\mathbb{R}^{n \times n} \ni \boldsymbol{K}_x := \boldsymbol{K}(\boldsymbol{X}, \boldsymbol{X}) = \boldsymbol{\Phi}(\boldsymbol{X})^\top \boldsymbol{\Phi}(\boldsymbol{X}), \tag{6.69}$$

where $\boldsymbol{\Phi}(\boldsymbol{X}) := [\boldsymbol{\phi}(\boldsymbol{x}_1), \ldots, \boldsymbol{\phi}(\boldsymbol{x}_n)] \in \mathbb{R}^{t \times n}$ is the pulled (mapped) data.

6.8.2 *One-Dimensional Subspace*

6.8.2.1 Scatters in Two-Class Case

Equation (6.13) in the feature space is:

$$\mathbb{R}^{t \times t} \ni \boldsymbol{\Phi}(\boldsymbol{S}_B) := \big(\boldsymbol{\phi}(\boldsymbol{\mu}_1) - \boldsymbol{\phi}(\boldsymbol{\mu}_2)\big)\big(\boldsymbol{\phi}(\boldsymbol{\mu}_1) - \boldsymbol{\phi}(\boldsymbol{\mu}_2)\big)^\top, \tag{6.70}$$

where the mean of the j-th class in the feature space is:

$$\mathbb{R}^t \ni \boldsymbol{\phi}(\boldsymbol{\mu}_j) := \frac{1}{n_j} \sum_{i=1}^{n_j} \boldsymbol{\phi}(\boldsymbol{x}_i^{(j)}). \tag{6.71}$$

According to representation theory [1] (see Chap. 3), any solution (direction) $\boldsymbol{\phi}(\boldsymbol{u}) \in \mathcal{H}$ must lie in the span of "all" the training vectors mapped to $\mathcal{H}$, i.e., $\boldsymbol{\Phi}(\boldsymbol{X}) = [\boldsymbol{\phi}(\boldsymbol{x}_1), \ldots, \boldsymbol{\phi}(\boldsymbol{x}_n)] \in \mathbb{R}^{t \times n}$ (usually $t \gg d$). Note that $\mathcal{H}$ denotes the Hilbert space (feature space). Therefore:

$$\mathbb{R}^t \ni \boldsymbol{\phi}(\boldsymbol{u}) = \sum_{i=1}^{n} \theta_i\, \boldsymbol{\phi}(\boldsymbol{x}_i) = \boldsymbol{\Phi}(\boldsymbol{X})\, \boldsymbol{\theta}, \tag{6.72}$$

where $\mathbb{R}^n \ni \boldsymbol{\theta} := [\theta_1, \ldots, \theta_n]^\top$ is the unknown vector of coefficients, and $\boldsymbol{\phi}(\boldsymbol{u}) \in \mathbb{R}^t$ is the pulled Fisher direction to the feature space. The pulled directions can be put together in $\mathbb{R}^{t \times p} \ni \boldsymbol{\Phi}(\boldsymbol{U}) := [\boldsymbol{\phi}(\boldsymbol{u}_1), \ldots, \boldsymbol{\phi}(\boldsymbol{u}_p)]$:

$$\mathbb{R}^{t \times p} \ni \boldsymbol{\Phi}(\boldsymbol{U}) = \boldsymbol{\Phi}(\boldsymbol{X})\, \boldsymbol{\Theta}, \tag{6.73}$$

where $\boldsymbol{\Theta} := [\boldsymbol{\theta}_1, \ldots, \boldsymbol{\theta}_p] \in \mathbb{R}^{n \times p}$.

The d_B in the feature space is:

$$\begin{aligned}
\mathbb{R} \ni d_B &:= \boldsymbol{\phi}(\boldsymbol{u})^\top \boldsymbol{\Phi}(\boldsymbol{S}_B)\, \boldsymbol{\phi}(\boldsymbol{u}) \\
&\overset{(a)}{=} \boldsymbol{\theta}^\top \boldsymbol{\Phi}(\boldsymbol{X})^\top \big(\boldsymbol{\phi}(\boldsymbol{\mu}_1) - \boldsymbol{\phi}(\boldsymbol{\mu}_2)\big)\big(\boldsymbol{\phi}(\boldsymbol{\mu}_1) - \boldsymbol{\phi}(\boldsymbol{\mu}_2)\big)^\top \boldsymbol{\Phi}(\boldsymbol{X})\, \boldsymbol{\theta},
\end{aligned} \tag{6.74}$$

where (a) is because of Eqs. (6.70). and (6.72). For the j-th class (here $j \in \{1, 2\}$):

$$\begin{aligned}
\boldsymbol{\theta}^\top \boldsymbol{\Phi}(\boldsymbol{X})^\top \boldsymbol{\phi}(\boldsymbol{\mu}_j) &\overset{(6.72)}{=} \sum_{i=1}^{n} \theta_i\, \boldsymbol{\phi}(\boldsymbol{x}_i)^\top \boldsymbol{\phi}(\boldsymbol{\mu}_j) \\
&\overset{(6.71)}{=} \frac{1}{n_j} \sum_{i=1}^{n} \sum_{k=1}^{n_j} \theta_i\, \boldsymbol{\phi}(\boldsymbol{x}_i)^\top \boldsymbol{\phi}(\boldsymbol{x}_k^{(j)}) = \frac{1}{n_j} \sum_{i=1}^{n} \sum_{k=1}^{n_j} \theta_i\, k(\boldsymbol{x}_i, \boldsymbol{x}_k^{(j)}) = \boldsymbol{\theta}^\top \boldsymbol{m}_j,
\end{aligned} \tag{6.75}$$

where $\boldsymbol{m}_j \in \mathbb{R}^n$ whose i-th entry is:

$$\boldsymbol{m}_j(i) := \frac{1}{n_j} \sum_{k=1}^{n_j} k(\boldsymbol{x}_i, \boldsymbol{x}_k^{(j)}). \tag{6.76}$$

Therefore, Eq. (6.74) becomes:

$$d_B \overset{(6.75)}{=} \boldsymbol{\theta}^\top (\boldsymbol{m}_1 - \boldsymbol{m}_2)(\boldsymbol{m}_1 - \boldsymbol{m}_2)^\top \boldsymbol{\theta} = \boldsymbol{\theta}^\top \boldsymbol{M} \boldsymbol{\theta}, \tag{6.77}$$

where:

$$\mathbb{R}^{n\times n} \ni \boldsymbol{M} := (\boldsymbol{m}_1 - \boldsymbol{m}_2)(\boldsymbol{m}_1 - \boldsymbol{m}_2)^\top, \tag{6.78}$$

is the *between-scatter* in kernel FDA. Therefore, Eq. (6.74) becomes:

$$d_B = \boldsymbol{\phi}(\boldsymbol{u})^\top \boldsymbol{\Phi}(\boldsymbol{S}_B)\, \boldsymbol{\phi}(\boldsymbol{u}) = \boldsymbol{\theta}^\top \boldsymbol{M}\boldsymbol{\theta}. \tag{6.79}$$

Equation (6.19) in the feature space is:

$$\mathbb{R}^{t\times t} \ni \boldsymbol{\Phi}(\boldsymbol{S}_W) := \sum_{j=1}^{c}\sum_{i=1}^{n_j} \big(\boldsymbol{\phi}(\boldsymbol{x}_i^{(j)}) - \boldsymbol{\phi}(\boldsymbol{\mu}_j)\big)\big(\boldsymbol{\phi}(\boldsymbol{x}_i^{(j)}) - \boldsymbol{\phi}(\boldsymbol{\mu}_j)\big)^\top. \tag{6.80}$$

The d_W in the feature space is:

$$\begin{aligned}
\mathbb{R} \ni d_W &:= \boldsymbol{\phi}(\boldsymbol{u})^\top \boldsymbol{\Phi}(\boldsymbol{S}_W)\, \boldsymbol{\phi}(\boldsymbol{u}) \\
&\overset{(a)}{=} \Big(\sum_{\ell=1}^{n} \theta_\ell\, \boldsymbol{\phi}(\boldsymbol{x}_\ell)^\top\Big)\Big(\sum_{j=1}^{c}\sum_{i=1}^{n_j} (\boldsymbol{\phi}(\boldsymbol{x}_i^{(j)}) - \boldsymbol{\phi}(\boldsymbol{\mu}_j))(\boldsymbol{\phi}(\boldsymbol{x}_i^{(j)}) - \boldsymbol{\phi}(\boldsymbol{\mu}_j))^\top\Big)\Big(\sum_{k=1}^{n} \theta_k\, \boldsymbol{\phi}(\boldsymbol{x}_k)\Big) \\
&= \sum_{j=1}^{c}\sum_{\ell=1}^{n}\sum_{i=1}^{n_j}\sum_{k=1}^{n} \Big(\theta_\ell\, \boldsymbol{\phi}(\boldsymbol{x}_\ell)^\top (\boldsymbol{\phi}(\boldsymbol{x}_i^{(j)}) - \boldsymbol{\phi}(\boldsymbol{\mu}_j))(\boldsymbol{\phi}(\boldsymbol{x}_i^{(j)}) - \boldsymbol{\phi}(\boldsymbol{\mu}_j))^\top \theta_k\, \boldsymbol{\phi}(\boldsymbol{x}_k)\Big) \\
&\overset{(6.71)}{=} \sum_{j=1}^{c}\sum_{\ell=1}^{n}\sum_{i=1}^{n_j}\sum_{k=1}^{n} \Big(\theta_\ell\, \boldsymbol{\phi}(\boldsymbol{x}_\ell)^\top (\boldsymbol{\phi}(\boldsymbol{x}_i^{(j)}) - \frac{1}{n_j}\sum_{e=1}^{n_j} \boldsymbol{\phi}(\boldsymbol{x}_e^{(j)})) \\
&\qquad (\boldsymbol{\phi}(\boldsymbol{x}_i^{(j)}) - \frac{1}{n_j}\sum_{z=1}^{n_j} \boldsymbol{\phi}(\boldsymbol{x}_z^{(j)}))^\top \theta_k\, \boldsymbol{\phi}(\boldsymbol{x}_k)\Big) \\
&= \sum_{j=1}^{c}\sum_{\ell=1}^{n}\sum_{i=1}^{n_j}\sum_{k=1}^{n} \Big(\theta_\ell\, k(\boldsymbol{x}_\ell, \boldsymbol{x}_i^{(j)}) - \frac{1}{n_j}\sum_{e=1}^{n_j} \theta_\ell\, k(\boldsymbol{x}_\ell, \boldsymbol{x}_e^{(j)})\Big) \\
&\qquad \Big(\theta_k\, k(\boldsymbol{x}_i^{(j)}, \boldsymbol{x}_k) - \frac{1}{n_j}\sum_{z=1}^{n_j} \theta_k\, k(\boldsymbol{x}_z^{(j)}, \boldsymbol{x}_k)\Big) \\
&\overset{(b)}{=} \sum_{j=1}^{c}\sum_{\ell=1}^{n}\sum_{i=1}^{n_j}\sum_{k=1}^{n} \Big(\theta_\ell\, k(\boldsymbol{x}_\ell, \boldsymbol{x}_i^{(j)}) - \frac{1}{n_j}\sum_{e=1}^{n_j} \theta_\ell\, k(\boldsymbol{x}_\ell, \boldsymbol{x}_e^{(j)})\Big) \\
&\qquad \Big(\theta_k\, k(\boldsymbol{x}_k, \boldsymbol{x}_i^{(j)}) - \frac{1}{n_j}\sum_{z=1}^{n_j} \theta_k\, k(\boldsymbol{x}_k, \boldsymbol{x}_z^{(j)})\Big) \\
&= \sum_{j=1}^{c}\sum_{\ell=1}^{n}\sum_{i=1}^{n_j}\sum_{k=1}^{n} \Big(\theta_\ell\, \theta_k\, k(\boldsymbol{x}_\ell, \boldsymbol{x}_i^{(j)})\, k(\boldsymbol{x}_k, \boldsymbol{x}_i^{(j)})
\end{aligned}$$

$$
\begin{aligned}
&- \frac{2\theta_\ell \theta_k}{n_j} \sum_{z=1}^{n_j} k(\boldsymbol{x}_\ell, \boldsymbol{x}_i^{(j)})\, k(\boldsymbol{x}_k, \boldsymbol{x}_z^{(j)}) + \frac{\theta_\ell \theta_k}{n_j^2} \sum_{e=1}^{n_j} \sum_{z=1}^{n_j} k(\boldsymbol{x}_\ell, \boldsymbol{x}_e^{(j)})\, k(\boldsymbol{x}_k, \boldsymbol{x}_z^{(j)})\Big) \\
&= \sum_{j=1}^{c} \sum_{\ell=1}^{n} \sum_{i=1}^{n_j} \sum_{k=1}^{n} \Big(\theta_\ell \theta_k\, k(\boldsymbol{x}_\ell, \boldsymbol{x}_i^{(j)})\, k(\boldsymbol{x}_k, \boldsymbol{x}_i^{(j)}) - \frac{\theta_\ell \theta_k}{n_j} \sum_{z=1}^{n_j} k(\boldsymbol{x}_\ell, \boldsymbol{x}_i^{(j)})\, k(\boldsymbol{x}_k, \boldsymbol{x}_z^{(j)})\Big) \\
&= \sum_{j=1}^{c} \Big(\sum_{\ell=1}^{n} \sum_{i=1}^{n_j} \sum_{k=1}^{n} \Big(\theta_\ell \theta_k\, k(\boldsymbol{x}_\ell, \boldsymbol{x}_i^{(j)})\, k(\boldsymbol{x}_k, \boldsymbol{x}_i^{(j)})\Big) \\
&\qquad - \sum_{\ell=1}^{n} \sum_{i=1}^{n_j} \sum_{k=1}^{n} \Big(\frac{\theta_\ell \theta_k}{n_j} \sum_{z=1}^{n_j} k(\boldsymbol{x}_\ell, \boldsymbol{x}_i^{(j)})\, k(\boldsymbol{x}_k, \boldsymbol{x}_z^{(j)})\Big)\Big) \\
&\overset{(c)}{=} \sum_{j=1}^{c} (\boldsymbol{\theta}^\top \boldsymbol{K}_j \boldsymbol{K}_j^\top \boldsymbol{\theta} - \boldsymbol{\theta}^\top \boldsymbol{K}_j \frac{1}{n_j} \boldsymbol{1}\boldsymbol{1}^\top \boldsymbol{K}_j^\top \boldsymbol{\theta}) \\
&= \sum_{j=1}^{c} \boldsymbol{\theta}^\top \boldsymbol{K}_j (\boldsymbol{I} - \frac{1}{n_j} \boldsymbol{1}\boldsymbol{1}^\top) \boldsymbol{K}_j^\top \boldsymbol{\theta} \overset{(d)}{=} \sum_{j=1}^{c} \boldsymbol{\theta}^\top \boldsymbol{K}_j \boldsymbol{H}_j \boldsymbol{K}_j^\top \boldsymbol{\theta} = \boldsymbol{\theta}^\top \Big(\sum_{j=1}^{c} \boldsymbol{K}_j \boldsymbol{H}_j \boldsymbol{K}_j^\top\Big) \boldsymbol{\theta},
\end{aligned}
$$

where (a) is because of Eqs. (6.80) and (6.72), (b) is because $k(\boldsymbol{x}_1, \boldsymbol{x}_2) = k(\boldsymbol{x}_2, \boldsymbol{x}_1) \in \mathbb{R}$, and (c) is because $\boldsymbol{K}_j \in \mathbb{R}^{n \times n_j}$ is the kernel matrix of the whole training data and the training data of the j-th class. The (a, b)-th element of $\boldsymbol{K}_j$ is:

$$
\boldsymbol{K}_j(a, b) := k(\boldsymbol{x}_a, \boldsymbol{x}_b^{(j)}). \tag{6.81}
$$

The equation (d) is because:

$$
\mathbb{R}^{n_j \times n_j} \ni \boldsymbol{H}_j := \boldsymbol{I} - \frac{1}{n_j} \boldsymbol{1}\boldsymbol{1}^\top, \tag{6.82}
$$

is the centering matrix (see Chap. 2).

$$
\mathbb{R}^{n \times n} \ni \boldsymbol{N} := \sum_{j=1}^{c} \boldsymbol{K}_j \boldsymbol{H}_j \boldsymbol{K}_j^\top, \tag{6.83}
$$

is described as the *within-scatter* in kernel FDA. Therefore, d_W becomes:

$$
d_W = \boldsymbol{\phi}(\boldsymbol{u})^\top \boldsymbol{\Phi}(\boldsymbol{S}_W)\, \boldsymbol{\phi}(\boldsymbol{u}) = \boldsymbol{\theta}^\top \boldsymbol{N} \boldsymbol{\theta}. \tag{6.84}
$$

The kernel Fisher criterion is:

$$
f(\boldsymbol{\theta}) := \frac{d_B(\boldsymbol{\theta})}{d_W(\boldsymbol{\theta})} = \frac{\boldsymbol{\phi}(\boldsymbol{u})^\top \boldsymbol{\Phi}(\boldsymbol{S}_B)\, \boldsymbol{\phi}(\boldsymbol{u})}{\boldsymbol{\phi}(\boldsymbol{u})^\top \boldsymbol{\Phi}(\boldsymbol{S}_W)\, \boldsymbol{\phi}(\boldsymbol{u})} = \frac{\boldsymbol{\theta}^\top \boldsymbol{M} \boldsymbol{\theta}}{\boldsymbol{\theta}^\top \boldsymbol{N} \boldsymbol{\theta}}, \tag{6.85}
$$

where $\boldsymbol{\theta} \in \mathbb{R}^n$ is the *kernel Fisher direction*. Similar to the solution of Eq. (6.29), the solution to maximize Eq. (6.85) is:

$$\boldsymbol{M}\boldsymbol{\theta} = \lambda\, \boldsymbol{N}\boldsymbol{\theta}, \tag{6.86}$$

which is a generalized eigenvalue problem $(\boldsymbol{M}, \boldsymbol{N})$, where $\boldsymbol{\theta}$ is the eigenvector with the largest eigenvalue (because the optimization is a maximization problem), and λ is the corresponding eigenvalue. $\boldsymbol{\theta}$ is the *kernel Fisher direction* or *kernel Fisher axis*. A possible solution to the generalized eigenvalue problem $(\boldsymbol{M}, \boldsymbol{N})$ is (see Chap. 2):

$$\boldsymbol{\theta} = \mathbf{eig}(\boldsymbol{N}^{-1}\boldsymbol{M}), \tag{6.87}$$

or (see Chap. 2):

$$\boldsymbol{\theta} = \mathbf{eig}((\boldsymbol{N} + \varepsilon \boldsymbol{I})^{-1}\boldsymbol{M}), \tag{6.88}$$

where $\mathbf{eig}(.)$ denotes the eigenvector of the matrix with the largest eigenvalue.

The projection and reconstruction of the training data point $\boldsymbol{x}_i$ and the out-of-sample data point $\boldsymbol{x}_t$ are:

$$\mathbb{R} \ni \phi(\widetilde{\boldsymbol{x}}_i) = \boldsymbol{\phi}(\boldsymbol{u})^\top \boldsymbol{\phi}(\boldsymbol{x}_i) \overset{(6.72)}{=} \boldsymbol{\theta}^\top \boldsymbol{\Phi}(\boldsymbol{X})^\top \boldsymbol{\phi}(\boldsymbol{x}_i) = \boldsymbol{\theta}^\top \boldsymbol{k}(\boldsymbol{X}, \boldsymbol{x}_i), \tag{6.89}$$

$$\mathbb{R}^t \ni \boldsymbol{\phi}(\widehat{\boldsymbol{x}}_i) = \boldsymbol{\phi}(\boldsymbol{u})\boldsymbol{\phi}(\boldsymbol{u})^\top \boldsymbol{\phi}(\boldsymbol{x}_i) \overset{(6.72)}{=} \boldsymbol{\Phi}(\boldsymbol{X})\, \boldsymbol{\theta}\boldsymbol{\theta}^\top \boldsymbol{k}(\boldsymbol{X}, \boldsymbol{x}_i), \tag{6.90}$$

$$\mathbb{R} \ni \phi(\widetilde{\boldsymbol{x}}_t) = \boldsymbol{\theta}^\top \boldsymbol{k}(\boldsymbol{X}, \boldsymbol{x}_t), \tag{6.91}$$

$$\mathbb{R}^t \ni \boldsymbol{\phi}(\widehat{\boldsymbol{x}}_t) = \boldsymbol{\Phi}(\boldsymbol{X})\, \boldsymbol{\theta}\boldsymbol{\theta}^\top \boldsymbol{k}(\boldsymbol{X}, \boldsymbol{x}_t). \tag{6.92}$$

However, in reconstruction expressions, the $\boldsymbol{\Phi}(\boldsymbol{X})$ is not necessarily available; therefore, in kernel FDA, similar to kernel PCA (see Chap. 5), *reconstruction cannot be done.* For the whole training and out-of-sample data, the projections are:

$$\mathbb{R}^{1 \times n} \ni \boldsymbol{\Phi}(\widetilde{\boldsymbol{X}}) = \boldsymbol{\theta}^\top \boldsymbol{K}(\boldsymbol{X}, \boldsymbol{X}), \tag{6.93}$$

$$\mathbb{R}^{1 \times n_t} \ni \boldsymbol{\Phi}(\widetilde{\boldsymbol{X}}_t) = \boldsymbol{\theta}^\top \boldsymbol{K}(\boldsymbol{X}, \boldsymbol{X}_t). \tag{6.94}$$

6.8.2.2 Scatters in Multiclass Case: Variant 1

In a kernel FDA multiclass case, the *within-scatter* is the same as in the two-class case, which is seen in Eq. (6.83) and d_W is also seen in Eq. (6.84). However, the between-scatter is different. The between-scatter, Eq. (6.16), in the feature space is:

$$\mathbb{R}^{t\times t} \ni \boldsymbol{\Phi}(\boldsymbol{S}_B) := \sum_{j=1}^{c} \big(\boldsymbol{\phi}(\boldsymbol{\mu}_j) - \boldsymbol{\phi}(\boldsymbol{\mu})\big)\big(\boldsymbol{\phi}(\boldsymbol{\mu}_j) - \boldsymbol{\phi}(\boldsymbol{\mu})\big)^\top, \tag{6.95}$$

where the total mean in the feature space is:

$$\mathbb{R}^t \ni \boldsymbol{\phi}(\boldsymbol{\mu}) := \frac{1}{\sum_{k=1}^{c} n_k} \sum_{j=1}^{c} n_j\, \boldsymbol{\phi}(\boldsymbol{\mu}_j) = \frac{1}{n} \sum_{i=1}^{n} \boldsymbol{\phi}(\boldsymbol{x}_i), \tag{6.96}$$

The d_B in the feature space is:

$$\begin{aligned}\mathbb{R} \ni d_B &:= \boldsymbol{\phi}(\boldsymbol{u})^\top \boldsymbol{\Phi}(\boldsymbol{S}_B)\, \boldsymbol{\phi}(\boldsymbol{u}) \\ &\overset{(a)}{=} \sum_{j=1}^{c} \boldsymbol{\theta}^\top \boldsymbol{\Phi}(\boldsymbol{X})^\top \big(\boldsymbol{\phi}(\boldsymbol{\mu}_j) - \boldsymbol{\phi}(\boldsymbol{\mu})\big)\big(\boldsymbol{\phi}(\boldsymbol{\mu}_j) - \boldsymbol{\phi}(\boldsymbol{\mu})\big)^\top \boldsymbol{\Phi}(\boldsymbol{X})\, \boldsymbol{\theta},\end{aligned} \tag{6.97}$$

where (a) is because of Eqs. (6.95) and (6.72). Therefore:

$$\begin{aligned}\boldsymbol{\theta}^\top \boldsymbol{\Phi}(\boldsymbol{X})^\top \boldsymbol{\phi}(\boldsymbol{\mu}) &\overset{(6.72)}{=} \sum_{i=1}^{n} \theta_i\, \boldsymbol{\phi}(\boldsymbol{x}_i)^\top \boldsymbol{\phi}(\boldsymbol{\mu}) \\ &\overset{(6.96)}{=} \frac{1}{n} \sum_{i=1}^{n} \sum_{k=1}^{n} \theta_i\, \boldsymbol{\phi}(\boldsymbol{x}_i)^\top \boldsymbol{\phi}(\boldsymbol{x}_k) = \frac{1}{n} \sum_{i=1}^{n} \sum_{k=1}^{n} \theta_i\, k(\boldsymbol{x}_i, \boldsymbol{x}_k) = \boldsymbol{\theta}^\top \boldsymbol{m}_*,\end{aligned} \tag{6.98}$$

where $\boldsymbol{m}_* \in \mathbb{R}^n$ whose i-th entry is:

$$\boldsymbol{m}_*(i) := \frac{1}{n} \sum_{k=1}^{n} k(\boldsymbol{x}_i, \boldsymbol{x}_k). \tag{6.99}$$

According to Eqs. (6.75) and (6.98), Eq. (6.97) becomes:

$$d_B = \boldsymbol{\theta}^\top \sum_{j=1}^{c} (\boldsymbol{m}_j - \boldsymbol{m}_*)(\boldsymbol{m}_j - \boldsymbol{m}_*)^\top \boldsymbol{\theta} = \boldsymbol{\theta}^\top \boldsymbol{M} \boldsymbol{\theta}, \tag{6.100}$$

where:

$$\mathbb{R}^{n\times n} \ni \boldsymbol{M} := \sum_{j=1}^{c} (\boldsymbol{m}_j - \boldsymbol{m}_*)(\boldsymbol{m}_j - \boldsymbol{m}_*)^\top, \tag{6.101}$$

is the *between-scatter* in kernel FDA. Similar to Eq. (6.18), it is possible to use the following instead:

$$\mathbb{R}^{n\times n} \ni \boldsymbol{M} := \sum_{j=1}^{c} n_j\,(\boldsymbol{m}_j - \boldsymbol{m}_*)(\boldsymbol{m}_j - \boldsymbol{m}_*)^\top. \tag{6.102}$$

Therefore, Eq. (6.97) becomes:

$$d_B = \boldsymbol{\phi}(\boldsymbol{u})^\top \boldsymbol{\Phi}(\boldsymbol{S}_B)\,\boldsymbol{\phi}(\boldsymbol{u}) = \boldsymbol{\theta}^\top \boldsymbol{M}\boldsymbol{\theta}, \tag{6.103}$$

where $\boldsymbol{M}$ is Eq. (6.101) or (6.102). The Fisher direction is Eq. (6.85) and the solution is the generalized eigenvalue problem $(\boldsymbol{M}, \boldsymbol{N})$ (see Chap. 2).

6.8.2.3 Scatters in Multiclass Case: Variant 2

In the second version of the multiclass case for kernel FDA, the *within-scatter* is the same as in the two-class case, which is Eq. (6.83) and d_W is also Eq. (6.84).

For the between scatter in the second version, the process starts with Eqs. (6.35) and (6.36). The first step is to kernelize the objective function of Eq. (6.36):

$$d_T := \boldsymbol{\phi}(\boldsymbol{u})^\top \boldsymbol{\Phi}(\boldsymbol{S}_T)\,\boldsymbol{\phi}(\boldsymbol{u}), \tag{6.104}$$

where total scatter, Eq. (6.23), is pulled as:

$$\mathbb{R}^{t\times t} \ni \boldsymbol{\Phi}(\boldsymbol{S}_T) := \sum_{k=1}^{n} \big(\boldsymbol{\phi}(\boldsymbol{x}_k) - \boldsymbol{\phi}(\boldsymbol{\mu})\big)\big(\boldsymbol{\phi}(\boldsymbol{x}_k) - \boldsymbol{\phi}(\boldsymbol{\mu})\big)^\top. \tag{6.105}$$

According to Eqs. (6.72), (6.104), and (6.105), there is:

$$d_T = \sum_{k=1}^{n} \boldsymbol{\theta}^\top \boldsymbol{\Phi}(\boldsymbol{X})^\top \big(\boldsymbol{\phi}(\boldsymbol{x}_k) - \boldsymbol{\phi}(\boldsymbol{\mu})\big)\big(\boldsymbol{\phi}(\boldsymbol{x}_k) - \boldsymbol{\phi}(\boldsymbol{\mu})\big)^\top \boldsymbol{\Phi}(\boldsymbol{X})\,\boldsymbol{\theta}.$$

According to Eq. (6.98), there is:

$$\boldsymbol{\theta}^\top \boldsymbol{\Phi}(\boldsymbol{X})^\top \boldsymbol{\phi}(\boldsymbol{\mu}) = \boldsymbol{\theta}^\top \boldsymbol{m}_*, \tag{6.106}$$

where $\boldsymbol{m}_*$ is Eq. (6.99). On the other hand:

$$\boldsymbol{\theta}^\top \boldsymbol{\Phi}(\boldsymbol{X})^\top \boldsymbol{\phi}(\boldsymbol{x}_k) \overset{(6.72)}{=} \sum_{i=1}^{n} \theta_i\,\boldsymbol{\phi}(\boldsymbol{x}_i)^\top \boldsymbol{\phi}(\boldsymbol{x}_k) = \sum_{i=1}^{n} \theta_i\,k(\boldsymbol{x}_i, \boldsymbol{x}_k) = \boldsymbol{\theta}^\top \boldsymbol{g}_k, \tag{6.107}$$

where $\boldsymbol{g}_k \in \mathbb{R}^n$ whose i-th entry is:

$$\boldsymbol{g}_k(i) := k(\boldsymbol{x}_i, \boldsymbol{x}_k). \tag{6.108}$$

Therefore:

$$d_T = \sum_{k=1}^{n} \boldsymbol{\theta}^\top (\boldsymbol{g}_k - \boldsymbol{m}_*)(\boldsymbol{g}_k - \boldsymbol{m}_*)^\top \boldsymbol{\theta} = \boldsymbol{\theta}^\top \boldsymbol{G}\, \boldsymbol{\theta}, \tag{6.109}$$

where:

$$\mathbb{R}^{n\times n} \ni \boldsymbol{G} := \sum_{k=1}^{n} (\boldsymbol{g}_k - \boldsymbol{m}_*)(\boldsymbol{g}_k - \boldsymbol{m}_*)^\top. \tag{6.110}$$

The denominator of the Fisher criterion in the feature space is Eq. (6.84).

The optimization will be similar to Eq. (6.36), but in the feature space:

$$\begin{aligned} &\underset{\boldsymbol{\theta}}{\text{maximize}} \quad \boldsymbol{\theta}^\top \boldsymbol{G}\, \boldsymbol{\theta} \\ &\text{subject to} \quad \boldsymbol{\theta}^\top \boldsymbol{N}\, \boldsymbol{\theta} = 1, \end{aligned} \tag{6.111}$$

whose solution is similarly obtained as:

$$\boldsymbol{G}\, \boldsymbol{\theta} = \lambda\, \boldsymbol{N}\, \boldsymbol{\theta}, \tag{6.112}$$

which is a generalized eigenvalue problem $(\boldsymbol{G}, \boldsymbol{N})$ (see Chap. 2).

6.8.3 Multidimensional Subspace

The previous section discussed the one-dimensional kernel Fisher subspace. In the multidimensional kernel Fisher subspace, the within- and between-scatters are the same, but the Fisher criterion is different. According to Eq. (6.73), d_B and d_W are:

$$d_B = \mathbf{tr}\big(\boldsymbol{\phi}(\boldsymbol{U})^\top \boldsymbol{\Phi}(\boldsymbol{S}_B)\, \boldsymbol{\phi}(\boldsymbol{U})\big) = \mathbf{tr}(\boldsymbol{\Theta}^\top \boldsymbol{M}\boldsymbol{\Theta}), \tag{6.113}$$

$$d_W = \mathbf{tr}\big(\boldsymbol{\phi}(\boldsymbol{U})^\top \boldsymbol{\Phi}(\boldsymbol{S}_W)\, \boldsymbol{\phi}(\boldsymbol{U})\big) = \mathbf{tr}(\boldsymbol{\Theta}^\top \boldsymbol{N}\boldsymbol{\Theta}), \tag{6.114}$$

where $\mathbb{R}^{n\times p} \ni \boldsymbol{\Theta} = [\boldsymbol{\theta}_1, \dots, \boldsymbol{\theta}_p]$ and $\boldsymbol{M} \in \mathbb{R}^{n\times n}$ and $\boldsymbol{N} \in \mathbb{R}^{n\times n}$ are the between- and within-scatters, respectively, determined for either the two-class or multiclass case.

The Fisher criterion becomes:

$$f(\boldsymbol{\Theta}) := \frac{d_B(\boldsymbol{\Theta})}{d_W(\boldsymbol{\Theta})} = \frac{\mathbf{tr}\big(\boldsymbol{\phi}(\boldsymbol{U})^\top \boldsymbol{\Phi}(\boldsymbol{S}_B)\, \boldsymbol{\phi}(\boldsymbol{U})\big)}{\mathbf{tr}\big(\boldsymbol{\phi}(\boldsymbol{U})^\top \boldsymbol{\Phi}(\boldsymbol{S}_W)\, \boldsymbol{\phi}(\boldsymbol{U})\big)} = \frac{\mathbf{tr}(\boldsymbol{\Theta}^\top \boldsymbol{M}\boldsymbol{\Theta})}{\mathbf{tr}(\boldsymbol{\Theta}^\top \boldsymbol{N}\boldsymbol{\Theta})}, \tag{6.115}$$

where the columns of $\boldsymbol{\Theta}$ are the *kernel Fisher directions*. Similar to Eq. (6.40), the solution to maximize this criterion is:

$$\boldsymbol{M}\boldsymbol{\Theta} = \boldsymbol{N}\boldsymbol{\Theta}\boldsymbol{\Lambda}, \tag{6.116}$$

which is the generalized eigenvalue problem $(\boldsymbol{M}, \boldsymbol{N})$ (see Chap. 2). The columns of $\boldsymbol{\Theta}$ are the eigenvectors sorted from the largest to smallest eigenvalues and the diagonal entries of $\boldsymbol{\Lambda}$ are the corresponding eigenvalues.

It is possible to have another variant of the kernel FDA for the multidimensional subspace, where the optimization is (similar to Eq. (6.111)):

$$\begin{aligned} &\underset{\boldsymbol{\Theta}}{\text{maximize}} \quad \mathbf{tr}(\boldsymbol{\Theta}^\top \boldsymbol{G}\, \boldsymbol{\Theta}) \\ &\text{subject to} \quad \boldsymbol{\Theta}^\top \boldsymbol{N}\, \boldsymbol{\Theta} = \boldsymbol{I}, \end{aligned} \tag{6.117}$$

whose solution is similarly obtained as:

$$\boldsymbol{G}\boldsymbol{\Theta} = \boldsymbol{N}\boldsymbol{\Theta}\boldsymbol{\Lambda}, \tag{6.118}$$

which is a generalized eigenvalue problem $(\boldsymbol{G}, \boldsymbol{N})$ (see Chap. 2).

As mentioned previously in Sect. 6.8.2.1 of this chapter, there is no reconstruction in kernel FDA. The projection of the training data point $\boldsymbol{x}_i$ and the out-of-sample data point $\boldsymbol{x}_t$ are:

$$\mathbb{R}^p \ni \boldsymbol{\phi}(\widetilde{\boldsymbol{x}}_i) = \boldsymbol{\Phi}(\boldsymbol{U})^\top \boldsymbol{\phi}(\boldsymbol{x}_i) \overset{(6.73)}{=} \boldsymbol{\Theta}^\top \boldsymbol{\Phi}(\boldsymbol{X})^\top \boldsymbol{\phi}(\boldsymbol{x}_i) = \boldsymbol{\Theta}^\top \boldsymbol{k}(\boldsymbol{X}, \boldsymbol{x}_i), \tag{6.119}$$

$$\mathbb{R}^p \ni \boldsymbol{\phi}(\widetilde{\boldsymbol{x}}_t) = \boldsymbol{\Theta}^\top \boldsymbol{k}(\boldsymbol{X}, \boldsymbol{x}_t). \tag{6.120}$$

For the whole training and out-of-sample data, the projections are:

$$\mathbb{R}^{p \times n} \ni \boldsymbol{\Phi}(\widetilde{\boldsymbol{X}}) = \boldsymbol{\Theta}^\top \boldsymbol{K}(\boldsymbol{X}, \boldsymbol{X}), \tag{6.121}$$

$$\mathbb{R}^{p \times n_t} \ni \boldsymbol{\Phi}(\widetilde{\boldsymbol{X}}_t) = \boldsymbol{\Theta}^\top \boldsymbol{K}(\boldsymbol{X}, \boldsymbol{X}_t). \tag{6.122}$$

6.8.4 Discussion of the Dimensionality of the Kernel Fisher Subspace

According to Eq. (6.83), the rank of $\boldsymbol{N}$ is at most $\min(n, c)$ because the matrix is $n \times n$ and its calculation includes c iterations. Therefore, the rank of $\boldsymbol{N}^{-1}$ is also at most $\min(n, c)$. According to Eq. (6.102), the rank of $\boldsymbol{M}$ is at most $\min(n, c-1)$ because the matrix is $n \times n$, there are c iterations in its calculation, and -1 is because of subtracting the mean (refer to the explanation in Sect. 6.3.3). In Eq. (6.87), there is $\boldsymbol{N}^{-1}\boldsymbol{M}$ whose rank is:

$$\begin{aligned} \mathbf{rank}(\boldsymbol{N}^{-1}\boldsymbol{M}) &\leq \min\big(\mathbf{rank}(\boldsymbol{N}^{-1}), \mathbf{rank}(\boldsymbol{M})\big) \\ &\leq \min\big(\min(n, c), \min(n, c-1)\big) = \min(n, c, c-1) \overset{(a)}{=} c - 1, \end{aligned} \tag{6.123}$$

where (a) is because there is usually $c < n$. Therefore, the rank of $\boldsymbol{N}^{-1}\boldsymbol{M}$ is limited because of the rank of $\boldsymbol{M}$, which is at most $c - 1$.

According to Eq. (6.87), the $c - 1$ leading eigenvalues will be valid, and the rest are zero or very small. Therefore, p, which is the dimensionality of the kernel Fisher subspace, is at most $c - 1$. The $c - 1$ leading eigenvectors are considered as the kernel Fisher directions, and the rest of eigenvectors are invalid and ignored.

6.9 Fisherfaces

This section introduces one of the most fundamental applications of PCA and its variants—facial recognition.

6.9.1 Projection Directions of Facial Images

FDA and kernel FDA can be trained using images of diverse faces, to learn the most discriminative facial features, which separate human subjects based on their facial pictures. A facial dataset, i.e., the AT&T (ORL) face dataset [19], is used to illustrate this concept. Here, four different classes of the AT&T (or ORL) facial dataset were used for training FDA, kernel FDA, PCA, and kernel PCA, where the used kernels were linear, Radial Basis Function (RBF), and cosine kernels. Since there are four classes, the number of FDA directions is three (because $c - 1 = 3$). The three FDA directions and the top ten PCA directions for the used dataset are shown in Fig. 6.2. The projection directions of a facial dataset are some facial features that are similar to ghost faces. That is why the facial projection directions are also referred to as *ghost faces*, while the ghost faces in FDA and PCA are also referred to as *Fisherfaces* [2, 4, 27] and *eigenfaces* [20, 21], respectively. In Fig. 6.2, the projection directions have captured different facial features, such as eyes, nose, cheeks, chin, lips, hair, and glasses, which discriminate the data with respect to the maximum variance in PCA and maximum class separation and minimum within-class scatter in FDA. PCA and FDA extracted features that are different because PCA extracts features for maximum variance between faces while FDA finds features that are different among the classes for their separation. Figure 6.2 does not include projection directions for kernel FDA and kernel PCA because in kernel FDA, the projection directions are n-dimensional and not d-dimensional, and in

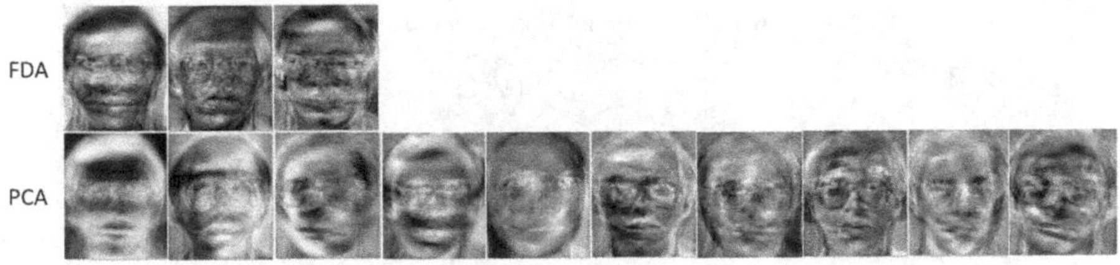

Fig. 6.2 The projection directions (ghost faces) of FDA and PCA for the first four classes of the facial AT&T dataset

kernel PCA, the projection directions are not available (see Chap. 5). Note that facial recognition using kernel FDA and kernel PCA are referred to as *kernel Fisherfaces* [14, 26] and *kernel eigenfaces* [25], respectively.

6.9.2 Projection of Facial Images

The projection of the images onto FDA and kernel FDA subspaces are demonstrated in Fig. 6.3. The projection of the images using PCA and kernel PCA are also depicted in Fig. 6.4. As can be seen, the FDA and kernel FDA subspaces have better separated the classes compared to the PCA and kernel PCA subspaces. This is because the FDA and kernel FDA make use of class labels to separate the classes in the subspace, while the PCA and kernel PCA only capture the variance (spread) of data regardless of class labels.

6.9.3 Reconstruction of Facial Images

Figure 6.5 illustrates the reconstruction of some training images. For reconstruction in this figure, FDA has used three projection directions (because $c - 1 = 3$), PCA once has used the top three PCA directions and has also used the whole d PCA directions. This figure demonstrates that the PCA reconstruction outperforms that of the FDA reconstruction. This makes sense because PCA is a linear method for reconstruction that has the least squared error (see Chap. 5). However, the primary responsibility of FDA is not reconstruction, but separation of the classes. Thus, the FDA directions try to separate the classes as much as possible and do not *necessarily* care for a good reconstruction. According to Fig. 6.1, even in some datasets, the FDA direction may be orthogonal to the PCA optimal direction for reconstruction. It is noteworthy that reconstruction cannot be performed in kernel FDA. However, it can be done in FDA for the out-of-sample data (for the sake of brevity, a simulation is not provided).

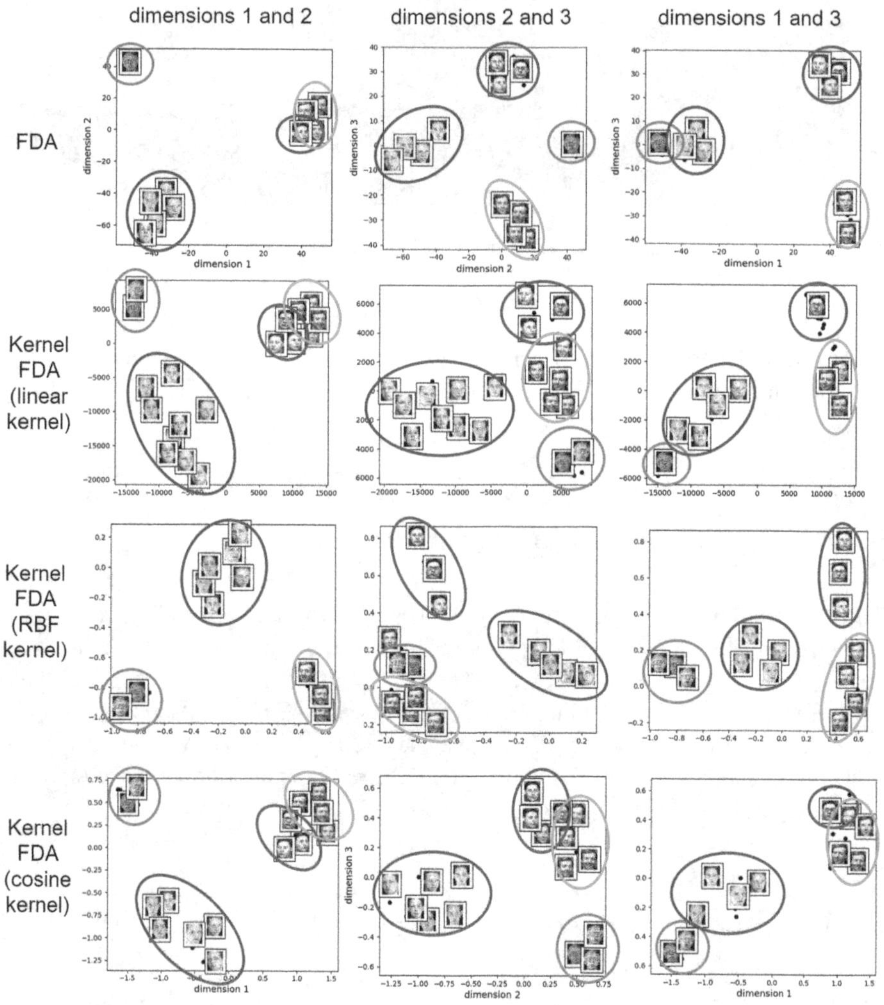

Fig. 6.3 The projection of the first four classes of the AT&T dataset onto the FDA and kernel FDA subspaces, where the kernels used were linear, RBF, and cosine kernels

6.9.4 Out-of-Sample Projection of Facial Images

The first six images of each of the first four subjects in the AT&T dataset were taken as training images, while the rest of the images were used as test (out-of-sample) images. The projection of the training and the out-of-sample images onto FDA and kernel FDA (using linear, RBF, and cosine kernels) are shown in Fig. 6.6. This figure demonstrates that the projection of the out-of-sample images has been properly carried out in FDA and kernel FDA. Therefore, FDA and kernel FDA are

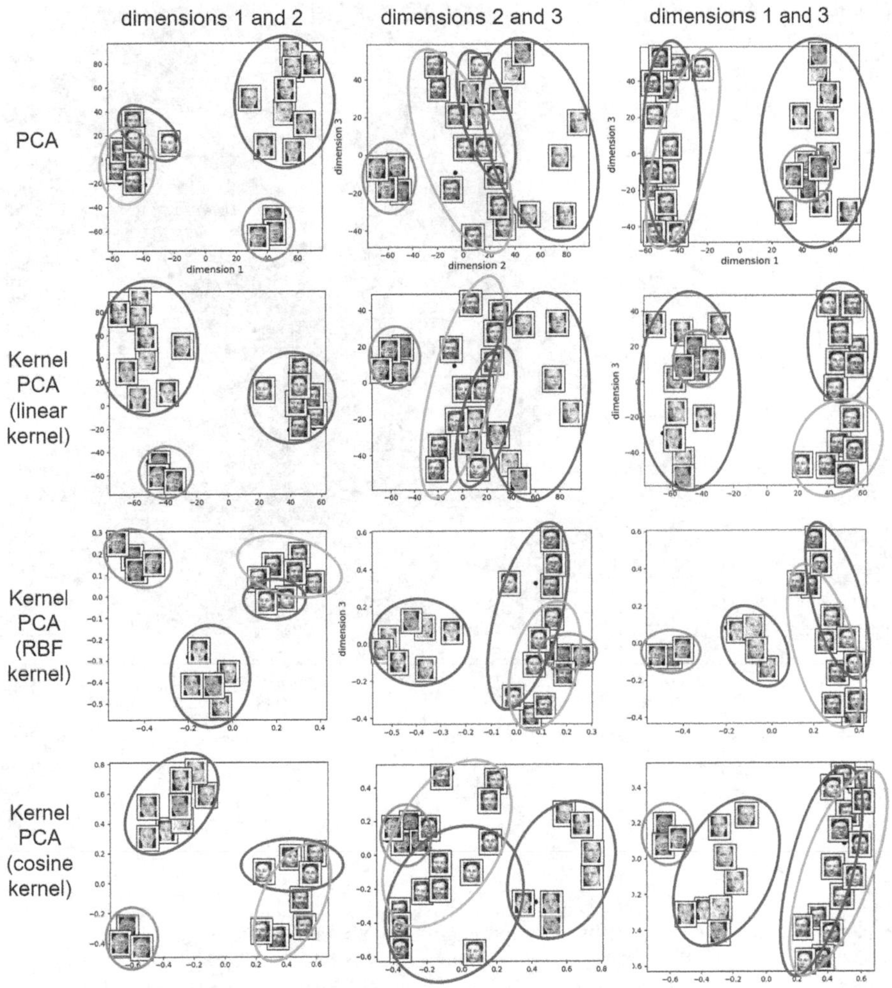

Fig. 6.4 The projection of the first four classes of the AT&T dataset onto the PCA and kernel PCA subspaces, where the kernels used were linear, RBF, and cosine kernels

able to generalize well to the out-of-sample data that are not introduced to the model during training.

6.10 Chapter Summary

FDA is the first proposed supervised spectral dimensionality reduction. This chapter introduced FDA and kernel FDA. The main approach of FDA, which is reducing the variance of every class and increasing the variance between classes for discrim-

Fig. 6.5 The reconstruction of four sample faces of AT&T datasets in FDA and PCA

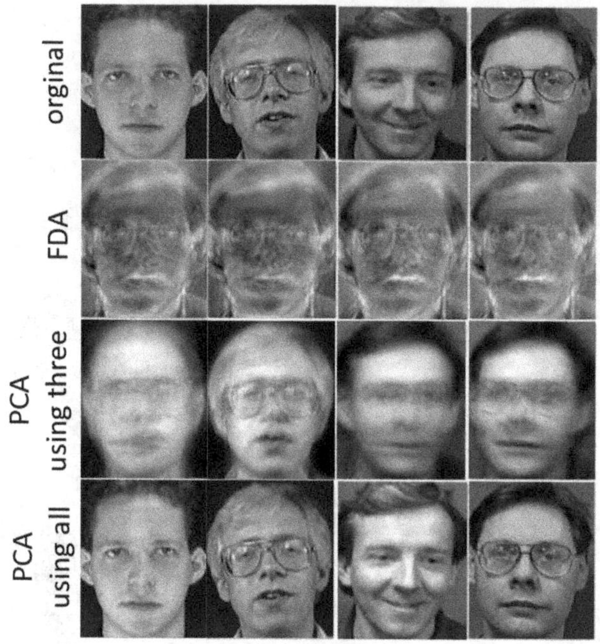

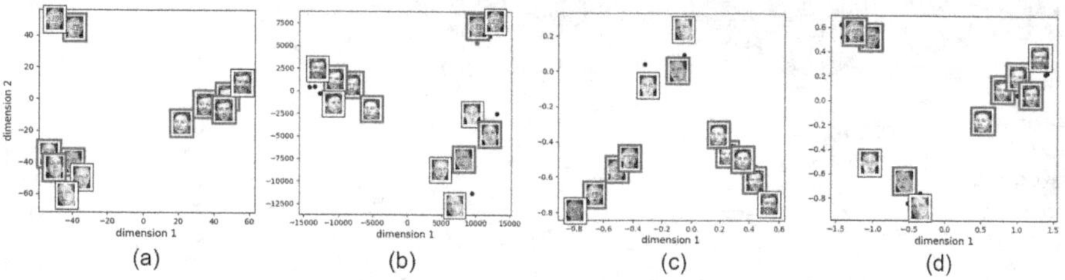

Fig. 6.6 The first two dimensions of the projection of both training and out-of-sample instances in the first four classes of the AT&T dataset onto subspaces of (**a**) FDA, (**b**) kernel FDA using a linear kernel, (**c**) kernel FDA using an RBF kernel, and (**d**) kernel FDA using a cosine kernel

ination of classes, is followed by almost all supervised dimensionality reduction methods. Another important spectral method is PCA, introduced in Chap. 5. The subspaces of FDA and PCA were compared with each other in this chapter to discuss the difference in their goals. PCA attempts to find the most variant direction of data, while FDA finds the direction for discrimination of classes. The chapter also showed that FDA and LDA subspaces are equivalent, justifying why LDA and FDA are used interchangeably in the literature. Discussions on the dimensionality of FDA and kernel FDA subspaces were provided to show that there are upper bounds on the dimensionality of subspace in FDA and kernel FDA, based on the ranks of matrices. The Fisherfaces method, introduced in this chapter, is one of the fundamental applications of FDA in computer vision. The Fisherfaces method was compared with the eigenfaces, introduced in Chap. 5.

References

1. Jonathan L Alperin. *Local representation theory: Modular representations as an introduction to the local representation theory of finite groups*. Vol. 11. Cambridge University Press, 1993.
2. Peter N Belhumeur, João P Hespanha, and David J Kriegman. “Eigenfaces vs. Fisherfaces: Recognition using class specific linear projection”. In: *IEEE Transactions on Pattern Analysis & Machine Intelligence* 7 (1997), pp. 711–720.
3. Weihong Deng et al. “Robust discriminant analysis of Gabor feature for face recognition”. In: *Fourth International Conference on Fuzzy Systems and Knowledge Discovery (FSKD 2007)*. Vol. 3. IEEE. 2007, pp. 248–252.
4. Kamran Etemad and Rama Chellappa. “Discriminant analysis for recognition of human face images”. In: *Journal of the Optical Society of America A* 14.8 (1997), pp. 1724–1733.
5. Ronald A Fisher. “The use of multiple measurements in taxonomic problems”. In: *Annals of eugenics* 7.2 (1936), pp. 179–188.
6. Ronald A Fisher. “XV.—The correlation between relatives on the supposition of Mendelian inheritance.” In: *Earth and Environmental Science Transactions of the Royal Society of Edinburgh* 52.2 (1919), pp. 399–433.
7. Ronald Aylmer Fisher. “Statistical methods for research workers”. In: *Breakthroughs in statistics*. Springer, 1992, pp. 66–70.
8. B Roy Frieden. *Science from Fisher information: a unification*. Cambridge University Press, 2004.
9. Jerome Friedman, Trevor Hastie, and Robert Tibshirani. *The elements of statistical learning*. Vol. 2. Springer series in statistics New York, NY, USA, 2009.
10. Benyamin Ghojogh and Mark Crowley. “Linear and Quadratic Discriminant Analysis: Tutorial”. In: *arXiv preprint arXiv:1906.02590* (2019).
11. Benyamin Ghojogh, Hoda Mohammadzade, and Mozhgan Mokari. “Fisherposes for human action recognition using Kinect sensor data”. In: *IEEE Sensors Journal* 18.4 (2017), pp. 1612–1627.
12. Amir Globerson and Sam T Roweis. “Metric learning by collapsing classes”. In: *Advances in neural information processing systems*. 2006, pp. 451–458.
13. Ming Guo and Zhelong Wang. “A feature extraction method for human action recognition using bodyworn inertial sensors”. In: *2015 IEEE 19th International Conference on Computer Supported Cooperative Work in Design (CSCWD)*. IEEE. 2015, pp. 576–581.
14. Qingshan Liu, Hanqing Lu, and Songde Ma. “Improving kernel Fisher discriminant analysis for face recognition”. In: *IEEE transactions on circuits and systems for video technology* 14.1 (2004), pp. 42–49.
15. Sebastian Mika et al. “Fisher discriminant analysis with kernels”. In: *Proceedings of the 1999 IEEE signal processing society workshop on Neural networks for signal processing IX*. IEEE. 1999, pp. 41–48.
16. Sebastian Mika et al. “Invariant feature extraction and classification in kernel spaces”. In: *Advances in neural information processing systems*. 2000, pp. 526–532.
17. Ali-Akbar Samadani, Ali Ghodsi, and Dana Kulić. “Discriminative functional analysis of human movements”. In: *Pattern Recognition Letters* 34.15 (2013), pp. 1829–1839.
18. Masashi Sugiyama. “Dimensionality reduction of multimodal labeled data by local fisher discriminant analysis”. In: *Journal of machine learning research* 8.May (2007), pp. 1027–1061.
19. The Digital Technology Group. *AT&T Laboratories Cambridge*. https://www.cl.cam.ac.uk/research/dtg/attarchive/facedatabase.html. Accessed: June 2017. 1994.
20. Matthew Turk and Alex Pentland. “Eigenfaces for recognition”. In: *Journal of cognitive neuroscience* 3.1 (1991), pp. 71–86.
21. Matthew A Turk and Alex P Pentland. “Face recognition using eigenfaces”. In: *Computer Vision and Pattern Recognition, 1991. Proceedings CVPR'91., IEEE Computer Society Conference on*. IEEE. 1991, pp. 586–591.

22. Yanxia Wang and Qiuqi Ruan. "Kernel fisher discriminant analysis for palmprint recognition". In: *18th International Conference on Pattern Recognition (ICPR'06)*. Vol. 4. IEEE. 2006, pp. 457–460.
23. Max Welling. *Fisher linear discriminant analysis*. Tech. rep. Department of Computer Science, University of Toronto, 2005.
24. Yong Xu and Guangming Lu. "Analysis on Fisher discriminant criterion and linear separability of feature space". In: *2006 International Conference on Computational Intelligence and Security*. Vol. 2. IEEE. 2006, pp. 1671–1676.
25. M-H Yang, Narendra Ahuja, and David Kriegman. "Face recognition using kernel eigenfaces". In: *Image processing, 2000. proceedings. 2000 international conference on*. Vol. 1. IEEE. 2000, pp. 37–40.
26. Ming-Hsuan Yang. "Kernel Eigenfaces vs. Kernel Fisherfaces: Face Recognition Using Kernel Methods". In: *Proceedings of the fifth IEEE international conference on automatic face and gesture recognition*. 2002, pp. 215–220.
27. Wenyi Zhao, Rama Chellappa, and P Jonathon Phillips. *Subspace linear discriminant analysis for face recognition*. Citeseer, 1999.

Chapter 7
Multidimensional Scaling, Sammon Mapping, and Isomap

7.1 Introduction

Multidimensional Scaling (MDS) was first proposed in [42] and is one of the earliest proposed dimensionality reduction methods [11]. It can be used for subspace learning, dimensionality reduction, and feature extraction. The idea of MDS is to preserve the similarity [41] or dissimilarity/distances [3] of points in the low-dimensional embedding space. Therefore, it fits the data locally to capture the global structure of the data [35]. MDS can be categorized into classical MDS, metric MDS, and nonmetric MDS [6, 11], which will be introduced in this chapter. The embedding results of these categories are different [24].

In later approaches, Sammon mapping [34], which is a special case of the distance-based metric MDS, was proposed. Sammon mapping can be considered the first proposed nonlinear dimensionality reduction method. The advantage of Sammon mapping is handling nonlinear data and its disadvantage is its iterative optimization solution, which makes this method slower than closed-form solutions such as classical MDS.

The classical MDS can be generalized to have a kernel classical MDS, in which any valid kernel can be used. Isomap [40] is a special case of the kernel classical MDS that uses a kernel constructed from geodesic distances between points. Because of the nonlinearity of the geodesic distance, Isomap is also a nonlinear manifold learning method.

MDS and its special cases, Sammon mapping and Isomap, have many applications [44]. For example, where MDS has been used for facial expression recognition [25, 33], kernel Isomap has been used for this application [45].

The goal of MDS is to embed the high-dimensional input data $\{\boldsymbol{x}_i\}_{i=1}^n$ into the lower dimensional embedded data $\{\boldsymbol{y}_i\}_{i=1}^n$, where n is the number of data points. The dimensionality of the input and embedding spaces is denoted by d and $p \leq d$, respectively, i.e., $\boldsymbol{x}_i \in \mathbb{R}^d$ and $\boldsymbol{y}_i \in \mathbb{R}^p$. Additionally, $\mathbb{R}^{d\times n} \ni \boldsymbol{X} := [\boldsymbol{x}_1, \dots, \boldsymbol{x}_n]$ and $\mathbb{R}^{p\times n} \ni \boldsymbol{Y} := [\boldsymbol{y}_1, \dots, \boldsymbol{y}_n]$.

B. Ghojogh et al., *Elements of Dimensionality Reduction and Manifold Learning*,
https://doi.org/10.1007/978-3-031-10602-6_7

7.2 Multidimensional Scaling

The following section introduces the three categories of MDS—the classical MDS, the metric MDS, and the nonmetric MDS.

7.2.1 *Classical Multidimensional Scaling*

7.2.1.1 Classical MDS with Euclidean Distance

The *classical MDS* is also referred to as *Principal Coordinates Analysis (PCoA)*, *Torgerson Scaling*, and *Torgerson–Gower scaling* [20]. The goal of classical MDS is to preserve the similarity of the data points in the embedding space as it was in the input space [41]. One way to measure similarity is to calculate the inner product. Therefore, it is possible to minimize the difference between similarities in the input and embedding spaces by:

$$\underset{\{\boldsymbol{y}_i\}_{i=1}^n}{\text{minimize}} \quad c_1 := \sum_{i=1}^{n}\sum_{j=1}^{n}(\boldsymbol{x}_i^\top \boldsymbol{x}_j - \boldsymbol{y}_i^\top \boldsymbol{y}_j)^2, \tag{7.1}$$

whose matrix form is:

$$\underset{\boldsymbol{Y}}{\text{minimize}} \quad c_1 = ||\boldsymbol{X}^\top \boldsymbol{X} - \boldsymbol{Y}^\top \boldsymbol{Y}||_F^2, \tag{7.2}$$

and where $\|\cdot\|_F$ denotes the Frobenius norm (see Chap. 2), and $\boldsymbol{X}^\top \boldsymbol{X}$ and $\boldsymbol{Y}^\top \boldsymbol{Y}$ are the Gram matrices of the original data $\boldsymbol{X}$ and the embedded data $\boldsymbol{Y}$, respectively.

The objective function, in Eq. (7.2), can be simplified to:

$$\begin{aligned}||\boldsymbol{X}^\top \boldsymbol{X} - \boldsymbol{Y}^\top \boldsymbol{Y}||_F^2 &= \mathbf{tr}\big[(\boldsymbol{X}^\top \boldsymbol{X} - \boldsymbol{Y}^\top \boldsymbol{Y})^\top (\boldsymbol{X}^\top \boldsymbol{X} - \boldsymbol{Y}^\top \boldsymbol{Y})\big] \\ &= \mathbf{tr}\big[(\boldsymbol{X}^\top \boldsymbol{X} - \boldsymbol{Y}^\top \boldsymbol{Y})(\boldsymbol{X}^\top \boldsymbol{X} - \boldsymbol{Y}^\top \boldsymbol{Y})\big] = \mathbf{tr}\big[(\boldsymbol{X}^\top \boldsymbol{X} - \boldsymbol{Y}^\top \boldsymbol{Y})^2\big],\end{aligned}$$

where $\mathbf{tr}(.)$ denotes the trace of the matrix. If $\boldsymbol{X}^\top \boldsymbol{X}$ and $\boldsymbol{Y}^\top \boldsymbol{Y}$ are decomposed using eigenvalue decomposition (see Chap. 2), the result is:

$$\boldsymbol{X}^\top \boldsymbol{X} = \boldsymbol{V}\boldsymbol{\Delta}\boldsymbol{V}^\top, \tag{7.3}$$

$$\boldsymbol{Y}^\top \boldsymbol{Y} = \boldsymbol{Q}\boldsymbol{\Psi}\boldsymbol{Q}^\top, \tag{7.4}$$

where the eigenvectors are sorted from leading (largest eigenvalue) to trailing (smallest eigenvalue). Note that, rather than eigenvalue decomposition of $\boldsymbol{X}^\top \boldsymbol{X}$ and $\boldsymbol{Y}^\top \boldsymbol{Y}$, it is possible to decompose $\boldsymbol{X}$ and $\boldsymbol{Y}$ using a Singular Value Decomposition (SVD) and by taking the right singular vectors of $\boldsymbol{X}$ and $\boldsymbol{Y}$ as $\boldsymbol{V}$ and $\boldsymbol{Q}$, respectively.

As shown in Chap. 2, the matrices $\boldsymbol{\Delta}$ and $\boldsymbol{\Psi}$ are obtained by squaring the singular values (to power 2).

The objective function can be further simplified as:

$$
\begin{aligned}
||\boldsymbol{X}^\top \boldsymbol{X} - \boldsymbol{Y}^\top \boldsymbol{Y}||_F^2 &= \mathbf{tr}\big[(\boldsymbol{X}^\top \boldsymbol{X} - \boldsymbol{Y}^\top \boldsymbol{Y})^2\big] = \mathbf{tr}\big[(\boldsymbol{V}\boldsymbol{\Delta}\boldsymbol{V}^\top - \boldsymbol{Q}\boldsymbol{\Psi}\boldsymbol{Q}^\top)^2\big] \\
&\overset{(a)}{=} \mathbf{tr}\big[(\boldsymbol{V}\boldsymbol{\Delta}\boldsymbol{V}^\top - \boldsymbol{V}\boldsymbol{V}^\top\boldsymbol{Q}\boldsymbol{\Psi}\boldsymbol{Q}^\top\boldsymbol{V}\boldsymbol{V}^\top)^2\big] = \mathbf{tr}\Big[\big(\boldsymbol{V}(\boldsymbol{\Delta} - \boldsymbol{V}^\top\boldsymbol{Q}\boldsymbol{\Psi}\boldsymbol{Q}^\top\boldsymbol{V})\boldsymbol{V}^\top\big)^2\Big] \\
&= \mathbf{tr}\Big[\boldsymbol{V}^2(\boldsymbol{\Delta} - \boldsymbol{V}^\top\boldsymbol{Q}\boldsymbol{\Psi}\boldsymbol{Q}^\top\boldsymbol{V})^2(\boldsymbol{V}^\top)^2\Big] \overset{(b)}{=} \mathbf{tr}\Big[(\boldsymbol{V}^\top)^2\boldsymbol{V}^2(\boldsymbol{\Delta} - \boldsymbol{V}^\top\boldsymbol{Q}\boldsymbol{\Psi}\boldsymbol{Q}^\top\boldsymbol{V})^2\Big] \\
&= \mathbf{tr}\Big[\underbrace{(\boldsymbol{V}^\top\boldsymbol{V})^2}_{\boldsymbol{I}}(\boldsymbol{\Delta} - \boldsymbol{V}^\top\boldsymbol{Q}\boldsymbol{\Psi}\boldsymbol{Q}^\top\boldsymbol{V})^2\Big] \overset{(c)}{=} \mathbf{tr}\Big[(\boldsymbol{\Delta} - \boldsymbol{V}^\top\boldsymbol{Q}\boldsymbol{\Psi}\boldsymbol{Q}^\top\boldsymbol{V})^2\Big],
\end{aligned}
$$

where (a) and (c) are for $\boldsymbol{V}^\top \boldsymbol{V} = \boldsymbol{V}\boldsymbol{V}^\top = \boldsymbol{I}$ because $\boldsymbol{V}$ is a nontruncated (square) orthogonal matrix (where $\boldsymbol{I}$ denotes the identity matrix). The reason for (b) is the cyclic property of the trace operator.

If $\mathbb{R}^{n\times n} \ni \boldsymbol{M} := \boldsymbol{V}^\top \boldsymbol{Q}$, then:

$$
||\boldsymbol{X}^\top \boldsymbol{X} - \boldsymbol{Y}^\top \boldsymbol{Y}||_F^2 = \mathbf{tr}\Big[(\boldsymbol{\Delta} - \boldsymbol{M}\boldsymbol{\Psi}\boldsymbol{M}^\top)^2\Big].
$$

Therefore:

$$
\underset{\boldsymbol{Y}}{\text{minimize}}\ ||\boldsymbol{X}^\top \boldsymbol{X} - \boldsymbol{Y}^\top \boldsymbol{Y}||_F^2 \equiv \underset{\boldsymbol{M},\boldsymbol{\Psi}}{\text{minimize}}\ \mathbf{tr}\Big[(\boldsymbol{\Delta} - \boldsymbol{M}\boldsymbol{\Psi}\boldsymbol{M}^\top)^2\Big].
$$

This means that the objective function is:

$$
\begin{aligned}
c_1 &= \mathbf{tr}\Big[(\boldsymbol{\Delta} - \boldsymbol{M}\boldsymbol{\Psi}\boldsymbol{M}^\top)^2\Big] = \mathbf{tr}(\boldsymbol{\Delta}^2 + (\boldsymbol{M}\boldsymbol{\Psi}\boldsymbol{M}^\top)^2 - 2\boldsymbol{\Delta}\boldsymbol{M}\boldsymbol{\Psi}\boldsymbol{M}^\top) \\
&= \mathbf{tr}(\boldsymbol{\Delta}^2) + \mathbf{tr}((\boldsymbol{M}\boldsymbol{\Psi}\boldsymbol{M}^\top)^2) - 2\,\mathbf{tr}(\boldsymbol{\Delta}\boldsymbol{M}\boldsymbol{\Psi}\boldsymbol{M}^\top).
\end{aligned}
$$

As the optimization problem is unconstrained and the objective function is the trace of a quadratic function, the minimum is nonnegative.

If the derivative with respect to the first objective variable, i.e., $\boldsymbol{M}$, is taken, then the result is:

$$
\begin{aligned}
&\mathbb{R}^{n\times n} \ni \frac{\partial c_1}{\partial \boldsymbol{M}} = 2(\boldsymbol{M}\boldsymbol{\Psi}\boldsymbol{M}^\top)\boldsymbol{M}\boldsymbol{\Psi} - 2\boldsymbol{\Delta}\boldsymbol{M}\boldsymbol{\Psi} \overset{\text{set}}{=} \boldsymbol{0} \implies (\boldsymbol{M}\boldsymbol{\Psi}\boldsymbol{M}^\top)(\boldsymbol{M}\boldsymbol{\Psi}) = (\boldsymbol{\Delta})(\boldsymbol{M}\boldsymbol{\Psi}) \\
&\overset{(a)}{\implies} \boldsymbol{M}\boldsymbol{\Psi}\boldsymbol{M}^\top = \boldsymbol{\Delta}, \qquad (7.5)
\end{aligned}
$$

where (a) is because $\boldsymbol{M}\boldsymbol{\Psi} \neq \boldsymbol{0}$.

For the derivative with respect to the second objective variable, i.e., $\boldsymbol{\Psi}$, the objective function can be simplified:

$$\begin{aligned} c_1 &= \mathbf{tr}(\boldsymbol{\Delta}^2) + \mathbf{tr}((\boldsymbol{M}\boldsymbol{\Psi}\boldsymbol{M}^\top)^2) - 2\,\mathbf{tr}(\boldsymbol{\Delta}\boldsymbol{M}\boldsymbol{\Psi}\boldsymbol{M}^\top) \\ &= \mathbf{tr}(\boldsymbol{\Delta}^2) + \mathbf{tr}(\boldsymbol{M}^2\boldsymbol{\Psi}^2\boldsymbol{M}^{\top 2}) - 2\,\mathbf{tr}(\boldsymbol{\Delta}\boldsymbol{M}\boldsymbol{\Psi}\boldsymbol{M}^\top) \\ &\overset{(a)}{=} \mathbf{tr}(\boldsymbol{\Delta}^2) + \mathbf{tr}(\boldsymbol{M}^{\top 2}\boldsymbol{M}^2\boldsymbol{\Psi}^2) - 2\,\mathbf{tr}(\boldsymbol{M}^\top\boldsymbol{\Delta}\boldsymbol{M}\boldsymbol{\Psi}) \\ &= \mathbf{tr}(\boldsymbol{\Delta}^2) + \mathbf{tr}((\boldsymbol{M}^\top\boldsymbol{M}\boldsymbol{\Psi})^2) - 2\,\mathbf{tr}(\boldsymbol{M}^\top\boldsymbol{\Delta}\boldsymbol{M}\boldsymbol{\Psi}), \end{aligned}$$

where (a) is because of the cyclic property of the trace.

Taking the derivative with respect to the second objective variable, i.e., $\boldsymbol{\Psi}$, results in:

$$\mathbb{R}^{n\times n} \ni \frac{\partial c_1}{\partial \boldsymbol{\Psi}} = 2\boldsymbol{M}^\top(\boldsymbol{M}\boldsymbol{\Psi}\boldsymbol{M}^\top)\boldsymbol{M} - 2\boldsymbol{M}^\top\boldsymbol{\Delta}\boldsymbol{M} \overset{\text{set}}{=} \boldsymbol{0} \implies \boldsymbol{M}^\top(\boldsymbol{M}\boldsymbol{\Psi}\boldsymbol{M}^\top)\boldsymbol{M} = \boldsymbol{M}^\top(\boldsymbol{\Delta})\boldsymbol{M}$$

$$\overset{(a)}{\implies} \boldsymbol{M}\boldsymbol{\Psi}\boldsymbol{M}^\top = \boldsymbol{\Delta}, \tag{7.6}$$

where (a) is because $\boldsymbol{M} \neq \boldsymbol{0}$. Both Eqs. (7.5) and (7.6) are:

$$\boldsymbol{M}\boldsymbol{\Psi}\boldsymbol{M}^\top = \boldsymbol{\Delta},$$

whose one possible solution is:

$$\boldsymbol{M} = \boldsymbol{I}, \tag{7.7}$$

$$\boldsymbol{\Psi} = \boldsymbol{\Delta}. \tag{7.8}$$

which means that the minimum value of the nonnegative objective function $\mathbf{tr}((\boldsymbol{\Delta} - \boldsymbol{M}\boldsymbol{\Psi}\boldsymbol{M}^\top)^2)$ is zero.

If there is $\boldsymbol{M} = \boldsymbol{V}^\top\boldsymbol{Q}$, then according to Eq. (7.7), there is:

$$\therefore \quad \boldsymbol{V}^\top\boldsymbol{Q} = \boldsymbol{I} \implies \boldsymbol{Q} = \boldsymbol{V}. \tag{7.9}$$

According to Eq. (7.4), there is:

$$\boldsymbol{Y}^\top\boldsymbol{Y} = \boldsymbol{Q}\boldsymbol{\Psi}\boldsymbol{Q}^\top \overset{(a)}{=} \boldsymbol{Q}\boldsymbol{\Psi}^{\frac{1}{2}}\boldsymbol{\Psi}^{\frac{1}{2}}\boldsymbol{Q}^\top \implies \boldsymbol{Y} = \boldsymbol{\Psi}^{\frac{1}{2}}\boldsymbol{Q}^\top \overset{(7.8),(7.9)}{\implies} \boldsymbol{Y} = \boldsymbol{\Delta}^{\frac{1}{2}}\boldsymbol{V}^\top, \tag{7.10}$$

where (a) is possible because $\boldsymbol{\Psi}$ does not include a negative entry because the Gram matrix $\boldsymbol{Y}^\top\boldsymbol{Y}$ is considered to be positive semidefinite by definition.

In summary, for embedding $\boldsymbol{X}$ using classical MDS, the eigenvalue decomposition of $\boldsymbol{X}^\top\boldsymbol{X}$ is obtained by using Eq. (7.3). Then, using Eq. (7.10), $\boldsymbol{Y} \in \mathbb{R}^{n\times n}$ is obtained. Truncating $\boldsymbol{Y}$ to $\boldsymbol{Y} \in \mathbb{R}^{p\times n}$, with the first (top) p rows, results in a p-dimensional embedding of the n points. Note that the leading p columns are used because singular values are sorted from largest to smallest in SVD, which can be used for Eq. (7.3).

7.2.1.2 Generalized Classical MDS (Kernel Classical MDS)

As shown in Chap. 3:

$$\boldsymbol{HGH} = \boldsymbol{HX}^\top \boldsymbol{XH} = -\frac{1}{2}\boldsymbol{HDH}, \tag{7.11}$$

where $\boldsymbol{H}$ is the centering matrix (see Chap. 2), $\mathbb{R}^{n\times n} \ni \boldsymbol{G} := \boldsymbol{X}^\top \boldsymbol{X}$ is the Gram matrix (see Chap. 3), and $\boldsymbol{D}$ is the distance matrix with a squared Euclidean distance as its elements. If data $\boldsymbol{X}$ are already centered, i.e., the mean has been removed ($\boldsymbol{X} \leftarrow \boldsymbol{XH}$), Eq. (7.11) becomes:

$$\boldsymbol{X}^\top \boldsymbol{X} = -\frac{1}{2}\boldsymbol{HDH}. \tag{7.12}$$

Corollary 7.1 *If Eq. (7.3) is used as the Gram matrix, the classical MDS uses the Euclidean distance as its metric. Due to using the Euclidean distance, the classical MDS using a Gram matrix is a* ***linear*** *subspace learning method.*

Proof In classical MDS, Eq. (7.3) is the eigenvalue decomposition of the Gram matrix $\boldsymbol{X}^\top \boldsymbol{X}$. According to Eq. (7.12), this Gram matrix can be restated to an expression based on the squared Euclidean distance. Therefore, the classical MDS with Eq. (7.3) uses Euclidean distance and is consequently linear. □

In Eq. (7.11) or (7.12), a general kernel matrix [22] can be written, rather than the double-centered Gram matrix, to have [11]:

$$\mathbb{R}^{n\times n} \ni \boldsymbol{K} = -\frac{1}{2}\boldsymbol{HDH}. \tag{7.13}$$

Note that the classical MDS with Eq. (7.3) uses a linear kernel $\boldsymbol{X}^\top \boldsymbol{X}$ for its kernel. This is another reason why classical MDS with Eq. (7.3) is a linear method. It is noteworthy that Eq. (7.13) can be used for unifying the spectral dimensionality reduction methods as special cases of kernel principal component analysis with different kernels. This will be explained in Chap. 10. See [4, 21] and [38, Table 2.1] for more details.

Comparing Eqs. (7.11), (7.12), and (7.13) with Eq. (7.3) shows that it is possible to use a general kernel matrix, such as a Radial Basis Function (RBF) kernel, in classical MDS to create a *generalized classical MDS*. In summary, for embedding $\boldsymbol{X}$ using classical MDS, the eigenvalue decomposition of the kernel matrix $\boldsymbol{K}$ is obtained in a similar way as Eq. (7.3):

$$\boldsymbol{K} = \boldsymbol{V\Delta V}^\top. \tag{7.14}$$

Then, by using Eq. (7.10), $\boldsymbol{Y} \in \mathbb{R}^{n\times n}$ is obtained. It is noteworthy that in this case, $\boldsymbol{X}^\top \boldsymbol{X}$ is being replaced with the kernel $\boldsymbol{K} = \boldsymbol{\Phi}(\boldsymbol{X})^\top \boldsymbol{\Phi}(\boldsymbol{X})$ and then, according to Eqs. (7.10) and (7.14), this results in:

$$\boldsymbol{K} = \boldsymbol{Y}^\top \boldsymbol{Y}. \tag{7.15}$$

Truncating $\boldsymbol{Y}$, obtained from Eq. (7.10), to have $\boldsymbol{Y} \in \mathbb{R}^{p \times n}$, with the first (top) p rows, results in the p-dimensional embedding of the n points. Interestingly, using the kernel in the generalized classical MDS can be called a *kernel classical MDS*.

7.2.1.3 Equivalence of PCA and Kernel PCA with Classical MDS and Generalized Classical MDS, Respectively

Proposition 7.1 *Classical MDS with the Euclidean distance is equivalent to Principal Component Analysis (PCA). Moreover, the generalized classical MDS is equivalent to kernel PCA.*

Proof On the one hand, Eq. (7.3) can be obtained by taking the SVD of $\boldsymbol{X}$, and the data projected onto a classical MDS subspace are obtained by Eq. (7.10), which is $\boldsymbol{\Delta}\boldsymbol{V}^\top$. On the other hand, according to Chap. 5, the data projected onto a PCA subspace are $\boldsymbol{\Delta}\boldsymbol{V}^\top$ where $\boldsymbol{\Delta}$ and $\boldsymbol{V}^\top$ are from the SVD of $\boldsymbol{X}$. Comparing these terms demonstrates that the classical MDS is equivalent to the PCA.

Moreover, Eq. (7.14) is the eigenvalue decomposition of the kernel matrix. The data projected onto the generalized classical MDS subspace are obtained through Eq. (7.10) which is $\boldsymbol{\Delta}\boldsymbol{V}^\top$. According to Chap. 5, the data projected onto the kernel PCA subspace are $\boldsymbol{\Delta}\boldsymbol{V}^\top$, where $\boldsymbol{\Delta}$ and $\boldsymbol{V}^\top$ are from the eigenvalue decomposition of the kernel matrix. Comparing these demonstrates that the generalized classical MDS is equivalent to the kernel PCA. □

7.2.2 *Metric Multidimensional Scaling*

Where classical MDS tries to preserve the similarities of points in the embedding space, later approaches, such as the *Metric MDS*, changed the cost function to preserve the distances, rather than the similarities [7, 28] in the embedding space [3]. *Metric MDS* minimizes the difference in distances of points in the input and embedding spaces [16]. The cost function in metric MDS is usually referred to as the *stress function* [12, 30]. This method is named metric MDS because it uses the distance metric in its optimization. The optimization in metric MDS is:

$$\underset{\{y_i\}_{i=1}^n}{\text{minimize}} \quad c_2 := \left(\frac{\sum_{i=1}^n \sum_{j=1, j<i}^n \big(d_x(\boldsymbol{x}_i, \boldsymbol{x}_j) - d_y(\boldsymbol{y}_i, \boldsymbol{y}_j)\big)^2}{\sum_{i=1}^n \sum_{j=1, j<i}^n d_x^2(\boldsymbol{x}_i, \boldsymbol{x}_j)} \right)^{\frac{1}{2}}, \tag{7.16}$$

or, if not using normalization factors:

$$\underset{\{y_i\}_{i=1}^n}{\text{minimize}} \quad c_2 := \Big(\sum_{i=1}^{n} \sum_{j=1, j<i}^{n} \big(d_x(\boldsymbol{x}_i, \boldsymbol{x}_j) - d_y(\boldsymbol{y}_i, \boldsymbol{y}_j)\big)^2\Big)^{\frac{1}{2}}, \tag{7.17}$$

where $d_x(.,.)$ and $d_y(.,.)$ denotes the distance metrics in the input and the embedded spaces, respectively.

Equations (7.16) and (7.17) use indices $j < i$, rather than $j \neq i$, because the distance metric is symmetric. In other words, it is not necessary to consider the distance of the j-th point from the i-th point when the distance of the i-th point from the j-point has already been considered. Note that in Eq. (7.16) and (7.17), d_y usually refers to the Euclidean distance, $d_y = \|\boldsymbol{y}_i - \boldsymbol{y}_j\|_2$, while d_x can be any valid metric distance, such as the Euclidean distance.

The optimization problem (7.16) can be solved using either the gradient descent or Newton's method (see Chap. 4). Note that the classical MDS is a linear method and has a closed-form solution; however, the metric and nonmetric MDS methods are *nonlinear*, but do *not have closed-form solutions*, and should be solved iteratively. In mathematics, whenever something is gained, something else is lost - give and take. Likewise, here, the method has become nonlinear, but in return its closed-form solution has become iterative.

Inspired by [34], the diagonal quasi-Newton's method can be used to solve this optimization problem. Considering the vectors componentwise, the diagonal quasi-Newton's method updates the solution to:

$$y_{i,k}^{(\nu+1)} := y_{i,k}^{(\nu)} - \eta \left|\frac{\partial^2 c_2}{\partial y_{i,k}^2}\right|^{-1} \frac{\partial c_2}{\partial y_{i,k}}, \tag{7.18}$$

where η is the learning rate, $y_{i,k}$ is the k-th element of the i-th embedded point $\mathbb{R}^p \ni \boldsymbol{y}_i = [y_{i,1}, \dots, y_{i,p}]^\top$, and $|\cdot|$ is the absolute value guaranteeing a move toward the minimum and not maximum in Newton's method [28]. If using the gradient descent for solving the optimization, the solution is updated as follows:

$$y_{i,k}^{(\nu+1)} := y_{i,k}^{(\nu)} - \eta \frac{\partial c_2}{\partial y_{i,k}}. \tag{7.19}$$

7.2.3 Nonmetric Multidimensional Scaling

In *nonmetric MDS*, rather than using a distance metric, $d_y(\boldsymbol{x}_i, \boldsymbol{x}_j)$, for the distances between points in the embedding space, $f(d_y(\boldsymbol{x}_i, \boldsymbol{x}_j))$ is used, where $f(.)$ is a nonparametric monotonic function. In other words, only the order of dissimilarities is important rather than the amount of dissimilarities [1, 24]:

$$d_y(\boldsymbol{y}_i, \boldsymbol{y}_j) \leq d_y(\boldsymbol{y}_k, \boldsymbol{y}_\ell) \iff f(d_y(\boldsymbol{y}_i, \boldsymbol{y}_j)) \leq f(d_y(\boldsymbol{x}_k, \boldsymbol{y}_\ell)). \tag{7.20}$$

The optimization in nonmetric MDS is [1]:

$$\underset{\{\boldsymbol{y}_i\}_{i=1}^n}{\text{minimize}} \quad c_3 := \left(\frac{\sum_{i=1}^n \sum_{j=1, j<i}^n \big(d_x(\boldsymbol{x}_i, \boldsymbol{x}_j) - f(d_y(\boldsymbol{y}_i, \boldsymbol{y}_j))\big)^2}{\sum_{i=1}^n \sum_{j=1, j<i}^n d_x^2(\boldsymbol{x}_i, \boldsymbol{x}_j)} \right)^{\frac{1}{2}}. \tag{7.21}$$

An example of a nonmetric MDS is the Smallest Space Analysis [36]. Another example is Kruskal's nonmetric MDS or Shepard-Kruskal Scaling (SKS) [26, 27]. In Kruskal's nonmetric MDS, the function $f(.)$ is the regression, where $f(d_y(\boldsymbol{y}_i, \boldsymbol{y}_j))$ is predicted from the regression, which preserves the order of dissimilarities [1, 23]. Equation (7.21) with $f(.)$ as the regression function, which is used in Kruskal's nonmetric MDS, is called the *Stress-1 formula* [1, 23, 24].

7.3 Sammon Mapping

Sammon mapping [34] is a special case of metric MDS, which is a *nonlinear* method. Looking at the history of algorithms, it is probably correct to call this method the first proposed nonlinear method of dimensionality reduction. This method has different names in the literature, such as *Sammon's nonlinear mapping*, *Sammon mapping*, and *Non-Linear Mapping (NLM)* [28]. Sammon originally named it NLM [34], although now it is most well-known as Sammon mapping.

The optimization problem in Sammon mapping is almost a weighted version of Eq. (7.16), formulated as:

$$\underset{\{\boldsymbol{y}_i\}_{i=1}^n}{\text{minimize}} \quad \frac{1}{a} \sum_{i=1}^{n} \sum_{j=1, j<i}^{n} w_{ij} \big(d_x(\boldsymbol{x}_i, \boldsymbol{x}_j) - d_y(\boldsymbol{y}_i, \boldsymbol{y}_j)\big)^2, \tag{7.22}$$

where w_{ij} is the weight and a is the normalizing factor. The $d_x(., .)$ can be any metric, but usually is considered to be the Euclidean distance for simplicity [28]. The $d_y(., .)$, however, is the Euclidean distance metric.

In Sammon mapping, the weights and the normalizing factor in Eq. (7.22) are:

$$w_{ij} = \frac{1}{d_x(\boldsymbol{x}_i, \boldsymbol{x}_j)}, \tag{7.23}$$

$$a = \sum_{i=1}^{n} \sum_{j=1, j<i}^{n} d_x(\boldsymbol{x}_i, \boldsymbol{x}_j). \tag{7.24}$$

The weight w_{ij} in Sammon mapping places more emphasis on the small distances (neighbour points), focusing on preserving the "local" structure of the manifold; therefore, it fits the manifold locally [35].

Substituting Eqs. (7.23) and (7.24) in Eq. (7.22) results in:

$$\underset{\boldsymbol{Y}}{\text{minimize}} \quad c_4 := \frac{1}{\sum_{i=1}^{n}\sum_{j=1,j<i}^{n} d_x(\boldsymbol{x}_i,\boldsymbol{x}_j)} \times \sum_{i=1}^{n}\sum_{j=1,j<i}^{n} \frac{\big(d_x(\boldsymbol{x}_i,\boldsymbol{x}_j) - d_y(\boldsymbol{y}_i,\boldsymbol{y}_j)\big)^2}{d_x(\boldsymbol{x}_i,\boldsymbol{x}_j)}. \tag{7.25}$$

Sammon used the diagonal quasi-Newton's method to solve this optimization problem [34], utilizing (7.18). The learning rate η (see Chap. 4) in the Sammon mapping is called the *magic factor* in [34]. To solve the optimization problem, both the gradient and second derivative are needed.

Note that, in practice, the classical MDS or PCA is used for the initialization of points in Sammon mapping optimization.

Proposition 7.2 *The gradient of the cost function c, with respect to $y_{i,k}$, is [28, 34]:*

$$\frac{\partial c_4}{\partial y_{i,k}} = \frac{-2}{a}\sum_{i=1}^{n}\sum_{j=1,j<i}^{n} \frac{d_x(\boldsymbol{x}_i,\boldsymbol{x}_j) - d_y(\boldsymbol{y}_i,\boldsymbol{y}_j)}{d_x(\boldsymbol{x}_i,\boldsymbol{x}_j)\, d_y(\boldsymbol{x}_i,\boldsymbol{x}_j)}(y_{i,k} - y_{j,k}). \tag{7.26}$$

Proof According to Lee and Verleysen [28], the proof is as follows. Utilizing the chain rule, the following is known:

$$\frac{\partial c_4}{\partial y_{i,k}} = \frac{\partial c_4}{\partial d_y(\boldsymbol{y}_i,\boldsymbol{y}_j)} \times \frac{\partial d_y(\boldsymbol{y}_i,\boldsymbol{y}_j)}{\partial y_{i,k}}.$$

The first term is:

$$\frac{\partial c_4}{\partial d_y(\boldsymbol{y}_i,\boldsymbol{y}_j)} = \frac{-2}{a}\sum_{i=1}^{n}\sum_{j=1,j<i}^{n} \frac{d_x(\boldsymbol{x}_i,\boldsymbol{x}_j) - d_y(\boldsymbol{y}_i,\boldsymbol{y}_j)}{d_x(\boldsymbol{x}_i,\boldsymbol{x}_j)},$$

and the second term is:

$$\frac{\partial d_y(\boldsymbol{y}_i,\boldsymbol{y}_j)}{\partial y_{i,k}} = \frac{\partial d_y(\boldsymbol{y}_i,\boldsymbol{y}_j)}{\partial d_y^2(\boldsymbol{y}_i,\boldsymbol{y}_j)} \times \frac{\partial d_y^2(\boldsymbol{y}_i,\boldsymbol{y}_j)}{\partial y_{i,k}}.$$

Therefore:

$$\frac{\partial d_y(\boldsymbol{y}_i,\boldsymbol{y}_j)}{\partial d_y^2(\boldsymbol{y}_i,\boldsymbol{y}_j)} = 1/\frac{\partial d_y^2(\boldsymbol{y}_i,\boldsymbol{y}_j)}{\partial d_y(\boldsymbol{y}_i,\boldsymbol{y}_j)} = 1/(2d_y(\boldsymbol{y}_i,\boldsymbol{y}_j)),$$

$$d_y^2(\boldsymbol{y}_i,\boldsymbol{y}_j) = ||\boldsymbol{y}_i - \boldsymbol{y}_j||_2^2 = \sum_{k=1}^{p}(y_{i,k} - y_{j,k})^2,$$

$$\frac{\partial d_y^2(\boldsymbol{y}_i, \boldsymbol{y}_j)}{\partial y_{i,k}} = 2\,(y_{i,k} - y_{j,k}),$$

$$\therefore \quad \frac{\partial d_y(\boldsymbol{y}_i, \boldsymbol{y}_j)}{\partial y_{i,k}} = \frac{y_{i,k} - y_{j,k}}{d_y(\boldsymbol{y}_i, \boldsymbol{y}_j)}. \tag{7.27}$$

Lastly:

$$\therefore \quad \frac{\partial c_4}{\partial y_{i,k}} = \frac{-2}{a} \sum_{i=1}^{n} \sum_{j=1, j<i}^{n} \frac{d_x(\boldsymbol{x}_i, \boldsymbol{x}_j) - d_y(\boldsymbol{y}_i, \boldsymbol{y}_j)}{d_x(\boldsymbol{x}_i, \boldsymbol{x}_j)\, d_y(\boldsymbol{x}_i, \boldsymbol{x}_j)} (y_{i,k} - y_{j,k}),$$

which is the gradient mentioned in the proposition. □

Proposition 7.3 *The second derivative of the cost function c, with respect to $y_{i,k}$, is [28, 34]:*

$$\frac{\partial^2 c_4}{\partial y_{i,k}^2} = \frac{-2}{a} \sum_{i=1}^{n} \sum_{j=1, j<i}^{n} \Big(\frac{d_x(\boldsymbol{x}_i, \boldsymbol{x}_j) - d_y(\boldsymbol{y}_i, \boldsymbol{y}_j)}{d_x(\boldsymbol{x}_i, \boldsymbol{x}_j)\, d_y(\boldsymbol{x}_i, \boldsymbol{x}_j)} - \frac{(y_{i,k} - y_{j,k})^2}{d_y^3(\boldsymbol{y}_i, \boldsymbol{y}_j)} \Big). \tag{7.28}$$

Proof Utilizing:

$$\frac{\partial^2 c_4}{\partial y_{i,k}^2} = \frac{\partial}{\partial y_{i,k}} \Big(\frac{\partial c_4}{\partial y_{i,k}} \Big),$$

where $\partial c_4 / \partial y_{i,k}$ is Eq. (7.26), results in:

$$\frac{\partial^2 c_4}{\partial y_{i,k}^2} = \frac{-2}{a} \sum_{i=1}^{n} \sum_{j=1, j<i}^{n} \frac{\partial}{\partial y_{i,k}} \Big(\frac{d_x(\boldsymbol{x}_i, \boldsymbol{x}_j) - d_y(\boldsymbol{y}_i, \boldsymbol{y}_j)}{d_x(\boldsymbol{x}_i, \boldsymbol{x}_j)\, d_y(\boldsymbol{x}_i, \boldsymbol{x}_j)} (y_{i,k} - y_{j,k}) \Big).$$

Therefore:

$$\frac{\partial}{\partial y_{i,k}} \Big(\frac{d_x(\boldsymbol{x}_i, \boldsymbol{x}_j) - d_y(\boldsymbol{y}_i, \boldsymbol{y}_j)}{d_x(\boldsymbol{x}_i, \boldsymbol{x}_j)\, d_y(\boldsymbol{x}_i, \boldsymbol{x}_j)} (y_{i,k} - y_{j,k}) \Big)$$

$$= (y_{i,k} - y_{j,k}) \frac{\partial}{\partial y_{i,k}} \Big(\frac{d_x(\boldsymbol{x}_i, \boldsymbol{x}_j) - d_y(\boldsymbol{y}_i, \boldsymbol{y}_j)}{d_x(\boldsymbol{x}_i, \boldsymbol{x}_j)\, d_y(\boldsymbol{x}_i, \boldsymbol{x}_j)} \Big) + \frac{d_x(\boldsymbol{x}_i, \boldsymbol{x}_j) - d_y(\boldsymbol{y}_i, \boldsymbol{y}_j)}{d_x(\boldsymbol{x}_i, \boldsymbol{x}_j)\, d_y(\boldsymbol{x}_i, \boldsymbol{x}_j)} \underbrace{\frac{\partial}{\partial y_{i,k}} (y_{i,k} - y_{j,k})}_{=1}.$$

Note that:

$$\frac{\partial}{\partial y_{i,k}} \Big(\frac{d_x(\boldsymbol{x}_i, \boldsymbol{x}_j) - d_y(\boldsymbol{y}_i, \boldsymbol{y}_j)}{d_x(\boldsymbol{x}_i, \boldsymbol{x}_j)\, d_y(\boldsymbol{x}_i, \boldsymbol{x}_j)} \Big) = \frac{1}{d_x(\boldsymbol{x}_i, \boldsymbol{x}_j)} \frac{\partial}{\partial y_{i,k}} \Big(\frac{d_x(\boldsymbol{x}_i, \boldsymbol{x}_j) - d_y(\boldsymbol{y}_i, \boldsymbol{y}_j)}{d_y(\boldsymbol{x}_i, \boldsymbol{x}_j)} \Big)$$

$$= \frac{1}{d_x(\boldsymbol{x}_i, \boldsymbol{x}_j)} \frac{\partial}{\partial y_{i,k}} \Big(\frac{d_x(\boldsymbol{x}_i, \boldsymbol{x}_j)}{d_y(\boldsymbol{x}_i, \boldsymbol{x}_j)} - 1 \Big) = \underbrace{\frac{d_x(\boldsymbol{x}_i, \boldsymbol{x}_j)}{d_x(\boldsymbol{x}_i, \boldsymbol{x}_j)}}_{=1} \frac{\partial}{\partial y_{i,k}} \Big(\frac{1}{d_y(\boldsymbol{x}_i, \boldsymbol{x}_j)} \Big) - \underbrace{\frac{\partial}{\partial y_{i,k}}(1)}_{=0}$$

$$= \frac{-1}{d_y^2(\boldsymbol{x}_i, \boldsymbol{x}_j)} \frac{\partial}{\partial y_{i,k}} (d_y(\boldsymbol{x}_i, \boldsymbol{x}_j)) \overset{(7.27)}{=} \frac{-1}{d_y^2(\boldsymbol{x}_i, \boldsymbol{x}_j)} \frac{y_{i,k} - y_{j,k}}{d_y(\boldsymbol{y}_i, \boldsymbol{y}_j)}.$$

Therefore:

$$\frac{\partial}{\partial y_{i,k}} \Big(\frac{d_x(\boldsymbol{x}_i, \boldsymbol{x}_j) - d_y(\boldsymbol{y}_i, \boldsymbol{y}_j)}{d_x(\boldsymbol{x}_i, \boldsymbol{x}_j)\, d_y(\boldsymbol{x}_i, \boldsymbol{x}_j)} (y_{i,k} - y_{j,k}) \Big) = \frac{-(y_{i,k} - y_{j,k})^2}{d_y^3(\boldsymbol{y}_i, \boldsymbol{y}_j)} + \frac{d_x(\boldsymbol{x}_i, \boldsymbol{x}_j) - d_y(\boldsymbol{y}_i, \boldsymbol{y}_j)}{d_x(\boldsymbol{x}_i, \boldsymbol{x}_j)\, d_y(\boldsymbol{x}_i, \boldsymbol{x}_j)}.$$

Therefore:

$$\frac{\partial^2 c_4}{\partial y_{i,k}^2} = \frac{-2}{a} \sum_{i=1}^{n} \sum_{j=1, j<i}^{n} \Big(\frac{d_x(\boldsymbol{x}_i, \boldsymbol{x}_j) - d_y(\boldsymbol{y}_i, \boldsymbol{y}_j)}{d_x(\boldsymbol{x}_i, \boldsymbol{x}_j)\, d_y(\boldsymbol{x}_i, \boldsymbol{x}_j)} - \frac{(y_{i,k} - y_{j,k})^2}{d_y^3(\boldsymbol{y}_i, \boldsymbol{y}_j)} \Big),$$

which is the derivative mentioned in the proposition. □

It is important to note that for better time complexity of the Sammon mapping, it is possible to use the k-Nearest Neighbours (kNN), rather than the whole data [17]:

$$\underset{\{\boldsymbol{y}_i\}_{i=1}^{n}}{\text{minimize}} \quad \frac{1}{a} \sum_{i=1}^{n} \sum_{j \in \mathcal{N}_i}^{n} w_{ij} \big(d_x(\boldsymbol{x}_i, \boldsymbol{x}_j) - d_y(\boldsymbol{y}_i, \boldsymbol{y}_j) \big)^2, \tag{7.29}$$

where $\mathcal{N}_i$ denotes the set of indices of kNN of the i-th point.

7.4 Isometric Mapping (Isomap)

7.4.1 Isomap

Isometric Mapping (*Isomap*) [40] is a special case of the generalized classical MDS, explained in Sect. 7.2.1.2. Rather than the Euclidean distance, Isomap uses an approximation of the geodesic distance. As was explained, the classical MDS is linear; therefore, it cannot capture the nonlinearity of the manifold. However, Isomap uses the geodesic distance to make the generalized classical MDS nonlinear.

7.4.1.1 Geodesic Distance

The *geodesic distance* is the length of the shortest path between two points on the possibly curvy (i.e., nonlinear) manifold. For computing the distance along a curvy

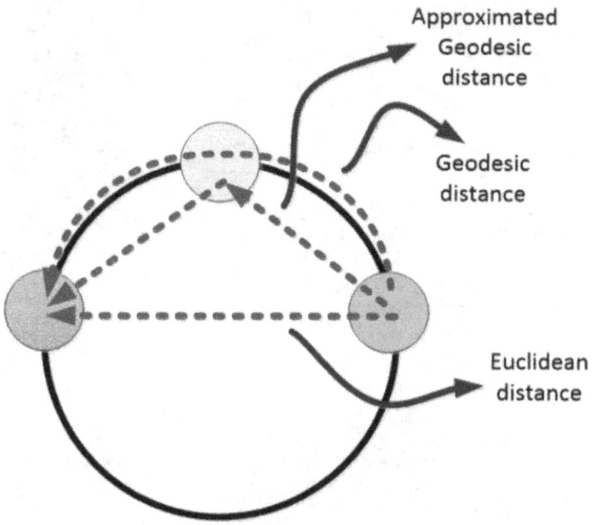

Fig. 7.1 An example of the Euclidean distance, geodesic distance, and approximated geodesic distance using piecewise Euclidean distances

nonlinear surface, it is ideal to use the geodesic distance. However, calculating the geodesic distance is very difficult since it needs to traverse from one point to another on the manifold, which requires differential geometry and Riemannian manifold calculations [2]. Therefore, Isomap approximates the geodesic distance by piecewise Euclidean distances, by finding the k-Nearest Neighbours (kNN) graph of the dataset. Then, the shortest path between two points, through their neighbours, is found using a shortest-path algorithm,[1] such as the Dijkstra algorithm or the Floyd-Warshal algorithm [10]. Note that the approximated geodesic distance is also referred to as the *curvilinear distance* [29]. The approximated geodesic distance can be formulated as [5]:

$$\boldsymbol{D}_{ij}^{(g)} := \min_{\boldsymbol{r}} \sum_{i=2}^{l} \|\boldsymbol{r}_i - \boldsymbol{r}_{i+1}\|_2, \tag{7.30}$$

where $l \geq 2$ is the length of the sequence of points $\boldsymbol{r}_i \in \{\boldsymbol{x}_i\}_{i=1}^n$ and $\boldsymbol{D}_{ij}^{(g)}$ denotes the (i, j)-th element of the geodesic distance matrix $\boldsymbol{D}^{(g)} \in \mathbb{R}^{n\times n}$.

An example of the Euclidean distance, geodesic distance, and the approximated geodesic distance using piecewise Euclidean distances is demonstrated in Fig. 7.1, using a girl's travel plans from Toronto to Athens. The most direct route would be to tunnel directly to Athens from Toronto (Euclidean distance). Although since that is not an option, she has two different routes available to her: (1) fly directly from Toronto to Athens, taking the shortest trip following the curve of the earth (geodesic

[1] A Scikit-learn function in python [31] for this is "graph_shortest_path" from the package "sklearn.utils.graph_shortest_path".

distance); and (2) fly from Toronto to London, London to Frankfurt, Frankfurt to Rome, and then Rome to Athens, thereby taking multiple short trips where every trip is almost on a straight line on the earth (approximated geodesic distance). Calculations of the lengths of paths in the approximated geodesic distance are much easier than those in the geodesic distance.

7.4.1.2 Isomap Formulation

As previously mentioned, Isomap is a special case of a generalized classical MDS with the geodesic distance used. Therefore, an Isomap uses Eq. (7.13) as:

$$\mathbb{R}^{n\times n} \ni \boldsymbol{K} = -\frac{1}{2}\boldsymbol{H}\boldsymbol{D}^{(g)}\boldsymbol{H}. \tag{7.31}$$

It then uses Eqs. (7.14) and (7.10) to embed the data. As Isomap uses the nonlinear geodesic distance in its kernel calculation, it is a *nonlinear* method.

7.4.2 *Kernel Isomap*

Consider $\boldsymbol{K}(\boldsymbol{D})$ to be Eq. (7.13). Consequently:

$$\mathbb{R}^{n\times n} \ni \boldsymbol{K}(\boldsymbol{D}^2) = -\frac{1}{2}\boldsymbol{H}\boldsymbol{D}^2\boldsymbol{H}, \tag{7.32}$$

where $\boldsymbol{D}$ is the geodesic distance matrix, defined by Eq. (7.30).

Define the following equation [11, Section 2.2.8]:

$$\mathbb{R}^{n\times n} \ni \boldsymbol{K}' := \boldsymbol{K}(\boldsymbol{D}^2) + 2c\boldsymbol{K}(\boldsymbol{D}) + \frac{1}{2}c^2\boldsymbol{H}. \tag{7.33}$$

According to [8], $\boldsymbol{K}'$ is guaranteed to be positive semidefinite for $c \geq c^*$, where c^* is the largest eigenvalue of the following matrix:

$$\begin{bmatrix} \boldsymbol{0} & 2\boldsymbol{K}(\boldsymbol{D}^2) \\ -\boldsymbol{I} & -4\boldsymbol{K}(\boldsymbol{D}) \end{bmatrix} \in \mathbb{R}^{2n\times 2n}. \tag{7.34}$$

Kernel Isomap [9] chooses a value $c \geq c^*$ and uses $\boldsymbol{K}'$ in Eq. (7.14), and then uses Eq. (7.10) for embedding the data.

7.5 Out-of-Sample Extensions for MDS and Isomap

Thus far, the training dataset $\{\boldsymbol{x}_i \in \mathbb{R}^d\}_{i=1}^n$ or $\boldsymbol{X} = [\boldsymbol{x}_1, \dots, \boldsymbol{x}_n] \in \mathbb{R}^{d\times n}$ has been embedded, where the embedding is denoted by $\{\boldsymbol{y}_i \in \mathbb{R}^p\}_{i=1}^n$ or $\boldsymbol{Y} = [\boldsymbol{y}_1, \dots, \boldsymbol{y}_n] \in \mathbb{R}^{p\times n}$. Assume an out-of-sample (test data) is denoted by $\{\boldsymbol{x}_i^{(t)} \in \mathbb{R}^d\}_{i=1}^{n_t}$ or $\boldsymbol{X}_t = [\boldsymbol{x}_1^{(t)}, \dots, \boldsymbol{x}_n^{(t)}] \in \mathbb{R}^{d\times n_t}$. The goal is to find its embedding $\{\boldsymbol{y}_i^{(t)} \in \mathbb{R}^p\}_{i=1}^{n_t}$ or $\boldsymbol{Y}_t = [\boldsymbol{y}_1^{(t)}, \dots, \boldsymbol{y}_n^{(t)}] \in \mathbb{R}^{p\times n_t}$ after the training phase. Several methods exist for out-of-sample extension of MDS and Isomap, such as extension by eigenfunctions [5] and kernel mapping [18, 19], which are explained in the following section. There are additional methods for the out-of-sample extension of MDS and Isomap [7, 37], which are not covered in this book.

7.5.1 *Out of Sample for Isomap and MDS Using Eigenfunctions*

Eigenfunctions were introduced in Chap. 3. Let the MDS or Isomap embedding of point $\boldsymbol{x}$ be $\mathbb{R}^p \ni \boldsymbol{y}(\boldsymbol{x}) = [y_1(\boldsymbol{x}), \dots, y_p(\boldsymbol{x})]^\top$. The k-th dimension of this embedding is (see Chap. 3):

$$y_k(\boldsymbol{x}) = \sqrt{\delta_k}\,\frac{f_k(\boldsymbol{x})}{\sqrt{n}} = \frac{1}{\sqrt{\delta_k}}\sum_{i=1}^{n} v_{ki}\,\breve{k}_t(\boldsymbol{x}_i, \boldsymbol{x}). \tag{7.35}$$

If a set of n_t out-of-sample data points is available, $\breve{k}_t(\boldsymbol{x}_i, \boldsymbol{x})$ is an element of the centered out-of-sample kernel (see Chap. 3), then:

$$\mathbb{R}^{n\times n_t} \ni \breve{\boldsymbol{K}}_t = \boldsymbol{K}_t - \frac{1}{n}\boldsymbol{1}_{n\times n}\boldsymbol{K}_t - \frac{1}{n}\boldsymbol{K}\boldsymbol{1}_{n\times n_t} + \frac{1}{n^2}\boldsymbol{1}_{n\times n}\boldsymbol{K}\boldsymbol{1}_{n\times n_t}, \tag{7.36}$$

where $\boldsymbol{1} := [1, 1, \dots, 1]^\top$, $\boldsymbol{K}_t \in \mathbb{R}^{n\times n_t}$ is not necessarily centered out-of-sample kernel, and $\boldsymbol{K} \in \mathbb{R}^{n\times n}$ is the training kernel. Equation (7.35) can be used to embed the i-th out-of-sample data point $\boldsymbol{x}_i^{(t)}$. For this purpose, $\boldsymbol{x}_i^{(t)}$ should be used in place of $\boldsymbol{x}$ in Eq. (7.35).

Note that Eq. (7.35) requires Eq. (7.36). In MDS and Isomap, $\boldsymbol{K}$ is obtained by the linear kernel, $\boldsymbol{X}^\top\boldsymbol{X}$, and Eq. (7.31), respectively. Additionally, the out-of-sample kernel $\boldsymbol{K}_t$ in MDS is obtained by the linear kernel between the training and out-of-sample data, i.e., $\boldsymbol{X}^\top\boldsymbol{X}_t$. In Isomap, the kernel $\boldsymbol{K}_t$ is obtained by centering the geodesic distance matrix (see Eq. (7.31)), where the geodesic distance matrix between the training and out-of-sample data is used. In calculation of this geodesic distance matrix, only the training data points, and not the test points, should be used as the intermediate points in the paths [5].

Bengio et al. [5, Corollary 1] showed that using the geodesic distance with only training data as intermediate points, for the sake of out-of-sample embedding in Isomap, is equivalent to the *landmark Isomap* method (which will be introduced in Sect. 7.6) [13]:

$$y_k(\boldsymbol{x}) = \frac{1}{2\sqrt{\delta_k}} \sum_{i=1}^{n} v_{ki} (\boldsymbol{D}^{(g)}_{\text{avg}} - \boldsymbol{D}^{(g)}_t(\boldsymbol{x}_i, \boldsymbol{x})), \tag{7.37}$$

where $\boldsymbol{D}^{(g)}_{\text{avg}}$ denotes the average geodesic distance between the training points and $\boldsymbol{D}^{(g)}_t$ is the geodesic distance between the i-th training point $\boldsymbol{x}_i$ and the out-of-sample point $\boldsymbol{x}$, in which the training set is used as intermediate points. Therefore, Eq. (7.37) can be used for out-of-sample embedding in Isomap.

7.5.2 *Out of Sample for Isomap, Kernel Isomap, and MDS Using Kernel Mapping*

Gisbrecht et al. developed a kernel mapping method [18, 19] to embed the out-of-sample data in Isomap, kernel Isomap, and MDS. Consider a mapping of any data point, defined as $\boldsymbol{x} \mapsto \boldsymbol{y}(\boldsymbol{x})$:

$$\mathbb{R}^p \ni \boldsymbol{y}(\boldsymbol{x}) := \sum_{j=1}^{n} \boldsymbol{\alpha}_j \frac{k(\boldsymbol{x}, \boldsymbol{x}_j)}{\sum_{\ell=1}^{n} k(\boldsymbol{x}, \boldsymbol{x}_\ell)}, \tag{7.38}$$

and $\boldsymbol{\alpha}_j \in \mathbb{R}^p$, and $\boldsymbol{x}_j$ and $\boldsymbol{x}_\ell$ denote the j-th and ℓ-th training data point, respectively. $k(\boldsymbol{x}, \boldsymbol{x}_j)$ is a kernel, such as the Gaussian kernel:

$$k(\boldsymbol{x}, \boldsymbol{x}_j) = \exp(\frac{-||\boldsymbol{x} - \boldsymbol{x}_j||_2^2}{2\sigma_j^2}), \tag{7.39}$$

where σ_j is calculated as [18]:

$$\sigma_j := \gamma \times \min_i(||\boldsymbol{x}_j - \boldsymbol{x}_i||_2), \tag{7.40}$$

where γ is a small positive number.

Assume the embedded training data points are using MDS (see Sect. 7.2), Isomap (see Sect. 7.4), or the kernel Isomap (see Sect. 7.4.2); therefore, the set $\{\boldsymbol{y}_i\}_{i=1}^n$ is available. If the training data points are mapped, the goal is to minimize the following least-squares cost function to obtain $\boldsymbol{y}(\boldsymbol{x}_i)$ close to $\boldsymbol{y}_i$ for the i-th training point:

$$\underset{\boldsymbol{\alpha}_j\text{'s}}{\text{minimize}} \quad \sum_{i=1}^{n} ||\boldsymbol{y}_i - \boldsymbol{y}(\boldsymbol{x}_i)||_2^2, \tag{7.41}$$

where the summation is over the training data points. This cost function can be written in the following matrix form:

$$\underset{\boldsymbol{A}}{\text{minimize}} \quad ||\boldsymbol{Y} - \boldsymbol{K}''\boldsymbol{A}||_F^2, \tag{7.42}$$

where $\mathbb{R}^{n\times p} \ni \boldsymbol{Y} := [\boldsymbol{y}_1, \ldots, \boldsymbol{y}_n]^\top$ and $\mathbb{R}^{n\times p} \ni \boldsymbol{A} := [\boldsymbol{\alpha}_1, \ldots, \boldsymbol{\alpha}_n]^\top$. $\boldsymbol{K}'' \in \mathbb{R}^{n\times n}$ is the kernel matrix, whose (i, j)-th element is defined as:

$$\boldsymbol{K}''(i, j) := \frac{k(\boldsymbol{x}_i, \boldsymbol{x}_j)}{\sum_{\ell=1}^{n} k(\boldsymbol{x}_i, \boldsymbol{x}_\ell)}. \tag{7.43}$$

Equation (7.42) is always nonnegative; thus, its smallest value is zero. Therefore, the solution to this equation is:

$$\boldsymbol{Y} - \boldsymbol{K}''\boldsymbol{A} = \boldsymbol{0} \implies \boldsymbol{Y} = \boldsymbol{K}''\boldsymbol{A} \overset{(a)}{\implies} \boldsymbol{A} = \boldsymbol{K}''^{\dagger}\,\boldsymbol{Y}, \tag{7.44}$$

where $\boldsymbol{K}''^{\dagger}$ is the pseudoinverse of $\boldsymbol{K}''$:

$$\boldsymbol{K}''^{\dagger} = (\boldsymbol{K}''^\top \boldsymbol{K}'')^{-1} \boldsymbol{K}''^\top, \tag{7.45}$$

and (a) is because $\boldsymbol{K}''^{\dagger}\,\boldsymbol{K}'' = \boldsymbol{I}$.

Lastly, the mapping, in Eq. (7.38), for the n_t the out-of-sample data points is:

$$\boldsymbol{Y}_t = \boldsymbol{K}_t''\,\boldsymbol{A}, \tag{7.46}$$

where the (i, j)-th element of the out-of-sample kernel matrix $\boldsymbol{K}_t'' \in \mathbb{R}^{n_t\times n}$ is:

$$\boldsymbol{K}_t''(i, j) := \frac{k(\boldsymbol{x}_i^{(t)}, \boldsymbol{x}_j)}{\sum_{\ell=1}^{n} k(\boldsymbol{x}_i^{(t)}, \boldsymbol{x}_\ell)}, \tag{7.47}$$

where $\boldsymbol{x}_i^{(t)}$ is the i-th out-of-sample data point, and $\boldsymbol{x}_j$ and $\boldsymbol{x}_\ell$ are the j-th and ℓ-th training data points, respectively.

7.6 Landmark MDS and Landmark Isomap for Big Data Embedding

MDS and Isomap cannot embed very large data. Landmark MDS and Isomap can be used for big data embedding. The landmark MDS can use kernel approximation or the distance matrix, which are explained in the following sections.

7.6.1 Using Kernel Approximation in Landmark MDS

The Nyström approximation, introduced in Chap. 3, can be used to make the spectral methods, such as MDS and Isomap, scalable, and suitable for big data embedding. Consider a kernel matrix:

$$\mathbb{R}^{n\times n} \ni \boldsymbol{K} = \left[\begin{array}{c|c} \boldsymbol{A} & \boldsymbol{B} \\ \hline \boldsymbol{B}^\top & \boldsymbol{C} \end{array}\right] = \begin{bmatrix} \boldsymbol{R}^\top \boldsymbol{R} & \boldsymbol{R}^\top \boldsymbol{S} \\ \boldsymbol{S}^\top \boldsymbol{R} & \boldsymbol{S}^\top \boldsymbol{S} \end{bmatrix} = \boldsymbol{O}^\top \boldsymbol{O} = \begin{bmatrix} \boldsymbol{R}^\top \\ \boldsymbol{S}^\top \end{bmatrix} [\boldsymbol{R}, \boldsymbol{S}], \tag{7.48}$$

where $\boldsymbol{A} \in \mathbb{R}^{m\times m}$, $\boldsymbol{B} \in \mathbb{R}^{m\times(n-m)}$, and $\boldsymbol{C} \in \mathbb{R}^{(n-m)\times(n-m)}$, in which $m \ll n$. According to Chap. 3, the following expressions hold approximately:

$$\boldsymbol{A} = \boldsymbol{U}\boldsymbol{\Sigma}\boldsymbol{U}^\top \tag{7.49}$$

$$\boldsymbol{R} = \boldsymbol{\Sigma}^{(1/2)}\boldsymbol{U}^\top. \tag{7.50}$$

$$\boldsymbol{S} = \boldsymbol{\Sigma}^{(-1/2)}\boldsymbol{U}^\top \boldsymbol{B}, \tag{7.51}$$

$$\boldsymbol{C} = \boldsymbol{B}^\top \boldsymbol{A}^{-1} \boldsymbol{B}, \tag{7.52}$$

where Eq. (7.49) is the eigenvalue decomposition of $\boldsymbol{A}$.

Consider Eq. (7.48) as the partitions of the kernel matrix $\boldsymbol{K}$. Note that the (Mercer) kernel matrix is positive semidefinite, which means that the Nyström approximation can be applied for kernels.

Recall that Eq. (7.14) decomposes the kernel matrix into eigenvectors and then Eq. (7.10) embeds data. However, for big data, the eigenvalue decomposition of the kernel matrix is intractable. Therefore, using Eq. (7.49), it is possible to decompose an $m \times m$ submatrix of the kernel. Comparing Eqs. (7.15) and (7.48) shows that:

$$\mathbb{R}^{n\times n} \ni \boldsymbol{Y} = [\boldsymbol{R}, \boldsymbol{S}] \overset{(a)}{=} [\boldsymbol{\Sigma}^{(1/2)}\boldsymbol{U}^\top, \boldsymbol{\Sigma}^{(-1/2)}\boldsymbol{U}^\top \boldsymbol{B}], \tag{7.53}$$

where (a) is because of Eqs. (7.50) and (7.51) and the terms $\boldsymbol{U}$ and $\boldsymbol{\Sigma}$ are obtained from Eq. (7.49). Equation (7.53) provides the approximately embedded data, with a good approximation. This is the embedding in *landmark MDS* [13, 14]. Truncating this matrix to have $\boldsymbol{Y} \in \mathbb{R}^{p\times n}$, with top p rows, results in the p-dimensional embedding of the n points.

Comparing Eq. (7.53) with Eq. (7.10) demonstrates that the formulas for embedding landmarks, $\boldsymbol{R}$, and the whole data (without the Nyström approximation) are similar, but one is completed with only landmarks and the other is completed with the whole data.

7.6.2 Using Distance Matrix in Landmark MDS

If $\boldsymbol{D}_{ij}$ denotes the (i, j)-th element of the distance matrix and $\boldsymbol{v}_j$ is the j-th element of a vector $\boldsymbol{v}$, Eq. (7.13) can be restated as [32]:

$$\boldsymbol{K} = \frac{-1}{2}\Big(\boldsymbol{D}_{ij}^2 - \mathbf{1}_j \sum_i \boldsymbol{c}_i \boldsymbol{D}_{ij}^2 - \mathbf{1}_i \sum_j \boldsymbol{c}_j \boldsymbol{D}_{ij}^2 + \sum_{i,j} \boldsymbol{c}_i \boldsymbol{c}_j \boldsymbol{D}_{ij}^2\Big), \tag{7.54}$$

where $\sum_i \boldsymbol{c}_i = 1$.

Let the partitions of the distance matrix be:

$$\mathbb{R}^{n\times n} \ni \boldsymbol{D} = \left[\begin{array}{c|c} \boldsymbol{E} & \boldsymbol{F} \\ \hline \boldsymbol{F}^\top & \boldsymbol{G} \end{array}\right], \tag{7.55}$$

where $\boldsymbol{E} \in \mathbb{R}^{m\times m}$, $\boldsymbol{F} \in \mathbb{R}^{m\times(n-m)}$, and $\boldsymbol{G} \in \mathbb{R}^{(n-m)\times(n-m)}$, in which $m \ll n$. Comparing Eqs. (7.48) and (7.55) demonstrates that the partitions of the kernel matrix can be obtained from the partitions of the distance matrix [32]:

$$\boldsymbol{A}_{ij} = \frac{-1}{2}\Big(\boldsymbol{E}_{ij}^2 - \mathbf{1}_i \frac{1}{m} \sum_p \boldsymbol{E}_{pj}^2 - \mathbf{1}_j \frac{1}{m} \sum_q \boldsymbol{E}_{iq}^2 + \frac{1}{m^2} \sum_{p,q} \boldsymbol{E}_{pq}^2\Big), \tag{7.56}$$

$$\boldsymbol{B}_{ij} = \frac{-1}{2}\Big(\boldsymbol{F}_{ij}^2 - \mathbf{1}_i \frac{1}{m} \sum_p \boldsymbol{F}_{qj}^2 - \mathbf{1}_j \frac{1}{m} \sum_q \boldsymbol{E}_{iq}^2\Big), \tag{7.57}$$

and $\boldsymbol{C}_{ij}$ can be obtained from Eq. (7.52).

In landmark MDS and landmark Isomap, the partitions (submatrices) $\boldsymbol{E}$ and $\boldsymbol{F}$ of the Euclidean and geodesic distance matrices are calculated, respectively (see Eq. (7.55)). Then, Eqs. (7.56), (7.57), and (7.52) results in the partitions of the kernel matrix, Eqs. 7.49) and (7.53) provide the embedded data.

It is noteworthy that the paper [32] demonstrates that different landmark MDS methods, such as *Landmark MDS (LMDS)* [13, 14], *FastMap* [15], and *MetricMap* [43] are reduced to the landmark MDS introduced here. The landmark MDS is also referred to as the *sparse MDS* [14]. Moreover, the *Landmark Isomap (L-Isomap)* [13] is reduced to the landmark Isomap method (see [5, Corollary 1] for proof). In other words, the large-scale dimensionality reduction methods make use of the Nyström approximation [39].

7.7 Chapter Summary

This chapter introduced MDS and its categories, which are the classical MDS, the metric MDS, and the nonmetric MDS. MDS is a fundamental method for both linear and nonlinear dimensionality reduction. Chapter 10 will demonstrate that most of the spectral dimensionality reduction methods can be unified as the kernel principal component analysis (kernel PCA) with different kernels. Many of the kernels used in this unified framework are in the form of Eq. (7.13). Therefore, the generalized classical MDS will be essential for the unified framework, introduced in Chap. 10. Sammon mapping was also introduced as a special case of the metric MDS. It is a very fundamental spectral method because it can be considered the first proposed nonlinear dimensionality reduction method. This chapter also explained Isomap as a special case of the generalized classical MDS. Isomap is another important spectral method because Isomap and LLE (introduced in Chap. 8) are the next oldest proposed methods, after Sammon mapping, for nonlinear dimensionality reduction.

References

1. Sameer Agarwal et al. "Generalized non-metric multidimensional scaling". In: *Artificial Intelligence and Statistics*. 2007, pp. 11–18.
2. Thierry Aubin. *A course in differential geometry*. Vol. 27. American Mathematical Society, Graduate Studies in Mathematics, 2001.
3. Richard Beals, David H Krantz, and Amos Tversky. "Foundations of multidimensional scaling." In: *Psychological review* 75.2 (1968), p. 127.
4. Yoshua Bengio et al. "Learning eigenfunctions links spectral embedding and kernel PCA". In: *Neural computation* 16.10 (2004), pp. 2197–2219.
5. Yoshua Bengio et al. "Out-of-sample extensions for LLE, Isomap, MDS, eigenmaps, and spectral clustering". In: *Advances in neural information processing systems*. 2004, pp. 177–184.
6. Ingwer Borg and Patrick JF Groenen. *Modern multidimensional scaling: Theory and applications*. Springer Science & Business Media, 2005.
7. Kerstin Bunte, Michael Biehl, and Barbara Hammer. "A general framework for dimensionality-reducing data visualization mapping". In: *Neural Computation* 24.3 (2012), pp. 771–804.
8. Francis Cailliez. "The analytical solution of the additive constant problem". In: *Psychometrika* 48.2 (1983), pp. 305–308.
9. Heeyoul Choi and Seungjin Choi. "Kernel Isomap". In: *Electronics letters* 40.25 (2004), pp. 1612–1613.
10. Thomas H Cormen et al. *Introduction to algorithms*. MIT press, 2009.
11. Michael AA Cox and Trevor F Cox. "Multidimensional scaling". In: *Handbook of data visualization*. Springer, 2008, pp. 315–347.
12. Jan De Leeuw. *Multidimensional scaling*. Tech. rep. University of California Los Angeles, 2011.
13. Vin De Silva and Joshua B Tenenbaum. "Global versus local methods in nonlinear dimensionality reduction". In: *Advances in neural information processing systems*. 2003, pp. 721–728.
14. Vin De Silva and Joshua B Tenenbaum. *Sparse multidimensional scaling using landmark points*. Tech. rep. Technical report, Stanford University, 2004.
15. Christos Faloutsos and King-Ip Lin. "FastMap: A fast algorithm for indexing, data-mining and visualization of traditional and multimedia datasets". In: *Proceedings of the 1995 ACM SIGMOD international conference on Management of data*. 1995, pp. 163–174.
16. Ali Ghodsi. *Dimensionality reduction a short tutorial*. Tech. rep. Department of Statistics and Actuarial Science, Univ. of Waterloo, Ontario, Canada, 2006.

17. Benyamin Ghojogh, Fakhri Karray, and Mark Crowley. "Quantile-Quantile Embedding for Distribution Transformation, Manifold Embedding, and Image Embedding with Choice of Embedding Distribution". In: *arXiv preprint arXiv:2006.11385* (2020).
18. Andrej Gisbrecht, Alexander Schulz, and Barbara Hammer. "Parametric nonlinear dimensionality reduction using kernel t-SNE". In: *Neurocomputing* 147 (2015), pp. 71–82.
19. Andrej Gisbrecht et al. "Out-of-sample kernel extensions for nonparametric dimensionality reduction." In: *European Symposium on Artificial Neural Networks, Computational Intelligence and Machine Learning*. Vol. 2012. 2012, pp. 531–536.
20. John C Gower. "Some distance properties of latent root and vector methods used in multivariate analysis". In: *Biometrika* 53.3-4 (1966), pp. 325–338.
21. Jihun Ham et al. "A kernel view of the dimensionality reduction of manifolds". In: *Proceedings of the twenty-first international conference on Machine learning*. 2004, p. 47.
22. Thomas Hofmann, Bernhard Schölkopf, and Alexander J Smola. "Kernel methods in machine learning". In: *The annals of statistics* (2008), pp. 1171–1220.
23. Steven M Holland. *Non-metric multidimensional scaling (MDS)*. Tech. rep. Department of Geology, University of Georgia, 2008.
24. Sungkyu Jung. "Lecture: Multidimensional scaling, Advanced Applied Multivariate Analysis". Lecture notes, Department of Statistics, University of Pittsburgh. 2013.
25. Mary Katsikitis. "The classification of facial expressions of emotion: A multidimensional-scaling approach". In: *Perception* 26.5 (1997), pp. 613–626.
26. J Kruskal. "Non-metric multidimensional scaling. A numerical method". In: *Psychometrika* 29.1 (1964), p. 1.
27. Joseph B Kruskal. "Multidimensional scaling by optimising goodness-of-fit to non-metric hypotheses". In: *Psychometrika* 29.1 (1964), pp. 115–29.
28. John A Lee and Michel Verleysen. *Nonlinear dimensionality reduction*. Springer Science & Business Media, 2007.
29. John Aldo Lee, Amaury Lendasse, Michel Verleysen, et al. "Curvilinear distance analysis versus Isomap". In: *European Symposium on Artificial Neural Networks*. Vol. 2. 2002, pp. 185–192.
30. Kanti V Mardia. "Some properties of clasical multi-dimesional scaling". In: *Communications in Statistics-Theory and Methods* 7.13 (1978), pp. 1233–1241.
31. Fabian Pedregosa et al. "Scikit-learn: Machine learning in Python". In: *Journal of machine learning research* 12.Oct (2011), pp. 2825–2830.
32. John Platt. "FastMap, MetricMap, and Landmark MDS are all Nystrom Algorithms." In: *AISTATS*. 2005.
33. James A Russell and Merry Bullock. "Multidimensional scaling of emotional facial expressions: similarity from preschoolers to adults." In: *Journal of personality and social psychology* 48.5 (1985), p. 1290.
34. John W Sammon. "A nonlinear mapping for data structure analysis". In: *IEEE Transactions on computers* 100.5 (1969), pp. 401–409.
35. Lawrence K Saul and Sam T Roweis. "Think globally, fit locally: unsupervised learning of low dimensional manifolds". In: *Journal of machine learning research* 4.Jun (2003), pp. 119–155.
36. ItzchakM Schlesinger and Louis Guttman. "Smallest space analysis of intelligence and achievement tests." In: *Psychological Bulletin* 71.2 (1969), p. 95.
37. Harry Strange and Reyer Zwiggelaar. "A generalised solution to the out-of-sample extension problem in manifold learning". In: *Proceedings of the Twenty-Fifth AAAI Conference on Artificial Intelligence*. 2011, pp. 471–476.
38. Harry Strange and Reyer Zwiggelaar. *Open Problems in Spectral Dimensionality Reduction*. Springer, 2014.
39. Ameet Talwalkar, Sanjiv Kumar, and Henry Rowley. "Large-scale manifold learning". In: *2008 IEEE Conference on Computer Vision and Pattern Recognition*. IEEE. 2008, pp. 1–8.
40. Joshua B Tenenbaum, Vin De Silva, and John C Langford. "A global geometric framework for nonlinear dimensionality reduction". In: *Science* 290.5500 (2000), pp. 2319–2323.

41. Warren S Torgerson. "Multidimensional scaling of similarity". In: *Psychometrika* 30.4 (1965), pp. 379–393.
42. Warren S Torgerson. "Multidimensional scaling: I. Theory and method". In: *Psychometrika* 17.4 (1952), pp. 401–419.
43. Jason Tsong-Li Wang et al. "Evaluating a class of distance-mapping algorithms for data mining and clustering". In: *Proceedings of the fifth ACM SIGKDD international conference on Knowledge discovery and data mining*. 1999, pp. 307–311.
44. Forrest W Young. *Multidimensional scaling: History, theory, and applications*. Psychology Press, 2013.
45. Xiaoming Zhao and Shiqing Zhang. "Facial expression recognition based on local binary patterns and kernel discriminant Isomap". In: *Sensors* 11.10 (2011), pp. 9573–9588.

Chapter 8
Locally Linear Embedding

8.1 Introduction

Locally Linear Embedding (LLE) [15, 53] is a nonlinear spectral dimensionality reduction method [57] that can be used for manifold embedding and feature extraction [23]. LLE tries to preserve the local structure of data in the embedding space. In other words, the close points in the high-dimensional input space should also be close to each other in the low-dimensional embedding space. For this goal, every point is reconstructed by a linear weighted combination of its neighbours in the high-dimensional input space; then, the same reconstruction weights are used in the lower-dimensional embedding space. LLE has many different applications, such as in medical areas [31, 46]. By this local fitting, hopefully the far points in the input space also fall far away from each other in the embedding space. This idea of fitting locally and thinking globally is the main idea of LLE [55, 56, 66, 68].

From a different perspective, the idea of local fitting by LLE is similar to the idea of piecewise spline regression [48]. LLE unfolds the nonlinear manifold by locally unfolding the manifold piece by piece, and it hopes that these local unfoldings result in a suitable total manifold unfolding (see Fig. 8.1).

Unsupervised dimensionality reduction methods usually have one of the following three approaches:

1. It is possible to have local fitting of data to preserve the global structure of data—fit locally, think globally [55]. In other words, similar points in the input space are embedded close to one another in the embedding space. This will hopefully embed the dissimilar points far away from each other in the embedding space.
2. It is also possible to have an approach, the other way around, i.e., global fitting of data to preserve the local structure of data—fit globally, think locally. In other words, the dissimilar points in the input space are embedded far away from one another in the embedding space. This will hopefully embed similar points close to each other in the embedding space.

B. Ghojogh et al., *Elements of Dimensionality Reduction and Manifold Learning*,
https://doi.org/10.1007/978-3-031-10602-6_8

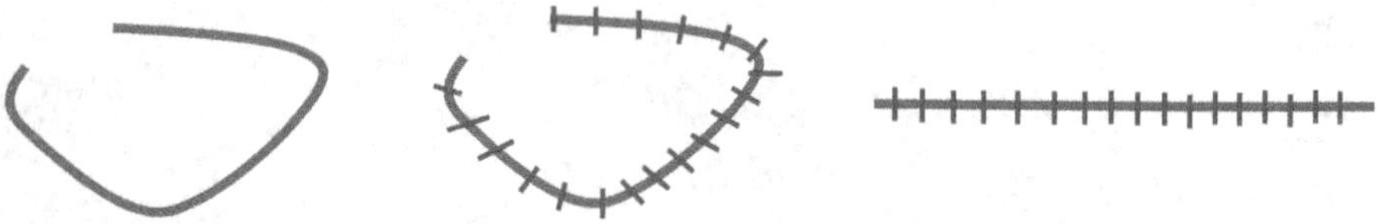

Fig. 8.1 Piecewise local unfolding of manifold by LLE (in this example from two dimensions to one intrinsic dimension). This local unfolding is expected to totally unfold the manifold properly

3. It is also possible to have both the first and second approaches together. In other words, similar points are embedded close to one another, and dissimilar points are embedded far away from each other, in the embedding space.

LLE has the first approach; in fact, it proposed the first approach for the first time [55]. This approach has been followed by most of the unsupervised dimensionality reduction methods. On the other hand, most of the supervised dimensionality reduction methods are based on increasing and decreasing the interclass and intraclass variances, respectively (see Chap. 6).

8.2 Locally Linear Embedding

LLE was first proposed in [53] and was developed in [54, 55]. The main idea of LLE is to use the same reconstruction weights in the lower dimensional embedding space and the high dimensional input space. There are three steps in LEE (see Fig. 8.2) [20]. First, it finds the k-Nearest neighbours (kNN) graph of all training points, before trying to find the weights for reconstructing every point by its neighbours, using linear combination. Lastly, using the same found weights, it embeds every point by a linear combination of its embedded neighbours.

8.2.1 *k-Nearest Neighbours*

The n data points in the input and feature spaces are denoted by $\{\boldsymbol{x}_i \in \mathbb{R}^d\}_{i=1}^n$ and $\{\boldsymbol{y}_i \in \mathbb{R}^p\}_{i=1}^n$, respectively, where usually there is $p \ll d$. As the first step in LLE, a kNN graph is formed using pairwise Euclidean distance between the data points. Therefore, every data point has k neighbours. Let $\boldsymbol{x}_{ij} \in \mathbb{R}^d$ denote the j-th neighbour of $\boldsymbol{x}_i$ and let the matrix $\mathbb{R}^{d\times k} \ni \boldsymbol{X}_i := [\boldsymbol{x}_{i1}, \dots, \boldsymbol{x}_{ik}]$ include the k neighbours of $\boldsymbol{x}_i$. It is assumed that k is large enough so that the kNN graph is connected.

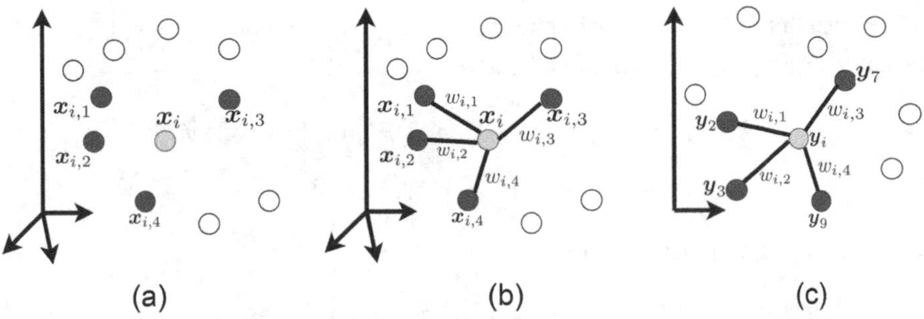

Fig. 8.2 Steps in LLE for embedding high dimensional data in a lower dimensional embedding space: (**a**) finding k-nearest neighbours, (**b**) linear reconstruction by the neighbours, and (**c**) linear embedding using the calculated weights. In this figure, it is assumed that $k = 4$, $\boldsymbol{x}_{i,1} = \boldsymbol{x}_2$, $\boldsymbol{x}_{i,2} = \boldsymbol{x}_3$, $\boldsymbol{x}_{i,3} = \boldsymbol{x}_7$, and $\boldsymbol{x}_{i,4} = \boldsymbol{x}_9$

8.2.2 Linear Reconstruction by the Neighbours

In the second step, the weights for linear reconstruction of every point by its kNN are found. The optimization for this linear reconstruction in the high dimensional input space is formulated as:

$$\begin{aligned} \underset{\widetilde{\boldsymbol{W}}}{\text{minimize}} \quad & \varepsilon(\widetilde{\boldsymbol{W}}) := \sum_{i=1}^{n} \Big|\Big|\boldsymbol{x}_i - \sum_{j=1}^{k} \widetilde{w}_{ij}\boldsymbol{x}_{ij}\Big|\Big|_2^2, \\ \text{subject to} \quad & \sum_{j=1}^{k} \widetilde{w}_{ij} = 1, \quad \forall i \in \{1, \ldots, n\}, \end{aligned} \tag{8.1}$$

where $\mathbb{R}^{n\times k} \ni \widetilde{\boldsymbol{W}} := [\widetilde{\boldsymbol{w}}_1, \ldots, \widetilde{\boldsymbol{w}}_n]^\top$ includes the weights, $\mathbb{R}^k \ni \widetilde{\boldsymbol{w}}_i := [\widetilde{w}_{i1}, \ldots, \widetilde{w}_{ik}]^\top$ includes the weights of linear reconstruction of the i-th data point using its k neighbours, and $\boldsymbol{x}_{ij} \in \mathbb{R}^d$ is the j-th neighbour of the i-th data point.

The constraint $\sum_{j=1}^{k} \widetilde{w}_{ij} = 1$ means that the weights of the linear reconstruction sum to one for every point. Note that some weights may be negative, which can cause the explosion of some weights due to large positive and negative weights that can cancel each other, summing to a total of one. However, this problem does not occur because the solution to this optimization problem has a closed form; thus, weights do not explode. If the solution was found iteratively, the weights would grow and explode gradually [22].

The objective $\varepsilon(\widetilde{\boldsymbol{W}})$ can be restated as:

$$\varepsilon(\widetilde{\boldsymbol{W}}) = \sum_{i=1}^{n} ||\boldsymbol{x}_i - \boldsymbol{X}_i\widetilde{\boldsymbol{w}}_i||_2^2. \tag{8.2}$$

The constraint $\sum_{j=1}^{k} \widetilde{w}_{ij} = 1$ implies that $\mathbf{1}^\top \widetilde{\boldsymbol{w}}_i = 1$; therefore, $\boldsymbol{x}_i = \boldsymbol{x}_i \mathbf{1}^\top \widetilde{\boldsymbol{w}}_i$. It is possible to simplify the term in $\varepsilon(\widetilde{\boldsymbol{W}})$ as:

$$\begin{aligned}||\boldsymbol{x}_i - \boldsymbol{X}_i \widetilde{\boldsymbol{w}}_i||_2^2 &= ||\boldsymbol{x}_i \mathbf{1}^\top \widetilde{\boldsymbol{w}}_i - \boldsymbol{X}_i \widetilde{\boldsymbol{w}}_i||_2^2 = ||(\boldsymbol{x}_i \mathbf{1}^\top - \boldsymbol{X}_i)\, \widetilde{\boldsymbol{w}}_i||_2^2 \\ &= \widetilde{\boldsymbol{w}}_i^\top (\boldsymbol{x}_i \mathbf{1}^\top - \boldsymbol{X}_i)^\top (\boldsymbol{x}_i \mathbf{1}^\top - \boldsymbol{X}_i)\, \widetilde{\boldsymbol{w}}_i = \widetilde{\boldsymbol{w}}_i^\top \boldsymbol{G}_i\, \widetilde{\boldsymbol{w}}_i,\end{aligned}$$

where $\boldsymbol{G}_i$ is a gram matrix defined as:

$$\mathbb{R}^{k \times k} \ni \boldsymbol{G}_i := (\boldsymbol{x}_i \mathbf{1}^\top - \boldsymbol{X}_i)^\top (\boldsymbol{x}_i \mathbf{1}^\top - \boldsymbol{X}_i). \tag{8.3}$$

Finally, Eq. (8.1) can be rewritten as:

$$\begin{aligned}&\underset{\{\widetilde{\boldsymbol{w}}_i\}_{i=1}^n}{\text{minimize}} \quad \sum_{i=1}^{n} \widetilde{\boldsymbol{w}}_i^\top \boldsymbol{G}_i\, \widetilde{\boldsymbol{w}}_i, \\ &\text{subject to} \quad \mathbf{1}^\top \widetilde{\boldsymbol{w}}_i = 1, \quad \forall i \in \{1, \dots, n\}.\end{aligned} \tag{8.4}$$

The Lagrangian for Eq. (8.4) is [11]:

$$\mathcal{L} = \sum_{i=1}^{n} \widetilde{\boldsymbol{w}}_i^\top \boldsymbol{G}_i\, \widetilde{\boldsymbol{w}}_i - \sum_{i=1}^{n} \lambda_i\, (\mathbf{1}^\top \widetilde{\boldsymbol{w}}_i - 1).$$

Setting the derivative of Lagrangian to zero results in:

$$\mathbb{R}^k \ni \frac{\partial \mathcal{L}}{\partial \widetilde{\boldsymbol{w}}_i} = 2\boldsymbol{G}_i \widetilde{\boldsymbol{w}}_i - \lambda_i \mathbf{1} \overset{\text{set}}{=} \mathbf{0} \implies \widetilde{\boldsymbol{w}}_i = \frac{1}{2} \boldsymbol{G}_i^{-1} \lambda_i \mathbf{1} = \frac{\lambda_i}{2} \boldsymbol{G}_i^{-1} \mathbf{1}. \tag{8.5}$$

$$\mathbb{R} \ni \frac{\partial \mathcal{L}}{\partial \lambda} = \mathbf{1}^\top \widetilde{\boldsymbol{w}}_i - 1 \overset{\text{set}}{=} 0 \implies \mathbf{1}^\top \widetilde{\boldsymbol{w}}_i = 1. \tag{8.6}$$

Using Eqs. (8.5) and (8.6) yields:

$$\frac{\lambda_i}{2} \mathbf{1}^\top \boldsymbol{G}_i^{-1} \mathbf{1} = 1 \implies \lambda_i = \frac{2}{\mathbf{1}^\top \boldsymbol{G}_i^{-1} \mathbf{1}}. \tag{8.7}$$

Using Eqs. (8.5) and (8.7) yields:

$$\widetilde{\boldsymbol{w}}_i = \frac{\lambda_i}{2} \boldsymbol{G}_i^{-1} \mathbf{1} = \frac{\boldsymbol{G}_i^{-1} \mathbf{1}}{\mathbf{1}^\top \boldsymbol{G}_i^{-1} \mathbf{1}}. \tag{8.8}$$

According to Eq. (8.3), the rank of the matrix $\boldsymbol{G}_i \in \mathbb{R}^{k \times k}$ is at most equal to $\min(k, d)$. If $d < k$, then $\boldsymbol{G}_i$ is singular and $\boldsymbol{G}_i$ should be replaced by $\boldsymbol{G}_i + \epsilon \boldsymbol{I}$, where ϵ is a small positive number. Usually, the data are high dimensional (so $k \ll d$) as in images, and thus, if $\boldsymbol{G}_i$ is full rank, there will not be a problem with inverting it. Strengthening the main diagonal of $\boldsymbol{G}$ is referred to as regularization in LLE [17]. This numerical technique is widely used in manifold and subspace learning (e.g., see [50]).

8.2.3 Linear Embedding

In the second step, the weights for linear reconstruction in the high-dimensional input space were found. In the third step, the data are embedded in the low-dimensional embedding space using the same weights as in the input space. This linear embedding can be formulated as the following optimization problem:

$$\begin{aligned}
\underset{\boldsymbol{Y}}{\text{minimize}} \quad & \sum_{i=1}^{n} \Big|\Big|\boldsymbol{y}_i - \sum_{j=1}^{n} w_{ij}\boldsymbol{y}_j\Big|\Big|_2^2, \\
\text{subject to} \quad & \frac{1}{n}\sum_{i=1}^{n} \boldsymbol{y}_i\boldsymbol{y}_i^\top = \boldsymbol{I}, \\
& \sum_{i=1}^{n} \boldsymbol{y}_i = \boldsymbol{0},
\end{aligned} \tag{8.9}$$

where $\boldsymbol{I}$ is the identity matrix, the rows of $\mathbb{R}^{n\times p} \ni \boldsymbol{Y} := [\boldsymbol{y}_1, \ldots, \boldsymbol{y}_n]^\top$ are the embedded data points (stacked rowwise), $\boldsymbol{y}_i \in \mathbb{R}^p$ is the i-th embedded data point, and w_{ij} is the weight obtained from the linear reconstruction if $\boldsymbol{x}_j$ is a neighbour of $\boldsymbol{x}_i$ and zero otherwise:

$$w_{ij} := \begin{cases} \widetilde{w}_{ij} & \text{if } \boldsymbol{x}_j \in k\text{NN}(\boldsymbol{x}_i) \\ 0 & \text{Otherwise.} \end{cases} \tag{8.10}$$

The second constraint in Eq. (8.9) ensures a zero mean of the embedded data points. The first and second constraints together satisfy having the unit covariance for the embedded points.

Suppose $\mathbb{R}^n \ni \boldsymbol{w}_i := [w_{i1}, \ldots, w_{in}]^\top$ and let $\mathbb{R}^n \ni \boldsymbol{1}_i := [0, \ldots, 1, \ldots, 0]^\top$ be the vector whose i-th element is one and all other elements are zero. The objective function in Eq. (8.9) can be restated as:

$$\sum_{i=1}^{n} \Big|\Big|\boldsymbol{y}_i - \sum_{j=1}^{n} w_{ij}\boldsymbol{y}_j\Big|\Big|_2^2 = \sum_{i=1}^{n} ||\boldsymbol{Y}^\top\boldsymbol{1}_i - \boldsymbol{Y}^\top\boldsymbol{w}_i||_2^2,$$

which can be stated in matrix form:

$$\sum_{i=1}^{n} ||\boldsymbol{Y}^\top \boldsymbol{1}_i - \boldsymbol{Y}^\top \boldsymbol{w}_i||_2^2 = ||\boldsymbol{Y}^\top \boldsymbol{I} - \boldsymbol{Y}^\top \boldsymbol{W}^\top||_F^2$$
$$= ||\boldsymbol{Y}^\top (\boldsymbol{I} - \boldsymbol{W})^\top||_F^2, \tag{8.11}$$

where the i-th row of $\mathbb{R}^{n\times n} \ni \boldsymbol{W} := [\boldsymbol{w}_1, \dots, \boldsymbol{w}_n]^\top$ includes the weights for the i-th data point and $||.||_F$ denotes the Frobenius norm of matrix. Equation (8.11) is simplified as:

$$||\boldsymbol{Y}^\top (\boldsymbol{I} - \boldsymbol{W})^\top||_F^2 = \mathbf{tr}\big((\boldsymbol{I} - \boldsymbol{W})\boldsymbol{Y}\boldsymbol{Y}^\top (\boldsymbol{I} - \boldsymbol{W})^\top\big)$$
$$= \mathbf{tr}\big(\boldsymbol{Y}^\top (\boldsymbol{I} - \boldsymbol{W})^\top (\boldsymbol{I} - \boldsymbol{W})\boldsymbol{Y}\big) = \mathbf{tr}(\boldsymbol{Y}^\top \boldsymbol{M}\boldsymbol{Y}), \tag{8.12}$$

where $\mathbf{tr}(.)$ denotes the trace of matrix and:

$$\mathbb{R}^{n\times n} \ni \boldsymbol{M} := (\boldsymbol{I} - \boldsymbol{W})^\top (\boldsymbol{I} - \boldsymbol{W}). \tag{8.13}$$

Note that $(\boldsymbol{I} - \boldsymbol{W})$ is the Laplacian of matrix $\boldsymbol{W}$ because the columns of $\boldsymbol{W}$, which are $\boldsymbol{w}_i$'s, add to one (for the constraint used in Eq. (8.1)). Therefore, according to Eq. (8.13), the matrix $\boldsymbol{M}$ can be considered as the Gram matrix over the Laplacian of the weight matrix.

Finally, Eq. (8.9) can be rewritten as:

$$\begin{aligned} \underset{\boldsymbol{Y}}{\text{minimize}} \quad & \mathbf{tr}(\boldsymbol{Y}^\top \boldsymbol{M}\boldsymbol{Y}), \\ \text{subject to} \quad & \frac{1}{n}\boldsymbol{Y}^\top \boldsymbol{Y} = \boldsymbol{I}, \\ & \boldsymbol{Y}^\top \boldsymbol{1} = \boldsymbol{0}, \end{aligned} \tag{8.14}$$

where the dimensionality of $\boldsymbol{1}$ and $\boldsymbol{0}$ are $\mathbb{R}^n$ and $\mathbb{R}^p$, respectively. Note that Sect. 8.8.6.2 will demonstrate that Eq. (8.14) can be interpreted as the maximized dependence between the input data $\boldsymbol{X}$ and the embedded data $\boldsymbol{Y}$. Proposition 8.1 will demonstrate that the second constraint will be satisfied implicitly. Therefore, if the second constraint is ignored, the Lagrangian for Eq. (8.14) is [11]:

$$\mathcal{L} = \mathbf{tr}(\boldsymbol{Y}^\top \boldsymbol{M}\boldsymbol{Y}) - \mathbf{tr}\big(\boldsymbol{\Lambda}^\top (\frac{1}{n}\boldsymbol{Y}^\top \boldsymbol{Y} - \boldsymbol{I})\big),$$

where $\boldsymbol{\Lambda} \in \mathbb{R}^{n\times n}$ is a diagonal matrix including the Lagrange multipliers. Equating the derivative of $\mathcal{L}$ to zero results in:

$$\mathbb{R}^{n\times p} \ni \frac{\partial \mathcal{L}}{\partial \boldsymbol{Y}} = 2\boldsymbol{M}\boldsymbol{Y} - \frac{2}{n}\boldsymbol{Y}\boldsymbol{\Lambda} \overset{\text{set}}{=} \boldsymbol{0} \implies \boldsymbol{M}\boldsymbol{Y} = \boldsymbol{Y}(\frac{1}{n}\boldsymbol{\Lambda}), \tag{8.15}$$

which is the eigenvalue problem for $\boldsymbol{M}$ (see Chap. 2). Therefore, the columns of $\boldsymbol{Y}$ are the eigenvectors of $\boldsymbol{M}$, where the eigenvalues are the diagonal elements of $(1/n)\boldsymbol{\Lambda}$.

Equation (8.14) is a minimization problem; therefore, the columns of $\boldsymbol{Y}$ should be sorted from the smallest to largest eigenvalues. Moreover, recall that $(\boldsymbol{I} - \boldsymbol{W})$ in $\boldsymbol{M}$ is the Laplacian matrix for the weights $\boldsymbol{W}$. It is well-known in linear algebra and graph theory that if a graph has k disjoint connected parts, its Laplacian matrix has k zero eigenvalues (see [47, Theorem 3.10] and [1, 52]). As the kNN graph, or $\boldsymbol{W}$, is a connected graph, $(\boldsymbol{I} - \boldsymbol{W})$ has one zero eigenvalue, whose eigenvector is $\boldsymbol{1} = [1, 1, \dots, 1]^\top$. After sorting the eigenvectors from smallest to largest eigenvalues, the first eigenvector having zero eigenvalue is ignored, and the p smallest eigenvectors of $\boldsymbol{M}$ with nonzero eigenvalues are taken as the columns of $\boldsymbol{Y} \in \mathbb{R}^{n \times p}$.

Proposition 8.1 *The eigenvector* $\boldsymbol{1}$ *with zero eigenvalue implicitly ensures that* $\sum_{i=1}^{n} \boldsymbol{y}_i = \boldsymbol{Y}^\top \boldsymbol{1} = \boldsymbol{0}$, *which was the second constraint.*

Proof Suppose the eigenvectors are sorted from the smallest to largest eigenvalues. Let $\boldsymbol{v}_i \in \mathbb{R}^n$ and $\lambda_i \in \mathbb{R}$ be the i-th eigenvector and eigenvalue, respectively. Therefore, in Eq. (8.15), if all of the eigenvectors and not just p of them are considered, then $\boldsymbol{Y} = [\boldsymbol{y}_1, \dots, \boldsymbol{y}_n]^\top = [\boldsymbol{v}_1, \dots, \boldsymbol{v}_n] \in \mathbb{R}^{n \times n}$. Eigenvectors are orthogonal by definition; therefore, $\boldsymbol{v}_1^\top \boldsymbol{v}_i = 0, \forall i \neq 1$. It is known that $\boldsymbol{v}_1 = \boldsymbol{1}$ with $\lambda_1 = 0$; therefore, $\boldsymbol{1}^\top \boldsymbol{v}_i = 0$, which means that the elements of every eigenvector, $\boldsymbol{v}_i, \forall i \neq 1$, add to zero. On the other hand, $[\boldsymbol{y}_1, \dots, \boldsymbol{y}_n]^\top = [\boldsymbol{v}_1, \dots, \boldsymbol{v}_n]$, meaning that the summation of a component among the $\boldsymbol{y}_i$'s (embedded data points) is zero. As the summation for 'every' component among $\boldsymbol{y}_i$'s is zero, then $\sum_{i=1}^{n} \boldsymbol{y}_i = \boldsymbol{0}$. This explanation can be summarized in this sentence: "Discarding this eigenvector enforces the constraint that the outputs have zero mean, since the components of other eigenvectors must sum to zero, by virtue of orthogonality with the bottom one (with smallest eigenvalue)" [55]. □

8.2.4 Additional Notes on LLE

8.2.4.1 Inverse Locally Linear Embedding

It is possible to have an inverse LLE where the data point $\boldsymbol{x}_i \in \mathbb{R}^d$ is in the input space for an embedding point $\boldsymbol{y}_i \in \mathbb{R}^p$ [55, Section 6.1]. If kNN is found in the embedding space, let $\boldsymbol{y}_{ij}$ denote the j-th neighbour of $\boldsymbol{y}_i$ in that embedding space. The following problem is solved to find the reconstruction weights, $\{\widetilde{w}_{ij}\}_{j=1}^{k}$, in the embedding space:

$$\begin{aligned} & \underset{\{\widetilde{w}_{ij}\}_{j=1}^k}{\text{minimize}} && \Big|\Big|\boldsymbol{y}_i - \sum_{j=1}^k \widetilde{w}_{ij}\, \boldsymbol{y}_{ij}\Big|\Big|_2^2, \\ & \text{subject to} && \sum_{j=1}^k \widetilde{w}_{ij} = 1, \end{aligned} \tag{8.16}$$

which is solved similarly to how Eq. (8.1) is solved. Thereafter, $\{w_{ij}\}_{j=1}^k$ are obtained from $\{\widetilde{w}_{ij}\}_{j=1}^k$, using Eq. (8.10). The original point in the input space is approximated using the obtained reconstruction weights:

$$\mathbb{R}^d \ni \boldsymbol{x}_i \approx \sum_{j=1}^k w_{ij}\, \boldsymbol{x}_j. \tag{8.17}$$

8.2.4.2 Feature Fusion in LLE

It is noteworthy that, in some cases, data points are represented by Q different features, i.e., $\{\boldsymbol{x}_i^q \,|\, i = 1, \dots, n, q = 1, \dots, Q\}$. In these cases, feature fusion using LLE [61], where Q weights obtained from LLE, denoted by $\boldsymbol{W}_1, \dots, \boldsymbol{W}_Q$, are needed. These weights can be combined as follows:

$$\bar{\boldsymbol{W}} := \frac{1}{Q} \sum_{q=1}^{Q} \boldsymbol{W}_q, \tag{8.18}$$

and $\bar{\boldsymbol{W}}$ is used in Eq. (8.13), rather than $\boldsymbol{W}$. The embedding optimization is used for finding the embeddings, although the data have Q features [61].

8.3 Kernel Locally Linear Embedding

It is possible to map data $\{\boldsymbol{x}_i \in \mathbb{R}^d\}_{i=1}^n$ to a higher-dimensional feature space, hoping to have the data fall close to a simpler-to-analyze manifold in the feature space. Suppose $\boldsymbol{\phi} : \boldsymbol{x} \to \mathcal{H}$ is the pulling function which maps data $\{\boldsymbol{x}_i\}_{i=1}^n$ to the feature space. In other words, $\boldsymbol{x}_i \mapsto \boldsymbol{\phi}(\boldsymbol{x}_i)$. Let t denote the dimensionality of the feature space, i.e., $\boldsymbol{\phi}(\boldsymbol{x}_i) \in \mathbb{R}^t$. Usually, there is $t \gg d$. The kernel of the data points $\boldsymbol{x}_1$ and $\boldsymbol{x}_2$ is $\boldsymbol{\phi}(\boldsymbol{x}_1)^\top \boldsymbol{\phi}(\boldsymbol{x}_2) \in \mathbb{R}$ [33]. The kernel matrix for the n data points is $\mathbb{R}^{n\times n} \ni \boldsymbol{K} := \boldsymbol{\Phi}(\boldsymbol{X})^\top \boldsymbol{\Phi}(\boldsymbol{X})$ where $\boldsymbol{\Phi}(\boldsymbol{X}) := [\boldsymbol{\phi}(\boldsymbol{x}_1), \dots, \boldsymbol{\phi}(\boldsymbol{x}_n)] \in \mathbb{R}^{t\times n}$. Kernel LLE [74] maps data to the feature space and performs the steps of kNN and linear reconstruction in the feature space.

8.3.1 k-Nearest Neighbours

As explained in Chap. 3, the Euclidean distance in the feature space is:

$$||\boldsymbol{\phi}(\boldsymbol{x}_i) - \boldsymbol{\phi}(\boldsymbol{x}_j)||_2 = \sqrt{k(\boldsymbol{x}_i, \boldsymbol{x}_i) - 2k(\boldsymbol{x}_i, \boldsymbol{x}_j) + k(\boldsymbol{x}_j, \boldsymbol{x}_j)}, \tag{8.19}$$

where $\mathbb{R} \ni k(\boldsymbol{x}_i, \boldsymbol{x}_j) = \boldsymbol{\phi}(\boldsymbol{x}_i)^\top \boldsymbol{\phi}(\boldsymbol{x}_j)$ is the (i, j)-th element of $\boldsymbol{K}$. Using the distances of the data points in the feature space, i.e. Eq. (8.19), it is possible to construct the kNN graph. Therefore, every data point has k neighbours in the feature space. Let the matrix $\mathbb{R}^{t \times k} \ni \boldsymbol{\Phi}(\boldsymbol{X}_i) := [\boldsymbol{\phi}(\boldsymbol{x}_{i1}), \dots, \boldsymbol{\phi}(\boldsymbol{x}_{ik})]$ include the neighbours of $\boldsymbol{x}_i$ in the feature space.

8.3.2 Linear Reconstruction by the Neighbours

Equation (8.1) in the feature space is:

$$\begin{aligned} \underset{\widetilde{\boldsymbol{W}}}{\text{minimize}} \quad & \varepsilon(\widetilde{\boldsymbol{W}}) := \sum_{i=1}^{n} \Big|\Big|\boldsymbol{\phi}(\boldsymbol{x}_i) - \sum_{j=1}^{k} \widetilde{w}_{ij} \boldsymbol{\phi}(\boldsymbol{x}_{ij})\Big|\Big|_2^2, \\ \text{subject to} \quad & \sum_{j=1}^{k} \widetilde{w}_{ij} = 1, \quad \forall i \in \{1, \dots, n\}. \end{aligned} \tag{8.20}$$

It is possible to restate $\varepsilon(\widetilde{\boldsymbol{W}})$ as follows:

$$\begin{aligned} \varepsilon(\widetilde{\boldsymbol{W}}) &= \sum_{i=1}^{n} \Big|\Big|\boldsymbol{\phi}(\boldsymbol{x}_i) - \sum_{j=1}^{k} \widetilde{w}_{ij} \boldsymbol{\phi}(\boldsymbol{x}_{ij})\Big|\Big|_2^2 \\ &\overset{(a)}{=} \sum_{i=1}^{n} \Big|\Big|\sum_{j=1}^{k} \widetilde{w}_{ij} \boldsymbol{\phi}(\boldsymbol{x}_i) - \sum_{j=1}^{k} \widetilde{w}_{ij} \boldsymbol{\phi}(\boldsymbol{x}_{ij})\Big|\Big|_2^2 = \sum_{i=1}^{n} \Big|\Big|\sum_{j=1}^{k} \widetilde{w}_{ij} \big(\boldsymbol{\phi}(\boldsymbol{x}_i) - \boldsymbol{\phi}(\boldsymbol{x}_{ij})\big)\Big|\Big|_2^2, \end{aligned}$$

where (a) is because $\sum_{j=1}^{k} \widetilde{w}_{ij} = 1$. The following is defined as:

$$\mathbb{R}^{t \times k} \ni \boldsymbol{P}_i = [\boldsymbol{p}_{i1}, \dots, \boldsymbol{p}_{ik}] := \big[\boldsymbol{\phi}(\boldsymbol{x}_i) - \boldsymbol{\phi}(\boldsymbol{x}_{i1}), \dots, \boldsymbol{\phi}(\boldsymbol{x}_i) - \boldsymbol{\phi}(\boldsymbol{x}_{ik})\big]. \tag{8.21}$$

Therefore:

$$\varepsilon(\widetilde{\boldsymbol{W}}) = \sum_{i=1}^{n} \Big|\Big|\sum_{j=1}^{k} \widetilde{w}_{ij} \big(\boldsymbol{\phi}(\boldsymbol{x}_i) - \boldsymbol{\phi}(\boldsymbol{x}_{ij})\big)\Big|\Big|_2^2 = \sum_{i=1}^{n} \Big|\Big|\sum_{j=1}^{k} \widetilde{w}_{ij} \boldsymbol{p}_{ij}\Big|\Big|_2^2 = \sum_{i=1}^{n} ||\boldsymbol{P}_i \widetilde{\boldsymbol{w}}_i||_2^2$$

$$= \sum_{i=1}^{n} (\boldsymbol{P}_i \widetilde{\boldsymbol{w}}_i)^\top (\boldsymbol{P}_i \widetilde{\boldsymbol{w}}_i) = \sum_{i=1}^{n} \widetilde{\boldsymbol{w}}_i^\top \boldsymbol{P}_i^\top \boldsymbol{P}_i \widetilde{\boldsymbol{w}}_i = \sum_{i=1}^{n} \widetilde{\boldsymbol{w}}_i^\top \boldsymbol{K}_i \widetilde{\boldsymbol{w}}_i, \tag{8.22}$$

where $\mathbb{R}^{k \times k} \ni \boldsymbol{K}_i := \boldsymbol{P}_i^\top \boldsymbol{P}_i$. The (a, b)-th element of $\boldsymbol{K}_i$ can be calculated as:

$$\begin{aligned}
\boldsymbol{K}_i(a, b) &= \boldsymbol{p}_{ia}^\top \boldsymbol{p}_{ib} = \big(\boldsymbol{\phi}(\boldsymbol{x}_i) - \boldsymbol{\phi}(\boldsymbol{x}_{ia})\big)^\top \big(\boldsymbol{\phi}(\boldsymbol{x}_i) - \boldsymbol{\phi}(\boldsymbol{x}_{ib})\big) \\
&= \boldsymbol{\phi}(\boldsymbol{x}_i)^\top \boldsymbol{\phi}(\boldsymbol{x}_i) - \boldsymbol{\phi}(\boldsymbol{x}_i)^\top \boldsymbol{\phi}(\boldsymbol{x}_{ia}) - \boldsymbol{\phi}(\boldsymbol{x}_i)^\top \boldsymbol{\phi}(\boldsymbol{x}_{ib}) + \boldsymbol{\phi}(\boldsymbol{x}_{ia})^\top \boldsymbol{\phi}(\boldsymbol{x}_{ib}) \\
&= k(\boldsymbol{x}_i, \boldsymbol{x}_i) - k(\boldsymbol{x}_i, \boldsymbol{x}_{ia}) - k(\boldsymbol{x}_i, \boldsymbol{x}_{ib}) + k(\boldsymbol{x}_{ia}, \boldsymbol{x}_{ib}).
\end{aligned} \tag{8.23}$$

Therefore, Eq. (8.20) is restated to:

$$\begin{aligned}
&\underset{\{\widetilde{\boldsymbol{w}}_i\}_{i=1}^n}{\text{minimize}} \quad \sum_{i=1}^{n} \widetilde{\boldsymbol{w}}_i^\top \boldsymbol{K}_i \, \widetilde{\boldsymbol{w}}_i, \\
&\text{subject to} \quad \boldsymbol{1}^\top \widetilde{\boldsymbol{w}}_i = 1, \quad \forall i \in \{1, \dots, n\}.
\end{aligned} \tag{8.24}$$

The Lagrangian for Eq. (8.24) is [11]:

$$\mathcal{L} = \widetilde{\boldsymbol{w}}_i^\top \boldsymbol{K}_i \, \widetilde{\boldsymbol{w}}_i - \sum_{i=1}^{n} \lambda_i \, (\boldsymbol{1}^\top \widetilde{\boldsymbol{w}}_i - 1).$$

Setting the derivative of the Lagrangian to zero results in:

$$\mathbb{R}^k \ni \frac{\partial \mathcal{L}}{\partial \widetilde{\boldsymbol{w}}_i} = 2\boldsymbol{K}_i \widetilde{\boldsymbol{w}}_i - \lambda_i \boldsymbol{1} \overset{\text{set}}{=} \boldsymbol{0} \implies \widetilde{\boldsymbol{w}}_i = \frac{1}{2} \boldsymbol{K}_i^{-1} \lambda_i \boldsymbol{1} = \frac{\lambda_i}{2} \boldsymbol{K}_i^{-1} \boldsymbol{1}. \tag{8.25}$$

$$\mathbb{R} \ni \frac{\partial \mathcal{L}}{\partial \lambda} = \boldsymbol{1}^\top \widetilde{\boldsymbol{w}}_i - 1 \overset{\text{set}}{=} 0 \implies \boldsymbol{1}^\top \widetilde{\boldsymbol{w}}_i = 1. \tag{8.26}$$

Using Eqs. (8.25) and (8.26) results in:

$$\frac{\lambda_i}{2} \boldsymbol{1}^\top \boldsymbol{K}_i^{-1} \boldsymbol{1} = 1 \implies \lambda_i = \frac{2}{\boldsymbol{1}^\top \boldsymbol{K}_i^{-1} \boldsymbol{1}}. \tag{8.27}$$

Using Eqs. (8.25) and (8.27) results in:

$$\widetilde{\boldsymbol{w}}_i = \frac{\lambda_i}{2} \boldsymbol{K}_i^{-1} \boldsymbol{1} = \frac{\boldsymbol{K}_i^{-1} \boldsymbol{1}}{\boldsymbol{1}^\top \boldsymbol{K}_i^{-1} \boldsymbol{1}}. \tag{8.28}$$

8.3.3 *Linear Embedding*

The linear embedding step in kernel LLE is the same as the linear embedding step in LLE (see Sect. 8.2.3).

8.4 Out-of-Sample Embedding in LLE

Suppose there are n_t out-of-sample (test) data points, i.e., $\mathbb{R}^{d \times n_t} \ni \boldsymbol{X}^{(t)} := [\boldsymbol{x}_1^{(t)}, \ldots, \boldsymbol{x}_{n_t}^{(t)}]$, which are not used for training. Let $\boldsymbol{x}_i^{(t)} \in \mathbb{R}^d$ denote the i-th out-of-sample data point. There are several approaches for the out-of-sample extension of LLE that can be used to find the low-dimensional embedding of out-of-sample points, denoted by $\{\boldsymbol{y}_i^{(t)} \in \mathbb{R}^p\}_{i=1}^{n_t}$ or $\boldsymbol{Y}_t = [\boldsymbol{y}_1^{(t)}, \ldots, \boldsymbol{y}_n^{(t)}] \in \mathbb{R}^{p \times n_t}$, after the training phase. Incremental LLE [40] and SLLEP [44], which are out-of-sample extensions of LLE methods, will be introduced in Sects. 8.5 and 8.8.3, respectively. For additional methods for out-of-sample extension of LEE that are not discussed in this section, see [12].

8.4.1 *Out-of-Sample Embedding Using Linear Reconstruction*

One way of extending LLE for out-of-sample data points is to use linear reconstruction [55]. For every out-of-sample data point $\boldsymbol{x}_i^{(t)}$, it is necessary to first find the kNN among the training points. Let $\boldsymbol{x}_{ij}^{(t)}$ denote the j-th training neighbour of $\boldsymbol{x}_i^{(t)}$ and let the matrix $\mathbb{R}^{d \times k} \ni \boldsymbol{X}_i^{(t)} := [\boldsymbol{x}_{i1}^{(t)}, \ldots, \boldsymbol{x}_{ik}^{(t)}]$ include the training neighbours of $\boldsymbol{x}_i^{(t)}$. The goal is to reconstruct every out-of-sample point by its training neighbours (note that out-of-sample points are not considered neighbours). Therefore, using an optimization problem similar to Eq. (8.1), results in:

$$
\begin{aligned}
\underset{\widetilde{\boldsymbol{W}}^{(t)}}{\text{minimize}} \quad & \varepsilon(\widetilde{\boldsymbol{W}}^{(t)}) := \sum_{i=1}^{n_t} \Big|\Big| \boldsymbol{x}_i^{(t)} - \sum_{j=1}^{k} \widetilde{w}_{ij}^{(t)} \boldsymbol{x}_{ij}^{(t)} \Big|\Big|_2^2, \\
\text{subject to} \quad & \sum_{j=1}^{k} \widetilde{w}_{ij}^{(t)} = 1, \quad \forall i \in \{1, \ldots, n_t\},
\end{aligned} \tag{8.29}
$$

where $\mathbb{R}^{n_t \times k} \ni \widetilde{\boldsymbol{W}}^{(t)} := [\widetilde{\boldsymbol{w}}_1^{(t)}, \ldots, \widetilde{\boldsymbol{w}}_{n_t}^{(t)}]^\top$ includes the weights and $\mathbb{R}^k \ni \widetilde{\boldsymbol{w}}_i^{(t)} := [\widetilde{w}_{i1}^{(t)}, \ldots, \widetilde{w}_{ik}^{(t)}]^\top$ includes the weights of the linear reconstruction of the m-th out-of-sample data point using its k training neighbours. Is it possible to restate the $\varepsilon(\widetilde{\boldsymbol{W}}^{(t)})$ as:

$$\varepsilon(\widetilde{\boldsymbol{W}}^{(t)}) = \sum_{i=1}^{n_t} ||\boldsymbol{x}_i^{(t)} - \boldsymbol{X}_i^{(t)} \widetilde{\boldsymbol{w}}_i^{(t)}||_2^2. \tag{8.30}$$

The constraint $\sum_{j=1}^{k} \widetilde{w}_{ij}^{(t)} = 1$ is restated as $\boldsymbol{1}^\top \widetilde{\boldsymbol{w}}_i^{(t)} = 1$; therefore, $\boldsymbol{x}_i^{(t)} = \boldsymbol{x}_i^{(t)} \boldsymbol{1}^\top \widetilde{\boldsymbol{w}}_i$. This term can be simplified in $\varepsilon(\widetilde{\boldsymbol{W}}^{(t)})$ as:

$$\begin{aligned} ||\boldsymbol{x}_i^{(t)} - \boldsymbol{X}_i^{(t)} \widetilde{\boldsymbol{w}}_i^{(t)}||_2^2 &= ||\boldsymbol{x}_i^{(t)} \boldsymbol{1}^\top \widetilde{\boldsymbol{w}}_i^{(t)} - \boldsymbol{X}_i^{(t)} \widetilde{\boldsymbol{w}}_i^{(t)}||_2^2 = ||(\boldsymbol{x}_i^{(t)} \boldsymbol{1}^\top - \boldsymbol{X}_i^{(t)})\, \widetilde{\boldsymbol{w}}_i^{(t)}||_2^2 \\ &= \widetilde{\boldsymbol{w}}_i^{(t)\top} (\boldsymbol{x}_i^{(t)} \boldsymbol{1}^\top - \boldsymbol{X}_i^{(t)})^\top (\boldsymbol{x}_i^{(t)} \boldsymbol{1}^\top - \boldsymbol{X}_i^{(t)})\, \widetilde{\boldsymbol{w}}_i^{(t)} = \widetilde{\boldsymbol{w}}_i^{(t)\top} \boldsymbol{G}_i^{(t)}\, \widetilde{\boldsymbol{w}}_i^{(t)}, \end{aligned}$$

where:

$$\mathbb{R}^{k \times k} \ni \boldsymbol{G}_i^{(t)} := (\boldsymbol{x}_i^{(t)} \boldsymbol{1}^\top - \boldsymbol{X}_i^{(t)})^\top (\boldsymbol{x}_i^{(t)} \boldsymbol{1}^\top - \boldsymbol{X}_i^{(t)}). \tag{8.31}$$

Equation (8.29) can be rewritten as:

$$\begin{aligned} \underset{\widetilde{\boldsymbol{W}}^{(t)}}{\text{minimize}} \quad & \sum_{i=1}^{n_t} \widetilde{\boldsymbol{w}}_i^{(t)\top} \boldsymbol{G}_i^{(t)}\, \widetilde{\boldsymbol{w}}_i^{(t)}, \\ \text{subject to} \quad & \boldsymbol{1}^\top \widetilde{\boldsymbol{w}}_i^{(t)} = 1, \quad \forall i \in \{1, \dots, n_t\}. \end{aligned} \tag{8.32}$$

This problem is solved similarly to the solution for Eq. (8.4). Therefore, similar to Eq. (8.8):

$$\widetilde{\boldsymbol{w}}_i^{(t)} = \frac{(\boldsymbol{G}_i^{(t)})^{-1} \boldsymbol{1}}{\boldsymbol{1}^\top (\boldsymbol{G}_i^{(t)})^{-1} \boldsymbol{1}}. \tag{8.33}$$

The embedding of the out-of-sample $\boldsymbol{x}_i^{(t)}$ is obtained by the linear combination (reconstruction) of the embedding of its k training neighbours:

$$\mathbb{R}^p \ni \boldsymbol{y}_i^{(t)} = \sum_{j=1}^{k} \widetilde{w}_{ij}^{(t)} \boldsymbol{y}_j. \tag{8.34}$$

8.4.2 Out-of-Sample Embedding Using Eigenfunctions

Another way of extending LLE for out-of-sample data points is to use eigenfunctions, which were introduced in Chap. 3. Let the MDS or Isomap embedding of the point $\boldsymbol{x}$ be $\mathbb{R}^p \ni \boldsymbol{y}(\boldsymbol{x}) = [y_1(\boldsymbol{x}), \dots, y_p(\boldsymbol{x})]^\top$. The k-th dimension of this embedding is (see Chap. 3):

$$y_k(\boldsymbol{x}) = \sqrt{\delta_k}\,\frac{f_k(\boldsymbol{x})}{\sqrt{n}} = \frac{1}{\sqrt{\delta_k}} \sum_{i=1}^{n} v_{ki}\, \breve{k}_t(\boldsymbol{x}_i, \boldsymbol{x}). \tag{8.35}$$

If a set of n_t out-of-sample data points exists, $\breve{k}_t(\boldsymbol{x}_i, \boldsymbol{x})$ is an element of the centered out-of-sample kernel (see Chap. 3):

$$\mathbb{R}^{n\times n_t} \ni \breve{\boldsymbol{K}}_t = \boldsymbol{K}_t - \frac{1}{n}\boldsymbol{1}_{n\times n}\boldsymbol{K}_t - \frac{1}{n}\boldsymbol{K}\boldsymbol{1}_{n\times n_t} + \frac{1}{n^2}\boldsymbol{1}_{n\times n}\boldsymbol{K}\boldsymbol{1}_{n\times n_t}, \tag{8.36}$$

where $\boldsymbol{1} := [1, 1, \dots, 1]^\top$, $\boldsymbol{K}_t \in \mathbb{R}^{n\times n_t}$ is the not-necessarily-centered out-of-sample kernel, and $\boldsymbol{K} \in \mathbb{R}^{n\times n}$ is the training kernel. Equation (8.35) can be used to embed the i-th out-of-sample data point $\boldsymbol{x}_i^{(t)}$. For this purpose, $\boldsymbol{x}_i^{(t)}$ should be used in place of $\boldsymbol{x}$ in Eq. (8.35).

Equation (8.35) can be used to embed the i-th out-of-sample data point $\boldsymbol{x}_i^{(t)}$. For this purpose, $\boldsymbol{x}_i^{(t)}$ should be used in place of $\boldsymbol{x}$ in Eq. (8.35). Note that Eq. (8.35) requires Eq. (8.36). A notion of kernel is required in LLE. LLE can be seen as a special case of kernel Principal Component Analysis (kernel PCA), where the inverse, or negative sign, of $\boldsymbol{M}$ can be interpreted as its kernel because Eq. (8.14) is a minimization, but kernel PCA optimization is a maximization problem (see Chap. 5). This will be explained more in Chap. 10. For more details on seeing LLE as a kernel PCA, see [7, 23, 29, 58] and [60, Table 2.1]. Therefore, the kernel in LLE can be [9]:

$$\begin{aligned} &\boldsymbol{M} \overset{(8.13)}{=} (\boldsymbol{I} - \boldsymbol{W})^\top(\boldsymbol{I} - \boldsymbol{W}) = \boldsymbol{I} - \boldsymbol{W} - \boldsymbol{W}^\top + \boldsymbol{W}^\top\boldsymbol{W}, \\ &\mathbb{R}^{n\times n} \ni \boldsymbol{K} := \mu\boldsymbol{I} - \boldsymbol{M}, \end{aligned} \tag{8.37}$$

$$\therefore \quad \boldsymbol{K}(i, j) = (\mu - 1)\,\delta_{ij} + w_{ij} + w_{ji} - \sum_{r=1}^{n} w_{ri}\, w_{rj}, \tag{8.38}$$

where δ_{ij} is the Kronecker delta, which is one if $i = j$ and is zero otherwise. If the hard similarity δ_{ij} is modified to a soft similarity $w_{ij}^{(t)}$, Eq. (8.38) can be slightly modified to [9]:

$$\boldsymbol{K}(\boldsymbol{x}_i, \boldsymbol{x}_j) = (\mu - 1)\, w_{ij}^{(t)} + w_{ij}^{(t)} + w_{ji}^{(t)} - \sum_{r=1}^{n} w_{ri}^{(t)}\, w_{rj}^{(t)}, \tag{8.39}$$

where either $\boldsymbol{x}_i$ or $\boldsymbol{x}_j$ (but not both) is an out-of-sample data point, i.e., there is either $\boldsymbol{x}_i^{(t)}$ or $\boldsymbol{x}_j^{(t)}$. The first and second terms in Eq. (8.39) are defined as [9]:

$$\boldsymbol{K}'_t(\boldsymbol{x}_i, \boldsymbol{x}_j) := w_{ij}^{(t)},$$

$$\boldsymbol{K}''_t(\boldsymbol{x}_i, \boldsymbol{x}_j) := w_{ij}^{(t)} + w_{ji}^{(t)} - \sum_{r=1}^{n} w_{ri}^{(t)} w_{rj}^{(t)},$$

respectively. Therefore, Eq. (8.39) can be restated as:

$$\boldsymbol{K}(\boldsymbol{x}_i, \boldsymbol{x}_j) = (\mu - 1)\, \boldsymbol{K}'_t(\boldsymbol{x}_i, \boldsymbol{x}_j) + \boldsymbol{K}''_t(\boldsymbol{x}_i, \boldsymbol{x}_j). \tag{8.40}$$

In LLE, the embeddings are the eigenvectors of $\boldsymbol{M}$. According to Eq. (8.37), the embeddings $\boldsymbol{y}$'s are the eigenvectors of the kernel, previously denoted by $\boldsymbol{v}$'s. Therefore, it is possible to employ Eq. (8.35), in which the kernel of LLE, Eq. (8.40), is used [9] (the dummy iterator[1] is changed from i to j):

$$y_r(\boldsymbol{x}) = \frac{1}{\sqrt{\delta_r}} \sum_{j=1}^{n} y_{jr} \Big((\mu - 1)\, \boldsymbol{K}'_t(\boldsymbol{x}_j, \boldsymbol{x}) + \boldsymbol{K}''_t(\boldsymbol{x}_j, \boldsymbol{x}) \Big). \tag{8.41}$$

Therefore, the r-th element of the out-of-sample embedding $\mathbb{R}^p \ni \boldsymbol{y}_i^{(t)} = [y_{i1}, \ldots, y_{ip}]^\top$ for $\boldsymbol{x}_i^{(t)} \in \mathbb{R}^d$ is:

$$y_{ir}^{(t)} = \frac{1}{\sqrt{\delta_r}} \sum_{j=1}^{n} y_{jr} \Big((\mu - 1)\, w_{ij}^{(t)} + \boldsymbol{K}''_t(\boldsymbol{x}_j, \boldsymbol{x}_i^{(t)}) \Big), \tag{8.42}$$

where y_{jr} denotes the r-th element of $\boldsymbol{y}_j \in \mathbb{R}^p$.

Corollary 8.1 *Out-of-sample embedding by linear reconstruction, i.e., Eq. (8.34), is a special case of out-of-sample embedding by eigenfunctions, i.e. Eq. (8.42), for* $\mu \to \infty$.

Proof On the one hand, inspired by Eq. (8.10), it is possible to restate Eq. (8.34) as:

$$\mathbb{R}^p \ni \boldsymbol{y}_i^{(t)} = \sum_{j=1}^{k} \widetilde{w}_{ij}^{(t)} \boldsymbol{y}_j = \sum_{j=1}^{n} w_{ij}^{(t)} \boldsymbol{y}_j,$$

whose elementwise expression is:

$$y_{ir}^{(t)} = \sum_{j=1}^{n} w_{ij}^{(t)} y_{jr}.$$

[1] A dummy iterator is a variable that can be renamed to any variable within the summation, without impacting the result of summation.

On the other hand, by $\mu \to \infty$, the first term in Eq. (8.42) dominates its second term as:

$$y_{ir}^{(t)} = \frac{\mu}{\sqrt{\delta_r}} \sum_{j=1}^{n} y_{jr}\, w_{ij}^{(t)}.$$

Scale is not important in manifold embedding; therefore, these two expressions are equivalent. □

8.4.3 Out-of-sample Embedding Using Kernel Mapping

There is a kernel mapping method [25, 26] to embed the out-of-sample data in LLE or kernel LLE. A mapping that maps any data point as $\boldsymbol{x} \mapsto \boldsymbol{y}(\boldsymbol{x})$ is defined, where:

$$\mathbb{R}^p \ni \boldsymbol{y}(\boldsymbol{x}) := \sum_{j=1}^{n} \boldsymbol{\alpha}_j \frac{k(\boldsymbol{x}, \boldsymbol{x}_j)}{\sum_{\ell=1}^{n} k(\boldsymbol{x}, \boldsymbol{x}_\ell)}, \tag{8.43}$$

and $\boldsymbol{\alpha}_j \in \mathbb{R}^p$, and $\boldsymbol{x}_j$ and $\boldsymbol{x}_\ell$ denote the j-th and ℓ-th training data points, respectively. $k(\boldsymbol{x}, \boldsymbol{x}_j)$ is a kernel, such as the Gaussian kernel:

$$k(\boldsymbol{x}, \boldsymbol{x}_j) = \exp\Big(\frac{-||\boldsymbol{x} - \boldsymbol{x}_j||_2^2}{2\sigma_j^2}\Big), \tag{8.44}$$

where σ_j is calculated as [25]:

$$\sigma_j := \gamma \times \min_i(||\boldsymbol{x}_j - \boldsymbol{x}_i||_2), \tag{8.45}$$

where γ is a small positive number.

If embedding the training data points using LLE or kernel LLE has already occurred, then the set $\{\boldsymbol{y}_i\}_{i=1}^n$ is available. If the training data points are mapped, the aim is to minimize the following least-squares cost function, to obtain $\boldsymbol{y}(\boldsymbol{x}_i)$ close to $\boldsymbol{y}_i$ for the i-th training point:

$$\underset{\boldsymbol{\alpha}_j\text{'s}}{\text{minimize}} \quad \sum_{i=1}^{n} ||\boldsymbol{y}_i - \boldsymbol{y}(\boldsymbol{x}_i)||_2^2, \tag{8.46}$$

where the summation is over the training data points. This cost function can be written in matrix form as:

$$\underset{\boldsymbol{A}}{\text{minimize}} \quad ||\boldsymbol{Y} - \boldsymbol{K}''\boldsymbol{A}||_F^2, \tag{8.47}$$

where $\mathbb{R}^{n\times p} \ni \boldsymbol{Y} := [\boldsymbol{y}_1, \dots, \boldsymbol{y}_n]^\top$ and $\mathbb{R}^{n\times p} \ni \boldsymbol{A} := [\boldsymbol{\alpha}_1, \dots, \boldsymbol{\alpha}_n]^\top$. $\boldsymbol{K}'' \in \mathbb{R}^{n\times n}$ is the kernel matrix whose (i, j)-th element is defined as:

$$\boldsymbol{K}''(i, j) := \frac{k(\boldsymbol{x}_i, \boldsymbol{x}_j)}{\sum_{\ell=1}^{n} k(\boldsymbol{x}_i, \boldsymbol{x}_\ell)}. \tag{8.48}$$

Equation (8.47) is always nonnegative; thus, its smallest value is zero. Therefore, the solution to this equation is:

$$\boldsymbol{Y} - \boldsymbol{K}''\boldsymbol{A} = \boldsymbol{0} \implies \boldsymbol{Y} = \boldsymbol{K}''\boldsymbol{A} \overset{(a)}{\implies} \boldsymbol{A} = \boldsymbol{K}''^{\dagger}\,\boldsymbol{Y}, \tag{8.49}$$

where $\boldsymbol{K}''^{\dagger}$ is the pseudoinverse of $\boldsymbol{K}''$:

$$\boldsymbol{K}''^{\dagger} = (\boldsymbol{K}''^{\top}\boldsymbol{K}'')^{-1}\boldsymbol{K}''^{\top}, \tag{8.50}$$

and (a) is because $\boldsymbol{K}''^{\dagger}\,\boldsymbol{K}'' = \boldsymbol{I}$.

Finally, the mapping of Eq. (8.43) for the n_t out-of-sample data points is:

$$\boldsymbol{Y}_t = \boldsymbol{K}_t''\,\boldsymbol{A}, \tag{8.51}$$

where the (i, j)-th element of the out-of-sample kernel matrix $\boldsymbol{K}_t'' \in \mathbb{R}^{n_t\times n}$ is:

$$\boldsymbol{K}_t''(i, j) := \frac{k(\boldsymbol{x}_i^{(t)}, \boldsymbol{x}_j)}{\sum_{\ell=1}^{n} k(\boldsymbol{x}_i^{(t)}, \boldsymbol{x}_\ell)}, \tag{8.52}$$

where $\boldsymbol{x}_i^{(t)}$ is the i-th out-of-sample data point, and $\boldsymbol{x}_j$ and $\boldsymbol{x}_\ell$ are the j-th and ℓ-th training data points, respectively.

8.5 Incremental LLE

Assume there is a stream of online data where the data points arrive gradually; therefore, data increments by time. Incremental LLE was proposed by Kouropteva et al. [40] to handle online data, by embedding newly arrived data using the already embedded data. In this sense, it can also be used for out-of-sample embedding.

Assume there are n data points; therefore, the embedding is obtained by Eq. (8.15). As the eigenvectors $\boldsymbol{Y}$ are orthonormal (so the matrix $\boldsymbol{Y}$ is orthogonal), Eq. (8.15) can be restated as:

$$\boldsymbol{Y}^\top \boldsymbol{M}\boldsymbol{Y} = (\frac{1}{n}\boldsymbol{\Lambda}). \tag{8.53}$$

Assume that $\boldsymbol{Y}$ has been truncated so that there are p eigenvalues, $\boldsymbol{Y} \in \mathbb{R}^{n\times p}$, $\boldsymbol{\Lambda} \in \mathbb{R}^{p\times p}$, and $\boldsymbol{M} \in \mathbb{R}^{n\times n}$.

If n_t new data points are received, then Eq. (8.53) becomes:

$$\boldsymbol{Y}_{\text{updated}}^\top \boldsymbol{M}_{\text{updated}} \boldsymbol{Y}_{\text{updated}} = (\frac{1}{n}\boldsymbol{\Lambda}_{\text{updated}}), \tag{8.54}$$

where $\boldsymbol{Y}_{\text{updated}} \in \mathbb{R}^{(n+n_t)\times p}$ and $\boldsymbol{M}_{\text{updated}} \in \mathbb{R}^{(n+n_t)\times(n+n_t)}$. Note that this considers the smallest eigenvalues when truncating. The eigenvalues in both $\boldsymbol{\Lambda}_{\text{updated}}$ and $\boldsymbol{\Lambda}$ are very small, meaning it is possible to say that $\boldsymbol{\Lambda}_{\text{updated}} \approx \boldsymbol{\Lambda}$. Therefore, considering Eq. (8.53) and the constraints in Eq. (8.14):

$$\begin{aligned}
\underset{\boldsymbol{Y}_{\text{updated}}}{\text{minimize}} \quad & \Big\|\boldsymbol{Y}_{\text{updated}}^\top \boldsymbol{M}_{\text{updated}} \boldsymbol{Y}_{\text{updated}} - \frac{1}{n}\boldsymbol{\Lambda}\Big\|_F^2, \\
\text{subject to} \quad & \frac{1}{n}\boldsymbol{Y}_{\text{updated}}^\top \boldsymbol{Y}_{\text{updated}} = \boldsymbol{I}, \\
& \boldsymbol{Y}_{\text{updated}}^\top \boldsymbol{1} = \boldsymbol{0}.
\end{aligned} \tag{8.55}$$

It is much more efficient than solving Eq. (8.14) for all $n + n_t$ data points, whose solution is Eq. (8.53) and is the eigenvalue problem for an $(n + n_t) \times (n + n_t)$ matrix $\boldsymbol{M}$. However, Eq. (8.55) is an optimization over the $p \times p$ matrix within the Frobenius norm. As $p \ll (n+n_t)$, it is much more efficient to use incremental LLE than regular LLE for the whole old and new data.

This optimization (8.55) can be solved using the interior point method [11]. After ignoring the second constraint, for the reason explained before, its Lagrangian is [11]:

$$\mathcal{L} = \Big\|\boldsymbol{Y}_{\text{updated}}^\top \boldsymbol{M}_{\text{updated}} \boldsymbol{Y}_{\text{updated}} - \frac{1}{n}\boldsymbol{\Lambda}\Big\|_F^2 - \mathbf{tr}\big(\boldsymbol{\Lambda}^\top(\frac{1}{n}\boldsymbol{Y}_{\text{updated}}^\top \boldsymbol{Y}_{\text{updated}} - \boldsymbol{I})\big).$$

According to the matrix derivatives and the chain rule, the derivative of this Lagrangian, with respect to $\boldsymbol{Y}_{\text{updated}}$, is:

$$\begin{aligned}
\frac{\partial \mathcal{L}}{\partial \boldsymbol{Y}_{\text{updated}}} &= 2\,(\boldsymbol{Y}_{\text{updated}}^\top \boldsymbol{M}_{\text{updated}} \boldsymbol{Y}_{\text{updated}} - \frac{1}{n}\boldsymbol{\Lambda})(\boldsymbol{M}_{\text{updated}} \boldsymbol{Y}_{\text{updated}} + \boldsymbol{M}_{\text{updated}}^\top \boldsymbol{Y}_{\text{updated}}) \\
&= 4\,(\boldsymbol{Y}_{\text{updated}}^\top \boldsymbol{M}_{\text{updated}} \boldsymbol{Y}_{\text{updated}} - \frac{1}{n}\boldsymbol{\Lambda})\boldsymbol{M}_{\text{updated}} \boldsymbol{Y}_{\text{updated}}.
\end{aligned}$$

The found $\boldsymbol{Y}_{\text{updated}} \in \mathbb{R}^{(n+n_t)\times p}$ through optimization contains the rowwise p-dimensional embeddings of both old and new data.

8.6 Landmark Locally Linear Embedding for Big Data Embedding

LLE is a spectral dimensionality reduction method [57] and its solution follows an eigenvalue problem; see Eq. (8.15). Therefore, it cannot handle big data, where $n \gg 1$. To tackle this issue, some landmark LLE methods exist that approximate the embedding of all points using the embedding of some landmarks. These methods are introduced in the following sections.

8.6.1 Landmark LLE Using Nyström Approximation

The Nyström approximation, introduced in Chap. 3, can be used to make the spectral methods, such as MDS and Isomap, scalable and suitable for big data embedding. Consider a kernel matrix:

$$\mathbb{R}^{n\times n} \ni \boldsymbol{K} = \left[\begin{array}{c|c} \boldsymbol{A} & \boldsymbol{B} \\ \hline \boldsymbol{B}^\top & \boldsymbol{C} \end{array}\right] = \begin{bmatrix} \boldsymbol{R}^\top \boldsymbol{R} & \boldsymbol{R}^\top \boldsymbol{S} \\ \boldsymbol{S}^\top \boldsymbol{R} & \boldsymbol{S}^\top \boldsymbol{S} \end{bmatrix} = \boldsymbol{O}^\top \boldsymbol{O} = \begin{bmatrix} \boldsymbol{R}^\top \\ \boldsymbol{S}^\top \end{bmatrix} [\boldsymbol{R}, \boldsymbol{S}], \tag{8.56}$$

where $\boldsymbol{A} \in \mathbb{R}^{m\times m}$, $\boldsymbol{B} \in \mathbb{R}^{m\times(n-m)}$, and $\boldsymbol{C} \in \mathbb{R}^{(n-m)\times(n-m)}$, in which $m \ll n$. According to Chap. 3, there are approximately:

$$\boldsymbol{A} = \boldsymbol{U}\boldsymbol{\Sigma}\boldsymbol{U}^\top \tag{8.57}$$

$$\boldsymbol{R} = \boldsymbol{\Sigma}^{(1/2)}\boldsymbol{U}^\top. \tag{8.58}$$

$$\boldsymbol{S} = \boldsymbol{\Sigma}^{(-1/2)}\boldsymbol{U}^\top \boldsymbol{B}, \tag{8.59}$$

$$\boldsymbol{C} = \boldsymbol{B}^\top \boldsymbol{A}^{-1} \boldsymbol{B}, \tag{8.60}$$

where Eq. (8.57) is the eigenvalue decomposition of $\boldsymbol{A}$. Equation (8.56) can be considered as the partitions of a kernel matrix $\boldsymbol{K}$. The (Mercer) kernel matrix is positive semidefinite (see Chap. 3), so the Nyström approximation can be applied for kernels.

Recall that LLE can be viewed as a special case of kernel PCA with the specified kernel in Eq. (8.37). Moreover, recall that according to eigenfunctions, introduced in Chap. 3, the eigenvectors of the kernel matrix can be utilized to embed data using Eq. (8.35). In other words, using the kernel defined by Eq. (8.37), it is possible to apply a kernel PCA and obtain the desired embedding of LLE. When the dataset is large, the eigenvalue decomposition of the kernel matrix is intractable. In this case, it is possible to decompose a $m \times m$ submatrix of the kernel, using Eq. (8.57). In the kernel PCA or generalized classical MDS, the kernel can be seen as the inner product of embeddings, i.e. (see Chap. 7):

$$\mathbb{R}^{n\times n} \ni \boldsymbol{K} = \boldsymbol{Y}'^\top \boldsymbol{Y}', \tag{8.61}$$

where $\mathbb{R}^{p\times n} \ni \boldsymbol{Y}' = \boldsymbol{Y}^\top$ because the embeddings are stacked rowwise in LLE, i.e., $\boldsymbol{Y} \in \mathbb{R}^{n\times p}$. Comparing Eqs. (8.61) and (8.56) demonstrates that:

$$\mathbb{R}^{n\times n} \ni \boldsymbol{Y} = [\boldsymbol{R}, \boldsymbol{S}] \overset{(a)}{=} [\boldsymbol{\Sigma}^{(1/2)}\boldsymbol{U}^\top, \boldsymbol{\Sigma}^{(-1/2)}\boldsymbol{U}^\top \boldsymbol{B}], \tag{8.62}$$

where (a) is due to Eqs. (8.58) and (8.59) and the terms $\boldsymbol{U}$ and $\boldsymbol{\Sigma}$ are obtained through Eq. (8.57). Equation (8.62) provides a good approximation of the embedded data, using the Nyström approximation. Truncating this matrix to have $\boldsymbol{Y}' \in \mathbb{R}^{p\times n}$, with top p rows, results in the p-dimensional embedding of the n points, $\mathbb{R}^{n\times p} \ni \boldsymbol{Y} = \boldsymbol{Y}'^\top$.

8.6.2 Landmark LLE Using Locally Linear Landmarks

Landmark LLE can handle big data by using Locally Linear Landmarks (LLL) [64]. This method maps the n embedded data $\boldsymbol{Y} \in \mathbb{R}^{n\times p}$ to m landmarks $\widetilde{\boldsymbol{Y}} \in \mathbb{R}^{m\times p}$, where $m \ll n$, using a projection matrix $\widetilde{\boldsymbol{U}} = [\widetilde{\boldsymbol{u}}_1, \dots, \widetilde{\boldsymbol{u}}_n]^\top \in \mathbb{R}^{n\times m}$:

$$\mathbb{R}^{n\times p} \ni \boldsymbol{Y} := \widetilde{\boldsymbol{U}}\widetilde{\boldsymbol{Y}}. \tag{8.63}$$

Assume that in some way, landmarks are chosen in the input space. For example, a subset of data points $\boldsymbol{X} \in \mathbb{R}^{d\times n}$ to have the landmarks $\widetilde{\boldsymbol{X}} \in \mathbb{R}^{d\times m}$ can be selected. In other words, $\mathbb{C}\text{ol}(\boldsymbol{X}) \subseteq \mathbb{C}\text{ol}(\widetilde{\boldsymbol{X}})$, where $\mathbb{C}\text{ol}(\cdot)$ denotes the column space of the matrix. Projecting to landmarks should additionally work for the input space as follows:

$$\mathbb{R}^{n\times d} \ni \boldsymbol{X}^\top := \widetilde{\boldsymbol{U}}\widetilde{\boldsymbol{X}}^\top. \tag{8.64}$$

By adding constraints, this goal can be written as an optimization problem:

$$\begin{aligned} &\underset{\widetilde{\boldsymbol{U}}}{\text{minimize}} && \sum_{i=1}^{n} ||\boldsymbol{x}_i - \widetilde{\boldsymbol{X}}\widetilde{\boldsymbol{u}}_i||_2^2., \\ &\text{subject to} && \boldsymbol{1}^\top \widetilde{\boldsymbol{u}}_i = 1, \quad \forall i \in \{1, \dots, n\}, \end{aligned} \tag{8.65}$$

which is in the form of Eq. (8.1). Therefore, its solution is similar to Eq. (8.8):

$$\mathbb{R}^{m} \ni \widetilde{\boldsymbol{u}}_i = \frac{\widetilde{\boldsymbol{G}}_i^{-1}\boldsymbol{1}}{\boldsymbol{1}^\top \widetilde{\boldsymbol{G}}_i^{-1}\boldsymbol{1}}, \tag{8.66}$$

where:

$$\mathbb{R}^{m\times m} \ni \widetilde{\boldsymbol{G}}_i := (\boldsymbol{x}_i\boldsymbol{1}^\top - \widetilde{\boldsymbol{X}})^\top(\boldsymbol{x}_i\boldsymbol{1}^\top - \widetilde{\boldsymbol{X}}). \tag{8.67}$$

According to Eq. (8.63), Eq. (8.14) becomes:

$$\begin{aligned} \underset{\widetilde{\boldsymbol{Y}}}{\text{minimize}} \quad & \textbf{tr}(\widetilde{\boldsymbol{Y}}^\top\widetilde{\boldsymbol{U}}^\top\boldsymbol{M}\widetilde{\boldsymbol{U}}\widetilde{\boldsymbol{Y}}), \\ \text{subject to} \quad & \frac{1}{n}\widetilde{\boldsymbol{Y}}^\top\widetilde{\boldsymbol{U}}^\top\widetilde{\boldsymbol{U}}\widetilde{\boldsymbol{Y}} = \boldsymbol{I}, \end{aligned} \tag{8.68}$$

whose second constraint is ignored because it has already been satisfied (see Proposition 8.1). Consider the following defined equation:

$$\mathbb{R}^{m\times m} \ni \widetilde{\boldsymbol{M}} := \widetilde{\boldsymbol{U}}^\top\boldsymbol{M}\widetilde{\boldsymbol{U}}. \tag{8.69}$$

Similar to the solution of Eq. (8.14), the solution to Eq. (8.68) is the eigenvalue problem for $\widetilde{\boldsymbol{M}}$ (see Chap. 2). In other words, the embeddings of the landmark points, $\widetilde{\boldsymbol{Y}}$, are the p smallest eigenvectors of $\widetilde{\boldsymbol{M}}$, after ignoring the eigenvector with zero eigenvalue. As the dimensionality of $\widetilde{\boldsymbol{M}}$ is $m \times m$, the landmark LLE, using LLL, is more efficient than LLE whose embeddings are the eigenvectors of $\boldsymbol{M} \in \mathbb{R}^{n\times n}$. The difference in efficiency is especially noticeable for big data, where $n \gg m$, because of the difference in the size of the matrices. Finally, using Eq. (8.63), the embeddings of all n points are approximated by the obtained embeddings of m landmarks.

8.7 Parameter Selection of the Number of Neighbours in LLE

The number of neighbours, k, is the hyperparameter of LLE. There are several different algorithms that can be used to find an optimal k, which include parameter selection using residual variance [41], Procrustes statistics [27], preservation neighbourhood error [3], and local neighbourhood selection [3]. These algorithms are introduced in the following sections.

8.7.1 Parameter Selection Using Residual Variance

The goal of parameter selection using a residual is to find the best k. Assume a candidate number of neighbours, denoted by $\{1, 2, \ldots, k_{\max}\}$. For every $k \in \{1, 2, \ldots, k_{\max}\}$, an LLE can be run and the embeddings $\boldsymbol{Y}$ for data $\boldsymbol{X}$ can be found. Let $\boldsymbol{D}_X$ and $\boldsymbol{D}_Y$ denote the Euclidean distance matrices over $\boldsymbol{X}$ and $\boldsymbol{Y}$, respectively. Let $\rho^2_{\boldsymbol{D}_X, \boldsymbol{D}_Y}$ be the standard linear correlation coefficient, i.e., $\rho^2_{\boldsymbol{D}_X, \boldsymbol{D}_Y} := S_{\boldsymbol{D}_X, \boldsymbol{D}_Y} / (S_{\boldsymbol{D}_X} S_{\boldsymbol{D}_Y})$, where $\rho^2_{\boldsymbol{D}_X, \boldsymbol{D}_Y}$ is the covariance of $\boldsymbol{D}_X$ and $\boldsymbol{D}_Y$ and $S_{\boldsymbol{D}_X}$ and $S_{\boldsymbol{D}_Y}$ are the standard deviations of $\boldsymbol{D}_X$ and $\boldsymbol{D}_Y$, respectively. The residual variance for the number of neighbours k is defined as [41]:

$$\sigma_k^2(\boldsymbol{D}_X, \boldsymbol{D}_Y) := 1 - \rho^2_{\boldsymbol{D}_X, \boldsymbol{D}_Y}. \tag{8.70}$$

The k value with the smallest residual variance value is the optimal number of neighbours because it maximizes the correlation between the distances in the input and embedding spaces. Therefore:

$$k := \arg\min_k \sigma_k^2(\boldsymbol{D}_X, \boldsymbol{D}_Y). \tag{8.71}$$

Instead of running an LLE for all $k \in \{1, 2, \dots, k_{\max}\}$, which is computationally expensive, it is possible to take a hierarchical approach [41]. In this approach, $\varepsilon(\widetilde{\boldsymbol{W}})$ is calculated by Eq. (8.1) for every value of $k \in \{1, 2, \dots, k_{\max}\}$. For the local minimums of $\varepsilon(\widetilde{\boldsymbol{W}})$ (whenever $\varepsilon(\widetilde{\boldsymbol{W}})$ for a k is smaller than that for $k-1$ and $k+1$), Eq. (8.71) can be calculated and the best k among the k's corresponding to the local minimums can be found.

8.7.2 Parameter Selection Using Procrustes Statistics

Goldberg and Ritov proposed a method for the parameter selection of k in LLE [27], which uses Procrustes statistics [59]. The Procrustes statistics between $\boldsymbol{X} = [\boldsymbol{x}_1, \dots, \boldsymbol{x}_n] \in \mathbb{R}^{d \times n}$ and their embeddings $\boldsymbol{Y} = [\boldsymbol{y}_1, \dots, \boldsymbol{y}_n]^\top \in \mathbb{R}^{n \times p}$ is [27, 59]:

$$P(X, Y) := \sum_{i=1}^{n} \|\boldsymbol{x}_i - \boldsymbol{y}_i \boldsymbol{A}^\top - \boldsymbol{b}\|_2^2 = \|\boldsymbol{H}_n(\boldsymbol{X}^\top - \boldsymbol{Y}\boldsymbol{A}^\top)\|_F^2, \tag{8.72}$$

with the orthogonal rotation matrix, i.e. $\boldsymbol{A}^\top \boldsymbol{A} = \boldsymbol{I}$, and the translation matrix $\boldsymbol{b} = \bar{\boldsymbol{x}} - \bar{\boldsymbol{y}}\boldsymbol{A}^\top$, where $\bar{\boldsymbol{x}}$ and $\bar{\boldsymbol{y}}$ are the means of samples $\boldsymbol{X}$ and $\boldsymbol{Y}$, respectively. The matrix $\mathbb{R}^{n \times n} \ni \boldsymbol{H}_n = \boldsymbol{I}_n - (1/n)\boldsymbol{1}\boldsymbol{1}^\top$ is the centering matrix. According to the Procrustes statistics [59], the rotation matrix can be computed by $\mathbb{R}^{d \times p} \ni \boldsymbol{A} = \boldsymbol{U}\boldsymbol{V}^\top$, where $\boldsymbol{U}\boldsymbol{\Sigma}\boldsymbol{V}^\top$ is the singular value decomposition of $\boldsymbol{X}\boldsymbol{H}_n\boldsymbol{Y} \in \mathbb{R}^{d \times p}$.

Let $\boldsymbol{X}_i \in \mathbb{R}^{d \times k}$ and $\boldsymbol{Y}_i \in \mathbb{R}^{k \times p}$ be the k neighbours of $\boldsymbol{x}_i$ in the input and embedding spaces, respectively. For every k, an LLE is applied on the dataset $\boldsymbol{X}$ to obtain an embedding $\boldsymbol{Y}$ and a neighbourhood graph. Normalized Procrustes statistics for the number of neighbours k is [27]:

$$R_k(\boldsymbol{X}, \boldsymbol{Y}) := \frac{1}{n} \sum_{i=1}^{n} \frac{P(\boldsymbol{X}_i, \boldsymbol{Y}_i)}{\|\boldsymbol{H}_k \boldsymbol{X}_i^\top\|_F^2}, \tag{8.73}$$

An optimal $k \in \{1. \dots, k_{\max}\}$ reduces the Procrustes statistics the most:

$$k := \arg\min_k R_k(\boldsymbol{X}, \boldsymbol{Y}). \tag{8.74}$$

A hierarchical approach, introduced in Sect. 8.7.1, can also be used here to determine the best value k using Eq. (8.74).

8.7.3 Parameter Selection Using Preservation Neighbourhood Error

Consider the data points $\{\boldsymbol{x}_i \in \mathbb{R}^d\}_{i=1}^n$ and their embeddings $\{\boldsymbol{y}_i \in \mathbb{R}^p\}_{i=1}^n$. For a point $\boldsymbol{x}_i$, let its k neighbours in the input space be denoted by $\{\boldsymbol{\eta}_i \in \mathbb{R}^d\}_{i=1}^k$. The embeddings of $\{\boldsymbol{\eta}_i\}_{i=1}^k$ are denoted by $\{\boldsymbol{\phi}_i \in \mathbb{R}^p\}_{i=1}^k$. Now, let the k neighbours of $\boldsymbol{y}_i$ in the embedding space be denoted by $\{\boldsymbol{\beta}_i \in \mathbb{R}^p\}_{i=1}^k$. The points that are among the k neighbours of $\boldsymbol{y}_i$, but not among the k neighbours of $\boldsymbol{x}_i$, are denoted by $\{\boldsymbol{\gamma}_i \in \mathbb{R}^p\}_{i=1}^{k_i'}$ in the embedding space, where the number of these points is denoted by k_i'. That is, $\{\gamma_i\}_{i=1}^{k_i'} = \{\boldsymbol{\beta}_i\}_{i=1}^k - \{\boldsymbol{\phi}_i\}_{i=1}^k$. The corresponding points to $\{\boldsymbol{\gamma}_i \in \mathbb{R}^p\}$ in the input space are denoted by $\{\boldsymbol{\theta}_i \in \mathbb{R}^d\}_{i=1}^{k_i'}$.

The Preservation Neighbourhood Error (PNE), for the number of neighbours k, is defined as [3]:

$$\text{PNE}_k(\boldsymbol{X}, \boldsymbol{Y})$$
$$:= \frac{1}{2n} \sum_{i=1}^{n} \Big(\sum_{j=1}^{k} \frac{(\|\boldsymbol{x}_i - \boldsymbol{\eta}_j\|_2 - \|\boldsymbol{y}_i - \boldsymbol{\phi}_j\|_2)^2}{k} + \sum_{j=1}^{k_i'} \frac{(\|\boldsymbol{x}_i - \boldsymbol{\theta}_j\|_2 - \|\boldsymbol{y}_i - \boldsymbol{\gamma}_j\|_2)^2}{k_i'} \Big). \tag{8.75}$$

The first term in the summation tries to preserve the local structure of points in the embedding space to be similar to the input space. The second term attempts to keep the points apart in the embedding space if they are far from each other in the input space; in other words, the second term avoids false folding of the manifold. The optimal $k \in \{1. \dots, k_{\max}\}$ reduces the PNE the most:

$$k := \arg\min_k \text{PNE}_k(\boldsymbol{X}, \boldsymbol{Y}). \tag{8.76}$$

A hierarchical approach, as introduced in Sect. 8.7.1, can be used to determine the best value k using Eq. (8.76).

8.7.4 Parameter Selection Using Local Neighbourhood Selection

The Local Neighbourhood Selection (LNS) [3] finds the optimal number of neighbours per each point $\boldsymbol{x}_i$; therefore, it allows a different number of neighbours for different points. The steps in this algorithm are as follows:

- The Euclidean and geodesic distance matrices, denoted by $\boldsymbol{D} \in \mathbb{R}^{n\times n}$ and $\boldsymbol{D}^{(g)} \in \mathbb{R}^{n\times n}$, respectively, are calculated first.
- $k_{\min}$ is initialized to one.
- The $k_{\min}$-NN graph is found using $\boldsymbol{D}$.
- To check if the $k_{\min}$-NN graph is connected, a Breadth First Search (BFS) [16] is used. If it is not connected, the $k_{\min}$ is incremented up by one until the graph becomes connected.
- $k_{\max} := n^2/(k_{\min} \times |E|)$ is set, where $|E|$ is the number of edges in the $k_{\min}$-NN graph.

The remainder of the algorithm is as follows. $\boldsymbol{k} = [\boldsymbol{k}(1), \dots, \boldsymbol{k}(k_{\max} - k_{\min})] := [k_{\min} + 1, \dots, k_{\max}] \in \mathbb{R}^{k_{\max}-k_{\min}}$ is defined. Let $\boldsymbol{\eta}_i^{D,k}$ and $\boldsymbol{\eta}_i^{D^{(g)},k}$ be the set of k-NN of $\boldsymbol{x}_i$ using the distance matrices $\boldsymbol{D}$ and $\boldsymbol{D}^{(g)}$, respectively, where $k \in \{k_{\min} + 1, \dots, k_{\max}\}$. If $|\cdot|$ and $\bar{\cdot}$ denote the cardinality and complement of the set, respectively. The (i, j)-th element of the linearity conservation matrix $\boldsymbol{V} \in \mathbb{R}^{n\times(k_{\max}-k_{\min})}$ is calculated as:

$$\boldsymbol{V}(i, j) := \frac{\overline{\big|(\boldsymbol{\eta}_i^{D,\boldsymbol{k}(j)} \cap \boldsymbol{\eta}_i^{D^{(g)},\boldsymbol{k}(j)})\big|}}{\boldsymbol{k}(j)}. \tag{8.77}$$

The smaller this quantity is, the closer the geodesic and Euclidean distances behave and the more local structure is preserved. For every row of the linearity conservation matrix (i.e., for every point $\boldsymbol{x}_i$), the ultimate number of neighbours is determined as:

$$k(\boldsymbol{x}_i) := \arg\min_{\boldsymbol{k}(j)} \boldsymbol{V}(i, j). \tag{8.78}$$

In the event of a tie, the largest value of $\boldsymbol{k}(j)$ is selected to better capture the neighbourhood structure.

8.8 Supervised and Semisupervised LLE

In supervised and semisupervised LLE, the class labels are used fully and partially, respectively. There are different versions of these methods which are explained in the following.

8.8.1 Supervised LLE

The Supervised LLE (SLLE) [19, 42, 43] can be used for both embedding and classification [18]. The main idea of SLLE is to artificially increase the interclass variance of data by adding to the distances between points from different classes. Assume the Euclidean distance matrix is denoted by $\boldsymbol{D} \in \mathbb{R}^{n\times n}$. In SLLE, the distance matrix is modified to [19]:

$$\mathbb{R}^{n\times n} \ni \boldsymbol{D}' := \boldsymbol{D} + \alpha\,(d_{\max})(\boldsymbol{1}\boldsymbol{1}^\top - \boldsymbol{\Delta}), \tag{8.79}$$

where $\boldsymbol{1}\boldsymbol{1}^\top \in \mathbb{R}^{n\times n}$ is the matrix with all elements equal to one, and $d_{\max} \in \mathbb{R}$ is the diameter of the data:

$$d_{\max} := \max_{i,j}(\|\boldsymbol{x}_i - \boldsymbol{x}_j\|_2), \tag{8.80}$$

and $\boldsymbol{\Delta}$ is a matrix whose (i, j)-th element is:

$$\boldsymbol{\Delta}(i, j) := \begin{cases} 1 \text{ if } c_i = c_j, \\ 0 \text{ Otherwise,} \end{cases} \tag{8.81}$$

where c_i denotes the class label of $\boldsymbol{x}_i$, and $\alpha \in [0, 1]$. When $\alpha = 0$, SLLE is reduced to LLE, which is unsupervised. When $\alpha = 1$, SLLE is fully supervised. This case is also named 1-SLLE [43]. SLEE is considered to be partially supervised when $\alpha \in (0, 1)$, also called α-SLLE [18]. Note that Eq. (8.79) does not change the distances between points belonging to the same class. After modifying the distance matrix, SLLE finds the kNN graph by using the modified distances, while the rest of the algorithm remains the same as in LLE.

8.8.2 Enhanced Supervised LLE

An Enhanced Supervised LLE (ESLLE) [71] artificially increases the interclass variances and artificially reduces the intraclass variances. Note that the idea of increasing and decreasing the interclass and intraclass variances, respectively, is common in supervised embedding, such as through a Fisher discriminant analysis (see Chap. 6). ESLLE modifies the distances to:

$$\boldsymbol{D}' := \begin{cases} \sqrt{1 - e^{-\boldsymbol{D}^2/\beta}} \text{ if } c_i = c_j, \\ \sqrt{e^{\boldsymbol{D}^2/\beta}} - \alpha \quad \text{Otherwise,} \end{cases} \tag{8.82}$$

where $\alpha \in [0, 1]$ and:

$$\beta := \text{average}_{i,j}(\|\boldsymbol{x}_i - \boldsymbol{x}_j\|_2). \tag{8.83}$$

In ESLLE, the distance of points from different classes grows exponentially, while the distances of points in the same class have a horizontal asymptote of one (see [71, Fig. 1]). Using the modified distances, a kNN graph is found and the remainder of the equation remains the same as that of LLE.

8.8.3 Supervised LLE Projection

It is possible to approximate the mapping of $\boldsymbol{X} \mapsto \boldsymbol{Y}$ by using a linear projection. Supervised LLE Projection (SLLEP) [44] finds a linear projection in the context of SLLE [19]. First, SLLEP finds the embedding of the training data, $\boldsymbol{Y}$, using SLLE (see Sect. 8.8.1). It then tries to approximate this embedding by using a linear projection $\boldsymbol{Y} = \boldsymbol{U}^\top \boldsymbol{X}$, where $\boldsymbol{U} = [\boldsymbol{u}_1, \dots, \boldsymbol{u}_p] \in \mathbb{R}^{d \times p}$ is the projection matrix. Let the embedding of point $\boldsymbol{x}_i$ be $\mathbb{R}^p \ni \boldsymbol{y}_i := [\boldsymbol{y}_i(1), \dots, \boldsymbol{y}_i(p)]^\top$. Additionally, let $\mathbb{R}^n \ni \boldsymbol{y}^j := [\boldsymbol{y}_1(j), \dots, \boldsymbol{y}_n(j)]^\top$. This approximation can be done using a least squares optimization:

$$\boldsymbol{u}_j = \arg\min_{\boldsymbol{u}} \sum_{i=1}^{n} (\boldsymbol{u}^\top \boldsymbol{x}_i - \boldsymbol{y}_i(j))^2, \quad \forall j \in \{1, \dots, p\}, \tag{8.84}$$

whose solution is similar to the solution for linear regression [30]:

$$\boldsymbol{u}_j = (\boldsymbol{X}\boldsymbol{X}^\top)^{-1} \boldsymbol{X} \boldsymbol{y}^j. \tag{8.85}$$

When $\boldsymbol{X}\boldsymbol{X}^\top$ is singular, regularized least squares optimization with the regularization parameter β can be used. In this case, the solution is similar to ridge regression[2]:

$$\boldsymbol{u}_j = (\boldsymbol{X}\boldsymbol{X}^\top + \beta \boldsymbol{I})^{-1} \boldsymbol{X} \boldsymbol{y}^j. \tag{8.86}$$

SLLEP can be used for the approximation of the out-of-sample embedding for new data $\boldsymbol{X}^{(t)}$ by $\boldsymbol{Y}^{(t)} = \boldsymbol{U}^\top \boldsymbol{X}^{(t)}$. It is noteworthy that the approximation used in SLLEP can also be used for approximating the unsupervised LLE with a linear projection.

[2] Ridge regression is the ℓ_2-norm regularized linear regression [30].

8.8.4 Probabilistic Supervised LLE

Probability-based LLE (PLLE) [73] is a supervised method for LLE that can handle out-of-sample data. For every training point $\boldsymbol{x}_i$, the probability of belonging to class c_i should be one; therefore, its one-hot encoding is:

$$\mathbb{R}^c \ni \boldsymbol{p}(\boldsymbol{x}_i) := \mathbf{1}_{c_i} = [0, \dots, 0, 1, 0, \dots, 0]^\top, \tag{8.87}$$

whose c_i-th element is one. However, for the out-of-sample data, the probability is found using logistic regression [38]. PLLE, first, applies an unsupervised LLE on both the training and out-of-sample data (see Sects. 8.2 and 8.4). Then, for the embedding of out-of-sample points, denoted by $\{\boldsymbol{y}_i^{(t)}\}_{i=1}^{n_t}$, it learns the logistic functions of all of the c classes:

$$\pi(\boldsymbol{y}_i^{(t)}; \boldsymbol{a}_\ell, \boldsymbol{b}_\ell) := \frac{e^{\boldsymbol{a}_\ell + \boldsymbol{b}_\ell^\top \boldsymbol{y}_i^{(t)}}}{1 + e^{\boldsymbol{a}_\ell + \boldsymbol{b}_\ell^\top \boldsymbol{y}_i^{(t)}}}, \quad \forall \ell \in \{1, \dots, c\}, \tag{8.88}$$

where the parameters $\boldsymbol{a}_\ell$ and $\boldsymbol{b}_\ell$ are found through logistic regression. Therefore, $\{\pi(\boldsymbol{y}_i^{(t)}; \boldsymbol{a}_\ell, \boldsymbol{b}_\ell)\}_{i=1}^{c}$. The probability of $\boldsymbol{x}_i$ belonging to every ℓ-th class is:

$$\mathbb{R} \ni p_\ell(\boldsymbol{x}_i^{(t)}) := \frac{\pi(\boldsymbol{y}_i^{(t)}; \boldsymbol{a}_\ell, \boldsymbol{b}_\ell)}{\sum_{\ell'=1}^{c} \pi(\boldsymbol{y}_i^{(t)}; \boldsymbol{a}_{\ell'}, \boldsymbol{b}_{\ell'})}. \tag{8.89}$$

Therefore, the probability vector for $\boldsymbol{x}_i^{(t)}$ is $\mathbb{R}^c \ni \boldsymbol{p}(\boldsymbol{x}_i^{(t)}) := [p_\ell(\boldsymbol{x}_1^{(t)}), \dots, p_\ell(\boldsymbol{x}_c^{(t)})]^\top$. PLLE uses Eq. (8.79) to modify the distances, but, since it can handle out-of-sample data as well, all training and out-of-sample points can be put together in this stage, to have $\boldsymbol{D}, \boldsymbol{D}', \boldsymbol{\Delta} \in \mathbb{R}^{(n+n_t)\times(n+n_t)}$. $\boldsymbol{\Delta}(i, j)$ is defined as follows:

$$\boldsymbol{\Delta}(i, j) := \begin{cases} 1 & \text{if } c_i = c_j, \\ \boldsymbol{p}(\boldsymbol{x}_i)^\top \boldsymbol{p}(\boldsymbol{x}_j) & \text{Otherwise}, \end{cases} \tag{8.90}$$

$\forall i, j \in \{1, \dots, n + n_t\}$. Using the modified distances, the kNN graph is found and the remainder of the equation remains the same as LLE.

8.8.5 Semi-Supervised LLE

If only some of the data have labels, a semisupervised LLE [70] can be used. Similar to Eq. (8.82), this method modifies the distances such that:

$$\boldsymbol{D}' := \begin{cases} \sqrt{1 - e^{-\boldsymbol{D}''^2/\beta}} - \alpha & \text{if } c_i = c_j, \\ \sqrt{1 - e^{-\boldsymbol{D}''^2/\beta}} & \text{if } \boldsymbol{x}_i \text{ or } \boldsymbol{x}_j \text{ is unlabeled}, \\ \sqrt{e^{\boldsymbol{D}''^2/\beta}} & \text{Otherwise}, \end{cases} \tag{8.91}$$

where:

$$\boldsymbol{D}''(i, j) := \frac{\boldsymbol{D}(i, j)}{\sqrt{m_i \times m_j}}, \tag{8.92}$$

and $\mathbb{R} \ni m_i := \text{average}_\ell(\|\boldsymbol{x}_i - \boldsymbol{x}_\ell)\|_2; \forall \ell \in \{1, \ldots, n\})$.

8.8.6 Supervised Guided LLE

The Guided LLE (GLLE) [2] uses the Hilbert-Schmidt Independence Criterion (HSIC) [28] (see Chap. 3) for utilizing labels for embedding. The following sections explain this method.

8.8.6.1 Seeing LLE as Kernel PCA

As mentioned in Sect. 8.4.2, LLE can be seen as a special case of kernel LLE, where the inverse, or negative sign of $\boldsymbol{M}$, can be interpreted as its kernel. The kernel of LLE in kernel PCA can be either Eq. (8.37) [8, 58] or [29]:

$$\mathbb{R}^{n \times n} \ni \boldsymbol{K} := \boldsymbol{M}^\dagger, \tag{8.93}$$

which is the pseudoinverse of the matrix $\boldsymbol{M}$.

8.8.6.2 Interpreting LLE Using HSIC

Recall the Hilbert-Schmidt Independence Criterion (HSIC) from Chap. 3. Suppose the kernel $\boldsymbol{K}_x$ in HSIC is Eq. (8.93) and its other kernel is a linear kernel, i.e., $\boldsymbol{K}_y := \boldsymbol{Y}\boldsymbol{Y}^\top$ (note that the embedded points are stacked in $\boldsymbol{Y}$ rowwise). The goal is to maximize the HSIC to have the largest dependence between the data $\boldsymbol{X}$ and their embedding $\boldsymbol{Y}$, so that the embedding of the data is related well enough to the data. This maximization can be modelled by a constrained optimization problem:

$$\begin{aligned}
&\underset{\boldsymbol{Y}}{\text{maximize}} \quad \mathbf{tr}(\boldsymbol{Y}^\top \boldsymbol{H}\boldsymbol{M}^\dagger \boldsymbol{H}\boldsymbol{Y}) \overset{(a)}{=} \mathbf{tr}(\boldsymbol{Y}^\top \boldsymbol{M}^\dagger \boldsymbol{Y}), \\
&\text{subject to} \quad \frac{1}{n}\boldsymbol{Y}^\top \boldsymbol{Y} = \boldsymbol{I},
\end{aligned} \tag{8.94}$$

where (a) is because the matrix $\boldsymbol{M}$ is already double-centered [2]. This maximization problem can be converted to a minimization problem by:

$$\begin{aligned}
&\underset{\boldsymbol{Y}}{\text{minimize}} \quad \mathbf{tr}(\boldsymbol{Y}^\top \boldsymbol{M}\boldsymbol{Y}), \\
&\text{subject to} \quad \frac{1}{n}\boldsymbol{Y}^\top \boldsymbol{Y} = \boldsymbol{I},
\end{aligned} \tag{8.95}$$

which is equivalent to Eq. (8.14), although the second constraint in Eq. (8.14) has already been satisfied. This shows that the optimization of embedding in LLE can be seen as maximizing the HSIC (or dependence) between the input and embedding data.

8.8.6.3 Guiding LLE Using Labels

For discriminating between classes, consider the maximization of the dependence between a linear kernel over embedding and a kernel over class labels (targets), denoted by $\boldsymbol{K}_t$:

$$\begin{aligned}
&\underset{\boldsymbol{Y}}{\text{maximize}} \quad \mathbf{tr}(\boldsymbol{Y}^\top \boldsymbol{H}\boldsymbol{K}_t \boldsymbol{H}\boldsymbol{Y}), \\
&\text{subject to} \quad \frac{1}{n}\boldsymbol{Y}^\top \boldsymbol{Y} = \boldsymbol{I},
\end{aligned} \tag{8.96}$$

which can also be converted to a minimization problem using the pseudoinverse of $\boldsymbol{K}_t$. The kernel over labels can be a delta kernel [5], introduced in Chap. 5. After converting Eq. (8.96) to a minimization problem, Eqs. (8.95) and (8.96) can be combined to:

$$\begin{aligned}
&\underset{\boldsymbol{Y}}{\text{minimize}} \quad \mathbf{tr}\big(\boldsymbol{Y}^\top ((1-\alpha)\boldsymbol{M} + \alpha \boldsymbol{K}_t)\boldsymbol{Y}\big), \\
&\text{subject to} \quad \frac{1}{n}\boldsymbol{Y}^\top \boldsymbol{Y} = \boldsymbol{I},
\end{aligned} \tag{8.97}$$

where $\alpha \in [0, 1]$. The solution to this optimization problem is the smallest p eigenvectors of $(1 - \alpha)\boldsymbol{M} + \alpha \boldsymbol{K}_t$ (see Chap. 2), after the first eigenvector with eigenvalue zero is ignored. Note that this optimization guides LLE to have an embedding with more discrimination between classes.

8.8.7 Other Supervised Algorithms

SLLEP [44] was not explained in this section, as a supervised LLE method because it was introduced in Sect. 8.8.3. Moreover, this section did not cover supervised LLE by adjusting weights [32] and Discriminant LLE [45] because they will be explained in Sects. 8.11.3 and 8.10.4, respectively.

8.9 Robust Locally Linear Embedding

In the presence of outliers and noise, LLE cannot sufficiently preserve the local structure of the manifold. This is because outliers can introduce bias through the reconstruction of the points [14]. The Robust LLE (RLLE) was proposed to resolve the problem caused by the presence of outliers in LLE.

8.9.1 Robust LLE Using the Least Squares Problem

One approach to conduct RLLE is by using the least squares problem to handle noise [14]. This RLLE uses an iterative optimization approach [36], where it iterates between Principal Component Analysis (PCA) and finding reliability weights. In every iteration, for every point $\boldsymbol{x}_i$, it minimizes the weighted reconstruction error using PCA by a least squares problem (note that PCA is the best method for linear reconstruction; see Chap. 5):

$$\underset{\boldsymbol{U}_i}{\text{minimize}} \quad \sum_{j=1}^{k} a_{ij}\, e_{ij} := \sum_{j=1}^{k} a_{ij}\, ||\boldsymbol{x}_{ij} - \boldsymbol{b}_i - \boldsymbol{U}_i\, \boldsymbol{y}_{ij}||_2^2, \tag{8.98}$$

where $\boldsymbol{b}_i \in \mathbb{R}^d$ and $\boldsymbol{U}_i \in \mathbb{R}^{d\times p}$ are the bias and PCA projection matrix, respectively, and $\boldsymbol{y}_{ij} \in \mathbb{R}^p$ is the embedding of $\boldsymbol{x}_{ij}$, and $\{a_{ij}\}_{j=1}^k$ are the reliability weights. The solution to this optimization problem is [14]:

$$\boldsymbol{b}_i := \frac{\sum_{j=1}^k a_{ij}\, \boldsymbol{x}_{ij}}{\sum_{j=1}^k a_{ij}}, \tag{8.99}$$

and the columns of $\boldsymbol{U}_i$ are the top p eigenvectors of the covariance matrix over the neighbours:

$$\boldsymbol{S}_i := \frac{1}{k} \sum_{j=1}^{k} a_{ij}\, (\boldsymbol{x}_{ij} - \boldsymbol{b}_i)(\boldsymbol{x}_{ij} - \boldsymbol{b}_i)^\top. \tag{8.100}$$

Then, the weights $\{a_j\}_{j=1}^k$ are obtained, inspired by the Huber function [14], through:

$$a_{ij} := \begin{cases} 1 & \text{if } e_{ij} \le c_i, \\ c_i/e_j & \text{if } e_{ij} > c_i, \end{cases} \tag{8.101}$$

where e_{ij} is defined in Eq. (8.98) and c_i is the mean error residual, i.e., $c_i := (1/k)\sum_{j=1}^k e_{ij}$. Using an iterative approach, named Iteratively Reweighted Least Squares (IRLS) [34], $\boldsymbol{b}_i$, $\boldsymbol{U}_i$, and $\{a_{ij}\}_{j=1}^k$ are fine tuned for the k neighbours of every point $\boldsymbol{x}_i$, by Eqs. (8.99), (8.100), and (8.101). In this way, the reliability weights $\{a_{ij}\}_{j=1}^k$ are calculated for each point. Let the mean reliability weights over the neighbours of a point determine the reliability weight of that point. This is calculated as $s_i := (1/k)\sum_{j=1}^k a_{ij}$. Then, RLLE weights the objective of Eq. (8.9) as [14]:

$$\underset{\boldsymbol{Y}}{\text{minimize}} \quad \sum_{i=1}^n s_i \Big|\Big| \boldsymbol{y}_i - \sum_{j=1}^n w_{ij}\boldsymbol{y}_j \Big|\Big|_2^2, \tag{8.102}$$

with the constraints in Eq. (8.9). Therefore, the embeddings are weighted to be robust to outliers.

8.9.2 *Robust LLE Using Penalty Functions*

RLLE can use the penalty function for regularized optimization [65]. In the presence of noise or outliers, some weights may explode, in the reconstruction of a point by its neighbours, because the distance between the outliers and other points is usually large. Winlaw et al. proposed two different penalty functions for RLLE [65]—ℓ_2 norm penalty and elastic-net penalty. These penalties are explained in the following subsection.

8.9.2.1 RLLE with ℓ_2 Norm Penalty

The penalty function can be the ℓ_2 norm. In RLLE, Eq. (8.4) is regularized, with the regularization parameter γ, as [65]:

$$\begin{aligned} \underset{\{\widetilde{\boldsymbol{w}}_i\}_{i=1}^n}{\text{minimize}} \quad & \sum_{i=1}^n \widetilde{\boldsymbol{w}}_i^\top \boldsymbol{G}_i\, \widetilde{\boldsymbol{w}}_i + \gamma\, \|\widetilde{\boldsymbol{w}}_i\|_2^2, \\ \text{subject to} \quad & \boldsymbol{1}^\top \widetilde{\boldsymbol{w}}_i = 1, \quad \forall i \in \{1, \dots, n\}. \end{aligned} \tag{8.103}$$

The Lagrangian for this optimization is [11]:

$$\mathcal{L} = \sum_{i=1}^{n} \widetilde{\boldsymbol{w}}_i^\top \boldsymbol{G}_i\, \widetilde{\boldsymbol{w}}_i + \gamma\, \|\widetilde{\boldsymbol{w}}_i\|_2^2 - \sum_{i=1}^{n} \lambda_i\, (\mathbf{1}^\top \widetilde{\boldsymbol{w}}_i - 1).$$

Setting the derivative of the Lagrangian to zero results in:

$$\mathbb{R}^k \ni \frac{\partial \mathcal{L}}{\partial \widetilde{\boldsymbol{w}}_i} = 2\boldsymbol{G}_i \widetilde{\boldsymbol{w}}_i + 2\gamma\, \widetilde{\boldsymbol{w}}_i - \lambda_i \mathbf{1} \overset{\text{set}}{=} \mathbf{0} \implies \widetilde{\boldsymbol{w}}_i = \frac{\lambda_i}{2}(\boldsymbol{G}_i + \gamma \boldsymbol{I})^{-1}\mathbf{1}. \tag{8.104}$$

$$\mathbb{R} \ni \frac{\partial \mathcal{L}}{\partial \lambda} = \mathbf{1}^\top \widetilde{\boldsymbol{w}}_i - 1 \overset{\text{set}}{=} 0 \implies \mathbf{1}^\top \widetilde{\boldsymbol{w}}_i = 1. \tag{8.105}$$

Using Eqs. (8.104) and (8.105), results in:

$$\frac{\lambda_i}{2}\mathbf{1}^\top(\boldsymbol{G}_i + \gamma \boldsymbol{I})^{-1}\mathbf{1} = 1 \implies \lambda_i = \frac{2}{\mathbf{1}^\top(\boldsymbol{G}_i + \gamma \boldsymbol{I})^{-1}\mathbf{1}}.$$

Therefore:

$$\widetilde{\boldsymbol{w}}_i = \frac{\lambda_i}{2}(\boldsymbol{G}_i + \gamma \boldsymbol{I})^{-1}\mathbf{1} = \frac{(\boldsymbol{G}_i + \gamma \boldsymbol{I})^{-1}\mathbf{1}}{\mathbf{1}^\top(\boldsymbol{G}_i + \gamma \boldsymbol{I})^{-1}\mathbf{1}}. \tag{8.106}$$

Note that, in addition to better handling of noise, this regularization solves the problem of possible singularity of the matrix $\boldsymbol{G}_i$ by strengthening its main diagonal.

8.9.2.2 RLLE with Elastic-Net Penalty

The elastic-net penalty function [76] incorporates sparsity in the regularization for the RLLE solution. This RLLE regularizes Eq. (8.4) as [65]:

$$\begin{aligned} \underset{\{\widetilde{\boldsymbol{w}}_i\}_{i=1}^n}{\text{minimize}} \quad & \sum_{i=1}^{n} \widetilde{\boldsymbol{w}}_i^\top \boldsymbol{G}_i\, \widetilde{\boldsymbol{w}}_i + \gamma(\alpha \|\widetilde{\boldsymbol{w}}_i\|_2^2 + (1-\alpha)\|\widetilde{\boldsymbol{w}}_i\|_1), \\ \text{subject to} \quad & \mathbf{1}^\top \widetilde{\boldsymbol{w}}_i = 1, \quad \forall i \in \{1, \dots, n\}, \end{aligned} \tag{8.107}$$

where $\alpha \in [0, 1]$. Note that $(\alpha \|\widetilde{\boldsymbol{w}}_i\|_2^2 + (1-\alpha)\|\widetilde{\boldsymbol{w}}_i\|_1)$ is the elastic-net function [76]. As the ℓ_1 norm, i.e., $\|\widetilde{\boldsymbol{w}}_i\|_1 = \sum_{j=1}^{k} |\widetilde{w}_{ij}|$, is not differentiable, $\widetilde{w}_{ij} := \widetilde{w}_{ij,+} - \widetilde{w}_{ij,-}$ can be used, where:

$$\begin{cases} \widetilde{w}_{ij,+} := |\widetilde{w}_{ij}|,\ \widetilde{w}_{ij,-} := 0 & \text{if } \widetilde{w}_{ij} \geq 0, \\ \widetilde{w}_{ij,+} := 0,\ \widetilde{w}_{ij,-} := -|\widetilde{w}_{ij}| & \text{if } \widetilde{w}_{ij} \geq 0. \end{cases} \tag{8.108}$$

Therefore $|\widetilde{w}_{ij}| := \widetilde{w}_{ij,+} + \widetilde{w}_{ij,-}$. Suppose $\mathbb{R}^k \ni \widetilde{\boldsymbol{w}}_{i,+} := [\widetilde{w}_{i1,+}, \dots, \widetilde{w}_{ik,+}]^\top$, $\mathbb{R}^k \ni \widetilde{\boldsymbol{w}}_{i,-} := [\widetilde{w}_{i1,-}, \dots, \widetilde{w}_{ik,-}]^\top$, $\mathbb{R}^{2k} \ni \widetilde{\boldsymbol{w}}_i^* := [\widetilde{\boldsymbol{w}}_{i,+}^\top, \widetilde{\boldsymbol{w}}_{i,-}^\top]^\top$, $\mathbb{R}^{d\times 2k} \ni \boldsymbol{X}_i^* := [\boldsymbol{X}_i, -\boldsymbol{X}_i]$, and $\mathbb{R}^{2k\times 2k} \ni \boldsymbol{G}_i^* := (\boldsymbol{x}_i \mathbf{1}_{2k\times 1}^\top - \boldsymbol{X}_i^*)^\top (\boldsymbol{x}_i \mathbf{1}_{2k\times 1}^\top - \boldsymbol{X}_i^*)$ are defined. Equation (8.107) can be restated as:

$$
\begin{aligned}
\underset{\{\widetilde{\boldsymbol{w}}_i^*\}_{i=1}^n}{\text{minimize}} \quad & \sum_{i=1}^n \widetilde{\boldsymbol{w}}_i^{*\top} \boldsymbol{G}_i^* \, \widetilde{\boldsymbol{w}}_i^* + \gamma(1-\alpha)\mathbf{1}_{2k\times 1}^\top \widetilde{\boldsymbol{w}}_i^*, \\
\text{subject to} \quad & \mathbf{1}_{k\times 1}^\top \widetilde{\boldsymbol{w}}_{i,+}^* - \mathbf{1}_{k\times 1}^\top \widetilde{\boldsymbol{w}}_{i,-}^* = 1, \\
& \widetilde{\boldsymbol{w}}_i^* \succeq 0, \quad \forall i \in \{1, \dots, n\}.
\end{aligned}
\tag{8.109}
$$

This optimization problem can be solved by sequential quadratic programming [10] (see Chap. 4).

8.10 Fusion of LLE with Other Dimensionality Reduction Methods

8.10.1 LLE with Geodesic Distances: Fusion of LLE with Isomap

ISOLLE [63] fuses LLE and Isomap [62], introduced in Chap. 7. Although LLE is a nonlinear manifold learning method, its *k*NN construction is linear because it uses the Euclidean distance. ISOLLE uses the geodesic distance in the LLE method; recall that the geodesic distance is also used in Isomap (see Chap. 7).

The *geodesic distance* is the length of the shortest path between two points on the possibly curvy manifold. Although ideal to use, calculating the geodesic distance is difficult since it requires traversing from one point to another on the manifold. This calculation requires differential geometry and Riemannian manifold calculations [4]. ISOLLE approximates the geodesic distance by using piecewise Euclidean distances. It finds the *k*-Nearest Neighbours (*k*NN) graph of the dataset. Then, the shortest path between two points, through their neighbours, is found using a shortest-path algorithm,[3] such as the Dijkstra algorithm or the Floyd-Warshal algorithm [16]. The approximated geodesic distance can be formulated as [9]:

$$
\boldsymbol{D}_{ij}^{(g)} := \min_{\boldsymbol{r}} \sum_{i=2}^{l} \|\boldsymbol{r}_i - \boldsymbol{r}_{i+1}\|_2, \tag{8.110}
$$

[3] A scikit-learn function in Python for this is “graph_shortest_path” from the package “sklearn.utils.graph_shortest_path”.

where $l \geq 2$ is the length of the sequence of points $\boldsymbol{r}_i \in \{\boldsymbol{x}_i\}_{i=1}^n$ and $\boldsymbol{D}_{ij}^{(g)}$ denotes the (i, j)-th element of the geodesic distance matrix $\boldsymbol{D}^{(g)} \in \mathbb{R}^{n \times n}$. ISOLLE uses the geodesic distance matrix $\boldsymbol{D}^{(g)}$, rather than the Euclidean distance matrix $\boldsymbol{D}$, for construction of the kNN graph. The rest of ISOLLE is the same as in LLE. For more information on geodesic distance, refer to Chap. 7.

8.10.2 Fusion of LLE with PCA

LLE is fused with Principal Component Analysis (PCA) (see Chap. 5) in the LLE-guided PCA (LLE-PCA) [37]. The centered data are denoted by $\mathbb{R}^{d \times n} \ni \breve{\boldsymbol{X}} := \boldsymbol{X}\boldsymbol{H}$, where $\boldsymbol{H} := \boldsymbol{I} - (1/n)\boldsymbol{1}\boldsymbol{1}^\top$ is the centering matrix (see Chap. 2). A PCA subspace can be found through Singular Value Decomposition (SVD) on the reconstructed data $\widehat{\boldsymbol{X}}$, i.e., $\widehat{\boldsymbol{X}} = \boldsymbol{U}\boldsymbol{\Sigma}\boldsymbol{V}$. According to the orthogonality of matrices in SVD, $\boldsymbol{U}^\top\boldsymbol{U} = \boldsymbol{I}$ and $\boldsymbol{V}\boldsymbol{U}^\top = \boldsymbol{I}$. Minimization of the reconstruction error is (see Chap. 5):

$$
\begin{aligned}
\underset{\boldsymbol{U},\boldsymbol{\Sigma},\boldsymbol{V}}{\text{minimize}} \quad & ||\breve{\boldsymbol{X}} - \widehat{\boldsymbol{X}}||_F^2 = ||\breve{\boldsymbol{X}} - \boldsymbol{U}\boldsymbol{\Sigma}\boldsymbol{V}||_F^2, \\
\text{subject to} \quad & \boldsymbol{U}^\top\boldsymbol{U} = \boldsymbol{I}, \\
& \boldsymbol{V}\boldsymbol{V}^\top = \boldsymbol{I}.
\end{aligned} \tag{8.111}
$$

$\boldsymbol{U}$ and $\boldsymbol{\Sigma}$ can be absorbed to have $\widehat{\boldsymbol{X}} = \boldsymbol{U}\boldsymbol{\Sigma}\boldsymbol{V} = \boldsymbol{R}\boldsymbol{V}$, where $\boldsymbol{R} = \boldsymbol{U}\boldsymbol{\Sigma}$. The embedded data or the projected data into the p-dimensional embedding space is $\mathbb{R}^{p \times n} \ni \boldsymbol{Y}^\top := \boldsymbol{U}^\top\breve{\boldsymbol{X}}$, where $\boldsymbol{U} \in \mathbb{R}^{d \times p}$ is the projection matrix. Therefore, the reconstructed data are $\widehat{\boldsymbol{X}} = \boldsymbol{U}\boldsymbol{U}^\top\breve{\boldsymbol{X}} = \boldsymbol{U}\boldsymbol{Y}^\top$. In summary, up to scale of singular values, the equality of $\boldsymbol{R}\boldsymbol{V}$ and $\boldsymbol{U}\boldsymbol{Y}^\top$ can be considered. Therefore, $\boldsymbol{Y} = \boldsymbol{V}$ and $||\breve{\boldsymbol{X}} - \boldsymbol{U}\boldsymbol{\Sigma}\boldsymbol{V}||_F^2 = ||\breve{\boldsymbol{X}} - \boldsymbol{R}\boldsymbol{Y}^\top||_F^2$ hold up to scale. Note that according to the first constraint in Eq. (8.14), there is $\boldsymbol{V}\boldsymbol{V}^\top = \boldsymbol{Y}^\top\boldsymbol{Y} = \boldsymbol{I}$ up to scale, meaning the second constraint in Eq. (8.111) is automatically satisfied. In conclusion, Eq. (8.111) is restated to:

$$
\underset{\boldsymbol{R}}{\text{minimize}} \quad ||\breve{\boldsymbol{X}} - \boldsymbol{R}\boldsymbol{Y}^\top||_F^2. \tag{8.112}
$$

The Lagrangian of this optimization is [11]:

$$
\begin{aligned}
\mathcal{L} = \frac{\partial ||\breve{\boldsymbol{X}} - \boldsymbol{R}\boldsymbol{Y}^\top||_F^2}{\partial \boldsymbol{R}} = 2(\breve{\boldsymbol{X}} - \boldsymbol{R}\boldsymbol{Y}^\top)\boldsymbol{Y} \overset{\text{set}}{=} \boldsymbol{0} \\
\implies \breve{\boldsymbol{X}} - \boldsymbol{R}\boldsymbol{Y}^\top = \boldsymbol{0} \implies \boldsymbol{R} = \breve{\boldsymbol{X}}(\boldsymbol{Y}^\top)^{-1} \overset{(a)}{=} \breve{\boldsymbol{X}}\boldsymbol{Y},
\end{aligned} \tag{8.113}
$$

where (a) is because $\boldsymbol{Y}$ is an orthogonal matrix, as $\boldsymbol{Y}^\top\boldsymbol{Y} = \boldsymbol{I}$. LLE-PCA [37] centers data first, then applies LLE to find the embedding $\boldsymbol{Y} \in \mathbb{R}^{n \times p}$. Lastly, it projects the data onto the PCA subspace:

$$\mathbb{R}^{d\times n} \ni \boldsymbol{Y}_{\text{LLE-PCA}} := \boldsymbol{R}\boldsymbol{Y}^\top \overset{(8.113)}{=} \breve{\boldsymbol{X}}\boldsymbol{Y}\boldsymbol{Y}^\top, \tag{8.114}$$

stacked columnwise. Considering the first p rows results in the p-dimensional embedding $\boldsymbol{Y}_{\text{LLE-PCA}} \in \mathbb{R}^{p\times n}$.

8.10.3 Fusion of LLE with FDA (or LDA)

Unified LLE and Linear Discriminant Analysis Algorithm (ULLELDA) [69] fuses LLE and FDA (see Chap. 6) (or linear discriminant analysis [21]). ULLELDA first applies LLE to the high-dimensional data to determine the embeddings $\{\boldsymbol{y}_i \in \mathbb{R}^p\}_{i=1}^n$ and the weights $\{w_{ij}\}_{i,j=1}^n$. These embeddings are projected onto the FDA subspace (see Chap. 6) to create new embeddings $\{\boldsymbol{z}_i \in \mathbb{R}^p\}_{i=1}^n$. The final embedding of $\boldsymbol{x}_i$ is obtained as:

$$\mathbb{R}^p \ni \boldsymbol{y}_{i,\text{ULLELDA}} := \sum_{j=1}^{n} w_{ij}\, \boldsymbol{z}_j. \tag{8.115}$$

This fusion of LLE with FDA makes LLE embeddings tuned for better separation of classes.

8.10.4 Fusion of LLE with FDA and Graph Embedding: Discriminant LLE

Discriminant LLE (DLLE) [45] is a supervised LLE method. Its scheme is to (I) take the kNN of every point only from the points in the same class as the point, and (II) maximize and minimize the interclass and intraclass variances of the data.

DLLE computes kNN of points from the points of their own classes and uses this kNN graph during optimization (8.1). Then, the weight matrix $\boldsymbol{W} = [w_{ij}] \in \mathbb{R}^{n\times n}$ is obtained through Eq. (8.10). A similarity matrix $\boldsymbol{S} \in \mathbb{R}^{n\times n}$ is defined using the obtained weight matrix:

$$\boldsymbol{S}(i, j) := \begin{cases} (\boldsymbol{W} + \boldsymbol{W}^\top - \boldsymbol{W}^\top\boldsymbol{W})(i, j) & \text{if } i = j, \\ 0 & \text{otherwise,} \end{cases} \tag{8.116}$$

inspired by graph embedding [67]. It also finds a kNN graph by considering the neighbours of a point from the different classes than the class of point. It defines a dissimilarity (or between-class) matrix $\boldsymbol{B} \in \mathbb{R}^{n\times n}$ by:

$$\boldsymbol{B}(i, j) := \begin{cases} 1/k & \text{if } c_i \neq c_j, \\ 0 & \text{Otherwise.} \end{cases} \tag{8.117}$$

Let the Laplacian matrices of $\boldsymbol{S}$ and $\boldsymbol{B}$ be denoted by $\boldsymbol{L}_S$ and $\boldsymbol{L}_B$, respectively. DLLE finds a projection matrix $\boldsymbol{U}$ for maximizing and minimizing the interclass and intraclass variances:

$$\underset{\boldsymbol{U}}{\text{maximize}} \quad \frac{\textbf{tr}(\boldsymbol{U}^\top \boldsymbol{X} \boldsymbol{L}_B \boldsymbol{X}^\top \boldsymbol{U})}{\textbf{tr}(\boldsymbol{U}^\top \boldsymbol{X} \boldsymbol{L}_S \boldsymbol{X}^\top \boldsymbol{U})}, \tag{8.118}$$

which is a Rayleigh-Ritz quotient, whose solution is a generalized eigenvalue problem $(\boldsymbol{X}\boldsymbol{L}_B\boldsymbol{X}^\top, \boldsymbol{X}\boldsymbol{L}_S\boldsymbol{X}^\top)$ (see Chap. 2). This optimization is inspired by Fisher discriminant analysis (see Chap. 6).

8.11 Weighted Locally Linear Embedding

Various works have been performed on the weighting of distances, reconstruction weights, and the embedding in LLE. Some examples include the weighted LLE for deformed distributed data, weighted LLE using probability of occurrence, supervised LLE by adjusting weights, the modified LLE method, and iterative LLE. These methods are introduced in the following sections.

8.11.1 Weighted LLE for Deformed Distributed Data

Weighted LLE [51] improves the LLE especially if the distribution of the data is deflated, in that it is different from a Gaussian distribution. A weighted LLE uses a weighted distance defined as [75]:

$$\text{dist}(\boldsymbol{x}_i, \boldsymbol{x}_j) := \frac{\|\boldsymbol{x}_i - \boldsymbol{x}_j\|_2}{a_i + b_i \frac{(\boldsymbol{x}_i - \boldsymbol{x}_j)^\top \boldsymbol{\tau}_i}{\|\boldsymbol{x}_i - \boldsymbol{x}_j\|_2}} = \frac{\|\boldsymbol{x}_i - \boldsymbol{x}_j\|_2}{(a_i + b_j \cos\theta)}, \tag{8.119}$$

where $\boldsymbol{v}_{ij} := \boldsymbol{x}_{ij} - \boldsymbol{x}_i$ is calculated using the kNN by the Euclidean distance and then [51]:

$$\boldsymbol{\tau}_i := \frac{\boldsymbol{g}_i}{\|\boldsymbol{g}_i\|_2}, \quad a_i := \frac{l_i}{c_2}, \quad b_i := \frac{\|\boldsymbol{g}_i\|_2}{c_1}, \tag{8.120}$$

$$\boldsymbol{g}_i := \frac{1}{k}\sum_{j=1}^{k} \boldsymbol{v}_{ij}, \quad l_i := \frac{1}{k}\sum_{j=1}^{k} \|\boldsymbol{v}_{ij}\|_2, \tag{8.121}$$

$$c_1 = \sqrt{2}\frac{\Gamma((d+1)/2)}{\Gamma(d/2)\, d}, \quad c_2 = \sqrt{2}\frac{\Gamma((d+1)/2)}{\Gamma(d/2)}, \tag{8.122}$$

where Γ is the Gamma function and d is the dimensionality of the input space. Using Eq. (8.119) as the distance, rather than the Euclidean distance, the kNN graph can be found. In the formulation of LLE, kNN is used, and the rest of the algorithm is the same as in LLE.

8.11.2 Weighted LLE Using Probability of Occurrence

The weighted LLE, which uses a probability of occurrence [49], assumes that data have a probability distribution; for example, a mixture distribution can be fitted to data using the expectation maximization algorithm [24]. Let the probability of the occurrence of the data point $\boldsymbol{x}_i$ be p_i. The distance used in this weighted LLE is weighted by the probability of the occurrence:

$$\text{dist}^2(\boldsymbol{x}_i, \boldsymbol{x}_j) := \frac{\|\boldsymbol{x}_i - \boldsymbol{x}_j\|_2^2}{p_i}. \tag{8.123}$$

Note that this weighting increases the distance of a point from its neighbours if its probability is low. This makes sense because an outlier or anomaly should be considered farther from other normal points. This weighting creates a more robust LLE when there are outliers.

Using this weighted distance rather than the Euclidean distance, the kNN graph can be calculated. Moreover, the Gram matrix, Eq. (8.3), is weighted by the probabilities of occurrence. If $\boldsymbol{G}_i(a, b)$ denotes the (a, b)-th element of $\boldsymbol{G}_i$, it is weighted as:

$$\boldsymbol{G}_i(a, b) := \sqrt{p_i\, p_j}\, \boldsymbol{G}_i(a, b). \tag{8.124}$$

The rest of the algorithm remains the same as in LLE. One of the applications of this method is in the field of facial recognition.

8.11.3 Supervised LLE by Adjusting Weights

The supervised LLE method uses labels to adjust the weights [32]. The obtained weights, by Eq. (8.8), in LLE are weighted using the class labels. If two points are in the same class, the reconstruction weight between them is strengthened because they are similar; otherwise, the weight is decreased:

$$\widetilde{w}_{ij} \leftarrow \begin{cases} \widetilde{w}_{ij} + \delta \text{ if } c_i = c_j, \\ \widetilde{w}_{ij} - \delta \text{ Otherwise.} \end{cases} \tag{8.125}$$

8.11.4 Modified Locally Linear Embedding

Modified LLE (MLLE) [72] modifies or adjusts the reconstruction weights, and defines the new weights as:

$$\mathbb{R}^k \ni \widetilde{\boldsymbol{w}}_i^{(l)} = (1 - \alpha_i)\, \widetilde{\boldsymbol{w}}_i + \boldsymbol{V}_i\, \boldsymbol{J}_i(:, l), \tag{8.126}$$

for $l \in \{1, \dots, s_i\}$, where $\boldsymbol{V}_i \in \mathbb{R}^{k \times s_i}$ is the matrix containing the s_i smallest right singular vectors of $\boldsymbol{G}_i$, $\alpha_i := (1/\sqrt{s_i})\|\boldsymbol{v}_i\|_2$, $\boldsymbol{v}_i := \boldsymbol{V}_i^\top \boldsymbol{1}_{k\times 1} \in \mathbb{R}^{s_i}$, and $\boldsymbol{J}_i$ is a Householder matrix [35] satisfying $\boldsymbol{H}_i \boldsymbol{V}_i^\top \boldsymbol{1}_{k\times 1} = \alpha_i \boldsymbol{1}_{s_i \times 1}$. MLLE uses $\widetilde{\boldsymbol{w}}_i^{(l)}$, rather than $\widetilde{\boldsymbol{w}}_i$, in Eq. (8.10), to obtain $w_{ij}^{(l)}$. This method slightly modifies the objective in optimization of Eq. (8.9):

$$\underset{\boldsymbol{Y}}{\text{minimize}} \quad \sum_{i=1}^{n} \sum_{l=1}^{s_i} \Big|\Big| \boldsymbol{y}_i - \sum_{j=1}^{n} w_{ij}^{(l)} \boldsymbol{y}_j \Big|\Big|_2^2, \tag{8.127}$$

with the same constraints as in Eq. (8.9). The rest of the algorithm is solved similarly to LLE but with this modified objective function.

8.11.5 Iterative Locally Linear Embedding

Iterative LLE [39] is an LLE-based method that has made several modifications to LLE. First, it restricts the weights to be nonnegative. Thereby, it changes Eq. (8.1) to:

$$\begin{aligned} \underset{\widetilde{\boldsymbol{W}}}{\text{minimize}} \quad & \varepsilon(\widetilde{\boldsymbol{W}}) := \sum_{i=1}^{n} \Big|\Big| \boldsymbol{x}_i - \sum_{j=1}^{k} \widetilde{w}_{ij} \boldsymbol{x}_{ij} \Big|\Big|_2^2, \\ \text{subject to} \quad & \widetilde{w}_{ij} \geq 0, \quad \forall i \in \{1, \dots, n\}. \end{aligned} \tag{8.128}$$

Additionally, iterative LLE adjusts and weights the embedded data $\boldsymbol{Y}$ by including the diagonal degree matrix $\boldsymbol{D} \in \mathbb{R}^{n\times n}$ to the constraint in Eq. (8.14):

$$\begin{aligned} \underset{\boldsymbol{Y}}{\text{minimize}} \quad & \mathbf{tr}(\boldsymbol{Y}^\top \boldsymbol{M} \boldsymbol{Y}), \\ \text{subject to} \quad & \frac{1}{n} \boldsymbol{Y}^\top \boldsymbol{D} \boldsymbol{Y} = \boldsymbol{I}, \\ & \boldsymbol{Y}^\top \boldsymbol{1} = \boldsymbol{0}, \end{aligned} \tag{8.129}$$

which has some relations with the spectral embedding [13] and the Laplacian embedding [6]. The iterative LLE [39] also iterates between the solutions of Eqs. (8.128) and (8.129) to improve the embedding of LLE.

8.12 Chapter Summary

This chapter introduced LLE and its variants, including kernel LLE, out-of-sample extension, incremental LLE, landmark LLE, supervised and semisupervised approaches, robust LLE, weighted LLE, and fusion of LLE with other methods. As explained in Sect. 8.1, the proposal of LLE was a breakthrough in dimensionality reduction since most of the unsupervised dimensionality reduction methods followed the main idea of LLE, which is local fitting while preserving the global structure.

References

1. Saeed Ahmadizadeh et al. "On eigenvalues of Laplacian matrix for a class of directed signed graphs". In: *Linear Algebra and its Applications* 523 (2017), pp. 281–306.
2. Babak Alipanahi and Ali Ghodsi. "Guided locally linear embedding". In: *Pattern recognition letters* 32.7 (2011), pp. 1029–1035.
3. Andrés Álvarez-Meza et al. "Global and local choice of the number of nearest neighbors in locally linear embedding". In: *Pattern Recognition Letters* 32.16 (2011), pp. 2171–2177.
4. Thierry Aubin. *A course in differential geometry*. Vol. 27. American Mathematical Society, Graduate Studies in Mathematics, 2001.
5. Elnaz Barshan et al. "Supervised principal component analysis: Visualization, classification and regression on subspaces and submanifolds". In: *Pattern Recognition* 44.7 (2011), pp. 1357–1371.
6. Mikhail Belkin and Partha Niyogi. "Laplacian eigenmaps for dimensionality reduction and data representation". In: *Neural computation* 15.6 (2003), pp. 1373–1396.
7. Yoshua Bengio et al. "Learning eigenfunctions links spectral embedding and kernel PCA". In: *Neural computation* 16.10 (2004), pp. 2197–2219.
8. Yoshua Bengio et al. *Learning eigenfunctions of similarity: linking spectral clustering and kernel PCA*. Tech. rep. Technical Report 1232, Departement d'Informatique et Recherche Oprationnelle ..., 2003.
9. Yoshua Bengio et al. "Out-of-sample extensions for LLE, Isomap, MDS, eigenmaps, and spectral clustering". In: *Advances in neural information processing systems*. 2004, pp. 177–184.
10. Paul T Boggs and Jon W Tolle. "Sequential quadratic programming". In: *Acta numerica* 4.1 (1995), pp. 1–51.
11. Stephen Boyd, Stephen P Boyd, and Lieven Vandenberghe. *Convex optimization*. Cambridge university press, 2004.
12. Kerstin Bunte, Michael Biehl, and Barbara Hammer. "A general framework for dimensionality-reducing data visualization mapping". In: *Neural Computation* 24.3 (2012), pp. 771–804.
13. Pak K Chan, Martine DF Schlag, and Jason Y Zien. "Spectral k-way ratio-cut partitioning and clustering". In: *IEEE Transactions on computer-aided design of integrated circuits and systems* 13.9 (1994), pp. 1088–1096.
14. Hong Chang and Dit-Yan Yeung. "Robust locally linear embedding". In: *Pattern recognition* 39.6 (2006), pp. 1053–1065.
15. Jing Chen and Yang Liu. "Locally linear embedding: a survey". In: *Artificial Intelligence Review* 36.1 (2011), pp. 29–48.
16. Thomas H Cormen et al. *Introduction to algorithms*. MIT press, 2009.

17. Genaro Daza-Santacoloma, Carlos D Acosta-Medina, and Germán Castellanos-Domınguez. "Regularization parameter choice in locally linear embedding". In: *neurocomputing* 73.10-12 (2010), pp. 1595–1605.
18. Dick De Ridder and Robert PW Duin. "Locally linear embedding for classification". In: *Pattern Recognition Group, Dept. of Imaging Science & Technology, Delft University of Technology, Delft, The Netherlands, Tech. Rep. PH-2002-01* (2002), pp. 1–12.
19. Dick De Ridder et al. "Supervised locally linear embedding". In: *Artificial Neural Networks and Neural Information Processing—ICANN/ICONIP 2003*. Springer, 2003, pp. 333–341.
20. Ali Ghodsi. *Dimensionality reduction a short tutorial*. Tech. rep. Department of Statistics and Actuarial Science, Univ. of Waterloo, Ontario, Canada, 2006.
21. Benyamin Ghojogh and Mark Crowley. "Linear and quadratic discriminant analysis: Tutorial". In: *arXiv preprint arXiv:1906.02590* (2019).
22. Benyamin Ghojogh, Fakhri Karray, and Mark Crowley. "Locally Linear Image Structural Embedding for Image Structure Manifold Learning". In: *International Conference on Image Analysis and Recognition*. Springer. 2019, pp. 126–138.
23. Benyamin Ghojogh et al. "Feature selection and feature extraction in pattern analysis: A literature review". In: *arXiv preprint arXiv:1905.02845* (2019).
24. Benyamin Ghojogh et al. "Fitting a mixture distribution to data: tutorial". In: *arXiv preprint arXiv:1901.06708* (2019).
25. Andrej Gisbrecht, Alexander Schulz, and Barbara Hammer. "Parametric nonlinear dimensionality reduction using kernel t-SNE". In: *Neurocomputing* 147 (2015), pp. 71–82.
26. Andrej Gisbrecht et al. "Out-of-sample kernel extensions for nonparametric dimensionality reduction." In: *European Symposium on Artificial Neural Networks, Computational Intelligence and Machine Learning*. 2012.
27. Yair Goldberg and Ya'acov Ritov. "Local procrustes for manifold embedding: a measure of embedding quality and embedding algorithms". In: *Machine learning* 77.1 (2009), pp. 1–25.
28. Arthur Gretton et al. "Measuring statistical dependence with Hilbert-Schmidt norms". In: *International conference on algorithmic learning theory*. Springer. 2005, pp. 63–77.
29. Jihun Ham et al. "A kernel view of the dimensionality reduction of manifolds". In: *Proceedings of the twenty-first international conference on Machine learning*. 2004, p. 47.
30. Trevor Hastie, Robert Tibshirani, and Jerome Friedman. *The elements of statistical learning: data mining, inference, and prediction*. Springer Science & Business Media, 2009.
31. Ping He et al. "Discriminative locally linear mapping for medical diagnosis". In: *Multimedia Tools and Applications* 79.21 (2020), pp. 14573–14591.
32. Ping He et al. "Nonlinear Manifold Classification Based on LLE". In: *Advances in Computer Communication and Computational Sciences*. Springer, 2019, pp. 227–234.
33. Thomas Hofmann, Bernhard Schölkopf, and Alexander J Smola. "Kernel methods in machine learning". In: *The annals of statistics* (2008), pp. 1171–1220.
34. Paul W Holland and Roy E Welsch. "Robust regression using iteratively reweighted least-squares". In: *Communications in Statistics-theory and Methods* 6.9 (1977), pp. 813–827.
35. Alston S Householder. *Principles of numerical analysis*. New York: McGraw-Hill, 1953, pp. 135–138.
36. Prateek Jain and Purushottam Kar. "Non-convex optimization for machine learning". In: *Foundations and Trends® in Machine Learning* 10.3-4 (2017), pp. 142–336.
37. Bo Jiang, Chris Ding, and Bin Luo. "Robust data representation using locally linear embedding guided PCA". In: *Neurocomputing* 275 (2018), pp. 523–532.
38. David G Kleinbaum et al. *Logistic regression*. Springer, 2002.
39. Deguang Kong et al. "An iterative locally linear embedding algorithm". In: *arXiv preprint arXiv:1206.6463* (2012).
40. Olga Kouropteva, Oleg Okun, and Matti Pietikäinen. "Incremental locally linear embedding". In: *Pattern recognition* 38.10 (2005), pp. 1764–1767.
41. Olga Kouropteva, Oleg Okun, and Matti Pietikäinen. "Selection of the optimal parameter value for the locally linear embedding algorithm." In: *FSKD* 2 (2002), pp. 359–363.

42. Olga Kouropteva, Oleg Okun, and Matti Pietikäinen. "Supervised locally linear embedding algorithm for pattern recognition". In: *Iberian Conference on Pattern Recognition and Image Analysis*. Springer. 2003, pp. 386–394.
43. Olga Kouropteva et al. *Beyond locally linear embedding algorithm*. Tech. rep. University of Oulu, 2002.
44. Benwei Li and Yun Zhang. "Supervised locally linear embedding projection (SLLEP) for machinery fault diagnosis". In: *Mechanical Systems and Signal Processing* 25.8 (2011), pp. 3125–3134.
45. Xuelong Li et al. "Discriminant locally linear embedding with high-order tensor data". In: *IEEE Transactions on Systems, Man, and Cybernetics, Part B (Cybernetics)* 38.2 (2008), pp. 342–352.
46. Xin Liu et al. "Locally linear embedding (LLE) for MRI based Alzheimer's disease classification". In: *Neuroimage* 83 (2013), pp. 148–157.
47. Anne Marsden. "Eigenvalues of the Laplacian and their relationship to the connectedness of a graph". In: *University of Chicago, REU* (2013).
48. Lawrence C Marsh and David R Cormier. *Spline regression models*. 137. Sage, 2001.
49. Nathan Mekuz, Christian Bauckhage, and John K Tsotsos. "Face recognition with weighted locally linear embedding". In: *The 2nd Canadian Conference on Computer and Robot Vision (CRV'05)*. IEEE. 2005, pp. 290–296.
50. Sebastian Mika et al. "Fisher discriminant analysis with kernels". In: *Neural networks for signal processing IX: Proceedings of the 1999 IEEE signal processing society workshop (cat. no. 98th8468)*. Ieee. 1999, pp. 41–48.
51. Yaozhang Pan, Shuzhi Sam Ge, and Abdullah Al Mamun. "Weighted locally linear embedding for dimension reduction". In: *Pattern Recognition* 42.5 (2009), pp. 798–811.
52. Marzia Polito and Pietro Perona. "Grouping and dimensionality reduction by locally linear embedding". In: *Advances in neural information processing systems*. 2002, pp. 1255–1262.
53. Sam T Roweis and Lawrence K Saul. "Nonlinear dimensionality reduction by locally linear embedding". In: *Science* 290.5500 (2000), pp. 2323–2326.
54. Lawrence K Saul and Sam T Roweis. *An introduction to locally linear embedding*. Tech. rep. 2000.
55. Lawrence K Saul and Sam T Roweis. "Think globally, fit locally: unsupervised learning of low dimensional manifolds". In: *Journal of machine learning research* 4.Jun (2003), pp. 119–155.
56. Lawrence K Saul and Sam T Roweis. *Think globally, fit locally: Unsupervised learning of nonlinear manifolds*. Tech. rep. Technical Report CIS-02-18, University of Pennsylvania, 2002.
57. Lawrence K Saul et al. "Spectral methods for dimensionality reduction." In: *Semi-supervised learning* 3 (2006).
58. Bernhard Schölkopf, Alexander J Smola, Francis Bach, et al. *Learning with kernels: support vector machines, regularization, optimization, and beyond*. MIT press, 2002.
59. Robin Sibson. "Studies in the robustness of multidimensional scaling: Procrustes statistics". In: *Journal of the Royal Statistical Society: Series B (Methodological)* 40.2 (1978), pp. 234–238.
60. Harry Strange and Reyer Zwiggelaar. *Open Problems in Spectral Dimensionality Reduction*. Springer, 2014.
61. Bing-Yu Sun et al. "Feature fusion using locally linear embedding for classification". In: *IEEE transactions on neural networks 21.1* (2009), pp. 163–168.
62. Joshua B Tenenbaum, Vin De Silva, and John C Langford. "A global geometric framework for nonlinear dimensionality reduction". In: *Science* 290.5500 (2000), pp. 2319–2323.
63. Claudio Varini, Andreas Degenhard, and Tim Nattkemper. "ISOLLE: Locally linear embedding with geodesic distance". In: *European Conference on Principles of Data Mining and Knowledge Discovery*. Springer. 2005, pp. 331–342.
64. Max Vladymyrov and Miguel Á Carreira-Perpinán. "Locally linear landmarks for large-scale manifold learning". In: *Joint European Conference on Machine Learning and Knowledge Discovery in Databases*. Springer. 2013, pp. 256–271.
65. Manda Winlaw, Leila Samimi Dehkordy, and Ali Ghodsi. "Robust locally linear embedding using penalty functions". In: *The 2011 International Joint Conference on Neural Networks*. IEEE. 2011, pp. 2305–2312.

66. Hau-Tieng Wu, Nan Wu, et al. "Think globally, fit locally under the manifold setup: Asymptotic analysis of locally linear embedding". In: *The Annals of Statistics* 46.6B (2018), pp. 3805–3837.
67. Shuicheng Yan et al. "Graph embedding: A general framework for dimensionality reduction". In: *2005 IEEE Computer Society Conference on Computer Vision and Pattern Recognition (CVPR'05)*. Vol. 2. IEEE. 2005, pp. 830–837.
68. Kamen Yotov, Keshav Pingali, and Paul Stodghill. "Think globally, search locally". In: *Proceedings of the 19th annual international conference on Supercomputing*. 2005, pp. 141–150.
69. Junping Zhang, Huanxing Shen, and Zhi-Hua Zhou. "Unified locally linear embedding and linear discriminant analysis algorithm (ULLELDA) for face recognition". In: *Chinese Conference on Biometric Recognition*. Springer. 2004, pp. 296–304.
70. Shanwen Zhang and Kwok-Wing Chau. "Dimension reduction using semi-supervised locally linear embedding for plant leaf classification". In: *International conference on intelligent computing*. Springer. 2009, pp. 948–955.
71. Shi-qing Zhang. "Enhanced supervised locally linear embedding". In: *Pattern Recognition Letters* 30.13 (2009), pp. 1208–1218.
72. Zhenyue Zhang and Jing Wang. "MLLE: Modified locally linear embedding using multiple weights". In: *Advances in neural information processing systems*. 2007, pp. 1593–1600.
73. Lingxiao Zhao and Zhenyue Zhang. "Supervised locally linear embedding with probability-based distance for classification". In: *Computers & Mathematics with Applications* 57.6 (2009), pp. 919–926.
74. Xiaoming Zhao and Shiqing Zhang. "Facial expression recognition using local binary patterns and discriminant kernel locally linear embedding". In: *EURASIP journal on Advances in signal processing* 2012.1 (2012), p. 20.
75. Chang Yin Zhou and Yan Qiu Chen. "Improving nearest neighbor classification with cam weighted distance". In: *Pattern Recognition* 39.4 (2006), pp. 635–645.
76. Hui Zou and Trevor Hastie. "Regularization and variable selection via the elastic net". In: *Journal of the royal statistical society: series B (statistical methodology)* 67.2 (2005), pp. 301–320.

Chapter 9
Laplacian-Based Dimensionality Reduction

9.1 Introduction

Spectral dimensionality reduction methods deal with the graph and geometry of data and usually reduce to an eigenvalue or generalized eigenvalue problem (see Chap. 2). A subfamily of the spectral dimensionality reduction methods is based on the Laplacian of the graph of data [64]. The importance of the Laplacian within spectral dimensionality reduction was discovered when researchers gradually proposed spectral clustering methods, which required the use of the Laplacian [70, 95]. Spectral clustering transforms data into a low-dimensional subspace to better separate the clusters. Then, within that discriminating subspace, a simple clustering algorithm is applied. Researchers have adopted the idea of spectral clustering for dimensionality reduction because it clusters data in a low-dimensional subspace. The result was the Laplacian eigenmap [7, 8]. In the meantime, out-of-sample extension techniques were proposed for spectral clustering and the Laplacian eigenmap to embed test data [12, 18, 37, 38]. Later, researchers decided to linearize the nonlinear Laplacian eigenmap using linear projection, creating Locality Preserving Projection (LPP) [46]. A kernel version of LPP was also proposed [26, 61] to apply LPP in the feature space rather than the input space. Afterwards, researchers determined that multiple spectral methods belonged to a family of methods, so they proposed a generalized spectral method using the Laplacian matrix. Therefore, graph embedding, with its various variants, was proposed [100, 101]. It was around that time that the diffusion map, which uses the Laplacian and kernel of data, was proposed, and revealed the underlying nonlinear manifold of data in a diffusion process [29, 30, 59]. This chapter will explore the theory and intuitions of these Laplacian-based methods in detail. Note that although Locally Linear Embedding (LLE) can be considered a Laplacian-based method, it is presented in Chap. 8.

B. Ghojogh et al., *Elements of Dimensionality Reduction and Manifold Learning*,
https://doi.org/10.1007/978-3-031-10602-6_9

9.2 Laplacian Matrix and Its Interpretation

Most of the methods in this chapter use the Laplacian matrix of the data graph. The following sections introduce the adjacency matrix to represent a dataset and then the Laplacian matrix of the adjacency matrix. The importance of Laplacian is interpreted and justified using the Laplace operator. Lastly, the eigenvalues of the Laplacian matrix and the convergence of Laplacian are briefly discussed.

9.2.1 Adjacency Matrix

Assume there is a dataset $\{\boldsymbol{x}_i\}_{i=1}^n$. In spectral clustering and related algorithms, a nearest neighbours graph is constructed. There are two ways of constructing the nearest neighbours graph [7]:

1. Graph using ϵ-neighbourhoods: nodes i and j are connected if $\|\boldsymbol{x}_i - \boldsymbol{x}_j\|_2^2 \leq \epsilon$
2. k-Nearest Neighbours (kNN) graph: node j is connected to node i if $\boldsymbol{x}_j$ is among the k nearest neighbours of $\boldsymbol{x}_i$

Using the nearest neighbours graph, it is possible to construct an adjacency matrix, also called a weight matrix or affinity matrix, which is a weighted version of the nearest neighbours graph. The weights of the adjacency matrix represent the similarity of nodes. The more similar two nodes are, the larger their weight is. Let $\boldsymbol{W} \in \mathbb{R}^{n \times n}$ denote the adjacency or weight matrix whose (i, j)-th element is [70]:

$$\boldsymbol{W}(i, j) := \begin{cases} w_{ij} & \text{if } \boldsymbol{x}_j \in \text{nearest neighbours of } \boldsymbol{x}_i, \\ w_{ij} = 0 & \text{if } i = j, \\ w_{ij} = 0 & \text{if } \boldsymbol{x}_j \notin \text{nearest neighbours of } \boldsymbol{x}_i. \end{cases} \tag{9.1}$$

Note that it is possible to choose a fully connected nearest neighbours graph where all points are among the neighbours of each other. In this case, the third condition in the above equation can be omitted [70].

An example of the weight function for the edge between $\boldsymbol{x}_i$ and $\boldsymbol{x}_j$ is the Radial Basis Function (RBF) or heat kernel [7]:

$$w_{ij} = \exp\Big(-\frac{\|\boldsymbol{x}_i - \boldsymbol{x}_j\|_2^2}{2\sigma^2}\Big), \tag{9.2}$$

whose parameter σ^2 can be set to 0.5 or one for simplicity. This weight function increases, or approaches one, when $\boldsymbol{x}_i$ and $\boldsymbol{x}_j$ are closer to each other. Another approach to select w_{ij} is the simple-minded approach [7]:

$$w_{ij} := \begin{cases} 1 \text{ if } i \text{ and } j \text{ are connected} \\ \qquad \text{in the nearest neighbours graph,} \\ 0 \text{ otherwise.} \end{cases} \tag{9.3}$$

9.2.2 Laplacian Matrix

The Laplacian matrix of the adjacency matrix, denoted by $\boldsymbol{L} \in \mathbb{R}^{n\times n}$, is defined as [64]:

$$\boldsymbol{L} := \boldsymbol{D} - \boldsymbol{W}, \tag{9.4}$$

where $\mathbb{R}^{n\times n} \ni \boldsymbol{D} = \text{diag}([d_1, \dots, d_n]^\top)$ is the degree matrix, with diagonal elements as:

$$d_i = \sum_{j=1}^{n} w_{ij}, \tag{9.5}$$

which is the summation of its corresponding row. Therefore, the row summation of the Laplacian matrix is zero:

$$\sum_{j=1}^{n} \boldsymbol{L}_{ij} = \boldsymbol{0}. \tag{9.6}$$

There are other variants of the Laplacian matrix, such as [70, 95]:

$$\boldsymbol{L} := \boldsymbol{D}^{-\alpha} \boldsymbol{W} \boldsymbol{D}^{-\alpha}, \tag{9.7}$$

where $\alpha \geq 0$ is a parameter. A common example is $\alpha = 0.5$. Some researchers have used [70, 95]:

$$\boldsymbol{L} := \boldsymbol{D}^{-1/2} \boldsymbol{W} \boldsymbol{D}^{-1/2}, \tag{9.8}$$

rather than Eq. (9.4) for the definition of the Laplacian. According to [95], Eq. (9.8) results in the (i, j)-th element of the Laplacian being:

$$\boldsymbol{L}(i, j) := \frac{\boldsymbol{W}(i, j)}{\sqrt{\boldsymbol{D}(i, i)\, \boldsymbol{D}(j, j)}}. \tag{9.9}$$

Note that Eq. (9.9) is similar to a technique used for kernel normalization [72]; therefore, the Laplacian is sometimes referred to as the *normalized kernel* matrix (e.g., diffusion map [29, 67], introduced in Sect. 9.7). Additionally, the adjacency matrix is a kernel, such as the RBF kernel in Eq. (9.2).

Equation (9.7) has an opposite effect compared to Eq. (9.4) because the matrix $\boldsymbol{W}$ is not negated in that equation. According to spectral graph theory [27], it is possible to use $(\boldsymbol{I} - \boldsymbol{L})$ or Eqs. (9.8) and (9.9), rather than $\boldsymbol{L}$, in optimization problems if the optimization is changed from minimization to maximization.

It is also noteworthy that Eq. (9.7) is similar to the *symmetric normalized Laplacian matrix* defined as [27, p. 2]:

$$\boldsymbol{L} \leftarrow \boldsymbol{D}^{-(1/2)} \boldsymbol{L} \boldsymbol{D}^{-(1/2)}. \tag{9.10}$$

9.2.3 Interpretation of Laplacian

Let $\boldsymbol{B}$ denote the incidence matrix of the data graph where the rows and columns of $\boldsymbol{B}$ correspond to the edges and vertices of the graph, respectively. The elements of this incidence matrix determine if a vertex is connected to an edge (it is zero if they are not connected). Therefore, this matrix shows if every two points (vertices) are connected or not.

Lemma 9.1 ([86, Section 5.6]) *The incidence matrix of a directed graph can be seen as a difference matrix (of graph nodes).*

Corollary 9.1 *As a gradient, denoted by* ∇, *is a measure of a small difference, it is possible to conclude from Lemma 9.1 that the incidence matrix is analogous to the gradient:*

$$\boldsymbol{B} \propto \nabla. \tag{9.11}$$

Lemma 9.2 ([55, Lemma 10]) *It is possible to factorize the Laplacian matrix of a graph using its incidence matrix:*

$$\boldsymbol{L} = \boldsymbol{B}^\top \boldsymbol{B}. \tag{9.12}$$

This also explains why the Laplacian matrix is positive semidefinite.

Corollary 9.2 *From Eqs. (9.11) and (9.12), it can be stated that:*

$$\boldsymbol{L} = \nabla^\top \nabla. \tag{9.13}$$

This operator, which is the inner produce of the gradients, is also referred to as the Laplace operator.

Therefore, the *Laplace operator* or *Laplacian* is a scalar operator, which is the inner product of two gradient vectors:

$$\begin{aligned}\boldsymbol{L} := \nabla \cdot \nabla = \nabla^2 &= \Big[\frac{\partial}{\partial x_1}, \dots, \frac{\partial}{\partial x_d}\Big] \begin{bmatrix} \frac{\partial}{\partial x_1} \\ \vdots \\ \frac{\partial}{\partial x_d} \end{bmatrix} \\ &= \frac{\partial^2}{\partial x_1 \partial x_1} + \dots + \frac{\partial^2}{\partial x_d \partial x_d} = \sum_{i=1}^{d} \frac{\partial^2}{\partial x_i \partial x_i}.\end{aligned} \tag{9.14}$$

This demonstrates that the Laplacian is the summation of the diagonal of the Hessian matrix (the Jacobian of the gradient); see Chap. 4 for the Hessian and Jacobian matrices. Therefore, it is possible to say that the Laplacian is the trace of the Hessian matrix. Therefore, it can also be seen as the summation of the eigenvalues of the Hessian matrix. The Laplacian captures the amount of divergence from a point. Therefore, the Laplacian measures whether the neighbouring points of a point are larger or smaller than it or are on a line. In other words, it shows how linear the neighbourhood of a point is. If a point is a local minimum/maximum, its Laplacian is positive or negative because its diverging neighbours are above or below it. In this case, the point is on a linear function, meaning its Laplacian is zero. Therefore, Laplacian shows how *bumpy* the neighbourhood of a point is.

The above explanation can be stated as a divergence:

$$\boldsymbol{L} \overset{(9.13)}{=} \nabla^\top \nabla = -\text{div}\nabla, \tag{9.15}$$

where div denotes the divergence operator. This explains that the Laplacian measures how bumpy the neighbours of every point are with respect to it in the nearest neighbours graph of data. Therefore, the Laplacian is used in spectral clustering and related algorithms because the Laplacian is a measure of the relationship between neighbouring points. Therefore, it can be used for inserting the information of the structure of data into the formulation of the algorithm. Similar to this discussion, there are some connections between Laplacian and the heat equations; refer to [9] for more information on the connections.

9.2.4 *Eigenvalues of Laplacian Matrix*

It is well known in linear algebra and graph theory that if a graph has k disjoint connected parts, its Laplacian matrix has k zero eigenvalues (see [3, Theorem 1], [63, Theorem 3.10], and [1, 73]). It can be assumed that the nearest neighbours graph of the data is a connected graph, and $\boldsymbol{L}$ has one zero eigenvalue whose eigenvector is $\boldsymbol{1} = [1, 1, \dots, 1]^\top$. In spectral clustering and related methods, the eigenvector corresponding to the zero eigenvalue is ignored. Laplacian and its eigenvalues are used in many spectral dimensionality reduction methods, including the methods described in this chapter and Chap. 8.

9.2.5 *Convergence of Laplacian*

Laplacian-based methods usually use either the Laplacian matrix in the Euclidean space or the Laplace–Beltrami operator on the Riemannian manifold [52]. It is shown in several works, such as [19], [89], and [33], that the eigenfunctions and eigenvalues of the Laplace–Beltrami operator can be approximated by the eigenvalues and eigenvectors of the Laplacian of a weight matrix, respectively. Moreover, it is shown in [51, 52] that the kernel-based averaging operators can approximate the weighted Laplacian pointwise, for epsilon-graph (i.e., not fully connected) weight matrices.

9.3 Spectral Clustering

Spectral clustering [70, 95] is a clustering method whose solution is reduced to an eigenvalue problem. In spectral clustering, it is aimed at transforming data from its input space to another low-dimensional subspace so that the clusters of data hopefully become more separated. Once the clusters of data have separated, a simple clustering algorithm in that discriminating subspace can be applied. The following section explains the theory of spectral clustering.

9.3.1 The Spectral Clustering Algorithm

9.3.1.1 Adjacency Matrix

As explained in Sect. 9.2, the adjacency matrix and its Laplacian are computed. These matrices are used in later sections for the spectral clustering algorithm.

9.3.1.2 The Cut

The graph of the dataset is defined as:

$$\mathcal{G} := (\mathcal{X}, \mathcal{W}), \tag{9.16}$$

where the data indices $\mathcal{X} := \{i\}_{i=1}^n$ are the vertices of the graph and the weights $\mathcal{W} := \{w_{ij}\}_{i,j=1}^n$ are the weights for the edges of the graph. If the data points $\mathcal{A}$ and $\mathcal{A}'$ are divided into two disjoint clusters, then:

$$\mathcal{X} = \mathcal{A} \cup \mathcal{A}', \quad \mathcal{A} \cap \mathcal{A}' = \varnothing. \tag{9.17}$$

The cut between two subgraphs is defined as [79, 80]:

$$\text{cut}(\mathcal{A}, \mathcal{A}') := \sum_{i \in \mathcal{A}} \sum_{j \in \mathcal{A}'} w_{ij}. \tag{9.18}$$

As the summations can be exchanged, the result is:

$$\mathrm{cut}(\mathcal{A}, \mathcal{A}') = \mathrm{cut}(\mathcal{A}', \mathcal{A}). \tag{9.19}$$

9.3.1.3 Optimization of Spectral Clustering

If the two clusters or the subgraphs are very different or far from each other, the summation of the weights between their points is small, meaning their cut is small. Therefore, the subgraphs that minimize the cut must be found to find the separated clusters:

$$\underset{\mathcal{A},\mathcal{A}'}{\text{minimize}} \quad \mathrm{cut}(\mathcal{A}, \mathcal{A}') = \underset{\mathcal{A},\mathcal{A}'}{\text{minimize}} \quad \sum_{i\in\mathcal{A}} \sum_{j\in\mathcal{A}'} w_{ij}. \tag{9.20}$$

However, this minimization is sensitive to outliers. This is because an outlier point can be considered to be a cluster itself in this optimization problem, as it is very far from other points. To overcome this issue, a normalized version of the cut, named the ratio cut, must be used, which is defined as [79, 80]:

$$\begin{aligned}\mathrm{ratioCut}(\mathcal{A}, \mathcal{A}') &= \frac{\mathrm{cut}(\mathcal{A}, \mathcal{A}')}{|\mathcal{A}|} + \frac{\mathrm{cut}(\mathcal{A}', \mathcal{A})}{|\mathcal{A}'|} \\ &\overset{(9.19)}{=} \frac{\mathrm{cut}(\mathcal{A}, \mathcal{A}')}{|\mathcal{A}|} + \frac{\mathrm{cut}(\mathcal{A}, \mathcal{A}')}{|\mathcal{A}'|},\end{aligned} \tag{9.21}$$

where $|\mathcal{A}|$ denotes the cardinality of set $\mathcal{A}$. Minimizing the ratio cut (Eq. (9.21)), rather than minimizing the cut (Eq. (9.18)), resolves the outlier problem because the cluster of outlier(s) is a small cluster. It is considered to be cardinality small, meaning the denominator of one of the fractions in the ratio cut becomes small while the ratio cut becomes large. Therefore, to minimize the ratio cut for clustering, the following optimization problem can be used:

$$\underset{\mathcal{A},\mathcal{A}'}{\text{minimize}} \quad \mathrm{ratioCut}(\mathcal{A}, \mathcal{A}'). \tag{9.22}$$

Proposition 9.1 ([79]) *Minimizing the ratio cut, i.e., Eq. (9.22), is equivalent to:*

$$\underset{\boldsymbol{f}}{\text{minimize}} \quad \sum_{i=1}^{n} \sum_{j=1}^{n} w_{ij}\, (f_i - f_j)^2, \tag{9.23}$$

where $\boldsymbol{f} := [f_1, \ldots, f_n]^\top \in \mathbb{R}^n$ *and:*

$$f_i := \begin{cases} \sqrt{\frac{|\mathcal{A}'|}{|\mathcal{A}|}} & \text{if } i \in \mathcal{A}, \\ -\sqrt{\frac{|\mathcal{A}|}{|\mathcal{A}'|}} & \text{if } i \in \mathcal{A}'. \end{cases} \tag{9.24}$$

Proof

$$\begin{aligned}
\sum_{i=1}^{n}\sum_{j=1}^{n} w_{ij}\,(f_i - f_j)^2 &= \sum_{i\in\mathcal{A}}\sum_{j\in\mathcal{A}} w_{ij}\underbrace{\Big(\sqrt{\frac{|\mathcal{A}'|}{|\mathcal{A}|}} - \sqrt{\frac{|\mathcal{A}'|}{|\mathcal{A}|}}\Big)^2}_{=0} \\
&+ \sum_{i\in\mathcal{A}'}\sum_{j\in\mathcal{A}'} w_{ij}\underbrace{\Big(-\sqrt{\frac{|\mathcal{A}|}{|\mathcal{A}'|}} + \sqrt{\frac{|\mathcal{A}|}{|\mathcal{A}'|}}\Big)^2}_{=0} + \sum_{i\in\mathcal{A}}\sum_{j\in\mathcal{A}'} w_{ij}\Big(\sqrt{\frac{|\mathcal{A}'|}{|\mathcal{A}|}} + \sqrt{\frac{|\mathcal{A}|}{|\mathcal{A}'|}}\Big)^2 \\
&+ \sum_{i\in\mathcal{A}'}\sum_{j\in\mathcal{A}} w_{ij}\Big(-\sqrt{\frac{|\mathcal{A}|}{|\mathcal{A}'|}} - \sqrt{\frac{|\mathcal{A}'|}{|\mathcal{A}|}}\Big)^2 \\
&= \sum_{i\in\mathcal{A}}\sum_{j\in\mathcal{A}'} w_{ij}\Big(\sqrt{\frac{|\mathcal{A}'|}{|\mathcal{A}|}} + \sqrt{\frac{|\mathcal{A}|}{|\mathcal{A}'|}}\Big)^2 + \sum_{i\in\mathcal{A}'}\sum_{j\in\mathcal{A}} w_{ij}\Big(\sqrt{\frac{|\mathcal{A}|}{|\mathcal{A}'|}} + \sqrt{\frac{|\mathcal{A}'|}{|\mathcal{A}|}}\Big)^2 \\
&= \sum_{i\in\mathcal{A}}\sum_{j\in\mathcal{A}'} w_{ij}\Big(\frac{|\mathcal{A}'|}{|\mathcal{A}|} + \frac{|\mathcal{A}|}{|\mathcal{A}'|} + 2\Big) + \sum_{i\in\mathcal{A}'}\sum_{j\in\mathcal{A}} w_{ij}\Big(\frac{|\mathcal{A}|}{|\mathcal{A}'|} + \frac{|\mathcal{A}'|}{|\mathcal{A}|} + 2\Big) \\
&= \Big(\frac{|\mathcal{A}|}{|\mathcal{A}'|} + \frac{|\mathcal{A}'|}{|\mathcal{A}|} + 2\Big)\Big(\sum_{i\in\mathcal{A}}\sum_{j\in\mathcal{A}'} w_{ij} + \sum_{i\in\mathcal{A}'}\sum_{j\in\mathcal{A}} w_{ij}\Big) \\
&\overset{(9.18)}{=} \Big(\frac{|\mathcal{A}|}{|\mathcal{A}'|} + \frac{|\mathcal{A}'|}{|\mathcal{A}|} + 2\Big)\big(\text{cut}(\mathcal{A},\mathcal{A}') + \text{cut}(\mathcal{A}',\mathcal{A})\big) \\
&\overset{(9.19)}{=} 2\Big(\frac{|\mathcal{A}|}{|\mathcal{A}'|} + \frac{|\mathcal{A}'|}{|\mathcal{A}|} + 2\Big)\,\text{cut}(\mathcal{A},\mathcal{A}') = 2\Big(\frac{|\mathcal{A}|}{|\mathcal{A}'|} + \frac{|\mathcal{A}'|}{|\mathcal{A}|} + \underbrace{\frac{|\mathcal{A}|}{|\mathcal{A}|} + \frac{|\mathcal{A}'|}{|\mathcal{A}'|}}_{=2}\Big)\,\text{cut}(\mathcal{A},\mathcal{A}') \\
&= 2\Big(\frac{|\mathcal{A}| + |\mathcal{A}'|}{|\mathcal{A}'|} + \frac{|\mathcal{A}'| + |\mathcal{A}|}{|\mathcal{A}|}\Big)\,\text{cut}(\mathcal{A},\mathcal{A}') = 2\Big(\frac{n}{|\mathcal{A}'|} + \frac{n}{|\mathcal{A}|}\Big)\,\text{cut}(\mathcal{A},\mathcal{A}') \\
&= 2n\Big(\frac{\text{cut}(\mathcal{A},\mathcal{A}')}{|\mathcal{A}'|} + \frac{\text{cut}(\mathcal{A},\mathcal{A}')}{|\mathcal{A}|}\Big) \overset{(9.21)}{=} 2n\,\text{ratioCut}(\mathcal{A},\mathcal{A}').
\end{aligned}$$

□

Proposition 9.2 ([7]) *Therefore:*

$$\frac{1}{2}\sum_{i=1}^{n}\sum_{j=1}^{n} w_{ij}\,(f_i - f_j)^2 = \boldsymbol{f}^\top \boldsymbol{L}\boldsymbol{f}, \tag{9.25}$$

where $\boldsymbol{L} \in \mathbb{R}^{n \times n}$ is the Laplacian matrix of the weight matrix.

Proof

$$\begin{aligned}
\boldsymbol{f}^\top \boldsymbol{L} \boldsymbol{f} &\overset{(9.4)}{=} \boldsymbol{f}^\top (\boldsymbol{D} - \boldsymbol{W}) \boldsymbol{f} = \boldsymbol{f}^\top \boldsymbol{D} \boldsymbol{f} - \boldsymbol{f}^\top \boldsymbol{W} \boldsymbol{f} \\
&\overset{(a)}{=} \sum_{i=1}^{n} d_i f_i^2 - \sum_{i=1}^{n}\sum_{j=1}^{n} w_{ij} f_i f_j = \frac{1}{2}(2\sum_{i=1}^{n} d_i f_i^2 - 2\sum_{i=1}^{n}\sum_{j=1}^{n} w_{ij} f_i f_j) \\
&\overset{(b)}{=} \frac{1}{2}(\sum_{i=1}^{n} d_i f_i^2 + \sum_{j=1}^{n} d_j f_j^2 - 2\sum_{i=1}^{n}\sum_{j=1}^{n} w_{ij} f_i f_j) \\
&\overset{(9.5)}{=} \frac{1}{2}(\sum_{i=1}^{n} f_i^2 \sum_{j=1}^{n} w_{ij} + \sum_{j=1}^{n} f_j^2 \sum_{i=1}^{n} w_{ji} - 2\sum_{i=1}^{n}\sum_{j=1}^{n} w_{ij} f_i f_j) \\
&\overset{(c)}{=} \frac{1}{2}(\sum_{i=1}^{n}\sum_{j=1}^{n} f_i^2 w_{ij} + \sum_{i=1}^{n}\sum_{j=1}^{n} f_j^2 w_{ij} - 2\sum_{i=1}^{n}\sum_{j=1}^{n} w_{ij} f_i f_j) \\
&= \frac{1}{2}\sum_{i=1}^{n}\sum_{j=1}^{n} w_{ij} (f_i - f_j)^2,
\end{aligned}$$

where (a) notices that $\boldsymbol{D}$ is a diagonal matrix, (b) is because the dummy index i is replaced with j, and (c) notices that $w_{ij} = w_{ji}$ as symmetric similarity measures are usually used. □

Therefore, $\boldsymbol{f}^\top \boldsymbol{L} \boldsymbol{f}$ can be minimized for spectral clustering. Therefore, the optimization in spectral clustering is:

$$\begin{aligned}
\underset{\boldsymbol{f}}{\text{minimize}} \quad & \boldsymbol{f}^\top \boldsymbol{L} \boldsymbol{f} \\
\text{subject to} \quad & f_i := \begin{cases} \sqrt{\frac{|\mathcal{A}'|}{|\mathcal{A}|}} & \text{if } i \in \mathcal{A}, \\ -\sqrt{\frac{|\mathcal{A}|}{|\mathcal{A}'|}} & \text{if } i \in \mathcal{A}'. \end{cases}, \quad \forall i.
\end{aligned} \tag{9.26}$$

This optimization problem is complicated to solve. The constraint of having an orthonormal vector $\boldsymbol{f}$ can be relaxed. Recall that according to Eq. (9.24) or the constraint, if $f_i > 0$, it is in the first cluster or $\mathcal{A}$ and if $f_i < 0$, it is in the second cluster or $\mathcal{A}'$. Therefore, in the modified constraint, the sign characteristic of Eq. (9.24) can be kept as:

$$f_i := \begin{cases} > 0 \text{ if } i \in \mathcal{A}, \\ < 0 \text{ if } i \in \mathcal{A}'. \end{cases} \tag{9.27}$$

This modification satisfies $\sum_{i=1}^{n}(f_i \times f_i) = 1$. Therefore, the optimization is relaxed to:

$$\begin{aligned} & \underset{\boldsymbol{f}}{\text{minimize}} && \boldsymbol{f}^\top \boldsymbol{L} \boldsymbol{f} \\ & \text{subject to} && \boldsymbol{f}^\top \boldsymbol{f} = 1. \end{aligned} \tag{9.28}$$

9.3.1.4 Solution to Optimization

The Lagrangian of Eq. (9.28) is:

$$\mathcal{L} = \boldsymbol{f}^\top \boldsymbol{L} \boldsymbol{f} - \lambda(\boldsymbol{f}^\top \boldsymbol{f} - 1),$$

with λ as the dual variable. Setting the derivative of the Lagrangian to zero results in:

$$\mathbb{R}^n \ni \frac{\partial \mathcal{L}}{\partial \boldsymbol{f}} = 2\boldsymbol{L}\boldsymbol{f} - 2\lambda \boldsymbol{f} \overset{\text{set}}{=} \boldsymbol{0} \implies \boldsymbol{L}\boldsymbol{f} = \lambda \boldsymbol{f}, \tag{9.29}$$

which is the eigenvalue problem for the Laplacian matrix $\boldsymbol{L}$ (see Chap. 2).

As Eq. (9.28) is a minimization problem, the eigenvectors should be sorted from the corresponding smallest to the largest eigenvalues. Moreover, note that $\boldsymbol{L}$ is the Laplacian matrix for the weights $\boldsymbol{W}$. Therefore, according to Sect. 9.2.4, after sorting the eigenvectors from smallest to largest eigenvalues, the first eigenvector having an eigenvalue of zero is ignored, and instead, the smallest eigenvector of $\boldsymbol{L}$, which has a nonzero eigenvalue as the solution $\boldsymbol{f} \in \mathbb{R}^n$, is taken. According to Eq. (9.27), after finding $\boldsymbol{f} = [f_1, \ldots, f_n]^\top$, the cluster of each point is defined as:

$$i := \begin{cases} \in \mathcal{A} \text{ if } f_i > 0, \\ \in \mathcal{A}' \text{ if } f_i < 0. \end{cases} \tag{9.30}$$

9.3.1.5 Extension of Spectral Clustering to Multiple Clusters

If there are more than binary clusters in the data, the optimization problem of Eq. (9.28) should become:

$$\begin{aligned} & \underset{\boldsymbol{F}}{\text{minimize}} && \mathbf{tr}(\boldsymbol{F}^\top \boldsymbol{L} \boldsymbol{F}) \\ & \text{subject to} && \boldsymbol{F}^\top \boldsymbol{F} = \boldsymbol{I}, \end{aligned} \tag{9.31}$$

where $\mathbf{tr}(.)$ denotes the trace of matrix and $\boldsymbol{I}$ is the identity matrix. The Lagrangian is therefore:

$$\mathcal{L} = \boldsymbol{F}^\top \boldsymbol{L}\boldsymbol{F} - \textbf{tr}(\boldsymbol{\Lambda}^\top(\boldsymbol{F}^\top \boldsymbol{F} - \boldsymbol{I})),$$

where $\boldsymbol{\Lambda}$ is the diagonal matrix with the dual variables. Setting the derivative of the Lagrangian to zero results in:

$$\mathbb{R}^{n\times n} \ni \frac{\partial \mathcal{L}}{\partial \boldsymbol{F}} = 2\boldsymbol{L}\boldsymbol{F} - 2\boldsymbol{F}\boldsymbol{\Lambda} \overset{\text{set}}{=} \boldsymbol{0} \implies \boldsymbol{L}\boldsymbol{F} = \boldsymbol{F}\boldsymbol{\Lambda}, \tag{9.32}$$

which is the eigenvalue problem for the Laplacian matrix $\boldsymbol{L}$, where columns of $\boldsymbol{F}$ are the eigenvectors (see Chap. 2). As this is a minimization problem, the columns of $\boldsymbol{F}$ are sorted from corresponding smallest to largest eigenvalues. Additionally, as it is an eigenvalue problem for the Laplacian matrix, the smallest eigenvector with an eigenvalue of zero is ignored (see Sect. 9.2.4). If there are c clusters, the sorted $\boldsymbol{F}$ is truncated to have the c smallest eigenvectors with nonzero eigenvalues. Therefore, the result is $\boldsymbol{F} \in \mathbb{R}^{n\times c}$.

After finding the matrix $\boldsymbol{F} \in \mathbb{R}^{n\times c}$, the rows of $\boldsymbol{F}$ are treated as new data points in a c-dimensional space. Another clustering algorithm, such as K-means, can be applied to the rows of $\boldsymbol{F}$ to cluster the n points into c clusters [70]. Note that regular clustering methods, such as K-means, are applied to the input space (space of $\boldsymbol{X}$), while spectral clustering is applied to the embedding space (space of $\boldsymbol{Y}$). In other words, spectral clustering first extracts features for better discrimination of the clusters and then applies clustering. This usually results in better clustering because the extracted features are better able to discriminate between the clusters. Recall from Sect. 9.2.2 that Eq. (9.8) can be used as the Laplacian if maximization, instead of minimization, is used.

9.3.1.6 Optimization Approach 2

In the original work of spectral clustering, the following optimization was used, rather than Eq. (9.31) [79]:

$$\begin{aligned} \underset{\boldsymbol{F}}{\text{minimize}} \quad & \textbf{tr}(\boldsymbol{F}^\top \boldsymbol{L}\boldsymbol{F}) \\ \text{subject to} \quad & \boldsymbol{F}^\top \boldsymbol{D}\boldsymbol{F} = \boldsymbol{I}, \end{aligned} \tag{9.33}$$

which makes the rotated $\boldsymbol{F}$, by the degree matrix, orthogonal. Note that, according to Eq. (9.5), the degree matrix is a natural measure for the data points, in the way that the larger d_i corresponds to the more important f_i because it has more similarity (weights) with its neighbours [46]. That is why having $\boldsymbol{D}$ in the constraint makes sense, as it places more weight on the importance of f_i's. The Lagrangian is:

$$\mathcal{L} = \boldsymbol{F}^\top \boldsymbol{L}\boldsymbol{F} - \textbf{tr}(\boldsymbol{\Lambda}^\top(\boldsymbol{F}^\top \boldsymbol{D}\boldsymbol{F} - \boldsymbol{I})),$$

where $\boldsymbol{\Lambda}$ is the diagonal matrix with the dual variables. Setting the derivative of the Lagrangian to zero results in:

$$\mathbb{R}^{n\times n} \ni \frac{\partial \mathcal{L}}{\partial \boldsymbol{F}} = 2\boldsymbol{L}\boldsymbol{F} - 2\boldsymbol{D}\boldsymbol{F}\boldsymbol{\Lambda} \overset{\text{set}}{=} \boldsymbol{0} \implies \boldsymbol{L}\boldsymbol{F} = \boldsymbol{D}\boldsymbol{F}\boldsymbol{\Lambda}, \tag{9.34}$$

which is the generalized eigenvalue problem $(\boldsymbol{L}, \boldsymbol{D})$ where columns of $\boldsymbol{F}$ are the eigenvectors (see Chap. 2). Sorting the matrix $\boldsymbol{F}$, ignoring the eigenvector with a zero eigenvalue (see Sect. 9.2.4), and the remainder of the algorithm is completed as explained in Sect. 9.3.1.5.

9.3.2 Other Improvements Over Spectral Clustering

Refer to [95] for a survey of initial work on spectral clustering approaches. A list of some initial works developing to result in the modern spectral clustering is [31, 71, 78–80]. There exist many different improvements over the basic spectral clustering. Some of these developments are distributed spectral clustering [25], consistency spectral clustering [92], correctional spectral clustering [16], spectral clustering by autoencoder [4], multiview spectral clustering [56, 57, 107], self-tuning spectral clustering [110], and fuzzy spectral clustering [103]. Some existing surveys and tutorials on spectral clustering are [43, 44, 69, 91]. The optimization setup of spectral clustering has also been successfully used in the field of hashing (mapping data of arbitrary size to fixed-size values). Spectral hashing [96] is a successful example. Moreover, spectral clustering has had various applications, e.g., in image segmentation [104].

9.4 Laplacian Eigenmap

9.4.1 Laplacian Eigenmap

The Laplacian eigenmap [7, 8] is a *nonlinear* dimensionality reduction method that is related to spectral clustering. As explained in Sect. 9.3.1.5, spectral clustering extracts features for better discriminative clusters and then applies clustering in the embedding space. Therefore, spectral clustering carries out some dimensionality reduction and feature extraction tasks. This guided researchers to propose the Laplacian eigenmap, which is based on the concepts of spectral clustering.

Consider the dataset matrix $\boldsymbol{X} := [\boldsymbol{x}_1, \ldots, \boldsymbol{x}_n] \in \mathbb{R}^{d\times n}$. p-dimensional embeddings of data points $\boldsymbol{Y} := [\boldsymbol{y}_1, \ldots, \boldsymbol{y}_n] \in \mathbb{R}^{n\times p}$ are desired, where $p \leq d$ and usually $p \ll d$. Note that in the Laplacian eigenmap, for simplicity of algebra, the embedding vectors are put rowwise in the matrix $\boldsymbol{Y}$.

9.4.1.1 Adjacency Matrix

Similar to spectral clustering, an adjacency matrix is created using Eq. (9.1). The elements w_{ij} can be determined by the RBF kernel, i.e., Eq. (9.2), or the simple-minded approach, i.e., Eq. (9.3).

9.4.1.2 Interpretation of Laplacian Eigenmap

Inspired by Eq. (9.23) in spectral clustering, the optimization problem of the Laplacian eigenmap is:

$$\underset{\boldsymbol{Y}}{\text{minimize}} \quad \sum_{i=1}^{n}\sum_{j=1}^{n} w_{ij}\, \|\boldsymbol{y}_i - \boldsymbol{y}_j\|_2^2. \tag{9.35}$$

Equation (9.35) can be interpreted as follows [8]:

- If $\boldsymbol{x}_i$ and $\boldsymbol{x}_j$ are close to each other, w_{ij} is large. Therefore, to minimize the objective function, the term $\|\boldsymbol{y}_i - \boldsymbol{y}_j\|_2^2$ should be minimized, which results in close $\boldsymbol{y}_i$ and $\boldsymbol{y}_j$. This is expected because $\boldsymbol{x}_i$ and $\boldsymbol{x}_j$ are close and their embeddings should be close as well.
- If $\boldsymbol{x}_i$ and $\boldsymbol{x}_j$ are far apart, w_{ij} is small. Therefore, the objective function is small because it is multiplied by the small weight w_{ij}. This means that $\boldsymbol{y}_i$ and $\boldsymbol{y}_j$ are less important as the objective is already small.

According to the above two notes, the Laplacian eigenmap fits data locally because it places more emphasis on the nearby points to ensure that the global structure is preserved by the local fitting [77]. This might be considered a weakness of the Laplacian eigenmap, since it preserves the locality of the data pattern. This connects this method to the locality preserving projection, which is explained later in Sect. 9.5.

9.4.1.3 Optimization Approach 1

According to Proposition 9.2, Eq. (9.35) can be restated by adding a constraint inspired by Eq. (9.31):

$$\begin{aligned} \underset{\boldsymbol{Y}}{\text{minimize}} \quad & \mathbf{tr}(\boldsymbol{Y}^\top \boldsymbol{L} \boldsymbol{Y}) \\ \text{subject to} \quad & \boldsymbol{Y}^\top \boldsymbol{Y} = \boldsymbol{I}, \end{aligned} \tag{9.36}$$

where $\boldsymbol{L}$ is the Laplacian of the adjacency matrix. The Lagrangian for this optimization problem is:

$$\mathcal{L} = \boldsymbol{Y}^\top \boldsymbol{L}\boldsymbol{Y} - \textbf{tr}(\boldsymbol{\Lambda}^\top(\boldsymbol{Y}^\top\boldsymbol{Y} - \boldsymbol{I})),$$

where $\boldsymbol{\Lambda}$ is the diagonal matrix with the dual variables. Setting the derivative of the Lagrangian to zero results in:

$$\mathbb{R}^{n\times n} \ni \frac{\partial \mathcal{L}}{\partial \boldsymbol{Y}} = 2\boldsymbol{L}\boldsymbol{Y} - 2\boldsymbol{Y}\boldsymbol{\Lambda} \overset{\text{set}}{=} \boldsymbol{0} \implies \boldsymbol{L}\boldsymbol{Y} = \boldsymbol{Y}\boldsymbol{\Lambda}, \tag{9.37}$$

which is the eigenvalue problem for the Laplacian matrix $\boldsymbol{L}$, where columns of $\boldsymbol{Y}$ are the eigenvectors (see Chap. 2). As this is a minimization problem, the columns of $\boldsymbol{Y}$ are sorted from corresponding smallest to largest eigenvalues. Additionally, as it is an eigenvalue problem for the Laplacian matrix, the smallest eigenvector with an eigenvalue of zero is ignored (see Sect. 9.2.4). For a p-dimensional embedding space, the sorted $\boldsymbol{Y}$ is truncated to have the p smallest eigenvectors with nonzero eigenvalues. Therefore, the result is $\boldsymbol{Y} \in \mathbb{R}^{n\times p}$. The rows of $\boldsymbol{Y}$ are the p-dimensional embedding vectors. Note that as the adjacency matrix can use any kernel, such as the RBF kernel of Eq. (9.2), some researchers have named it the *kernel Laplacian eigenmap* [45].

9.4.1.4 Optimization Approach 2

Another optimization approach for the Laplacian eigenmap is to use the following optimization, rather than Eq. (9.36) [7]:

$$\begin{aligned} \underset{\boldsymbol{Y}}{\text{minimize}} \quad & \textbf{tr}(\boldsymbol{Y}^\top \boldsymbol{L}\boldsymbol{Y}) \\ \text{subject to} \quad & \boldsymbol{Y}^\top \boldsymbol{D}\boldsymbol{Y} = \boldsymbol{I}, \end{aligned} \tag{9.38}$$

where $\boldsymbol{D}$ is the degree matrix defined by Eq. (9.5). In this optimization, rather than having orthonormal embeddings, rotated embeddings by the degree matrix are preferred to be orthonormal. Note that, according to Eq. (9.5), the degree matrix is a natural measure for the data points, in the way that the larger d_i corresponds to the more important $\boldsymbol{x}_i$ because it has more similarity (weights) with its neighbours [46]. That is why having $\boldsymbol{D}$ in the constraint makes sense as it places more weight on $\boldsymbol{y}_i$'s. The Lagrangian for this optimization problem is:

$$\mathcal{L} = \boldsymbol{Y}^\top \boldsymbol{L}\boldsymbol{Y} - \textbf{tr}(\boldsymbol{\Lambda}^\top(\boldsymbol{Y}^\top\boldsymbol{D}\boldsymbol{Y} - \boldsymbol{I})),$$

where $\boldsymbol{\Lambda}$ is the diagonal matrix with the dual variables. Setting the derivative of Lagrangian to zero results in:

$$\mathbb{R}^{n\times n} \ni \frac{\partial \mathcal{L}}{\partial \boldsymbol{Y}} = 2\boldsymbol{L}\boldsymbol{Y} - 2\boldsymbol{D}\boldsymbol{Y}\boldsymbol{\Lambda} \overset{\text{set}}{=} \boldsymbol{0} \implies \boldsymbol{L}\boldsymbol{Y} = \boldsymbol{D}\boldsymbol{Y}\boldsymbol{\Lambda}, \tag{9.39}$$

which is the generalized eigenvalue problem $(\boldsymbol{L}, \boldsymbol{D})$, where columns of $\boldsymbol{Y}$ are the eigenvectors (see Chap. 2). Sorting the matrix $\boldsymbol{Y}$ and ignoring the eigenvector with a zero eigenvalue (see Sect. 9.2.4) is completed as previously explained in Sect. 9.3.1.5. Weiss demonstrated that the solution to Eq. (9.39) is equivalent to the solution of the eigenvalue problem for Eq. (9.9) [95, Normalization Lemma 1]. In other words, the eigenvectors of Eq. (9.9) are equivalent to the generalized eigenvectors of Eq. (9.39).

9.4.2 *Out-of-Sample Extension for Laplacian Eigenmap*

The training dataset $\{\boldsymbol{x}_i \in \mathbb{R}^d\}_{i=1}^n$ or $\boldsymbol{X} = [\boldsymbol{x}_1, \dots, \boldsymbol{x}_n] \in \mathbb{R}^{d\times n}$ has been embedded to have their embedding $\{\boldsymbol{y}_i \in \mathbb{R}^p\}_{i=1}^n$ or $\boldsymbol{Y} = [\boldsymbol{y}_1, \dots, \boldsymbol{y}_n]^\top \in \mathbb{R}^{n\times p}$. Assume there are some out-of-sample (test) data, denoted by $\{\boldsymbol{x}_i^{(t)} \in \mathbb{R}^d\}_{i=1}^{n_t}$ or $\boldsymbol{X}_t = [\boldsymbol{x}_1^{(t)}, \dots, \boldsymbol{x}_n^{(t)}] \in \mathbb{R}^{d\times n_t}$. The objective is to find their embedding $\{\boldsymbol{y}_i^{(t)} \in \mathbb{R}^p\}_{i=1}^{n_t}$ or $\boldsymbol{Y}_t = [\boldsymbol{y}_1^{(t)}, \dots, \boldsymbol{y}_n^{(t)}]^\top \in \mathbb{R}^{n_t\times p}$ after the training phase.

9.4.2.1 Embedding Using Eigenfunctions

Recall the definition of eigenfunctions and the following proposition from Chap. 3.

Proposition 9.3 *If v_{ki} is the i-th element of the n-dimensional vector $\boldsymbol{v}_k$ and $k(\boldsymbol{x}, \boldsymbol{x}_i)$ is the kernel between vectors $\boldsymbol{x}$ and $\boldsymbol{x}_i$, the eigenfunction for the point $\boldsymbol{x}$ and the i-th training point $\boldsymbol{x}_i$ are:*

$$f_k(\boldsymbol{x}) = \frac{\sqrt{n}}{\delta_k} \sum_{i=1}^{n} v_{ki}\, \breve{k}_t(\boldsymbol{x}_i, \boldsymbol{x}), \tag{9.40}$$

$$f_k(\boldsymbol{x}_i) = \sqrt{n}\, v_{ki}, \tag{9.41}$$

respectively, where $\breve{k}_t(\boldsymbol{x}_i, \boldsymbol{x})$ is the centered kernel between the training set and the out-of-sample point $\boldsymbol{x}$.

Let the embedding of the point $\boldsymbol{x}$ be $\mathbb{R}^p \ni \boldsymbol{y}(\boldsymbol{x}) = [y_1(\boldsymbol{x}), \dots, y_p(\boldsymbol{x})]^\top$. The k-th dimension of this embedding is:

$$y_k(\boldsymbol{x}) = \sqrt{\delta_k}\,\frac{f_k(\boldsymbol{x})}{\sqrt{n}} = \frac{1}{\sqrt{\delta_k}} \sum_{i=1}^{n} v_{ki}\, \breve{k}_t(\boldsymbol{x}_i, \boldsymbol{x}). \tag{9.42}$$

Proof This proposition is taken from [13, Proposition 1]. For proof, refer to [10, Proposition 1], [15, Proposition 1], and [14, Proposition 1 and Theorem 1]. More complete proofs can be found in [11]. □

The adjacency matrix can be seen as a kernel (e.g. see Eq. (9.2) where the RBF kernel is used for the adjacency matrix). This fact was also explained in Sect. 9.2.2, where it was explained why the Laplacian can be seen as a normalized kernel matrix. Therefore, here, the kernel $k_t(\boldsymbol{x}_i, \boldsymbol{x})$ refers to the adjacency matrix.

According to Eq. (9.9) and Eq. (9.32), the solution for the spectral clustering is the eigenvectors of the Laplacian in Eq. (9.9). Moreover, as explained in Sect. 9.4.1.4, the solution of the Laplacian eigenmap is also the eigenvectors of the Laplacian in Eq. (9.9). Therefore, eigenfunctions can be used for out-of-sample extension.

If there is a set of n_t out-of-sample data points, the kernel $\boldsymbol{K}_t \in \mathbb{R}^{n \times n_t}$ between training and out-of-sample data can be computed. The kernel matrix can be normalized as follows [72]. The $\breve{k}_t(\boldsymbol{x}_i, \boldsymbol{x})$, used in Eq. (9.42), is an element of the normalized kernel (adjacency) matrix [12]:

$$\breve{k}_t(\boldsymbol{x}_i, \boldsymbol{x}) := \frac{1}{n} \frac{k_t(\boldsymbol{x}_i, \boldsymbol{x})}{\sqrt{\mathbb{E}[\boldsymbol{K}_t(\boldsymbol{x}_i, .)]\ \mathbb{E}[\boldsymbol{K}_t(., \boldsymbol{x})]}}, \tag{9.43}$$

where the expectations are computed as:

$$\mathbb{E}[\boldsymbol{K}_t(\boldsymbol{x}_i, .)] \approx \frac{1}{n_t} \sum_{j=1}^{n_t} k_t(\boldsymbol{x}_i, \boldsymbol{x}_j^{(t)}), \tag{9.44}$$

$$\mathbb{E}[\boldsymbol{K}_t(., \boldsymbol{x})] \approx \frac{1}{n} \sum_{i=1}^{n} k_t(\boldsymbol{x}_i, \boldsymbol{x}). \tag{9.45}$$

Noting Eq. (9.5) for $\boldsymbol{D}$ and comparing Eq. (9.9) to Eqs. (9.43), (9.44), and (9.45) demonstrate the approximate equivalency of Eqs. (9.9) and (9.43), particularly if $k_t \approx k$ and $n_t \approx n$ are assumed. Therefore, Eq. (9.43) is valid to be used in Eq. (9.42).

9.4.2.2 Out-of-Sample Embedding

Equation (9.42) can be used to embed the i-th out-of-sample data point $\boldsymbol{x}_i^{(t)}$. Thus, $\boldsymbol{x}_i^{(t)}$ should be used in place of $\boldsymbol{x}$ in Eq. (9.42). This out-of-sample extension technique can be used for both the Laplacian eigenmap and spectral clustering [13]. In addition to this technique, there are other methods for out-of-sample extension of the Laplacian eigenmap, such as kernel mapping [37, 38] and optimization [18] - which are not covered in this book for brevity.

9.4.3 *Other Improvements Over the Laplacian Eigenmap*

There have been recent developments since the basic Laplacian eigenmap. Some examples are the Laplacian Eigenmaps Latent Variable Model (LELVM) [21], robust Laplacian eigenmap [76], and Laplacian forest [62]. Convergence analysis of the Laplacian eigenmap methods can be found in several papers [6, 83]. Inspired by eigenfaces [90] and Fisherfaces [5], which have been proposed based on principal component analysis (see Chap. 5) and Fisher discriminant analysis (see Chap. 6), respectively, *Laplacianfaces* [47] has been proposed for facial recognition using Laplacian eigenmaps. Finally, a recent survey on the Laplacian eigenmap is [60, 97].

9.5 Locality Preserving Projection

Locality Preserving Projection (LPP) [46] is a *linear* dimensionality reduction method, which is a linear approximation of the Laplacian eigenmap [7, 8]. It uses linear projection in the formulation of the Laplacian eigenmap. Note that this method is called locality preserving projection because it approximates the Laplacian eigenmap by linear projection, where as explained in Sect. 9.4.1.2, the Laplacian eigenmap fits data locally and preserves the local structure of data in the embedding space.

9.5.1 *Locality Preserving Projection*

LPP assumes that the embeddings of the data points are obtained by a linear projection (see Chap. 2 for definition of linear projection), which approximates the Laplacian eigenmap linearly. The goal is to find the embeddings of data points $\boldsymbol{X} = [\boldsymbol{x}_1, \dots, \boldsymbol{x}_n] \in \mathbb{R}^{d\times n}$ using linear projection.

9.5.1.1 One-Dimensional Subspace

For the derivation of LPP optimization, $p = 1$ is assumed first, meaning that the projection is in line with the projection vector $\boldsymbol{u} \in \mathbb{R}^d$. In this case, the vector of the projections of all of the points is:

$$\mathbb{R}^{1\times n} \ni \boldsymbol{Y} = [y_1, y_2, \dots, y_n] := \boldsymbol{u}^\top \boldsymbol{X}. \tag{9.46}$$

Therefore, the one-dimensional embedding of every point is:

$$\mathbb{R} \ni y_i = \boldsymbol{u}^\top \boldsymbol{x}_i, \quad \forall i \in \{1, \dots, n\}. \tag{9.47}$$

Inspired by Eq. (9.35) for the Laplacian eigenmap, the optimization for obtaining the one-dimensional embedding is:

$$\underset{\boldsymbol{Y}}{\text{minimize}} \quad \sum_{i=1}^{n}\sum_{j=1}^{n} w_{ij}\,(y_i - y_j)^2, \tag{9.48}$$

the same as the interpretation in the Laplacian eigenmap.

Proposition 9.4 ([46]) *The following expression holds:*

$$\frac{1}{2}\sum_{i=1}^{n}\sum_{j=1}^{n} w_{ij}(y_i - y_j)^2 = \boldsymbol{u}^\top \boldsymbol{X}\boldsymbol{L}\boldsymbol{X}^\top \boldsymbol{u}. \tag{9.49}$$

Proof The proof is similar to the proof of Proposition 9.2.

$$\boldsymbol{u}^\top \boldsymbol{X}\boldsymbol{L}\boldsymbol{X}^\top \boldsymbol{u} \overset{(9.4)}{=} \boldsymbol{u}^\top \boldsymbol{X}(\boldsymbol{D}-\boldsymbol{W})\boldsymbol{X}^\top \boldsymbol{u} = \boldsymbol{u}^\top \boldsymbol{X}\boldsymbol{D}\boldsymbol{X}^\top \boldsymbol{u} - \boldsymbol{u}^\top \boldsymbol{X}\boldsymbol{W}\boldsymbol{X}^\top \boldsymbol{u}$$

$$\overset{(a)}{=} \sum_{i=1}^{n} \boldsymbol{u}^\top \boldsymbol{x}_i\, d_i\, \boldsymbol{x}_i^\top \boldsymbol{u} - \sum_{i=1}^{n}\sum_{j=1}^{n} \boldsymbol{u}^\top \boldsymbol{x}_i\, w_{ij}\, \boldsymbol{x}_j^\top \boldsymbol{u}$$

$$= \frac{1}{2}\Big(2\sum_{i=1}^{n} \boldsymbol{u}^\top \boldsymbol{x}_i\, d_i\, \boldsymbol{x}_i^\top \boldsymbol{u} - 2\sum_{i=1}^{n}\sum_{j=1}^{n} \boldsymbol{u}^\top \boldsymbol{x}_i\, w_{ij}\, \boldsymbol{x}_j^\top \boldsymbol{u}\Big)$$

$$\overset{(b)}{=} \frac{1}{2}\Big(2\sum_{i=1}^{n} (\boldsymbol{u}^\top \boldsymbol{x}_i)^2 d_i - 2\sum_{i=1}^{n}\sum_{j=1}^{n} (\boldsymbol{u}^\top \boldsymbol{x}_i)(\boldsymbol{u}^\top \boldsymbol{x}_j) w_{ij}\Big)$$

$$\overset{(c)}{=} \frac{1}{2}\Big(\sum_{i=1}^{n} (\boldsymbol{u}^\top \boldsymbol{x}_i)^2 d_i + \sum_{j=1}^{n} (\boldsymbol{u}^\top \boldsymbol{x}_j)^2 d_j - 2\sum_{i=1}^{n}\sum_{j=1}^{n} (\boldsymbol{u}^\top \boldsymbol{x}_i)(\boldsymbol{u}^\top \boldsymbol{x}_j) w_{ij}\Big)$$

$$\overset{(9.5)}{=} \frac{1}{2}\Big(\sum_{i=1}^{n} (\boldsymbol{u}^\top \boldsymbol{x}_i)^2 \sum_{j=1}^{n} w_{ij} + \sum_{j=1}^{n} (\boldsymbol{u}^\top \boldsymbol{x}_j)^2 \sum_{i=1}^{n} w_{ji} - 2\sum_{i=1}^{n}\sum_{j=1}^{n} (\boldsymbol{u}^\top \boldsymbol{x}_i)(\boldsymbol{u}^\top \boldsymbol{x}_j) w_{ij}\Big)$$

$$\overset{(d)}{=} \frac{1}{2}\Big(\sum_{i=1}^{n} (\boldsymbol{u}^\top \boldsymbol{x}_i)^2 \sum_{j=1}^{n} w_{ij} + \sum_{j=1}^{n} (\boldsymbol{u}^\top \boldsymbol{x}_j)^2 \sum_{i=1}^{n} w_{ij} - 2\sum_{i=1}^{n}\sum_{j=1}^{n} (\boldsymbol{u}^\top \boldsymbol{x}_i)(\boldsymbol{u}^\top \boldsymbol{x}_j) w_{ij}\Big)$$

$$= \frac{1}{2}\sum_{i=1}^{n}\sum_{j=1}^{n} w_{ij}(\boldsymbol{u}^\top \boldsymbol{x}_i - \boldsymbol{u}^\top \boldsymbol{x}_j)^2 \overset{(9.47)}{=} \frac{1}{2}\sum_{i=1}^{n}\sum_{j=1}^{n} w_{ij}(y_i - y_j)^2,$$

where (a) notices that $\boldsymbol{D}$ is a diagonal matrix, (b) is because $\boldsymbol{u}^\top \boldsymbol{x}_i = \boldsymbol{x}_i^\top \boldsymbol{u}_i \in \mathbb{R}$, (c) is because the dummy index i is replaced with j, and (d) notices that $w_{ij} = w_{ji}$ as symmetric similarity measures are usually used. □

According to Proposition 9.4, Eq. (9.48) can be restated by adding a constraint inspired by Eq. (9.38):

$$\begin{aligned} & \underset{\boldsymbol{u}}{\text{minimize}} \quad \boldsymbol{u}^\top \boldsymbol{X}\boldsymbol{L}\boldsymbol{X}^\top \boldsymbol{u} \\ & \text{subject to} \quad \boldsymbol{u}^\top \boldsymbol{X}\boldsymbol{D}\boldsymbol{X}^\top \boldsymbol{u} = 1, \end{aligned} \tag{9.50}$$

where $\boldsymbol{L}$ and $\boldsymbol{D}$ are the Laplacian and degree matrix of the adjacency matrix, respectively. Note that the objective variable has been changed to the projection vector $\boldsymbol{u}$. The Lagrangian for this optimization problem is:

$$\mathcal{L} = \boldsymbol{u}^\top \boldsymbol{X}\boldsymbol{L}\boldsymbol{X}^\top \boldsymbol{u} - \lambda^\top (\boldsymbol{u}^\top \boldsymbol{X}\boldsymbol{D}\boldsymbol{X}^\top \boldsymbol{u} - 1),$$

where λ is the dual variable. Setting the derivative of the Lagrangian to zero results in:

$$\begin{aligned} \mathbb{R}^d \ni \frac{\partial \mathcal{L}}{\partial \boldsymbol{u}} &= 2\boldsymbol{X}\boldsymbol{L}\boldsymbol{X}^\top \boldsymbol{u} - 2\lambda \boldsymbol{X}\boldsymbol{D}\boldsymbol{X}^\top \boldsymbol{u} \overset{\text{set}}{=} \boldsymbol{0} \\ &\implies \boldsymbol{X}\boldsymbol{L}\boldsymbol{X}^\top \boldsymbol{u} = \lambda \boldsymbol{X}\boldsymbol{D}\boldsymbol{X}^\top \boldsymbol{u}, \end{aligned} \tag{9.51}$$

which is the generalized eigenvalue problem $(\boldsymbol{X}\boldsymbol{L}\boldsymbol{X}^\top, \boldsymbol{X}\boldsymbol{D}\boldsymbol{X}^\top)$, where $\boldsymbol{u}$ is the eigenvector (see Chap. 2). The embedded points are obtained using Eq. (9.46).

9.5.1.2 Multidimensional Subspace

Section 9.5.1.1 explained LPP with linear projection onto a one-dimensional subspace. It is possible to extend LPP to linear projection onto a span of p basis vectors. In other words, it is extended to have a p-dimensional subspace. The embedding of $\boldsymbol{x}_i$ is denoted by $\boldsymbol{y}_i \in \mathbb{R}^p$. In LPP, the embedding vectors are placed columnwise in the matrix $\boldsymbol{Y} = [\boldsymbol{y}_1, \boldsymbol{y}_2, \ldots, \boldsymbol{y}_n] \in \mathbb{R}^{p\times n}$. In this case, the projection is:

$$\mathbb{R}^{p\times n} \ni \boldsymbol{Y} := \boldsymbol{U}^\top \boldsymbol{X}, \tag{9.52}$$

where $\boldsymbol{U} = [\boldsymbol{u}_1, \ldots, \boldsymbol{u}_p] \in \mathbb{R}^{d\times p}$ is the projection matrix onto its column space. Equation (9.36) can be extended to the multidimensional subspace as:

$$\begin{aligned} & \underset{\boldsymbol{U}}{\text{minimize}} \quad \mathbf{tr}(\boldsymbol{U}^\top \boldsymbol{X}\boldsymbol{L}\boldsymbol{X}^\top \boldsymbol{U}) \\ & \text{subject to} \quad \boldsymbol{U}^\top \boldsymbol{X}\boldsymbol{D}\boldsymbol{X}^\top \boldsymbol{U} = \boldsymbol{I}. \end{aligned} \tag{9.53}$$

In optimization, it is assumed that the dimensionality of $\boldsymbol{U}$ is $d \times d$, but the explanation in this section will truncate it to $d \times p$ after solving the optimization. The Lagrangian for this optimization problem is:

$$\mathcal{L} = \mathbf{tr}(\boldsymbol{U}^\top \boldsymbol{X}\boldsymbol{L}\boldsymbol{X}^\top \boldsymbol{U}) - \mathbf{tr}(\boldsymbol{\Lambda}^\top (\boldsymbol{U}^\top \boldsymbol{X}\boldsymbol{D}\boldsymbol{X}^\top \boldsymbol{U} - \boldsymbol{I})),$$

where $\boldsymbol{\Lambda}$ is the diagonal matrix with the dual variables. Setting the derivative of the Lagrangian to zero results in:

$$\mathbb{R}^{d\times d} \ni \frac{\partial \mathcal{L}}{\partial \boldsymbol{D}} = 2\boldsymbol{X}\boldsymbol{L}\boldsymbol{X}^\top \boldsymbol{U} - 2\boldsymbol{X}\boldsymbol{D}\boldsymbol{X}^\top \boldsymbol{U}\boldsymbol{\Lambda} \overset{\text{set}}{=} \boldsymbol{0}$$
$$\implies \boldsymbol{X}\boldsymbol{L}\boldsymbol{X}^\top \boldsymbol{U} = \boldsymbol{X}\boldsymbol{D}\boldsymbol{X}^\top \boldsymbol{U}\boldsymbol{\Lambda}, \tag{9.54}$$

which is the generalized eigenvalue problem $(\boldsymbol{X}\boldsymbol{L}\boldsymbol{X}^\top, \boldsymbol{X}\boldsymbol{D}\boldsymbol{X}^\top)$, where columns of $\boldsymbol{U}$ are the eigenvectors (see Chap. 2). As it is a minimization problem, the columns of $\boldsymbol{U}$, which are the eigenvectors, are sorted from the corresponding smallest to largest eigenvalues. Moreover, as this is a Laplacian matrix, there is one eigenvector with an eigenvalue of zero, which should be ignored (see Sect. 9.2.4). The embedded points are obtained using Eq. (9.52).

9.5.2 *Kernel Locality Preserving Projection*

Kernel LPP, also called kernel supervised LPP, [26, 61] performs LPP in the feature space. Assume $\boldsymbol{\phi}(.)$ denotes the pulling function to the feature space. According to representation theory [2], any pulled solution (direction) $\boldsymbol{\phi}(\boldsymbol{u}) \in \mathcal{H}$ must lie in the span of all the training vectors pulled to $\mathcal{H}$, i.e., $\boldsymbol{\Phi}(\boldsymbol{X}) = [\boldsymbol{\phi}(\boldsymbol{x}_1), \dots, \boldsymbol{\phi}(\boldsymbol{x}_n)] \in \mathbb{R}^{t\times n}$. Therefore, $\mathbb{R}^t \ni \boldsymbol{\phi}(\boldsymbol{u}) = \sum_{i=1}^n \theta_i\, \boldsymbol{\phi}(\boldsymbol{x}_i) = \boldsymbol{\Phi}(\boldsymbol{X})\,\boldsymbol{\theta}$ where $\mathbb{R}^n \ni \boldsymbol{\theta} = [\theta_1, \dots, \theta_n]^\top$ is the unknown vector of coefficients. If considering multiple projection directions, then:

$$\mathbb{R}^{t\times p} \ni \boldsymbol{\Phi}(\boldsymbol{U}) = \boldsymbol{\Phi}(\boldsymbol{X})\,\boldsymbol{\Theta}, \tag{9.55}$$

where $\mathbb{R}^{n\times p} \ni \boldsymbol{\Theta} = [\boldsymbol{\theta}_1, \dots, \boldsymbol{\theta}_p]$. Therefore, the objective function in Eq. (9.53) is converted to:

$$\mathbf{tr}(\boldsymbol{\Phi}(\boldsymbol{U})^\top \boldsymbol{\Phi}(\boldsymbol{X})\boldsymbol{L}\boldsymbol{\Phi}(\boldsymbol{X})^\top \boldsymbol{\Phi}(\boldsymbol{U})) \overset{(9.55)}{=} \mathbf{tr}(\boldsymbol{\Theta}^\top \boldsymbol{\Phi}(\boldsymbol{X})^\top \boldsymbol{\Phi}(\boldsymbol{X})\boldsymbol{L}\boldsymbol{\Phi}(\boldsymbol{X})^\top \boldsymbol{\Phi}(\boldsymbol{X})\,\boldsymbol{\Theta})$$
$$= \mathbf{tr}(\boldsymbol{\Theta}^\top \boldsymbol{K}_x \boldsymbol{L}\boldsymbol{K}_x\,\boldsymbol{\Theta}),$$

where:

$$\mathbb{R}^{n\times n} \ni \boldsymbol{K}_x := \boldsymbol{\Phi}(\boldsymbol{X})^\top \boldsymbol{\Phi}(\boldsymbol{X}), \tag{9.56}$$

is the kernel of data [53]. Similarly, the constraint of Eq. (9.53) is converted to $\boldsymbol{\Theta}^\top \boldsymbol{K}_x \boldsymbol{D}\boldsymbol{K}_x\,\boldsymbol{\Theta}$. Finally, Eq. (9.53) is changed to:

$$\begin{aligned} \underset{\boldsymbol{\Theta}}{\text{minimize}} \quad & \textbf{tr}(\boldsymbol{\Theta}^\top \boldsymbol{K}_x \boldsymbol{L} \boldsymbol{K}_x \boldsymbol{\Theta}) \\ \text{subject to} \quad & \boldsymbol{\Theta}^\top \boldsymbol{K}_x \boldsymbol{D} \boldsymbol{K}_x \boldsymbol{\Theta} = \boldsymbol{I}, \end{aligned} \tag{9.57}$$

where $\boldsymbol{\Theta}$ is the optimization variable. The solution to this problem is similar to the solution for Eq. (9.53). Therefore, its solution is the generalized eigenvalue problem $(\boldsymbol{K}_x \boldsymbol{L} \boldsymbol{K}_x^\top, \boldsymbol{K}_x \boldsymbol{D} \boldsymbol{K}_x^\top)$, where columns of $\boldsymbol{\Theta}$ are the eigenvectors (see Chap. 2). Sorting the matrix $\boldsymbol{\Theta}$ and ignoring the eigenvector with a zero eigenvalue is completed as explained before (see Sect. 9.2.4). The p smallest eigenvectors form $\boldsymbol{\Theta} \in \mathbb{R}^{n \times p}$. The embedded data are obtained as follows:

$$\mathbb{R}^{p \times n} \ni \boldsymbol{Y} \overset{(9.52)}{=} \boldsymbol{\Phi}(\boldsymbol{U})^\top \boldsymbol{\Phi}(\boldsymbol{X}) \overset{(9.55)}{=} \boldsymbol{\Theta}^\top \boldsymbol{\Phi}(\boldsymbol{X})^\top \boldsymbol{\Phi}(\boldsymbol{X}) \overset{(9.56)}{=} \boldsymbol{\Theta}^\top \boldsymbol{K}_x. \tag{9.58}$$

9.5.3 Other Improvements Over Locality Preserving Projection

There have been many improvements over the basic LPP, which include the supervised LPP [98], extended LPP [81], multiview uncorrelated LPP [108], graph-optimized LPP [111], and LPP for Grassmann manifold [94]. For image recognition purposes, a two-dimensional (2D) LPP, [23] is proposed whose robust version is [24] and some of its applications are [54, 99]. A survey on LPP variants can be found in [82]. LPP has been shown to be effective for different applications, such as document clustering [20], facial recognition [48, 105, 109], and speech recognition [87].

9.6 Graph Embedding

Graph Embedding (GE) [100, 101] is a generalized dimensionality reduction method, where the graph of the data is embedded in the lower dimensional space [22]. Note that the terminology of graph embedding is used for two different families of methods in the literature. The first family, including GE [100, 101], refers to embedding the graph of the data in the lower dimensional space. The second family refers to embedding every node of the data graph in a subspace [41]. This section is on embedding the whole data graph.[1]

[1] Note that there are other generalized subspace learning methods, such as Roweis discriminant analysis [35], in the literature.

9.6.1 Direct Graph Embedding

Consider the high dimensional data points $\{\boldsymbol{x}_i \in \mathbb{R}^d\}_{i=1}^n$. The aim is to reduce the dimensionality of the data from d to $p \leq d$, which results in $\{\boldsymbol{y}_i \in \mathbb{R}^p\}_{i=1}^n$, where usually $p \ll d$. The following are denoted as $\boldsymbol{X} := [\boldsymbol{x}_1, \dots, \boldsymbol{x}_n] \in \mathbb{R}^{d\times n}$ and $\boldsymbol{Y} := [\boldsymbol{y}_1, \dots, \boldsymbol{y}_n] \in \mathbb{R}^{p\times n}$. Similar to spectral clustering, an adjacency matrix is created using Eq. (9.1). The elements w_{ij} can be determined by the RBF kernel, i.e., Eq. (9.2), or the simple-minded approach, i.e., Eq. (9.3).

The objective function of optimization in a direct GE [100, 101] is inspired by Eq. (9.23) or Eq. (9.33) in spectral clustering:

$$\begin{aligned} \underset{\boldsymbol{Y}}{\text{minimize}} \quad & \sum_{i=1}^{n}\sum_{j=1}^{n} w_{ij}\, \|\boldsymbol{y}_i - \boldsymbol{y}_j\|_2^2 \\ \text{subject to} \quad & \boldsymbol{Y}^\top \boldsymbol{B}\boldsymbol{Y} = \boldsymbol{I}, \end{aligned} \tag{9.59}$$

where $\boldsymbol{B} \succeq 0$ is called the constraint matrix. According to Proposition 9.2, this optimization problem is equivalent to:

$$\begin{aligned} \underset{\boldsymbol{Y}}{\text{minimize}} \quad & \boldsymbol{Y}^\top \boldsymbol{L}\boldsymbol{Y} \\ \text{subject to} \quad & \boldsymbol{Y}^\top \boldsymbol{B}\boldsymbol{Y} = \boldsymbol{I}, \end{aligned} \tag{9.60}$$

where $\boldsymbol{L}$ is the Laplacian of the graph of data. The Lagrangian for this optimization problem is:

$$\mathcal{L} = \boldsymbol{Y}^\top \boldsymbol{L}\boldsymbol{Y} - \mathbf{tr}(\boldsymbol{\Lambda}^\top(\boldsymbol{Y}^\top \boldsymbol{B}\boldsymbol{Y} - \boldsymbol{I})),$$

where $\boldsymbol{\Lambda}$ is the diagonal matrix with the dual variables. Setting the derivative of the Lagrangian to zero results in:

$$\mathbb{R}^{n\times n} \ni \frac{\partial \mathcal{L}}{\partial \boldsymbol{Y}} = 2\boldsymbol{L}\boldsymbol{Y} - 2\boldsymbol{B}\boldsymbol{Y}\boldsymbol{\Lambda} \overset{\text{set}}{=} \boldsymbol{0} \implies \boldsymbol{L}\boldsymbol{Y} = \boldsymbol{B}\boldsymbol{Y}\boldsymbol{\Lambda}, \tag{9.61}$$

which is the generalized eigenvalue problem $(\boldsymbol{L}, \boldsymbol{B})$, where columns of $\boldsymbol{Y}$ are the eigenvectors (see Chap. 2). Sorting the matrix $\boldsymbol{Y}$ and ignoring the eigenvector with a zero eigenvalue is completed as previously explained (see Sect. 9.2.4).

9.6.2 Linearized Graph Embedding

Similar to LPP, linearized GE [100, 101] assumes that the embedded data can be obtained by a linear projection onto the column space of a projection matrix $\boldsymbol{U} \in \mathbb{R}^{d\times p}$. Therefore, linearized GE uses a similar optimization to Eq. (9.53):

$$
\begin{aligned}
& \underset{U}{\text{minimize}} \quad \mathbf{tr}(\boldsymbol{U}^\top \boldsymbol{X}\boldsymbol{L}\boldsymbol{X}^\top \boldsymbol{U}) \\
& \text{subject to} \quad \boldsymbol{U}^\top \boldsymbol{X}\boldsymbol{B}\boldsymbol{X}^\top \boldsymbol{U} = \boldsymbol{I},
\end{aligned} \tag{9.62}
$$

whose solution is the generalized eigenvalue problem $(\boldsymbol{X}\boldsymbol{L}\boldsymbol{X}^\top, \boldsymbol{X}\boldsymbol{B}\boldsymbol{X}^\top)$, similar to Eq. (9.54).

Another approach to the linearized GE is a modification to Eq. (9.62) which is [101]:

$$
\begin{aligned}
& \underset{U}{\text{minimize}} \quad \mathbf{tr}(\boldsymbol{U}^\top \boldsymbol{X}\boldsymbol{L}\boldsymbol{X}^\top \boldsymbol{U}) \\
& \text{subject to} \quad \boldsymbol{U}^\top \boldsymbol{U} = \boldsymbol{I},
\end{aligned} \tag{9.63}
$$

whose solution is the eigenvalue decomposition of $\boldsymbol{X}\boldsymbol{L}\boldsymbol{X}^\top$.

9.6.3 Kernelized Graph Embedding

Kernelized GE [100, 101] performs the linearized GE in the feature space. It uses a similar optimization problem as Eq. (9.57):

$$
\begin{aligned}
& \underset{\Theta}{\text{minimize}} \quad \mathbf{tr}(\boldsymbol{\Theta}^\top \boldsymbol{K}_x \boldsymbol{L}\boldsymbol{K}_x \boldsymbol{\Theta}) \\
& \text{subject to} \quad \boldsymbol{\Theta}^\top \boldsymbol{K}_x \boldsymbol{B}\boldsymbol{K}_x \boldsymbol{\Theta} = \boldsymbol{I},
\end{aligned} \tag{9.64}
$$

whose solution is the generalized eigenvalue problem $(\boldsymbol{K}_x\boldsymbol{L}\boldsymbol{K}_x^\top, \boldsymbol{K}_x\boldsymbol{B}\boldsymbol{K}_x^\top)$, where columns of $\boldsymbol{\Theta}$ are the eigenvectors (see Chap. 2). Sorting the matrix $\boldsymbol{\Theta}$ and ignoring the eigenvector with a zero eigenvalue is completed as previously explained (see Sect. 9.2.4). The embedding is then obtained using Eq. (9.58).

Another approach for linearized GE is a slight revision to Eq. (9.64). This approach is the kernelized version of Eq. (9.63) and is [101]:

$$
\begin{aligned}
& \underset{\Theta}{\text{minimize}} \quad \mathbf{tr}(\boldsymbol{\Theta}^\top \boldsymbol{K}_x \boldsymbol{L}\boldsymbol{K}_x \boldsymbol{\Theta}) \\
& \text{subject to} \quad \boldsymbol{\Theta}^\top \boldsymbol{K}_x \boldsymbol{\Theta} = \boldsymbol{I},
\end{aligned} \tag{9.65}
$$

whose solution is the generalized eigenvalue problem $(\boldsymbol{K}_x\boldsymbol{L}\boldsymbol{K}_x^\top, \boldsymbol{K}_x)$.

9.6.4 Special Cases of Graph Embedding

GE is a generalized subspace learning method, whose special cases are some of the well-known dimensionality reduction methods. These special cases are cases of either direct, linearized, or kernelized GE. Some of these special cases are Laplacian eigenmap, LPP, kernel LPP, Principal Component Analysis (PCA), kernel PCA, Fisher Discriminant Analysis (FDA), kernel FDA, Multidimensional Scaling (MDS), Isomap, and Locally Linear Embedding (LLE). The following sections explain why these methods are special cases of graph embedding.

9.6.4.1 Laplacian Eigenmap, LPP, and Kernel LPP

On the one hand, comparing Eq. (9.36) of the Laplacian eigenmap with Eq. (9.60), it is discernible that $\boldsymbol{B} = \boldsymbol{I}$. Additionally, comparing Eq. (9.38), as the second approach of the Laplacian eigenmap, with Eq. (9.60) demonstrates that $\boldsymbol{B} = \boldsymbol{D}$. Therefore, the Laplacian eigenmap is a special case of direct GE. On the other hand, comparing Eq. (9.53) with Eq. (9.62) and comparing Eq. (9.57) with Eq. (9.64) show that LPP and kernel LPP are special cases of linearized GE and kernelized GE, respectively.

9.6.4.2 PCA and Kernel PCA

The optimization of PCA is as follows (see Chap. 5):

$$\begin{aligned} \underset{\boldsymbol{U}}{\text{minimize}} \quad & \mathbf{tr}(\boldsymbol{U}^\top \boldsymbol{S}\boldsymbol{U}) \overset{(a)}{=} \mathbf{tr}(\boldsymbol{U}^\top \boldsymbol{X}\boldsymbol{X}^\top \boldsymbol{U}) \\ \text{subject to} \quad & \boldsymbol{U}^\top \boldsymbol{U} = \boldsymbol{I}, \end{aligned} \tag{9.66}$$

where $\boldsymbol{U}$ is the projection matrix, $\boldsymbol{S}$ is the covariance matrix of the data, and (a) is because $\boldsymbol{S} = \boldsymbol{X}\boldsymbol{X}^\top$, assuming that the data $\boldsymbol{X}$ are already centered. Moreover, the solution to a kernel PCA is the eigenvalue decomposition of the kernel of the data (see Chap. 5). Therefore, optimization of the kernel PCA is the kernelized version of Eq. (9.66). Comparing Eq. (9.66) with Eq. (9.63) demonstrates that PCA is a special case of the linearized GE with $\boldsymbol{L} = \boldsymbol{I}$. Moreover, the kernel PCA is a special case of the kernelized GE with $\boldsymbol{L} = \boldsymbol{I}$.

9.6.4.3 FDA and Kernel FDA

The optimization of the FDA is [34] (see Chap. 6):

$$\begin{aligned} &\underset{U}{\text{maximize}} && \textbf{tr}(U^\top S_B\, U) \\ &\text{subject to} && U^\top S_W\, U = I, \end{aligned} \tag{9.67}$$

where S_B and S_W are the between- and within-class scatters, respectively. Additionally, [106]:

$$S = S_B + S_W \implies S_B = S - S_W, \tag{9.68}$$

where S denotes the total covariance of the data. Therefore, the Fisher criterion, to be maximized, can be restated as:

$$\frac{\textbf{tr}(U^\top S_B\, U)}{\textbf{tr}(U^\top S_W\, U)} = \frac{\textbf{tr}(U^\top S\, U)}{\textbf{tr}(U^\top S_W\, U)} - 1, \tag{9.69}$$

whose second constant term can be ignored in the maximization problem. Therefore, Eq. (9.67) is restated to [35, 101]:

$$\begin{aligned} &\underset{U}{\text{maximize}} && \textbf{tr}(U^\top S\, U) \overset{(a)}{=} \textbf{tr}(U^\top X X^\top U) \\ &\text{subject to} && U^\top S_W\, U = I, \end{aligned} \tag{9.70}$$

where (a) is because $S = XX^\top$, assuming that the data X are already centered.

Let c_i denote the class to which point x_i belongs and μ_{c_i} be the mean of class c_i. Let c denote the number of classes and n_j be the sample size of class j. For every class j, $e_j = [e_{j1}, \ldots, e_{jn}]^\top \in \mathbb{R}^n$ is defined as:

$$e_{ji} = \begin{cases} 1 \text{ if } c_i = j, \\ 0 \text{ otherwise.} \end{cases} \tag{9.71}$$

Note that the within scatter can be obtained through [101]:

$$S_W = \sum_{i=1}^{n} (x_i - \mu_{c_i})(x_i - \mu_{c_i})^\top = X(I - \sum_{j=1}^{c} \frac{1}{n_j} e_j e_j^\top) X^\top. \tag{9.72}$$

Comparing Eq. (9.70) with Eq. (9.62) and noticing Eq. (9.72) demonstrate that FDA is a special case of linearized GE with $L = I$ and $B = I - \sum_{j=1}^{c} (1/n_j) e_j e_j^\top$. Kernel FDA [65] (see Chap. 6) is the kernelized version of Eq. (9.70) so it is a special case of kernelized GE with the mentioned L and B matrices.

9.6.4.4 MDS and Isomap

Kernel classical MDS uses a kernel as follows (see Chap. 7):

$$\mathbb{R}^{n\times n} \ni \boldsymbol{K} = -\frac{1}{2}\boldsymbol{H}\boldsymbol{D}\boldsymbol{H}, \tag{9.73}$$

where $\boldsymbol{D} \in \mathbb{R}^{n\times n}$ is a matrix with the squared Euclidean distance between points and $\mathbb{R}^{n\times n} \ni \boldsymbol{H} := \boldsymbol{I} - (1/n)\boldsymbol{1}\boldsymbol{1}^\top$ is the centering matrix. Isomap also applies multidimensional scaling with a geodesic kernel, which uses piecewise Euclidean distance for computing $\boldsymbol{D}$ [88] (see Chap. 7). The row summation of this kernel matrix is [100, 101]:

$$\begin{aligned}
\sum_{j=1}^{n} \boldsymbol{K}_{ij} &\overset{(9.73)}{=} \sum_{j=1}^{n}(-\frac{1}{2}\boldsymbol{H}\boldsymbol{D}\boldsymbol{H})_{ij} = \sum_{j=1}^{n}\Big(-\frac{1}{2}(\boldsymbol{I} - \frac{1}{n}\boldsymbol{1}\boldsymbol{1}^\top)\boldsymbol{D}(\boldsymbol{I} - \frac{1}{n}\boldsymbol{1}\boldsymbol{1}^\top)\Big)_{ij} \\
&= \frac{1}{2}\sum_{j=1}^{n}\Big(-\boldsymbol{D}_{ij} + \frac{1}{n}\sum_{i'=1}^{n}\boldsymbol{D}_{i'j} + \frac{1}{n}\sum_{j'=1}^{n}(\boldsymbol{D}_{ij'} - \frac{1}{n}\sum_{k'=1}^{n}\boldsymbol{D}_{k'j'})\Big) \\
&= (-\frac{1}{2}\sum_{j=1}^{n}\boldsymbol{D}_{ij} + \frac{1}{2n}\sum_{j=1}^{n}\sum_{i'=1}^{n}\boldsymbol{D}_{i'j}) \\
&+ (\frac{1}{2n}\sum_{j=1}^{n}\sum_{j'=1}^{n}\boldsymbol{D}_{ij'} - \frac{1}{2n^2}\sum_{j=1}^{n}\sum_{j'=1}^{n}\sum_{k'=1}^{n}\boldsymbol{D}_{k'j'}) = \boldsymbol{0} + \boldsymbol{0} = \boldsymbol{0}.
\end{aligned}$$

According to Eq. (9.6), the kernel used in kernel classical MDS or the geodesic kernel used in Isomap can be interpreted as the Laplacian matrix of the data graph as it satisfies its row-sum property. The optimization of MDS or Isomap is (see Chap. 7):

$$\begin{aligned}
&\underset{\boldsymbol{Y}}{\text{minimize}} \quad \boldsymbol{Y}^\top\boldsymbol{K}\boldsymbol{Y} \\
&\text{subject to} \quad \boldsymbol{Y}^\top\boldsymbol{Y} = \boldsymbol{I},
\end{aligned} \tag{9.74}$$

with the kernel of Eq. (9.73). Comparing Eq. (9.74) with Eq. (9.60) demonstrates that kernel classical MDS and Isomap are special cases of direct GE with $\boldsymbol{L} = \boldsymbol{K}$ and $\boldsymbol{B} = \boldsymbol{I}$.

9.6.4.5 LLE

Let $\boldsymbol{W} \in \mathbb{R}^{n\times n}$ be the reconstruction weight matrix in LLE [75] (see Chap. 8). Let $\mathbb{R}^{n\times n} \ni \boldsymbol{M} := (\boldsymbol{I}-\boldsymbol{W})(\boldsymbol{I}-\boldsymbol{W})^\top$. The row summation of the matrix $\boldsymbol{M}$ is [100, 101]:

$$\begin{aligned}
\sum_{j=1}^{n} \boldsymbol{M}_{ij} &= \sum_{j=1}^{n} \big((\boldsymbol{I} - \boldsymbol{W})(\boldsymbol{I} - \boldsymbol{W})^\top\big)_{ij} \\
&= \sum_{j=1}^{n} \big(\boldsymbol{I}_{ij} - \boldsymbol{W}_{ij} - \boldsymbol{W}_{ji} + (\boldsymbol{W}\boldsymbol{W}^\top)_{ij}\big) \\
&= \mathbf{1} - \sum_{j=1}^{n} (\boldsymbol{W}_{ij} + \boldsymbol{W}_{ji}) + \sum_{j=1}^{n}\sum_{k=1}^{n} (\boldsymbol{W}_{ij}\boldsymbol{W}_{jk}) \\
&= \mathbf{1} - \sum_{j=1}^{n} (\boldsymbol{W}_{ij} + \boldsymbol{W}_{ji}) + \sum_{j=1}^{n} (\boldsymbol{W}_{ij} \sum_{k=1}^{n} \boldsymbol{W}_{jk}) \\
&\overset{(a)}{=} \mathbf{1} - \sum_{j=1}^{n} \boldsymbol{W}_{ij} - \sum_{j=1}^{n} \boldsymbol{W}_{ji} + \sum_{j=1}^{n} \boldsymbol{W}_{ij} = \mathbf{0},
\end{aligned}$$

where (a) is because the row summation of the reconstruction matrix is one, i.e. $\sum_{k=1}^{n} \boldsymbol{W}_{jk} = \mathbf{1}$, according to the constraint in the linear reconstruction step of LLE (see Chap. 8 for more details). According to Eq. (9.6), the matrix $\boldsymbol{M}$ used in LLE can be interpreted as the Laplacian matrix of the data graph as it satisfies its row-sum property. The optimization of LLE is as follows (see Chap. 8):

$$\begin{aligned}
\underset{\boldsymbol{Y}}{\text{minimize}} \quad & \mathbf{tr}(\boldsymbol{Y}^\top \boldsymbol{M} \boldsymbol{Y}), \\
\text{subject to} \quad & \frac{1}{n} \boldsymbol{Y}^\top \boldsymbol{Y} = \boldsymbol{I}, \\
& \boldsymbol{Y}^\top \mathbf{1} = \mathbf{0},
\end{aligned} \tag{9.75}$$

whose second constraint can be ignored because it is automatically satisfied (see Chap. 8). Comparing Eq. (9.75), without its second constraint, to Eq. (9.60) demonstrates that LLE is a special case of direct GE with $\boldsymbol{L} = \boldsymbol{M}$ and $\boldsymbol{B} = (1/n)\boldsymbol{I}$.

9.6.5 Other Improvements Over Graph embedding

Some of the improved variants of graph embedding are kernel eigenmaps for graph embedding with side information [17], fuzzy graph embedding [93], graph embedding with extreme learning machines [102], and deep dynamic graph embedding [39, 42]. Graph embedding has also been used for domain adaptation [49, 50]. A Python package for graph embedding toolbox is available [40].

9.7 Diffusion Map

The diffusion map, proposed in [29, 30, 59], is a nonlinear dimensionality reduction method that makes use of the Laplacian of data. Note that the literature on diffusion maps usually refers to the Laplacian as the normalized kernel (see Sect. 9.2.2 on why Laplacian is sometimes called the normalized kernel). A diffusion map reveals the nonlinear underlying manifold of the data through the diffusion process, where the structure of the manifold becomes increasingly obvious by iterations of the algorithm (i.e., by passing time). Its algorithm includes several concepts explained in the following sections.

9.7.1 Discrete Time Markov Chain

Consider the graph of the data, denoted by $\mathcal{X} = \{\boldsymbol{x}_1, \ldots, \boldsymbol{x}_n\}$, where $\boldsymbol{x}_i \in \mathbb{R}^d$ is the i-th data point in the dataset. Consider a kNN graph, or an adjacency matrix, of the data. Starting from any point in the graph, it is possible to randomly traverse the data graph to move to another point. This process is known as a random walk on the graph of data [32]. Let $\mathbb{P}(\boldsymbol{x}_i, \boldsymbol{x}_j)$ denote the probability of moving from point $\boldsymbol{x}_i$ to point $\boldsymbol{x}_j$. Obviously, if the two points are closer to one another in the graph or adjacency matrix, i.e. if they are more similar, this probability is larger. Movements from one point to another point in this graph can be modelled as a discrete time Markov chain[2] [74]. Let $\boldsymbol{M}$ be the transition matrix of a Markov chain on $\mathcal{X}$, where its elements are the probabilities of transition [36]. This transition can be repeated any number of times. The transition matrix at the t-th step is obtained as $\boldsymbol{M}^t$ [74].

Let $\boldsymbol{W} \in \mathbb{R}^{n\times n}$ be the adjacency matrix. The diffusion map usually uses the RBF kernel, Eq. (9.2), for the adjacency matrix [67, 68]. According to Eq. (9.7), the Laplacian is calculated for a value of α:

$$\mathbb{R}^{n\times n} \ni \boldsymbol{L}^{(\alpha)} := \boldsymbol{D}^{-\alpha}\boldsymbol{W}\boldsymbol{D}^{-\alpha}. \tag{9.76}$$

Let $\boldsymbol{D}^{(\alpha)} \in \mathbb{R}^{n\times n}$ denote the diagonal degree matrix for $\boldsymbol{L}^{(\alpha)}$, which is calculated similarly to Eq. (9.5), where $\boldsymbol{W}$ is replaced by $\boldsymbol{L}^{(\alpha)}$ in that equation. Applying the so-called *random-walk graph Laplacian normalization* to the Laplacian matrix results in [27]:

$$\mathbb{R}^{n\times n} \ni \boldsymbol{M} := (\boldsymbol{D}^{(\alpha)})^{-1}\boldsymbol{L}^{(\alpha)}, \tag{9.77}$$

[2] A discrete time Markov chain is a sequence of random variables, where the value of the next random variable depends only on the value of the current random value and not the previous random variables in the sequence.

which is the transition matrix of the Markov chain. Note that Eq. (9.77) is a stochastic transition matrix with nonnegative rows, which sum to one [17]. According to the definition of a transition matrix, the (i, j)-th element of the transition matrix at time step t, i.e., $\boldsymbol{M}^t$, is:

$$\mathbb{P}(\boldsymbol{x}_j, t \mid \boldsymbol{x}_i) = \boldsymbol{M}^t(i, j). \tag{9.78}$$

9.7.2 The Optimization Problem

At time step t, the diffusion map calculates the embeddings $\boldsymbol{Y} \in \mathbb{R}^{n \times d}$ by maximizing $\mathbf{tr}(\boldsymbol{Y}^\top \boldsymbol{M}^t \boldsymbol{Y})$, where it assumes the embeddings have scaled the identity covariance, i.e. $\boldsymbol{Y}^\top \boldsymbol{Y} = \boldsymbol{I}$. The term $\mathbf{tr}(\boldsymbol{Y}^\top \boldsymbol{M}^t \boldsymbol{Y})$ demonstrates the transition matrix of the data, at time t, after projection onto the column space of $\boldsymbol{Y}$ (see Chap. 2). The optimization problem is:

$$\begin{aligned} & \underset{\boldsymbol{Y}}{\text{maximize}} && \mathbf{tr}(\boldsymbol{Y}^\top \boldsymbol{M}^t \boldsymbol{Y}) \\ & \text{subject to} && \boldsymbol{Y}^\top \boldsymbol{Y} = \boldsymbol{I}, \end{aligned} \tag{9.79}$$

whose solution is the eigenvalue problem of $\boldsymbol{M}^t$:

$$\boldsymbol{M}^t \boldsymbol{Y} = \boldsymbol{Y}\boldsymbol{\Lambda}, \tag{9.80}$$

where columns of $\boldsymbol{Y}$ are the eigenvectors (see Chap. 2) and diagonal elements of $\boldsymbol{\Lambda}$ are the eigenvalues of $\boldsymbol{M}^t$. As it is a maximization problem, the eigenvectors are sorted from the corresponding largest to smallest eigenvalues. Then, the solution is truncated by taking the p leading eigenvectors to obtain the p-dimensional embedding $\boldsymbol{Y} \in \mathbb{R}^{n \times p}$, where $p \leq d$. Note that the diffusion map usually scales the embedding using eigenvalues. Further explanation is found in the following section.

9.7.3 Diffusion Distance

A diffusion distance can be defined, measuring how dissimilar two points are with regard to their probability of random walk from their neighbours to them. In other words, this distance at time step t is defined as [67, 68]:

$$d^{(t)}(\boldsymbol{x}_i, \boldsymbol{x}_j) := \sqrt{\sum_{\ell=1}^{n} \frac{\big(\mathbb{P}(\boldsymbol{x}_\ell, t|\boldsymbol{x}_i) - \mathbb{P}(\boldsymbol{x}_\ell, t|\boldsymbol{x}_j)\big)^2}{\boldsymbol{\psi}_1(\ell)}}, \tag{9.81}$$

where $\boldsymbol{\psi}_1 = [\boldsymbol{\psi}_1(1), \dots, \boldsymbol{\psi}_1(n)] \in \mathbb{R}^n$ is the leading eigenvector with the largest eigenvalue, i.e. $\boldsymbol{Y} = [\boldsymbol{\psi}_1, \dots, \boldsymbol{\psi}_p]$. The denominator is for the sake of normalization. It is demonstrated by Nadler et al. [67] that this distance is equivalent to the original diffusion distance proposed in [29]:

$$d^{(t)}(\boldsymbol{x}_i, \boldsymbol{x}_j) := \sqrt{\sum_{\ell=1}^{n} \lambda_\ell^{2t} \big(\boldsymbol{\psi}_\ell(i) - \boldsymbol{\psi}_\ell(j)\big)^2}, \tag{9.82}$$

where λ_ℓ denotes the ℓ-th largest eigenvalue.

If, at time t, the diffusion distance of two points is small, it implies that the two points are similar and close to each other in the embedding of step t. In other words, a small diffusion map between two points means those points probably belong to the same cluster in the embedding. Note that, as was explained in Sect. 9.7.2, the diffusion map, i.e. the embedding, is computed as $\boldsymbol{Y} = [\boldsymbol{y}_1, \dots, \boldsymbol{y}_n]^\top \in \mathbb{R}^{n \times p}$, where [29]:

$$\mathbb{R}^p \ni \boldsymbol{y}_i^{(t)} := [\lambda_1^t \boldsymbol{\psi}_1(i), \dots, \lambda_p^t \boldsymbol{\psi}_p(i)]^\top, \tag{9.83}$$

for time step t. Note that the diffusion map scales the embedding using the corresponding eigenvalues to the power of the time step. Therefore, the diffusion map becomes larger by moving forward in time. That is where the diffusion map receives its name from because it diffuses in time, where t is the time step and acts as a parameter for the scale.

Moreover, Nadler et al. have demonstrated that [67]:

$$d^{(t)}(\boldsymbol{x}_i, \boldsymbol{x}_j) = \|\boldsymbol{y}_i^{(t)} - \boldsymbol{y}_j^{(t)}\|_2. \tag{9.84}$$

Therefore, the diffusion distance is equivalent to the distance of embedded points in the diffusion map.

9.7.4 Other Improvements Over Diffusion maps

An improvement over the diffusion map is the vector diffusion map [85], which is a generalization of the diffusion map and several other nonlinear dimensionality reduction methods. The diffusion map has also been used in several applications, such as data fusion [58], nonlinear independent component analysis [84], and changing data [28]. The Fokker-Planck operator, which describes the time evolution of a probability density function, has also been used in some variants of diffusion maps [66].

9.8 Chapter Summary

This chapter introduced the Laplacian-based dimensionality reduction methods, including spectral clustering, Laplacian eigenmap, LPP, graph embedding, and diffusion map. These methods form a family of methods in the spectral category of dimensionality reduction. They all use the graph of data, so the topology of data points matters to them. Therefore, they are important algorithms in the spectral dimensionality reduction, addressing with the geometry of data. As was discussed in Sect. 9.4.1.2, the Laplacian-based methods fit data locally. This local fitting is similar to the main idea of locally linear embedding, introduced in Chap. 8. Note that a survey on the Laplacian-based methods is [97].

References

1. Saeed Ahmadizadeh et al. "On eigenvalues of Laplacian matrix for a class of directed signed graphs". In: *Linear Algebra and its Applications* 523 (2017), pp. 281–306.
2. Jonathan L Alperin. *Local representation theory: Modular representations as an introduction to the local representation theory of finite groups* Vol. 11. Cambridge University Press, 1993.
3. William N Anderson Jr and Thomas D Morley. "Eigenvalues of the Laplacian of a graph". In: *Linear and multilinear algebra* 18.2 (1985), pp. 141–145.
4. Ershad Banijamali and Ali Ghodsi. "Fast spectral clustering using autoencoders and landmarks". In: *International Conference Image Analysis and Recognition* Springer. 2017, pp. 380–388.
5. Peter N. Belhumeur, Joao P Hespanha, and David J. Kriegman. "Eigenfaces vs. Fisherfaces: Recognition using class specific linear projection". In: *IEEE Transactions on pattern analysis and machine intelligence* 19.7 (1997), pp. 711–720.
6. Mikhail Belkin and Partha Niyogi. "Convergence of Laplacian eigenmaps". In: *Advances in neural information processing systems* 19 (2006), pp. 129–136.
7. Mikhail Belkin and Partha Niyogi. "Laplacian eigenmaps and spectral techniques for embedding and clustering". In: *Advances in neural information processing systems* 14 (2001), pp. 585–591.
8. Mikhail Belkin and Partha Niyogi. "Laplacian eigenmaps for dimensionality reduction and data representation". In: *Neural computation* 15.6 (2003), pp. 1373–1396.
9. Mikhail Belkin and Partha Niyogi. "Towards a theoretical foundation for Laplacian-based manifold methods". In: *International Conference on Computational Learning Theory* Springer. 2005, pp. 486–500.
10. Yoshua Bengio et al. "Learning eigenfunctions links spectral embedding and kernel PCA". In: *Neural computation* 16.10 (2004), pp. 2197–2219.
11. Yoshua Bengio et al. *Learning eigenfunctions of similarity: linking spectral clustering and kernel PCA* Tech. rep. Departement d'Informatique et Recherche Operationnelle, 2003.
12. Yoshua Bengio et al. "Out-of-sample extensions for LLE, Isomap MDS, eigenmaps, and spectral clustering". In: *Advances in neural information processing systems* 16 (2003), pp. 177–184.
13. Yoshua Bengio et al. "Out-of-sample extensions for LLE, Isomap MDS, eigenmaps, and spectral clustering". In: *Advances in neural information processing systems* 2004, pp. 177–184.
14. Yoshua Bengio et al. *Spectral clustering and kernel PCA are learning eigenfunctions* Vol. 1239. Citeseer, 2003.

15. Yoshua Bengio et al. "Spectral dimensionality reduction". In: *Feature Extraction* Springer, 2006, pp. 519–550.
16. Matthew B Blaschko and Christoph H Lampert. "Correlational spectral clustering". In: *2008 IEEE Conference on Computer Vision and Pattern Recognition* IEEE. 2008, pp. 1–8.
17. Matthew Brand. "Continuous nonlinear dimensionality reduction by kernel eigenmaps". In: *International Joint Conference on Artificial Intelligence* 2003, pp. 547–554.
18. Kerstin Bunte, Michael Biehl, and Barbara Hammer. "A general framework for dimensionality-reducing data visualization mapping". In: *Neural Computation* 24.3 (2012), pp. 771–804.
19. Dmitri Burago, Sergei Ivanov, and Yaroslav Kurylev. "A graph discretization of the Laplace–Beltrami operator". In: *Journal of Spectral Theory* 4.4 (2015), pp. 675–714.
20. Deng Cai, Xiaofei He, and Jiawei Han. "Document clustering using locality preserving indexing". In: *IEEE Transactions on Knowledge and Data Engineering* 17.12 (2005), pp. 1624–1637.
21. Miguel A Carreira-Perpinán and Zhengdong Lu. "The Laplacian eigenmaps latent variable model". In: *Artificial Intelligence and Statistics* 2007, pp. 59–66.
22. Yale Chang. *Graph Embedding and Extensions: A General Framework for Dimensionality Reduction* Tech. rep. Department of ECE, Northeastern University, 2014.
23. Sibao Chen et al. "2D-LPP: A two-dimensional extension of locality preserving projections". In: *Neurocomputing* 70.4-6 (2007), pp. 912–921.
24. Wei-Jie Chen et al. "2DRLPP: Robust two-dimensional locality preserving projection with regularization". In: *Knowledge-Based Systems* 169 (2019), pp. 53–66.
25. Wen-Yen Chen et al. "Parallel spectral clustering in distributed systems". In: *IEEE transactions on pattern analysis and machine intelligence* 33.3 (2010), pp. 568–586.
26. Jian Cheng et al. "Supervised kernel locality preserving projections for face recognition". In: *Neurocomputing* 67 (2005), pp. 443–449.
27. Fan RK Chung. *Spectral graph theory* 92. American Mathematical Soc., 1997.
28. Ronald R Coifman and Matthew J Hirn. "Diffusion maps for changing data". In: *Applied and computational harmonic analysis* 36.1 (2014), pp. 79–107.
29. Ronald R Coifman and Stéphane Lafon. "Diffusion maps". In: *Applied and computational harmonic analysis* 21.1 (2006), pp. 5–30.
30. Ronald R Coifman et al. "Geometric diffusions as a tool for harmonic analysis and structure definition of data: Diffusion maps". In: *Proceedings of the national academy of sciences* 102.21 (2005), pp. 7426–7431.
31. Joao Costeira and Takeo Kanade. "A multi-body factorization method for motion analysis". In: *Proceedings of IEEE International Conference on Computer Vision* IEEE. 1995, pp. 1071–1076.
32. J De la Porte et al. "An introduction to diffusion maps". In: *Proceedings of the 19th Symposium of the Pattern Recognition Association of South Africa (PRASA 2008), Cape Town, South Africa* 2008, pp. 15–25.
33. David B Dunson, Hau-Tieng Wu, and Nan Wu. "Spectral convergence of graph Laplacian and Heat kernel reconstruction in L^∞ from random samples". In: *Applied and Computational Harmonic Anal- ysis* (2021).
34. Ronald A Fisher. "The use of multiple measurements in taxonomic problems". In: *Annals of eugenics* 7.2 (1936), pp. 179–188.
35. Benyamin Ghojogh, Fakhri Karray, and Mark Crowley. "Generalized Subspace Learning by Roweis Discriminant Analysis". In: *International Conference on Image Analysis and Recognition* Springer. 2020, pp. 328–342.
36. Benyamin Ghojogh, Fakhri Karray, and Mark Crowley. "Hidden Markov Model: Tutorial". In: *engrXiv* (2019).
37. Andrej Gisbrecht, Alexander Schulz, and Barbara Hammer. "Parametric nonlinear dimensionality reduction using kernel t-SNE". In: *Neurocomputing* 147 (2015), pp. 71–82.
38. Andrej Gisbrecht et al. "Out-of-sample kernel extensions for nonparametric dimensionality reduction". In: *European Symposium on Artificial Neural Networks, Computational Intelligence and Machine Learning* Vol. 2012. 2012, pp. 531–536.

39. Palash Goyal, Sujit Rokka Chhetri, and Arquimedes Canedo. “dyngraph2vec: Capturing network dynamics using dynamic graph representation learning”. In: *Knowledge-Based Systems* 187 (2020), p. 104816.
40. Palash Goyal and Emilio Ferrara. “GEM: a Python package for graph embedding methods”. In: *Journal of Open Source Software* 3.29 (2018), p. 876.
41. Palash Goyal and Emilio Ferrara. “Graph embedding techniques, applications, and performance: A survey”. In: *Knowledge-Based Systems* 151 (2018), pp. 78–94.
42. Palash Goyal et al. “DynGEM: Deep embedding method for dynamic graphs”. In: *arXiv preprint arXiv:1805.11273* (2018).
43. CAI Xiao-yan DAI Guan-zhong and YANG Li-bin. “Survey on Spectral Clustering Algorithms [J]”. In: *Computer Science* 7.005 (2008).
44. Cuimei Guo et al. “A survey on spectral clustering”. In: *World Automation Congress 2012* IEEE. 2012, pp. 53–56.
45. Yi Guo, Junbin Gao, and Paul WH Kwan. “Kernel Laplacian eigenmaps for visualization of nonvectorial data”. In: *Australasian Joint Conference on Artificial Intelligence* Springer. 2006, pp. 1179–1183.
46. Xiaofei He and Partha Niyogi. “Locality preserving projections”. In: *Advances in neural information processing systems* 2004, pp. 153–160.
47. Xiaofei He et al. “Face recognition using Laplacianfaces”. In: *IEEE transactions on pattern analysis and machine intelligence* 27.3 (2005), pp. 328–340.
48. Xiaofei He et al. “Learning a locality preserving subspace for visual recognition”. In: *Proceedings Ninth IEEE International Conference on Computer Vision* IEEE. 2003, pp. 385–392.
49. Lukas Hedegaard, Omar Ali Sheikh-Omar, and Alexandros Iosifidis. “Supervised domain adaptation using graph embedding”. In: *arXiv preprint arXiv:2003.04063* (2020).
50. Lukas Hedegaard, Omar Ali Sheikh-Omar, and Alexandros Iosifidis. “Supervised Domain Adaptation: A Graph Embedding Perspective and a Rectified Experimental Protocol”. In: *arXiv e-prints* (2020), arXiv–2004.
51. Matthias Hein, Jean-Yves Audibert, and Ulrike von Luxburg. “Graph Laplacians and their convergence on random neighborhood graphs”. In: *Journal of Machine Learning Research* 8.6 (2007).
52. Matthias Hein, Jean-Yves Audibert, and Ulrike Von Luxburg. “From graphs to manifolds–weak and strong pointwise consistency of graph Laplacians”. In: *International Conference on Computational Learning Theory* Springer. 2005, pp. 470–485.
53. Thomas Hofmann, Bernhard Schölkopf, and Alexander J Smola. “Kernel methods in machine learning”. In: *The annals of statistics* (2008), pp. 1171–1220.
54. Dewen Hu, Guiyu Feng, and Zongtan Zhou. “Two-dimensional locality preserving projections (2DLPP) with its application to palmprint recognition”. In: *Pattern recognition* 40.1 (2007), pp. 339–342.
55. Jonathan Kelner. *An Algorithm's Toolkit: Properties of the Laplacian, Positive Semidefinite Matrices, Spectra of Common Graphs, and Connection to the Continuous Laplacian* Tech. rep. Department of Mathematics and CSAIL, MIT University, 2007.
56. Abhishek Kumar and Hal Daumé. “A co-training approach for multi-view spectral clustering”. In: *Proceedings of the 28th international conference on machine learning* 2011, pp. 393–400.
57. Abhishek Kumar, Piyush Rai, and Hal Daume. “Co-regularized multi-view spectral clustering”. In: *Advances in neural information processing systems* 2011, pp. 1413–1421.
58. Stephane Lafon, Yosi Keller, and Ronald R Coifman. “Data fusion and multicue data matching by diffusion maps”. In: *IEEE Transactions on pattern analysis and machine intelligence* 28.11 (2006), pp. 1784–1797.
59. Stéphane S Lafon. “Diffusion maps and geometric harmonics”. PhD thesis. Yale University, 2004.
60. Bo Li, Yan-Rui Li, and Xiao-Long Zhang. “A survey on Laplacian eigenmaps based manifold learning methods”. In: *Neurocomputing* 335 (2019), pp. 336–351.

61. Jun-Bao Li, Jeng-Shyang Pan, and Shu-Chuan Chu. "Kernel class-wise locality preserving projection". In: *Information Sciences* 178.7 (2008), pp. 1825–1835.
62. Herve Lombaert et al. "Laplacian forests: semantic image segmentation by guided bagging". In: *International Conference on Medical Image Computing and Computer-Assisted Intervention* Springer. 2014, pp. 496–504.
63. Anne Marsden. "Eigenvalues of the Laplacian and their relationship to the connectedness of a graph". In: *University of Chicago, REU* (2013).
64. Russell Merris. "Laplacian matrices of graphs: a survey". In: *Linear algebra and its applications* 197 (1994), pp. 143–176.
65. Sebastian Mika et al. "Fisher discriminant analysis with kernels". In: *Proceedings of the 1999 IEEE signal processing society workshop on Neural networks for signal processing IX* IEEE. 1999, pp. 41–48.
66. Boaz Nadler et al. "Diffusion maps, spectral clustering and eigenfunctions of Fokker-Planck operators". In: *Advances in neural information processing systems* 18 (2005).
67. Boaz Nadler et al. "Diffusion maps, spectral clustering and eigenfunctions of Fokker-Planck operators". In: *Advances in neural information processing systems* 2006, pp. 955–962.
68. Boaz Nadler et al. "Diffusion maps, spectral clustering and reaction coordinates of dynamical systems". In: *Applied and Computational Harmonic Analysis* 21.1 (2006), pp. 113–127.
69. Maria CV Nascimento and Andre CPLF De Carvalho. "Spectral methods for graph clustering–a survey". In: *European Journal of Operational Research* 211.2 (2011), pp. 221–231.
70. Andrew Ng, Michael Jordan, and Yair Weiss. "On spectral clustering: Analysis and an algorithm". In: *Advances in neural information processing systems* 14 (2001), pp. 849–856.
71. Pietro Perona and William Freeman. "A factorization approach to grouping". In: *European Conference on Computer Vision* Springer. 1998, pp. 655–670.
72. Julien Ah-Pine. "Normalized kernels as similarity indices". In: *Pacific-Asia Conference on Knowledge Discovery and Data Mining* Springer. 2010, pp. 362–373.
73. Marzia Polito and Pietro Perona. "Grouping and dimensionality reduction by locally linear embedding". In: *Advances in neural information processing systems* 2002, pp. 1255–1262.
74. Sheldon M Ross. *Introduction to probability models* Academic press, 2014.
75. Sam T Roweis and Lawrence K Saul. "Nonlinear dimensionality reduction by locally linear embedding". In: *Science* 290.5500 (2000), pp. 2323–2326.
76. Shounak Roychowdhury. "Robust Laplacian Eigenmaps using global information". In: *2009 AAAI Fall Symposium Series* Citeseer. 2009.
77. Lawrence K Saul and Sam T Roweis. "Think globally fit locally: unsupervised learning of low dimensional manifolds". In: *Journal of machine learning research* 4.Jun (2003), pp. 119–155.
78. Guy L Scott and Hugh Christopher Longuet-Higgins. "Feature grouping by "relocalisation" of eigenvectors of the proximity matrix." In: *BMVC* 1990, pp. 1–6.
79. Jianbo Shi and Jitendra Malik. "Normalized cuts and image segmentation". In: *Proceedings of IEEE computer society conference on computer vision and pattern recognition* IEEE. 1997, pp. 731–737.
80. Jianbo Shi and Jitendra Malik. "Normalized cuts and image segmentation". In: *IEEE Transactions on pattern analysis and machine intelligence* 22.8 (2000), pp. 888–905.
81. Gitam Shikkenawis and Suman K Mitra. "Improving the locality preserving projection for dimensionality reduction". In: *2012 Third International Conference on Emerging Applications of Information Technology* IEEE. 2012, pp. 161–164.
82. Gitam Shikkenawis and Suman K Mitra. "On some variants of locality preserving projection". In: *Neurocomputing* 173 (2016), pp. 196–211.
83. Amit Singer. "From graph to manifold Laplacian: The convergence rate". In: *Applied and Computational Harmonic Analysis* 21.1 (2006), pp. 128–134.
84. Amit Singer and Ronald R Coifman. "Non-linear independent component analysis with diffusion maps". In: *Applied and Computational Harmonic Analysis* 25.2 (2008), pp. 226–239.

85. Amit Singer and H-T Wu. “Vector diffusion maps and the connection Laplacian”. In: *Communications on pure and applied mathematics* 65.8 (2012), pp. 1067–1144.
86. Gilbert Strang. *Differential equations and linear algebra* Wellesley-Cambridge Press Wellesley, 2014.
87. Yun Tang and Richard Rose. “A study of using locality preserving projections for feature extraction in speech recognition”. In: *2008 IEEE International Conference on Acoustics, Speech and Signal Processing* IEEE. 2008, pp. 1569–1572.
88. Joshua B Tenenbaum, Vin De Silva, and John C Langford. “A global geometric framework for nonlinear dimensionality reduction”. In: *Science* 290.5500 (2000), pp. 2319–2323.
89. Nicolás Garcıa Trillos et al. “Error estimates for spectral convergence of the graph Laplacian on random geometric graphs toward the Laplace–Beltrami operator”. In: *Foundations of Computational Mathematics* 20.4 (2020), pp. 827–887.
90. Matthew Turk and Alex Pentland. “Eigenfaces for recognition”. In: *Journal of cognitive neuroscience* 3.1 (1991), pp. 71–86.
91. Ulrike Von Luxburg. “A tutorial on spectral clustering”. In: *Statistics and computing* 17.4 (2007), pp. 395–416.
92. Ulrike Von Luxburg, Mikhail Belkin, and Olivier Bousquet. “Consistency of spectral clustering”. In: *The Annals of Statistics* (2008), pp. 555–586.
93. Minghua Wan et al. “Local graph embedding based on maximum margin criterion via fuzzy set”. In: *Fuzzy Sets and Systems* 318 (2017), pp. 120–131.
94. Boyue Wang et al. “Locality preserving projections for Grassmann manifold”. In: *arXiv preprint arXiv:1704.08458* (2017).
95. Yair Weiss. “Segmentation using eigenvectors: a unifying view”. In: *Proceedings of the seventh IEEE international conference on computer vision* Vol. 2. IEEE. 1999, pp. 975–982.
96. Yair Weiss, Antonio Torralba, and Rob Fergus. “Spectral hashing”. In: *Advances in neural information processing systems* 21 (2008), pp. 1753–1760.
97. Laurenz Wiskott and Fabian Schönfeld. “Laplacian matrix for dimensionality reduction and clustering”. In: *European Big Data Management and Analytics Summer School* Springer. 2019, pp. 93–119.
98. Wai Keung Wong and HT Zhao. “Supervised optimal locality preserving projection”. In: *Pattern Recognition* 45.1 (2012), pp. 186–197.
99. Yong Xu, Ge Feng, and Yingnan Zhao. “One improvement to two-dimensional locality preserving projection method for use with face recognition”. In: *Neurocomputing* 73.1-3 (2009), pp. 245–249.
100. Shuicheng Yan et al. “Graph embedding and extensions: A general framework for dimensionality reduction”. In: *IEEE transactions on pattern analysis and machine intelligence* 29.1 (2006), pp. 40–51.
101. Shuicheng Yan et al. “Graph embedding: A general framework for dimensionality reduction”. In: *2005 IEEE Computer Society Conference on Computer Vision and Pattern Recognition (CVPR'05)* Vol. 2. IEEE. 2005, pp. 830–837.
102. Le Yang et al. “Graph Embedding-Based Dimension Reduction With Extreme Learning Machine”. In: *IEEE Transactions on Systems, Man, and Cybernetics: Systems* (2019).
103. Yifang Yang and Yuping Wang. “Fuzzy Partition based Similarity Measure for Spectral Clustering”. In: *International Journal of Signal Processing, Image Processing and Pattern Recognition* 9.10 (2016), pp. 417–428.
104. Yifang Yang, Yuping Wang, and Xingsi Xue. “A novel spectral clustering method with superpixels for image segmentation”. In: *Optik* 127.1 (2016), pp. 161–167.
105. Yifang Yang, Yuping Wang, and Xingsi Xue. “Discriminant sparse locality preserving projection for face recognition”. In: *Multimedia Tools and Applications* 76.2 (2017), pp. 2697–2712.
106. Jieping Ye. “Least squares linear discriminant analysis”. In: *Proceedings of the 24th international conference on machine learning* ACM. 2007, pp. 1087–1093.
107. Hongwei Yin et al. “Multi-view clustering via spectral embedding fusion”. In: *Soft Computing* 23.1 (2019), pp. 343–356.

108. Jun Yin and Shiliang Sun. "Multiview uncorrelated locality preserving projection". In: *IEEE transactions on neural networks and learning systems* (2019).
109. Weiwei Yu, Xiaolong Teng, and Chongqing Liu. "Face recognition using discriminant locality preserving projections". In: *Image and Vision computing* 24.3 (2006), pp. 239–248.
110. Lihi Zelnik-Manor and Pietro Perona. "Self-tuning spectral clustering". In: *Advances in neural information processing systems* 17 (2004), pp. 1601–1608.
111. Limei Zhang, Lishan Qiao, and Songcan Chen. "Graph-optimized locality preserving projections". In: *Pattern Recognition* 43.6 (2010), pp. 1993–2002.

Chapter 10
Unified Spectral Framework and Maximum Variance Unfolding

10.1 Introduction

Various spectral methods have been proposed over the past few decades. Some of the most well-known spectral methods include Principal Component Analysis (PCA), Multidimensional Scaling (MDS), Isomap, spectral clustering, Laplacian eigenmap, diffusion map, and Locally Linear Embedding (LLE). After the development of these methods, it was discovered that most of the spectral methods are learning eigenfunctions and can be reduced to kernel PCA with different kernels [4–8], where the kernels are constructed from various distance matrices [21]. Recall from Chap. 3 that a kernel can be written in terms of the distance matrix. This kernel-based unified framework for spectral dimensionality reduction encouraged researchers to obtain the best kernel matrix for each specific dataset.

At the same time as the discovery of spectral methods being learning eigenfunctions, Lanckriet et al. [25] determined that the kernel matrix can also be learned using Semidefinite Programming (SDP) [40] (see Chap. 4 for SDP). This kernel learning was proposed for the goal of transduction, i.e., learning the labels of an unlabelled part of data. The fact that the kernel can be learned through SDP inspired researchers to use SDP for learning the best kernel for dimensionality reduction and manifold unfolding to its maximum variance. This resulted in Semidefinite Embedding (SDE) [44, 46–48] being proposed, which was later renamed to Maximum Variance Unfolding (MVU) [45]. MVU unfolds a manifold to its maximum variance in its intrinsic dimensionality; see Fig. 10.1. To better understand manifold unfolding to its maximum variance, see the depiction in Fig. 10.2. Following the proposal of MVU, various versions of MVU were developed, such as supervised MVU [26, 36, 43, 52], landmark MVU [44], action respecting embedding [9], out-of-sample extensions [13], relaxed MVU [22], etc. This chapter first explains the unified framework of spectral methods as kernel PCA and then introduces MVU for finding the best kernel in dimensionality reduction for each dataset.

B. Ghojogh et al., *Elements of Dimensionality Reduction and Manifold Learning*,
https://doi.org/10.1007/978-3-031-10602-6_10

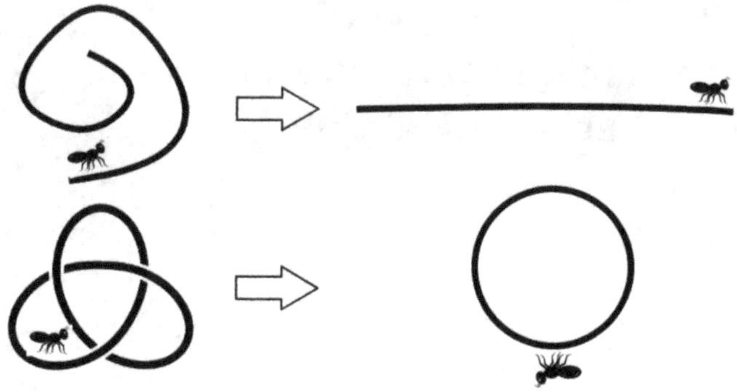

Fig. 10.1 The intrinsic dimensionality for a Swiss roll (above) and a trefoil knot (below). Suppose a small ant, which cannot see the whole manifold together and can only see what is in front of it, travels across the manifold once completely. The ant reaches the end point in the Swiss roll, so it thinks that it is a 1D line. Where in the trefoil the ant returns to where it started, so it thinks it is on a 2D circle. The dimensionality that the ant feels on its travels is the intrinsic dimensionality

Fig. 10.2 Iterative unfolding of a nonlinear Swiss roll manifold using maximum variance unfolding

10.2 Unified Framework for Spectral Methods

After the development of many spectral dimensionality reduction methods, it was found by researchers that these methods all can be unified as kernel Principal Component Analysis (PCA) (see Chap. 5) with different kernels. This is mostly because the spectral methods reduce to an eigenvalue or a generalized eigenvalue problem (see Chap. 2). This unification can be analyzed from different perspectives, including eigenfunction learning, having kernels in kernel PCA, and generalized embedding. The following section explains these points of view.

10.2.1 Learning Eigenfunctions

Recall eigenfunctions and the following lemma and theorem from Chap. 3.

Lemma 10.1 (Relation of Eigenfunctions and Eigenvectors for Kernel [6, Proposition 1], [7, Theorem 1]) *Consider a training dataset* $\{\boldsymbol{x}_i \in \mathbb{R}^d\}_{i=1}^n$ *and the eigenvalue problem:*

$$\boldsymbol{K}\boldsymbol{v}_k = \delta_k \boldsymbol{v}_k, \quad \forall k \in \{1, \dots, n\}, \tag{10.1}$$

where $\boldsymbol{v}_k \in \mathbb{R}^n$ *and* δ_k *are the* k*-th eigenvector and eigenvalue of kernel matrix* $\boldsymbol{K} \in \mathbb{R}^{n\times n}$*. If* v_{ki} *is the* i*-th element of the vector* $\boldsymbol{v}_k$*, the eigenfunction for the point* $\boldsymbol{x}$ *and the* i*-th training point* $\boldsymbol{x}_i$ *are:*

$$f_k(\boldsymbol{x}) = \frac{\sqrt{n}}{\delta_k} \sum_{i=1}^{n} v_{ki}\, \breve{k}(\boldsymbol{x}_i, \boldsymbol{x}), \tag{10.2}$$

$$f_k(\boldsymbol{x}_i) = \sqrt{n}\, v_{ki}, \tag{10.3}$$

respectively, where $\breve{k}(\boldsymbol{x}_i, \boldsymbol{x})$ *is the centered kernel. If* $\boldsymbol{x}$ *is a training point,* $\breve{k}(\boldsymbol{x}_i, \boldsymbol{x})$ *is the centered kernel over the training data and if* $\boldsymbol{x}$ *is an out-of-sample point, then* $\breve{k}(\boldsymbol{x}_i, \boldsymbol{x}) = \breve{k}_t(\boldsymbol{x}_i, \boldsymbol{x})$ *is between the training set and the out-of-sample point.*

Theorem 10.4 (Embedding from Eigenfunctions of Kernel Operator [6, Proposition 1], [7, Section 4]) *Consider a dimensionality reduction algorithm that embeds data into a low-dimensional embedding space. Let the embedding of the point* $\boldsymbol{x}$ *be* $\mathbb{R}^p \ni \boldsymbol{y}(\boldsymbol{x}) = [y_1(\boldsymbol{x}), \dots, y_p(\boldsymbol{x})]^\top$*, where* $p \leq n$*. The* k*-th dimension of this embedding is:*

$$y_k(\boldsymbol{x}) = \sqrt{\delta_k}\, \frac{f_k(\boldsymbol{x})}{\sqrt{n}} = \frac{1}{\sqrt{\delta_k}} \sum_{i=1}^{n} v_{ki}\, \breve{k}(\boldsymbol{x}_i, \boldsymbol{x}), \tag{10.4}$$

where $\breve{k}(\boldsymbol{x}_i, \boldsymbol{x})$ *is the centered training or the out-of-sample kernel depending on whether* $\boldsymbol{x}$ *is a training or an out-of-sample point.*

Theorem 10.4 has been widely used for out-of-sample (test data) embedding in many spectral dimensionality reduction algorithms [6]. This theorem also explains why spectral methods can all be seen as kernel PCA for learning eigenfunctions. The following section discusses the justifications behind the unification of the spectral methods.

10.2.2 Unified Framework as Kernel PCA

Many spectral dimensionality reduction methods can be reduced to kernel PCA, where the eigenvectors of the kernel matrix or eigenfunctions of the kernel operator are used for embedding, as stated in Lemma 10.1 and Theorem 10.4. This unification was analyzed in the following two categories of papers:

- papers for eigenfunction learning: [4–8]
- papers for unification by kernels constructed from distance matrices: [21] and [38, Table 2.1]

In the following, the unification of spectral methods as kernel PCA is explained using both approaches together.

Principal Component analysis (PCA) As kernel PCA makes use of the kernel trick for kernelization, it is considered to be equivalent to PCA when a linear kernel is used. Therefore, PCA (see Chap. 5) is equivalent to kernel PCA by the linear kernel:

$$\boldsymbol{K} = \boldsymbol{X}^\top \boldsymbol{X}. \tag{10.5}$$

Multidimensional Scaling (MDS) According to Chaps. 3 and 7, MDS [15] is equivalent to kernel PCA with the kernel:

$$\boldsymbol{K} = -\frac{1}{2}\boldsymbol{H}\boldsymbol{D}\boldsymbol{H}, \tag{10.6}$$

where $\boldsymbol{D}$ is the squared Euclidean distance matrix and $\boldsymbol{H} := \boldsymbol{I} - (1/n)\boldsymbol{1}\boldsymbol{1}^\top$ is the centering matrix.

Spectral Clustering Let $\boldsymbol{W} \in \mathbb{R}^{n\times n}$ be the adjacency matrix of points (see Chap. 9). Spectral clustering [29, 49] uses the normalized adjacency matrix as its kernel matrix. Suppose $\boldsymbol{D}_{i,i} := \sum_{j=1}^{n} \boldsymbol{W}_{ij}$ is the (i, i)-th element of the diagonal degree matrix of the adjacency matrix. Therefore, the (i, j)-th element of the kernel is [6, 49]:

$$\boldsymbol{K}_{ij} = \frac{\boldsymbol{W}_{ij}}{\sqrt{\boldsymbol{D}_{ii}\,\boldsymbol{D}_{jj}}}, \tag{10.7}$$

where $\boldsymbol{K}_{ij} = k(\boldsymbol{x}_i, \boldsymbol{x}_j)$ for the kernel function $k(.,.)$. This relation of the kernel matrix with the adjacency matrix makes sense because the kernel is a notion of similarity (see Chap. 3).

Laplacian Eigenmap The solution of the Laplacian eigenmap [3] (see Chap. 9) is the generalized eigenvalue problem $(\boldsymbol{L}, \boldsymbol{D})$ (see Chap. 2), where $\boldsymbol{L} := \boldsymbol{D} - \boldsymbol{W}$ is the Laplacian of the adjacency matrix $\boldsymbol{W}$. According to [49, Normalization Lemma 1], the eigenvectors of the generalized eigenvalue problem $(\boldsymbol{L}, \boldsymbol{D})$ are equivalent to the eigenvectors of the normalized kernel in Eq. (10.7).

In addition to the above analysis [6], there is another analysis for the Laplacian eigenmap [21]. Consider the Laplacian matrix of the data graph. It can, for example, be the Laplacian of the RBF adjacency matrix. It was mentioned above that the solution of the Laplacian eigenmap is the generalized eigenvalue problem $(\boldsymbol{L}, \boldsymbol{D})$. However, there exists another form of optimization for the Laplacian eigenmap whose solution is the eigenvalue problem for $\boldsymbol{L}$ (see Chap. 9). As the optimization of the Laplacian eigenmap is a minimization problem, its eigenvectors are sorted from the smallest to largest eigenvalues (see Chap. 2). However, the optimization in PCA and kernel PCA is a maximization problem and the eigenvectors are sorted from the largest to smallest eigenvalues. Therefore, to have an equivalency with kernel PCA, a modification from a minimization problem to a maximization problem is needed. One way to do this is to replace $\boldsymbol{L}$ with [21]:

$$\mathbb{R}^{n \times n} \ni \boldsymbol{K} = \boldsymbol{L}^{\dagger}, \tag{10.8}$$

where † denotes the pseudoinverse of the matrix. This replacement changes the minimization problem to a maximization problem. This is because the Laplacian matrix is positive semidefinite (see Chap. 9), and the eigenvalues of the inverse of a positive semidefinite matrix are equivalent to the reciprocal of the eigenvalues of that matrix, while the eigenvectors remain the same. Therefore, the order of the eigenvalues and eigenvectors become reversed. It is also important to note that the pseudoinverse of the Laplacian used as the kernel should be double-centered because the kernel is double-centered in kernel PCA (see Chap. 5). It is interesting that Eq. (10.8) is already double-centered [21]:

$$\boldsymbol{L}\boldsymbol{1} = \boldsymbol{L}^{\dagger}\boldsymbol{1} = \boldsymbol{0} \implies \boldsymbol{H}\boldsymbol{L}^{\dagger}\boldsymbol{H} = \boldsymbol{L}^{\dagger},$$

where $\boldsymbol{H} := \boldsymbol{I} - (1/n)\boldsymbol{1}\boldsymbol{1}^{\top}$ is the centering matrix. Note that $\boldsymbol{L}\boldsymbol{1} = \boldsymbol{L}^{\dagger}\boldsymbol{1} = \boldsymbol{0}$ holds because the row summation of the Laplacian matrix is zero (see Chap. 9).

Isomap The kernel in Isomap is [39] (see Chap. 7):

$$\boldsymbol{K} = -\frac{1}{2}\boldsymbol{H}\boldsymbol{D}^{(g)}\boldsymbol{H}, \tag{10.9}$$

where $\boldsymbol{D}^{(g)}$ is the geodesic distance matrix whose elements are the (squared) approximation of the geodesic distances using piecewise Euclidean distances. Therefore, it can be seen as a kernel PCA with the abovementioned kernel [21].

Locally Linear Embedding (LLE) The solution of LLE is the eigenvectors of $\boldsymbol{M} := (\boldsymbol{I} - \boldsymbol{W})^{\top}(\boldsymbol{I} - \boldsymbol{W})$, where $\boldsymbol{I}$ is the identity matrix and $\boldsymbol{W}$ is the weight matrix obtained from the linear reconstruction step in LLE [31] (see Chap. 8). As the optimization of LLE is a minimization problem, its eigenvalues are sorted from smallest to largest (see Chap. 2). However, PCA and kernel PCA have maximization problems in their formulation. Therefore, to have equivalency with the kernel PCA, the minimization problem should be converted to a maximization problem. One way to do this is to replace $\boldsymbol{M}$ with [8, 34]:

$$\mathbb{R}^{n \times n} \ni \boldsymbol{K} = \mu\boldsymbol{I} - \boldsymbol{M}, \tag{10.10}$$

where $\mu > 0$. It is suggested to set μ to the largest eigenvalue $\lambda_{\max}$ of $\boldsymbol{M}$ so that the kernel $\boldsymbol{K}$ becomes positive definite [21]. This replacement of $\boldsymbol{M}$ with the provided kernel changes the minimization problem to a maximization problem because:

$$\arg\max(\boldsymbol{K}) \overset{(10.10)}{=} \arg\max(-\boldsymbol{M}) = \arg\min(\boldsymbol{M}).$$

Another replacement option exists where $\boldsymbol{M}$ is replaced with its pseudoinverse [1, Section 2.1]:

$$\mathbb{R}^{n\times n} \ni \boldsymbol{K} = \boldsymbol{M}^{\dagger}, \tag{10.11}$$

whose justification is similar to the justification of the pseudoinverse of the Laplacian for the Laplacian eigenmap.

Diffusion Map A diffusion map [14] is a Laplacian-based method (see Chap. 9) that can be seen as a special case of the kernel PCA with a specific kernel [21], [38, Chapter 2]. Let $\boldsymbol{M}$ denote the random-walk graph Laplacian normalization of the Laplacian, i.e., $\boldsymbol{M} := (\boldsymbol{D}^{(\alpha)})^{-1}\boldsymbol{L}^{(\alpha)}$ where α is a parameter. At time t, is can be said that $\boldsymbol{M}^t$ is the probability of going from $\boldsymbol{x}_i$ to $\boldsymbol{x}_j$. The solution of the diffusion map at time t is the eigenvectors of $\boldsymbol{M}^t$ sorted from the largest to smallest eigenvalues. Therefore, the following kernel can be considered for the equivalency of the diffusion map and kernel PCA:

$$\mathbb{R}^{n\times n} \ni \boldsymbol{K} = \boldsymbol{M}^{t}, \tag{10.12}$$

because the kernel PCA also has a maximization problem in its formulation.

10.2.3 Summary of Kernels in Spectral Methods

Kernels in some algorithms, such as PCA and MDS, are closed-form and can be computed quickly. However, some methods, such as Isomap, require several steps of the algorithm (i.e., a piece of code) to calculate their kernels. Table 10.1 provides a summary of the kernels used for unification of the spectral dimensionality reduction methods as kernel PCA. A similar summary exists in Chap. 2 of another book on open problems in spectral dimensionality reduction [38, Table 2.1].

10.2.4 Generalized Embedding

It was demonstrated above that many of the spectral dimensionality reduction methods belong to a unified framework. Therefore, there can be generalized embed-

Table 10.1 Summary of kernels used for unification of the spectral methods as kernel PCA

Method	Kernel
PCA	$\boldsymbol{X}^\top \boldsymbol{X}$
MDS	$-\frac{1}{2}\boldsymbol{H}\boldsymbol{D}\boldsymbol{H}$
Spectral clustering	$\frac{\boldsymbol{W}_{ij}}{\sqrt{\boldsymbol{D}_{ii}\,\boldsymbol{D}_{jj}}}$
Laplacian eigenmap	$\boldsymbol{L}^\dagger$ or $\frac{\boldsymbol{W}_{ij}}{\sqrt{\boldsymbol{D}_{ii}\,\boldsymbol{D}_{jj}}}$
Isomap	$-\frac{1}{2}\boldsymbol{H}\boldsymbol{D}^{(g)}\boldsymbol{H}$
LLE	$\boldsymbol{M}^\dagger$ or $\lambda_{\max}\boldsymbol{I} - \boldsymbol{M}$
Diffusion map	$\boldsymbol{M}^t$

ding methods, which generalize the spectral methods to broader algorithms. Graph Embedding (GE) [50, 51] (see Chap. 9) demonstrates that many spectral methods, including the Laplacian eigenmap, locality preserving projection, PCA and kernel PCA, Fisher Discriminant Analysis (FDA) and kernel FDA, MDS, Isomap, and LLE, are special cases of a model named graph embedding. Another generalized subspace learning method, named Roweis Discriminant Analysis (RDA), also generalized PCA, supervised PCA, and FDA as unified methods. These generalized methods also justify why spectral methods can be unified.

10.3 Kernel Learning for Transduction

Kernel learning by Semidefinite Programming (SDP) was initially proposed in [25] with the goal of transduction. Transduction is a task in which the labelling of an incompletely labelled dataset is completed. In other words, using the labelled part of the data, the embedding for the unlabelled part of the data is also calculated. It seems that paper [25] has inspired the authors of MVU or SDE [47] to use SDP in kernel learning for the task of dimensionality reduction. As kernel learning for transduction is not completely related to dimensionality reduction, it is briefly introduced. Refer to [25] or its summary in [23] for additional information.

Consider a dataset that is partially labelled. The labelled and unlabelled sets of data can be the training and test sets, respectively. Let the training and test sets be denoted by $\{(\boldsymbol{x}_i, y_i)\}_{i=1}^{n_{\text{tr}}}$ and $\{\boldsymbol{x}_i\}_{i=n_{\text{tr}}+1}^{n_{\text{te}}}$, respectively, where n_{tr} is the number of labelled training points and n_{te} is the number of unlabelled test points and $n := n_{\text{tr}} + n_{\text{te}}$. The kernel matrix has the following submatrices:

$$\mathbb{R}^{n\times n} \ni \boldsymbol{K} = \begin{bmatrix} \boldsymbol{K}_{\text{tr,tr}} & \boldsymbol{K}_{\text{tr,te}} \\ \boldsymbol{K}_{\text{te,tr}} & \boldsymbol{K}_{\text{te,te}}, \end{bmatrix}, \tag{10.13}$$

where $\boldsymbol{K}_{\text{tr,tr}} \in \mathbb{R}^{n_{\text{tr}}\times n_{\text{tr}}}$, $\boldsymbol{K}_{\text{tr,te}} \in \mathbb{R}^{n_{\text{tr}}\times n_{\text{te}}}$, $\boldsymbol{K}_{\text{te,tr}} \in \mathbb{R}^{n_{\text{te}}\times n_{\text{tr}}}$, and $\boldsymbol{K}_{\text{te,te}} \in \mathbb{R}^{n_{\text{te}}\times n_{\text{te}}}$. As discussed later in Sect. 10.4, MVU [47] learns the optimal kernel matrix for manifold unfolding. However, in contrast to kernel learning in MVU, which learns the kernel matrix through optimization, the kernel learning proposed by Lanckriet et al. [25] finds the optimal kernel among a set of available kernels. Let $\mathcal{K} := \{\boldsymbol{K}_1, \dots, \boldsymbol{K}_m\}$ be the set of kernels whose elements can be different kernels, such as linear kernel, RBF kernel, Laplacian kernel, etc. (see Chap. 3).

It is possible to determine the kernel among the sets of kernels by the following optimization problem [25, Theorem 16]:

$$\begin{aligned}
\underset{\boldsymbol{K},\boldsymbol{\nu},\boldsymbol{\delta},\lambda,t}{\text{minimize}} \quad & t \\
\text{subject to} \quad & \mathbf{tr}(\boldsymbol{K}) = c_1, \\
& \boldsymbol{K} \in \mathcal{K}, \\
& \boldsymbol{\nu} \geq \mathbf{0}, \\
& \boldsymbol{\delta} \geq \mathbf{0}, \\
& \begin{bmatrix} G(\boldsymbol{K}_{\text{tr}}) + \tau \boldsymbol{I} & \mathbf{1} + \boldsymbol{\nu} - \boldsymbol{\delta} + \lambda \boldsymbol{y} \\ (\mathbf{1} + \boldsymbol{\nu} - \boldsymbol{\delta} + \lambda \boldsymbol{y})^\top & t - 2\, c_2\, \boldsymbol{\delta}^\top \mathbf{1} \end{bmatrix} \succeq \mathbf{0},
\end{aligned} \tag{10.14}$$

where c_1, c_2, and τ are constants, and the (i, j)-th element of $G(\boldsymbol{K}_{\text{tr}})$ is defined as $G_{ij}(\boldsymbol{K}_{\text{tr}}) := y_i y_j \boldsymbol{K}_{ij}$. Note that $\boldsymbol{K}_{ij}$ is the (i, j)-th element of the kernel matrix and a binary classification is considered, so the labels are $y_i \in \{-1, 1\}, \forall i$. The derivation of this problem is available in [23, Appendix B]. This optimization problem is in the form of an SDP problem [40]. See Chap. 4 for more information on SDP and how it is solved. As it is an SDP, it is a convex problem and has only one local optimum [11].

After learning the kernel $\boldsymbol{K} \in \mathcal{K}$, it can be used to predict the labels of the unlabelled part of the data, i.e., the test dataset. Lanckriet et al. [25] use a kernel Support Vector Machine (SVM) [41] for predicting labels using the learned kernel. Using kernelization techniques (see Chap. 3), the predictor of the labels in a kernel SVM becomes:

$$f(\boldsymbol{x}) = \sum_{i=1}^{n} \alpha_i\, k(\boldsymbol{x}_i, \boldsymbol{x}) + b, \tag{10.15}$$

where $k(.,.)$ is the kernel function that determines the elements of the kernel matrix $\boldsymbol{K}_{ij} = k(\boldsymbol{x}_i, \boldsymbol{x}_j)$. Additionally, $\boldsymbol{\alpha} = [\alpha_1, \dots, \alpha_n]^\top$ and b is the bias. By using the learned kernel $\boldsymbol{K}$ in Eq. (10.15), it is possible to predict the labels of unlabelled test data. As mentioned at the start of this section, this kernel learning may have inspired the authors of MVU (SDE) [47] to propose kernel learning for dimensionality reduction.

10.4 Maximum Variance Unfolding (or Semidefinite Embedding)

The kernel can be learned using Semidefinite Programming (SDP) (see Chap. 4) for dimensionality reduction. The following section introduces kernel learning using SDP for the goal of dimensionality reduction and manifold unfolding.

10.4.1 Intuitions and Comparison with Kernel PCA

As seen in Sect. 10.2, most of the spectral dimensionality reduction methods can be unified as kernel PCA with different kernels. Therefore, the aim can be to learn the best kernel in dimensionality reduction for each specific dataset. Semidefinite Embedding (SDE) [44, 46–48], which was renamed later to Maximum Variance Unfolding (MVU) [45], aims to identify the best kernel that unfolds the manifold of the data to its maximum variance. It learns the best kernel for manifold unfolding using semidefinite programming. An example manifold unfolding, where MVU gradually unfolds the nonlinear manifold to its extreme variance by iterations of semidefinite programming, is shown in Fig. 10.2.

Using MVU, the embedding of the data has its maximum variance in the embedding space. In this sense, the goal is similar to the goal of PCA (see Chap. 5), but there are major differences between MVU and PCA and kernel PCA. These differences and similarities include the following:

- MVU performs embedding in the feature space or the so-called Reproducing Kernel Hilbert Space (RKHS), as does kernel PCA. However, PCA performs in the input space.
- MVU learns the optimal kernel, while kernel PCA uses a ready kernel such as the Radial Basis Function (RBF) kernel.
- MVU is a nonlinear method. Kernel PCA transforms data to the feature space and then applies a linear PCA to it. See Chap. 3 for an explanation of the difference between these two approaches.
- MVU is an iterative algorithm because it solves the semidefinite programming iteratively (see Chap. 4). However, kernel PCA and PCA are almost closed form. The word "almost" is used because solving the eigenvalue decomposition (EVD) or singular value decomposition (SVD) requires some algorithm iterations, such as the power method or Jordan's method [37]. However, the solution of EVD or SVD can be seen as a black box. The iterative solution of semidefinite programming in MVU is much more complicated and time consuming, although the task completed by MVU is more promising in manifold unfolding.

Assume there is a d-dimensional dataset $\mathcal{X} := \{\boldsymbol{x}_i \in \mathbb{R}^d\}_{i=1}^n$. The goal is to find a p-dimensional embedding of this dataset, denoted by $\mathcal{Y} := \{\boldsymbol{y}_i \in \mathbb{R}^p\}_{i=1}^n$ where $p \leq d$ and usually $p \ll d$. Let $\boldsymbol{X} := [\boldsymbol{x}_1, \ldots, \boldsymbol{x}_n] \in \mathbb{R}^{d \times n}$ and $\boldsymbol{Y} :=$

$[\boldsymbol{y}_1, \ldots, \boldsymbol{y}_n] \in \mathbb{R}^{p \times n}$. The embedding $\boldsymbol{Y}$ is supposed to be the maximum variance unfolding of the manifold of data. As mentioned before, SDE [44, 46–48] or MVU [45] performs this task. MVU embeds data in the feature space or RKHS; in other words, the embedding space is the feature space (RKHS). Therefore:

$$\boldsymbol{y}_i = \boldsymbol{\phi}(\boldsymbol{x}_i), \quad \forall i \in \{1, \ldots, n\}. \tag{10.16}$$

10.4.2 Local Isometry

Definition 10.22 (Isometric Manifolds [48]) Two Riemannian manifolds are isometric if there is a diffeomorphism, such that the metric on one of them pulls back to the metric on the other. In other words, isometry is a smooth invertible mapping, that locally resembles an affine transformation, i.e., a rotation and a translation. Therefore, isometry preserves the local distances on the manifold.

The notion of isometry between the data $\mathcal{X}$ and their embedding $\mathcal{Y}$ is used. In other words, the local structure of the data should be preserved in the embedding space [32]. The datasets $\mathcal{X}$ and $\mathcal{Y}$ are locally isometric if they have similar rotation and translation relations between neighbouring points. Let $\boldsymbol{x}_j$ and $\boldsymbol{x}_l$ be neighbours of $\boldsymbol{x}_i$ so that they form a triangle. This triangle should also exist in the low-dimensional embedding space with some rotation and translation. Therefore, for isometry, there should be equal relative angles of points:

$$(\boldsymbol{y}_i - \boldsymbol{y}_j)^\top (\boldsymbol{y}_i - \boldsymbol{y}_l) = (\boldsymbol{x}_i - \boldsymbol{x}_j)^\top (\boldsymbol{x}_i - \boldsymbol{x}_l), \tag{10.17}$$

because the inner product is proportional to the cosine of the angle. A special case of Eq. (10.17) is $l = j$:

$$\|\boldsymbol{y}_i - \boldsymbol{y}_j\|_2^2 = \|\boldsymbol{x}_i - \boldsymbol{x}_j\|_2^2. \tag{10.18}$$

The Gram (kernel) matrices of points in the input and embedding spaces are denoted by $\mathbb{R}^{n \times n} \ni \boldsymbol{G} := \boldsymbol{X}^\top \boldsymbol{X}$ and $\mathbb{R}^{n \times n} \ni \boldsymbol{K} := \boldsymbol{Y}^\top \boldsymbol{Y}$, respectively. Let $\boldsymbol{G}_{ij}$ and $\boldsymbol{K}_{ij}$ denote the (i, j)-th element of $\boldsymbol{G}$ and $\boldsymbol{K}$, respectively. Then:

$$\|\boldsymbol{x}_i - \boldsymbol{x}_j\|_2^2 = (\boldsymbol{x}_i - \boldsymbol{x}_j)^\top (\boldsymbol{x}_i - \boldsymbol{x}_j) = \boldsymbol{x}_i^\top \boldsymbol{x}_i + \boldsymbol{x}_j^\top \boldsymbol{x}_j - 2\boldsymbol{x}_i^\top \boldsymbol{x}_j = \boldsymbol{G}_{ii} + \boldsymbol{G}_{jj} - 2\boldsymbol{G}_{ij}.$$

Likewise, in the embedding space:

$$\begin{aligned}\|\boldsymbol{y}_i - \boldsymbol{y}_j\|_2^2 &= (\boldsymbol{y}_i - \boldsymbol{y}_j)^\top (\boldsymbol{y}_i - \boldsymbol{y}_j) = \boldsymbol{y}_i^\top \boldsymbol{y}_i + \boldsymbol{y}_j^\top \boldsymbol{y}_j - 2\boldsymbol{y}_i^\top \boldsymbol{y}_j \\ &\overset{(10.16)}{=} \boldsymbol{\phi}(\boldsymbol{x}_i)^\top \boldsymbol{\phi}(\boldsymbol{x}_i) + \boldsymbol{\phi}(\boldsymbol{x}_j)^\top \boldsymbol{\phi}(\boldsymbol{x}_j) - 2\boldsymbol{\phi}(\boldsymbol{x}_i)^\top \boldsymbol{\phi}(\boldsymbol{x}_j) \overset{(a)}{=} \boldsymbol{K}_{ii} + \boldsymbol{K}_{jj} - 2\boldsymbol{K}_{ij},\end{aligned} \tag{10.19}$$

where (a) is because of the kernel trick. Note that Eq. (10.19) is in fact the distance of points in RKHS equipped with kernel k [33] (see Chap. 3):

$$\|\boldsymbol{y}_i - \boldsymbol{y}_j\|_2^2 \overset{(10.16)}{=} \|\boldsymbol{\phi}(\boldsymbol{x}_i) - \boldsymbol{\phi}(\boldsymbol{x}_j)\|_k^2 = \boldsymbol{K}_{ii} + \boldsymbol{K}_{jj} - 2\boldsymbol{K}_{ij}.$$

Therefore, Eq. (10.18) is simplified to:

$$\boldsymbol{K}_{ii} + \boldsymbol{K}_{jj} - 2\,\boldsymbol{K}_{ij} = \boldsymbol{G}_{ii} + \boldsymbol{G}_{jj} - 2\,\boldsymbol{G}_{ij}. \tag{10.20}$$

A version of MVU uses the k-Nearest Neighbours (kNN) in the local isometry [47]. This version forms a kNN graph between the points of the training data points. Therefore, the k neighbours of every point in the dataset are known. Let τ_{ij} be one if $\boldsymbol{x}_j$ is a neighbour of $\boldsymbol{x}_i$ and zero otherwise:

$$\tau_{ij} := \begin{cases} 1 & \boldsymbol{x}_j \in k\mathrm{NN}(\boldsymbol{x}_i) \\ 0 & \text{Otherwise.} \end{cases} \tag{10.21}$$

Considering the neighbouring points only to have local isometry modifies Eq. (10.20) slightly:

$$\tau_{ij}(\boldsymbol{K}_{ii} + \boldsymbol{K}_{jj} - 2\,\boldsymbol{K}_{ij}) = \tau_{ij}(\boldsymbol{G}_{ii} + \boldsymbol{G}_{jj} - 2\,\boldsymbol{G}_{ij}). \tag{10.22}$$

10.4.3 Centering

It is possible to have a mean of zero for the embeddings of the points:

$$\sum_{i=1}^{n} \boldsymbol{y}_i = \boldsymbol{0}. \tag{10.23}$$

This removes the translational degree of freedom. According to Eq. (10.16), centered embedding data means centered pulled data in the feature space. According to Chap. 3, this is equivalent to double centering the kernel matrix, resulting in:

$$\frac{1}{n}\sum_{i=1}^{n}\sum_{j=1}^{n} \boldsymbol{K}_{ij} = 0 \implies \sum_{i=1}^{n}\sum_{j=1}^{n} \boldsymbol{K}_{ij} = 0. \tag{10.24}$$

Another justification for this is as follows. From Eq. (10.23):

$$\Big|\sum_{i=1}^{n} \boldsymbol{y}_i\Big|^2 = \sum_{i=1}^{n}\sum_{j=1}^{n} \boldsymbol{y}_i^\top \boldsymbol{y}_j \overset{(10.16)}{=} \sum_{i=1}^{n}\sum_{j=1}^{n} \boldsymbol{\phi}(\boldsymbol{x}_i)^\top \boldsymbol{\phi}(\boldsymbol{x}_j) \overset{(a)}{=} \sum_{i=1}^{n}\sum_{j=1}^{n} \boldsymbol{K}_{ij} \overset{(10.23)}{=} \boldsymbol{0},$$

where (a) is because of the kernel trick (see Chap. 3).

10.4.4 Positive Semidefiniteness

Thus far, kernels are being used to express embedding because it was assumed that the embedding space is the feature space. Therefore, optimization can be performed over the kernel matrix, rather than the embedding points. However, a valid Mercer kernel should be symmetric and positive semidefinite (see Chap. 3). Therefore, the kernel matrix should be constrained to belong to the cone of the semidefinite matrices:

$$\boldsymbol{K} \in \mathbb{S}^n_+, \quad \text{or} \quad \boldsymbol{K} \succeq \boldsymbol{0}. \tag{10.25}$$

10.4.5 Manifold Unfolding

The goal of MVU is unfolding the manifold of the data with its maximum variance, as shown in Fig. 10.2. According to the definition of variance, MVU maximizes the following quantitative:

$$\mathcal{T}(\boldsymbol{Y}) := \frac{1}{2n} \sum_{i=1}^{n} \sum_{j=1}^{n} \|\boldsymbol{y}_i - \boldsymbol{y}_j\|_2^2. \tag{10.26}$$

Lemma 10.2 (Boundedness of Variance of Embedding [46]) *The value of $\mathcal{T}(\boldsymbol{Y})$ is bounded.*

Proof Suppose $\eta_{ij} = 1$ if $\boldsymbol{x}_j$ is one of the k-Nearest Neighbours (kNN) of $\boldsymbol{x}_i$; otherwise, $\eta_{ij} = 0$. Let:

$$\tau := \max_{i,j} \Big(\eta_{ij} \, \|\boldsymbol{x}_i - \boldsymbol{x}_j\|_2\Big) \overset{(a)}{<} \infty, \tag{10.27}$$

where (a) is because $\|\boldsymbol{x}_i - \boldsymbol{x}_j\|_2 < \infty$. Assuming that the kNN graph is connected, the longest path is at most $n\tau$, i.e., $\|\boldsymbol{y}_i - \boldsymbol{y}_j\|_2 \leq n\tau$. Therefore, an upper bound of Eq. (10.26) is:

$$\mathcal{T}(\boldsymbol{Y}) \leq \frac{1}{2n} \sum_{i=1}^{n} \sum_{j=1}^{n} (n\tau)^2 = \frac{1}{2n} n^2 (n\tau)^2 = \frac{n^3 \tau^2}{2} \overset{(10.27)}{<} \infty.$$

□

According to Lemma 10.2, the variance of embedding is bounded so it can be maximized. Equation (10.26) can be simplified to:

$$
\begin{aligned}
\mathcal{T}(\boldsymbol{Y}) &:= \frac{1}{2n}\sum_{i=1}^{n}\sum_{j=1}^{n}\|\boldsymbol{y}_i - \boldsymbol{y}_j\|_2^2 \overset{(10.19)}{=} \frac{1}{2n}\sum_{i=1}^{n}\sum_{j=1}^{n}(\boldsymbol{K}_{ii} + \boldsymbol{K}_{jj} - 2\boldsymbol{K}_{ij}) \\
&= \frac{1}{2n}\Big[\sum_{i=1}^{n}\boldsymbol{K}_{ii} + \sum_{j=1}^{n}\boldsymbol{K}_{jj} - 2\sum_{i=1}^{n}\sum_{j=1}^{n}\boldsymbol{K}_{ij}\Big] \\
&\overset{(10.24)}{=} \frac{1}{2n}\Big[\sum_{i=1}^{n}\boldsymbol{K}_{ii} + \sum_{j=1}^{n}\boldsymbol{K}_{jj}\Big] = \frac{1}{2n}\Big[2\sum_{i=1}^{n}\boldsymbol{K}_{ii}\Big] = \frac{1}{n}\sum_{i=1}^{n}\boldsymbol{K}_{ii} \propto \sum_{i=1}^{n}\boldsymbol{K}_{ii} \overset{(a)}{=} \mathbf{tr}(\boldsymbol{K}),
\end{aligned}
\tag{10.28}
$$

where $\mathbf{tr}(.)$ denotes the trace of the matrix and (a) is because the trace of a matrix is equivalent to the summation of its diagonal elements. Equation (10.28) makes sense because the kernel is a measure of the similarity of points, so it is related to the distance of the points and the variance of unfolding.

In summary, MVU maximizes the variance of embedding, i.e. Eq. (10.28), with the constraints of Eqs. (10.20), (10.24), and (10.25). Therefore, it solves the following optimization problem:

$$
\begin{aligned}
& \underset{\boldsymbol{K}}{\text{maximize}} \quad \mathbf{tr}(\boldsymbol{K}) \\
& \text{subject to} \quad \boldsymbol{K}_{ii} + \boldsymbol{K}_{jj} - 2\,\boldsymbol{K}_{ij} = \boldsymbol{G}_{ii} + \boldsymbol{G}_{jj} - 2\,\boldsymbol{G}_{ij}, \\
& \qquad\qquad\qquad\qquad \forall i, j \in \{1, \dots, n\}, \\
& \qquad\quad \sum_{i=1}^{n}\sum_{j=1}^{n}\boldsymbol{K}_{ij} = 0, \\
& \qquad\quad \boldsymbol{K} \succeq \boldsymbol{0},
\end{aligned}
\tag{10.29}
$$

which is a semidefinite programming problem [40] (see Chap. 4). Equation (10.29) is a convex optimization problem so it has only one local optimum, which is the global optimum [11].

Equation (10.29) uses Eq. (10.20) for local isometry. Some versions of MVU use Eq. (10.22) as the local isometry constraint:

$$
\begin{aligned}
& \underset{\boldsymbol{K}}{\text{maximize}} \quad \mathbf{tr}(\boldsymbol{K}) \\
& \text{subject to} \quad \tau_{ij}(\boldsymbol{K}_{ii} + \boldsymbol{K}_{jj} - 2\,\boldsymbol{K}_{ij}) = \tau_{ij}(\boldsymbol{G}_{ii} + \boldsymbol{G}_{jj} - 2\,\boldsymbol{G}_{ij}), \\
& \qquad\qquad\qquad\qquad \forall i, j \in \{1, \dots, n\}, \\
& \qquad\quad \sum_{i=1}^{n}\sum_{j=1}^{n}\boldsymbol{K}_{ij} = 0, \\
& \qquad\quad \boldsymbol{K} \succeq \boldsymbol{0},
\end{aligned}
\tag{10.30}
$$

Note that if the entire dataset is seen as one class and there is $k = n$, Eq. (10.30) becomes equivalent to Eq. (10.29). Using kNN in Eq. (10.30) makes the optimization slightly more efficient because it does not compute the kernel between all points, although computation of the kNN graph can be time-consuming.

10.4.6 Spectral Embedding

After the kernel is found by solving optimization problem (10.29), it is possible to calculate the eigenvalues and eigenvectors of the kernel, and then the embedding of every point is obtained through Eq. (10.31) in the following lemma. The obtained embedding dataset has the maximum variance in the embedding space. To find the intrinsic dimensionality of a manifold, denoted by $p \leq n$, the eigenvalues are sorted from largest to smallest; then, a large separation between two successive eigenvalues indicates a good cut-off for the number of required dimensions. A scree plot can be used to visualize this cut-off. For better understanding of intrinsic dimensionality, see the examples in Fig. 10.1.

Lemma 10.3 (Embedding from Eigenvectors of Kernel Matrix) *Let $\boldsymbol{v}_k = [v_{k1}, \dots, v_{kn}]^\top$ and δ_k be the k-th eigenvector and eigenvalue of the kernel matrix, respectively. It is possible to compute the embedding of point $\boldsymbol{x}$, denoted by $\boldsymbol{y}(\boldsymbol{x}) = [y_1(\boldsymbol{x}), \dots, y_p(\boldsymbol{x})]^\top$ (where $p \leq n$) using the eigenvector of the kernel as:*

$$y_k(\boldsymbol{x}) = \sqrt{\delta_k}\, v_{ki}. \tag{10.31}$$

Proof See Chap. 3 for proof. □

An additional way to justify Eq. (10.31) is by using the eigenvalue decomposition of the kernel matrix (see Chaps. 2 and 3) [46]:

$$\boldsymbol{K} = \boldsymbol{V}\boldsymbol{\Delta}\boldsymbol{V}^\top \implies \boldsymbol{K}_{ij} = \sum_{k=1}^{n} \delta_k\, v_{ki}\, v_{kj}, \quad \forall i, j \in \{1, \dots, n\}, \tag{10.32}$$

where the columns of $\boldsymbol{V}$ are the eigenvectors and the diagonal elements of $\boldsymbol{\Delta}$ are eigenvalues. Additionally, the kernel can be stated as follows (see Chap. 3):

$$\boldsymbol{K}_{ij} = \boldsymbol{\phi}(\boldsymbol{x}_i)^\top \boldsymbol{\phi}(\boldsymbol{x}_j) \overset{(10.16)}{=} \boldsymbol{y}_i^\top \boldsymbol{y}_j. \tag{10.33}$$

Considering both Eqs. (10.32) and (10.33) results in:

$$\boldsymbol{y}_i^\top \boldsymbol{y}_j = \sum_{k=1}^{n} \delta_k\, v_{ki}\, v_{kj} \overset{(a)}{=} \sum_{k=1}^{n} \sqrt{\delta_k}\, v_{ki} \sqrt{\delta_k}\, v_{kj} \implies \boldsymbol{y}_i(k) = \sqrt{\delta_k}\, v_{ki},$$

where $\boldsymbol{y}_i(k)$ is the k-th element of $\boldsymbol{y}_i$ and (a) is allowed because the kernel matrix is positive semidefinite (see Eq. (10.25)), meaning its eigenvalues, δ_k's, are nonnegative. This equation is equal to Eq. (10.31).

10.5 Supervised Maximum Variance Unfolding

MVU is an unsupervised manifold learning method. Several variants exist for supervised MVU or supervised SDE that make use of class labels. In the following section, these methods are introduced.

10.5.1 Supervised MVU Using kNN Within Classes

One of the methods for supervised MVU or supervised SDE was proposed by Ma et al. [26, 52], which uses k-Nearest Neighbours (kNN) within the classes. It modifies Eq. (10.21) in a way that the kNN graph is formed between points of every class and not among all of the data points. Therefore, the k neighbours of every point in each class are known. Let τ_{ij} be one, if $\boldsymbol{x}_i$ and $\boldsymbol{x}_i$ belong to the same class and $\boldsymbol{x}_j$ is a neighbour of $\boldsymbol{x}_i$; otherwise, it is zero. The optimization problem of MVU is the same as Eq. (10.30), where τ_{ij} has been computed differently.

10.5.2 Supervised MVU by Classwise Unfolding

SMVU1, proposed in [43], is a supervised MVU method that unfolds the manifold classwise. Let G_c denote the set of points belonging to class c and $\bar{\boldsymbol{x}}_c$ be the mean of class c. For the class c, the representative $\boldsymbol{x}_c$ is defined as being the closest point of the class to the mean of the class:

$$\boldsymbol{x}_c := \min_{\boldsymbol{x}_i \in G_c} \|\boldsymbol{x}_i - \bar{\boldsymbol{x}}_c\|_2^2. \tag{10.34}$$

Note that, rather than the above definition, $\boldsymbol{x}_c$ can be defined to be the medoid of the class, which is the closest point to all of the points of the class; however, computation of medoid can be more time-consuming.

Assume there are C classes denoted by $\{c_1, \ldots, c_C\}$. Recall that the local isometry yielded $\boldsymbol{K}_{ii} + \boldsymbol{K}_{jj} - 2\,\boldsymbol{K}_{ij} = \|\boldsymbol{x}_i - \boldsymbol{x}_j\|_2^2$. For classwise local isometry, it is possible to have:

$$\boldsymbol{K}_{c_i c_i} + \boldsymbol{K}_{c_j c_j} - 2\,\boldsymbol{K}_{c_i c_j} = \alpha^2 \|\boldsymbol{x}_{c_i} - \boldsymbol{x}_{c_j}\|_2^2, \tag{10.35}$$

$\forall i, j \in \{1, \dots, C\}$, where $\alpha > 1$ is a hyperparameter [43] and $\boldsymbol{K}_{c_i c_j} := \boldsymbol{\phi}(\boldsymbol{x}_{c_i})^\top \boldsymbol{\phi}(\boldsymbol{x}_{c_j})$ is the kernel between the representatives of classes i and j. Consider the following pairs of classes [43]:

$$\boldsymbol{K}_{c_i c_i} + \boldsymbol{K}_{c_{i+1} c_{i+1}} - 2\,\boldsymbol{K}_{c_i c_{i+1}} = \alpha^2 \|\boldsymbol{x}_{c_i} - \boldsymbol{x}_{c_j}\|_2^2,$$
$$\boldsymbol{K}_{c_i c_i} + \boldsymbol{K}_{c_C c_C} - 2\,\boldsymbol{K}_{c_i c_C} = \alpha^2 \|\boldsymbol{x}_{c_i} - \boldsymbol{x}_{C}\|_2^2,$$

$\forall i, j \in \{1, \dots, C\}$. We define:

$$\Gamma_c = \frac{1}{2} \sum_{i=1}^{n_c} \sum_{j=1}^{n_c} (\boldsymbol{K}_{ii} + \boldsymbol{K}_{jj} - 2\boldsymbol{K}_{ij}), \ \forall \boldsymbol{x}_i, \boldsymbol{x}_j \in G_c, \tag{10.36}$$

$$\Gamma := \sum_{c=1}^{C} \frac{\Gamma_c}{n_c}. \tag{10.37}$$

where n_c denotes the number of points in class c. This term is maximized to maximize the variance of unfolding for each class. The SDP optimization for kernel learning is [43]:

$$\begin{aligned}
& \underset{\boldsymbol{K}}{\text{maximize}} \quad \sum_{c=1}^{C} \frac{\Gamma_c}{n_c} \\
& \text{subject to} \\
& \tau_{ij}(\boldsymbol{K}_{ii} + \boldsymbol{K}_{jj} - 2\,\boldsymbol{K}_{ij}) = \tau_{ij}(\boldsymbol{G}_{ii} + \boldsymbol{G}_{jj} - 2\,\boldsymbol{G}_{ij}), \\
& \qquad\qquad\qquad\qquad \forall i, j \in \{1, \dots, n\}, \\
& \boldsymbol{K}_{c_i c_i} + \boldsymbol{K}_{c_{i+1} c_{i+1}} - 2\,\boldsymbol{K}_{c_i c_{i+1}} = \alpha^2 \|\boldsymbol{x}_{c_i} - \boldsymbol{x}_{c_j}\|_2^2, \\
& \boldsymbol{K}_{c_i c_i} + \boldsymbol{K}_{c_C c_C} - 2\,\boldsymbol{K}_{c_i c_C} = \alpha^2 \|\boldsymbol{x}_{c_i} - \boldsymbol{x}_{C}\|_2^2, \\
& \qquad\qquad\qquad\qquad \forall i, j \in \{1, \dots, C\}, \\
& \sum_{i=1}^{n} \sum_{j=1}^{n} \boldsymbol{K}_{ij} = 0, \\
& \boldsymbol{K} \succeq \boldsymbol{0},
\end{aligned} \tag{10.38}$$

where τ_{ij} is the same as defined before in Eq. (10.21).

10.5.3 Supervised MVU by Fisher Criterion

The method named SMVU2 proposed in [43] is a supervised MVU method, which unfolds manifold by the Fisher criterion. Therefore, this method is named the Fisher-MVU. The Fisher criterion maximizes the between-class scatter and minimizes the within-class scatter. The within-class scatter in the embedding space (or feature space equipped with kernel k) is:

$$\sigma_W := \sum_{c=1}^{C} \frac{1}{n_c} \sum_{\boldsymbol{x}_i \in G_c} \|\boldsymbol{\phi}(\boldsymbol{x}_i) - \boldsymbol{\phi}(\boldsymbol{x}_c)\|_k^2 \overset{(a)}{=} \sum_{c=1}^{C} \frac{1}{n_c} \sum_{\boldsymbol{x}_i \in G_c} (\boldsymbol{K}_{ii} + \boldsymbol{K}_{cc} - 2\boldsymbol{K}_{ci}), \tag{10.39}$$

where $\boldsymbol{K}_{ii} := \boldsymbol{\phi}(\boldsymbol{x}_i)^\top \boldsymbol{\phi}(\boldsymbol{x}_i)$, $\boldsymbol{K}_{cc} := \boldsymbol{\phi}(\boldsymbol{x}_c)^\top \boldsymbol{\phi}(\boldsymbol{x}_c)$, and $\boldsymbol{K}_{ci} := \boldsymbol{\phi}(\boldsymbol{x}_c)^\top \boldsymbol{\phi}(\boldsymbol{x}_i)$, and (a) is because of the distance in the feature space [33] (see Chap. 3). The between-class scatter in the embedding space (or feature space equipped with kernel k) is:

$$\sigma_B := \sum_{c_i=1}^{C} \sum_{c_j=1}^{C} \|\boldsymbol{\phi}(\boldsymbol{x}_{c_i}) - \boldsymbol{\phi}(\boldsymbol{x}_{c_j})\|_k^2 \overset{(a)}{=} \sum_{c_i=1}^{C} \sum_{c_j=1}^{C} (\boldsymbol{K}_{c_i c_i} + \boldsymbol{K}_{c_j c_j} - 2\boldsymbol{K}_{c_i c_j}). \tag{10.40}$$

where (a) is because of the distance in the feature space [33]. Note that the between-class scatter has another form that uses the mean of classes. However, as the mean of a class is not necessarily one of the points, its embedding (in the feature space) is not available. Therefore, the scatter of all points from the classes is used for computation of the between-class scatter. One of the forms of the Fisher criterion is [17]:

$$\Gamma = C \times (\sigma_B - \sigma_W), \tag{10.41}$$

which should be maximized. The SDP optimization for kernel learning is [43]:

$$\begin{aligned}
\underset{\boldsymbol{K}}{\text{maximize}} \quad & C(\sigma_B - \sigma_W) \\
\text{subject to} \quad & \tau_{ij}(\boldsymbol{K}_{ii} + \boldsymbol{K}_{jj} - 2\,\boldsymbol{K}_{ij}) \\
& \qquad = \tau_{ij}(\boldsymbol{G}_{ii} + \boldsymbol{G}_{jj} - 2\,\boldsymbol{G}_{ij}), \\
& \qquad\qquad \forall i, j \in \{1, \dots, n\}, \\
& \sum_{i=1}^{n} \sum_{j=1}^{n} \boldsymbol{K}_{ij} = 0, \\
& \boldsymbol{K} \succeq \boldsymbol{0}.
\end{aligned} \tag{10.42}$$

10.5.4 Supervised MVU by Coloured MVU

Coloured MVU [36], one of the supervised approaches for MVU, uses some side information such as labels. In its formulation, it uses the Hilbert-Schmidt Independence Criterion (HSIC) [20] between the embedded data points $\{\boldsymbol{y}_i = \boldsymbol{\phi}(\boldsymbol{x}_i)\}_{i=1}^n$ and the labels $\{l_i\}_{i=1}^n$ (see Chap. 3):

$$\text{HSIC} := \frac{1}{(n-1)^2}\,\mathbf{tr}(\boldsymbol{K}_l \boldsymbol{H}\boldsymbol{K}\boldsymbol{H}) \overset{(a)}{=} \frac{1}{(n-1)^2}\,\mathbf{tr}(\boldsymbol{H}\boldsymbol{K}\boldsymbol{H}\boldsymbol{K}_l), \tag{10.43}$$

where (a) is because of the cyclic property of the trace, $\boldsymbol{K} \in \mathbb{R}^{n\times n}$ and $\boldsymbol{K}_l \in \mathbb{R}^{n\times n}$ are the kernel matrices over the embedded points and labels, respectively, and $\mathbb{R}^{n\times n} \ni \boldsymbol{H} := \boldsymbol{I} - (1/n)\mathbf{1}\mathbf{1}^\top$ is the centering matrix. HSIC is a measure of the dependence between two random variables. Coloured MVU maximizes the HSIC between the embedded points and the labels; in other words, the dependence of the embedding and labels is maximized to be supervised. Therefore, the coloured MVU maximizes Eq. (10.43), i.e., $\mathbf{tr}(\boldsymbol{H}\boldsymbol{K}\boldsymbol{H}\boldsymbol{K}_l)$ or $\mathbf{tr}(\boldsymbol{K}\boldsymbol{H}\boldsymbol{K}_l\boldsymbol{H})$, rather than maximizing $\mathbf{tr}(\boldsymbol{K})$, which is done in MVU [36]:

$$\begin{aligned}
\underset{\boldsymbol{K}}{\text{maximize}} \quad & \mathbf{tr}(\boldsymbol{H}\boldsymbol{K}\boldsymbol{H}\boldsymbol{K}_l) \\
\text{subject to} \quad & \tau_{ij}(\boldsymbol{K}_{ii} + \boldsymbol{K}_{jj} - 2\,\boldsymbol{K}_{ij}) \\
& \quad = \tau_{ij}(\boldsymbol{G}_{ii} + \boldsymbol{G}_{jj} - 2\,\boldsymbol{G}_{ij}), \\
& \qquad \forall i, j \in \{1, \dots, n\}, \\
& \sum_{i=1}^n \sum_{j=1}^n \boldsymbol{K}_{ij} = 0, \\
& \boldsymbol{K} \succeq \mathbf{0}.
\end{aligned} \tag{10.44}$$

Note that as the kernel is a soft measure of similarity, the labels or side information $\{l_i\}_{i=1}^n$ can be soft labels (e.g., regression labels) or hard labels (e.g., classification labels). In the case of hard labels, one of the best choices for the $\boldsymbol{K}_l$ is the delta kernel [2], where the (i, j)-th element of the kernel is:

$$\boldsymbol{K}_l(i, j) = \delta_{l_i, l_j} := \begin{cases} 1 \text{ if } \boldsymbol{l}_i = \boldsymbol{l}_j, \\ 0 \text{ if } \boldsymbol{l}_i \neq \boldsymbol{l}_j, \end{cases} \tag{10.45}$$

where δ_{l_i, l_j} is the Kronecker delta, which is one if the $\boldsymbol{x}_i$ and $\boldsymbol{x}_j$ belong to the same class.

10.6 Out-of-Sample Extension of MVU

There are several approaches for the out-of-sample extension of MVU. The following section introduces some of these approaches, which use eigenfunctions [13] and kernel mapping [18]. Additional methods can be used for out-of-sample extension, such as paper [12], which are not covered here.

10.6.1 Out-of-Sample Extension Using Eigenfunctions

One method for out-of-sample extension of MVU is using eigenfunctions [13]. Recall Eq. (10.2) which relates the eigenvectors of the kernel function and eigenfunctions of the kernel operator. According to Schwaighofer et al. [35], this equation can be slightly modified to:

$$f_k(\boldsymbol{x}) = \sum_{i=1}^{n} b_{ki}\, r(\boldsymbol{x}_i, \boldsymbol{x}) = \sum_{i=1}^{n} b_{ki}\, r(\boldsymbol{x}, \boldsymbol{x}_i), \tag{10.46}$$

where $r(.,.)$ is an auxiliary smoothing kernel, such as the RBF kernel, and b_{ki} is the i-th element of:

$$\mathbb{R}^n \ni \boldsymbol{b}_k := (\boldsymbol{R} + \eta \boldsymbol{I})^{-1} \boldsymbol{v}_k, \tag{10.47}$$

where $\boldsymbol{v}_k$ is the k-th eigenvector of the kernel matrix $\boldsymbol{K} \in \mathbb{R}^{n\times n}$ over the training data, and $\boldsymbol{R} \in \mathbb{R}^{n\times n}$ is the smoothing kernel matrix on the n training data points using the kernel $r(.,.)$, and $\eta > 0$ is a regularization parameter for the stable inverse of the matrix.

According to the Mercer's theorem (see Chap. 3), the kernel can be written as [13]:

$$k(\boldsymbol{x}, \boldsymbol{x}_i) = \sum_{j=1}^{n} \lambda_j \psi_j(\boldsymbol{x}) \psi_j(\boldsymbol{x}_i) \overset{(a)}{=} \boldsymbol{r}(\boldsymbol{x})^\top (\boldsymbol{R} + \eta \boldsymbol{I})^{-1} \boldsymbol{K} (\boldsymbol{R} + \eta \boldsymbol{I})^{-1} \boldsymbol{r}(\boldsymbol{x}_i) \tag{10.48}$$

where (a) is because of Eqs. (10.46) and (10.47), $\boldsymbol{r}(\boldsymbol{x}) := [r(\boldsymbol{x}_1, \boldsymbol{x}), \dots, r(\boldsymbol{x}_n, \boldsymbol{x})]^\top \in \mathbb{R}^n$, $\{\psi_j\}_{j=1}^n$ and $\{\lambda_j\}_{j=1}^n$ are the eigenfunctions and eigenvalues of the kernel operator k, respectively. According to Eq. (10.4):

$$y_k(\boldsymbol{x}) = \frac{1}{\sqrt{\delta_k}} \sum_{i=1}^{n} v_{ki}\, k(\boldsymbol{x}, \boldsymbol{x}_i) \overset{(10.48)}{=} \boldsymbol{p}_k\, \boldsymbol{r}(\boldsymbol{x}), \tag{10.49}$$

where δ_k is the k-th eigenvalue of the kernel matrix $\boldsymbol{K}$ and [13]:

$$\mathbb{R}^{1\times n} \ni \boldsymbol{p}_k := \delta_k^{-1/2} \boldsymbol{v}_k^\top \boldsymbol{R}(\boldsymbol{R} + \eta \boldsymbol{I})^{-1} \boldsymbol{K}(\boldsymbol{R} + \eta \boldsymbol{I})^{-1}. \tag{10.50}$$

Therefore, for a point $\boldsymbol{x}$, which can be an out-of-sample point, the k-th dimension of the embedding is calculated using Eq. (10.49). Considering the top p dimensions of the embedding, where $p \leq n$, results in $\mathbb{R}^p \ni \boldsymbol{y}(\boldsymbol{x}) = [y_1(\boldsymbol{x}), \ldots, y_p(\boldsymbol{x})]^\top$.

10.6.2 Out-of-Sample Extension Using Kernel Mapping

The kernel mapping method [18] can embed the out-of-sample data in MVU. A map that maps any data point as $\boldsymbol{x} \mapsto \boldsymbol{y}(\boldsymbol{x})$ can be defined as:

$$\mathbb{R}^p \ni \boldsymbol{y}(\boldsymbol{x}) := \sum_{j=1}^{n} \boldsymbol{\alpha}_j \frac{k(\boldsymbol{x}, \boldsymbol{x}_j)}{\sum_{\ell=1}^{n} k(\boldsymbol{x}, \boldsymbol{x}_\ell)}, \tag{10.51}$$

and $\boldsymbol{\alpha}_j \in \mathbb{R}^p$, and $\boldsymbol{x}_j$ and $\boldsymbol{x}_\ell$ denote the j-th and ℓ-th training data points, respectively. $k(\boldsymbol{x}, \boldsymbol{x}_j)$ is a kernel, such as the Gaussian kernel:

$$k(\boldsymbol{x}, \boldsymbol{x}_j) = \exp\Big(\frac{-||\boldsymbol{x} - \boldsymbol{x}_j||_2^2}{2\sigma_j^2}\Big), \tag{10.52}$$

where σ_j is calculated as [18]:

$$\sigma_j := \gamma \times \min_i(||\boldsymbol{x}_j - \boldsymbol{x}_i||_2), \tag{10.53}$$

where γ is a small positive number.

Assume the training data points have already been embedded using MVU; therefore, the set $\{\boldsymbol{y}_i\}_{i=1}^n$ is available. If the training data points are mapped, the aim is to minimize the following least-squares cost function to obtain $\boldsymbol{y}(\boldsymbol{x}_i)$ close to $\boldsymbol{y}_i$ for the i-th training point:

$$\underset{\boldsymbol{\alpha}_j\text{'s}}{\text{minimize}} \quad \sum_{i=1}^{n} ||\boldsymbol{y}_i - \boldsymbol{y}(\boldsymbol{x}_i)||_2^2, \tag{10.54}$$

where the summation is over the training data points. The cost function can be written in the matrix form:

$$\underset{\boldsymbol{A}}{\text{minimize}} \quad ||\boldsymbol{Y} - \boldsymbol{K}''\boldsymbol{A}||_F^2, \tag{10.55}$$

where $\mathbb{R}^{n\times p} \ni \boldsymbol{Y} := [\boldsymbol{y}_1, \ldots, \boldsymbol{y}_n]^\top$ and $\mathbb{R}^{n\times p} \ni \boldsymbol{A} := [\boldsymbol{\alpha}_1, \ldots, \boldsymbol{\alpha}_n]^\top$. The $\boldsymbol{K}'' \in \mathbb{R}^{n\times n}$ is the kernel matrix, whose (i, j)-th element is defined as:

$$\boldsymbol{K}''(i, j) := \frac{k(\boldsymbol{x}_i, \boldsymbol{x}_j)}{\sum_{\ell=1}^{n} k(\boldsymbol{x}_i, \boldsymbol{x}_\ell)}. \tag{10.56}$$

Equation (10.55) is always nonnegative; thus, its smallest value is zero. Therefore, the solution to this equation is:

$$\boldsymbol{Y} - \boldsymbol{K}''\boldsymbol{A} = \boldsymbol{0} \implies \boldsymbol{Y} = \boldsymbol{K}''\boldsymbol{A} \overset{(a)}{\implies} \boldsymbol{A} = \boldsymbol{K}''^{\dagger}\,\boldsymbol{Y}, \tag{10.57}$$

where $\boldsymbol{K}''^{\dagger}$ is the pseudoinverse of $\boldsymbol{K}''$:

$$\boldsymbol{K}''^{\dagger} = (\boldsymbol{K}''^{\top}\boldsymbol{K}'')^{-1}\boldsymbol{K}''^{\top}, \tag{10.58}$$

and (a) is because $\boldsymbol{K}''^{\dagger}\,\boldsymbol{K}'' = \boldsymbol{I}$.

Lastly, the mapping of Eq. (10.51) for the n_t out-of-sample data points is:

$$\boldsymbol{Y}_t = \boldsymbol{K}_t''\,\boldsymbol{A}, \tag{10.59}$$

where the (i, j)-th element of the out-of-sample kernel matrix $\boldsymbol{K}_t'' \in \mathbb{R}^{n_t \times n}$ is:

$$\boldsymbol{K}_t''(i, j) := \frac{k(\boldsymbol{x}_i^{(t)}, \boldsymbol{x}_j)}{\sum_{\ell=1}^{n} k(\boldsymbol{x}_i^{(t)}, \boldsymbol{x}_\ell)}, \tag{10.60}$$

where $\boldsymbol{x}_i^{(t)}$ is the i-th out-of-sample data point, and $\boldsymbol{x}_j$ and $\boldsymbol{x}_\ell$ are the j-th and ℓ-th training data points.

10.7 Other Variants of Maximum Variance Unfolding

10.7.1 Action Respecting Embedding

Most dimensionality reduction methods, including MVU, do not consider the order of points. However, some data may consist of temporal information, for example, the frames of a video where the order of images matters. In this case, a variant of MVU is required that considers the temporal information when unfolding the manifold of data. Action Respecting Embedding (ARE) [9] is an MVU variant that places importance on the order of points. It has various applications in reinforcement learning and robotics [10]. ARE has the same constraints in the optimization of MVU, plus an additional constraint of temporal information. Assume two actions (e.g., rotation or transformation of an image or a combination of rotation and transformation), denoted by a_i and a_j, are applied on the data points $\boldsymbol{x}_i$ and $\boldsymbol{x}_j$ which results in $\boldsymbol{x}_{i+1}$ and $\boldsymbol{x}_{j+1}$, respectively:

$$\boldsymbol{x}_i \xrightarrow{a_i} \boldsymbol{x}_{i+1},$$

$$\boldsymbol{x}_j \xrightarrow{a_j} \boldsymbol{x}_{j+1}.$$

Consider the same actions in the embedding space of MVU, i.e. the feature space:

$$\boldsymbol{\phi}(\boldsymbol{x}_i) \xrightarrow{a_i} \boldsymbol{\phi}(\boldsymbol{x}_{i+1}),$$

$$\boldsymbol{\phi}(\boldsymbol{x}_j) \xrightarrow{a_j} \boldsymbol{\phi}(\boldsymbol{x}_{j+1}).$$

If the two actions a_i and a_j are equal, the distances of the embedded points should remain the same before and after the action:

$$\begin{aligned} a_i = a_j &\implies \|\boldsymbol{\phi}(\boldsymbol{x}_i) - \boldsymbol{\phi}(\boldsymbol{x}_j)\|_k^2 = \|\boldsymbol{\phi}(\boldsymbol{x}_{i+1}) - \boldsymbol{\phi}(\boldsymbol{x}_{j+1})\|_k^2 \\ &\overset{(a)}{\implies} \boldsymbol{K}_{ii} + \boldsymbol{K}_{jj} - 2\boldsymbol{K}_{ij} = \boldsymbol{K}_{i+1,i+1} + \boldsymbol{K}_{j+1,j+1} - 2\boldsymbol{K}_{i+1,j+1}, \end{aligned} \tag{10.61}$$

where (a) is because of the distance in the feature space [33] (see Chap. 3). Lastly, given actions $\{a_i\}_{i=1}^n$, the optimization in ARE is:

$$\begin{aligned} &\underset{\boldsymbol{K}}{\text{maximize}} \quad \mathbf{tr}(\boldsymbol{K}) \\ &\text{subject to} \\ &\boldsymbol{K}_{ii} + \boldsymbol{K}_{jj} - 2\,\boldsymbol{K}_{ij} = \boldsymbol{G}_{ii} + \boldsymbol{G}_{jj} - 2\,\boldsymbol{G}_{ij}, \\ &\qquad\qquad\qquad\qquad \forall i, j \in \{1, \dots, n\}, \\ &\boldsymbol{K}_{ii}+\boldsymbol{K}_{jj}- 2\boldsymbol{K}_{ij} = \boldsymbol{K}_{i+1,i+1}+\boldsymbol{K}_{j+1,j+1}- 2\boldsymbol{K}_{i+1,j+1}, \\ &\qquad\qquad\qquad\qquad \forall i, j : a_i = a_j, \\ &\sum_{i=1}^{n}\sum_{j=1}^{n} \boldsymbol{K}_{ij} = 0, \\ &\boldsymbol{K} \succeq \boldsymbol{0}, \end{aligned} \tag{10.62}$$

which is an SDP problem. The solution of this problem results in a kernel for unfolding the manifold of data where the temporal information of the actions is taken into account. After finding the kernel from the optimization, the unfolded embedding is calculated by Eq. (10.31).

10.7.2 Relaxed Maximum Variance Unfolding

Two problems exist in MVU—short circuits in the kNN graph and rescaling local distances. These problems are addressed and resolved in the relaxed MVU [22]. The following sections explain these problems and how relaxed MVU resolves them.

10.7.2.1 Short Circuits in the kNN Graph

For some k values, short circuits may occur in the kNN graph. As seen in Fig. 10.3, these short circuits result in the incorrect unfolding of the manifold. Let $k\text{NN}(\boldsymbol{x}_i)$ denote the set of k nearest neighbour points of $\boldsymbol{x}_i$. Consider a kNN of data where $v(i, j)$ is an edge between $\boldsymbol{x}_i$ and $\boldsymbol{x}_j$ in this graph. The deviation of an edge $v(i, j)$ is defined as [22]:

$$d(v(i,j)) := \frac{1}{|k\text{NN}(\boldsymbol{x}_i) \cup k\text{NN}(\boldsymbol{x}_j)|} \times \sum_{\boldsymbol{x}_l \in k\text{NN}(\boldsymbol{x}_i) \cup k\text{NN}(\boldsymbol{x}_j)} \|\boldsymbol{x}_l - \boldsymbol{x}_{ij}^{(\text{mid})}\|_2^2, \tag{10.63}$$

where $\boldsymbol{x}_{ij}^{(\text{mid})} := (\boldsymbol{x}_i + \boldsymbol{x}_j)/2$ and $|.|$ denotes the size of the set. This deviation is related to the density of the points; that is, the lower the density is, the larger the deviation is. The deviations of points are sorted from smallest to largest and the points from kNN, which have larger deviation than a threshold, can be discarded. A scree plot can be used to find the suitable threshold.

10.7.2.2 Rescaling Local Distances

In some cases, the mapping is conformal but not isometric. Conformal maps are locally isometric, but up to a scale. Let $s(\boldsymbol{x}_i)$ denote the average distance of $\boldsymbol{x}_i$ to its k nearest neighbours. Assuming that the original sampling in the input space is

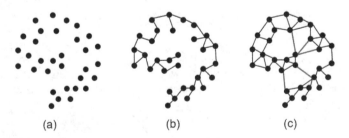

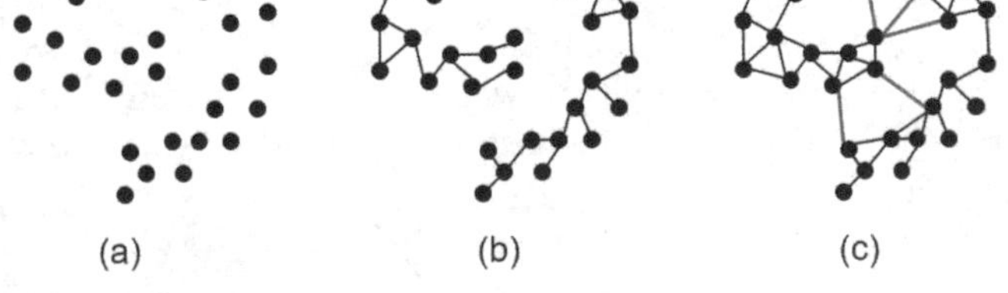

Fig. 10.3 Visualizing the possible problem of short circuit: (**a**) data points lying on a Swiss roll manifold, (**b**) Correct kNN graph for the manifold, (**c**) and incorrect kNN graph having short circuits (shown by red edges) by some larger value for k

uniform, the conformal scaling factor between points $\boldsymbol{x}_i$ and $\boldsymbol{x}_j$ is $(s(\boldsymbol{x}_i)s(\boldsymbol{x}_j))^{1/2}$ [16]. Relaxed MVU is robust to violation of this assumption [22]. Relaxed MVU scales the distances of points in the local isometry constraint:

$$\tau_{ij}(\boldsymbol{K}_{ii} + \boldsymbol{K}_{jj} - 2\,\boldsymbol{K}_{ij}) = \tau_{ij}\|(s(\boldsymbol{x}_i)s(\boldsymbol{x}_j))^{1/2}(\boldsymbol{x}_i - \boldsymbol{x}_j)\|_2^2. \tag{10.64}$$

10.7.3 Landmark Maximum Variance Unfolding for Big Data

As explained in Chap. 4, MVU uses an interior-point method, for solving SPD, which is slow especially for big data. Landmark SDE or Landmark MVU [44] uses randomly selected landmarks from the data points. Let n and $m \ll n$ denote the total number of points and the number of landmarks, respectively, and let $\{\boldsymbol{\mu}_\alpha\}_{\alpha=1}^m \subset \{\boldsymbol{x}_i\}_{i=1}^n$ be the landmarks. Every point can be reconstructed linearly by the landmarks:

$$\widehat{\boldsymbol{x}}_i = \sum_{\alpha=1}^{m} q_{i\alpha}\,\boldsymbol{\mu}_\alpha, \tag{10.65}$$

where $q_{i\alpha}, \forall i, \alpha$ are the reconstruction weights. Inspired by Locally Linear embedding (LLE) (see Chap. 8), every embedded point should also be reconstructed from the embedded landmarks, denoted by $\{\boldsymbol{\ell}_\alpha\}_{\alpha=1}^m = \{\boldsymbol{\phi}(\boldsymbol{\mu}_\alpha)\}_{\alpha=1}^m$, with the same reconstruction weights:

$$\widehat{\boldsymbol{y}}_i = \sum_{\alpha=1}^{m} q_{i\alpha}\,\boldsymbol{\ell}_\alpha. \tag{10.66}$$

The kernel can be stated as the inner product of the points in the feature space (see Chap. 3):

$$\begin{aligned}
&\boldsymbol{K}_{ij} = \boldsymbol{\phi}(\boldsymbol{x}_i)^\top\boldsymbol{\phi}(\boldsymbol{x}_j) \overset{(10.16)}{=} \boldsymbol{y}_i^\top\boldsymbol{y}_j \implies \boldsymbol{K}_{ij} \approx \widehat{\boldsymbol{y}}_i^\top\widehat{\boldsymbol{y}}_j.\\
&\overset{(10.66)}{\implies} \boldsymbol{K}_{ij} \approx q_{i\alpha}\boldsymbol{\ell}_\alpha^\top\boldsymbol{\ell}_\beta q_{i\beta} \implies \boldsymbol{K} = \boldsymbol{V}\boldsymbol{L}\boldsymbol{V}^\top,
\end{aligned} \tag{10.67}$$

where $\boldsymbol{Q}_{i\alpha} = q_{i\alpha}$, $\boldsymbol{Q} \in \mathbb{R}^{n\times m}$, and $\boldsymbol{L} = \boldsymbol{\ell}_\alpha^\top\boldsymbol{\ell}_\beta \in \mathbb{R}^{m\times m}$. This decomposition of $\boldsymbol{L}$ demonstrates that $\boldsymbol{L} \succeq \boldsymbol{0}$ which will be used as one of the constraints in optimization. Now, consider linear reconstruction of the points by all of the points in the training set:

$$\min_{w_{ij}} \quad \sum_{i=1}^{n}\|\boldsymbol{x}_i - \sum_{j=1}^{n} w_{ij}\boldsymbol{x}_j\|_2^2, \tag{10.68}$$

where w_{ij}'s are the reconstruction weights. This optimization also exists in LLE and can be restated as (see Chap. 8):

$$\min_{\boldsymbol{M}} \quad \sum_{i=1}^{n}\sum_{j=1}^{n} \boldsymbol{M}_{ij}\boldsymbol{x}_i\boldsymbol{x}_j, \tag{10.69}$$

where $\mathbb{R}^{n\times n} \ni \boldsymbol{M} := (\boldsymbol{I}-\boldsymbol{W})^\top(\boldsymbol{I}-\boldsymbol{W})$ where $\boldsymbol{W}_{ij} = w_{ij}$. The matrix $\boldsymbol{M}$ can be found by solving the least squares problem in Eq. (10.69). Considering the $m \ll n$ landmarks, the matrix $\boldsymbol{M}$ can be decomposed to:

$$\boldsymbol{M} \approx \left[\begin{array}{c|c} \boldsymbol{M}^{ll} \in \mathbb{R}^{m\times m} & \boldsymbol{M}^{lu} \in \mathbb{R}^{m\times(n-m)} \\ \hline \boldsymbol{M}^{ul} \in \mathbb{R}^{(n-m)\times m} & \boldsymbol{M}^{uu} \in \mathbb{R}^{(n-m)\times(n-m)} \end{array}\right]. \tag{10.70}$$

As the reconstruction weights $q_{i\alpha}, \forall i, \alpha$ can be seen as a subset of reconstruction weights w_{ij}, $\boldsymbol{Q}$ can be written using the parts of the matrix $\boldsymbol{M}$ [44]:

$$\mathbb{R}^{n\times m} \ni \boldsymbol{Q} = \begin{bmatrix} \boldsymbol{I}_{m\times m} \\ (\boldsymbol{M}^{uu})^{-1}\boldsymbol{M}^{ul} \end{bmatrix}. \tag{10.71}$$

Note that this usage of a small part of the matrix (because $m \ll n$) is inspired by the Nyström method (see Chap. 3).

Landmark MVU solves the following SDP optimization problem [44]:

$$\begin{aligned} \underset{\boldsymbol{L}}{\text{maximize}} \quad & \mathbf{tr}(\boldsymbol{Q}\boldsymbol{L}\boldsymbol{Q}^\top) \\ \text{subject to} \quad & \tau_{ij}((\boldsymbol{Q}\boldsymbol{L}\boldsymbol{Q}^\top)_{ii} + (\boldsymbol{Q}\boldsymbol{L}\boldsymbol{Q}^\top)_{jj} - 2(\boldsymbol{Q}\boldsymbol{L}\boldsymbol{Q}^\top)_{ij}) \\ & \quad \leq \tau_{ij}(\boldsymbol{G}_{ii} + \boldsymbol{G}_{jj} - 2\boldsymbol{G}_{ij}), \\ & \qquad \forall i, j \in \{1, \dots, n\}, \\ & \sum_{i=1}^{n}\sum_{j=1}^{n} (\boldsymbol{Q}\boldsymbol{L}\boldsymbol{Q}^\top)_{ij} = 0, \\ & \boldsymbol{L} \succeq \boldsymbol{0}, \end{aligned} \tag{10.72}$$

where the optimization variable is the small matrix $\boldsymbol{L} \in \mathbb{R}^{m\times m}$, rather than the large matrix $\boldsymbol{K} \in \mathbb{R}^{n\times n}$; therefore, a large problem is reduced to an efficient small one. Note that Weinberger et al. [44] converted the equality of the local isometry constraint to an inequality. This has made the constraint more restricted and harder than the equality constraint. After solving the optimization problem to find the optimal $\boldsymbol{L}$, Eqs. (10.71) and (10.67) are used to calculate $\boldsymbol{K}$ and then embedding is calculated using Eq. (10.31).

10.7.4 Other Improvements Over Maximum Variance Unfolding and Kernel Learning

There have been other improvements over MVU and kernel learning by SDP whose details are not covered here. The improvements here are for further reading if desired. An application of MVU in nonlinear process monitoring can be found in [27]. SDP has also been used for kernel matrix completion [19] and low-rank kernel learning [24]. Maximum covariance unfolding [28] has been proposed for bimodal manifold unfolding. MVU can also be interpreted as a regularized shortest path problem on the graph of the data [30]; therefore, it can be related to the Isomap algorithm (see Chap. 7). An existing book chapter on MVU for interested readers in further studies is [42, Chapter 9].

10.8 Chapter Summary

This chapter explained that most of the spectral dimensionality reduction methods, including PCA, MDS, spectral clustering, Laplacian eigenmap, Isomap, LLE, and diffusion map, can be unified. These methods can be considered special cases of kernel PCA (introduced in Chap. 5) with different kernel functions because they are learning eigenfunctions. This unification was encouraging to find the best kernel function for every specific dataset. This led to MVU, also known as SDE, which learns the optimal kernel function for unfolding the manifold of data to its maximum variance. MVU is formulated as semidefinite programming, which was introduced in Chap. 4.

References

1. Babak Alipanahi and Ali Ghodsi. "Guided locally linear embedding". In: *Pattern recognition letters* 32.7 (2011), pp. 1029–1035.
2. Elnaz Barshan et al. "Supervised principal component analysis: Visualization, classification and regression on subspaces and submanifolds". In: *Pattern Recognition* 44.7 (2011), pp. 1357–1371.
3. Mikhail Belkin and Partha Niyogi. "Laplacian eigenmaps and spectral techniques for embedding and clustering." In: *Nips* Vol. 14. 14. 2001, pp. 585–591.
4. Yoshua Bengio et al. "Learning eigenfunctions links spectral embedding and kernel PCA". In: *Neural computation* 16.10 (2004), pp. 2197–2219.
5. Yoshua Bengio et al. *Learning eigenfunctions of similarity: linking spectral clustering and kernel PCA* Tech. rep. Departement d'Informatique et Recherche Operationnelle, 2003.
6. Yoshua Bengio et al. "Out-of-sample extensions for LLE, Isomap, MDS, eigenmaps, and spectral clustering". In: *Advances in neural information processing systems* 16 (2003), pp. 177–184.
7. Yoshua Bengio et al. *Spectral clustering and kernel PCA are learning eigenfunctions* Tech. rep. Departement d'Informatique et Recherche Operationnelle, Technical Report 1239, 2003.

8. Yoshua Bengio et al "Spectral dimensionality reduction". In: *Feature Extraction* Springer, 2006, pp. 519–550.
9. Michael Bowling, Ali Ghodsi, and Dana Wilkinson. "Action respecting embedding". In: *Proceedings of the 22nd international conference on Machine learning* 2005, pp. 65–72.
10. Michael Bowling et al. "Subjective localization with action respecting embedding". In: *Robotics Research* Springer, 2007, pp. 190–202.
11. Stephen Boyd, Stephen P Boyd, and Lieven Vandenberghe. *Convex optimization* Cambridge university press, 2004.
12. Kerstin Bunte, Michael Biehl, and Barbara Hammer. "A general framework for dimensionality-reducing data visualization mapping". In: *Neural Computation* 24.3 (2012), pp. 771–804.
13. Tat-Jun Chin and David Suter. "Out-of-sample extrapolation of learned manifolds". In: *IEEE Transactions on Pattern Analysis and Machine Intelligence* 30.9 (2008), pp. 1547–1556.
14. Ronald R Coifman and Stéphane Lafon. "Diffusion maps". In: *Applied and computational harmonic analysis* 21.1 (2006), pp. 5–30.
15. Michael AA Cox and Trevor F Cox. "Multidimensional scaling". In: *Handbook of data visualization* Springer, 2008, pp. 315–347.
16. Vin De Silva and Joshua B Tenenbaum. "Global versus local methods in nonlinear dimensionality reduction". In: *Advances in neural information processing systems* (2003), pp. 721–728.
17. Keinosuke Fukunaga. *Introduction to statistical pattern recognition* Academic Press, 1990.
18. Andrej Gisbrecht, Alexander Schulz, and Barbara Hammer. "Parametric nonlinear dimensionality reduction using kernel t-SNE". In: *Neurocomputing* 147 (2015), pp. 71–82.
19. Thore Graepel. "Kernel matrix completion by semidefinite programming". In: *International Conference on Artificial Neural Networks* Springer. 2002, pp. 694–699.
20. Arthur Gretton et al. "Measuring statistical dependence with Hilbert-Schmidt norms". In: *International conference on algorithmic learning theory* Springer. 2005, pp. 63–77.
21. Jihun Ham et al. "A kernel view of the dimensionality reduction of manifolds". In: *Proceedings of the twenty-first international conference on Machine learning* 2004, p. 47.
22. Chenping Hou et al. "Relaxed maximum-variance unfolding". In: *Optical Engineering* 47.7 (2008), p. 077202.
23. Amir-Hossein Karimi. "A Summary Of The Kernel Matrix, And How To Learn It Effectively Using Semidefinite Programming". In: *arXiv preprint arXiv:1709.06557* (2017).
24. Brian Kulis, Mátyás Sustik, and Inderjit Dhillon. "Learning low-rank kernel matrices". In: *Proceedings of the 23rd international conference on Machine learning* 2006, pp. 505–512.
25. Gert RG Lanckriet et al. "Learning the kernel matrix with semidefinite programming". In: *Journal of Machine learning research* 5.Jan (2004), pp. 27–72.
26. Ning Liu et al. "Supervised semi-definite embedding for email data cleaning and visualization". In: *Asia-Pacific Web Conference* Springer. 2005, pp. 972–982.
27. Yuan-Jui Liu, Tao Chen, and Yuan Yao. "Nonlinear process monitoring and fault isolation using extended maximum variance unfolding". In: *Journal of process control* 24.6 (2014), pp. 880–891.
28. Vijay Mahadevan et al. "Maximum covariance unfolding: Manifold learning for bimodal data". In: *Advances in Neural Information Processing Systems* 24 (2011), pp. 918–926.
29. Andrew Ng, Michael Jordan, and Yair Weiss. "On spectral clustering: Analysis and an algorithm". In: *Advances in neural information processing systems* 14 (2001), pp. 849–856.
30. Alexander Paprotny and Jochen Garcke. "On a connection between maximum variance unfolding, shortest path problems and Isomap". In: *Artificial Intelligence and Statistics* PMLR. 2012, pp. 859–867.
31. Sam T Roweis and Lawrence K Saul. "Nonlinear dimensionality reduction by locally linear embedding". In: *Science* 290.5500 (2000), pp. 2323–2326.
32. Lawrence K Saul and Sam T Roweis. "Think globally fit locally: unsupervised learning of low dimensional manifolds". In: *Journal of Machine Learning Research* 4 (2003), pp. 119–155.
33. Bernhard Schölkopf. "The kernel trick for distances". In: *Advances in neural information processing systems* (2001), pp. 301–307.

34. Bernhard Schölkopf, Alexander J Smola, and Francis Bach. *Learning with kernels: support vector machines, regularization, optimization, and beyond* MIT press, 2002.
35. Anton Schwaighofer, Volker Tresp, and Kai Yu. "Learning Gaussian process kernels via hierarchical Bayes". In: *Advances in neural information processing systems* 2005, pp. 1209–1216.
36. Le Song et al. "Colored Maximum Variance Unfolding." In: *Nips* Citeseer. 2007, pp. 1385–1392.
37. Gilbert W Stewart. "On the early history of the singular value decomposition". In: *SIAM review* 35.4 (1993), pp. 551–566.
38. Harry Strange and Reyer Zwiggelaar. *Open Problems in Spectral Dimensionality Reduction* Springer, 2014.
39. Joshua B Tenenbaum, Vin De Silva, and John C Langford. "A global geometric framework for nonlinear dimensionality reduction". In: *Science* 290.5500 (2000), pp. 2319–2323.
40. Lieven Vandenberghe and Stephen Boyd. "Semidefinite programming". In: *SIAM review* 38.1 (1996), pp. 49–95.
41. Vladimir Vapnik. *The nature of statistical learning theory* Springer science & business media, 1995.
42. Jianzhong Wang. *Geometric structure of high-dimensional data and dimensionality reduction* Vol. 5. Springer, 2012.
43. Chihang Wei, Junghui Chen, and Zhihuan Song. "Developments of two supervised maximum variance unfolding algorithms for process classification". In: *Chemometrics and Intelligent Laboratory Systems* 159 (2016), pp. 31–44.
44. Kilian Q Weinberger, Benjamin Packer, and Lawrence K Saul. "Nonlinear dimensionality reduction by semidefinite programming and kernel matrix factorization". In: *AISTATS* 2005.
45. Kilian Q Weinberger and Lawrence K Saul. "An introduction to nonlinear dimensionality reduction by maximum variance unfolding". In: *Proceedings of the AAAI Conference on Artificial Intelligence* Vol. 6. 2006, pp. 1683–1686.
46. Kilian Q Weinberger and Lawrence K Saul. "Unsupervised learning of image manifolds by semidefinite programming". In: *International journal of computer vision* 70.1 (2006), pp. 77–90.
47. Kilian Q Weinberger, Fei Sha, and Lawrence K Saul. "Learning a kernel matrix for nonlinear dimensionality reduction". In: *Proceedings of the twenty-first international conference on Machine learning* 2004, p. 106.
48. KQ Weinberger and LK Saul. "Unsupervised learning of image manifolds by semidefinite programming". In: *Proceedings of the 2004 IEEE Computer Society Conference on Computer Vision and Pattern Recognition* Vol. 2. IEEE. 2004.
49. Yair Weiss. "Segmentation using eigenvectors: a unifying view". In: *Proceedings of the seventh IEEE international conference on computer vision* Vol. 2. IEEE. 1999, pp. 975–982.
50. Shuicheng Yan et al. "Graph embedding and extensions: A general framework for dimensionality reduction". In: *IEEE transactions on pattern analysis and machine intelligence* 29.1 (2006), pp. 40–51.
51. Shuicheng Yan et al. "Graph embedding: A general framework for dimensionality reduction". In: *2005 IEEE Computer Society Conference on Computer Vision and Pattern Recognition (CVPR'05)* Vol. 2. IEEE. 2005, pp. 830–837.
52. Benyu Zhang et al. "Supervised semi-definite embedding for image manifolds". In: *2005 IEEE International Conference on Multimedia and Expo* IEEE. 2005, 4–pp.

Chapter 11
Spectral Metric Learning

11.1 Introduction

A family of dimensionality reduction methods known as metric learning learns a distance metric in an embedding space to separate dissimilar points and bring together similar points. In supervised metric learning, the aim is to discriminate classes by learning an appropriate metric. As discussed in Chap. 1, dimensionality reduction methods can be divided into spectral, probabilistic, and deep learning methods. Spectral methods use a geometrical approach and are usually reduced to generalized eigenvalue problems (see Chap. 2). Probabilistic methods are based on probability distributions. Deep learning methods use neural networks for learning. In each of these categories, several metric learning methods exist. This chapter introduces spectral metric learning algorithms. Probabilistic metric learning and deep metric learning will be introduced in Chaps. 13 and 19, respectively.

11.2 Generalized Mahalanobis Distance Metric

11.2.1 Distance Metric

Definition 11.1 (Distance Metric) Consider a metric space $\mathcal{X}$. A distance metric is a mapping $d : \mathcal{X} \times \mathcal{X} \rightarrow [0, \infty)$, which satisfies the following properties:

- nonnegativity: $d(\boldsymbol{x}_i, \boldsymbol{x}_j) \geq 0$
- identity: $d(\boldsymbol{x}_i, \boldsymbol{x}_j) = 0 \iff \boldsymbol{x}_i = \boldsymbol{x}_j$
- symmetry: $d(\boldsymbol{x}_i, \boldsymbol{x}_j) = d(\boldsymbol{x}_j, \boldsymbol{x}_i)$
- triangle inequality: $d(\boldsymbol{x}_i, \boldsymbol{x}_j) \leq d(\boldsymbol{x}_i, \boldsymbol{x}_k) + d(\boldsymbol{x}_k, \boldsymbol{x}_j)$

where $\boldsymbol{x}_i, \boldsymbol{x}_j, \boldsymbol{x}_k \in \mathcal{X}$.

B. Ghojogh et al., *Elements of Dimensionality Reduction and Manifold Learning*,
https://doi.org/10.1007/978-3-031-10602-6_11

An example of a distance metric is the Euclidean distance:

$$\|\boldsymbol{x}_i - \boldsymbol{x}_j\|_2 := \sqrt{(\boldsymbol{x}_i - \boldsymbol{x}_j)^\top (\boldsymbol{x}_i - \boldsymbol{x}_j)}. \tag{11.1}$$

11.2.2 Mahalanobis Distance

The Mahalanobis distance is a distance metric that was originally proposed by Mahalanobis [28].

Definition 11.2 (Mahalanobis Distance [28]) Consider a d-dimensional metric space $\mathcal{X}$. Let two clouds or sets of points $\mathcal{X}_1$ and $\mathcal{X}_2$ be part of the data, i.e., $\mathcal{X}_1, \mathcal{X}_2 \in \mathcal{X}$. A point is considered in each set, i.e., $\boldsymbol{x}_i \in \mathcal{X}_1$ and $\boldsymbol{x}_j \in \mathcal{X}_2$. The Mahalanobis distance between the two points is:

$$\|\boldsymbol{x}_i - \boldsymbol{x}_j\|_{\boldsymbol{\Sigma}} := \sqrt{(\boldsymbol{x}_i - \boldsymbol{x}_j)^\top \boldsymbol{\Sigma}^{-1} (\boldsymbol{x}_i - \boldsymbol{x}_j)}, \tag{11.2}$$

where $\boldsymbol{\Sigma} \in \mathbb{R}^{d \times d}$ is the covariance matrix of the data in the two sets $\mathcal{X}_1$ and $\mathcal{X}_2$.

If the points $\boldsymbol{x}_i$ and $\boldsymbol{x}_j$ are the means of the sets $\mathcal{X}_1$ and $\mathcal{X}_2$, respectively, as the representatives of the sets, this Mahalanobis distance is a good measure of the distance of the sets [29]:

$$\|\boldsymbol{\mu}_1 - \boldsymbol{\mu}_2\|_{\boldsymbol{\Sigma}} := \sqrt{(\boldsymbol{\mu}_1 - \boldsymbol{\mu}_2)^\top \boldsymbol{\Sigma}^{-1} (\boldsymbol{\mu}_1 - \boldsymbol{\mu}_2)}, \tag{11.3}$$

where $\boldsymbol{\mu}_1$ and $\boldsymbol{\mu}_2$ are the means of the sets $\mathcal{X}_1$ and $\mathcal{X}_2$, respectively.

Let $\mathcal{X}_1 := \{\boldsymbol{x}_{1,i}\}_{i=1}^{n_1}$ and $\mathcal{X}_2 := \{\boldsymbol{x}_{2,i}\}_{i=1}^{n_2}$. The unbiased sample covariance matrices of these two sets are:

$$\boldsymbol{\Sigma}_1 := \frac{1}{n_1 - 1} \sum_{i=1}^{n_1} (\boldsymbol{x}_{1,i} - \boldsymbol{\mu}_1)(\boldsymbol{x}_{1,i} - \boldsymbol{\mu}_1)^\top,$$

and $\boldsymbol{\Sigma}_2$ similarly. The covariance matrix $\boldsymbol{\Sigma}$ can be an unbiased sample covariance matrix [29]:

$$\boldsymbol{\Sigma} := \frac{1}{n_1 + n_2 - 2}\Big((n_1 - 1)\boldsymbol{\Sigma}_1 + (n_2 - 1)\boldsymbol{\Sigma}_2\Big).$$

The Mahalanobis distance can also be defined between a point $\boldsymbol{x}$ and a cloud or set of points $\mathcal{X}$ [17]. Let $\boldsymbol{\mu}$ and $\boldsymbol{\Sigma}$ be the mean and the (sample) covariance matrix of the set $\mathcal{X}$. The Mahalanobis distance of $\boldsymbol{x}$ and $\mathcal{X}$ is:

$$\|\boldsymbol{x} - \boldsymbol{\mu}\|_{\boldsymbol{\Sigma}} := \sqrt{(\boldsymbol{x} - \boldsymbol{\mu})^\top \boldsymbol{\Sigma}^{-1} (\boldsymbol{x} - \boldsymbol{\mu})}. \tag{11.4}$$

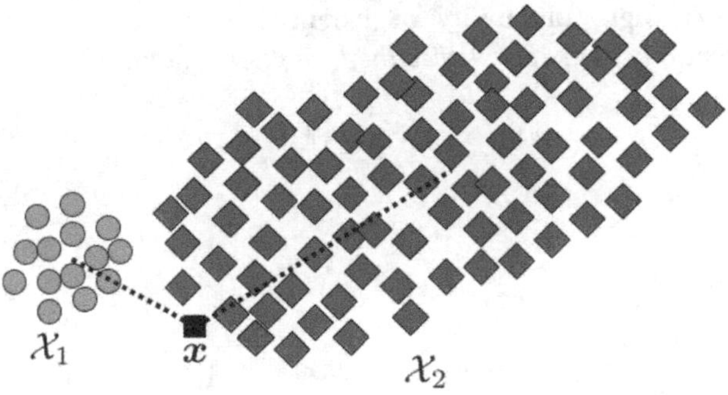

Fig. 11.1 An example for comparison of the Euclidean and Mahalanobis distances

Remark 11.5 (Justification of the Mahalanobis Distance [17]) Consider two clouds of data, $\mathcal{X}_1$ and $\mathcal{X}_2$, depicted in Fig. 11.1. The aim is to compute the distance of a point $\boldsymbol{x}$ from these two data clouds to see which cloud this point is closer to. The Euclidean distance ignores the scatter/variance of the clouds and only measures the distances of the point from the means of the clouds. Therefore, in this example, $\boldsymbol{x}$ belongs to $\mathcal{X}_1$ because it is closer to the mean of $\mathcal{X}_1$ compared to $\mathcal{X}_2$. However, the Mahalanobis distance takes the variance of clouds into account and says that $\boldsymbol{x}$ belongs to $\mathcal{X}_2$ because it is closer to its scatter compared to $\mathcal{X}_1$. Visually, humans also say $\boldsymbol{x}$ belongs to $\mathcal{X}_2$; therefore, the Mahalanobis distance has performed better than the Euclidean distance by considering the variances of the data.

11.2.3 Generalized Mahalanobis Distance

Definition 11.3 (Generalized Mahalanobis Distance) In the Mahalanobis distance, i.e., Eq. (11.2), the covariance matrix $\boldsymbol{\Sigma}$ and its inverse $\boldsymbol{\Sigma}^{-1}$ are positive semidefinite. It is possible to replace $\boldsymbol{\Sigma}^{-1}$ with a positive semidefinite weight matrix $\boldsymbol{W} \succeq \boldsymbol{0}$ in the squared Mahalanobis distance. This distance is named the generalized Mahalanobis distance:

$$
\begin{aligned}
&\|\boldsymbol{x}_i - \boldsymbol{x}_j\|_{\boldsymbol{W}} := \sqrt{(\boldsymbol{x}_i - \boldsymbol{x}_j)^\top \boldsymbol{W} (\boldsymbol{x}_i - \boldsymbol{x}_j)}. \\
&\therefore \quad \|\boldsymbol{x}_i - \boldsymbol{x}_j\|_{\boldsymbol{W}}^2 := (\boldsymbol{x}_i - \boldsymbol{x}_j)^\top \boldsymbol{W} (\boldsymbol{x}_i - \boldsymbol{x}_j).
\end{aligned} \tag{11.5}
$$

The generalized Mahalanobis norm is defined as:

$$
\|\boldsymbol{x}\|_{\boldsymbol{W}} := \sqrt{\boldsymbol{x}^\top \boldsymbol{W} \boldsymbol{x}}. \tag{11.6}
$$

Lemma 11.1 (Triangle Inequality of Norm) *Let $\|.\|$ be a norm. Using the Cauchy-Schwarz inequality, it satisfies the triangle inequality:*

$$\|\boldsymbol{x}_i + \boldsymbol{x}_j\| \leq \|\boldsymbol{x}_i\| + \|\boldsymbol{x}_j\|. \tag{11.7}$$

Proof

$$\begin{aligned}\|\boldsymbol{x}_i + \boldsymbol{x}_j\|^2 &= (\boldsymbol{x}_i + \boldsymbol{x}_j)^\top(\boldsymbol{x}_i + \boldsymbol{x}_j) = \|\boldsymbol{x}_i\|^2 + \|\boldsymbol{x}_j\|^2 + 2\boldsymbol{x}_i^\top \boldsymbol{x}_j \\ &\overset{(a)}{\leq} \|\boldsymbol{x}_i\|^2 + \|\boldsymbol{x}_j\|^2 + 2\|\boldsymbol{x}_i\|\|\boldsymbol{x}_j\| = (\|\boldsymbol{x}_i\| + \|\boldsymbol{x}_j\|)^2,\end{aligned}$$

where (a) is because of the Cauchy-Schwarz inequality, i.e., $\boldsymbol{x}_i^\top \boldsymbol{x}_j \leq \|\boldsymbol{x}_i\|\|\boldsymbol{x}_j\|$. Taking the second root from the sides results in Eq. (11.7). □

Proposition 11.1 *The generalized Mahalanobis distance is a valid distance metric (as stated in Definition 11.1).*

Proof The characteristics in Definition 11.1 are satisfied through:

- As $\boldsymbol{W} \succeq \boldsymbol{0}$, Eq. (11.5) is nonnegative.
- identity: if $\|\boldsymbol{x}_i - \boldsymbol{x}_j\|_{\boldsymbol{W}} = 0$, according to Eq. (11.5), the following holds: $\boldsymbol{x}_i - \boldsymbol{x}_j = 0 \implies \boldsymbol{x}_i = \boldsymbol{x}_j$. If $\boldsymbol{x}_i = \boldsymbol{x}_j$, then $\|\boldsymbol{x}_i - \boldsymbol{x}_j\|_{\boldsymbol{W}} = 0$ according to Eq. (11.5).
- symmetry: $\|\boldsymbol{x}_i - \boldsymbol{x}_j\|_{\boldsymbol{W}} = \sqrt{(\boldsymbol{x}_i - \boldsymbol{x}_j)^\top \boldsymbol{W}(\boldsymbol{x}_i - \boldsymbol{x}_j)} = \sqrt{(\boldsymbol{x}_j - \boldsymbol{x}_i)^\top \boldsymbol{W}(\boldsymbol{x}_j - \boldsymbol{x}_i)} = \|\boldsymbol{x}_j - \boldsymbol{x}_i\|_{\boldsymbol{W}}$.
- triangle inequality: $\|\boldsymbol{x}_i - \boldsymbol{x}_j\|_{\boldsymbol{W}} = \|\boldsymbol{x}_i - \boldsymbol{x}_k + \boldsymbol{x}_k - \boldsymbol{x}_j\|_{\boldsymbol{W}} \overset{(11.7)}{\leq} \|\boldsymbol{x}_i - \boldsymbol{x}_k\|_{\boldsymbol{W}} + \|\boldsymbol{x}_k - \boldsymbol{x}_j\|_{\boldsymbol{W}}$. □

Remark 11.6 It is noteworthy that $\boldsymbol{W} \succeq \boldsymbol{0}$ is required so that the generalized Mahalanobis distance is convex and satisfies the triangle inequality.

Remark 11.7 The weight matrix $\boldsymbol{W}$ in Eq. (11.5) weighs the dimensions and determines that there is a degree of correlation between the dimensions of the data points. In other words, it changes the space in such a way that the scatters of clouds are considered.

Remark 11.8 The Euclidean distance is a special case of the Mahalanobis distance where the weight matrix is the identity matrix, i.e., $\boldsymbol{W} = \boldsymbol{I}$ (cf. Eqs. (11.1) and (11.5)). In other words, the Euclidean distance does not change the space for computing the distance.

Proposition 11.2 (Projection in Metric Learning) *Consider the eigenvalue decomposition of the weight matrix $\boldsymbol{W}$ in the generalized Mahalanobis distance with $\boldsymbol{V}$ and $\boldsymbol{\Lambda}$ as the matrix of the eigenvectors and the diagonal matrix of eigenvalues of* *the weight, respectively. Let* $\boldsymbol{U} := \boldsymbol{V}\boldsymbol{\Lambda}^{(1/2)}$. *The generalized Mahalanobis distance can be seen as the Euclidean distance after applying a linear projection onto the column space of* $\boldsymbol{U}$:

$$\|\boldsymbol{x}_i - \boldsymbol{x}_j\|_{\boldsymbol{W}}^2 = (\boldsymbol{U}^\top \boldsymbol{x}_i - \boldsymbol{U}^\top \boldsymbol{x}_j)^\top (\boldsymbol{U}^\top \boldsymbol{x}_i - \boldsymbol{U}^\top \boldsymbol{x}_j) = \|\boldsymbol{U}^\top \boldsymbol{x}_i - \boldsymbol{U}^\top \boldsymbol{x}_j\|_2^2. \tag{11.8}$$

If $\boldsymbol{U} \in \mathbb{R}^{d \times p}$ *with* $p \leq d$, *the column space of the projection matrix* $\boldsymbol{U}$ *is a* p*-dimensional subspace.*

Proof Using the eigenvalue decomposition of $\boldsymbol{W}$:

$$\boldsymbol{W} = \boldsymbol{V}\boldsymbol{\Lambda}\boldsymbol{V}^\top \overset{(a)}{=} \boldsymbol{V}\boldsymbol{\Lambda}^{(1/2)}\boldsymbol{\Lambda}^{(1/2)}\boldsymbol{V}^\top \overset{(b)}{=} \boldsymbol{U}\boldsymbol{U}^\top, \tag{11.9}$$

where (a) is because $\boldsymbol{W}$ is positive semidefinite so all its eigenvalues are nonnegative and can be written as the multiplication of its second roots. Additionally, (b) is because $\boldsymbol{U} := \boldsymbol{V}\boldsymbol{\Lambda}^{(1/2)}$ is defined. Substituting Eq. (11.9) into Eq. (11.5) results in:

$$\begin{aligned}\|\boldsymbol{x}_i - \boldsymbol{x}_j\|_{\boldsymbol{W}}^2 &= (\boldsymbol{x}_i - \boldsymbol{x}_j)^\top \boldsymbol{U}\boldsymbol{U}^\top (\boldsymbol{x}_i - \boldsymbol{x}_j) \\ &= (\boldsymbol{U}^\top \boldsymbol{x}_i - \boldsymbol{U}^\top \boldsymbol{x}_j)^\top (\boldsymbol{U}^\top \boldsymbol{x}_i - \boldsymbol{U}^\top \boldsymbol{x}_j) = \|\boldsymbol{U}^\top \boldsymbol{x}_i - \boldsymbol{U}^\top \boldsymbol{x}_j\|_2^2.\end{aligned}$$

□

It is noteworthy that Eq. (11.9) can also be obtained using singular value decomposition rather than eigenvalue decomposition. In that case, the matrices of the right and left singular vectors are equal because of the symmetry of $\boldsymbol{W}$. □

11.2.4 The Main Idea of Metric Learning

Consider a d-dimensional dataset $\{\boldsymbol{x}_i\}_{i=1}^n \subset \mathbb{R}^d$ of size n. Assume some data points are similar. For example, they have similar patterns or the same characteristics. Therefore, there is a set of similar pair points, denoted by $\mathcal{S}$. In contrast, there can be dissimilar points that are different in pattern or characteristics. Let the set of dissimilar pair points be denoted by $\mathcal{D}$. In summary:

$$\begin{aligned}&(\boldsymbol{x}_i, \boldsymbol{x}_j) \in \mathcal{S} \text{ if } \boldsymbol{x}_i \text{ and } \boldsymbol{x}_j \text{ are similar,} \\ &(\boldsymbol{x}_i, \boldsymbol{x}_j) \in \mathcal{D} \text{ if } \boldsymbol{x}_i \text{ and } \boldsymbol{x}_j \text{ are dissimilar.}\end{aligned} \tag{11.10}$$

The measure of similarity and dissimilarity can belong to the same or different classes, if class labels are available for the dataset. In this case:

$$\begin{aligned}&(\boldsymbol{x}_i, \boldsymbol{x}_j) \in \mathcal{S} \text{ if } \boldsymbol{x}_i \text{ and } \boldsymbol{x}_j \text{ are in the same class,} \\ &(\boldsymbol{x}_i, \boldsymbol{x}_j) \in \mathcal{D} \text{ if } \boldsymbol{x}_i \text{ and } \boldsymbol{x}_j \text{ are in different classes.}\end{aligned} \tag{11.11}$$

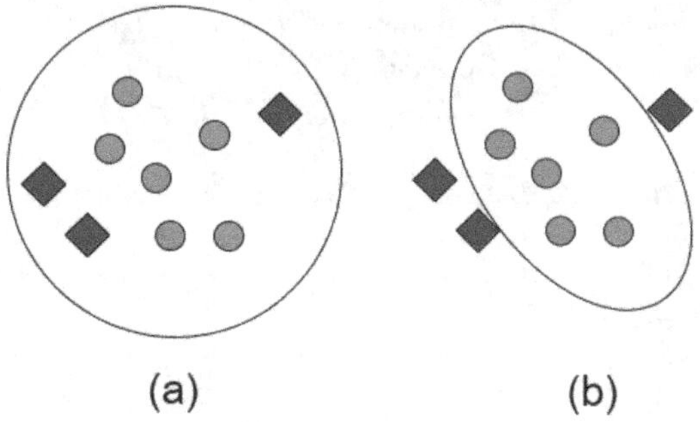

Fig. 11.2 Visualizing metric learning in 2D: (**a**) the contour of the Euclidean distance that does not properly discriminate classes, and (**b**) the contour of the Euclidean distance that is better at discriminating between classes

In metric learning, the weight matrix is determined so that the distances between similar points decrease and the distances between dissimilar points increase. In this way, the variance of similar and dissimilar points decreases and increases, respectively. A 2D visualization of metric learning is depicted in Fig. 11.2. If the class labels are available, metric learning attempts to make the intraclass and interclass variances decrease and increase, respectively. This is the same idea as the idea of Fisher Discriminant Analysis (FDA) [19] (see Chap. 6).

11.3 Spectral Methods Using Scatters

11.3.1 The First Spectral Method

The first metric learning method was proposed in [45]. In this method, the distances between similar points are minimized using the weight matrix $\boldsymbol{W}$, where this matrix is positive semidefinite:

$$\begin{aligned} \underset{\boldsymbol{W}}{\text{minimize}} \quad & \sum_{(\boldsymbol{x}_i, \boldsymbol{x}_j) \in \mathcal{S}} \|\boldsymbol{x}_i - \boldsymbol{x}_j\|_{\boldsymbol{W}}^2 \\ \text{subject to} \quad & \boldsymbol{W} \succeq \boldsymbol{0}. \end{aligned}$$

However, the solution to this optimization problem is trivial, i.e., $\boldsymbol{W} = \boldsymbol{0}$. Therefore, a constraint is added to the dissimilar points to increase the distances by a positive amount:

$$\begin{aligned}
&\underset{\boldsymbol{W}}{\text{minimize}} \quad \sum_{(\boldsymbol{x}_i,\boldsymbol{x}_j)\in\mathcal{S}} \|\boldsymbol{x}_i - \boldsymbol{x}_j\|_{\boldsymbol{W}}^2 \\
&\text{subject to} \quad \sum_{(\boldsymbol{x}_i,\boldsymbol{x}_j)\in\mathcal{D}} \|\boldsymbol{x}_i - \boldsymbol{x}_j\|_{\boldsymbol{W}} \geq \alpha, \\
&\qquad\qquad\quad \boldsymbol{W} \succeq \boldsymbol{0}.
\end{aligned} \tag{11.12}$$

where $\alpha > 0$ is a positive number, such as $\alpha = 1$.

Lemma 11.2 ([45]) *If the constraint in Eq. (11.12) is squared, i.e., $\sum_{(\boldsymbol{x}_i,\boldsymbol{x}_j)\in\mathcal{D}} \|\boldsymbol{x}_i - \boldsymbol{x}_j\|_{\boldsymbol{W}}^2 \geq \alpha$, the solution to the optimization problem will have a rank of* 1. *Therefore, a nonsquared constraint is used in this optimization problem.*

Proof If the constraint in Eq. (11.12) is squared, the problem is equivalent to maximizing the following Rayleigh-Ritz quotient (see Chap. 2):

$$\underset{\boldsymbol{W}}{\text{maximize}} \quad \frac{\sum_{(\boldsymbol{x}_i,\boldsymbol{x}_j)\in\mathcal{D}} \|\boldsymbol{x}_i - \boldsymbol{x}_j\|_{\boldsymbol{W}}^2}{\sum_{(\boldsymbol{x}_i,\boldsymbol{x}_j)\in\mathcal{S}} \|\boldsymbol{x}_i - \boldsymbol{x}_j\|_{\boldsymbol{W}}^2}.$$

It is possible to restate $\|\boldsymbol{x}_i - \boldsymbol{x}_j\|_{\boldsymbol{W}}^2$ as:

$$\begin{aligned}
&\sum_{(\boldsymbol{x}_i,\boldsymbol{x}_j)\in\mathcal{S}} \|\boldsymbol{x}_i - \boldsymbol{x}_j\|_{\boldsymbol{W}}^2 = \textbf{tr}(\boldsymbol{W}\boldsymbol{\Sigma}_{\mathcal{S}}), \\
&\sum_{(\boldsymbol{x}_i,\boldsymbol{x}_j)\in\mathcal{D}} \|\boldsymbol{x}_i - \boldsymbol{x}_j\|_{\boldsymbol{W}}^2 = \textbf{tr}(\boldsymbol{W}\boldsymbol{\Sigma}_{\mathcal{D}}),
\end{aligned} \tag{11.13}$$

where $\textbf{tr}(.)$ denotes the trace of the matrix and:

$$\begin{aligned}
&\boldsymbol{\Sigma}_{\mathcal{S}} := \sum_{(\boldsymbol{x}_i,\boldsymbol{x}_j)\in\mathcal{S}} (\boldsymbol{x}_i - \boldsymbol{x}_j)(\boldsymbol{x}_i - \boldsymbol{x}_j)^\top, \\
&\boldsymbol{\Sigma}_{\mathcal{D}} := \sum_{(\boldsymbol{x}_i,\boldsymbol{x}_j)\in\mathcal{D}} (\boldsymbol{x}_i - \boldsymbol{x}_j)(\boldsymbol{x}_i - \boldsymbol{x}_j)^\top.
\end{aligned} \tag{11.14}$$

Therefore:

$$\begin{aligned}
\frac{\sum_{(\boldsymbol{x}_i,\boldsymbol{x}_j)\in\mathcal{D}} \|\boldsymbol{x}_i - \boldsymbol{x}_j\|_{\boldsymbol{W}}^2}{\sum_{(\boldsymbol{x}_i,\boldsymbol{x}_j)\in\mathcal{S}} \|\boldsymbol{x}_i - \boldsymbol{x}_j\|_{\boldsymbol{W}}^2} &= \frac{\textbf{tr}(\boldsymbol{W}\boldsymbol{\Sigma}_{\mathcal{D}})}{\textbf{tr}(\boldsymbol{W}\boldsymbol{\Sigma}_{\mathcal{S}})} \overset{(11.9)}{=} \frac{\textbf{tr}(\boldsymbol{U}\boldsymbol{U}^\top\boldsymbol{\Sigma}_{\mathcal{D}})}{\textbf{tr}(\boldsymbol{U}\boldsymbol{U}^\top\boldsymbol{\Sigma}_{\mathcal{S}})} \\
&\overset{(a)}{=} \frac{\textbf{tr}(\boldsymbol{U}^\top\boldsymbol{\Sigma}_{\mathcal{D}}\boldsymbol{U})}{\textbf{tr}(\boldsymbol{U}^\top\boldsymbol{\Sigma}_{\mathcal{S}}\boldsymbol{U})} = \frac{\sum_{i=1}^d \boldsymbol{u}^\top\boldsymbol{\Sigma}_{\mathcal{D}}\boldsymbol{u}}{\sum_{i=1}^d \boldsymbol{u}^\top\boldsymbol{\Sigma}_{\mathcal{S}}\boldsymbol{u}},
\end{aligned}$$

where (a) is the result of the cyclic property of the trace and (b) is the result of $\boldsymbol{U} = [\boldsymbol{u}_1, \ldots, \boldsymbol{u}_d]$. Maximizing this Rayleigh-Ritz quotient results in the following generalized eigenvalue problem (see Chap. 2):

$$\boldsymbol{\Sigma}_{\mathcal{D}} \boldsymbol{u}_1 = \lambda \boldsymbol{\Sigma}_{\mathcal{S}} \boldsymbol{u}_1,$$

where $\boldsymbol{u}_1$ is the eigenvector with the largest eigenvalue and the other eigenvectors $\boldsymbol{u}_2, \ldots, \boldsymbol{u}_d$ are zero vectors. □

Equation (11.12) can be restated as a maximization problem:

$$\begin{aligned} \underset{\boldsymbol{W}}{\text{maximize}} \quad & \sum_{(\boldsymbol{x}_i, \boldsymbol{x}_j) \in \mathcal{D}} \|\boldsymbol{x}_i - \boldsymbol{x}_j\|_{\boldsymbol{W}} \\ \text{subject to} \quad & \sum_{(\boldsymbol{x}_i, \boldsymbol{x}_j) \in \mathcal{S}} \|\boldsymbol{x}_i - \boldsymbol{x}_j\|_{\boldsymbol{W}}^2 \leq \alpha, \\ & \boldsymbol{W} \succeq \boldsymbol{0}. \end{aligned} \tag{11.15}$$

This problem can be solved using the projected gradient method (see Chap. 4) where a step of gradient ascent is followed by a projection onto the two constraint sets:

$$\boldsymbol{W} := \boldsymbol{W} + \eta \frac{\partial}{\partial \boldsymbol{W}} \Big(\sum_{(\boldsymbol{x}_i, \boldsymbol{x}_j) \in \mathcal{D}} \|\boldsymbol{x}_i - \boldsymbol{x}_j\|_{\boldsymbol{W}} \Big),$$

$$\boldsymbol{W} := \arg\min_{\boldsymbol{Q}} \Big(\|\boldsymbol{Q} - \boldsymbol{W}\|_F^2 \text{ s.t. } \sum_{(\boldsymbol{x}_i, \boldsymbol{x}_j) \in \mathcal{S}} \|\boldsymbol{x}_i - \boldsymbol{x}_j\|_{\boldsymbol{Q}}^2 \leq \alpha \Big),$$

$$\boldsymbol{W} := \boldsymbol{V} \, \textbf{diag}(\max(\lambda_1, 0), \ldots, \max(\lambda_d, 0)) \, \boldsymbol{V}^\top,$$

where $\eta > 0$ is the learning rate and $\boldsymbol{V}$ and $\boldsymbol{\Lambda} = \textbf{diag}(\lambda_1, \ldots, \lambda_d)$ are the eigenvectors and eigenvalues of $\boldsymbol{W}$, respectively (see Eq. (11.9)).

11.3.1.1 Formulating as Semidefinite Programming

Ghodsi et al.'s metric learning method [21] minimizes the distances between similar points and maximizes the distances between dissimilar points. For this, the distances between similar points and the negation of the distances between dissimilar points are minimized. The weight matrix should be positive semidefinite to satisfy the triangle inequality and convexity of the distance metric. The trace of the weight matrix is also set to a constant to eliminate the trivial solution $\boldsymbol{W} = \boldsymbol{0}$. The optimization problem is:

$$\begin{aligned} &\underset{W}{\text{minimize}} \quad \frac{1}{|\mathcal{S}|} \sum_{(x_i,x_j)\in\mathcal{S}} \|x_i - x_j\|_W^2 - \frac{1}{|\mathcal{D}|} \sum_{(x_i,x_j)\in\mathcal{D}} \|x_i - x_j\|_W^2 \\ &\text{subject to} \quad W \succeq \mathbf{0}, \\ &\qquad\qquad\quad \mathbf{tr}(W) = 1, \end{aligned} \tag{11.16}$$

where $|.|$ denotes the cardinality of the set.

Lemma 11.3 ([21]) *The objective function can be simplified to:*

$$\begin{aligned} &\frac{1}{|\mathcal{S}|} \sum_{(x_i,x_j)\in\mathcal{S}} \|x_i - x_j\|_W^2 - \frac{1}{|\mathcal{D}|} \sum_{(x_i,x_j)\in\mathcal{D}} \|x_i - x_j\|_W \\ &= \mathbf{vec}(W)^\top \Big(\frac{1}{|\mathcal{S}|} \sum_{(x_i,x_j)\in\mathcal{S}} \mathbf{vec}\big((x_i - x_j)(x_i - x_j)^\top\big) \\ &\qquad - \frac{1}{|\mathcal{D}|} \sum_{(x_i,x_j)\in\mathcal{D}} \mathbf{vec}\big((x_i - x_j)(x_i - x_j)^\top\big)\Big), \end{aligned} \tag{11.17}$$

where $\mathbf{vec}(.)$ *vectorizes the matrix to a vector.*

Proof See [21, Section 2.1] for proof. □

According to Lemma 11.3, Eq. (11.16) is a Semidefinite Programming (SDP) problem. It can be solved iteratively using the interior-point method (see Chap. 4).

11.3.1.2 Relevant to Fisher Discriminant Analysis

Alipanahi et al.'s metric learning method [2] takes two approaches. The relation of metric learning with Fisher discriminant analysis [19] (see Chap. 6) has been discussed in that paper [2]. This relation is explained in this section.

Approach 1 As $W \succeq \mathbf{0}$, the weight matrix can be decomposed as demonstrated in Eq. (11.9), namely, $W = UU^\top$. Therefore:

$$\begin{aligned} \|x_i - x_j\|_W^2 &\overset{(11.5)}{=} (x_i - x_j)^\top W (x_i - x_j) \overset{(a)}{=} \mathbf{tr}\big((x_i - x_j)^\top W (x_i - x_j)\big) \\ &\overset{(11.9)}{=} \mathbf{tr}\big((x_i - x_j)^\top U U^\top (x_i - x_j)\big) \overset{(b)}{=} \mathbf{tr}\big(U^\top (x_i - x_j)(x_i - x_j)^\top U\big), \end{aligned} \tag{11.18}$$

where (a) is because a scalar is equal to its trace and (b) is because of the cyclic property of that trace. Equation (11.18) can be substituted into Eq. (11.16) to obtain the optimization problem:

$$\begin{aligned}
\underset{\boldsymbol{U}}{\text{minimize}} \quad & \frac{1}{|\mathcal{S}|} \sum_{(\boldsymbol{x}_i, \boldsymbol{x}_j) \in \mathcal{S}} \mathbf{tr}\big(\boldsymbol{U}^\top (\boldsymbol{x}_i - \boldsymbol{x}_j)(\boldsymbol{x}_i - \boldsymbol{x}_j)^\top \boldsymbol{U}\big) \\
& - \frac{1}{|\mathcal{D}|} \sum_{(\boldsymbol{x}_i, \boldsymbol{x}_j) \in \mathcal{D}} \mathbf{tr}\big(\boldsymbol{U}^\top (\boldsymbol{x}_i - \boldsymbol{x}_j)(\boldsymbol{x}_i - \boldsymbol{x}_j)^\top \boldsymbol{U}\big) \\
\text{subject to} \quad & \mathbf{tr}(\boldsymbol{U}\boldsymbol{U}^\top) = 1,
\end{aligned} \tag{11.19}$$

whose objective variable is $\boldsymbol{U}$. Note that the constraint $\boldsymbol{W} \succeq \mathbf{0}$ is implicitly satisfied because of the decomposition $\boldsymbol{W} = \boldsymbol{U}\boldsymbol{U}^\top$. The following expressions are defined as:

$$\begin{aligned}
\mathbf{\Sigma}'_{\mathcal{S}} &:= \frac{1}{|\mathcal{S}|} \sum_{(\boldsymbol{x}_i, \boldsymbol{x}_j) \in \mathcal{S}} (\boldsymbol{x}_i - \boldsymbol{x}_j)(\boldsymbol{x}_i - \boldsymbol{x}_j)^\top \overset{(11.13)}{=} \frac{1}{|\mathcal{S}|} \mathbf{\Sigma}_{\mathcal{S}}, \\
\mathbf{\Sigma}'_{\mathcal{D}} &:= \frac{1}{|\mathcal{D}|} \sum_{(\boldsymbol{x}_i, \boldsymbol{x}_j) \in \mathcal{D}} (\boldsymbol{x}_i - \boldsymbol{x}_j)(\boldsymbol{x}_i - \boldsymbol{x}_j)^\top \overset{(11.13)}{=} \frac{1}{|\mathcal{D}|} \mathbf{\Sigma}_{\mathcal{D}}.
\end{aligned} \tag{11.20}$$

Therefore, Eq. (11.19) can be restated as:

$$\begin{aligned}
\underset{\boldsymbol{U}}{\text{minimize}} \quad & \mathbf{tr}(\boldsymbol{U}^\top (\mathbf{\Sigma}'_{\mathcal{S}} - \mathbf{\Sigma}'_{\mathcal{D}}) \boldsymbol{U}) \\
\text{subject to} \quad & \mathbf{tr}(\boldsymbol{U}\boldsymbol{U}^\top) = 1,
\end{aligned} \tag{11.21}$$

whose Lagrangian is (see Chap. 4):

$$\mathcal{L} = \mathbf{tr}(\boldsymbol{U}^\top (\mathbf{\Sigma}'_{\mathcal{S}} - \mathbf{\Sigma}'_{\mathcal{D}}) \boldsymbol{U}) - \lambda(\mathbf{tr}(\boldsymbol{U}\boldsymbol{U}^\top) - 1).$$

Setting the derivative of the Lagrangian to zero results in:

$$\frac{\partial \mathcal{L}}{\partial \boldsymbol{U}} = 2(\mathbf{\Sigma}'_{\mathcal{S}} - \mathbf{\Sigma}'_{\mathcal{D}})\boldsymbol{U} - 2\lambda \boldsymbol{U} \overset{\text{set}}{=} \mathbf{0} \implies (\mathbf{\Sigma}'_{\mathcal{S}} - \mathbf{\Sigma}'_{\mathcal{D}})\boldsymbol{U} = \lambda \boldsymbol{U}, \tag{11.22}$$

which is the eigenvalue problem for $(\mathbf{\Sigma}'_{\mathcal{S}} - \mathbf{\Sigma}'_{\mathcal{D}})$ (see Chap. 2). Therefore, $\boldsymbol{U}$ is the eigenvector of $(\mathbf{\Sigma}'_{\mathcal{S}} - \mathbf{\Sigma}'_{\mathcal{D}})$ with the smallest eigenvalue because Eq. (11.19) is a minimization problem.

Approach 2 The constraint in Eq. (11.21) can be altered to have an orthogonal projection matrix, i.e., $\boldsymbol{U}^\top \boldsymbol{U} = \boldsymbol{I}$. Rather, it is possible to make the rotation of the projection matrix, by the matrix $\mathbf{\Sigma}'_{\mathcal{S}}$, orthogonal, i.e., $\boldsymbol{U}^\top \mathbf{\Sigma}'_{\mathcal{S}} \boldsymbol{U} = \boldsymbol{I}$. Therefore, the optimization problem becomes:

$$\begin{aligned} &\underset{\boldsymbol{U}}{\text{minimize}} \quad \mathbf{tr}(\boldsymbol{U}^\top(\boldsymbol{\Sigma}'_{\mathcal{S}} - \boldsymbol{\Sigma}'_{\mathcal{D}})\boldsymbol{U}) \\ &\text{subject to} \quad \boldsymbol{U}^\top \boldsymbol{\Sigma}'_{\mathcal{S}} \boldsymbol{U} = \boldsymbol{I}, \end{aligned} \tag{11.23}$$

whose Lagrangian is (see Chap. 4):

$$\begin{aligned} &\mathcal{L} = \mathbf{tr}(\boldsymbol{U}^\top(\boldsymbol{\Sigma}'_{\mathcal{S}} - \boldsymbol{\Sigma}'_{\mathcal{D}})\boldsymbol{U}) - \mathbf{tr}(\boldsymbol{\Lambda}^\top(\boldsymbol{U}^\top \boldsymbol{\Sigma}'_{\mathcal{S}} \boldsymbol{U} - \boldsymbol{I})). \\ &\frac{\partial \mathcal{L}}{\partial \boldsymbol{U}} = 2(\boldsymbol{\Sigma}'_{\mathcal{S}} - \boldsymbol{\Sigma}'_{\mathcal{D}})\boldsymbol{U} - 2\boldsymbol{\Sigma}'_{\mathcal{S}} \boldsymbol{U}\boldsymbol{\Lambda} \overset{\text{set}}{=} \boldsymbol{0} \implies (\boldsymbol{\Sigma}'_{\mathcal{S}} - \boldsymbol{\Sigma}'_{\mathcal{D}})\boldsymbol{U} = \boldsymbol{\Sigma}'_{\mathcal{S}} \boldsymbol{U}\boldsymbol{\Lambda}, \end{aligned} \tag{11.24}$$

which is the generalized eigenvalue problem for $(\boldsymbol{\Sigma}'_{\mathcal{S}} - \boldsymbol{\Sigma}'_{\mathcal{D}}, \boldsymbol{\Sigma}'_{\mathcal{S}})$ (see Chap. 2). Therefore, $\boldsymbol{U}$ is a matrix whose columns are the eigenvectors sorted from the smallest to largest eigenvalues.

This optimization problem is similar to the optimization of Fisher discriminant analysis (FDA) [19] (see Chap. 6) where $\boldsymbol{\Sigma}'_{\mathcal{S}}$ and $\boldsymbol{\Sigma}'_{\mathcal{D}}$ are replaced with the intraclass and interclass covariance matrices of the data, respectively. This demonstrates the relation between this method and FDA, which is evident because both metric learning and FDA have the same goal—decreasing and increasing the variances of similar and dissimilar points, respectively.

11.3.1.3 Relevant Component Analysis (RCA)

Suppose the n data points can be divided into c clusters, or so-called chunklets. These chunklets can be classes if the class labels are available. If $\mathcal{X}_l$ denotes the data of the l-th cluster and $\boldsymbol{\mu}_l$ is the mean of $\mathcal{X}_l$, the summation of the intracluster scatters is:

$$\mathbb{R}^{d \times d} \ni \boldsymbol{S}_w := \frac{1}{n} \sum_{l=1}^{c} \sum_{\boldsymbol{x}_i \in \mathcal{X}_l} (\boldsymbol{x}_i - \boldsymbol{\mu}_l)(\boldsymbol{x}_i - \boldsymbol{\mu}_l)^\top. \tag{11.25}$$

Relevant Component Analysis (RCA) [36] is a metric learning method. In this method, Principal Component Analysis (PCA) (see Chap. 5) is first applied to the data using the total scatter of the data. Let the projection matrix of PCA be denoted by $\boldsymbol{U}$. After projection onto the PCA subspace, the summation of the intracluster scatters is $\widehat{\boldsymbol{S}}_w := \boldsymbol{U}^\top \boldsymbol{S}_w \boldsymbol{U}$ because of the quadratic characteristic of the covariance. RCA uses $\widehat{\boldsymbol{S}}_w$ as the covariance matrix in the Mahalanobis distance, i.e., Eq. (11.2). According to Eq. (11.8), the subspace of RDA is obtained through the eigenvalue (or singular value) decomposition of $\widehat{\boldsymbol{S}}_w^{-1}$ (see Eq. (11.9)).

11.3.1.4 Discriminative Component Analysis (DCA)

Discriminative Component Analysis (DCA) [26] is a spectral metric learning method based on the scatters of clusters/classes. Consider the c clusters, chunklets, or classes of data. The intraclass scatter is the same as in Eq. (11.25). The interclass scatter is:

$$\begin{aligned}\mathbb{R}^{d\times d} \ni \boldsymbol{S}_b &:= \frac{1}{n}\sum_{l=1}^{c}\sum_{j=1}^{c}(\boldsymbol{\mu}_l - \boldsymbol{\mu}_j)(\boldsymbol{\mu}_l - \boldsymbol{\mu}_j)^\top, \text{ or} \\ \mathbb{R}^{d\times d} \ni \boldsymbol{S}_b &:= \frac{1}{n}\sum_{l=1}^{c}(\boldsymbol{\mu}_l - \boldsymbol{\mu})(\boldsymbol{\mu}_l - \boldsymbol{\mu})^\top,\end{aligned} \tag{11.26}$$

where $\boldsymbol{\mu}_l$ is the mean of the l-th class and $\boldsymbol{\mu}$ is the total mean of the data. According to Proposition 11.2, metric learning can be seen as a Euclidean distance after projection onto the column space of a projection matrix $\boldsymbol{U}$, where $\boldsymbol{W} = \boldsymbol{U}\boldsymbol{U}^\top$. Similar to Fisher discriminant analysis [19] (see Chap. 6), DCA maximizes the interclass variance and minimizes the intraclass variance after projection. Therefore, its optimization is:

$$\underset{\boldsymbol{U}}{\text{maximize}} \quad \frac{\textbf{tr}(\boldsymbol{U}^\top \boldsymbol{S}_b \boldsymbol{U})}{\textbf{tr}(\boldsymbol{U}^\top \boldsymbol{S}_w \boldsymbol{U})}, \tag{11.27}$$

which is a generalized Rayleigh-Ritz quotient. The solution $\boldsymbol{U}$ to this optimization problem is the generalized eigenvalue problem $(\boldsymbol{S}_b, \boldsymbol{S}_w)$ (see Chap. 2). According to Eq. (11.9), the weight matrix of the generalized Mahalanobis distance can be set to $\boldsymbol{W} = \boldsymbol{U}\boldsymbol{U}^\top$, where $\boldsymbol{U}$ is the matrix of the eigenvectors.

11.3.1.5 High-Dimensional Discriminative Component Analysis

High-dimensional discriminative component analysis is a spectral method for metric learning [44], which minimizes and maximizes the intraclass and interclass variances, respectively, by the same optimization problem as Eq. (11.27) with an additional constraint on the orthogonality of the projection matrix, i.e., $\boldsymbol{U}^\top\boldsymbol{U} = \boldsymbol{I}$. This problem can be restated by posing a penalty on the denominator:

$$\begin{aligned}&\underset{\boldsymbol{U}}{\text{maximize}} && \textbf{tr}(\boldsymbol{U}^\top(\boldsymbol{S}_b - \lambda \boldsymbol{S}_w)\boldsymbol{U}) \\ &\text{subject to} && \boldsymbol{U}^\top\boldsymbol{U} = \boldsymbol{I},\end{aligned} \tag{11.28}$$

where $\lambda > 0$ is the regularization parameter. The solution to this problem is the eigenvalue problem for $\boldsymbol{S}_b - \lambda \boldsymbol{S}_w$. The eigenvectors are the columns of $\boldsymbol{U}$ and the weight matrix of the generalized Mahalanobis is obtained using Eq. (11.9).

If the dimensionality of data is large, computing the eigenvectors of $(\boldsymbol{S}_b - \lambda \boldsymbol{S}_w) \in \mathbb{R}^{d\times d}$ is extremely time-consuming. According to [44, Theorem 3], the optimization problem (11.28) can be solved in the orthogonal complement space of the null space of $\boldsymbol{S}_b + \boldsymbol{S}_w$, without loss of any information (see [44, Appendix A] for proof). Therefore, if $d \gg 1$, then $\boldsymbol{U}$ is found as follows. Let $\boldsymbol{X} := [\boldsymbol{x}_1, \ldots, \boldsymbol{x}_n] \in \mathbb{R}^{d\times n}$ be the matrix of the data. Let $\boldsymbol{A}_w$ and $\boldsymbol{A}_b$ be the adjacency matrices for the sets $\mathcal{S}$ and $\mathcal{D}$, respectively. For example, if $(\boldsymbol{x}_i, \boldsymbol{x}_j) \in \mathcal{S}$, then $\boldsymbol{A}_w(i, j) = 1$; otherwise, $\boldsymbol{A}_w(i, j) = 0$. If $\boldsymbol{L}_w$ and $\boldsymbol{L}_b$ are the Laplacian matrices of $\boldsymbol{A}_w$ and $\boldsymbol{A}_b$, respectively, then $\boldsymbol{S}_w = 0.5\boldsymbol{X}\boldsymbol{L}_w\boldsymbol{X}^\top$ and $\boldsymbol{S}_b = 0.5\boldsymbol{X}\boldsymbol{L}_b\boldsymbol{X}^\top$ (see [7] and Chap. 9 for proof). There is $\mathbf{tr}(\boldsymbol{S}_w + \boldsymbol{S}_b) = \mathbf{tr}(\boldsymbol{X}(0.5\boldsymbol{L}_w + 0.5\boldsymbol{L}_b)\boldsymbol{X}^\top) = \mathbf{tr}(\boldsymbol{X}^\top\boldsymbol{X}(0.5\boldsymbol{L}_w + 0.5\boldsymbol{L}_b))$ because of the cyclic property of the trace. If the rank of $\boldsymbol{L} := \boldsymbol{X}^\top\boldsymbol{X}(0.5\boldsymbol{L}_w + 0.5\boldsymbol{L}_b) \in \mathbb{R}^{n\times n}$ is $r \leq n$, it has r nonzero eigenvalues whose corresponding eigenvectors are computed. These eigenvectors are stacked in a matrix to have $\boldsymbol{V} \in \mathbb{R}^{d\times r}$. The projected intraclass and interclass variances after projection onto the column space of $\boldsymbol{V}$ are $\boldsymbol{S}'_w := \boldsymbol{V}^\top\boldsymbol{S}_w\boldsymbol{V}$ and $\boldsymbol{S}'_b := \boldsymbol{V}^\top\boldsymbol{S}_b\boldsymbol{V}$, respectively. Then, $\boldsymbol{S}'_w$ and $\boldsymbol{S}'_b$ are used in Eq. (11.28) and the weight matrix of the generalized Mahalanobis is obtained using Eq. (11.9).

11.3.1.6 Regularization by Locally Linear Embedding

The spectral metric learning methods using scatters can be modelled as a maximization problem of the following Rayleigh–Ritz quotient [5] (see Chap. 2):

$$\begin{aligned} \underset{\boldsymbol{U}}{\text{maximize}} \quad & \frac{\sum_{(\boldsymbol{x}_i,\boldsymbol{x}_j)\in\mathcal{S}} \|\boldsymbol{x}_i - \boldsymbol{x}_j\|_{\boldsymbol{W}}}{\sum_{(\boldsymbol{x}_i,\boldsymbol{x}_j)\in\mathcal{D}} \|\boldsymbol{x}_i - \boldsymbol{x}_j\|_{\boldsymbol{W}} + \lambda\Omega(\boldsymbol{U})}, \\ \text{subject to} \quad & \boldsymbol{U}^\top\boldsymbol{U} = \boldsymbol{I}, \end{aligned} \tag{11.29}$$

where $\boldsymbol{W} = \boldsymbol{U}\boldsymbol{U}^\top$ (see Eq. (11.9)), $\lambda > 0$ is the regularization parameter, and $\Omega(\boldsymbol{U})$ is a penalty or regularization term on the projection matrix $\boldsymbol{U}$. This optimization maximizes and minimizes the distances of the similar and dissimilar points, respectively. According to Sect. 11.3.1.2, Eq. (11.29) can be restated as:

$$\begin{aligned} \underset{\boldsymbol{U}}{\text{maximize}} \quad & \frac{\mathbf{tr}(\boldsymbol{U}^\top\boldsymbol{S}_b\boldsymbol{U})}{\mathbf{tr}(\boldsymbol{U}^\top\boldsymbol{S}_w\boldsymbol{U}) + \lambda\Omega(\boldsymbol{U})}, \\ \text{subject to} \quad & \boldsymbol{U}^\top\boldsymbol{U} = \boldsymbol{I}. \end{aligned} \tag{11.30}$$

As discussed in Proposition 11.2, metric learning can be seen as projection onto a subspace. The regularization term can be a linear reconstruction of every projected point by its k Nearest Neighbors (kNN), using the same reconstruction weights as before the projection [5]. The weights used for linear reconstruction in the input space can be found as in locally linear embedding [33] (see Chap. 8). If s_{ij} denotes the weight of $\boldsymbol{x}_j$ in the reconstruction of $\boldsymbol{x}_i$ and $\mathcal{N}(\boldsymbol{x}_i)$ is the set of kNN for $\boldsymbol{x}_i$, then:

$$\begin{aligned}
&\underset{s_{ij}}{\text{minimize}} && \sum_{i=1}^{n} \Big\| \boldsymbol{x}_i - \sum_{\boldsymbol{x}_j \in \mathcal{N}(\boldsymbol{x}_i)} s_{ij} \boldsymbol{x}_j \Big\|_2^2, \\
&\text{subject to} && \sum_{\boldsymbol{x}_j \in \mathcal{N}(\boldsymbol{x}_i)} s_{ij} = 1.
\end{aligned}$$

The solution to this optimization problem is (see Chap. 8):

$$s_{ij}^* = \frac{\boldsymbol{G}_i^{-1}\mathbf{1}}{\mathbf{1}^\top \boldsymbol{G}_i^{-1}\mathbf{1}},$$

where $\boldsymbol{G}_i := (\boldsymbol{x}_i\mathbf{1}^\top - \boldsymbol{X}_i)^\top(\boldsymbol{x}_i\mathbf{1}^\top - \boldsymbol{X}_i)$ in which $\boldsymbol{X}_i \in \mathbb{R}^{d\times k}$ denotes the stack of kNN for $\boldsymbol{x}_i$. The following is defined as $\boldsymbol{S}^* := [s_{ij}^*] \in \mathbb{R}^{n\times n}$. The regularization term can be the reconstruction in the subspace using the same reconstruction weights as in the input space [5]:

$$\Omega(\boldsymbol{U}) := \sum_{i=1}^{n} \Big\| \boldsymbol{U}^\top \boldsymbol{x} - \sum_{\boldsymbol{x}_j \in \mathcal{N}(\boldsymbol{x}_i)} s_{ij}^* \boldsymbol{U}^\top \boldsymbol{x}_j \Big\|_2^2 = \textbf{tr}(\boldsymbol{U}^\top \boldsymbol{X}\boldsymbol{E}\boldsymbol{X}^\top \boldsymbol{U}), \tag{11.31}$$

where $\boldsymbol{X} = [\boldsymbol{x}_1, \dots, \boldsymbol{x}_n] \in \mathbb{R}^{d\times n}$ and $\mathbb{R}^{n\times n} \ni \boldsymbol{E} := (\boldsymbol{I} - \boldsymbol{S}^*)^\top(\boldsymbol{I} - \boldsymbol{S}^*)$. Using Eq. (11.31) into Eq. (11.30) results in:

$$\begin{aligned}
&\underset{\boldsymbol{U}}{\text{maximize}} && \frac{\textbf{tr}(\boldsymbol{U}^\top \boldsymbol{S}_b \boldsymbol{U})}{\textbf{tr}\big(\boldsymbol{U}^\top(\boldsymbol{S}_w + \lambda \boldsymbol{X}\boldsymbol{E}\boldsymbol{X}^\top)\boldsymbol{U}\big)}, \\
&\text{subject to} && \boldsymbol{U}^\top \boldsymbol{U} = \boldsymbol{I}.
\end{aligned} \tag{11.32}$$

The solution to this optimization problem is the generalized eigenvalue problem $(\boldsymbol{S}_b, \boldsymbol{S}_w + \lambda \boldsymbol{X}\boldsymbol{E}\boldsymbol{X}^\top)$, where $\boldsymbol{U}$ has the eigenvectors as its columns (see Chap. 2). According to Eq. (11.9), the weight matrix of the metric is $\boldsymbol{W} = \boldsymbol{U}\boldsymbol{U}^\top$.

11.3.1.7 Fisher-HSIC Multiview Metric Learning (FISH-MML)

Fisher-HSIC Multiview Metric Learning (FISH-MML) [50] is a metric learning method used for multiview data. In multiview data, there are different types (views) of features for every data point. For example, an image dataset, which has a descriptive caption for every image, is considered to be multiview data. Let $\boldsymbol{X}^{(r)} := \{\boldsymbol{x}_i^{(r)}\}_{i=1}^n$ be the features of the data points in the r-th view, c be the number of classes/clusters, and v be the number of views. According to Proposition 11.2, metric learning is the Euclidean distance after projection with $\boldsymbol{U}$. The interclass scatter of the data, in the r-th view, is denoted by $\boldsymbol{S}_b^{(r)}$ and calculated using

Eqs. (11.26). The total scatter of the data, in the r-th view, is denoted by $\boldsymbol{S}_t^{(r)}$ and is the covariance of the data in that view.

Inspired by Fisher discriminant analysis [19] (see Chap. 6), the interclass variances of the projected data are maximized, $\sum_{r=1}^{v} \textbf{tr}(\boldsymbol{U}^\top \boldsymbol{S}_b^{(r)} \boldsymbol{U})$, to discriminate the classes after projection. Additionally, inspired by principal component analysis (see Chap. 5), the total scatter of the projected data is maximized, $\sum_{r=1}^{v} \textbf{tr}(\boldsymbol{U}^\top \boldsymbol{S}_t^{(r)} \boldsymbol{U})$, for expressiveness. Moreover, the dependence of the projected data is maximized in all views because various views of a point should be related. A measure of dependence between two random variables X and Y is the Hilbert-Schmidt Independence Criterion (HSIC) [22], whose empirical estimation is $\text{HSIC}(X, Y) = 1/((n-1)^2)\textbf{tr}(\boldsymbol{K}_x \boldsymbol{H} \boldsymbol{K}_y \boldsymbol{H})$, where $\boldsymbol{K}_x$ and $\boldsymbol{K}_y$ are kernel matrices over the X and Y variables, respectively, and $\boldsymbol{H} := \boldsymbol{I} - (1/n)\boldsymbol{1}\boldsymbol{1}^\top$ is the centering matrix (see Chap. 3). The HSIC between the projection of the two views $\boldsymbol{X}^{(r)}$ and $\boldsymbol{X}^{(w)}$ is:

$$\begin{aligned}
&\text{HSIC}(\boldsymbol{U}^\top \boldsymbol{X}^{(r)}, \boldsymbol{U}^\top \boldsymbol{X}^{(w)}) \propto \textbf{tr}(\boldsymbol{K}^{(r)} \boldsymbol{H} \boldsymbol{K}^{(w)} \boldsymbol{H}) \\
&\overset{(a)}{=} \textbf{tr}(\boldsymbol{X}^{(r)\top} \boldsymbol{U}\boldsymbol{U}^\top \boldsymbol{X}^{(r)} \boldsymbol{H} \boldsymbol{K}^{(w)} \boldsymbol{H}) \overset{(b)}{=} \textbf{tr}(\boldsymbol{U}^\top \boldsymbol{X}^{(r)} \boldsymbol{H} \boldsymbol{K}^{(w)} \boldsymbol{H} \boldsymbol{X}^{(r)\top} \boldsymbol{U}),
\end{aligned}$$

where (a) is because the linear kernel is used for $\boldsymbol{U}^\top \boldsymbol{X}^{(r)}$, i.e., $\boldsymbol{K}^{(r)} := (\boldsymbol{U}^\top \boldsymbol{X}^{(r)})^\top \boldsymbol{U}^\top \boldsymbol{X}^{(r)}$ and (b) is because of the cyclic property of the trace.

In summary, the summation of interclass scatter, total scatter, and the dependence of views are maximized, which is:

$$\begin{aligned}
&\sum_{r=1}^{v} \Big(\textbf{tr}(\boldsymbol{U}^\top \boldsymbol{S}_b^{(r)} \boldsymbol{U}) + \lambda_1 \textbf{tr}(\boldsymbol{U}^\top \boldsymbol{S}_t^{(r)} \boldsymbol{U}) + \lambda_2 \textbf{tr}(\boldsymbol{U}^\top \boldsymbol{X}^{(r)} \boldsymbol{H} \boldsymbol{K}^{(w)} \boldsymbol{H} \boldsymbol{X}^{(r)\top} \boldsymbol{U})\Big) \\
&= \sum_{r=1}^{v} \textbf{tr}\big(\boldsymbol{U}^\top (\boldsymbol{S}_b^{(r)} + \lambda_1 \boldsymbol{S}_t^{(r)} + \lambda_2 \boldsymbol{X}^{(r)} \boldsymbol{H} \boldsymbol{K}^{(w)} \boldsymbol{H} \boldsymbol{X}^{(r)\top}) \boldsymbol{U}\big),
\end{aligned}$$

where $\lambda_1 > 0$ and $\lambda_2 > 0$ are the regularization parameters. The optimization problem is:

$$\begin{aligned}
&\underset{\boldsymbol{U}}{\text{maximize}} \quad \sum_{r=1}^{v} \textbf{tr}\big(\boldsymbol{U}^\top (\boldsymbol{S}_b^{(r)} + \lambda_1 \boldsymbol{S}_t^{(r)} + \lambda_2 \boldsymbol{X}^{(r)} \boldsymbol{H} \boldsymbol{K}^{(w)} \boldsymbol{H} \boldsymbol{X}^{(r)\top}) \boldsymbol{U}\big) \\
&\text{subject to} \quad \boldsymbol{U}^\top \boldsymbol{U} = \boldsymbol{I},
\end{aligned} \tag{11.33}$$

whose solution is the eigenvalue problem for $\boldsymbol{S}_b^{(r)} + \lambda_1 \boldsymbol{S}_t^{(r)} + \lambda_2 \boldsymbol{X}^{(r)} \boldsymbol{H} \boldsymbol{K}^{(w)} \boldsymbol{H} \boldsymbol{X}^{(r)\top}$, where $\boldsymbol{U}$ has the eigenvectors as its columns (see Chap. 2).

11.3.2 *Spectral Methods Using Hinge Loss*

11.3.2.1 Large-Margin Metric Learning

k-Nearest Neighbors (kNN) classification is highly impacted by the metric used for measuring distances between points. Therefore, metric learning can be used to improve the performance of kNN classification [42, 43]. Let $y_{ij} = 1$ if $(\boldsymbol{x}_i, \boldsymbol{x}_j) \in \mathcal{S}$ and $y_{ij} = 0$ if $(\boldsymbol{x}_i, \boldsymbol{x}_j) \in \mathcal{D}$. Moreover, kNN is considered for similar points, where the nearest neighbors of every point are found among its similar points. Let $\eta_{ij} = 1$ if $(\boldsymbol{x}_i, \boldsymbol{x}_j) \in \mathcal{S}$ and $\boldsymbol{x}_j$ is among kNN of $\boldsymbol{x}_i$; otherwise, $\eta_{ij} = 0$. The optimization problem for finding the optimal weight matrix in the metric can be [42, 43]:

$$\begin{aligned} \underset{\boldsymbol{W}}{\text{minimize}} \quad & \sum_{i=1}^{n}\sum_{j=1}^{n} \eta_{ij} \|\boldsymbol{x}_i - \boldsymbol{x}_j\|_{\boldsymbol{W}}^2 \\ & + \lambda \sum_{i=1}^{n}\sum_{j=1}^{n}\sum_{l=1}^{n} \eta_{ij}(1 - y_{il})\Big[1 + \|\boldsymbol{x}_i - \boldsymbol{x}_j\|_{\boldsymbol{W}}^2 - \|\boldsymbol{x}_i - \boldsymbol{x}_l\|_{\boldsymbol{W}}^2\Big]_+, \\ \text{subject to} \quad & \boldsymbol{W} \succeq \boldsymbol{0}, \end{aligned} \tag{11.34}$$

where $\lambda > 0$ is the regularization parameter, and $[.]_+ := \max(., 0)$ is the standard Hinge loss.

The first term in Eq. (11.34) pushes the similar neighbors close to each other. The second term in this equation is the triplet loss [35], which pushes the similar neighbors to each other and pulls the dissimilar points away from one another. This is because minimizing $\|\boldsymbol{x}_i - \boldsymbol{x}_j\|_{\boldsymbol{W}}^2$ for $\eta_{ij} = 1$ decreases the distances between similar neighbors. Moreover, minimizing $-\|\boldsymbol{x}_i - \boldsymbol{x}_l\|_{\boldsymbol{W}}^2$ for $1 - y_{il} = 1$ (i.e., $y_{il} = 0$) is equivalent to maximize $\|\boldsymbol{x}_i - \boldsymbol{x}_l\|_{\boldsymbol{W}}^2$, which maximizes the distances between dissimilar points. Minimizing the whole second term forces the distances between dissimilar points to be at least greater than the distances between similar points up to a threshold (or margin) of one. The margin can be altered by changing 1 in this term with a different positive number. According to this discussion, this loss is closely related to the triplet loss for neural networks [35] (see Chap. 19).

Equation (11.34) can be restated using the slack variables $\xi_{ijl}, \forall i, j, l \in \{1, \dots, n\}$. The Hinge loss in term $[1 + \|\boldsymbol{x}_i - \boldsymbol{x}_j\|_{\boldsymbol{W}}^2 - \|\boldsymbol{x}_i - \boldsymbol{x}_l\|_{\boldsymbol{W}}^2]_+$ requires that:

$$1 + \|\boldsymbol{x}_i - \boldsymbol{x}_j\|_{\boldsymbol{W}}^2 - \|\boldsymbol{x}_i - \boldsymbol{x}_l\|_{\boldsymbol{W}}^2 \geq 0 \implies \|\boldsymbol{x}_i - \boldsymbol{x}_l\|_{\boldsymbol{W}}^2 - \|\boldsymbol{x}_i - \boldsymbol{x}_j\|_{\boldsymbol{W}}^2 \leq 1.$$

If $\xi_{ijl} \geq 0$, then the term $\|\boldsymbol{x}_i - \boldsymbol{x}_l\|_{\boldsymbol{W}}^2 - \|\boldsymbol{x}_i - \boldsymbol{x}_j\|_{\boldsymbol{W}}^2$ can be sandwiched to be minimized:

$$1 - \xi_{ijl} \leq \|\boldsymbol{x}_i - \boldsymbol{x}_l\|_{\boldsymbol{W}}^2 - \|\boldsymbol{x}_i - \boldsymbol{x}_j\|_{\boldsymbol{W}}^2 \leq 1.$$

Therefore, the term of Hinge loss can be replaced with the slack variable. Therefore, Eq. (11.34) can be restated as [42, 43]:

$$
\begin{aligned}
\underset{\boldsymbol{W}, \{\xi_{ijl}\}}{\text{minimize}} \quad & \sum_{i=1}^{n}\sum_{j=1}^{n} \eta_{ij} \|\boldsymbol{x}_i - \boldsymbol{x}_j\|_{\boldsymbol{W}}^2 + \lambda \sum_{i=1}^{n}\sum_{j=1}^{n}\sum_{l=1}^{n} \eta_{ij}(1 - y_{il})\, \xi_{ijl} \\
\text{subject to} \quad & \|\boldsymbol{x}_i - \boldsymbol{x}_l\|_{\boldsymbol{W}}^2 - \|\boldsymbol{x}_i - \boldsymbol{x}_j\|_{\boldsymbol{W}}^2 \geq 1 - \xi_{ijl}, \\
& \qquad \forall (\boldsymbol{x}_i, \boldsymbol{x}_j) \in \mathcal{S}, \eta_{ij} = 1, (\boldsymbol{x}_i, \boldsymbol{x}_l) \in \mathcal{D}, \\
& \xi_{ijl} \geq 0, \\
& \boldsymbol{W} \succeq \boldsymbol{0}.
\end{aligned}
\tag{11.35}
$$

This optimization problem is a semidefinite programming problem, which can be solved iteratively using the interior-point method (see Chap. 4).

This problem uses triplets of similar and dissimilar points, i.e., $\{\boldsymbol{x}_i, \boldsymbol{x}_j, \boldsymbol{x}_l\}$, where $(\boldsymbol{x}_i, \boldsymbol{x}_j) \in \mathcal{S}$, $\eta_{ij} = 1$, $(\boldsymbol{x}_i, \boldsymbol{x}_l) \in \mathcal{D}$. Therefore, the triplets should be extracted randomly from the dataset for metric learning. Solving semidefinite programming is usually slow and time-consuming, especially for large datasets. Triplet minimizing can be used to find the best triplets for metric learning [31]. For example, among all combinations of similar and dissimilar points, only the similar and dissimilar points with the smallest and/or largest distances can be used to limit the number of triplets [37].

11.3.2.2 Imbalanced Metric Learning (IML)

Imbalanced Metric Learning (IML) [20] is a spectral metric learning method that handles imbalanced classes by further decomposition of the similar set $\mathcal{S}$ and dissimilar set $\mathcal{D}$. Suppose the dataset is composed of two classes c_0 and c_1. Let $\mathcal{S}_0$ and $\mathcal{S}_1$ denote the similarity between sets for classes c_0 and c_1, respectively. Suppose the following similarity and dissimilarity sets are defined based on the pairs of points taken randomly from the sets $\mathcal{S}_0$ and $\mathcal{S}_1$ [20]:

$$
\begin{aligned}
\text{Sim}_0 \subseteq \mathcal{S}_0 \times \mathcal{S}_0, \quad & \text{Sim}_1 \subseteq \mathcal{S}_1 \times \mathcal{S}_1, \\
\text{Dis}_0 \subseteq \mathcal{S}_0 \times \mathcal{S}_1, \quad & \text{Dis}_1 \subseteq \mathcal{S}_1 \times \mathcal{S}_0.
\end{aligned}
$$

The optimization problem of IML is:

$$
\begin{aligned}
\underset{\boldsymbol{W}}{\text{minimize}} \quad & \frac{\lambda}{4|\text{Sim}_0|} \sum_{(\boldsymbol{x}_i, \boldsymbol{x}_j) \in \text{Sim}_0} \big[\|\boldsymbol{x}_i - \boldsymbol{x}_j\|_{\boldsymbol{W}}^2 - 1\big]_+ \\
& + \frac{\lambda}{4|\text{Sim}_1|} \sum_{(\boldsymbol{x}_i, \boldsymbol{x}_j) \in \text{Sim}_1} \big[\|\boldsymbol{x}_i - \boldsymbol{x}_j\|_{\boldsymbol{W}}^2 - 1\big]_+
\end{aligned}
$$

$$+ \frac{1-\lambda}{4|\text{Dis}_0|} \sum_{(\boldsymbol{x}_i, \boldsymbol{x}_j) \in \text{Dis}_0} \big[-\|\boldsymbol{x}_i - \boldsymbol{x}_j\|_{\boldsymbol{W}}^2 + 1 + m\big]_+$$

$$+ \frac{1-\lambda}{4|\text{Dis}_1|} \sum_{(\boldsymbol{x}_i, \boldsymbol{x}_j) \in \text{Dis}_1} \big[-\|\boldsymbol{x}_i - \boldsymbol{x}_j\|_{\boldsymbol{W}}^2 + 1 + m\big]_+ + \gamma \|\boldsymbol{W} - \boldsymbol{I}\|_F^2$$

$$\text{subject to} \quad \boldsymbol{W} \succeq \boldsymbol{0}, \tag{11.36}$$

where $|.|$ denotes the cardinality of set, $[.]_+ := \max(., 0)$ is the standard Hinge loss, $m > 0$ is the desired margin between classes, and $\lambda \in [0, 1]$ and $\gamma > 0$ are the regularization parameters. This optimization pulls similar points that are separated by a distance less than 1 and pushes dissimilar points away if the distance between them is more than $m + 1$. Additionally, the regularization term $\|\boldsymbol{W} - \boldsymbol{I}\|_F^2$ attempts to make the weight matrix, in the generalized Mahalanobis distance, close to identity for simplicity of metric. In this way, the metric becomes close to the Euclidean distance, preventing overfitting, while satisfying the desired margins between distances.

11.3.3 Locally Linear Metric Adaptation (LLMA)

Another method for metric learning is Locally Linear Metric Adaptation (LLMA) [12], which performs nonlinear and linear transformations globally and locally, respectively. For every point $\boldsymbol{x}_l$, its k nearest (similar) neighbors are considered. The local linear transformation for every point $\boldsymbol{x}_l$ is:

$$\mathbb{R}^d \ni \boldsymbol{y}_l := \boldsymbol{x}_l + \boldsymbol{B}\boldsymbol{\pi}_i, \tag{11.37}$$

where $\boldsymbol{B} \in \mathbb{R}^{d \times k}$ is the matrix of biases, $\mathbb{R}^k \ni \boldsymbol{\pi}_i = [\pi_{i1}, \ldots, \pi_{ik}]^\top$, and $\pi_{ij} := \exp(-\|\boldsymbol{x}_i - \boldsymbol{x}_j\|_2^2 / 2w^2)$ is a Gaussian measure of similarity between $\boldsymbol{x}_i$ and $\boldsymbol{x}_j$. The variables $\boldsymbol{B}$ and w are found through optimization.

In this method, the distances between the linearly transformed similar points are minimized, while attempting to preserve the distances between similar points after the transformation:

$$\underset{\{\boldsymbol{y}_i\}_{i=1}^n, \boldsymbol{B}, w, \sigma}{\text{minimize}} \quad \sum_{(\boldsymbol{y}_i, \boldsymbol{y}_j) \in \mathcal{S}} \|\boldsymbol{y}_i - \boldsymbol{y}_j\|_2^2 + \lambda \sum_{i=1}^n \sum_{j=1}^n (q_{ij} - d_{ij})^2 \exp(\frac{-d_{ij}^2}{\sigma^2}), \tag{11.38}$$

where $\lambda > 0$ is the regularization parameter, σ_2^2 is the variance to be optimized, and $d_{ij} := \|\boldsymbol{x}_i - \boldsymbol{x}_j\|_2$ and $q_{ij} := \|\boldsymbol{y}_i - \boldsymbol{y}_j\|_2$. This objective function is optimized iteratively until convergence.

11.3.4 Relevant to Support Vector Machine

Inspired by ν-Support Vector Machine (ν-SVM) [34], the weight matrix in the generalized Mahalanobis distance can be obtained through [40]:

$$
\begin{aligned}
\underset{\boldsymbol{W},\gamma,\{\xi_{il}\}}{\text{minimize}} \quad & \frac{1}{2}\|\boldsymbol{W}\|_2^2 + \frac{\lambda_1}{|\mathcal{S}|}\sum_{(\boldsymbol{x}_i,\boldsymbol{x}_j)\in\mathcal{S}} \|\boldsymbol{x}_i - \boldsymbol{x}_j\|_{\boldsymbol{W}}^2 \\
& + \lambda_2\Big(\nu\gamma + \frac{1}{|\mathcal{D}|}\sum_{(\boldsymbol{x}_i,\boldsymbol{x}_l)\in\mathcal{D}} \xi_{il}\Big) \\
\text{subject to} \quad & \boldsymbol{W} \succeq \boldsymbol{0}, \\
& \gamma \geq 0, \\
& \|\boldsymbol{x}_i - \boldsymbol{x}_j\|_{\boldsymbol{W}}^2 - \|\boldsymbol{x}_i - \boldsymbol{x}_l\|_{\boldsymbol{W}}^2 \geq \gamma - \xi_{il}, \\
& \qquad\qquad \forall(\boldsymbol{x}_i,\boldsymbol{x}_j)\in\mathcal{S}, (\boldsymbol{x}_i,\boldsymbol{x}_l)\in\mathcal{D}, \\
& \xi_{il} \geq 0, \quad \forall(\boldsymbol{x}_i,\boldsymbol{x}_l)\in\mathcal{D},
\end{aligned}
\tag{11.39}
$$

where $\lambda_1 > 0$ and $\lambda_2 > 0$ are regularization parameters. Using KKT conditions and Lagrange multipliers (see Chap. 4), the dual optimization problem is (see [40] for derivation):

$$
\begin{aligned}
\underset{\{\alpha_{ij}\}}{\text{maximize}} \quad & \sum_{(\boldsymbol{x}_i,\boldsymbol{x}_j)\in\mathcal{D}} \alpha_{ij}(\boldsymbol{x}_i - \boldsymbol{x}_j)^\top \boldsymbol{W}(\boldsymbol{x}_i - \boldsymbol{x}_j) \\
& - \frac{1}{2}\sum_{(\boldsymbol{x}_i,\boldsymbol{x}_j)\in\mathcal{D}}\sum_{(\boldsymbol{x}_k,\boldsymbol{x}_l)\in\mathcal{D}} \alpha_{ij}\alpha_{kl}((\boldsymbol{x}_i - \boldsymbol{x}_j)^\top(\boldsymbol{x}_k - \boldsymbol{x}_l))^2 \\
& + \frac{\lambda_1}{|\mathcal{S}|}\sum_{(\boldsymbol{x}_i,\boldsymbol{x}_j)\in\mathcal{D}}\sum_{(\boldsymbol{x}_k,\boldsymbol{x}_l)\in\mathcal{S}} \alpha_{ij}((\boldsymbol{x}_i - \boldsymbol{x}_j)^\top(\boldsymbol{x}_k - \boldsymbol{x}_l))^2 \\
\text{subject to} \quad & \frac{1}{\lambda_2}\sum_{(\boldsymbol{x}_i,\boldsymbol{x}_j)\in\mathcal{D}} \alpha_{ij} \geq \nu, \\
& \alpha_{ij} \in [0, \frac{\lambda_2}{|\mathcal{D}|}],
\end{aligned}
\tag{11.40}
$$

where $\{\alpha_{ij}\}$ are the dual variables. This problem is a quadratic programming problem and can be solved using optimization solvers.

11.3.5 Relevant to Multidimensional Scaling

Multidimensional Scaling (MDS) attempts to preserve the distance after projection onto its subspace [16] (see Chap. 7). Proposition 11.2 demonstrated that metric learning can be seen as a projection onto the column space of $\boldsymbol{U}$, where $\boldsymbol{W} = \boldsymbol{U}\boldsymbol{U}^\top$. Inspired by MDS, it is possible to learn a metric that preserves the distances between points after projection onto the subspace of the metric [51]:

$$\begin{aligned} \underset{\boldsymbol{W}}{\text{minimize}} \quad & \sum_{i=1}^{n}\sum_{j=1}^{n}(\|\boldsymbol{x}_i - \boldsymbol{x}_j\|_2^2 - \|\boldsymbol{x}_i - \boldsymbol{x}_j\|_{\boldsymbol{W}}^2)^2 \\ \text{subject to} \quad & \boldsymbol{W} \succeq \boldsymbol{0}. \end{aligned} \tag{11.41}$$

It can be solved using any optimization method, such as the projected gradient method (see Chap. 4).

11.3.6 Kernel Spectral Metric Learning

Let $k(\boldsymbol{x}_i, \boldsymbol{x}_j) := \boldsymbol{\phi}(\boldsymbol{x}_i)^\top \boldsymbol{\phi}(\boldsymbol{x}_j)$ be the kernel function over the data points $\boldsymbol{x}_i$ and $\boldsymbol{x}_j$, where $\boldsymbol{\phi}(.)$ is the pulling function to the Reproducing Kernel Hilbert Space (RKHS) (see Chap. 3). Let $\mathbb{R}^{n\times n} \ni \boldsymbol{K} := \boldsymbol{\Phi}(\boldsymbol{X})^\top \boldsymbol{\Phi}(\boldsymbol{X})$ be the kernel matrix of the data. The following sections introduce several kernel spectral metric learning methods.

11.3.6.1 Using Eigenvalue Decomposition of the Kernel

Eigenvalue Decomposition of Kernel is considered to be a kernel method for spectral metric learning [48]. There are two approaches to using the eigenvalue decomposition of the kernel; this section reviews one of these approaches. The eigenvalue decomposition of the kernel matrix is:

$$\boldsymbol{K} = \sum_{r=1}^{p} \beta_r^2 \boldsymbol{\alpha}_r \boldsymbol{\alpha}_r^\top \overset{(a)}{=} \sum_{r=1}^{p} \beta_r^2 \boldsymbol{K}_r \tag{11.42}$$

where p is the rank of kernel matrix, β_r^2 is the nonnegative r-th eigenvalue (because $\boldsymbol{K} \succeq \boldsymbol{0}$), $\boldsymbol{\alpha}_r \in \mathbb{R}^n$ is the r-th eigenvector, and (a) is for the definition of $\boldsymbol{K}_r := \boldsymbol{\alpha}_r \boldsymbol{\alpha}_r^\top$. Consider $\{\beta_r^2\}_{r=1}^p$ to be learnable parameters and not the eigenvalues. Therefore, $\{\beta_r^2\}_{r=1}^p$ should be learned for the sake of metric learning. The distance metric of the pulled data points to RKHS is (see Chap. 3):

$$\|\boldsymbol{\phi}(\boldsymbol{x}_i) - \boldsymbol{\phi}(\boldsymbol{x}_j)\|_2^2 = k(\boldsymbol{x}_i, \boldsymbol{x}_i) + k(\boldsymbol{x}_j, \boldsymbol{x}_j) - 2k(\boldsymbol{x}_i, \boldsymbol{x}_j). \tag{11.43}$$

In metric learning, the goal is to decrease the distances between similar points. Therefore:

$$\sum_{(x_i,x_j)\in\mathcal{S}} \|\boldsymbol{\phi}(\boldsymbol{x}_i) - \boldsymbol{\phi}(\boldsymbol{x}_j)\|_2^2 = \sum_{(x_i,x_j)\in\mathcal{S}} k(\boldsymbol{x}_i, \boldsymbol{x}_i) + k(\boldsymbol{x}_j, \boldsymbol{x}_j) - 2k(\boldsymbol{x}_i, \boldsymbol{x}_j)$$

$$\overset{(11.42)}{=} \sum_{r=1}^{p} \beta_r^2 \sum_{(x_i,x_j)\in\mathcal{S}} k_r(\boldsymbol{x}_i, \boldsymbol{x}_i) + k_r(\boldsymbol{x}_j, \boldsymbol{x}_j) - 2k_r(\boldsymbol{x}_i, \boldsymbol{x}_j)$$

$$\overset{(a)}{=} \sum_{r=1}^{p} \beta_r^2 \sum_{(x_i,x_j)\in\mathcal{S}} (\boldsymbol{e}_i - \boldsymbol{e}_j)^\top \boldsymbol{K}_r(\boldsymbol{e}_i - \boldsymbol{e}_j) \overset{(b)}{=} \sum_{r=1}^{p} \beta_r^2 f_r \overset{(c)}{=} \boldsymbol{\beta}^\top \boldsymbol{D}_{\mathcal{S}} \boldsymbol{\beta},$$

where (a) is because $\boldsymbol{e}_i$ is the vector whose i-th element is one and the other elements are zero, (b) is because the following is defined as $f_r := \sum_{(x_i,x_j)\in\mathcal{S}} (\boldsymbol{e}_i - \boldsymbol{e}_j)^\top \boldsymbol{K}_r(\boldsymbol{e}_i - \boldsymbol{e}_j)$, and (c) is because the following is defined as $\boldsymbol{D}_{\mathcal{S}} := \textbf{diag}([f_1, \ldots, f_p]^\top)$ and $\boldsymbol{\beta} := [\beta_1, \ldots, \beta_p]^\top$. By adding a constraint on the summation of $\{\beta_r^2\}_{r=1}^p$, the optimization problem for metric learning becomes:

$$\begin{aligned} &\underset{\boldsymbol{\beta}}{\text{minimize}} \quad \boldsymbol{\beta}^\top \boldsymbol{D}_{\mathcal{S}} \boldsymbol{\beta} \\ &\text{subject to} \quad \mathbf{1}^\top \boldsymbol{\beta} = 1. \end{aligned} \tag{11.44}$$

This optimization is similar to one of the optimization problems in locally linear embedding [33] (see Chap. 8). The Lagrangian for this problem is:

$$\mathcal{L} = \boldsymbol{\beta}^\top \boldsymbol{D}_{\mathcal{S}} \boldsymbol{\beta} - \lambda(\mathbf{1}^\top \boldsymbol{\beta} - 1),$$

where λ is the dual variable. Taking the derivative of the Lagrangian with respect to the variables and setting it to zero results in:

$$\frac{\partial \mathcal{L}}{\partial \boldsymbol{\beta}} = 2\boldsymbol{D}_{\mathcal{S}}\boldsymbol{\beta} - \lambda\mathbf{1} \overset{\text{set}}{=} 0 \implies \boldsymbol{\beta} = \frac{\lambda}{2}\boldsymbol{D}_{\mathcal{S}}^{-1}\mathbf{1},$$

$$\frac{\partial \mathcal{L}}{\partial \lambda} = \mathbf{1}^\top \boldsymbol{\beta} - 1 \overset{\text{set}}{=} 0 \implies \mathbf{1}^\top \boldsymbol{\beta} = 1 \implies \frac{\lambda}{2}\mathbf{1}^\top \boldsymbol{D}_{\mathcal{S}}^{-1}\mathbf{1} = 1 \implies \lambda = \frac{2}{\mathbf{1}^\top \boldsymbol{D}_{\mathcal{S}}^{-1}\mathbf{1}}$$

$$\implies \boldsymbol{\beta} = \frac{\boldsymbol{D}_{\mathcal{S}}^{-1}\mathbf{1}}{\mathbf{1}^\top \boldsymbol{D}_{\mathcal{S}}^{-1}\mathbf{1}}.$$

Therefore, the optimal $\boldsymbol{\beta}$ is obtained for metric learning in the RKHS, where the distances of similar points are smaller than in the input Euclidean space.

11.3.6.2 Regularization by Locally Linear Embedding

Baghshah and Shouraki's metric learning method [5], which was introduced in Sect. 11.3.1.6, can be kernelized. Recall that this method used locally linear embedding for regularization. According to representation theory, the solution in the RKHS can be represented as a linear combination of all pulled data points to RKHS (see Chap. 3):

$$\boldsymbol{\Phi}(\boldsymbol{U}) = \boldsymbol{\Phi}(\boldsymbol{X})\boldsymbol{T}, \tag{11.45}$$

where $\boldsymbol{X} = [\boldsymbol{x}_1, \dots, \boldsymbol{x}_n]$, and $\boldsymbol{T} \in \mathbb{R}^{n \times p}$ (p is the dimensionality of subspace) is the coefficients.

The similarity and dissimilarity adjacency matrices are defined as:

$$\begin{aligned}
\boldsymbol{A}_S(i,j) &:= \begin{cases} 1 \text{ if } (\boldsymbol{x}_i, \boldsymbol{x}_j) \in \mathcal{S}, \\ 0 \text{ otherwise.} \end{cases} \\
\boldsymbol{A}_D(i,j) &:= \begin{cases} 1 \text{ if } (\boldsymbol{x}_i, \boldsymbol{x}_j) \in \mathcal{D}, \\ 0 \text{ otherwise.} \end{cases}
\end{aligned} \tag{11.46}$$

Let $\boldsymbol{L}_w$ and $\boldsymbol{L}_b$ denote the Laplacian matrices of these adjacency matrices (see Chap. 9):

$$\boldsymbol{L}_w := \boldsymbol{D}_S - \boldsymbol{A}_S(i,j), \quad \boldsymbol{L}_b := \boldsymbol{D}_D - \boldsymbol{A}_D(i,j),$$

where $\boldsymbol{D}_S(i,i) := \sum_{j=1}^n \boldsymbol{A}_S(i,j)$ and $\boldsymbol{D}_D(i,i) := \sum_{j=1}^n \boldsymbol{A}_D(i,j)$ are diagonal matrices. The terms in the objective of Eq. (11.32) can be restated using the Laplacian of the adjacency matrices, rather than the scatters:

$$\begin{aligned}
\underset{\boldsymbol{U}}{\text{maximize}} \quad & \frac{\textbf{tr}(\boldsymbol{U}^\top \boldsymbol{L}_b \boldsymbol{U})}{\textbf{tr}\big(\boldsymbol{U}^\top (\boldsymbol{L}_w + \lambda \boldsymbol{X}\boldsymbol{E}\boldsymbol{X}^\top)\boldsymbol{U}\big)}, \\
\text{subject to} \quad & \boldsymbol{U}^\top \boldsymbol{U} = \boldsymbol{I}.
\end{aligned} \tag{11.47}$$

According to representation theory, the pulled Laplacian matrices to RKHS are $\boldsymbol{\Phi}(\boldsymbol{L}_b) = \boldsymbol{\Phi}(\boldsymbol{X})\boldsymbol{L}_b\boldsymbol{\Phi}(\boldsymbol{X})^\top$ and $\boldsymbol{\Phi}(\boldsymbol{L}_w) = \boldsymbol{\Phi}(\boldsymbol{X})\boldsymbol{L}_w\boldsymbol{\Phi}(\boldsymbol{X})^\top$. Therefore, the numerator of Eq. (11.32) in RKHS becomes:

$$\begin{aligned}
&\textbf{tr}(\boldsymbol{\Phi}(\boldsymbol{U})^\top \boldsymbol{\Phi}(\boldsymbol{X})\boldsymbol{L}_b\boldsymbol{\Phi}(\boldsymbol{X})^\top \boldsymbol{\Phi}(\boldsymbol{U})) \\
&= \textbf{tr}\big(\boldsymbol{T}^\top \boldsymbol{\Phi}(\boldsymbol{X})^\top \boldsymbol{\Phi}(\boldsymbol{X})\boldsymbol{L}_b\boldsymbol{\Phi}(\boldsymbol{X})^\top \boldsymbol{\Phi}(\boldsymbol{X})\boldsymbol{T}\big) \overset{(a)}{=} \textbf{tr}\big(\boldsymbol{T}^\top \boldsymbol{K}_x \boldsymbol{L}_b \boldsymbol{K}_x \boldsymbol{T}\big),
\end{aligned}$$

where (a) is because of the kernel trick (see Chap. 3), i.e.,

$$\boldsymbol{K}_x := \boldsymbol{\Phi}(\boldsymbol{X})^\top \boldsymbol{\Phi}(\boldsymbol{X}). \tag{11.48}$$

Similarly, the denominator of Eq. (11.32) in RKHS becomes:

$$\begin{aligned}
&\mathbf{tr}\big(\boldsymbol{\Phi}(\boldsymbol{U})^\top(\boldsymbol{\Phi}(\boldsymbol{X})\boldsymbol{L}_w\boldsymbol{\Phi}(\boldsymbol{X})^\top + \lambda\boldsymbol{\Phi}(\boldsymbol{X})\boldsymbol{E}\boldsymbol{\Phi}(\boldsymbol{X})^\top)\boldsymbol{\Phi}(\boldsymbol{U})\big)\\
&\overset{(11.45)}{=} \mathbf{tr}\big(\boldsymbol{T}^\top\boldsymbol{\Phi}(\boldsymbol{X})^\top(\boldsymbol{\Phi}(\boldsymbol{X})\boldsymbol{L}_w\boldsymbol{\Phi}(\boldsymbol{X})^\top + \lambda\boldsymbol{\Phi}(\boldsymbol{X})\boldsymbol{E}\boldsymbol{\Phi}(\boldsymbol{X})^\top)\boldsymbol{\Phi}(\boldsymbol{X})\boldsymbol{T}\big)\\
&\overset{(11.48)}{=} \mathbf{tr}\big(\boldsymbol{T}^\top\boldsymbol{K}_x(\boldsymbol{L}_w + \lambda\boldsymbol{E})\boldsymbol{K}_x\boldsymbol{T}\big).
\end{aligned}$$

The constraint in RKHS becomes:

$$\boldsymbol{\Phi}(\boldsymbol{U})^\top\boldsymbol{\Phi}(\boldsymbol{U}) \overset{(11.45)}{=} \boldsymbol{T}^\top\boldsymbol{\Phi}(\boldsymbol{X})^\top\boldsymbol{\Phi}(\boldsymbol{X})\boldsymbol{T} \overset{(11.48)}{=} \boldsymbol{T}^\top\boldsymbol{K}_x\boldsymbol{T}.$$

Hence, Eq. (11.32) in RKHS is:

$$\begin{aligned}
&\underset{\boldsymbol{T}}{\text{maximize}} \quad \frac{\mathbf{tr}\big(\boldsymbol{T}^\top\boldsymbol{K}_x\boldsymbol{L}_b\boldsymbol{K}_x\boldsymbol{T}\big)}{\mathbf{tr}\big(\boldsymbol{T}^\top\boldsymbol{K}_x(\boldsymbol{L}_w + \lambda\boldsymbol{E})\boldsymbol{K}_x\boldsymbol{T}\big)},\\
&\text{subject to} \quad \boldsymbol{T}^\top\boldsymbol{K}_x\boldsymbol{T} = \boldsymbol{I}.
\end{aligned} \tag{11.49}$$

It can be solved using the projected gradient method (see Chap. 4) to find the optimal $\boldsymbol{T}$. Then, the data projected onto the subspace of the metric are found as:

$$\boldsymbol{\Phi}(\boldsymbol{U})^\top\boldsymbol{\Phi}(\boldsymbol{X}) \overset{(11.45)}{=} \boldsymbol{T}^\top\boldsymbol{\Phi}(\boldsymbol{X})^\top\boldsymbol{\Phi}(\boldsymbol{X}) \overset{(11.48)}{=} \boldsymbol{T}^\top\boldsymbol{K}_x. \tag{11.50}$$

11.3.6.3 Regularization by Laplacian

Regularization by Laplacian is another kernel spectral metric learning method [4] whose optimization is in the form:

$$\begin{aligned}
&\underset{\boldsymbol{\Phi}(X)}{\text{minimize}} \quad \frac{1}{|\mathcal{S}|}\sum_{(\boldsymbol{x}_i,\boldsymbol{x}_j)\in\mathcal{S}} \|\boldsymbol{\phi}(\boldsymbol{x}_i) - \boldsymbol{\phi}(\boldsymbol{x}_j)\|_2^2 + \lambda\Omega(\boldsymbol{\Phi}(X)),\\
&\text{subject to} \quad \|\boldsymbol{\phi}(\boldsymbol{x}_i) - \boldsymbol{\phi}(\boldsymbol{x}_j)\|_2^2 \geq c, \quad \forall(\boldsymbol{x}_i, \boldsymbol{x}_j) \in \mathcal{D},
\end{aligned} \tag{11.51}$$

where $c > 0$ is a hyperparameter and $\lambda > 0$ is the regularization parameter. Consider the kNN graph of data with an adjacency matrix $\boldsymbol{A} \in \mathbb{R}^{n\times n}$ whose (i, j)-th element is one if $\boldsymbol{x}_i$ and $\boldsymbol{x}_j$ are neighbors and is zero otherwise. Let the Laplacian matrix of this adjacency matrix be denoted by $\boldsymbol{L}$.

In this method, the regularization term $\Omega(\boldsymbol{\Phi}(X))$ can be the objective of the Laplacian eigenmap (see Chap. 9):

$$\begin{aligned}
\Omega(\boldsymbol{\Phi}(X)) &:= \frac{1}{2n}\sum_{i=1}^{n}\sum_{j=1}^{n} \|\boldsymbol{\phi}(\boldsymbol{x}_i) - \boldsymbol{\phi}(\boldsymbol{x}_j)\|_2^2 A(i, j)\\
&\overset{(a)}{=} \mathbf{tr}(\boldsymbol{\Phi}(\boldsymbol{X})\boldsymbol{L}\boldsymbol{\Phi}(\boldsymbol{X})^\top) \overset{(b)}{=} \mathbf{tr}(\boldsymbol{L}\boldsymbol{\Phi}(\boldsymbol{X})^\top\boldsymbol{\Phi}(\boldsymbol{X})) \overset{(11.48)}{=} \mathbf{tr}(\boldsymbol{L}\boldsymbol{K}_x),
\end{aligned}$$

where (a) is according to [6] (see Chap. 9 for proof) and (b) is because of the cyclic property of the trace. Moreover, according to Eq. (11.43), the distance in RKHS is $\|\boldsymbol{\phi}(\boldsymbol{x}_i)-\boldsymbol{\phi}(\boldsymbol{x}_j)\|_2^2 = k(\boldsymbol{x}_i,\boldsymbol{x}_i)+k(\boldsymbol{x}_j,\boldsymbol{x}_j)-2k(\boldsymbol{x}_i,\boldsymbol{x}_j)$. This term can be simplified in Eq. (11.51) as:

$$\frac{1}{|\mathcal{S}|}\sum_{(\boldsymbol{x}_i,\boldsymbol{x}_j)\in\mathcal{S}}\|\boldsymbol{\phi}(\boldsymbol{x}_i)-\boldsymbol{\phi}(\boldsymbol{x}_j)\|_2^2 \overset{(11.43)}{=} \frac{1}{|\mathcal{S}|}\sum_{(\boldsymbol{x}_i,\boldsymbol{x}_j)\in\mathcal{S}} k(\boldsymbol{x}_i,\boldsymbol{x}_i)+k(\boldsymbol{x}_j,\boldsymbol{x}_j)-2k(\boldsymbol{x}_i,\boldsymbol{x}_j)$$

$$= \frac{1}{|\mathcal{S}|}\sum_{(\boldsymbol{x}_i,\boldsymbol{x}_j)\in\mathcal{S}}(\boldsymbol{e}_i-\boldsymbol{e}_j)^\top\boldsymbol{K}_x(\boldsymbol{e}_i-\boldsymbol{e}_j) \overset{(a)}{=} \mathbf{tr}(\boldsymbol{E}_{\mathcal{S}}\boldsymbol{K}_x),$$

where (a) is because the scalar is equal to its trace and the cyclic property of the trace is used, i.e., $(\boldsymbol{e}_i-\boldsymbol{e}_j)^\top\boldsymbol{K}_x(\boldsymbol{e}_i-\boldsymbol{e}_j) = \mathbf{tr}((\boldsymbol{e}_i-\boldsymbol{e}_j)^\top\boldsymbol{K}_x(\boldsymbol{e}_i-\boldsymbol{e}_j)) = \mathbf{tr}((\boldsymbol{e}_i-\boldsymbol{e}_j)(\boldsymbol{e}_i-\boldsymbol{e}_j)^\top\boldsymbol{K}_x)$, and then the following is defined as $\boldsymbol{E}_{\mathcal{S}} := (1/|\mathcal{S}|)\sum_{(\boldsymbol{x}_i,\boldsymbol{x}_j)\in\mathcal{S}}(\boldsymbol{e}_i-\boldsymbol{e}_j)(\boldsymbol{e}_i-\boldsymbol{e}_j)^\top$.

Therefore, Eq. (11.51) can be restated as:

$$\begin{aligned}
&\underset{\boldsymbol{K}_x}{\text{minimize}} && \mathbf{tr}(\boldsymbol{E}_{\mathcal{S}}\boldsymbol{K}_x)+\lambda\,\mathbf{tr}(\boldsymbol{L}\boldsymbol{K}_x),\\
&\text{subject to} && k(\boldsymbol{x}_i,\boldsymbol{x}_i)+k(\boldsymbol{x}_j,\boldsymbol{x}_j)-2k(\boldsymbol{x}_i,\boldsymbol{x}_j)\ge c,\\
& && \quad\forall(\boldsymbol{x}_i,\boldsymbol{x}_j)\in\mathcal{D},\\
& && \boldsymbol{K}_x\succeq\boldsymbol{0},
\end{aligned} \tag{11.52}$$

noticing that the kernel matrix is positive semidefinite. This problem is a Semidefinite Programming (SDP) problem and can be solved using the interior point method (see Chap. 4). The optimal kernel matrix can be decomposed using eigenvalue decomposition to find the embedding of the data in RKHS, i.e., $\boldsymbol{\Phi}(\boldsymbol{X})$:

$$\boldsymbol{K}_x = \boldsymbol{V}^\top\boldsymbol{\Sigma}\boldsymbol{V} = \boldsymbol{V}^\top\boldsymbol{\Sigma}^{(1/2}\boldsymbol{\Sigma}^{(1/2)}\boldsymbol{V} \overset{(11.48)}{=} \boldsymbol{\Phi}(\boldsymbol{X})^\top\boldsymbol{\Phi}(\boldsymbol{X}),$$

where $\boldsymbol{V}$ and $\boldsymbol{\Sigma}$ are the eigenvectors and eigenvalues, (a) is because $\boldsymbol{K}_x\succeq\boldsymbol{0}$, so its eigenvalues are nonnegative and can be taken the second root of, and (b) is because $\boldsymbol{\Phi}(\boldsymbol{X}) := \boldsymbol{\Sigma}^{(1/2)}\boldsymbol{V}$ is taken.

11.3.6.4 Kernel Discriminative Component Analysis

This section explores the kernel version of DCA [26] which was introduced in Sect. 11.3.1.4.

Lemma 11.4 *The generalized Mahalanobis distance metric in RKHS, with the pulled weight matrix to RKHS denoted by $\boldsymbol{\Phi}(\boldsymbol{W})$, can be seen as measuring the Euclidean distance in RKHS after projection onto the column subspace of $\boldsymbol{T}$, where*

$\boldsymbol{T}$ is the coefficient of the matrix in Eq. (11.45). In other words:

$$\|\boldsymbol{\phi}(\boldsymbol{x}_i)-\boldsymbol{\phi}(\boldsymbol{x}_j)\|^2_{\boldsymbol{\Phi}(\boldsymbol{W})} = \|\boldsymbol{k}_i - \boldsymbol{k}_j\|^2_{\boldsymbol{T}\boldsymbol{T}^\top} = (\boldsymbol{k}_i - \boldsymbol{k}_j)^\top \boldsymbol{T}\boldsymbol{T}^\top (\boldsymbol{k}_i - \boldsymbol{k}_j), \tag{11.53}$$

where $\boldsymbol{k}_i := \boldsymbol{k}(\boldsymbol{X}, \boldsymbol{x}_i) = \boldsymbol{\Phi}(\boldsymbol{X})^\top \boldsymbol{\phi}(\boldsymbol{x}_i) = [k(\boldsymbol{x}_1, \boldsymbol{x}_i), \dots, k(\boldsymbol{x}_n, \boldsymbol{x}_i)]^\top \in \mathbb{R}^n$ is the kernel vector between $\boldsymbol{X}$ and $\boldsymbol{x}_i$.

Proof The decomposition of the weight matrix, i.e. Eq. (11.9), in RKHS is:

$$\boldsymbol{\Phi}(\boldsymbol{W}) = \boldsymbol{\Phi}(\boldsymbol{U})\boldsymbol{\Phi}(\boldsymbol{U})^\top. \tag{11.54}$$

The generalized Mahalanobis distance metric in RKHS is:

$$\begin{aligned}
&\|\boldsymbol{\phi}(\boldsymbol{x}_i) - \boldsymbol{\phi}(\boldsymbol{x}_j)\|^2_{\boldsymbol{\Phi}(\boldsymbol{W})} \overset{(11.9)}{=} (\boldsymbol{\phi}(\boldsymbol{x}_i) - \boldsymbol{\phi}(\boldsymbol{x}_j))^\top \boldsymbol{\Phi}(\boldsymbol{U})\boldsymbol{\Phi}(\boldsymbol{U})^\top (\boldsymbol{\phi}(\boldsymbol{x}_i) - \boldsymbol{\phi}(\boldsymbol{x}_j)) \\
&= \big(\boldsymbol{\Phi}(\boldsymbol{U})^\top \boldsymbol{\phi}(\boldsymbol{x}_i) - \boldsymbol{\Phi}(\boldsymbol{U})^\top \boldsymbol{\phi}(\boldsymbol{x}_j)\big)^\top \big(\boldsymbol{\Phi}(\boldsymbol{U})^\top \boldsymbol{\phi}(\boldsymbol{x}_i) - \boldsymbol{\Phi}(\boldsymbol{U})^\top \boldsymbol{\phi}(\boldsymbol{x}_j)\big) \\
&\overset{(11.45)}{=} \big(\boldsymbol{T}^\top \boldsymbol{\Phi}(\boldsymbol{X})^\top \boldsymbol{\phi}(\boldsymbol{x}_i) - \boldsymbol{T}^\top \boldsymbol{\Phi}(\boldsymbol{X})^\top \boldsymbol{\phi}(\boldsymbol{x}_j)\big)^\top \big(\boldsymbol{T}^\top \boldsymbol{\Phi}(\boldsymbol{X})^\top \boldsymbol{\phi}(\boldsymbol{x}_i) - \boldsymbol{T}^\top \boldsymbol{\Phi}(\boldsymbol{X})^\top \boldsymbol{\phi}(\boldsymbol{x}_j)\big) \\
&\overset{(a)}{=} \big(\boldsymbol{T}^\top \boldsymbol{k}_i - \boldsymbol{T}^\top \boldsymbol{k}_j\big)^\top \big(\boldsymbol{T}^\top \boldsymbol{k}_i - \boldsymbol{T}^\top \boldsymbol{k}_j\big) = \big(\boldsymbol{k}_i - \boldsymbol{k}_j\big)^\top \boldsymbol{T}\boldsymbol{T}^\top \big(\boldsymbol{k}_i - \boldsymbol{k}_j\big) = \|\boldsymbol{k}_i - \boldsymbol{k}_j\|^2_{\boldsymbol{T}\boldsymbol{T}^\top},
\end{aligned}$$

where (a) is because of the kernel trick, i.e., $\boldsymbol{k}(\boldsymbol{X}, \boldsymbol{x}_i) = \boldsymbol{\Phi}(\boldsymbol{X})^\top \boldsymbol{\phi}(\boldsymbol{x}_i)$. □

Let $\boldsymbol{v}_l := [\frac{1}{n_l}\sum_{i=1}^{n_l} k(\boldsymbol{x}_1, \boldsymbol{x}_i), \dots, \frac{1}{n_l}\sum_{i=1}^{n_l} k(\boldsymbol{x}_n, \boldsymbol{x}_i)]^\top \in \mathbb{R}^n$, where n_l denotes the cardinality of the l-th class. Let $\boldsymbol{K}_w$ and $\boldsymbol{K}_b$ be the kernelized versions of $\boldsymbol{S}_w$ and $\boldsymbol{S}_b$, respectively (see Eqs. (11.25) and (11.26)). If $\mathcal{X}_l$ denotes the l-th class, then:

$$\mathbb{R}^{n \times n} \ni \boldsymbol{K}_w := \frac{1}{n}\sum_{l=1}^{c}\sum_{\boldsymbol{x}_i \in \mathcal{X}_l} (\boldsymbol{k}_i - \boldsymbol{v}_l)(\boldsymbol{k}_i - \boldsymbol{v}_l)^\top \tag{11.55}$$

$$\mathbb{R}^{n \times n} \ni \boldsymbol{K}_b := \frac{1}{n}\sum_{l=1}^{c}\sum_{j=1}^{c} (\boldsymbol{v}_l - \boldsymbol{v}_j)(\boldsymbol{v}_l - \boldsymbol{v}_j)^\top. \tag{11.56}$$

The metric in RKHS can be seen as the projection onto a subspace with the projection matrix $\boldsymbol{T}$. Therefore, Eq. (11.27) in RKHS becomes [26]:

$$\underset{\boldsymbol{T}}{\text{maximize}} \quad \frac{\mathbf{tr}(\boldsymbol{T}^\top \boldsymbol{K}_b \boldsymbol{T})}{\mathbf{tr}(\boldsymbol{T}^\top \boldsymbol{K}_w \boldsymbol{T})}, \tag{11.57}$$

which is a generalized Rayleigh-Ritz quotient. The solution $\boldsymbol{T}$ to this optimization problem is the generalized eigenvalue problem $(\boldsymbol{K}_b, \boldsymbol{K}_w)$ (see Chap. 2). The weight matrix of the generalized Mahalanobis distance is obtained through Eqs. (11.45) and (11.54).

11.3.6.5 Relevant to Kernel Fisher Discriminant Analysis

This section explains the kernel version of the metric learning method [2] which was introduced in Sect. 11.3.1.2.

According to Eq. (11.53):

$$\|\boldsymbol{\phi}(\boldsymbol{x}_i) - \boldsymbol{\phi}(\boldsymbol{x}_j)\|_{\boldsymbol{\Phi}(\boldsymbol{W})}^2 = (\boldsymbol{k}_i - \boldsymbol{k}_j)^\top \boldsymbol{T}\boldsymbol{T}^\top (\boldsymbol{k}_i - \boldsymbol{k}_j)$$
$$\overset{(a)}{=} \mathbf{tr}\big((\boldsymbol{k}_i - \boldsymbol{k}_j)^\top \boldsymbol{T}\boldsymbol{T}^\top (\boldsymbol{k}_i - \boldsymbol{k}_j)\big) \overset{(b)}{=} \mathbf{tr}\big(\boldsymbol{T}^\top (\boldsymbol{k}_i - \boldsymbol{k}_j)(\boldsymbol{k}_i - \boldsymbol{k}_j)^\top \boldsymbol{T}\big),$$

where (a) is because a scalar is equal to its trace and (b) is because of the cyclic property of that trace. Therefore, Eq. (11.20) in RKHS becomes:

$$\frac{1}{|\mathcal{S}|} \sum_{(\boldsymbol{x}_i,\boldsymbol{x}_j)\in\mathcal{S}} \mathbf{tr}\big(\boldsymbol{T}^\top (\boldsymbol{k}_i - \boldsymbol{k}_j)(\boldsymbol{k}_i - \boldsymbol{k}_j)^\top \boldsymbol{T}\big)$$
$$= \mathbf{tr}\Big(\boldsymbol{T}^\top \big(\frac{1}{|\mathcal{S}|} \sum_{(\boldsymbol{x}_i,\boldsymbol{x}_j)\in\mathcal{S}} (\boldsymbol{k}_i - \boldsymbol{k}_j)(\boldsymbol{k}_i - \boldsymbol{k}_j)^\top \boldsymbol{T}\big)\Big) = \mathbf{tr}(\boldsymbol{T}^\top \boldsymbol{\Sigma}_{\mathcal{S}}^{\phi} \boldsymbol{T}),$$

and likewise:

$$\frac{1}{|\mathcal{D}|} \sum_{(\boldsymbol{x}_i,\boldsymbol{x}_j)\in\mathcal{D}} \mathbf{tr}\big(\boldsymbol{T}^\top (\boldsymbol{k}_i - \boldsymbol{k}_j)(\boldsymbol{k}_i - \boldsymbol{k}_j)^\top \boldsymbol{T}\big) = \mathbf{tr}(\boldsymbol{T}^\top \boldsymbol{\Sigma}_{\mathcal{D}}^{\phi} \boldsymbol{T}),$$

where:

$$\boldsymbol{\Sigma}_{\mathcal{S}}^{\phi} := \frac{1}{|\mathcal{S}|} \sum_{(\boldsymbol{x}_i,\boldsymbol{x}_j)\in\mathcal{S}} (\boldsymbol{k}_i - \boldsymbol{k}_j)(\boldsymbol{k}_i - \boldsymbol{k}_j)^\top,$$
$$\boldsymbol{\Sigma}_{\mathcal{D}}^{\phi} := \frac{1}{|\mathcal{D}|} \sum_{(\boldsymbol{x}_i,\boldsymbol{x}_j)\in\mathcal{D}} (\boldsymbol{k}_i - \boldsymbol{k}_j)(\boldsymbol{k}_i - \boldsymbol{k}_j)^\top.$$

Therefore, in RKHS, the objective of the optimization problem (11.23) becomes $\mathbf{tr}(\boldsymbol{T}^\top(\boldsymbol{\Sigma}_{\mathcal{S}}^{\phi} - \boldsymbol{\Sigma}_{\mathcal{D}}^{\phi})\boldsymbol{T}^\top)$. The constraint in Eq. (11.23) can be changed to $\boldsymbol{U}^\top\boldsymbol{U} = \boldsymbol{I}$. In RKHS, this constraint becomes:

$$\boldsymbol{\Phi}(\boldsymbol{U})^\top \boldsymbol{\Phi}(\boldsymbol{U}) \overset{(11.45)}{=} \boldsymbol{T}^\top \boldsymbol{\Phi}(\boldsymbol{X})^\top \boldsymbol{\Phi}(\boldsymbol{X}) \boldsymbol{T} \overset{(11.48)}{=} \boldsymbol{T}^\top \boldsymbol{K}_x \boldsymbol{T} \overset{\text{set}}{=} \boldsymbol{I},$$

Lastly, (11.23) in RKHS becomes:

$$\begin{aligned} \underset{\boldsymbol{T}}{\text{minimize}} \quad & \mathbf{tr}(\boldsymbol{T}^\top (\boldsymbol{\Sigma}_{\mathcal{S}}^{\phi} - \boldsymbol{\Sigma}_{\mathcal{D}}^{\phi})\boldsymbol{T}) \\ \text{subject to} \quad & \boldsymbol{T}^\top \boldsymbol{K}_x \boldsymbol{T} = \boldsymbol{I}, \end{aligned} \tag{11.58}$$

whose solution is a generalized eigenvalue problem $(\boldsymbol{\Sigma}_{\mathcal{S}}^{\phi} - \boldsymbol{\Sigma}_{\mathcal{D}}^{\phi}, \boldsymbol{K}_x)$, where $\boldsymbol{T}$ is the matrix of the eigenvectors. The weight matrix of the generalized Mahalanobis distance is obtained through Eqs. (11.45) and (11.54). This is relevant to kernel Fisher discriminant analysis (see Chap. 6) which minimizes and maximizes the intraclass and interclass variances in RKHS.

11.3.6.6 Relevant to Kernel Support Vector Machine

This section explains the kernel version of the metric learning method [40], which was introduced in Sect. 11.3.4. It is relevant to kernel SVM because it has a similar optimization problem. Using the kernel trick (see Chap. 3) and Eq. (11.53), Eq. (11.40) can be kernelized as [40]:

$$
\begin{aligned}
\underset{\{\alpha_{ij}\}}{\text{maximize}} \quad & \sum_{(\boldsymbol{x}_i,\boldsymbol{x}_j)\in\mathcal{D}} \alpha_{ij}\boldsymbol{T}^\top(k_{ii} + k_{jj} - 2k_{ij}) \\
& - \frac{1}{2}\sum_{(\boldsymbol{x}_i,\boldsymbol{x}_j)\in\mathcal{D}}\sum_{(\boldsymbol{x}_k,\boldsymbol{x}_l)\in\mathcal{D}} \alpha_{ij}\alpha_{kl}(k_{ik} - k_{il} - k_{jk} + k_{jl})^2 \\
& + \frac{\lambda_1}{|\mathcal{S}|}\sum_{(\boldsymbol{x}_i,\boldsymbol{x}_j)\in\mathcal{D}}\sum_{(\boldsymbol{x}_k,\boldsymbol{x}_l)\in\mathcal{S}} \alpha_{ij}(k_{ik} - k_{il} - k_{jk} + k_{jl})^2 \\
\text{subject to} \quad & \frac{1}{\lambda_2}\sum_{(\boldsymbol{x}_i,\boldsymbol{x}_j)\in\mathcal{D}} \alpha_{ij} \geq \nu, \\
& \alpha_{ij} \in [0, \frac{\lambda_2}{|\mathcal{D}|}],
\end{aligned}
\tag{11.59}
$$

which is a quadratic programming problem and can be solved by optimization solvers.

11.3.7 Geometric Spectral Metric Learning

Some spectral metric learning methods are considered geometric methods, which use Riemannian manifolds. This section introduces several well-known geometric methods. There are some other geometric methods, such as [25], which are not covered for brevity.

11.3.7.1 Geometric Mean Metric Learning

One of the geometric spectral metric learning methods is the Geometric Mean Metric Learning (GMML) [49]. Let $\boldsymbol{W}$ be the weight matrix in the generalized Mahalanobis distance for similar points.

Regular GMML In GMML, the inverse of weight matrix, i.e. $\boldsymbol{W}^{-1}$, is used for the dissimilar points. The optimization problem of GMML is [49]:

$$\begin{aligned} \underset{\boldsymbol{W}}{\text{minimize}} \quad & \sum_{(\boldsymbol{x}_i,\boldsymbol{x}_j)\in\mathcal{S}} \|\boldsymbol{x}_i - \boldsymbol{x}_j\|_{\boldsymbol{W}}^2 + \sum_{(\boldsymbol{x}_i,\boldsymbol{x}_j)\in\mathcal{D}} \|\boldsymbol{x}_i - \boldsymbol{x}_j\|_{\boldsymbol{W}^{-1}}^2 \\ \text{subject to} \quad & \boldsymbol{W} \succeq \boldsymbol{0}. \end{aligned} \tag{11.60}$$

According to Eq. (11.13), this problem can be restated as:

$$\begin{aligned} \underset{\boldsymbol{W}}{\text{minimize}} \quad & \textbf{tr}(\boldsymbol{W}\boldsymbol{\Sigma}_{\mathcal{S}}) + \textbf{tr}(\boldsymbol{W}^{-1}\boldsymbol{\Sigma}_{\mathcal{D}}) \\ \text{subject to} \quad & \boldsymbol{W} \succeq \boldsymbol{0}, \end{aligned} \tag{11.61}$$

where $\boldsymbol{\Sigma}_{\mathcal{S}}$ and $\boldsymbol{\Sigma}_{\mathcal{D}}$ are defined in Eq. (11.14). Taking the derivative of the objective function with respect to $\boldsymbol{W}$ and setting it to zero results in:

$$\frac{\partial}{\partial \boldsymbol{W}}\big(\textbf{tr}(\boldsymbol{W}\boldsymbol{\Sigma}_{\mathcal{S}}) + \textbf{tr}(\boldsymbol{W}^{-1}\boldsymbol{\Sigma}_{\mathcal{D}})\big) = \boldsymbol{\Sigma}_{\mathcal{S}} - \boldsymbol{W}^{-1}\boldsymbol{\Sigma}_{\mathcal{D}}\boldsymbol{W}^{-1} \overset{\text{set}}{=} \boldsymbol{0} \implies \boldsymbol{\Sigma}_{\mathcal{D}} = \boldsymbol{W}\boldsymbol{\Sigma}_{\mathcal{S}}\boldsymbol{W}. \tag{11.62}$$

This Eq. (11.62) is called the Riccati equation [32] and its solution is the midpoint of the geodesic connecting $\boldsymbol{\Sigma}_{\mathcal{S}}^{-1}$ and $\boldsymbol{\Sigma}_{\mathcal{D}}$ [10, Section 1.2.13].

Lemma 11.5 ([10, Chapter 6]) *The geodesic curve connecting two points $\boldsymbol{\Sigma}_1$ and $\boldsymbol{\Sigma}_2$ on the Symmetric Positive Definite (SPD) Riemannian manifold is denoted by $\boldsymbol{\Sigma}_1 \sharp_t \boldsymbol{\Sigma}_2$ and is computed as:*

$$\boldsymbol{\Sigma}_1 \sharp_t \boldsymbol{\Sigma}_2 := \boldsymbol{\Sigma}_1^{(1/2)}\big(\boldsymbol{\Sigma}_1^{(-1/2)}\boldsymbol{\Sigma}_2\boldsymbol{\Sigma}_1^{(-1/2)}\big)^t\boldsymbol{\Sigma}_1^{(1/2)}, \tag{11.63}$$

where $t \in [0, 1]$.

Therefore, the solution of Eq. (11.62) is:

$$\boldsymbol{W} = \boldsymbol{\Sigma}_{\mathcal{S}}^{-1} \sharp_{(1/2)} \boldsymbol{\Sigma}_{\mathcal{D}} \overset{(11.63)}{=} \boldsymbol{\Sigma}_{\mathcal{S}}^{(-1/2)}\big(\boldsymbol{\Sigma}_{\mathcal{S}}^{(1/2)}\boldsymbol{\Sigma}_{\mathcal{D}}\boldsymbol{\Sigma}_{\mathcal{S}}^{(1/2)}\big)^{(1/2)}\boldsymbol{\Sigma}_{\mathcal{S}}^{(-1/2)}. \tag{11.64}$$

The proof of Eq. (11.64) is as follows [23]:

$$\boldsymbol{\Sigma}_{\mathcal{D}} \overset{(11.62)}{=} \boldsymbol{W}\boldsymbol{\Sigma}_{\mathcal{S}}\boldsymbol{W} \implies \boldsymbol{\Sigma}_{\mathcal{S}}^{(1/2)}\boldsymbol{\Sigma}_{\mathcal{D}}\boldsymbol{\Sigma}_{\mathcal{S}}^{(1/2)} = \boldsymbol{\Sigma}_{\mathcal{S}}^{(1/2)}\boldsymbol{W}\boldsymbol{\Sigma}_{\mathcal{S}}\boldsymbol{W}\boldsymbol{\Sigma}_{\mathcal{S}}^{(1/2)}$$

$$
\begin{aligned}
&\implies (\boldsymbol{\Sigma}_{\mathcal{S}}^{(1/2)} \boldsymbol{\Sigma}_{\mathcal{D}} \boldsymbol{\Sigma}_{\mathcal{S}}^{(1/2)})^{(1/2)} = (\boldsymbol{\Sigma}_{\mathcal{S}}^{(1/2)} \boldsymbol{W} \boldsymbol{\Sigma}_{\mathcal{S}} \boldsymbol{W} \boldsymbol{\Sigma}_{\mathcal{S}}^{(1/2)})^{(1/2)} \\
&\implies (\boldsymbol{\Sigma}_{\mathcal{S}}^{(1/2)} \boldsymbol{\Sigma}_{\mathcal{D}} \boldsymbol{\Sigma}_{\mathcal{S}}^{(1/2)})^{(1/2)} \overset{(a)}{=} ((\boldsymbol{\Sigma}_{\mathcal{S}}^{(1/2)} \boldsymbol{W} \boldsymbol{\Sigma}_{\mathcal{S}}^{(1/2)})(\boldsymbol{\Sigma}_{\mathcal{S}}^{(1/2)} \boldsymbol{W} \boldsymbol{\Sigma}_{\mathcal{S}}^{(1/2)}))^{(1/2)} \\
&\qquad\qquad\qquad\qquad\quad = (\boldsymbol{\Sigma}_{\mathcal{S}}^{(1/2)} \boldsymbol{W} \boldsymbol{\Sigma}_{\mathcal{S}}^{(1/2)}) \\
&\implies \boldsymbol{\Sigma}_{\mathcal{S}}^{(-1/2)} (\boldsymbol{\Sigma}_{\mathcal{S}}^{(1/2)} \boldsymbol{\Sigma}_{\mathcal{D}} \boldsymbol{\Sigma}_{\mathcal{S}}^{(1/2)})^{(1/2)} \boldsymbol{\Sigma}_{\mathcal{S}}^{(-1/2)} = \boldsymbol{\Sigma}_{\mathcal{S}}^{(-1/2)} (\boldsymbol{\Sigma}_{\mathcal{S}}^{(1/2)} \boldsymbol{W} \boldsymbol{\Sigma}_{\mathcal{S}}^{(1/2)}) \boldsymbol{\Sigma}_{\mathcal{S}}^{(-1/2)} = \boldsymbol{W},
\end{aligned}
$$

where (a) is because $\boldsymbol{\Sigma}_{\mathcal{S}} \succeq \mathbf{0}$ so its eigenvalues are nonnegative, and the matrix of the eigenvalues can be decomposed by the second root in its eigenvalue decomposition to have $\boldsymbol{\Sigma}_{\mathcal{S}} = \boldsymbol{\Sigma}_{\mathcal{S}}^{(1/2)} \boldsymbol{\Sigma}_{\mathcal{S}}^{(1/2)}$.

Regularized GMML The matrix $\boldsymbol{\Sigma}_{\mathcal{S}}$ might be singular or near singular and consequently noninvertible. Therefore, Eq. (11.61) is regularized to ensure that the weight matrix is close to a prior known positive definite matrix $\boldsymbol{W}_0$.

$$
\begin{aligned}
\underset{\boldsymbol{W}}{\text{minimize}} \quad & \mathbf{tr}(\boldsymbol{W}\boldsymbol{\Sigma}_{\mathcal{S}}) + \mathbf{tr}(\boldsymbol{W}^{-1}\boldsymbol{\Sigma}_{\mathcal{D}}) \\
& + \lambda\big(\mathbf{tr}(\boldsymbol{W}\boldsymbol{W}_0^{-1}) + \mathbf{tr}(\boldsymbol{W}^{-1}\boldsymbol{W}_0) - 2d\big), \\
\text{subject to} \quad & \boldsymbol{W} \succeq \mathbf{0},
\end{aligned}
\tag{11.65}
$$

where $\lambda > 0$ is the regularization parameter. The regularization term is the symmetrized log-determinant divergence between $\boldsymbol{W}$ and $\boldsymbol{W}_0$. Taking the derivative of the objective function with respect to $\boldsymbol{W}$ and setting it to zero results in:

$$
\begin{aligned}
&\frac{\partial}{\partial \boldsymbol{W}}\big(\mathbf{tr}(\boldsymbol{W}\boldsymbol{\Sigma}_{\mathcal{S}}) + \mathbf{tr}(\boldsymbol{W}^{-1}\boldsymbol{\Sigma}_{\mathcal{D}}) + \lambda\mathbf{tr}(\boldsymbol{W}\boldsymbol{W}_0^{-1}) + \lambda\mathbf{tr}(\boldsymbol{W}^{-1}\boldsymbol{W}_0) - 2\lambda d\big) \\
&= \boldsymbol{\Sigma}_{\mathcal{S}} - \boldsymbol{W}^{-1}\boldsymbol{\Sigma}_{\mathcal{D}}\boldsymbol{W}^{-1} + \lambda\boldsymbol{W}_0^{-1} + \lambda\boldsymbol{W}^{-1}\boldsymbol{W}_0\boldsymbol{W}^{-1} \overset{\text{set}}{=} \mathbf{0} \\
&\implies \boldsymbol{\Sigma}_{\mathcal{D}} + \lambda\boldsymbol{W}_0 = \boldsymbol{W}(\boldsymbol{\Sigma}_{\mathcal{S}} + \lambda\boldsymbol{W}_0^{-1})\boldsymbol{W},
\end{aligned}
$$

which is again a Riccati equation [32], whose solution is the midpoint of the geodesic connecting $(\boldsymbol{\Sigma}_{\mathcal{S}} + \lambda\boldsymbol{W}_0^{-1})^{-1}$ and $(\boldsymbol{\Sigma}_{\mathcal{D}} + \lambda\boldsymbol{W}_0)$:

$$
\boldsymbol{W} = (\boldsymbol{\Sigma}_{\mathcal{S}} + \lambda\boldsymbol{W}_0^{-1})^{-1} \sharp_{(1/2)} (\boldsymbol{\Sigma}_{\mathcal{D}} + \lambda\boldsymbol{W}_0). \tag{11.66}
$$

Weighted GMML Equation (11.61) can be restated as:

$$
\begin{aligned}
\underset{\boldsymbol{W}}{\text{minimize}} \quad & \delta^2(\boldsymbol{W}, \boldsymbol{\Sigma}_{\mathcal{S}}^{-1}) + \delta^2(\boldsymbol{W}, \boldsymbol{\Sigma}_{\mathcal{D}}) \\
\text{subject to} \quad & \boldsymbol{W} \succeq \mathbf{0},
\end{aligned}
\tag{11.67}
$$

where $\delta(.,.)$ is the Riemannian distance (or Fréchet mean) on the SPD manifold [3, Eq 1.1]:

$$\delta(\boldsymbol{\Sigma}_1, \boldsymbol{\Sigma}_2) := \| \log(\boldsymbol{\Sigma}_2^{(-1/2)} \boldsymbol{\Sigma}_1 \boldsymbol{\Sigma}_2^{(-1/2)}) \|_F,$$

where $\|.\|_F$ is the Frobenius norm. The objective in Eq. (11.67) can be weighted as:

$$\begin{aligned} \underset{\boldsymbol{W}}{\text{minimize}} \quad & (1-t)\delta^2(\boldsymbol{W}, \boldsymbol{\Sigma}_{\mathcal{S}}^{-1}) + t\delta^2(\boldsymbol{W}, \boldsymbol{\Sigma}_{\mathcal{D}}) \\ \text{subject to} \quad & \boldsymbol{W} \succeq \boldsymbol{0}, \end{aligned} \tag{11.68}$$

where $t \in [0, 1]$ is a hyperparameter. The solution to this problem is the weighted version of Eq. (11.66):

$$\boldsymbol{W} = (\boldsymbol{\Sigma}_{\mathcal{S}} + \lambda \boldsymbol{W}_0^{-1})^{-1} \sharp_t (\boldsymbol{\Sigma}_{\mathcal{D}} + \lambda \boldsymbol{W}_0). \tag{11.69}$$

11.3.7.2 Low-Rank Geometric Mean Metric Learning

A low-rank weight matrix in GMML [11], where the rank of weight matrix is set to be $p \ll d$, can be determined by:

$$\begin{aligned} \underset{\boldsymbol{W}}{\text{minimize}} \quad & \mathbf{tr}(\boldsymbol{W}\boldsymbol{\Sigma}_{\mathcal{S}}) + \mathbf{tr}(\boldsymbol{W}^{-1}\boldsymbol{\Sigma}_{\mathcal{D}}) \\ \text{subject to} \quad & \boldsymbol{W} \succeq \boldsymbol{0}, \\ & \mathbf{rank}(\boldsymbol{W}) = p. \end{aligned} \tag{11.70}$$

It is possible to decompose it using eigenvalue decomposition, which was done in Eq. (11.9), i.e., $\boldsymbol{W} = \boldsymbol{V}\boldsymbol{\Lambda}\boldsymbol{V}^\top = \boldsymbol{U}\boldsymbol{U}^\top$, where only p eigenvectors and p eigenvalues are known. Therefore, the sizes of the matrices are $\boldsymbol{V} \in \mathbb{R}^{d\times p}$, $\boldsymbol{\Lambda} \in \mathbb{R}^{p\times p}$, and $\boldsymbol{U} \in \mathbb{R}^{d\times p}$. By this decomposition, the objective function in Eq. (11.70 can be restated as:

$$\begin{aligned} &\mathbf{tr}(\boldsymbol{V}\boldsymbol{\Lambda}\boldsymbol{V}^\top\boldsymbol{\Sigma}_{\mathcal{S}}) + \mathbf{tr}(\boldsymbol{V}\boldsymbol{\Lambda}^{-1}\boldsymbol{V}^\top\boldsymbol{\Sigma}_{\mathcal{D}}) \\ &\overset{(a)}{=} \mathbf{tr}(\boldsymbol{\Lambda}\boldsymbol{V}^\top\boldsymbol{\Sigma}_{\mathcal{S}}\boldsymbol{V}) + \mathbf{tr}(\boldsymbol{\Lambda}^{-1}\boldsymbol{V}^\top\boldsymbol{\Sigma}_{\mathcal{D}}\boldsymbol{V}) \overset{(b)}{=} \mathbf{tr}(\boldsymbol{\Lambda}\widetilde{\boldsymbol{\Sigma}}_{\mathcal{S}}) + \mathbf{tr}(\boldsymbol{\Lambda}^{-1}\widetilde{\boldsymbol{\Sigma}}_{\mathcal{D}}), \end{aligned}$$

where $(\boldsymbol{V}^\top)^{-1} = \boldsymbol{V}$ because it is orthogonal, (a) is because of the cyclic property of the trace, and (b) is because $\widetilde{\boldsymbol{\Sigma}}_{\mathcal{S}} := \boldsymbol{V}^\top\boldsymbol{\Sigma}_{\mathcal{S}}\boldsymbol{V}$ and $\widetilde{\boldsymbol{\Sigma}}_{\mathcal{D}} := \boldsymbol{V}^\top\boldsymbol{\Sigma}_{\mathcal{D}}\boldsymbol{V}$ are defined. Noting that the matrix of eigenvectors $\boldsymbol{V}$ is orthogonal, Eq. (11.70) is restated to:

$$
\begin{aligned}
\underset{\boldsymbol{\Lambda}, \boldsymbol{V}}{\text{minimize}} \quad & \mathbf{tr}(\boldsymbol{\Lambda}\widetilde{\boldsymbol{\Sigma}}_{\mathcal{S}}) + \mathbf{tr}(\boldsymbol{\Lambda}^{-1}\widetilde{\boldsymbol{\Sigma}}_{\mathcal{D}}) \\
\text{subject to} \quad & \boldsymbol{\Lambda} \succeq \mathbf{0}, \\
& \boldsymbol{V}^\top \boldsymbol{V} = \boldsymbol{I},
\end{aligned}
\tag{11.71}
$$

where $\mathbf{rank}(\boldsymbol{W}) = p$ is automatically satisfied by taking $\boldsymbol{V} \in \mathbb{R}^{d\times p}$ and $\boldsymbol{\Lambda} \in \mathbb{R}^{p\times p}$ in the decomposition. This problem can be solved by alternative optimization (see Chap. 4). If variable $\boldsymbol{V}$ is fixed, the minimization problem with respect to $\boldsymbol{\Lambda}$ is similar to problem (11.61). Therefore, its solution is similar to Eq. (11.64), i.e., $\boldsymbol{\Lambda} = \widetilde{\boldsymbol{\Sigma}}_{\mathcal{S}}^{-1} \sharp_{(1/2)} \widetilde{\boldsymbol{\Sigma}}_{\mathcal{D}}$ (see Eq. (11.63) for the definition of $\sharp_t$). If $\boldsymbol{\Lambda}$ is fixed, the orthogonality constraint $\boldsymbol{V}^\top \boldsymbol{V} = \boldsymbol{I}$ can be modelled by $\boldsymbol{V}$, belonging to the Grassmannian manifold $G(p, d)$, which is the set of p-dimensional subspaces of $\mathbb{R}^d$. In summary, the alternative optimization is:

$$
\begin{aligned}
&\boldsymbol{\Lambda}^{(\tau+1)} = (\boldsymbol{V}^{(\tau)\top}\boldsymbol{\Sigma}_{\mathcal{S}}\boldsymbol{V}^{(\tau)})^{-1} \sharp_{(1/2)} (\boldsymbol{V}^{(\tau)\top}\boldsymbol{\Sigma}_{\mathcal{D}}\boldsymbol{V}^{(\tau)}), \\
&\boldsymbol{V}^{(\tau+1)} := \arg \min_{\boldsymbol{V}\in G(p,d)} \Big(\mathbf{tr}(\boldsymbol{\Lambda}^{(\tau+1)}\boldsymbol{V}^\top\boldsymbol{\Sigma}_{\mathcal{S}}\boldsymbol{V}) + \mathbf{tr}((\boldsymbol{\Lambda}^{(\tau+1)})^{-1}\boldsymbol{V}^\top\boldsymbol{\Sigma}_{\mathcal{D}}\boldsymbol{V})\Big),
\end{aligned}
$$

where τ is the iteration index. The optimization of $\boldsymbol{V}$ can be solved by Riemannian optimization [1].

11.3.7.3 Geometric Mean Metric Learning for Partial Labels

Partial label learning [15] refers to when a set of candidate labels is available for every data point. GMML can be modified to be used for partial label learning [52]. Let $\mathcal{Y}_i$ denote the set of candidate labels for $\boldsymbol{x}_i$. If there are q candidate labels in total, then $\boldsymbol{y}_i = [y_{i1}, \dots, y_{iq}]^\top \in \{0, 1\}^q$ is denoted, where y_{ij} is one if the j-th label is a candidate label for $\boldsymbol{x}_i$ and zero otherwise. The following terms are defined as $\boldsymbol{X}_i^+ := \{\boldsymbol{x}_j | j = 1, \dots, n, j \neq i, \mathcal{Y}_i \cap \mathcal{Y}_j \neq \varnothing\}$ and $\boldsymbol{X}_i^- := \{\boldsymbol{x}_j | j = 1, \dots, n, \mathcal{Y}_i \cap \mathcal{Y}_j = \varnothing\}$. In other words, $\boldsymbol{X}_i^+$ and $\boldsymbol{X}_i^-$ are the data points that share and do not share any candidate labels with $\boldsymbol{x}_i$, respectively. Let $\mathcal{N}_i^+$ be the indices of the k nearest neighbors of $\boldsymbol{x}_i$ among $\boldsymbol{X}_i^+$. Additionally, let $\mathcal{N}_i^-$ be the indices of points in $\boldsymbol{X}_i^-$, whose distances from $\boldsymbol{x}_i$ are smaller than the distance of the furthest point in $\mathcal{N}_i^+$ from $\boldsymbol{x}_i$. That is, $\mathcal{N}_i^- := \{j | j = 1, \dots, n, \boldsymbol{x}_j \in \boldsymbol{X}_i^-, \|\boldsymbol{x}_i - \boldsymbol{x}_j\|_2 \leq \max_{t\in\mathcal{N}_i^+} \|\boldsymbol{x}_i - \boldsymbol{x}_t\|_2\}$.

Let $\boldsymbol{w}_i^{(1)} = [w_{i,t}^{(1)}, \forall t \in \mathcal{N}_i^+]^\top \in \mathbb{R}^k$ contain the probabilities that each of the k neighbors of $\boldsymbol{x}_i$ share the same label as $\boldsymbol{x}_i$. The vector $\boldsymbol{w}_i^{(1)}$ can be estimated by linear reconstruction of $\boldsymbol{y}_i$ by the neighbor $\boldsymbol{y}_t$'s:

$$\begin{aligned}
\underset{\boldsymbol{w}_i^{(1)}}{\text{minimize}} \quad & \frac{1}{q} \| \boldsymbol{y}_i - \sum_{t \in \mathcal{N}_i^+} w_{i,t}^{(1)} \boldsymbol{y}_t \|_2^2 + \frac{\lambda_1}{k} \sum_{t \in \mathcal{N}_i^+} (w_{i,t}^{(1)})^2 \\
\text{subject to} \quad & w_{i,t}^{(1)} \geq 0, \quad t \in \mathcal{N}_i^+,
\end{aligned}$$

where $\lambda_1 > 0$ is the regularization parameter. Let $\boldsymbol{w}_i^{(2)} = [w_{i,t}^{(2)}, \forall t \in \mathcal{N}_i^+]^\top \in \mathbb{R}^k$ denote the coefficients for linear reconstruction of $\boldsymbol{x}_i$ by its k nearest neighbors. It is obtained by:

$$\begin{aligned}
\underset{\boldsymbol{w}_i^{(2)}}{\text{minimize}} \quad & \| \boldsymbol{x}_i - \sum_{t \in \mathcal{N}_i^+} w_{i,t}^{(2)} \boldsymbol{x}_t \|_2^2 \\
\text{subject to} \quad & w_{i,t}^{(2)} \geq 0, \quad t \in \mathcal{N}_i^+.
\end{aligned}$$

These two optimization problems are in the form of quadratic programming and can be solved using the interior point method (see Chap. 4).

The main optimization problem for GMML dealing with partial labels is [52]:

$$\begin{aligned}
\underset{\boldsymbol{W}}{\text{minimize}} \quad & \mathbf{tr}(\boldsymbol{W} \boldsymbol{\Sigma}'_{\mathcal{S}}) + \mathbf{tr}(\boldsymbol{W}^{-1} \boldsymbol{\Sigma}'_{\mathcal{D}}) \\
\text{subject to} \quad & \boldsymbol{W} \succeq \mathbf{0},
\end{aligned} \tag{11.72}$$

where:

$$\begin{aligned}
\boldsymbol{\Sigma}'_{\mathcal{S}} := \sum_{i=1}^{n} \Bigg(& \frac{\sum_{t \in \mathcal{N}_i^+} w_{i,t}^{(1)} (\boldsymbol{x}_i - \boldsymbol{x}_t)(\boldsymbol{x}_i - \boldsymbol{x}_t)^\top}{\sum_{t \in \mathcal{N}_i^+} w_{i,t}^{(1)}} \\
& + \lambda \Big(\boldsymbol{x}_i - \sum_{t \in \mathcal{N}_i^+} w_{i,t}^{(2)} \boldsymbol{x}_t\Big)\Big(\boldsymbol{x}_i - \sum_{t \in \mathcal{N}_i^+} w_{i,t}^{(2)} \boldsymbol{x}_t\Big)^\top \Bigg), \\
\boldsymbol{\Sigma}'_{\mathcal{D}} := \sum_{i=1}^{n} & \sum_{t \in \mathcal{N}_i^-} (\boldsymbol{x}_i - \boldsymbol{x}_t)(\boldsymbol{x}_i - \boldsymbol{x}_t)^\top.
\end{aligned}$$

Minimizing the first term of $\boldsymbol{\Sigma}'_{\mathcal{S}}$ in $\mathbf{tr}(\boldsymbol{W}\boldsymbol{\Sigma}'_{\mathcal{S}})$ decreases the distance between similar points that share any candidate labels. Minimizing the second term of $\boldsymbol{\Sigma}'_{\mathcal{S}}$ in $\mathbf{tr}(\boldsymbol{W}\boldsymbol{\Sigma}'_{\mathcal{S}})$ tries to preserve the linear reconstruction of $\boldsymbol{x}_i$ by its neighbors after projection onto the subspace of the metric. Minimizing $\mathbf{tr}(\boldsymbol{W}^{-1}\boldsymbol{\Sigma}'_{\mathcal{D}})$ increases the distance between dissimilar points that do not share any candidate labels. Problem (11.72) is similar to problem (11.61); therefore, its solution is similar to Eq. (11.64), i.e., $\boldsymbol{W} = \boldsymbol{\Sigma}'^{-1}_{\mathcal{S}} \sharp_{(1/2)} \boldsymbol{\Sigma}'_{\mathcal{D}}$ (see Eq. (11.63) for the definition of $\sharp_t$).

11.3.7.4 Geometric Mean Metric Learning on SPD and Grassmannian Manifolds

The GMML method [49], introduced in Sect. 11.3.7.1, can be implemented on Symmetric Positive Definite (SPD) and Grassmannian manifolds [53]. If $\boldsymbol{X}_i, \boldsymbol{X}_j \in \mathcal{S}_{++}^d$ is a point on the SPD manifold, the distance metric on this manifold is [53]:

$$d_{\boldsymbol{W}}(\boldsymbol{T}_i, \boldsymbol{T}_j) := \textbf{tr}\big(\boldsymbol{W}(\boldsymbol{T}_i - \boldsymbol{T}_j)(\boldsymbol{T}_i - \boldsymbol{T}_j)\big), \tag{11.73}$$

where $\boldsymbol{W} \in \mathbb{R}^{d\times d}$ is the weight matrix of the metric and $\boldsymbol{T}_i := \log(\boldsymbol{X}_i)$ is the logarithm operation on the SPD manifold. The Grassmannian manifold $Gr(k, d)$ is the k-dimensional subspaces of the d-dimensional vector space. A point in $Gr(k, d)$ is a linear subspace spanned by a full-rank $\boldsymbol{X}_i \in \mathbb{R}^{d\times k}$, which is orthogonal, i.e., $\boldsymbol{X}_i^\top \boldsymbol{X}_i = \boldsymbol{I}$. If $\boldsymbol{M} \in \mathbb{R}^{d\times r}$ is any matrix, then $\boldsymbol{X}_i'$ is defined in a way that $\boldsymbol{M}^\top \boldsymbol{X}_i'$ is the orthogonal components of $\boldsymbol{M}^\top \boldsymbol{X}_i$. If $\mathbb{R}^{d\times d} \ni \boldsymbol{T}_{ij} := \boldsymbol{X}_i' \boldsymbol{X}_i'^\top - \boldsymbol{X}_j' \boldsymbol{X}_j'^\top$, the distance on the Grassmannian manifold is [53]:

$$d_{\boldsymbol{W}}(\boldsymbol{T}_{ij}) := \textbf{tr}\big(\boldsymbol{W}\boldsymbol{T}_{ij}\boldsymbol{T}_{ij}\big), \tag{11.74}$$

$\boldsymbol{W} \in \mathbb{R}^{d\times d}$ is the weight matrix of the metric.

Similar to GMML's optimization problem, i.e. Eq. (11.60), the following problem can be solved for the SPD manifold:

$$\begin{aligned}
\underset{\boldsymbol{W}}{\text{minimize}} \quad & \sum_{(\boldsymbol{T}_i,\boldsymbol{T}_j)\in\mathcal{S}} \textbf{tr}\big(\boldsymbol{W}(\boldsymbol{T}_i - \boldsymbol{T}_j)(\boldsymbol{T}_i - \boldsymbol{T}_j)\big) \\
& + \sum_{(\boldsymbol{T}_i,\boldsymbol{T}_j)\in\mathcal{D}} \textbf{tr}\big(\boldsymbol{W}^{-1}(\boldsymbol{T}_i - \boldsymbol{T}_j)(\boldsymbol{T}_i - \boldsymbol{T}_j)\big) \\
\text{subject to} \quad & \boldsymbol{W} \succeq \boldsymbol{0}.
\end{aligned} \tag{11.75}$$

Likewise, the Grassmannian manifold optimization problem is:

$$\begin{aligned}
\underset{\boldsymbol{W}}{\text{minimize}} \quad & \sum_{(\boldsymbol{T}_i,\boldsymbol{T}_j)\in\mathcal{S}} \textbf{tr}\big(\boldsymbol{W}\boldsymbol{T}_{ij}\boldsymbol{T}_{ij}\big) + \sum_{(\boldsymbol{T}_i,\boldsymbol{T}_j)\in\mathcal{D}} \textbf{tr}\big(\boldsymbol{W}^{-1}\boldsymbol{T}_{ij}\boldsymbol{T}_{ij}\big) \\
\text{subject to} \quad & \boldsymbol{W} \succeq \boldsymbol{0}.
\end{aligned} \tag{11.76}$$

Suppose that the following terms are defined for the SPD manifold:

$$\boldsymbol{\Sigma}'_{\mathcal{S}} := \sum_{(\boldsymbol{T}_i,\boldsymbol{T}_j)\in\mathcal{S}} (\boldsymbol{T}_i - \boldsymbol{T}_j)(\boldsymbol{T}_i - \boldsymbol{T}_j),$$

$$\boldsymbol{\Sigma}'_{\mathcal{D}} := \sum_{(\boldsymbol{T}_i,\boldsymbol{T}_j)\in\mathcal{D}} (\boldsymbol{T}_i - \boldsymbol{T}_j)(\boldsymbol{T}_i - \boldsymbol{T}_j).$$

and, the following terms are defined for the Grassmannian manifold:

$$\boldsymbol{\Sigma}'_{\mathcal{S}} := \sum_{(\boldsymbol{T}_i,\boldsymbol{T}_j)\in\mathcal{S}} \boldsymbol{T}_{ij}\boldsymbol{T}_{ij},$$

$$\boldsymbol{\Sigma}'_{\mathcal{D}} := \sum_{(\boldsymbol{T}_i,\boldsymbol{T}_j)\in\mathcal{D}} \boldsymbol{T}_{ij}\boldsymbol{T}_{ij}.$$

Therefore, for either an SPD or a Grassmannian manifold, the optimization problem becomes Eq. (11.61) in which $\boldsymbol{\Sigma}_{\mathcal{S}}$ and $\boldsymbol{\Sigma}_{\mathcal{D}}$ are replaced with $\boldsymbol{\Sigma}'_{\mathcal{S}}$ and $\boldsymbol{\Sigma}'_{\mathcal{D}}$, respectively.

11.3.7.5 Metric Learning on Stiefel and SPD Manifolds

According to Eq. (11.9), the weight matrix in the metric can be decomposed to $\boldsymbol{W} = \boldsymbol{V}\boldsymbol{\Lambda}\boldsymbol{V}^\top$. If $\boldsymbol{V}$ and $\boldsymbol{\Lambda}$ are not restricted to be the matrices of the eigenvectors and eigenvalues, as in Eq. (11.9), it is possible to determine $\boldsymbol{V} \in \mathbb{R}^{d\times p}$ and $\boldsymbol{\Lambda} \in \mathbb{R}^{p\times p}$ through optimization [24]. The optimization problem in this method is:

$$\begin{aligned} \underset{\boldsymbol{V},\boldsymbol{\Lambda}}{\text{minimize}} \quad & \sum_{(\boldsymbol{x}_i,\boldsymbol{x}_j)\in\mathcal{S}} \log(1+q_{ij}) + \sum_{(\boldsymbol{x}_i,\boldsymbol{x}_j)\in\mathcal{D}} \log(1+q_{ij}^{-1}) \\ & + \lambda\Big(\mathbf{tr}(\boldsymbol{\Lambda}\boldsymbol{\Lambda}_0^{-1}) - \log\big(\mathbf{det}(\boldsymbol{\Lambda}\boldsymbol{\Lambda}_0^{-1})\big) - p\Big) \qquad (11.77) \\ \text{subject to} \quad & \boldsymbol{V}^\top\boldsymbol{V} = \boldsymbol{I}, \\ & \boldsymbol{\Lambda} \succeq \boldsymbol{0}, \end{aligned}$$

where $\lambda > 0$ is the regularization parameter, $\mathbf{det}(.)$ denotes the determinant of the matrix, and q_{ij} models the Gaussian distribution with the generalized Mahalanobis distance metric:

$$q_{ij} := \exp(\|\boldsymbol{x}_i - \boldsymbol{x}_j\|_{\boldsymbol{V}\boldsymbol{\Lambda}\boldsymbol{V}^\top}).$$

The constraint $\boldsymbol{V}^\top\boldsymbol{V} = \boldsymbol{I}$ means that the matrix $\boldsymbol{V}$ belongs to the Stiefel manifold $\mathrm{St}(p,d) := \{\boldsymbol{V} \in \mathbb{R}^{d\times p} | \boldsymbol{V}^\top\boldsymbol{V} = \boldsymbol{I}\}$, and the constraint $\boldsymbol{\Lambda} \succeq \boldsymbol{0}$ means $\boldsymbol{\Lambda}$ belongs to the SPD manifold $\mathcal{S}^p_{++}$. This means that these two variables belong to the product manifold $\mathrm{St}(p,d) \times \mathcal{S}^p_{++}$. Therefore, it is possible to solve this optimization problem using Riemannian optimization methods [1]. This method can also be kernelized (see [24, Section 4] for further information on its kernelized version).

11.3.7.6 Curvilinear Distance Metric Learning (CDML)

Lemma 11.6 ([14]) *The generalized Mahalanobis distance can be restated as:*

$$\|\boldsymbol{x}_i - \boldsymbol{x}_j\|_{\boldsymbol{W}}^2 = \sum_{l=1}^{p} \|\boldsymbol{u}_l\|_2^2 \Big(\int_{T_l(\boldsymbol{x}_i)}^{T_l(\boldsymbol{x}_j)} \|\boldsymbol{u}_l\|_2 \, dt \Big)^2, \tag{11.78}$$

where $\boldsymbol{u}_l \in \mathbb{R}^d$ is the l-th column of $\boldsymbol{U}$ in Eq. (11.9), $t \in \mathbb{R}$, and $T_l(\boldsymbol{x}) \in \mathbb{R}$ is the projection of $\boldsymbol{x}$ satisfying $(\boldsymbol{u}_l T_l(\boldsymbol{x}) - \boldsymbol{x})^\top \boldsymbol{u}_l = 0$.

Proof

$$\|\boldsymbol{x}_i - \boldsymbol{x}_j\|_{\boldsymbol{W}}^2 = (\boldsymbol{x}_i - \boldsymbol{x}_j)^\top \boldsymbol{W} (\boldsymbol{x}_i - \boldsymbol{x}_j) \overset{(11.9)}{=} (\boldsymbol{x}_i - \boldsymbol{x}_j)^\top \boldsymbol{U}\boldsymbol{U}^\top (\boldsymbol{x}_i - \boldsymbol{x}_j)$$

$$= \|\boldsymbol{U}^\top (\boldsymbol{x}_i - \boldsymbol{x}_j)\|_2^2 = \|[\boldsymbol{u}_1^\top (\boldsymbol{x}_i - \boldsymbol{x}_j), \dots, \boldsymbol{u}_p^\top (\boldsymbol{x}_i - \boldsymbol{x}_j)]^\top\|_2^2$$

$$= \sum_{l=1}^{p} \big(\boldsymbol{u}_l^\top (\boldsymbol{x}_i - \boldsymbol{x}_j)\big)^2 \overset{(a)}{=} \sum_{l=1}^{p} \|\boldsymbol{u}_l\|_2^2 \|\boldsymbol{x}_i - \boldsymbol{x}_j\|_2^2 \cos^2(\boldsymbol{u}_l, \boldsymbol{x}_i - \boldsymbol{x}_j)$$

$$\overset{(b)}{=} \sum_{l=1}^{p} \|\boldsymbol{u}_l\|_2^2 \|\boldsymbol{u}_l T_l(\boldsymbol{x}_i) - \boldsymbol{u}_l T_l(\boldsymbol{x}_j)\|_2^2,$$

where (a) is because of the law of cosines and (b) is because of $(\boldsymbol{u}_l T_l(\boldsymbol{x}) - \boldsymbol{x})^\top \boldsymbol{u}_l = 0$. The distance $\|\boldsymbol{u}_l T_l(\boldsymbol{x}_i) - \boldsymbol{u}_l T_l(\boldsymbol{x}_j)\|_2$ can be replaced by the length of the arc between $T_l(\boldsymbol{x}_i)$ and $T_l(\boldsymbol{x}_j)$ on the straight line $\boldsymbol{u}_l t$ for $t \in \mathbb{R}$. This results in Eq. (11.78). □

The condition $(\boldsymbol{u}_l T_l(\boldsymbol{x}) - \boldsymbol{x})^\top \boldsymbol{u}_l = 0$ is equivalent to finding the nearest neighbor to the line $\boldsymbol{u}_l t$, $\forall t \in \mathbb{R}$, i.e., $T_l(\boldsymbol{x}) := \arg\min_{t \in \mathbb{R}} \|\boldsymbol{u}_l t - \boldsymbol{x}\|_2^2$ [14]. This equation can be generalized to find the nearest neighbor to the geodesic curve $\boldsymbol{\theta}_l(t)$, rather than the line $\boldsymbol{u}_l t$:

$$T_{\boldsymbol{\theta}_l}(\boldsymbol{x}) := \arg\min_{t \in \mathbb{R}} \|\boldsymbol{\theta}_l(t) - \boldsymbol{x}\|_2^2. \tag{11.79}$$

Therefore, it is possible to replace the arc length of the straight line in Eq. (11.78) with the arc length of the curve:

$$\|\boldsymbol{x}_i - \boldsymbol{x}_j\|_{\boldsymbol{W}}^2 = \sum_{l=1}^{p} \alpha_l \Big(\int_{T_{\boldsymbol{\theta}_l}(\boldsymbol{x}_i)}^{T_{\boldsymbol{\theta}_l}(\boldsymbol{x}_j)} \|\boldsymbol{\theta}_l'(t)\|_2 \, dt \Big)^2, \tag{11.80}$$

where $\boldsymbol{\theta}_l'(t)$ is the derivative of $\boldsymbol{\theta}_l(t)$ with respect to t and $\alpha_l := (\int_0^1 \|\boldsymbol{\theta}_l'(t)\|_2 \, dt)^2$ is the scale factor. Curvilinear Distance Metric Learning (CDML) [14] uses this approximation of the distance metric by the above curvy geodesic on the manifold, i.e., Eq. (11.80). The optimization problem in CDML is:

$$\underset{\boldsymbol{\Theta}}{\text{minimize}} \quad \frac{1}{n}\sum_{i=1}^{n} \mathcal{L}(\|\boldsymbol{x}_i - \boldsymbol{x}_j\|_{\boldsymbol{W}}^2; y_{ij}) + \lambda\Omega(\boldsymbol{\Theta}), \tag{11.81}$$

where n is the number of points, $\boldsymbol{\Theta} := [\boldsymbol{\theta}_1, \dots, \boldsymbol{\theta}_p]$, $y_{ij} = 1$ if $(\boldsymbol{x}_i, \boldsymbol{x}_j) \in \mathcal{S}$ and $y_{ij} = 0$ if $(\boldsymbol{x}_i, \boldsymbol{x}_j) \in \mathcal{D}$, $\|\boldsymbol{x}_i - \boldsymbol{x}_j\|_{\boldsymbol{W}}^2$ is defined in Eq. (11.80), $\lambda > 0$ is the regularization parameter, $\mathcal{L}(.)$ is a loss function, and $\Omega(\boldsymbol{\Theta})$ is a penalty term. The optimal $\boldsymbol{\Theta}$, obtained from Eq. (11.81), can be used in Eq. (11.80) to determine the optimal distance metric.

11.3.8 Adversarial Metric Learning (AML)

Adversarial Metric Learning (AML) [13] uses adversarial learning for metric learning. In adversarial learning, two models compete by attempting to fool the other, forcing each other to gradually become stronger throughout the competition (see Chap. 21). Likewise, there are two stages in the AML that are trained simultaneously during the competition. The distinguishing stage attempts to discriminate between dissimilar points and push similar points close together, while the confusion or adversarial stage attempts to fool the metric learning method by pulling the dissimilar points close to each other and pushing the similar points away.

From the dataset, some random pairs $\mathcal{X} := \{(\boldsymbol{x}_i, \boldsymbol{x}_i')\}_{i=1}^{n/2}$ are formed. If $\boldsymbol{x}_i$ and $\boldsymbol{x}_i'$ are similar points, then $y_i = 1$, and if they are dissimilar, then $y_i = -1$. Additional new points are randomly generated in pairs $\mathcal{X}^g := \{(\boldsymbol{x}_i^g, \boldsymbol{x}_i^{g'})\}_{i=1}^{n/2}$. The generated points are updated iteratively through optimization of the confusion stage to fool the metric. The loss functions for both stages are Eq. (11.60) used in geometric mean metric learning (see Sect. 11.3.7.1).

The alternative optimization (see Chap. 4) used in AML is:

$$\begin{aligned}
\boldsymbol{W}^{(t+1)} := \arg\min_{\boldsymbol{W}} \Big(& \sum_{y_i=1} \|\boldsymbol{x}_i - \boldsymbol{x}_i'\|_{\boldsymbol{W}}^2 + \sum_{y_i=-1} \|\boldsymbol{x}_i - \boldsymbol{x}_i'\|_{\boldsymbol{W}^{-1}}^2 \\
& + \lambda_1 \big(\sum_{y_i=1} \|\boldsymbol{x}_i^{g(t)} - \boldsymbol{x}_i^{g'(t)}\|_{\boldsymbol{W}}^2 + \sum_{y_i=-1} \|\boldsymbol{x}_i^{g(t)} - \boldsymbol{x}_i^{g'(t)}\|_{\boldsymbol{W}^{-1}}^2 \big) \Big), \\
\mathcal{X}^{g(t+1)} := \arg\min_{\mathcal{X}'} \Big(& \sum_{y_i=-1} \|\boldsymbol{x}_i - \boldsymbol{x}_i'\|_{\boldsymbol{W}^{(t+1)}}^2 + \sum_{y_i=1} \|\boldsymbol{x}_i - \boldsymbol{x}_i'\|_{(\boldsymbol{W}^{(t+1)})^{-1}}^2 \\
& + \lambda_2 \big(\sum_{i=1}^{n/2} \|\boldsymbol{x}_i - \boldsymbol{x}_i^g\|_{\boldsymbol{W}^{(t+1)}}^2 + \sum_{i=1}^{n/2} \|\boldsymbol{x}_i' - \boldsymbol{x}_i^{g'}\|_{\boldsymbol{W}^{(t+1)}}^2 \big) \Big),
\end{aligned}$$

until convergence, where $\lambda_1, \lambda_2 > 0$ are the regularization parameters. Updating $\boldsymbol{W}$ and $\mathcal{X}^g$ are the distinguishing and confusion stages, respectively. In the distinguishing stage, a weight matrix $\boldsymbol{W}$ is found to minimize the distances between similar points in both $\mathcal{X}$ and $\mathcal{X}^g$ and maximize the distances between dissimilar points in both $\mathcal{X}$ and $\mathcal{X}^g$. In the confusion stage, new points $\mathcal{X}^g$ are generated to adversarially maximize the distances between similar points in $\mathcal{X}$, and adversarially minimize the distances between dissimilar points in $\mathcal{X}$. In this stage, points $\boldsymbol{x}_i^g$ and $\boldsymbol{x}_i^{g'}$ are also made similar to their corresponding points $\boldsymbol{x}_i$ and $\boldsymbol{x}'_i$, respectively.

11.4 Chapter Summary

Metric learning is a family of dimensionality reduction methods that learns the optimal distance metric or embedding space for the better representation of data or discrimination between classes. It can be divided into three categories of spectral, probabilistic, and deep metric learning. This chapter introduced spectral metric learning, while the probabilistic and deep metric learning are going to be introduced in Chaps. 13 and 19, respectively. Consequently, many of the preliminaries introduced in this chapter will also be used in Chaps. 13 and 19. Note that there are additional surveys on metric learning, such as [8, 27, 38, 41, 46, 47]. Further reading on metric learning can be found in [9]. Python toolboxes for metric learning are available through [18, 30, 39].

References

1. P-A Absil, Robert Mahony, and Rodolphe Sepulchre. *Optimization algorithms on matrix manifolds* Princeton University Press, 2009.
2. Babak Alipanahi, Michael Biggs, and Ali Ghodsi. "Distance metric learning vs. Fisher discriminant analysis". In: *Proceedings of the 23rd national conference on Artificial intelligence* Vol. 2. 2008, pp. 598–603.
3. Vincent Arsigny et al. "Geometric means in a novel vector space structure on symmetric positivedefinite matrices". In: *SIAM journal on matrix analysis and applications* 29.1 (2007), pp. 328–347.
4. Mahdieh Soleymani Baghshah and Saeed Bagheri Shouraki. "Kernel-based metric learning for semisupervised clustering". In: *Neurocomputing* 73.7-9 (2010), pp. 1352–1361.
5. Mahdieh Soleymani Baghshah and Saeed Bagheri Shouraki. "Semi-supervised metric learning using pairwise constraints". In: *Twenty-First International Joint Conference on Artificial Intelligence* 2009.
6. Mikhail Belkin and Partha Niyogi. "Laplacian eigenmaps and spectral techniques for embedding and clustering". In: *Advances in neural information processing systems* Vol. 14. 14. 2001, pp. 585–591.
7. Mikhail Belkin and Partha Niyogi. "Laplacian eigenmaps and spectral techniques for embedding and clustering". In: *Advances in neural information processing systems* 2002, pp. 585–591.
8. Aurélien Bellet, Amaury Habrard, and Marc Sebban. "A survey on metric learning for feature vectors and structured data". In: *arXiv preprint arXiv:1306.6709* (2013).
9. Aurélien Bellet, Amaury Habrard, and Marc Sebban. "Metric learning". In: *Synthesis Lectures on Artificial Intelligence and Machine Learning* 9.1 (2015), pp. 1–151.

10. Rajendra Bhatia. *Positive definite matrices* Princeton university press, 2007.
11. Mukul Bhutani et al. "Low-rank geometric mean metric learning". In: *arXiv preprint arXiv:1806.05454* (2018).
12. Hong Chang and Dit-Yan Yeung. "Locally linear metric adaptation for semi-supervised clustering". In: *Proceedings of the twenty-first international conference on Machine learning* 2004, p. 20.
13. Shuo Chen et al. "Adversarial metric learning". In: *arXiv preprint arXiv:1802.03170* (2018).
14. Shuo Chen et al. "Curvilinear distance metric learning". In: *Advances in Neural Information Processing Systems* 32 (2019).
15. Timothee Cour, Ben Sapp, and Ben Taskar. "Learning from partial labels". In: *The Journal of Machine Learning Research* 12 (2011), pp. 1501–1536.
16. Michael AA Cox and Trevor F Cox. "Multidimensional scaling". In: *Handbook of data visualization* Springer, 2008, pp. 315–347.
17. Roy De Maesschalck, Delphine Jouan-Rimbaud, and Désiré L Massart. "The Mahalanobis distance". In: *Chemometrics and intelligent laboratory systems* 50.1 (2000), pp. 1–18.
18. William De Vazelhes et al. "metric-learn: Metric Learning Algorithms in Python". In: *Journal of Machine Learning Research* 21 (2020), pp. 138–1.
19. Ronald A Fisher. "The use of multiple measurements in taxonomic problems". In: *Annals of eugenics* 7.2 (1936), pp. 179–188.
20. Léo Gautheron et al. "Metric learning from imbalanced data". In: *2019 IEEE 31st International Conference on Tools with Artificial Intelligence (ICTAI)* IEEE. 2019, pp. 923–930.
21. Ali Ghodsi, Dana F Wilkinson, and Finnegan Southey. "Improving Embeddings by Flexible Exploitation of Side Information". In: *IJCAI* 2007, pp. 810–816.
22. Arthur Gretton et al. "Measuring statistical dependence with Hilbert-Schmidt norms". In: *International conference on algorithmic learning theory*. Springer. 2005, pp. 63–77.
23. Hamideh Hajiabadi et al. "Layered Geometric Learning". In: *International Conference on Artificial Intelligence and Soft Computing* Springer. 2019, pp. 571–582.
24. Mehrtash Harandi, Mathieu Salzmann, and Richard Hartley. "Joint dimensionality reduction and metric learning: A geometric take". In: *International Conference on Machine Learning* PMLR. 2017, pp. 1404–1413.
25. Søren Hauberg, Oren Freifeld, and Michael J Black. "A Geometric take on Metric Learning". In: *Advances in neural information processing systems* Vol. 25. 2012, pp. 2033–2041.
26. Steven CH Hoi et al. "Learning distance metrics with contextual constraints for image retrieval". In: *2006 IEEE Computer Society Conference on Computer Vision and Pattern Recognition (CVPR'06)* Vol. 2. IEEE. 2006, pp. 2072–2078.
27. Brian Kulis. "Metric learning: A survey". In: *Foundations and Trends® in Machine Learning* 5.4 (2013), pp. 287–364.
28. Prasanta Chandra Mahalanobis. "On tests and measures of group divergence". In: *Journal of the Asiatic Society of Bengal* 26 (1930), pp. 541–588.
29. Goeffrey J McLachlan. "Mahalanobis distance". In: *Resonance* 4.6 (1999), pp. 20–26.
30. Kevin Musgrave, Serge Belongie, and Ser-Nam Lim. "Pytorch metric learning". In: *arXiv preprint arXiv:2008.09164* (2020).
31. Parisa Abdolrahim Poorheravi et al. "Acceleration of large margin metric learning for nearest neighbor classification using triplet mining and stratified sampling". In: *Journal of Computational Vision and Imaging Systems* 6.1 (2020).
32. Jacobo Riccati. "Animadversiones in aequationes differentiales secundi gradus". In: *Actorum Eruditorum Supplementa* 8.1724 (1724), pp. 66–73.
33. Sam T Roweis and Lawrence K Saul. "Nonlinear dimensionality reduction by locally linear embedding". In: *Science* 290.5500 (2000), pp. 2323–2326.
34. Bernhard Schölkopf et al. "New support vector algorithms". In: *Neural computation* 12.5 (2000), pp. 1207–1245.
35. Florian Schroff, Dmitry Kalenichenko, and James Philbin. "FaceNet: A unified embedding for face recognition and clustering". In: *Proceedings of the IEEE conference on computer vision and pattern recognition* 2015, pp. 815–823.

36. Noam Shental et al. "Adjustment learning and relevant component analysis". In: *European conference on computer vision* Springer. 2002, pp. 776–790.
37. Milad Sikaroudi et al. "Offline versus online triplet mining based on extreme distances of histopathology patches". In: *International Symposium on Visual Computing* Springer. 2020, pp. 333–345.
38. Juan Luis Suárez, Salvador Garcıa, and Francisco Herrera. "A tutorial on distance metric learning: Mathematical foundations, algorithms, experimental analysis, prospects and challenges". In: *Neurocomputing* 425 (2021), pp. 300–322.
39. Juan-Luis Suárez, Salvador Garcıa, and Francisco Herrera. "pyDML: A Python Library for Distance Metric Learning". In: *Journal of Machine Learning Research* 21 (2020), pp. 96–1.
40. Ivor W Tsang et al. "Distance metric learning with kernels". In: *Proceedings of the International Conference on Artificial Neural Networks* 2003, pp. 126–129.
41. Fei Wang and Jimeng Sun. "Survey on distance metric learning and dimensionality reduction in data mining". In: *Data mining and knowledge discovery* 29.2 (2015), pp. 534–564.
42. Kilian Q Weinberger, John Blitzer, and Lawrence K Saul. "Distance metric learning for large margin nearest neighbor classification". In: *Advances in neural information processing systems* 2006, pp. 1473–1480.
43. Kilian Q Weinberger and Lawrence K Saul. "Distance metric learning for large margin nearest neighbor classification". In: *Journal of machine learning research* 10.2 (2009).
44. Shiming Xiang, Feiping Nie, and Changshui Zhang. "Learning a Mahalanobis distance metric for data clustering and classification". In: *Pattern recognition* 41.12 (2008), pp. 3600–3612.
45. Eric Xing et al. "Distance metric learning with application to clustering with side-information". In: *Advances in neural information processing systems* 15 (2002), pp. 521–528.
46. Liu Yang. "An overview of distance metric learning". In: *Proceedings of the computer vision and pattern recognition conference* 2007.
47. Liu Yang and Rong Jin. "Distance metric learning: A comprehensive survey". In: *Michigan State Universiy* 2.2 (2006), p. 4.
48. Dit-Yan Yeung and Hong Chang. "A kernel approach for semisupervised metric learning". In: *IEEE Transactions on Neural Networks* 18.1 (2007), pp. 141–149.
49. Pourya Zadeh, Reshad Hosseini, and Suvrit Sra. "Geometric mean metric learning". In: *International conference on machine learning* 2016, pp. 2464–2471.
50. Changqing Zhang et al. "FISH-MML: Fisher-HSIC multi-view metric learning." In: *IJCAI* 2018, pp. 3054–3060.
51. Zhihua Zhang, James T Kwok, and Dit-Yan Yeung. "Parametric distance metric learning with label information". In: *IJCAI* Vol. 1450. 2003.
52. Yu Zhou and Hong Gu. "Geometric mean metric learning for partial label data". In: *Neurocomputing* 275 (2018), pp. 394–402.
53. Pengfei Zhu et al. "Towards Generalized and Efficient Metric Learning on Riemannian Manifold". In: *IJCAI* 2018, pp. 3235–3241.

Part III
Probabilistic Dimensionality Reduction

Chapter 12
Factor Analysis and Probabilistic Principal Component Analysis

12.1 Introduction

Learning models can be divided into discriminative and generative models [3, 17]. Discriminative models discriminate the classes of data for better separation of classes while the generative models learn a latent space that generates the data points. This chapter introduces generative models.

Variational inference is a technique that finds the lower bound on the log-likelihood of the data and maximizes it, rather than the log-likelihood, in the Maximum Likelihood Estimation (MLE). This lower bound is usually referred to as the Evidence Lower Bound (ELBO). Learning the parameters of the latent space can be done using Expectation Maximization (EM) [2].

Factor analysis assumes that every data point is generated from a latent factor/variable, where some noise may have been added to the data in the data space. Using EM, the ELBO is maximized and the parameters of the latent space are learned iteratively. Probabilistic Principal Component Analysis (PPCA), a special case of factor analysis, restricts the noise of the dimensions to be uncorrelated and assumes the variance of the noise to be equal in all dimensions. This restriction makes PPCA's solution closed-form and simple.

12.2 Variational Inference

Consider a dataset $\{\boldsymbol{x}_i\}_{i=1}^n$. Assume that every data point $\boldsymbol{x}_i \in \mathbb{R}^d$ is generated from a latent variable $\boldsymbol{z}_i \in \mathbb{R}^p$. This latent variable has a prior distribution $\mathbb{P}(\boldsymbol{z}_i)$. According to the Bayes' rule:

$$\mathbb{P}(\boldsymbol{z}_i \mid \boldsymbol{x}_i) = \frac{\mathbb{P}(\boldsymbol{x}_i \mid \boldsymbol{z}_i)\, \mathbb{P}(\boldsymbol{z}_i)}{\mathbb{P}(\boldsymbol{x}_i)}. \tag{12.1}$$

B. Ghojogh et al., *Elements of Dimensionality Reduction and Manifold Learning*,
https://doi.org/10.1007/978-3-031-10602-6_12

Let $\mathbb{P}(z_i)$ be an arbitrary distribution denoted by $q(z_i)$. Suppose the parameter of the conditional distribution of z_i on $\boldsymbol{x}_i$ is denoted by $\boldsymbol{\theta}$, which means that $\mathbb{P}(z_i \,|\, \boldsymbol{x}_i) = \mathbb{P}(z_i \,|\, \boldsymbol{x}_i, \boldsymbol{\theta})$. Therefore:

$$\mathbb{P}(z_i \,|\, \boldsymbol{x}_i, \boldsymbol{\theta}) = \frac{\mathbb{P}(\boldsymbol{x}_i \,|\, z_i, \boldsymbol{\theta})\, \mathbb{P}(z_i \,|\, \boldsymbol{\theta})}{\mathbb{P}(\boldsymbol{x}_i \,|\, \boldsymbol{\theta})}. \tag{12.2}$$

12.2.1 Evidence Lower Bound (ELBO)

Consider the Kullback–Leibler (KL) divergence [14] between the prior probability of the latent variable and the posterior of the latent variable:

$$\begin{aligned}
&\text{KL}\big(q(z_i) \,\|\, \mathbb{P}(z_i \,|\, \boldsymbol{x}_i, \boldsymbol{\theta})\big) \overset{(a)}{=} \int q(z_i) \log\big(\frac{q(z_i)}{\mathbb{P}(z_i \,|\, \boldsymbol{x}_i, \boldsymbol{\theta})}\big) dz_i \\
&= \int q(z_i)\big(\log(q(z_i)) - \log(\mathbb{P}(z_i \,|\, \boldsymbol{x}_i, \boldsymbol{\theta}))\big) dz_i \\
&\overset{(12.2)}{=} \int q(z_i)\big(\log(q(z_i)) - \log(\mathbb{P}(\boldsymbol{x}_i \,|\, z_i, \boldsymbol{\theta})) - \log(\mathbb{P}(z_i \,|\, \boldsymbol{\theta})) + \log(\mathbb{P}(\boldsymbol{x}_i \,|\, \boldsymbol{\theta}))\big)\, dz_i \\
&\overset{(b)}{=} \log(\mathbb{P}(\boldsymbol{x}_i \,|\, \boldsymbol{\theta})) + \int q(z_i)\big(\log(q(z_i)) - \log(\mathbb{P}(\boldsymbol{x}_i \,|\, z_i, \boldsymbol{\theta})) - \log(\mathbb{P}(z_i \,|\, \boldsymbol{\theta}))\big)\, dz_i \\
&= \log(\mathbb{P}(\boldsymbol{x}_i \,|\, \boldsymbol{\theta})) + \int q(z_i) \log\big(\frac{q(z_i)}{\mathbb{P}(\boldsymbol{x}_i \,|\, z_i, \boldsymbol{\theta})\mathbb{P}(z_i \,|\, \boldsymbol{\theta})}\big)\, dz_i \\
&= \log(\mathbb{P}(\boldsymbol{x}_i \,|\, \boldsymbol{\theta})) + \int q(z_i) \log\big(\frac{q(z_i)}{\mathbb{P}(\boldsymbol{x}_i, z_i \,|\, \boldsymbol{\theta})}\big)\, dz_i \\
&= \log(\mathbb{P}(\boldsymbol{x}_i \,|\, \boldsymbol{\theta})) + \text{KL}\big(q(z_i) \,\|\, \mathbb{P}(\boldsymbol{x}_i, z_i \,|\, \boldsymbol{\theta})\big),
\end{aligned}$$

where (a) is for definition of KL divergence and (b) is because $\log(\mathbb{P}(\boldsymbol{x}_i \,|\, \boldsymbol{\theta}))$ is independent of z_i and comes out of integral and $\int dz_i = 1$. Therefore:

$$\log(\mathbb{P}(\boldsymbol{x}_i \,|\, \boldsymbol{\theta})) = \text{KL}\big(q(z_i) \,\|\, \mathbb{P}(z_i \,|\, \boldsymbol{x}_i, \boldsymbol{\theta})\big) - \text{KL}\big(q(z_i) \,\|\, \mathbb{P}(\boldsymbol{x}_i, z_i \,|\, \boldsymbol{\theta})\big). \tag{12.3}$$

The *Evidence Lower Bound (ELBO)* is defined as:

$$\mathcal{L}(q, \boldsymbol{\theta}) := -\text{KL}\big(q(z_i) \,\|\, \mathbb{P}(\boldsymbol{x}_i, z_i \,|\, \boldsymbol{\theta})\big). \tag{12.4}$$

So:

$$\log(\mathbb{P}(\boldsymbol{x}_i \,|\, \boldsymbol{\theta})) = \text{KL}\big(q(z_i) \,\|\, \mathbb{P}(z_i \,|\, \boldsymbol{x}_i, \boldsymbol{\theta})\big) + \mathcal{L}(q, \boldsymbol{\theta}).$$

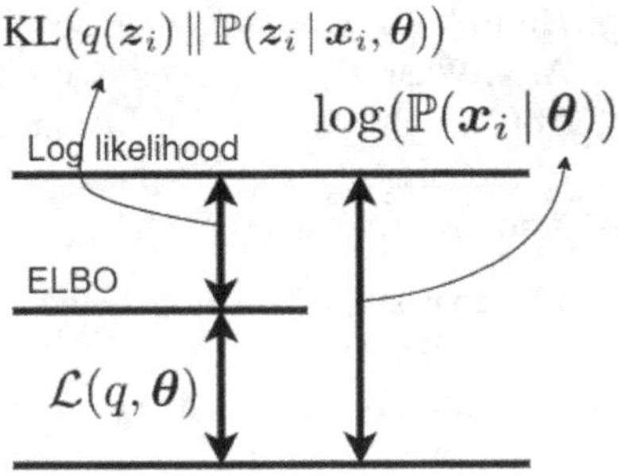

Fig. 12.1 Depiction of ELBO as the lower bound on log likelihood. The image is inspired by [2]

Therefore:

$$\mathcal{L}(q, \boldsymbol{\theta}) = \log(\mathbb{P}(\boldsymbol{x}_i \,|\, \boldsymbol{\theta})) - \underbrace{\mathrm{KL}\big(q(z_i) \,\|\, \mathbb{P}(z_i \,|\, \boldsymbol{x}_i, \boldsymbol{\theta})\big)}_{\geq 0}. \tag{12.5}$$

As the second term is negative with its minus, the log likelihood of the data is lower bounded by the ELBO:

$$\mathcal{L}(q, \boldsymbol{\theta}) \leq \log(\mathbb{P}(\boldsymbol{x}_i \,|\, \boldsymbol{\theta})). \tag{12.6}$$

The likelihood $\mathbb{P}(\boldsymbol{x}_i \,|\, \boldsymbol{\theta})$ is also referred to as the *evidence*. Note that this lower bound tightens when:

$$\begin{aligned}
\mathcal{L}(q, \boldsymbol{\theta}) \approx \log(\mathbb{P}(\boldsymbol{x}_i \,|\, \boldsymbol{\theta})) &\implies 0 \leq \mathrm{KL}\big(q(z_i) \,\|\, \mathbb{P}(z_i \,|\, \boldsymbol{x}_i, \boldsymbol{\theta})\big) \overset{\text{set}}{=} 0 \\
&\implies q(z_i) = \mathbb{P}(z_i \,|\, \boldsymbol{x}_i, \boldsymbol{\theta}).
\end{aligned} \tag{12.7}$$

This lower bound is depicted in Fig. 12.1.

12.2.2 Expectation Maximization

12.2.2.1 Background on Expectation Maximization

Sometimes, the data are not fully observable. For example, the values of data are not known, but the data points are known to be either zero or greater than zero. In this case, the Maximum Likelihood Expectation (MLE) cannot be directly applied because it is not possible to access complete information and some data are missing. Instead of MLE, Expectation Maximization (EM) is useful in such cases. The main idea behind EM can be summarized in this short friendly conversation:

- What shall we do? Some of the data are missing! The log-likelihood is not known completely, meaning MLE cannot be used!

- Hmm, we can probably replace the missing data with something.
- Aha! What if we replace it with its mean?
- You're right! We can take the mean of the log-likelihood over the possible values of the missing data. Then everything in the log-likelihood will be known.
- And then we can do MLE!

EM consists of two steps—the E-step (expectation step) and the M-step (maximization step). In the E-step, the expectation of log-likelihood, with respect to the missing data, is calculated to determine its mean estimation. In the M-step, the MLE approach is used, where the log-likelihood is replaced with its expectation. These two steps are iteratively repeated until convergence of the estimated parameters.

12.2.2.2 Expectation Maximization in Variational Inference

The aim of MLE is to maximize the log-likelihood of data. According to Eq. (12.6), maximizing the ELBO will also maximize the log-likelihood. Equation (12.6) holds for any prior distribution q. The goal is to find the optimal distribution to maximize the lower bound. Therefore, EM for variational inference is performed iteratively:

$$\text{E-step:} \quad q^{(t)} := \arg\max_{q} \quad \mathcal{L}(q, \boldsymbol{\theta}^{(t-1)}), \tag{12.8}$$

$$\text{M-step:} \quad \boldsymbol{\theta}^{(t)} := \arg\max_{\boldsymbol{\theta}} \quad \mathcal{L}(q^{(t)}, \boldsymbol{\theta}), \tag{12.9}$$

where t denotes the iteration index.

E-step in EM for Variational Inference The E-step is:

$$\begin{aligned} \max_{q} \mathcal{L}(q, \boldsymbol{\theta}^{(t-1)}) &\overset{(12.5)}{=} \max_{q} \log(\mathbb{P}(\boldsymbol{x}_i \,|\, \boldsymbol{\theta}^{(t-1)})) + \max_{q} \big(-\text{KL}\big(q(\boldsymbol{z}_i) \,\|\, \mathbb{P}(\boldsymbol{z}_i \,|\, \boldsymbol{x}_i, \boldsymbol{\theta}^{(t-1)})\big)\big) \\ &= \max_{q} \log(\mathbb{P}(\boldsymbol{x}_i \,|\, \boldsymbol{\theta}^{(t-1)})) + \min_{q} \text{KL}\big(q(\boldsymbol{z}_i) \,\|\, \mathbb{P}(\boldsymbol{z}_i \,|\, \boldsymbol{x}_i, \boldsymbol{\theta}^{(t-1)})\big). \end{aligned}$$

The second term is always nonnegative; therefore, its minimum is zero:

$$\text{KL}\big(q(\boldsymbol{z}_i) \,\|\, \mathbb{P}(\boldsymbol{z}_i \,|\, \boldsymbol{x}_i, \boldsymbol{\theta}^{(t-1)})\big) \overset{\text{set}}{=} 0 \implies q(\boldsymbol{z}_i) = \mathbb{P}(\boldsymbol{z}_i \,|\, \boldsymbol{x}_i, \boldsymbol{\theta}^{(t-1)}),$$

which was already determined in Eq. (12.7). Thus, the E-step assigns:

$$q^{(t)}(\boldsymbol{z}_i) \leftarrow \mathbb{P}(\boldsymbol{z}_i \,|\, \boldsymbol{x}_i, \boldsymbol{\theta}^{(t-1)}). \tag{12.10}$$

In other words, as demonstrated in Fig. 12.1, it pushes the middle line toward the above line by maximizing the ELBO.

M-step in EM for Variational Inference The M-step is:

$$\begin{aligned}
\max_{\boldsymbol{\theta}} \mathcal{L}(q^{(t)}, \boldsymbol{\theta}) &\overset{(12.4)}{=} \max_{\boldsymbol{\theta}} \big(-\text{KL}\big(q^{(t)}(z_i) \,\|\, \mathbb{P}(\boldsymbol{x}_i, z_i \,|\, \boldsymbol{\theta})\big)\big) \\
&\overset{(a)}{=} \max_{\boldsymbol{\theta}} \Big[-\int q^{(t)}(z_i) \log\big(\frac{q^{(t)}(z_i)}{\mathbb{P}(\boldsymbol{x}_i, z_i \,|\, \boldsymbol{\theta})}\big)\, dz_i\Big] \\
&= \max_{\boldsymbol{\theta}} \int q^{(t)}(z_i) \log(\mathbb{P}(\boldsymbol{x}_i, z_i \,|\, \boldsymbol{\theta}))\, dz_i - \max_{\boldsymbol{\theta}} \int q^{(t)}(z_i) \log(q^{(t)}(z_i))\, dz_i,
\end{aligned}$$

where (a) is for the definition of KL divergence. The second term is constant with respect to $\boldsymbol{\theta}$. Therefore:

$$\begin{aligned}
\max_{\boldsymbol{\theta}} \mathcal{L}(q^{(t)}, \boldsymbol{\theta}) &= \max_{\boldsymbol{\theta}} \int q^{(t)}(z_i) \log(\mathbb{P}(\boldsymbol{x}_i, z_i \,|\, \boldsymbol{\theta}))\, dz_i \\
&\overset{(a)}{=} \max_{\boldsymbol{\theta}} \mathbb{E}_{\sim q^{(t)}(z_i)}\big[\log \mathbb{P}(\boldsymbol{x}_i, z_i \,|\, \boldsymbol{\theta})\big],
\end{aligned}$$

where (a) is because of the definition of the expectation operator. Thus, the M-step assigns:

$$\boldsymbol{\theta}^{(t)} \leftarrow \arg\max_{\boldsymbol{\theta}} \mathbb{E}_{\sim q^{(t)}(z_i)}\big[\log \mathbb{P}(\boldsymbol{x}_i, z_i \,|\, \boldsymbol{\theta})\big]. \tag{12.11}$$

In other words, as demonstrated in Fig. 12.1, it pushes the above line higher. The E-step and M-step together play a game, where the E-step tries to reach the middle line (or the ELBO) to the log-likelihood and the M-step tries to increase the above line (or the log-likelihood). This procedure is performed repeatedly so that the two steps help each other push to higher values.

To summarize, the EM in variational inference is:

$$q^{(t)}(z_i) \leftarrow \mathbb{P}(z_i \,|\, \boldsymbol{x}_i, \boldsymbol{\theta}^{(t-1)}), \tag{12.12}$$

$$\boldsymbol{\theta}^{(t)} \leftarrow \arg\max_{\boldsymbol{\theta}} \mathbb{E}_{\sim q^{(t)}(z_i)}\big[\log \mathbb{P}(\boldsymbol{x}_i, z_i \,|\, \boldsymbol{\theta})\big]. \tag{12.13}$$

It is noteworthy that, in variational inference, sometimes, the parameter $\boldsymbol{\theta}$ is absorbed into the latent variable z_i. According to the chain rule:

$$\mathbb{P}(\boldsymbol{x}_i, z_i, \boldsymbol{\theta}) = \mathbb{P}(\boldsymbol{x}_i \,|\, z_i, \boldsymbol{\theta})\, \mathbb{P}(z_i \,|\, \boldsymbol{\theta})\, \mathbb{P}(\boldsymbol{\theta}).$$

By considering the term $\mathbb{P}(z_i \,|\, \boldsymbol{\theta})\, \mathbb{P}(\boldsymbol{\theta})$ as one probability term, the above equation becomes:

$$\mathbb{P}(\boldsymbol{x}_i, z_i) = \mathbb{P}(\boldsymbol{x}_i \,|\, z_i)\, \mathbb{P}(z_i),$$

where the parameter $\boldsymbol{\theta}$ disappears because it is absorbed.

12.3 Factor Analysis

12.3.1 Background on Marginal Multivariate Gaussian Distribution

Consider two random variables $\boldsymbol{x}_i \in \mathbb{R}^d$ and $\boldsymbol{z}_i \in \mathbb{R}^p$, and let $\boldsymbol{y}_i := [\boldsymbol{x}_i^\top, \boldsymbol{z}_i^\top]^\top \in \mathbb{R}^{d+p}$. Assume that $\boldsymbol{x}_i$ and $\boldsymbol{z}_i$ are jointly multivariate Gaussian; therefore, the variable $\boldsymbol{y}_i$ has a multivariate Gaussian distribution, i.e., $\boldsymbol{y}_i \sim \mathcal{N}(\boldsymbol{\mu}_y, \boldsymbol{\Sigma}_y)$. The mean and covariance can be decomposed to:

$$\boldsymbol{\mu}_y = [\boldsymbol{\mu}^\top, \boldsymbol{\mu}_0^\top]^\top \in \mathbb{R}^{d+p}, \tag{12.14}$$

$$\boldsymbol{\Sigma}_y = \begin{bmatrix} \boldsymbol{\Sigma}_{11} & \boldsymbol{\Sigma}_{12} \\ \boldsymbol{\Sigma}_{21} & \boldsymbol{\Sigma}_{22} \end{bmatrix} \in \mathbb{R}^{(d+p)\times(d+p)}, \tag{12.15}$$

where $\boldsymbol{\mu} \in \mathbb{R}^d$, $\boldsymbol{\mu}_0 \in \mathbb{R}^p$, $\boldsymbol{\Sigma}_{11} \in \mathbb{R}^{d\times d}$, $\boldsymbol{\Sigma}_{22} \in \mathbb{R}^{p\times p}$, $\boldsymbol{\Sigma}_{12} \in \mathbb{R}^{d\times p}$, and $\boldsymbol{\Sigma}_{21} = \boldsymbol{\Sigma}_{12}^\top \in \mathbb{R}^{p\times d}$.

The marginal distributions for $\boldsymbol{x}_i$ and $\boldsymbol{z}_i$ are Gaussian distributions, where $\mathbb{E}[\boldsymbol{x}_i] = \boldsymbol{\mu}$ and $\mathbb{E}[\boldsymbol{z}_i] = \boldsymbol{\mu}_0$ [16]. The covariance matrix of the joint distribution can be simplified to [16]:

$$\begin{aligned}\boldsymbol{\Sigma} &= \mathbb{E}[(\boldsymbol{y}_i - \boldsymbol{\mu}_y)(\boldsymbol{y}_i - \boldsymbol{\mu}_y)^\top] = \mathbb{E}\Bigg[\begin{bmatrix} \boldsymbol{x}_i - \boldsymbol{\mu} \\ \boldsymbol{z}_i - \boldsymbol{\mu}_0 \end{bmatrix}\begin{bmatrix} \boldsymbol{x}_i - \boldsymbol{\mu} \\ \boldsymbol{z}_i - \boldsymbol{\mu}_0 \end{bmatrix}^\top\Bigg] \\ &= \mathbb{E}\Bigg[\begin{bmatrix} (\boldsymbol{x}_i - \boldsymbol{\mu})(\boldsymbol{x}_i - \boldsymbol{\mu})^\top, (\boldsymbol{x}_i - \boldsymbol{\mu})(\boldsymbol{z}_i - \boldsymbol{\mu}_0)^\top \\ (\boldsymbol{z}_i - \boldsymbol{\mu}_0)(\boldsymbol{x}_i - \boldsymbol{\mu})^\top, (\boldsymbol{z}_i - \boldsymbol{\mu}_0)(\boldsymbol{z}_i - \boldsymbol{\mu}_0)^\top \end{bmatrix}\Bigg].\end{aligned} \tag{12.16}$$

This demonstrates that the marginal distributions are:

$$\boldsymbol{x}_i \sim \mathcal{N}(\boldsymbol{\mu}, \boldsymbol{\Sigma}_{11}), \tag{12.17}$$

$$\boldsymbol{z}_i \sim \mathcal{N}(\boldsymbol{\mu}_0, \boldsymbol{\Sigma}_{22}). \tag{12.18}$$

According to the definition of the multivariate Gaussian distribution, the conditional distribution is also a Gaussian distribution, i.e., $\boldsymbol{x}_i | \boldsymbol{z}_i \sim \mathcal{N}(\boldsymbol{\mu}_{x|z}, \boldsymbol{\Sigma}_{x|z})$ where [16]:

$$\mathbb{R}^d \ni \boldsymbol{\mu}_{x|z} := \boldsymbol{\mu} + \boldsymbol{\Sigma}_{12}\boldsymbol{\Sigma}_{22}^{-1}(\boldsymbol{z}_i - \boldsymbol{\mu}_0), \tag{12.19}$$

$$\mathbb{R}^{d\times d} \ni \boldsymbol{\Sigma}_{x|z} := \boldsymbol{\Sigma}_{11} - \boldsymbol{\Sigma}_{12}\boldsymbol{\Sigma}_{22}^{-1}\boldsymbol{\Sigma}_{21}, \tag{12.20}$$

and likewise for $\boldsymbol{z}_i | \boldsymbol{x}_i \sim \mathcal{N}(\boldsymbol{\mu}_{z|x}, \boldsymbol{\Sigma}_{z|x})$:

$$\mathbb{R}^p \ni \boldsymbol{\mu}_{z|x} := \boldsymbol{\mu}_0 + \boldsymbol{\Sigma}_{21}\boldsymbol{\Sigma}_{11}^{-1}(\boldsymbol{x}_i - \boldsymbol{\mu}), \tag{12.21}$$

$$\mathbb{R}^{p\times p} \ni \boldsymbol{\Sigma}_{z|x} := \boldsymbol{\Sigma}_{22} - \boldsymbol{\Sigma}_{21}\boldsymbol{\Sigma}_{11}^{-1}\boldsymbol{\Sigma}_{12}. \tag{12.22}$$

12.3.2 The Main Idea Behind Factor Analysis

Factor analysis [4–6, 11] is one of the simplest and most fundamental generative models. Although its theoretical derivations are complicated, its main idea is very simple. Factor analysis assumes that every data point $\boldsymbol{x}_i \in \mathbb{R}^d$ is generated from a latent variable $\boldsymbol{z}_i \in \mathbb{R}^p$. The latent variable is also referred to as the latent factor; therefore, the name of factor analysis comes from the fact that it analyzes the latent factors.

In factor analysis, it is assumed that the data point $\boldsymbol{x}_i$ is obtained through the following steps: (1) by linear projection of the p-dimensional $\boldsymbol{z}_i$ onto a d-dimensional space by the projection matrix $\boldsymbol{\Lambda} \in \mathbb{R}^{d\times p}$, then (2) applying a linear translation, and finally (3) adding a Gaussian noise $\boldsymbol{\epsilon} \in \mathbb{R}^d$ with covariance matrix $\boldsymbol{\Psi} \in \mathbb{R}^{d\times d}$. Note that as the noises in the different dimensions are independent, the covariance matrix $\boldsymbol{\Psi}$ is diagonal. Factor analysis can be illustrated as a graphical model [7], where the visible data variable is conditioned on the latent variable and the noise random variable. Figure 12.2 shows this graphical model.

12.3.3 The Factor Analysis Model

For simplicity, the prior distribution of the latent variable can be assumed to be a multivariate Gaussian distribution:

$$\mathbb{P}(\boldsymbol{z}_i) = \mathcal{N}(\boldsymbol{z}_i \mid \boldsymbol{\mu}_0, \boldsymbol{\Sigma}_0) = \frac{1}{\sqrt{(2\pi)^p |\boldsymbol{\Sigma}_0|}} \exp\Big(-\frac{(\boldsymbol{z}_i - \boldsymbol{\mu}_0)^\top \boldsymbol{\Sigma}_0^{-1} (\boldsymbol{z}_i - \boldsymbol{\mu}_0)}{2}\Big), \tag{12.23}$$

where $\boldsymbol{\mu}_0 \in \mathbb{R}^p$ and $\boldsymbol{\Sigma}_0 \in \mathbb{R}^{p\times p}$ are the mean and the covariance matrix of $\boldsymbol{z}_i$ and $|.|$ is the determinant of the matrix. As explained in Sect. 12.3.2, $\boldsymbol{x}_i$ is obtained through (1) the linear projection of $\boldsymbol{z}_i$ by $\boldsymbol{\Lambda} \in \mathbb{R}^{d\times p}$, (2) applying a linear translation, and (3) adding Gaussian noise $\boldsymbol{\epsilon} \in \mathbb{R}^d$ with covariance $\boldsymbol{\Psi} \in \mathbb{R}^{d\times d}$. Therefore, the data point

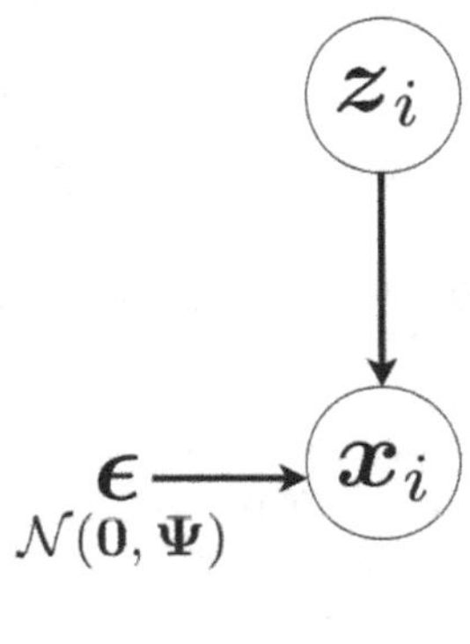

Fig. 12.2 The graphical model for factor analysis. The image is inspired by [7]

$\boldsymbol{x}_i$ has a conditional multivariate Gaussian distribution given the latent variable; its conditional likelihood is:

$$\mathbb{P}(\boldsymbol{x}_i \mid z_i) = \mathbb{P}(\boldsymbol{x}_i \mid z_i, \boldsymbol{\Lambda}, \boldsymbol{\mu}, \boldsymbol{\Psi}) = \mathcal{N}(\boldsymbol{\Lambda} z_i + \boldsymbol{\mu}, \boldsymbol{\Psi}), \tag{12.24}$$

where $\boldsymbol{\mu}$, which is the translation vector, is the mean of the data $\{\boldsymbol{x}_i\}_{i=1}^n$:

$$\mathbb{R}^d \ni \boldsymbol{\mu} := \frac{1}{n}\sum_{i=1}^{n} \boldsymbol{x}_i. \tag{12.25}$$

The marginal distribution of $\boldsymbol{x}_i$ is:

$$\mathbb{P}(\boldsymbol{x}_i) = \int \mathbb{P}(\boldsymbol{x}_i \mid z_i)\, \mathbb{P}(z_i)\, dz_i \implies$$

$$\mathbb{P}(\boldsymbol{x}_i \mid \boldsymbol{\Lambda}, \boldsymbol{\mu}, \boldsymbol{\Psi}) = \int \mathbb{P}(\boldsymbol{x}_i \mid z_i, \boldsymbol{\Lambda}, \boldsymbol{\mu}, \boldsymbol{\Psi})\, \mathbb{P}(z_i \mid \boldsymbol{\mu}_0, \boldsymbol{\Sigma}_0)\, dz_i$$

$$\overset{(a)}{=} \mathcal{N}(\boldsymbol{\Lambda}\boldsymbol{\mu}_0 + \boldsymbol{\mu}, \boldsymbol{\Psi} + \boldsymbol{\Lambda}\boldsymbol{\Sigma}_0\boldsymbol{\Lambda}^\top) = \mathcal{N}(\widehat{\boldsymbol{\mu}}, \boldsymbol{\Psi} + \widehat{\boldsymbol{\Lambda}}\widehat{\boldsymbol{\Lambda}}^\top), \tag{12.26}$$

where $\mathbb{R}^d \ni \widehat{\boldsymbol{\mu}} := \boldsymbol{\Lambda}\boldsymbol{\mu}_0 + \boldsymbol{\mu}$, $\mathbb{R}^{d\times d} \ni \widehat{\boldsymbol{\Lambda}} := \boldsymbol{\Lambda}\boldsymbol{\Sigma}_0^{(1/2)}$, and (a) is because the mean is linear and the variance is quadratic, which means that the mean and variance of the projection are applied linearly and quadratically, respectively. As the mean $\widehat{\boldsymbol{\mu}}$ and covariance $\widehat{\boldsymbol{\Lambda}}$ should be determined, $\boldsymbol{\mu}_0$ and $\boldsymbol{\Sigma}_0$ can be absorbed into $\boldsymbol{\mu}$ and $\boldsymbol{\Lambda}$ by assuming that $\boldsymbol{\mu}_0 = \boldsymbol{0}$ and $\boldsymbol{\Sigma}_0 = \boldsymbol{I}$.

In summary, factor analysis assumes that every data point $\boldsymbol{x}_i \in \mathbb{R}^d$ is obtained by projecting a latent variable $z_i \in \mathbb{R}^p$ onto a d-dimensional space, by the projection matrix $\boldsymbol{\Lambda} \in \mathbb{R}^{d\times p}$ and translating it by $\boldsymbol{\mu} \in \mathbb{R}^d$, and finally adding additional Gaussian noise $\boldsymbol{\epsilon} \in \mathbb{R}^d$ (whose dimensions are independent):

$$\boldsymbol{x}_i := \boldsymbol{\Lambda} z_i + \boldsymbol{\mu} + \boldsymbol{\epsilon}, \tag{12.27}$$

$$\mathbb{P}(z_i) = \mathcal{N}(\boldsymbol{0}, \boldsymbol{I}), \tag{12.28}$$

$$\mathbb{P}(\boldsymbol{\epsilon}) = \mathcal{N}(\boldsymbol{0}, \boldsymbol{\Psi}). \tag{12.29}$$

12.3.4 The Joint and Marginal Distributions in Factor Analysis

The joint distribution of $\boldsymbol{x}_i$ and z_i is:

$$\boldsymbol{y}_i := \begin{bmatrix} \boldsymbol{x}_i \\ z_i \end{bmatrix} \sim \mathcal{N}(\boldsymbol{\mu}_y, \boldsymbol{\Sigma}_y). \tag{12.30}$$

The expectation of $\boldsymbol{x}_i$ is:

$$\mathbb{E}[\boldsymbol{x}_i] \overset{(12.27)}{=} \mathbb{E}[\boldsymbol{\Lambda} z_i + \boldsymbol{\mu} + \boldsymbol{\epsilon}] = \boldsymbol{\Lambda}\mathbb{E}[z_i] + \boldsymbol{\mu} + \mathbb{E}[\boldsymbol{\epsilon}] \overset{(a)}{=} \boldsymbol{\mu}, \tag{12.31}$$

where (a) is because of Eqs. (12.28) and (12.29). Therefore:

$$\boldsymbol{\mu}_y := \begin{bmatrix} \boldsymbol{\mu}_x \\ \boldsymbol{\mu}_z \end{bmatrix} \overset{(a)}{=} \begin{bmatrix} \boldsymbol{\mu} \\ \mathbf{0} \end{bmatrix}, \tag{12.32}$$

where (a) is because of Eqs. (12.28) and (12.31). Consider Eq. (12.15). According to Eq. (12.28), $\boldsymbol{\Sigma}_{22} = \boldsymbol{\Sigma}_z = \boldsymbol{I}$. According to Eq. (12.27):

$$\begin{aligned}
\boldsymbol{\Sigma}_{11} = \boldsymbol{\Sigma}_x &= \mathbb{E}[(\boldsymbol{x}_i - \boldsymbol{\mu})(\boldsymbol{x}_i - \boldsymbol{\mu})^\top] \\
&= \mathbb{E}[(\boldsymbol{\Lambda} z_i + \boldsymbol{\mu} + \boldsymbol{\epsilon} - \boldsymbol{\mu})(\boldsymbol{\Lambda} z_i + \boldsymbol{\mu} + \boldsymbol{\epsilon} - \boldsymbol{\mu})^\top] \\
&= \mathbb{E}[\boldsymbol{\Lambda} z_i z_i^\top \boldsymbol{\Lambda}^\top + \boldsymbol{\epsilon} z_i^\top \boldsymbol{\Lambda}^\top + \boldsymbol{\Lambda} z_i \boldsymbol{\epsilon}^\top + \boldsymbol{\epsilon}\boldsymbol{\epsilon}^\top] \\
&= \boldsymbol{\Lambda}\mathbb{E}[z_i z_i^\top]\boldsymbol{\Lambda}^\top + \mathbb{E}[\boldsymbol{\epsilon}]\mathbb{E}[z_i]^\top \boldsymbol{\Lambda}^\top + \boldsymbol{\Lambda}\mathbb{E}[z_i]\mathbb{E}[\boldsymbol{\epsilon}]^\top + \mathbb{E}[\boldsymbol{\epsilon}\boldsymbol{\epsilon}^\top] \\
&\overset{(a)}{=} \boldsymbol{\Lambda}\boldsymbol{I}\boldsymbol{\Lambda}^\top + \mathbf{0} + \mathbf{0} + \boldsymbol{\Psi} = \boldsymbol{\Lambda}\boldsymbol{\Lambda}^\top + \boldsymbol{\Psi},
\end{aligned} \tag{12.33}$$

where (a) is because of Eqs. (12.28) and (12.29). Moreover:

$$\begin{aligned}
\boldsymbol{\Sigma}_{12} = \boldsymbol{\Sigma}_{xz} &= \mathbb{E}[(\boldsymbol{x}_i - \boldsymbol{\mu})(z_i - \boldsymbol{\mu}_0)^\top] \\
&\overset{(a)}{=} \mathbb{E}[(\boldsymbol{\Lambda} z_i + \boldsymbol{\mu} + \boldsymbol{\epsilon} - \boldsymbol{\mu})(z_i - \mathbf{0})^\top] \\
&\overset{(b)}{=} \boldsymbol{\Lambda}\mathbb{E}[z_i z_i^\top] + \mathbb{E}[\boldsymbol{\epsilon}]\mathbb{E}[z_i^\top] = \boldsymbol{\Lambda}\boldsymbol{I} + (\mathbf{0}\mathbf{0}^\top) = \boldsymbol{\Lambda},
\end{aligned} \tag{12.34}$$

where (a) is because of Eqs. (12.27) and (12.28) and (b) is because z_i and $\boldsymbol{\epsilon}$ are independent. Additionally, $\boldsymbol{\Sigma}_{21} = \boldsymbol{\Sigma}_{12}^\top = \boldsymbol{\Lambda}^\top$. Therefore:

$$\begin{bmatrix} \boldsymbol{x}_i \\ z_i \end{bmatrix} \sim \mathcal{N}\Big(\begin{bmatrix} \boldsymbol{\mu} \\ \mathbf{0} \end{bmatrix}, \begin{bmatrix} \boldsymbol{\Lambda}\boldsymbol{\Lambda}^\top + \boldsymbol{\Psi} & \boldsymbol{\Lambda} \\ \boldsymbol{\Lambda}^\top & \boldsymbol{I} \end{bmatrix}\Big). \tag{12.35}$$

Therefore, the marginal distribution of data point $\boldsymbol{x}_i$ is:

$$\mathbb{P}(\boldsymbol{x}_i) = \mathbb{P}(\boldsymbol{x}_i \,|\, \boldsymbol{\Lambda}, \boldsymbol{\mu}, \boldsymbol{\Psi}) = \mathcal{N}(\boldsymbol{\mu}, \boldsymbol{\Lambda}\boldsymbol{\Lambda}^\top + \boldsymbol{\Psi}). \tag{12.36}$$

According to Eqs. (12.21) and (12.22), the posterior or the conditional distribution of the latent variable given the data is:

$$q(z_i) \overset{(12.12)}{=} \mathbb{P}(z_i \,|\, \boldsymbol{x}_i) = \mathbb{P}(z_i \,|\, \boldsymbol{x}_i, \boldsymbol{\Lambda}, \boldsymbol{\mu}, \boldsymbol{\Psi}) = \mathcal{N}(\boldsymbol{\mu}_{z|x}, \boldsymbol{\Sigma}_{z|x}), \tag{12.37}$$

where:

$$\mathbb{R}^p \ni \boldsymbol{\mu}_{z|x} := \boldsymbol{\Lambda}^\top(\boldsymbol{\Lambda}\boldsymbol{\Lambda}^\top + \boldsymbol{\Psi})^{-1}(\boldsymbol{x}_i - \boldsymbol{\mu}), \tag{12.38}$$

$$\mathbb{R}^{p\times p} \ni \boldsymbol{\Sigma}_{z|x} := \boldsymbol{I} - \boldsymbol{\Lambda}^\top(\boldsymbol{\Lambda}\boldsymbol{\Lambda}^\top + \boldsymbol{\Psi})^{-1}\boldsymbol{\Lambda}. \tag{12.39}$$

Recall that the conditional distribution of the data given the latent variable, $\mathbb{P}(\boldsymbol{x}_i \,|\, \boldsymbol{z}_i)$, was introduced in Eq. (12.24).

If the data $\{\boldsymbol{x}_i\}_{i=1}^n$ are centered, i.e., $\boldsymbol{\mu} = \boldsymbol{0}$, then the marginal of the data, Eq. (12.36), and the likelihood of the data, Eq. (12.24), become:

$$\mathbb{P}(\boldsymbol{x}_i \,|\, \boldsymbol{\Lambda}, \boldsymbol{\Psi}) = \mathcal{N}(\boldsymbol{0}, \boldsymbol{\Psi} + \boldsymbol{\Lambda}\boldsymbol{\Lambda}^\top), \tag{12.40}$$

$$\mathbb{P}(\boldsymbol{x}_i \,|\, \boldsymbol{z}_i, \boldsymbol{\Lambda}, \boldsymbol{\Psi}) = \mathcal{N}(\boldsymbol{\Lambda}\boldsymbol{z}_i, \boldsymbol{\Psi}), \tag{12.41}$$

respectively. In some works, the data are centered as a preprocessing to factor analysis.

12.3.5 Expectation Maximization in Factor Analysis

12.3.5.1 Maximization of Joint Likelihood

Factor analysis has two parameters $\boldsymbol{\theta}$ of variational inference, namely $\boldsymbol{\Lambda}$ and $\boldsymbol{\Psi}$. As seen in Eq. (12.13), consider maximization of the joint likelihood, which reduces to the likelihood of the data, over all n data points:

$$\max_{\boldsymbol{\Lambda},\boldsymbol{\Psi}} \sum_{i=1}^n \mathbb{E}_{\sim q^{(t)}(\boldsymbol{z}_i)}\big[\log \mathbb{P}(\boldsymbol{x}_i, \boldsymbol{z}_i \,|\, \boldsymbol{\Lambda}, \boldsymbol{\Psi})\big]$$

$$\overset{(a)}{=} \max_{\boldsymbol{\Lambda},\boldsymbol{\Psi}} \sum_{i=1}^n \Big(\mathbb{E}_{\sim q^{(t)}(\boldsymbol{z}_i)}\big[\log \mathbb{P}(\boldsymbol{x}_i \,|\, \boldsymbol{z}_i, \boldsymbol{\Lambda}, \boldsymbol{\Psi})\big] + \mathbb{E}_{\sim q^{(t)}(\boldsymbol{z}_i)}\big[\log \mathbb{P}(\boldsymbol{z}_i)\big]\Big),$$

$$\overset{(b)}{=} \max_{\boldsymbol{\Lambda},\boldsymbol{\Psi}} \sum_{i=1}^n \mathbb{E}_{\sim q^{(t)}(\boldsymbol{z}_i)}\big[\log \mathbb{P}(\boldsymbol{x}_i \,|\, \boldsymbol{z}_i, \boldsymbol{\Lambda}, \boldsymbol{\Psi})\big]$$

$$\overset{(12.24)}{=} \max_{\boldsymbol{\Lambda},\boldsymbol{\Psi}} \sum_{i=1}^n \mathbb{E}_{\sim q^{(t)}(\boldsymbol{z}_i)}\big[\log \mathcal{N}(\boldsymbol{\Lambda}\boldsymbol{z}_i + \boldsymbol{\mu}, \boldsymbol{\Psi})\big]$$

$$= \max_{\boldsymbol{\Lambda},\boldsymbol{\Psi}} \sum_{i=1}^n \mathbb{E}_{\sim q^{(t)}(\boldsymbol{z}_i)}\Big[\log \big(\frac{1}{(2\pi)^{p/2}|\boldsymbol{\Psi}|^{1/2}}$$

$$\exp\Big(-\frac{(\boldsymbol{x}_i - \boldsymbol{\Lambda}\boldsymbol{z}_i - \boldsymbol{\mu})^\top \boldsymbol{\Psi}^{-1}(\boldsymbol{x}_i - \boldsymbol{\Lambda}\boldsymbol{z}_i - \boldsymbol{\mu})}{2}\Big)\Big)\Big]$$

$$= \max_{\boldsymbol{\Lambda},\boldsymbol{\Psi}} \Big(\underbrace{-\frac{dn}{2}\log(2\pi)}_{\text{constant}} - \frac{n}{2}\log|\boldsymbol{\Psi}|$$

$$- \sum_{i=1}^{n} \mathbb{E}_{\sim q^{(t)}(\boldsymbol{z}_i)}\big[\frac{1}{2}(\boldsymbol{x}_i - \boldsymbol{\Lambda}\boldsymbol{z}_i - \boldsymbol{\mu})^\top \boldsymbol{\Psi}^{-1}(\boldsymbol{x}_i - \boldsymbol{\Lambda}\boldsymbol{z}_i - \boldsymbol{\mu})\big]\Big) \tag{12.42}$$

where (a) is because of the chain rule $\mathbb{P}(\boldsymbol{x}_i, \boldsymbol{z}_i \,|\, \boldsymbol{\Lambda}, \boldsymbol{\Psi}) = \mathbb{P}(\boldsymbol{x}_i \,|\, \boldsymbol{z}_i, \boldsymbol{\Lambda}, \boldsymbol{\Psi})\,\mathbb{P}(\boldsymbol{z}_i)$, and (b) is because the second term is zero due to the zero mean of the prior on $\boldsymbol{z}_i$ (see Eq. (12.28)).

12.3.5.2 The E-Step in EM for Factor Analysis

There are two expectation terms that need to be computed in the E-step (it will be seen later in the M-step of EM that those expectations are required). These expectations, which are over the $q(\boldsymbol{z}_i)$ distribution, are $\mathbb{E}_{\sim q^{(t)}(\boldsymbol{z}_i)}[\boldsymbol{z}_i]$ and $\mathbb{E}_{\sim q^{(t)}(\boldsymbol{z}_i)}[\boldsymbol{z}_i\boldsymbol{z}_i^\top]$. According to Eq. (12.12), $q(\boldsymbol{z}_i) = \mathbb{P}(\boldsymbol{z}_i \,|\, \boldsymbol{x}_i)$. Therefore, according to Eqs. (12.37), (12.38), and (12.39):

$$\mathbb{E}_{\sim q^{(t)}(\boldsymbol{z}_i)}[\boldsymbol{z}_i] = \boldsymbol{\mu}_{z|x} := \boldsymbol{\Lambda}^\top(\boldsymbol{\Lambda}\boldsymbol{\Lambda}^\top + \boldsymbol{\Psi})^{-1}(\boldsymbol{x}_i - \boldsymbol{\mu}), \tag{12.43}$$

$$\mathbb{E}_{\sim q^{(t)}(\boldsymbol{z}_i)}[\boldsymbol{z}_i\boldsymbol{z}_i^\top] = \boldsymbol{\Sigma}_{z|x} := \boldsymbol{I} - \boldsymbol{\Lambda}^\top(\boldsymbol{\Lambda}\boldsymbol{\Lambda}^\top + \boldsymbol{\Psi})^{-1}\boldsymbol{\Lambda}. \tag{12.44}$$

12.3.5.3 The M-Step in EM for Factor Analysis

There are two variables $\boldsymbol{\Lambda}$ and $\boldsymbol{\Psi}$, meaning that the maximization problem is solved with respect to these variables.

Finding Parameter $\boldsymbol{\Lambda}$

$$\mathbb{R}^{d\times p} \ni \frac{\partial\, \text{Eq. (12.42)}}{\partial \boldsymbol{\Lambda}}$$

$$= -\sum_{i=1}^{n} \frac{\partial}{\partial \boldsymbol{\Lambda}} \mathbb{E}_{\sim q^{(t)}(\boldsymbol{z}_i)}\Big[\frac{1}{2}\mathbf{tr}(\boldsymbol{z}_i^\top \boldsymbol{\Lambda}^\top \boldsymbol{\Psi}^{-1}\boldsymbol{\Lambda}\boldsymbol{z}_i) - \mathbf{tr}(\boldsymbol{z}_i^\top \boldsymbol{\Lambda}^\top \boldsymbol{\Psi}^{-1}(\boldsymbol{x}_i - \boldsymbol{\mu}))\Big]$$

$$\overset{(a)}{=} -\sum_{i=1}^{n} \frac{\partial}{\partial \boldsymbol{\Lambda}} \mathbb{E}_{\sim q^{(t)}(\boldsymbol{z}_i)}\Big[\frac{1}{2}\mathbf{tr}(\boldsymbol{\Lambda}^\top \boldsymbol{\Psi}^{-1}\boldsymbol{\Lambda}\boldsymbol{z}_i\boldsymbol{z}_i^\top) - \mathbf{tr}(\boldsymbol{\Lambda}^\top \boldsymbol{\Psi}^{-1}(\boldsymbol{x}_i - \boldsymbol{\mu})\boldsymbol{z}_i^\top)\Big]$$

$$= -\sum_{i=1}^{n} \mathbb{E}_{\sim q^{(t)}(z_i)}\Big[\boldsymbol{\Psi}^{-1}\boldsymbol{\Lambda} z_i z_i^\top - \boldsymbol{\Psi}^{-1}(\boldsymbol{x}_i - \boldsymbol{\mu}) z_i^\top\Big]$$

$$= -\sum_{i=1}^{n} \Big[\boldsymbol{\Psi}^{-1}\boldsymbol{\Lambda}\mathbb{E}_{\sim q^{(t)}(z_i)}[z_i z_i^\top] - \boldsymbol{\Psi}^{-1}(\boldsymbol{x}_i - \boldsymbol{\mu})\mathbb{E}_{\sim q^{(t)}(z_i)}[z_i]^\top\Big],$$

where (a) is because of the cyclic property of the trace. Setting this derivative to zero results in the optimal $\boldsymbol{\Lambda}$:

$$\sum_{i=1}^{n} \boldsymbol{\Psi}^{-1}\boldsymbol{\Lambda}\mathbb{E}_{\sim q^{(t)}(z_i)}[z_i z_i^\top] = \sum_{i=1}^{n} \boldsymbol{\Psi}^{-1}(\boldsymbol{x}_i - \boldsymbol{\mu})\mathbb{E}_{\sim q^{(t)}(z_i)}[z_i]^\top$$

$$\implies \boldsymbol{\Lambda} = \Big(\sum_{i=1}^{n} \boldsymbol{\Psi}^{-1}(\boldsymbol{x}_i - \boldsymbol{\mu})\mathbb{E}_{\sim q^{(t)}(z_i)}[z_i]^\top\Big)\Big(\sum_{i=1}^{n} \boldsymbol{\Psi}^{-1}\boldsymbol{\Lambda}\mathbb{E}_{\sim q^{(t)}(z_i)}[z_i z_i^\top]\Big)^{-1}. \tag{12.45}$$

Finding Parameter $\boldsymbol{\Psi}$ Now, consider maximization with respect to $\boldsymbol{\Psi}$. Equation (12.42) can be restated as [18]:

$$\max_{\boldsymbol{\Lambda},\boldsymbol{\Psi}} \Big(\underbrace{-\frac{d\,n}{2}\log(2\pi)}_{\text{constant}} - \frac{n}{2}\log|\boldsymbol{\Psi}| - \frac{n}{2}\mathbf{tr}(\boldsymbol{\Psi}^{-1}\boldsymbol{S})\Big), \tag{12.46}$$

where $\boldsymbol{S} \in \mathbb{R}^{d\times d}$ is a sample covariance matrix defined as:

$$\boldsymbol{S} := \frac{1}{n}\sum_{i=1}^{n} \mathbb{E}_{\sim q^{(t)}(z_i)}[(\boldsymbol{x}_i - \boldsymbol{\Lambda} z_i - \boldsymbol{\mu})(\boldsymbol{x}_i - \boldsymbol{\Lambda} z_i - \boldsymbol{\mu})^\top]$$

$$= \frac{1}{n}\sum_{i=1}^{n} \Big((\boldsymbol{x}_i - \boldsymbol{\mu})(\boldsymbol{x}_i - \boldsymbol{\mu})^\top - 2\boldsymbol{\Lambda}\mathbb{E}_{\sim q^{(t)}(z_i)}[z_i](\boldsymbol{x}_i - \boldsymbol{\mu})^\top + \boldsymbol{\Lambda}\mathbb{E}_{\sim q^{(t)}(z_i)}[z z^\top]\boldsymbol{\Lambda}^\top\Big). \tag{12.47}$$

The maximization results in [18]:

$$\mathbb{R}^{d\times d} \ni \frac{\partial\, \text{Eq. (12.46)}}{\partial \boldsymbol{\Psi}^{-1}} = \frac{n}{2}\boldsymbol{\Psi} - \frac{n}{2}\boldsymbol{S} \overset{\text{set}}{=} \boldsymbol{0} \implies \boldsymbol{\Psi} = \boldsymbol{S}.$$

Note that as the dimensions of the noise $\boldsymbol{\epsilon} \in \mathbb{R}^d$ are independent, the covariance matrix of the noise, $\boldsymbol{\Psi}$, is a diagonal matrix. Therefore:

$$\boldsymbol{\Psi} = \mathbf{diag}(S) \overset{(12.47)}{=} \frac{1}{n}\mathbf{diag}\bigg(\sum_{i=1}^{n}\Big[(\boldsymbol{x}_i - \boldsymbol{\mu})(\boldsymbol{x}_i - \boldsymbol{\mu})^\top$$
$$- 2\boldsymbol{\Lambda}\mathbb{E}_{\sim q^{(t)}(\boldsymbol{z}_i)}[\boldsymbol{z}_i](\boldsymbol{x}_i - \boldsymbol{\mu})^\top + \boldsymbol{\Lambda}\mathbb{E}_{\sim q^{(t)}(\boldsymbol{z}_i)}[\boldsymbol{z}\boldsymbol{z}^\top]\boldsymbol{\Lambda}^\top\Big]\bigg). \tag{12.48}$$

12.3.5.4 Summary of Factor Analysis Algorithm

According to the derived Eqs. (12.43), (12.44), (12.45), and (12.48), the EM algorithm in factor analysis is summarized as follows. The mean of the data, $\boldsymbol{\mu}$, is computed. Then, the aim is to iteratively solve for every data point $\boldsymbol{x}_i$:

$$\mathbb{E}_{\sim q^{(t)}(\boldsymbol{z}_i)}[\boldsymbol{z}_i] \leftarrow \boldsymbol{\Lambda}^{(t)\top}(\boldsymbol{\Lambda}^{(t)}\boldsymbol{\Lambda}^{(t)\top} + \boldsymbol{\Psi}^{(t)})^{-1}(\boldsymbol{x}_i - \boldsymbol{\mu}),$$
$$\mathbb{E}_{\sim q^{(t)}(\boldsymbol{z}_i)}[\boldsymbol{z}_i\boldsymbol{z}_i^\top] \leftarrow \boldsymbol{I} - \boldsymbol{\Lambda}^{(t)\top}(\boldsymbol{\Lambda}^{(t)}\boldsymbol{\Lambda}^{(t)\top} + \boldsymbol{\Psi}^{(t)})^{-1}\boldsymbol{\Lambda}^{(t)},$$
$$\boldsymbol{\Lambda}^{(t+1)} \leftarrow \Big(\sum_{i=1}^{n}(\boldsymbol{\Psi}^{(t)})^{-1}(\boldsymbol{x}_i - \boldsymbol{\mu})\mathbb{E}_{\sim q^{(t)}(\boldsymbol{z}_i)}[\boldsymbol{z}_i]^\top\Big)\Big(\sum_{i=1}^{n}(\boldsymbol{\Psi}^{(t)})^{-1}\boldsymbol{\Lambda}^{(t)}\mathbb{E}_{\sim q^{(t)}(\boldsymbol{z}_i)}[\boldsymbol{z}_i\boldsymbol{z}_i^\top]\Big)^{-1}.$$
$$\boldsymbol{\Psi}^{(t+1)} \leftarrow \frac{1}{n}\mathbf{diag}\bigg(\sum_{i=1}^{n}\Big[(\boldsymbol{x}_i - \boldsymbol{\mu})(\boldsymbol{x}_i - \boldsymbol{\mu})^\top - 2\boldsymbol{\Lambda}^{(t+1)}\mathbb{E}_{\sim q^{(t)}(\boldsymbol{z}_i)}[\boldsymbol{z}_i](\boldsymbol{x}_i - \boldsymbol{\mu})^\top$$
$$+ \boldsymbol{\Lambda}^{(t+1)}\mathbb{E}_{\sim q^{(t)}(\boldsymbol{z}_i)}[\boldsymbol{z}\boldsymbol{z}^\top]\boldsymbol{\Lambda}^{(t+1)}\Big]\bigg).$$

Note that if the data are centered as a preprocessing to factor analysis, i.e., $\boldsymbol{\mu} = \boldsymbol{0}$, then the algorithm of the factor analysis is simplified to:

$$\mathbb{E}_{\sim q^{(t)}(\boldsymbol{z}_i)}[\boldsymbol{z}_i] \leftarrow \boldsymbol{\Lambda}^{(t)\top}(\boldsymbol{\Lambda}^{(t)}\boldsymbol{\Lambda}^{(t)\top} + \boldsymbol{\Psi}^{(t)})^{-1}\boldsymbol{x}_i,$$
$$\mathbb{E}_{\sim q^{(t)}(\boldsymbol{z}_i)}[\boldsymbol{z}_i\boldsymbol{z}_i^\top] \leftarrow \boldsymbol{I} - \boldsymbol{\Lambda}^{(t)\top}(\boldsymbol{\Lambda}^{(t)}\boldsymbol{\Lambda}^{(t)\top} + \boldsymbol{\Psi}^{(t)})^{-1}\boldsymbol{\Lambda}^{(t)},$$
$$\boldsymbol{\Lambda}^{(t+1)} \leftarrow \Big(\sum_{i=1}^{n}(\boldsymbol{\Psi}^{(t)})^{-1}\boldsymbol{x}_i\,\mathbb{E}_{\sim q^{(t)}(\boldsymbol{z}_i)}[\boldsymbol{z}_i]^\top\Big)\Big(\sum_{i=1}^{n}(\boldsymbol{\Psi}^{(t)})^{-1}\boldsymbol{\Lambda}^{(t)}\mathbb{E}_{\sim q^{(t)}(\boldsymbol{z}_i)}[\boldsymbol{z}_i\boldsymbol{z}_i^\top]\Big)^{-1}.$$
$$\boldsymbol{\Psi}^{(t+1)} \leftarrow \frac{1}{n}\mathbf{diag}\bigg(\sum_{i=1}^{n}\Big[\boldsymbol{x}_i\boldsymbol{x}_i^\top - 2\boldsymbol{\Lambda}^{(t+1)}\mathbb{E}_{\sim q^{(t)}(\boldsymbol{z}_i)}[\boldsymbol{z}_i]\,\boldsymbol{x}_i^\top$$
$$+ \boldsymbol{\Lambda}^{(t+1)}\mathbb{E}_{\sim q^{(t)}(\boldsymbol{z}_i)}[\boldsymbol{z}\boldsymbol{z}^\top]\boldsymbol{\Lambda}^{(t+1)}\Big]\bigg).$$

As demonstrated above, factor analysis does not have a closed-form solution, and its solution, which are the projection matrix $\mathbf{\Lambda}$ and the noise covariance matrix $\mathbf{\Psi}$, are found iteratively until convergence. It is noteworthy that a mixture of factor analysis [7] also exists in the literature, which considers a mixture distribution for the factor analysis and trains the parameters of the mixture using EM [8].

12.4 Probabilistic Principal Component Analysis

12.4.1 Main Idea of Probabilistic PCA

Probabilistic PCA (PPCA) [19, 21] is a special case of factor analysis, where the variance of the noise is equal in all dimensions of the data space, with covariance between dimensions, i.e.:

$$\mathbf{\Psi} = \sigma^2 \boldsymbol{I}. \tag{12.49}$$

In other words, PPCA considers isotropic noise in its formulation. Therefore, Eq. (12.29) can be simplified to:

$$\mathbb{P}(\boldsymbol{\epsilon}) = \mathcal{N}(\boldsymbol{0}, \sigma^2 \boldsymbol{I}). \tag{12.50}$$

As there is zero covariance of noise between the different dimensions, PPCA assumes that the data points are independent of each other given the latent variables. As depicted in Fig. 12.3, PPCA can be illustrated as a graphical model, where the visible data variable is conditioned on the latent variable and the isotropic noise random variable.

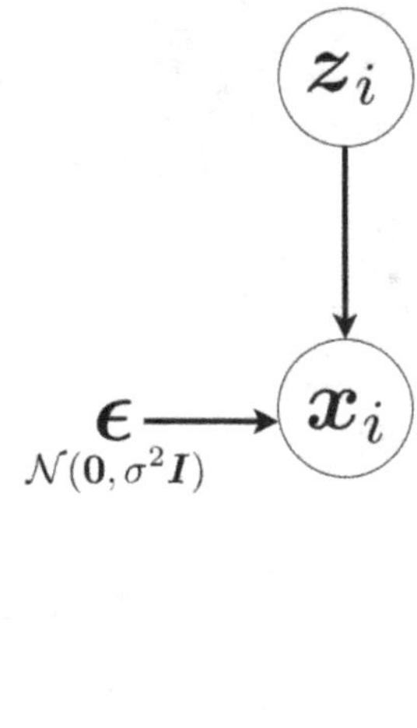

Fig. 12.3 The graphical model for PPCA

12.4.2 MLE for Probabilistic PCA

As PPCA is a special case of factor analysis, it can also be iteratively solved using EM [19]. However, it is also possible to find a closed-form solution to its EM approach [21]. Therefore, by restricting the noise covariance to be isotropic, its solution becomes simpler and closed-form. The iterative approach is the same as in factor analysis, explained in Sect. 12.3.5. In this section, the closed-form solution is described.

Consider the likelihood or the marginal distribution of the data points $\{\boldsymbol{x}_i \in \mathbb{R}^d\}_{i=1}^n$, which is Eq. (12.36). The log-likelihood of the data is:

$$
\begin{aligned}
\sum_{i=1}^n \log \mathbb{P}(\boldsymbol{x}_i) &\overset{(12.36)}{=} \sum_{i=1}^n \log \mathcal{N}(\boldsymbol{\mu}, \boldsymbol{\Lambda}\boldsymbol{\Lambda}^\top + \sigma^2 \boldsymbol{I}) \\
&= \sum_{i=1}^n \Big[\log \Big(\frac{1}{(2\pi)^{d/2} |\boldsymbol{\Lambda}\boldsymbol{\Lambda}^\top + \sigma^2 \boldsymbol{I}|^{1/2}} \exp\Big(- \frac{(\boldsymbol{x}_i - \boldsymbol{\mu})^\top (\boldsymbol{\Lambda}\boldsymbol{\Lambda}^\top + \sigma^2 \boldsymbol{I})^{-1} (\boldsymbol{x}_i - \boldsymbol{\mu})}{2} \Big) \Big) \Big] \\
&= \underbrace{-\frac{d\,n}{2} \log(2\pi)}_{\text{constant}} - \frac{n}{2} \log |\boldsymbol{\Lambda}\boldsymbol{\Lambda}^\top + \sigma^2 \boldsymbol{I}| - \sum_{i=1}^n \frac{1}{2} (\boldsymbol{x}_i - \boldsymbol{\mu})^\top (\boldsymbol{\Lambda}\boldsymbol{\Lambda}^\top + \sigma^2 \boldsymbol{I})^{-1} (\boldsymbol{x}_i - \boldsymbol{\mu}) \\
&= \underbrace{-\frac{d\,n}{2} \log(2\pi)}_{\text{constant}} - \frac{n}{2} \log |\boldsymbol{\Lambda}\boldsymbol{\Lambda}^\top + \sigma^2 \boldsymbol{I}| - \frac{n}{2} \mathbf{tr}\big((\boldsymbol{\Lambda}\boldsymbol{\Lambda}^\top + \sigma^2 \boldsymbol{I})^{-1} \boldsymbol{S}_x \big),
\end{aligned}
$$

where $\boldsymbol{S}_x \in \mathbb{R}^{d \times d}$ is the sample covariance matrix of the data:

$$
\boldsymbol{S}_x := \frac{1}{n} \sum_{i=1}^n (\boldsymbol{x}_i - \boldsymbol{\mu})(\boldsymbol{x}_i - \boldsymbol{\mu})^\top. \tag{12.51}
$$

MLE can be used where the variables of the maximization problem are the projection matrix $\boldsymbol{\Lambda}$ and the noise variance σ:

$$
\max_{\boldsymbol{\Lambda}, \sigma} \Big(\underbrace{-\frac{d\,n}{2} \log(2\pi)}_{\text{constant}} - \frac{n}{2} \log |\boldsymbol{\Lambda}\boldsymbol{\Lambda}^\top + \sigma^2 \boldsymbol{I}| - \frac{n}{2} \mathbf{tr}\big((\boldsymbol{\Lambda}\boldsymbol{\Lambda}^\top + \sigma^2 \boldsymbol{I})^{-1} \boldsymbol{S}_x \big) \Big). \tag{12.52}
$$

It is noteworthy that the literature usually defines the following:

$$
\mathbb{R}^{d \times d} \ni \boldsymbol{C} := (\boldsymbol{\Lambda}\boldsymbol{\Lambda}^\top + \sigma^2 \boldsymbol{I}_{d \times d}), \tag{12.53}
$$

$$
\mathbb{R}^{p \times p} \ni \boldsymbol{M} := (\boldsymbol{\Lambda}^\top \boldsymbol{\Lambda} + \sigma^2 \boldsymbol{I}_{p \times p}). \tag{12.54}
$$

According to the matrix inversion lemma, the inversion of $\boldsymbol{C}$ is:

$$\boldsymbol{C}^{-1} = \sigma^{-1}\boldsymbol{I}_{d\times d} - \sigma^{-2}\boldsymbol{\Lambda}\boldsymbol{M}^{-1}\boldsymbol{\Lambda}^\top. \tag{12.55}$$

This inversion is interesting because the inverse of a $(d \times d)$ matrix $\boldsymbol{C}$ is reduced to the inversion of a $(p \times p)$ matrix $\boldsymbol{M}$, which is much simpler because $p \ll d$.

12.4.2.1 MLE for Determining Λ

Taking the derivative of Eq. (12.52) with respect to $\boldsymbol{\Lambda} \in \mathbb{R}^{d\times p}$ and setting it to zero results in:

$$\frac{\partial\, \text{Eq.(12.52)}}{\partial \boldsymbol{\Lambda}} = -n\big((\boldsymbol{\Lambda}\boldsymbol{\Lambda}^\top + \sigma^2\boldsymbol{I})^{-1}\boldsymbol{S}_x(\boldsymbol{\Lambda}\boldsymbol{\Lambda}^\top + \sigma^2\boldsymbol{I})^{-1}\boldsymbol{\Lambda} - (\boldsymbol{\Lambda}\boldsymbol{\Lambda}^\top + \sigma^2\boldsymbol{I})^{-1}\boldsymbol{\Lambda}\big) \stackrel{\text{set}}{=} \boldsymbol{0}$$
$$\implies (\boldsymbol{\Lambda}\boldsymbol{\Lambda}^\top + \sigma^2\boldsymbol{I})^{-1}\boldsymbol{S}_x(\boldsymbol{\Lambda}\boldsymbol{\Lambda}^\top + \sigma^2\boldsymbol{I})^{-1}\boldsymbol{\Lambda} = (\boldsymbol{\Lambda}\boldsymbol{\Lambda}^\top + \sigma^2\boldsymbol{I})^{-1}\boldsymbol{\Lambda}$$
$$\implies \boldsymbol{S}_x(\boldsymbol{\Lambda}\boldsymbol{\Lambda}^\top + \sigma^2\boldsymbol{I})^{-1}\boldsymbol{\Lambda} = \boldsymbol{\Lambda},$$

whose trivial solutions are $\boldsymbol{\Lambda} = \boldsymbol{0}$ and $\boldsymbol{S}_x = (\boldsymbol{\Lambda}\boldsymbol{\Lambda}^\top + \sigma^2\boldsymbol{I})$, which are not valid. For the nontrivial solution, consider the Singular Value Decomposition (SVD) $\mathbb{R}^{d\times p} \ni \boldsymbol{\Lambda} = \boldsymbol{U}\boldsymbol{L}\boldsymbol{V}^\top$ where $\boldsymbol{U} \in \mathbb{R}^{d\times p}$ and $\boldsymbol{V} \in \mathbb{R}^{p\times p}$ contain the left and right singular vectors, respectively, and $\boldsymbol{L} \in \mathbb{R}^{p\times p}$ is the diagonal matrix containing the singular values denoted by $\{l_j\}_{j=1}^p$ (see Chap. 2 for more information on SVD). Moreover, note that $\mathbf{tr}(\boldsymbol{\Lambda}\boldsymbol{\Lambda}^\top) = \mathbf{tr}(\boldsymbol{U}\boldsymbol{L}\boldsymbol{V}^\top\boldsymbol{V}\boldsymbol{L}\boldsymbol{U}^\top) = \mathbf{tr}(\boldsymbol{U}\boldsymbol{L}\boldsymbol{L}\boldsymbol{U}^\top) = \mathbf{tr}(\boldsymbol{U}^\top\boldsymbol{U}\boldsymbol{L}^2) = \mathbf{tr}(\boldsymbol{L}^2)$ because $\boldsymbol{U}$ and $\boldsymbol{V}$ are orthogonal matrices.

From the previous calculations [12]:

$$\boldsymbol{S}_x\boldsymbol{U}\boldsymbol{L}(\boldsymbol{L}^2 + \sigma^2\boldsymbol{I})^{-1}\boldsymbol{V}^\top = \boldsymbol{U}\boldsymbol{L}\boldsymbol{V}^\top \implies \boldsymbol{S}_x\boldsymbol{U}\boldsymbol{L}(\boldsymbol{L}^2 + \sigma^2\boldsymbol{I})^{-1} = \boldsymbol{U}\boldsymbol{L}$$
$$\implies \boldsymbol{S}_x\boldsymbol{U}\boldsymbol{L} = \boldsymbol{U}(\boldsymbol{L}^2 + \sigma^2\boldsymbol{I})\boldsymbol{L} \implies \boldsymbol{S}_x\boldsymbol{U} = \boldsymbol{U}(\boldsymbol{L}^2 + \sigma^2\boldsymbol{I}), \tag{12.56}$$

which is an eigenvalue problem (see Chap. 2) for the covariance matrix $\boldsymbol{S}_x$, where the columns of $\boldsymbol{U}$ are the eigenvectors of $\boldsymbol{S}_x$ and the eigenvalues are $\sigma^2 + l_j^2$. Recall that σ is the variance of noise in different dimensions and l_j is the j-th singular value of $\boldsymbol{\Lambda}$ (sorted from largest to smallest). The j-th eigenvalue of the covariance matrix $\boldsymbol{S}_x$ is denoted by:

$$\delta_j := \sigma^2 + l_j^2 \implies l_j = (\delta_j - \sigma^2)^{(1/2)}. \tag{12.57}$$

Only the top p singular values l_j and the top p eigenvalues δ_j are considered; substituting the singular values in the SVD of the projection matrix $\boldsymbol{\Lambda}$ results in:

$$\boldsymbol{\Lambda} = \boldsymbol{U}\boldsymbol{L}\boldsymbol{V}^\top = \boldsymbol{U}(\boldsymbol{\Delta}_p - \sigma^2\boldsymbol{I})^{(1/2)}\boldsymbol{V}^\top,$$

where $\boldsymbol{\Delta}_p := \mathbf{diag}(\delta_1, \ldots, \delta_p)$. However, as Eq. (12.56) does not include any $\boldsymbol{V}$, $\boldsymbol{V}^\top$ can be replaced with any arbitrary orthogonal matrix $\boldsymbol{R} \in \mathbb{R}^{p \times p}$ [21]:

$$\boldsymbol{\Lambda} = \boldsymbol{U}(\boldsymbol{\Delta}_p - \sigma^2 \boldsymbol{I})^{(1/2)} \boldsymbol{R}. \tag{12.58}$$

The arbitrary orthogonal matrix $\boldsymbol{R}$ is a rotation matrix that rotates data in projection. This is arbitrary because rotation is not important in dimensionality reduction (the relative distances of the embedded data points do not change if all of the embedded points are rotated). A simple choice for this rotation matrix is $\boldsymbol{R} = \boldsymbol{I}$, which results in:

$$\boldsymbol{\Lambda} = \boldsymbol{U}(\boldsymbol{\Delta}_p - \sigma^2 \boldsymbol{I})^{(1/2)}. \tag{12.59}$$

12.4.2.2 MLE for Determining σ

If Eq. (12.58) is substituted into the log-likelihood, Eq. (12.52), the log-likelihood becomes [12]:

$$\max_{\sigma} \quad -\frac{n}{2}\Big(\underbrace{d \log(2\pi)}_{\text{constant}} + \sum_{j=1}^{p} \log(\delta_j) + \sum_{j=p+1}^{d} \delta_j + (d-p)\log((\sigma^2)^{-1}) + p\Big). \tag{12.60}$$

Note that there are d eigenvalues $\{\delta_j\}_{j=1}^d$ because the covariance matrix $\boldsymbol{S}_x$ is a $(d \times d)$ matrix. However, as there are only p singular values $\{l_j\}_{j=1}^p$, the eigenvalues $\{\delta_j\}_{j=p+1}^d$ are very small.

Taking the derivative of Eq. (12.60) with respect to σ^2 and setting it to zero [21] results in:

$$\frac{\partial\, \text{Eq.(12.60)}}{\partial \sigma^2} = -\frac{n}{2}\big(0 + \sum_{j=p+1}^{d} \delta_j + (d-p)\sigma^2 + 0\big) \overset{\text{set}}{=} 0$$

$$\implies \sigma^2 = \frac{1}{d-p} \sum_{j=p+1}^{d} \delta_j. \tag{12.61}$$

12.4.2.3 Summary of MLE Formulas

In summary, the MLE estimations for the variables of PPCA are:

$$\sigma^2 = \frac{1}{d-p} \sum_{j=p+1}^{d} \delta_j, \tag{12.62}$$

$$\boldsymbol{\Lambda} = \boldsymbol{U}(\boldsymbol{\Delta}_p - \sigma^2 \boldsymbol{I})^{(1/2)} \boldsymbol{R} = \boldsymbol{U}(\boldsymbol{\Delta}_p - \sigma^2 \boldsymbol{I})^{(1/2)}. \tag{12.63}$$

Note that Eq. (12.62) is a measure of the variance lost in the projection by the projection matrix $\boldsymbol{\Lambda}$. Equation (12.63) is the projection or mapping matrix from the latent space to the data space.

12.4.3 Zero Noise Limit: PCA Is a Special Case of Probabilistic PCA

Recall the posterior, i.e. Eq. (12.37). According to Eqs. (12.38), (12.39), and (12.49), the posterior in PPCA is:

$$\begin{aligned} q(z_i) &= \mathbb{P}(z_i \,|\, \boldsymbol{x}_i) \\ &= \mathcal{N}\Big(\boldsymbol{\Lambda}^\top(\boldsymbol{\Lambda}\boldsymbol{\Lambda}^\top + \sigma^2 \boldsymbol{I})^{-1}(\boldsymbol{x}_i - \boldsymbol{\mu}), \boldsymbol{I} - \boldsymbol{\Lambda}^\top(\boldsymbol{\Lambda}\boldsymbol{\Lambda}^\top + \sigma^2 \boldsymbol{I})^{-1}\boldsymbol{\Lambda}\Big). \end{aligned} \tag{12.64}$$

Consider a zero noise limit, where the variance of the noise goes to zero:

$$\lim_{\sigma^2 \to 0} \mathbb{P}(\boldsymbol{\epsilon}) = \lim_{\sigma^2 \to 0} \mathcal{N}(\boldsymbol{0}, \sigma^2 \boldsymbol{I}) = \mathcal{N}(\boldsymbol{0}, \lim_{\sigma^2 \to 0} (\sigma^2)\boldsymbol{I}). \tag{12.65}$$

In this case, the uncertainty of the PPCA almost disappears. The following section demonstrates that in a zero noise limit, PPCA is reduced to PCA [13] (introduced in Chap. 5), which explains why the PPCA method is a probabilistic approach to PCA.

In the zero noise limit, the posterior becomes:

$$\begin{aligned} \lim_{\sigma^2 \to 0} q(z_i) &= \lim_{\sigma^2 \to 0} \mathbb{P}(z_i \,|\, \boldsymbol{x}_i) \\ &\overset{(12.64)}{=} \mathcal{N}\Big(\boldsymbol{\Lambda}^\top(\boldsymbol{\Lambda}\boldsymbol{\Lambda}^\top)^{-1}(\boldsymbol{x}_i - \boldsymbol{\mu}), \boldsymbol{I} - \boldsymbol{\Lambda}^\top(\boldsymbol{\Lambda}\boldsymbol{\Lambda}^\top)^{-1}\boldsymbol{\Lambda}\Big) \\ &\overset{(a)}{=} \mathcal{N}\Big((\boldsymbol{\Lambda}^\top\boldsymbol{\Lambda})^{-1}\boldsymbol{\Lambda}^\top(\boldsymbol{x}_i - \boldsymbol{\mu}), \boldsymbol{I} - (\boldsymbol{\Lambda}^\top\boldsymbol{\Lambda})^{-1}\boldsymbol{\Lambda}^\top\boldsymbol{\Lambda}\Big). \end{aligned} \tag{12.66}$$

where (a) is because according to [19, footnote 4]:

$$\boldsymbol{\Lambda}^\top(\boldsymbol{\Lambda}\boldsymbol{\Lambda}^\top)^{-1} = (\boldsymbol{\Lambda}^\top\boldsymbol{\Lambda})^{-1}\boldsymbol{\Lambda}^\top. \tag{12.67}$$

On the other hand, according to [21, Appendix C], PCA minimizes the reconstruction error:

$$\underset{\mathbf{\Lambda}}{\text{minimize}} \quad ||(\boldsymbol{X} - \boldsymbol{\mu}) - \mathbf{\Lambda}\mathbf{\Lambda}^\top(\boldsymbol{X} - \boldsymbol{\mu})||_F^2, \tag{12.68}$$

where $\|.\|_F$ denotes the Frobenius norm of the matrix (see Chap. 4). See Chap. 5 for more details on minimizing the reconstruction error using PCA.

Instead of minimizing the reconstruction error, it is possible to minimize the reconstruction error for the mean of the posterior [21, Appendix C]:

$$\underset{\mathbf{\Lambda}}{\text{minimize}} \quad ||(\boldsymbol{X} - \boldsymbol{\mu}) - \mathbf{\Lambda}(\mathbf{\Lambda}^\top\mathbf{\Lambda})^{-1}\mathbf{\Lambda}^\top(\boldsymbol{X} - \boldsymbol{\mu})||_F^2. \tag{12.69}$$

This optimization minimizes the reconstruction error after projection by $(\mathbf{\Lambda}^\top\mathbf{\Lambda})^{-1}\mathbf{\Lambda}^\top$. According to the posterior in the zero noise limit (see Eq. (12.66)), this is equivalent to PPCA in the zero noise limit model. Therefore, PCA is a deterministic special case of PPCA, where the variance of noise goes to zero.

12.4.4 Other Variants of Probabilistic PCA

There are other variants of PPCA. PPCA has a hyperparameter that is the dimensionality of latent space p. Bayesian PCA [1] models this hyperparameter as another latent random variable, which is learned during EM training.

According to Eqs. (12.27), (12.28), and (12.50), PPCA uses Gaussian distributions. A PPCA with a Student-t distribution [23] has been proposed that uses t distributions. This change may improve the embedding of PPCA because of the heavier tails of the Student-t distribution compared to the Gaussian distribution. This avoids the crowding problem, which motivated the proposal of t-SNE (see Chap. 16).

Sparse PPCA [10, 15] has inserted sparsity into PPCA. Supervised PPCA [22] makes use of class labels in the formulation of PPCA. Mixture of PPCA [20] trains the parameters of a mixture distribution using EM [8] in the formulation of PPCA. Generalized PPCA for correlated data is another recent variant of PPCA [9].

12.5 Chapter Summary

Probabilistic dimensionality reduction uses probabilities, Bayesian analysis, and variational inference. This chapter introduced the variational inference and Bayesian inference through maximum likelihood estimation and likelihood maximization using the evidence lower bound. Using the introduced fundamental concepts, it introduced the factor analysis and probabilistic PCA as its special case. These algorithms are fundamental and important methods in the probabilistic category of dimensionality reduction. Chapter 20 will use the expectation maximization introduced in this chapter for variational inference, to implement it in an autoencoder structure.

References

1. Christopher M Bishop. “Bayesian PCA”. In: *Advances in neural information processing systems*. 1999, pp. 382–388.
2. Christopher M Bishop. *Pattern recognition and machine learning*. Springer, 2006.
3. Guillaume Bouchard and Bill Triggs. “The tradeoff between generative and discriminative classifiers”. In: *16th IASC International Symposium on Computational Statistics*. 2004.
4. Raymond B Cattell. “A biometrics invited paper. Factor analysis: An introduction to essentials I. The purpose and underlying models”. In: *Biometrics* 21.1 (1965), pp. 190–215.
5. Dennis Child. The *essentials of factor analysis*. Cassell Educational, 1990.
6. Benjamin Fruchter. *Introduction to factor analysis*. Van Nostrand, 1954.
7. Zoubin Ghahramani and Geoffrey E Hinton. *The EM algorithm for mixtures of factor analyzers*. Tech. rep. Technical Report CRG-TR-96-1, University of Toronto, 1996.
8. Benyamin Ghojogh et al. “Fitting a mixture distribution to data: tutorial”. *In: arXiv preprint arXiv:1901.06708* (2019).
9. Mengyang Gu and Weining Shen. “Generalized probabilistic principal component analysis of correlated data.” In: *Journal of Machine Learning Research* 21.13 (2020), pp. 1–41.
10. Yue Guan and Jennifer Dy. “Sparse probabilistic principal component analysis”. In: *Artificial Intelligence and Statistics*. 2009, pp. 185–192.
11. Harry H Harman. *Modern factor analysis*. University of Chicago press, 1976.
12. Milos Hauskrecht. *CS3750 Lecture notes for Probabilistic Principal Component Analysis and the E-M algorithm*. Tech. rep. University of Pittsburgh, 2007.
13. Ian T Jolliffe and Jorge Cadima. “Principal component analysis: a review and recent developments”. In: *Philosophical Transactions of the Royal Society A: Mathematical, Physical and Engineering Sciences* 374.2065 (2016).
14. Solomon Kullback and Richard A Leibler. “On information and sufficiency”. In: *The annals of mathematical statistics* 22.1 (1951), pp. 79–86.
15. Pierre-Alexandre Mattei, Charles Bouveyron, and Pierre Latouche. “Globally sparse probabilistic PCA”. In: *Artificial Intelligence and Statistics*. 2016, pp. 976–984.
16. Andrew Ng. *CS229 Lecture notes for Factor Analysis*. Tech. rep. Stanford University, 2018.
17. Andrew Y Ng and Michael I Jordan. “On discriminative vs. generative classifiers: A comparison of logistic regression and naive Bayes”. In: *Advances in neural information processing systems*. 2002, pp. 841–848.
18. Leibny Paola Garcia. *Lecture notes for Factor Analysis*. Tech. rep. Carnegie Mellon University, 2018.
19. Sam Roweis. “EM algorithms for PCA and SPCA”. In: *Advances in neural information processing systems* 10 (1997), pp. 626–632.
20. Michael E Tipping and Christopher M Bishop. “Mixtures of probabilistic principal component analyzers”. In: *Neural computation* 11.2 (1999), pp. 443–482.
21. Michael E Tipping and Christopher M Bishop. “Probabilistic principal component analysis”. In: *Journal of the Royal Statistical Society: Series B (Statistical Methodology)* 61.3 (1999), pp. 611–622.
22. Shipeng Yu et al. “Supervised probabilistic principal component analysis”. In: *Proceedings of the 12th ACM SIGKDD international conference on Knowledge discovery and data mining*. 2006, pp. 464–473.
23. J Zhao and Q Jiang. “Probabilistic PCA for t distributions”. In: *Neurocomputing* 69.16–18 (2006), pp. 2217–2226.

Chapter 13
Probabilistic Metric Learning

13.1 Introduction

It was mentioned in Chap. 11 that metric learning can be divided into three types of learning—spectral, probabilistic and deep metric learning. This chapter introduces and reviews the probabilistic metric learning methods, which take a probabilistic approach to learning a generalized Mahalanobis distance metric. It is recommended that the reader studies Sect. 11.2 in Chap. 11 prior to reading this chapter to ensure a thorough understanding of the generalized Mahalanobis distance metric.

Recall Proposition 11.2 from Chap. 11:

Proposition 13.1 (Projection in Metric Learning) *Consider the eigenvalue decomposition of the weight matrix* $\boldsymbol{W}$ *in the generalized Mahalanobis distance, with* $\boldsymbol{V}$ *and* $\boldsymbol{\Lambda}$ *as the matrix of the eigenvectors and the diagonal matrix of the eigenvalues of the weight, respectively. Let* $\boldsymbol{U} := \boldsymbol{V}\boldsymbol{\Lambda}^{(1/2)}$. *The generalized Mahalanobis distance can be seen as the Euclidean distance, after applying a linear projection onto the column space of* $\boldsymbol{U}$:

$$\begin{aligned}\|\boldsymbol{x}_i - \boldsymbol{x}_j\|_{\boldsymbol{W}}^2 &= (\boldsymbol{U}^\top \boldsymbol{x}_i - \boldsymbol{U}^\top \boldsymbol{x}_j)^\top (\boldsymbol{U}^\top \boldsymbol{x}_i - \boldsymbol{U}^\top \boldsymbol{x}_j) \\ &= \|\boldsymbol{U}^\top \boldsymbol{x}_i - \boldsymbol{U}^\top \boldsymbol{x}_j\|_2^2.\end{aligned} \tag{13.1}$$

If $\boldsymbol{U} \in \mathbb{R}^{d \times p}$ *with* $p \leq d$, *the column space of the projection matrix* $\boldsymbol{U}$ *is a* p-*dimensional subspace.*

Additionally, recall the eigenvalue decomposition of the weight matrix $\boldsymbol{W}$ in metric (see Eq. (11.9) in Chap. 11)):

$$\boldsymbol{W} = \boldsymbol{V}\boldsymbol{\Lambda}\boldsymbol{V}^\top = \boldsymbol{V}\boldsymbol{\Lambda}^{(1/2)}\boldsymbol{\Lambda}^{(1/2)}\boldsymbol{V}^\top = \boldsymbol{U}\boldsymbol{U}^\top. \tag{13.2}$$

B. Ghojogh et al., *Elements of Dimensionality Reduction and Manifold Learning*,
https://doi.org/10.1007/978-3-031-10602-6_13

This chapter also demonstrated that, according to representation theory (see Chap. 3), the projection matrix in the RKHS can be represented as a linear combination of all of the pulled data points to RKHS (see Eq. (11.45) in Chap. 11):

$$\boldsymbol{\Phi}(\boldsymbol{U}) = \boldsymbol{\Phi}(\boldsymbol{X})\boldsymbol{T}, \tag{13.3}$$

where $\boldsymbol{X} = [\boldsymbol{x}_1, \ldots, \boldsymbol{x}_n]$, and $\boldsymbol{T} \in \mathbb{R}^{n \times p}$ (p is the dimensionality of the subspace) is the coefficients. Therefore, the distance metric in the RKHS becomes (see Eq. (11.53) in Chap. 11):

$$\|\boldsymbol{\phi}(\boldsymbol{x}_i) - \boldsymbol{\phi}(\boldsymbol{x}_j)\|_{\boldsymbol{\Phi}(\boldsymbol{W})}^2 = \|\boldsymbol{k}_i - \boldsymbol{k}_j\|_{\boldsymbol{T}\boldsymbol{T}^\top}^2 = (\boldsymbol{k}_i - \boldsymbol{k}_j)^\top \boldsymbol{T}\boldsymbol{T}^\top (\boldsymbol{k}_i - \boldsymbol{k}_j), \tag{13.4}$$

where $\boldsymbol{k}_i \in \mathbb{R}^n$ is the kernel of all of the training points with point $\boldsymbol{x}_i$.

Probabilistic methods for metric learning learn the weight matrix in the generalized Mahalanobis distance using probability distributions. These methods define some probability distribution for each point to accept other points as neighbours. Of course, the closer points have a higher probability of being neighbours. The following sections introduce the important probabilistic metric learning methods.

13.2 Collapsing Classes

A probabilistic method for metric learning is collapsing similar points to the same class, while pushing dissimilar points away from one another [6]. The probability distribution between points can be a Gaussian distribution to describe the probability of points being neighbours. A generalized Mahalanobis distance is used as the metric in the Gaussian distribution. The distribution for $\boldsymbol{x}_i$ to take $\boldsymbol{x}_j$ as its neighbour is [7]:

$$p_{ij}^W := \frac{\exp(-\|\boldsymbol{x}_i - \boldsymbol{x}_j\|_{\boldsymbol{W}}^2)}{\sum_{k \neq i} \exp(-\|\boldsymbol{x}_i - \boldsymbol{x}_k\|_{\boldsymbol{W}}^2)}, \quad j \neq i, \tag{13.5}$$

where the normalization factor, also called the partition function, is defined as $Z_i := \sum_{k \neq i} \exp(-\|\boldsymbol{x}_i - \boldsymbol{x}_k\|_{\boldsymbol{W}}^2)$. This factor makes the summation of distribution one. Equation (13.5) is a Gaussian distribution whose covariance matrix is $\boldsymbol{W}^{-1}$ because it is equivalent to:

$$p_{ij}^W := \frac{1}{Z_i} \exp\big(-(\boldsymbol{x}_i - \boldsymbol{x}_j)^\top \boldsymbol{W} (\boldsymbol{x}_i - \boldsymbol{x}_j)\big).$$

The aim is to have similar points to collapse to the same point after projection onto the subspace of the metric (see Proposition 13.1). Therefore, the desired neighbourhood distribution is defined as a bilevel distribution [6]:

$$p_{ij}^0 := \begin{cases} 1 \text{ if } (\boldsymbol{x}_i, \boldsymbol{x}_j) \in \mathcal{S} \\ 0 \text{ if } (\boldsymbol{x}_i, \boldsymbol{x}_j) \in \mathcal{D}. \end{cases} \tag{13.6}$$

This makes all similar points of a group/class the same point after projection.

13.2.1 Collapsing Classes in the Input Space

To make p_{ij}^W close to the desired distribution p_{ij}^0, the KL-divergence is minimized between them [6]:

$$\begin{aligned} \underset{\boldsymbol{W}}{\text{minimize}} \quad & \sum_{i=1}^{n} \sum_{j=1, j\neq i}^{n} \text{KL}(p_{ij}^0 \parallel p_{ij}^W) \\ \text{subject to} \quad & \boldsymbol{W} \succeq \boldsymbol{0}. \end{aligned} \tag{13.7}$$

Lemma 13.1 ([6]) *Let the objective function in Eq. (13.7) be denoted by* $c := \sum_{i=1}^{n} \sum_{j=1, j\neq i}^{n} KL(p_{ij}^0 \parallel p_{ij}^W)$*. The gradient of this function with respect to* $\boldsymbol{W}$ *is:*

$$\frac{\partial c}{\partial \boldsymbol{W}} = \sum_{i=1}^{n} \sum_{j=1, j\neq i}^{n} (p_{ij}^0 - p_{ij}^W)(\boldsymbol{x}_i - \boldsymbol{x}_j)(\boldsymbol{x}_i - \boldsymbol{x}_j)^\top. \tag{13.8}$$

Proof The derivation is similar to the derivation of the gradient in Stochastic Neighbour Embedding (SNE) and t-SNE (see Chap. 16). Let:

$$\mathbb{R} \ni r_{ij} := d_{ij}^2 = ||\boldsymbol{x}_i - \boldsymbol{x}_j||_{\boldsymbol{W}}^2. \tag{13.9}$$

Changing $\boldsymbol{x}_i$ impacts only d_{ij} and d_{ji} (or r_{ij} and r_{ji}) for all j's. According to the chain rule in derivatives:

$$\frac{\partial c}{\partial \boldsymbol{W}} = \sum_{i,j} \Big(\frac{\partial c}{\partial r_{ij}} \frac{\partial r_{ij}}{\partial \boldsymbol{W}} + \frac{\partial c}{\partial r_{ji}} \frac{\partial r_{ji}}{\partial \boldsymbol{W}} \Big).$$

According to Eq. (13.9):

$$r_{ij} = ||\boldsymbol{x}_i - \boldsymbol{x}_j||_{\boldsymbol{W}}^2 = \textbf{tr}((\boldsymbol{x}_i - \boldsymbol{x}_j)^\top \boldsymbol{W} (\boldsymbol{x}_i - \boldsymbol{x}_j)) \overset{(a)}{=} \textbf{tr}((\boldsymbol{x}_i - \boldsymbol{x}_j)(\boldsymbol{x}_i - \boldsymbol{x}_j)^\top \boldsymbol{W})$$

$$\implies \frac{\partial r_{ij}}{\partial \boldsymbol{W}} = (\boldsymbol{x}_i - \boldsymbol{x}_j)(\boldsymbol{x}_i - \boldsymbol{x}_j)^\top,$$

$$r_{ji} = ||\boldsymbol{x}_j - \boldsymbol{x}_i||_{\boldsymbol{W}}^2 = ||\boldsymbol{x}_i - \boldsymbol{x}_j||_{\boldsymbol{W}}^2 = r_{ij} \implies \frac{\partial r_{ji}}{\partial \boldsymbol{W}} = (\boldsymbol{x}_i - \boldsymbol{x}_j)(\boldsymbol{x}_i - \boldsymbol{x}_j)^\top,$$

where (a) is due to the cyclic property of the trace. Therefore:

$$\therefore \quad \frac{\partial c}{\partial \boldsymbol{W}} = 2 \sum_{i,j} (\frac{\partial c}{\partial r_{ij}})(\boldsymbol{x}_i - \boldsymbol{x}_j)(\boldsymbol{x}_i - \boldsymbol{x}_j)^\top. \tag{13.10}$$

The dummy variables in the cost function can be rewritten as:

$$\begin{aligned} c &= \sum_k \sum_{l \neq k} p_O(l|k) \log(\frac{p_O(l|k)}{p_W(l|k)}) = \sum_{k \neq l} p_O(l|k) \log(\frac{p_O(l|k)}{p_W(l|k)}) \\ &= \sum_{k \neq l} \big(p_O(l|k) \log(p_O(l|k)) - p_O(l|k) \log(p_W(l|k))\big), \end{aligned}$$

whose first term is a constant with respect to $p_W(l|k)$ and thus to $\boldsymbol{W}$. The derivative is:

$$\mathbb{R} \ni \frac{\partial c}{\partial r_{ij}} = - \sum_{k \neq l} p_O(l|k) \frac{\partial (\log(p_W(l|k)))}{\partial r_{ij}}.$$

According to Eqs. (13.5) and (13.9), the $p_W(l|k)$ is:

$$p_W(l|k) := \frac{\exp(-d_{kl}^2)}{\sum_{k \neq f} \exp(-d_{kf}^2)} = \frac{\exp(-r_{kl})}{\sum_{k \neq f} \exp(-r_{kf})}.$$

The denominator of $p_W(l|k)$ is denoted as follows:

$$\beta := \sum_{k \neq f} \exp(-d_{kf}^2) = \sum_{k \neq f} \exp(-r_{kf}). \tag{13.11}$$

It can be said that $\log(p_W(l|k)) = \log(p_W(l|k)) + \log \beta - \log \beta = \log(p_W(l|k)\, \beta) - \log \beta$. Therefore:

$$\begin{aligned} \therefore \quad \frac{\partial c}{\partial r_{ij}} &= - \sum_{k \neq l} p_O(l|k) \frac{\partial \big(\log(p_W(l|k)\beta) - \log \beta\big)}{\partial r_{ij}} \\ &= - \sum_{k \neq l} p_O(l|k) \Big[\frac{\partial \big(\log(p_W(l|k)\beta)\big)}{\partial r_{ij}} - \frac{\partial \big(\log \beta\big)}{\partial r_{ij}} \Big] \\ &= - \sum_{k \neq l} p_O(l|k) \Big[\frac{1}{p_W(l|k)\beta} \frac{\partial \big(p_W(l|k)\beta\big)}{\partial r_{ij}} - \frac{1}{\beta} \frac{\partial \beta}{\partial r_{ij}} \Big]. \end{aligned}$$

The $p_W(l|k)\beta$ is:

$$p_W(l|k)\beta = \frac{\exp(-r_{kl})}{\sum_{f\neq k}\exp(-r_{kf})} \times \sum_{k\neq f}\exp(-r_{kf}) = \exp(-r_{kl}).$$

Therefore:

$$\therefore \quad \frac{\partial c}{\partial r_{ij}} = -\sum_{k\neq l} p_O(l|k)\Big[\frac{1}{p_W(l|k)\beta}\frac{\partial\big(\exp(-r_{kl})\big)}{\partial r_{ij}} - \frac{1}{\beta}\frac{\partial \beta}{\partial r_{ij}}\Big].$$

The $\partial\big(\exp(-r_{kl})\big)/\partial r_{ij}$ is nonzero for only $k = i$ and $l = j$; therefore:

$$\frac{\partial\big(\exp(-r_{ij})\big)}{\partial r_{ij}} = -\exp(-r_{ij}),$$

$$\frac{\partial \beta}{\partial r_{ij}} = \frac{\partial \sum_{k\neq f}\exp(-r_{kf})}{\partial r_{ij}} = \frac{\partial \exp(-r_{ij})}{\partial r_{ij}} = -\exp(-r_{ij}).$$

Therefore:

$$\therefore \quad \frac{\partial c}{\partial r_{ij}} = -\Big(p_{ij}^0\Big[\frac{-1}{p_{ij}^W\beta}\exp(-r_{ij})\Big] + 0 + \cdots + 0\Big) - \sum_{k\neq l} p_O(l|k)\Big[\frac{1}{\beta}\exp(-r_{ij})\Big].$$

$\sum_{k\neq l} p_O(l|k) = 1$ holds because the summation of all possible probabilities is one. Thus:

$$\frac{\partial c}{\partial r_{ij}} = -p_{ij}^0\Big[\frac{-1}{p_{ij}^W\beta}\exp(-r_{ij})\Big] - \Big[\frac{1}{\beta}\exp(-r_{ij})\Big] = \underbrace{\frac{\exp(-r_{ij})}{\beta}}_{=p_{ij}^W}\Big[\frac{p_{ij}^0}{p_{ij}^W} - 1\Big] = p_{ij}^0 - p_{ij}^W. \tag{13.12}$$

Substituting the obtained derivative in Eq. (13.10) results in Eq. (13.8). □

Optimization problem (13.7) is considered convex; therefore, it has a unique solution. It can be solved using any optimization method, such as the projected gradient method, where after every gradient descent step, the solution is projected onto the positive semidefinite cone (see Chap. 4):

$$\boldsymbol{W} := \boldsymbol{W} - \eta\frac{\partial c}{\partial \boldsymbol{W}},$$

$$\boldsymbol{W} := \boldsymbol{V}\,\mathbf{diag}(\max(\lambda_1, 0), \ldots, \max(\lambda_d, 0))\,\boldsymbol{V}^\top,$$

where $\eta > 0$ is the learning rate and $\boldsymbol{V}$ and $\boldsymbol{\Lambda} = \mathbf{diag}(\lambda_1, \ldots, \lambda_d)$ are the eigenvectors and eigenvalues of $\boldsymbol{W}$, respectively (see Eq. (13.2)).

13.2.2 Collapsing Classes in the Feature Space

According to Eq. (13.4), the distance in the feature space can be stated using kernels as $\|\boldsymbol{k}_i - \boldsymbol{k}_j\|^2_{\boldsymbol{T}\boldsymbol{T}^\top}$, where $\boldsymbol{k}_i \in \mathbb{R}^n$ is the kernel vector between dataset $\boldsymbol{X}$ and point $\boldsymbol{x}_i$. The following is defined as $\boldsymbol{R} := \boldsymbol{T}\boldsymbol{T}^\top \in \mathbb{R}^{n\times n}$. Therefore, in the feature space, Eq. (13.5) becomes:

$$p^R_{ij} := \frac{\exp(-\|\boldsymbol{k}_i - \boldsymbol{k}_j\|^2_{\boldsymbol{R}})}{\sum_{k\neq i}\exp(-\|\boldsymbol{k}_i - \boldsymbol{k}_k\|^2_{\boldsymbol{R}})}, \quad j \neq i. \tag{13.13}$$

The gradient in Eq. (13.8) becomes:

$$\frac{\partial c}{\partial \boldsymbol{R}} = \sum_{i=1}^{n}\sum_{j=1, j\neq i}^{n} (p^0_{ij} - p^R_{ij})(\boldsymbol{k}_i - \boldsymbol{k}_j)(\boldsymbol{k}_i - \boldsymbol{k}_j)^\top. \tag{13.14}$$

Again, the optimal $\boldsymbol{R}$ can be found using the projected gradient method. This results in the optimal metric for collapsing classes in the feature space [6]. Note that it is possible to regularize the objective function, using the trace operator or Frobenius norm, to avoid overfitting.

13.3 Neighbourhood Component Analysis Methods

Neighbourhood Component Analysis (NCA) is one of the most well-known probabilistic metric learning methods. The following section introduces different variants of NCA.

13.3.1 Neighbourhood Component Analysis (NCA)

In the original NCA [7], Eq. (13.5) is used for the probability that $\boldsymbol{x}_j$ takes $\boldsymbol{x}_i$ as its neighbour, where $p^W_{ii} = 0$ is assumed by convention:

$$p^W_{ij} := \begin{cases} \frac{\exp(-\|\boldsymbol{x}_i - \boldsymbol{x}_j\|^2_{\boldsymbol{W}})}{\sum_{k\neq i}\exp(-\|\boldsymbol{x}_i - \boldsymbol{x}_k\|^2_{\boldsymbol{W}})} & \text{if } j \neq i \\ 0 & \text{if } j = i. \end{cases} \tag{13.15}$$

Consider the decomposition of the weight matrix of the metric as seen in Eq. (13.2), i.e., $\boldsymbol{W} = \boldsymbol{U}\boldsymbol{U}^\top$. Let $\mathcal{S}_i$ denote a set of similar points to $\boldsymbol{x}_i$, where $(\boldsymbol{x}_i, \boldsymbol{x}_j) \in \mathcal{S}$. The goal of NCA's optimization problem is to find a $\boldsymbol{U}$ to maximize the probability distribution for similar points [7]:

$$\underset{U}{\text{maximize}} \quad \sum_{(x_i,x_j)\in\mathcal{S}} p_{ij}^W = \sum_{i=1}^{n} \sum_{x_j\in\mathcal{S}_i} p_{ij}^W = \sum_{i=1}^{n} p_i^W, \tag{13.16}$$

where:

$$p_i^W := \sum_{x_j\in\mathcal{S}_i} p_{ij}^W. \tag{13.17}$$

Note that the required constraint $W \succeq \mathbf{0}$ is already satisfied because of the decomposition in Eq. (13.5).

Lemma 13.2 ([7]) *Suppose the objective function of Eq. (13.16) is denoted by c. The gradient of this cost function with respect to* U *is:*

$$\frac{\partial c}{\partial U} = 2\sum_{i=1}^{n}\Big(p_i^W \sum_{k=1}^{n} p_{ik}^W (x_i - x_k)(x_i - x_k)^\top - \sum_{x_j\in\mathcal{S}_i} p_{ij}^W (x_i - x_j)(x_i - x_j)^\top\Big)U. \tag{13.18}$$

The derivation of this gradient is similar to the approach in the proof of Lemma 13.1. The gradient ascent can be used to solve this optimization problem (see Chap. 4).

Another approach is to maximize the log-likelihood of the neighbourhood probability [7]:

$$\underset{U}{\text{maximize}} \quad \sum_{i=1}^{n} \log\Big(\sum_{x_j\in\mathcal{S}_i} p_{ij}^W\Big), \tag{13.19}$$

whose gradient is [7]:

$$\frac{\partial c}{\partial U} = 2\sum_{i=1}^{n}\Big(\sum_{k=1}^{n} p_{ik}^W (x_i - x_k)(x_i - x_k)^\top - \frac{\sum_{x_j\in\mathcal{S}_i} p_{ij}^W (x_i - x_j)(x_i - x_j)^\top}{\sum_{x_j\in\mathcal{S}_i} p_{ij}^W}\Big)U. \tag{13.20}$$

Again, gradient ascent can provide the optimal U. As explained in Proposition 13.1, the subspace of the metric is the column space of U, and the projection of points onto this subspace reduces the dimensionality of the data.

13.3.2 Regularized Neighbourhood Component Analysis

NCA has been found by experiments to overfit to the training data, specifically for high-dimensional data [18]. Therefore, it can be regularized to avoid overfitting. In the regularized NCA [18], the log-posterior of the matrix U is used, which is equal to:

$$\mathbb{P}(\boldsymbol{U}|\boldsymbol{x}_i, \mathcal{S}_i) = \frac{\mathbb{P}(\boldsymbol{x}_i, \mathcal{S}_i|\boldsymbol{U})\,\mathbb{P}(\boldsymbol{U})}{\mathbb{P}(\boldsymbol{x}_i, \mathcal{S}_i)}, \tag{13.21}$$

according to Bayes' rule. A Gaussian distribution can be used for the prior:

$$\mathbb{P}(\boldsymbol{U}) = \prod_{k=1}^{d}\prod_{l=1}^{d} c\, \exp(-\lambda(\boldsymbol{U}(k,l))^2), \tag{13.22}$$

where $c > 0$ is a constant factor including the normalization factor, $\lambda > 0$ is the inverse of the variance, and $\boldsymbol{U}(k,l)$ is the (k,l)-th element of $\boldsymbol{U} \in \mathbb{R}^{d\times d}$. Note that it is possible to have $\boldsymbol{U} \in \mathbb{R}^{d\times p}$, if it is truncated to have p leading eigenvectors of $\boldsymbol{W}$ (see Eq. (13.2)). The likelihood

$$\mathbb{P}(\boldsymbol{x}_i, \mathcal{S}_i|\boldsymbol{U}) \propto \exp\Big(\sum_{(\boldsymbol{x}_i,\boldsymbol{x}_j)\in\mathcal{S}} p_{ij}^W\Big). \tag{13.23}$$

The regularized NCA maximizes the log-posterior [18]:

$$\begin{aligned}\log \mathbb{P}(\boldsymbol{U}|\boldsymbol{x}_i, \mathcal{S}_i) &\overset{(13.21)}{=} \log \mathbb{P}(\boldsymbol{x}_i, \mathcal{S}_i|\boldsymbol{U}) + \log \mathbb{P}(\boldsymbol{U}) - \underbrace{\log \mathbb{P}(\boldsymbol{x}_i, \mathcal{S}_i)}_{\text{constant with respect to } \boldsymbol{U}}\\ &\overset{(a)}{=} \sum_{(\boldsymbol{x}_i,\boldsymbol{x}_j)\in\mathcal{S}} p_{ij}^W - \lambda\|\boldsymbol{U}\|_F^2,\end{aligned}$$

where (a) is the result of Eqs. (13.22) and (13.23) and $\|.\|_F$ denotes the Frobenius norm. Therefore, the optimization problem of a regularized NCA is [18]:

$$\underset{\boldsymbol{U}}{\text{maximize}} \quad \sum_{(\boldsymbol{x}_i,\boldsymbol{x}_j)\in\mathcal{S}} p_{ij}^W - \lambda\|\boldsymbol{U}\|_F^2, \tag{13.24}$$

where $\lambda > 0$ can be seen as the regularization parameter. The gradient is similar to Eq. (13.18), but with addition of the derivative of the regularization term, which is $-2\lambda\boldsymbol{U}$.

13.3.3 Fast Neighbourhood Component Analysis

13.3.3.1 Fast NCA

The fast NCA [16] accelerates NCA by using k-Nearest Neighbours (kNN), rather than using all points for computing the neighbourhood distribution of every point. Let $\mathcal{N}_i$ and $\mathcal{M}_i$ denote the kNN of $\boldsymbol{x}_i$ among the similar points to $\boldsymbol{x}_i$ (denoted by

$\mathcal{S}_i$) and dissimilar points (denoted by $\mathcal{D}_i$), respectively. Fast NCA uses the following probability distribution for $\boldsymbol{x}_i$ to take $\boldsymbol{x}_i$ as its neighbour [16]:

$$p_{ij}^W := \begin{cases} \frac{\exp(-\|\boldsymbol{x}_i - \boldsymbol{x}_j\|_{\boldsymbol{W}})}{\sum_{\boldsymbol{x}_k \in \mathcal{N}_i \cup \mathcal{M}_i} \exp(-\|\boldsymbol{x}_i - \boldsymbol{x}_k\|_{\boldsymbol{W}})} & \text{if } \boldsymbol{x}_k \in \mathcal{N}_i \cup \mathcal{M}_i \\ 0 & \text{otherwise.} \end{cases} \tag{13.25}$$

The optimization problem of the fast NCA is similar to Eq. (13.24):

$$\underset{\boldsymbol{U}}{\text{maximize}} \quad \sum_{i=1}^{n} \sum_{\boldsymbol{x}_j \in \mathcal{M}_i} p_{ij}^W - \lambda \|\boldsymbol{U}\|_F^2, \tag{13.26}$$

where p_{ij}^W is Eq. (13.25), and $\boldsymbol{U}$ is the matrix in the decomposition of $\boldsymbol{W}$ (see Eq. (13.2)).

Lemma 13.3 ([16]) *Suppose the objective function of Eq. (13.26) is denoted by c. The gradient of this cost function with respect to* $\boldsymbol{U}$ *is:*

$$\begin{aligned} \frac{\partial c}{\partial \boldsymbol{U}} = \sum_{i=1}^{n} \Big(p_i^W \sum_{\boldsymbol{x}_k \in \mathcal{N}_i} p_{ik}^W (\boldsymbol{x}_i - \boldsymbol{x}_k)(\boldsymbol{x}_i - \boldsymbol{x}_k)^\top \\ + (p_i^W - 1) \sum_{\boldsymbol{x}_j \in \mathcal{M}_i} p_{ij}^W (\boldsymbol{x}_i - \boldsymbol{x}_j)(\boldsymbol{x}_i - \boldsymbol{x}_j)^\top \Big) \boldsymbol{U} - 2\lambda \boldsymbol{U}. \end{aligned} \tag{13.27}$$

This is similar to Eq. (13.18). The gradient ascent can be used to solve this optimization problem (see Chap. 4).

13.3.3.2 Kernel Fast NCA

According to Eq. (13.4), the distance in the feature space is $\|\boldsymbol{k}_i - \boldsymbol{k}_j\|_{\boldsymbol{T}\boldsymbol{T}^\top}^2$, where $\boldsymbol{k}_i \in \mathbb{R}^n$ is the kernel vector between dataset $\boldsymbol{X}$ and point $\boldsymbol{x}_i$. This distance metric can be used in Eq. (13.25) to have a kernel fast NCA [16]. Therefore, the gradient of the kernel fast NCA is similar to Eq. (13.27):

$$\begin{aligned} \frac{\partial c}{\partial \boldsymbol{T}} = \sum_{i=1}^{n} \Big(p_i^W \sum_{\boldsymbol{x}_k \in \mathcal{N}_i} p_{ik}^W (\boldsymbol{k}_i - \boldsymbol{k}_k)(\boldsymbol{k}_i - \boldsymbol{k}_k)^\top \\ + (p_i^W - 1) \sum_{\boldsymbol{x}_j \in \mathcal{M}_i} p_{ij}^W (\boldsymbol{k}_i - \boldsymbol{k}_j)(\boldsymbol{k}_i - \boldsymbol{k}_j)^\top \Big) \boldsymbol{T} - 2\lambda \boldsymbol{T}. \end{aligned} \tag{13.28}$$

Again, the optimal $\boldsymbol{T}$ can be found using the gradient ascent. Note that the same technique can be used to kernelize the original NCA.

13.4 Bayesian Metric Learning Methods

This section introduces the Bayesian metric learning methods, which use variational inference (see Chaps. 12 and 20) for metric learning. A distribution for the distance metric between every two points is learned in the Bayesian metric learning, and the pairwise distances are sampled from these learned distributions.

According to Eq. (13.2), it is possible to decompose the weight matrix in the metric using the eigenvalue decomposition. Accordingly, it is possible to approximate this matrix by:

$$\boldsymbol{W} \approx \boldsymbol{V}_x \boldsymbol{\Lambda} \boldsymbol{V}_x^\top, \tag{13.29}$$

where $\boldsymbol{V}_x$ contains the eigenvectors of $\boldsymbol{X}\boldsymbol{X}^\top$ and $\boldsymbol{\Lambda} = \textbf{diag}([\lambda_1, \dots, \lambda_d]^\top)$ is the diagonal matrix of the eigenvalues, which are learned through Bayesian metric learning. Let X and Y denote the random variables for data and labels, respectively, and let $\boldsymbol{\lambda} = [\lambda_1, \dots, \lambda_d]^\top \in \mathbb{R}^d$ denote the learnable eigenvalues. Let $\boldsymbol{v}_x^l \in \mathbb{R}^d$ denote the l-th column of $\boldsymbol{V}_x$. Therefore, the following is defined as $\boldsymbol{w}_{ij} = [w_{ij}^1, \dots, w_{ij}^d]^\top := [((\boldsymbol{v}_x^1)^\top(\boldsymbol{x}_i - \boldsymbol{x}_j))^2, \dots, ((\boldsymbol{v}_x^d)^\top(\boldsymbol{x}_i - \boldsymbol{x}_j))^2]^\top \in \mathbb{R}^d$. The reader should not confuse $\boldsymbol{w}_{ij}$ with $\boldsymbol{W}$, which is the weight matrix of the metric in the notations used.

13.4.1 Bayesian Metric Learning Using the Sigmoid Function

One of the Bayesian metric learning methods is Yang et al.'s method [14], explained in the following. The following is defined as:

$$y_{ij} := \begin{cases} 1 & \text{if } (\boldsymbol{x}_i, \boldsymbol{x}_j) \in \mathcal{S} \\ -1 & \text{if } (\boldsymbol{x}_i, \boldsymbol{x}_j) \in \mathcal{D}. \end{cases} \tag{13.30}$$

A sigmoid function can be considered for the likelihood [14]:

$$\mathbb{P}(Y|X, \boldsymbol{\Lambda}) = \frac{1}{1 + \exp(y_{ij}(\sum_{l=1}^{d} \lambda_l w_{ij}^l - \mu))}, \tag{13.31}$$

where $\mu > 0$ is a threshold. It is also possible to derive an evidence lower bound for $\mathbb{P}(\mathcal{S}, \mathcal{D})$; this section does not provide the derivation for brevity (see [14] for the derivation of the lower bound). As in the variational inference, this lower bound is maximized for likelihood maximization (see Eq. (13.2)). Assume a Gaussian distribution with mean $\boldsymbol{m}_\lambda \in \mathbb{R}^d$ and covariance $\boldsymbol{V}_\lambda \in \mathbb{R}^{d\times d}$ for the distribution $\mathbb{P}(\boldsymbol{\lambda})$. By maximizing the lower bound, it is possible to estimate these parameters as [14]:

$$V_T := \Big(\delta I + 2 \sum_{(x_i,x_j)\in\mathcal{S}} \frac{\tanh(\xi_{ij}^s)}{4\xi_{ij}^s} w_{ij} w_{ij}^\top + 2 \sum_{(x_i,x_j)\in\mathcal{D}} \frac{\tanh(\xi_{ij}^d)}{4\xi_{ij}^d} w_{ij} w_{ij}^\top\Big)^{-1}, \tag{13.32}$$

$$m_T := V_T\Big(\delta \gamma_0 - \frac{1}{2} \sum_{(x_i,x_j)\in\mathcal{S}} w_{ij} + \frac{1}{2} \sum_{(x_i,x_j)\in\mathcal{D}} w_{ij}\Big), \tag{13.33}$$

where $\delta > 0$ and γ_0 are hyperparameters related to the priors on the weight matrix of the metric and the threshold. The following variational parameter [14] is defined as:

$$\xi_{ij}^s := \sqrt{(m_T^\top w_{ij})^2 + w_{ij}^\top V_T w_{ij}}, \tag{13.34}$$

for $(x_i, x_j) \in \mathcal{S}$. Similarly, the variational parameter ξ_{ij}^d is defined for $(x_i, x_j) \in \mathcal{D}$. The variables V_T, m_T, ξ_{ij}^s, and ξ_{ij}^d are updated iteratively by Eqs. (13.36), (13.37), and (13.34), respectively, until convergence. After these parameters are learned, it is possible to sample the eigenvalues from the posterior, $\lambda \sim \mathcal{N}(m_T, V_T)$. These eigenvalues can be used in Eq. (13.29) to obtain the weight matrix in the metric. Note that Bayesian metric learning can also be used for active learning (see [14] for details).

13.4.2 Bayesian Neighbourhood Component Analysis

Bayesian NCA [12] uses variational inference (see Chaps. 12 and 20) in the NCA formulation. If $\mathcal{N}_{im}$ denotes the dataset index of the m-th nearest neighbour of x_i, then $W_i^j := [w_{ij} - w_{i\mathcal{N}_{i1}}, \dots, w_{ij} - w_{i\mathcal{N}_{ik}}] \in \mathbb{R}^{d\times k}$ is defined. As in the variational inference, an evidence lower-bound on the log-likelihood is considered:

$$\log(\mathbb{P}(Y|X, \Lambda)) > \sum_{i=1}^{n} \sum_{x_j\in\mathcal{N}_i} \Big(-\frac{1}{2}\lambda^\top W_i^j H (W_i^j)^\top \lambda + b_{ij}^\top (W_i^j)^\top \lambda - c_{ij}\Big),$$

where $\mathcal{N}_i$ was defined before in Sect. 13.3.3, $H := \frac{1}{2}(I - \frac{1}{k+1}\mathbf{1}\mathbf{1}^\top) \in \mathbb{R}^{k\times k}$ is the centering matrix, and:

$$\mathbb{R}^k \ni b_{ij} := H\psi_{ij} - \exp\Big(\psi_{ij} - \log\Big(1 + \sum_{x_t\in\mathcal{N}_i} \exp\big((w_{ij} - w_{it})^\top \lambda\big)\Big)\Big), \tag{13.35}$$

in which $\boldsymbol{\psi}_{ij} \in \mathbb{R}^k$ is the learnable variational parameter. See [12] for the derivation of this lower-bound. The sketch of this derivation uses Eq. (13.5), but for the kNN among the similar points, i.e., $\mathcal{N}_i$. Then, the lower-bound is obtained by a logarithm inequality, as well as Bohning's quadratic bound [10].

Assume a Gaussian distribution for the prior of $\boldsymbol{\lambda}$ with a mean $\boldsymbol{m}_0 \in \mathbb{R}^d$ and a covariance $\boldsymbol{V}_0 \in \mathbb{R}^{d\times d}$. This prior is assumed to be known. Likewise, assume a Gaussian distribution, with a mean $\boldsymbol{m}_T \in \mathbb{R}^d$ and covariance $\boldsymbol{V}_T \in \mathbb{R}^{d\times d}$, for the posterior $\mathbb{P}(X, \boldsymbol{\Lambda}|Y)$. Using Bayes' rule and the above lower bound on the likelihood, it is possible to estimate these parameters [12]:

$$\boldsymbol{V}_T := \Big(\boldsymbol{V}_0^{-1} + \sum_{i=1}^{n} \sum_{\boldsymbol{x}_j \in \mathcal{N}_i} \boldsymbol{W}_i^j \boldsymbol{H} (\boldsymbol{W}_i^j)^\top\Big)^{-1}, \tag{13.36}$$

$$\boldsymbol{m}_T := \boldsymbol{V}_T\Big(\boldsymbol{V}_0^{-1}\boldsymbol{m}_0 + \sum_{i=1}^{n} \sum_{\boldsymbol{x}_j \in \mathcal{N}_i} \boldsymbol{W}_i^j \boldsymbol{b}_{ij}\Big). \tag{13.37}$$

The variational parameter can also be obtained by [12]:

$$\boldsymbol{\psi}_{ij} := (\boldsymbol{W}_i^j)^\top \boldsymbol{m}_T. \tag{13.38}$$

The variables $\boldsymbol{b}_{ij}$, $\boldsymbol{V}_T$, $\boldsymbol{m}_T$, and $\boldsymbol{\psi}_{ij}$ are updated iteratively by Eqs. (13.35), (13.36), (13.37), and (13.38), respectively, until convergence.

After these parameters are learned, the eigenvalues can be sampled from the posterior, $\boldsymbol{\lambda} \sim \mathcal{N}(\boldsymbol{m}_T, \boldsymbol{V}_T)$. These eigenvalues can be used in Eq. (13.29) to obtain the weight matrix in the metric. Alternatively, it is possible to directly sample the distance metric from the following distribution:

$$\|\boldsymbol{x}_i - \boldsymbol{x}_j\|_{\boldsymbol{W}}^2 \sim \mathcal{N}(\boldsymbol{w}_{ij}^\top \boldsymbol{m}_T, \boldsymbol{w}_{ij}^\top \boldsymbol{V}_T \boldsymbol{w}_{ij}). \tag{13.39}$$

13.4.3 Local Distance Metric (LDM)

Let the set of similar and dissimilar points for point $\boldsymbol{x}_i$ be denoted by $\mathcal{S}_i$ and $\mathcal{D}_i$, respectively. In Local Distance Metric (LDM) [15], the following is considered for the likelihood:

$$\begin{aligned}\mathbb{P}(y_i|\boldsymbol{x}_i) = &\sum_{\boldsymbol{x}_j \in \mathcal{S}_i} \exp(-\|\boldsymbol{x}_i - \boldsymbol{x}_j\|_{\boldsymbol{W}}^2) \\ &\times \Big(\sum_{\boldsymbol{x}_j \in \mathcal{S}_i} \exp(-\|\boldsymbol{x}_i - \boldsymbol{x}_j\|_{\boldsymbol{W}}^2) + \sum_{\boldsymbol{x}_j \in \mathcal{D}_i} \exp(-\|\boldsymbol{x}_i - \boldsymbol{x}_j\|_{\boldsymbol{W}}^2)\Big)^{-1}.\end{aligned} \tag{13.40}$$

If Eq. (13.29) is considered for decomposition of the weight matrix, the log-likelihood becomes:

$$\sum_{i=1}^{n} \log(\mathbb{P}(y_i|\boldsymbol{x}_i, \boldsymbol{\Lambda})) = \sum_{i=1}^{n} \log\Big(\sum_{\boldsymbol{x}_j \in \mathcal{S}_i} \exp\big(-\sum_{l=1}^{d} \lambda_l w_{ij}^l\big)\Big) \times$$
$$\sum_{i=1}^{n} \log\Big(\sum_{\boldsymbol{x}_j \in \mathcal{S}_i} \exp\big(-\sum_{l=1}^{d} \lambda_l w_{ij}^l\big) + \sum_{\boldsymbol{x}_j \in \mathcal{D}_i} \exp\big(-\sum_{l=1}^{d} \lambda_l w_{ij}^l\big)\Big).$$

The aim is to maximize this log-likelihood for learning the variables $\{\lambda_1, \dots, \lambda_d\}$. An evidence lower bound on this log-likelihood can be [15]:

$$\sum_{i=1}^{n} \log(\mathbb{P}(y_i|\boldsymbol{x}_i, \boldsymbol{\Lambda})) \geq \sum_{i=1}^{n} \sum_{\boldsymbol{x}_j \in \mathcal{S}_i} \phi_{ij} \sum_{l=1}^{d} \lambda_l w_{ij}^l$$
$$- \sum_{i=1}^{n} \log\Big(\sum_{\boldsymbol{x}_j \in \mathcal{S}_i} \exp\big(-\sum_{l=1}^{d} \lambda_l w_{ij}^l\big) + \sum_{\boldsymbol{x}_j \in \mathcal{D}_i} \exp\big(-\sum_{l=1}^{d} \lambda_l w_{ij}^l\big)\Big), \tag{13.41}$$

where ϕ_{ij} is the variational parameter:

$$\phi_{ij} := \frac{\exp\big(-\sum_{l=1}^{d} \lambda_l w_{ij}^l\big)}{\sum_{\boldsymbol{x}_j \in \mathcal{S}_i} \exp\big(-\sum_{l=1}^{d} \lambda_l w_{ij}^l\big)} \times \Big(1 + \frac{\exp\big(-\sum_{l=1}^{d} \lambda_l w_{ij}^l\big)}{\sum_{\boldsymbol{x}_j \in \mathcal{S}_i} \exp\big(-\sum_{l=1}^{d} \lambda_l w_{ij}^l\big)}\Big)^{-1}. \tag{13.42}$$

The derivation of the lower bound can be found in [15] for interested readers. Iteratively, the lower bound, i.e., Eq. (13.41), is maximized, and ϕ_{ij} is updated by Eq. (13.42). The learned parameters $\{\lambda_1, \dots, \lambda_d\}$ can be used in Eq. (13.29) to obtain the weight matrix in the metric.

13.5 Information Theoretic Metric Learning

There exist information theoretic approaches for metric learning, where KL-divergence (relative entropy) or mutual information is used.

13.5.1 Information Theoretic Metric Learning with a Prior Weight Matrix

One of the information theoretic methods for metric learning uses a prior weight matrix [3], where a known weight matrix $\boldsymbol{W}_0$ is considered to be the regularizer and the aim is to minimize the KL-divergence between the distributions with $\boldsymbol{W}$ and $\boldsymbol{W}_0$:

$$\text{KL}(p_{ij}^{\boldsymbol{W}_0} \| p_{ij}^{\boldsymbol{W}}) := \sum_{i=1}^{n} \sum_{j=1}^{n} p_{ij}^{\boldsymbol{W}_0} \log \Big(\frac{p_{ij}^{\boldsymbol{W}_0}}{p_{ij}^{\boldsymbol{W}}} \Big). \tag{13.43}$$

There are both offline and online approaches for metric learning using batch and streaming data, respectively.

13.5.1.1 Offline Information Theoretic Metric Learning

Consider a Gaussian distribution, Eq. (13.5), for the probability of $\boldsymbol{x}_i$ taking $\boldsymbol{x}_j$ as its neighbour, i.e. $p_{ij}^{\boldsymbol{W}}$. While the weight matrix is made similar to the prior weight matrix through KL-divergence, a weight matrix is found to make all the distances between similar points less than an upper bound $u > 0$ and all the distances between dissimilar points larger than a lower bound l (where $l > u$). Note that, for Gaussian distributions, the KL divergence is related to LogDet $D_{ld}(.,.)$ between covariance matrices [4]; therefore:

$$\begin{aligned}\text{KL}(p_{ij}^{\boldsymbol{W}_0} \| p_{ij}^{\boldsymbol{W}}) &= \frac{1}{2} D_{ld}(\boldsymbol{W}_0^{-1}, \boldsymbol{W}^{-1}) = \frac{1}{2} D_{ld}(\boldsymbol{W}, \boldsymbol{W}_0) \\ &\overset{(a)}{=} \textbf{tr}(\boldsymbol{W}\boldsymbol{W}_0^{-1}) - \log(\det(\boldsymbol{W}\boldsymbol{W}_0^{-1})) - n,\end{aligned}$$

where (a) is because of the definition of LogDet. Therefore, the optimization problem can be [3]:

$$\begin{aligned} \underset{\boldsymbol{W}}{\text{minimize}} \quad & D_{ld}(\boldsymbol{W}, \boldsymbol{W}_0) \\ \text{subject to} \quad & \|\boldsymbol{x}_i - \boldsymbol{x}_j\|_{\boldsymbol{W}}^2 \leq u, \quad \forall (\boldsymbol{x}_i, \boldsymbol{x}_j) \in \mathcal{S}, \\ & \|\boldsymbol{x}_i - \boldsymbol{x}_j\|_{\boldsymbol{W}}^2 \geq l, \quad \forall (\boldsymbol{x}_i, \boldsymbol{x}_j) \in \mathcal{D}. \end{aligned} \tag{13.44}$$

13.5.1.2 Online Information Theoretic Metric Learning

Online information theoretic metric learning [3] is suitable for streaming data. For this, the offline approach can be used, where the known weight matrix $\boldsymbol{W}_0$ is the weight matrix learned by the data that have been received thus far. Consider the time slot t, where some data have been accumulated previously and some new data points are received at this time. The optimization problem is stated in Eq. (13.44), where $\boldsymbol{W}_0 = \boldsymbol{W}_t$, which is the learned weight matrix thus far at time t. Note that if there is some label information available, it can be incorporated into the optimization problem as a regularizer.

13.5.2 Information Theoretic Metric Learning for Imbalanced Data

The Distance Metric by Balancing KL-divergence (DMBK) [5] can be used for imbalanced data, where the cardinality of classes are different. Assume the classes have Gaussian distributions, where $\boldsymbol{\mu}_i \in \mathbb{R}^d$ and $\boldsymbol{\Sigma}_i \in \mathbb{R}^{d\times d}$ denote the mean and covariance of the i-th class. Recall the projection matrix $\boldsymbol{U}$ in Eq. (13.2) and Proposition 13.1. The KL-divergence between the probabilities of the i-th and j-th classes after projection onto the subspace of the metric is [5]:

$$\begin{aligned} \mathrm{KL}(p_i \| p_j) = \frac{1}{2}\Big(& \log\big(\det(\boldsymbol{U}^\top \boldsymbol{\Sigma}_j \boldsymbol{U})\big) - \log\big(\det(\boldsymbol{U}^\top \boldsymbol{\Sigma}_i \boldsymbol{U})\big) \\ & + \mathbf{tr}\big((\boldsymbol{U}^\top \boldsymbol{\Sigma}_j \boldsymbol{U})^{-1} \boldsymbol{U}^\top (\boldsymbol{\Sigma}_i + \boldsymbol{D}_{ij}) \boldsymbol{U}\big)\Big), \end{aligned} \tag{13.45}$$

where $\boldsymbol{D}_{ij} := (\boldsymbol{\mu}_i - \boldsymbol{\mu}_j)(\boldsymbol{\mu}_i - \boldsymbol{\mu}_j)^\top$. To cancel the effect of the cardinality of classes in imbalanced data, the normalized divergence of classes can be used:

$$e_{ij} := \frac{n_i n_j \mathrm{KL}(p_i \| p_j)}{\sum_{1\leq k<l\leq c} n_k n_l \mathrm{KL}(p_k \| p_l)}, \tag{13.46}$$

where n_i and c denote the cardinality of the i-th class and the number of classes, respectively. The geometric mean of this divergence is maximized between pairs of classes to separate classes after projection onto the subspace of the metric. A regularization term is used to increase the distances of dissimilar points, and a constraint is used to decrease the similar points [5]:

$$\begin{aligned} \underset{\boldsymbol{W}}{\text{maximize}} \quad & \log\Big(\Big(\prod_{1\leq i<j\leq c} e_{ij}\Big)^{\frac{1}{c(c-1)}}\Big) + \lambda \sum_{(\boldsymbol{x}_i,\boldsymbol{x}_j)\in\mathcal{D}} \|\boldsymbol{x}_i - \boldsymbol{x}_j\|_{\boldsymbol{W}} \\ \text{subject to} \quad & \sum_{(\boldsymbol{x}_i,\boldsymbol{x}_j)\in\mathcal{S}} \|\boldsymbol{x}_i - \boldsymbol{x}_j\|_{\boldsymbol{W}}^2 \leq 1, \\ & \boldsymbol{W} \succeq \boldsymbol{0}, \end{aligned} \tag{13.47}$$

where $\lambda > 0$ is the regularization parameter. This problem can be solved using the projected gradient method (see Chap. 4).

13.5.3 Probabilistic Relevant Component Analysis Methods

Recall the Relevant Component Analysis (RCA method) [11] which was introduced in Sect. 11.3.1.3 of Chap. 11. This section introduces probabilistic RCA [1, 2], which uses information theory. Suppose the n data points can be divided into c clusters, or so-called chunklets. Let $\mathcal{X}_l$ denote the data of the l-th chunklet and $\boldsymbol{\mu}_l$ be the mean of $\mathcal{X}_l$. Consider Eq. (13.2) for the decomposition of the weight matrix in the metric, where the column space of $\boldsymbol{U}$ is the subspace of the metric. Let the projection of data onto this subspace be denoted by $\boldsymbol{Y} = \boldsymbol{U}^\top \boldsymbol{X}$, the projected data in the l-th chunklet be $\mathcal{Y}_l$, and $\boldsymbol{\mu}_l^y$ be the mean of $\mathcal{Y}_l$.

In the probabilistic RCA, the mutual information between data and the projected data is maximized, while the aim is to have the summation of the distances between the points in a chunklet and the mean of the chunklet be less than a threshold or margin $m > 0$. The mutual information is related to the entropy as $I(X, Y) := H(Y) - H(Y|X)$; therefore, it is possible to maximize the entropy of the projected data $H(Y)$, rather than the mutual information. Since $\boldsymbol{Y} = \boldsymbol{U}^\top \boldsymbol{X}$, then $H(Y) \propto \det(\boldsymbol{U})$. According to Eq. (13.2), $\det(\boldsymbol{U}) \propto \det(\boldsymbol{W})$. Therefore, the optimization problem can be described as [1, 2]:

$$\begin{aligned} \underset{\boldsymbol{W}}{\text{maximize}} \quad & \det(\boldsymbol{W}) \\ \text{subject to} \quad & \sum_{l=1}^{c} \sum_{\boldsymbol{y}_i \in \mathcal{Y}_l} \|\boldsymbol{y}_i - \boldsymbol{\mu}_l^y\|_{\boldsymbol{W}}^2 \leq m, \\ & \boldsymbol{W} \succeq \boldsymbol{0}. \end{aligned} \tag{13.48}$$

This preserves the information of the data after projection, while the interchunklet variances are upper bounded by a margin.

If a Gaussian distribution is assumed for each chunklet with the covariance matrix $\boldsymbol{\Sigma}_l$ for the l-th chunklet, then $\det(\boldsymbol{W}) \propto \log(\det(\boldsymbol{U}^\top \boldsymbol{\Sigma}_l \boldsymbol{U}))$ because of the quadratic characteristic of the covariance. In this case, the optimization problem becomes:

$$\begin{aligned} \underset{\boldsymbol{U}}{\text{maximize}} \quad & \sum_{l=1}^{c} \log(\det(\boldsymbol{U}^\top \boldsymbol{\Sigma}_l \boldsymbol{U})) \\ \text{subject to} \quad & \sum_{l=1}^{c} \sum_{\boldsymbol{y}_i \in \mathcal{Y}_l} \|\boldsymbol{y}_i - \boldsymbol{\mu}_l^y\|_{\boldsymbol{U}\boldsymbol{U}^\top}^2 \leq m, \end{aligned} \tag{13.49}$$

where $\boldsymbol{W} \succeq \boldsymbol{0}$ is already satisfied because of Eq. (13.2).

13.5.4 Metric Learning by Information Geometry

An information theoretic method for metric learning uses information geometry, in which kernels on data and labels are used [13]. Let $\boldsymbol{L} \in \mathbb{R}^{c \times n}$ denote the one-hot encoded labels of n data points with c classes and let $\boldsymbol{X} \in \mathbb{R}^{d \times n}$ be the data points. The kernel matrix of the labels is $\boldsymbol{K}_L = \boldsymbol{Y}^\top \boldsymbol{Y} + \lambda \boldsymbol{I}$ whose main diagonal is strengthened by a small positive number λ to have a full rank. Recall Proposition 13.1 and Eq. (13.2), where $\boldsymbol{U}$ is the projection matrix onto the subspace of the metric. The kernel matrix over the projected data, $\boldsymbol{Y} = \boldsymbol{U}^\top \boldsymbol{X}$, is:

$$\boldsymbol{K}_Y = \boldsymbol{Y}^\top \boldsymbol{Y} = (\boldsymbol{U}^\top \boldsymbol{X})^\top (\boldsymbol{U}^\top \boldsymbol{X}) = \boldsymbol{X}^\top \boldsymbol{U}\boldsymbol{U}^\top \boldsymbol{X} \overset{(a)}{=} \boldsymbol{X}^\top \boldsymbol{W}\boldsymbol{X}, \tag{13.50}$$

where (a) is because of Eq. (13.2). The KL-divergence between the distributions of kernels $\boldsymbol{K}_Y$ and $\boldsymbol{K}_L$ can be minimized [13]:

$$\begin{aligned} &\underset{\boldsymbol{W}}{\text{minimize}} && \text{KL}(\boldsymbol{K}_Y \| \boldsymbol{K}_L) \\ &\text{subject to} && \boldsymbol{W} \succeq \boldsymbol{0}. \end{aligned} \tag{13.51}$$

For simplicity, assume Gaussian distributions for the kernels. The KL divergence between the distributions of two matrices, $\boldsymbol{K}_Y \in \mathbb{R}^{n\times n}$ and $\boldsymbol{K}_L \in \mathbb{R}^{n\times n}$, with Gaussian distributions is simplified to [13, Theorem 1]:

$$\begin{aligned} \text{KL}(\boldsymbol{K}_Y \| \boldsymbol{K}_L) &= \frac{1}{2}\Big(\textbf{tr}(\boldsymbol{K}_L^{-1}\boldsymbol{K}_Y) + \log(\det(\boldsymbol{K}_L)) - \log(\det(\boldsymbol{K}_Y)) - n\Big) \\ &\overset{(13.50)}{\propto} \frac{1}{2}\Big(\textbf{tr}(\boldsymbol{K}_L^{-1}\boldsymbol{X}^\top \boldsymbol{W}\boldsymbol{X}) + \log(\det(\boldsymbol{K}_L)) - \log(\det(\boldsymbol{W})) - n\Big). \end{aligned}$$

After ignoring the constant terms with respect to $\boldsymbol{W}$, Eq. (13.51) can be restated to:

$$\begin{aligned} &\underset{\boldsymbol{W}}{\text{minimize}} && \textbf{tr}(\boldsymbol{K}_L^{-1}\boldsymbol{X}^\top \boldsymbol{W}\boldsymbol{X}) - \log(\det(\boldsymbol{W})) \\ &\text{subject to} && \boldsymbol{W} \succeq \boldsymbol{0}. \end{aligned} \tag{13.52}$$

If the derivative of the objective function in Eq. (13.52) is set to zero, then:

$$\frac{\partial c}{\partial \boldsymbol{W}} = \boldsymbol{X}\boldsymbol{K}_L^{-1}\boldsymbol{X}^\top - \boldsymbol{W}^{-1} \overset{\text{set}}{=} \boldsymbol{0} \implies \boldsymbol{W} = (\boldsymbol{X}\boldsymbol{K}_L^{-1}\boldsymbol{X}^\top)^{-1}. \tag{13.53}$$

Note that the constraint $\boldsymbol{W} \succeq \boldsymbol{0}$ is already satisfied by the solution, i.e., Eq. (13.53).

Although this method has used kernels, it can be kernelized further by using Eq. (13.4) as the generalized Mahalanobis distance in the feature space, where $\boldsymbol{T}$ (defined in Eq. (13.3)) is the projection matrix for the metric. Using this in Eqs. (13.52) and (13.53) results in the kernel version of this method. See [13] for additional information.

13.6 Empirical Risk Minimization in Metric Learning

By minimizing some empirical risk, it is possible to learn the metric. This section introduces several metric learning metric learning methods, which use risk minimization.

13.6.1 Metric Learning Using the Sigmoid Function

Guillamin et al. developed a metric learning method using risk minimization [8]. The distribution for $\boldsymbol{x}_i$ to take $\boldsymbol{x}_j$ as its neighbour can be stated using a sigmoid function:

$$p_{ij}^W := \frac{1}{1 + \exp(\|\boldsymbol{x}_i - \boldsymbol{x}_j\|_{\boldsymbol{W}}^2 - b)}, \tag{13.54}$$

where $b > 0$ is a bias, because close-by points should have a larger probability. It is possible to maximize and minimize this probability for similar and dissimilar points, respectively:

$$\begin{aligned} \underset{\boldsymbol{W}}{\text{maximize}} \quad & \sum_{i=1}^{n}\sum_{j=1}^{n} y_{ij} \log(p_{ij}^W) + (1 - y_{ij}) \log(1 - p_{ij}^W) \\ \text{subject to} \quad & \boldsymbol{W} \succeq \boldsymbol{0}, \end{aligned} \tag{13.55}$$

where y_{ij} is defined in Eq. (13.30). This can be solved using the projected gradient method (see Chap. 4). This optimization can be seen as the minimization of the empirical risk, where close-by points are pushed toward each other and dissimilar points are pushed away to decrease the error.

13.6.2 Pairwise Constrained Component Analysis (PCCA)

Pairwise Constrained Component Analysis (PCCA) [9] minimizes the following empirical risk to minimize and maximize the distances between similar points and dissimilar points, respectively:

$$\underset{\boldsymbol{U}}{\text{minimize}} \quad \sum_{i=1}^{n}\sum_{j=1}^{n} \log\Big(1 + \exp\big(y_{ij}(\|\boldsymbol{x}_i - \boldsymbol{x}_j\|_{\boldsymbol{U}\boldsymbol{U}^\top}^2 - b)\big)\Big), \tag{13.56}$$

where y_{ij} is defined in Eq. (13.30), $b > 0$ is a bias, and $\boldsymbol{W} \succeq \boldsymbol{0}$ has been satisfied in Eq. (13.2). This can be solved using the projected gradient method (see Chap. 4) with the gradient [9]:

$$\frac{\partial c}{\partial \boldsymbol{U}} = 2 \sum_{i=1}^{n} \sum_{j=1}^{n} \frac{y_{ij}}{1 + \exp\big(y_{ij}(\|\boldsymbol{x}_i - \boldsymbol{x}_j\|_{\boldsymbol{U}\boldsymbol{U}^\top}^2 - b)\big)} \times (\boldsymbol{x}_i - \boldsymbol{x}_j)(\boldsymbol{x}_i - \boldsymbol{x}_j)^\top \boldsymbol{U}. \tag{13.57}$$

Note that a kernel PCCA can be obtained by using Eq. (13.4). In other words, replace $\|\boldsymbol{x}_i - \boldsymbol{x}_j\|_{\boldsymbol{U}\boldsymbol{U}^\top}^2$ and $(\boldsymbol{x}_i - \boldsymbol{x}_j)(\boldsymbol{x}_i - \boldsymbol{x}_j)^\top \boldsymbol{U}$ with $\|\boldsymbol{k}_i - \boldsymbol{k}_j\|_{\boldsymbol{T}\boldsymbol{T}^\top}^2$ and $(\boldsymbol{k}_i - \boldsymbol{k}_j)(\boldsymbol{k}_i - \boldsymbol{k}_j)^\top \boldsymbol{T}$, respectively, to have PCCA in the feature space.

13.6.3 Metric Learning for Privileged Information

In some applications, there is a dataset with privileged information, where for every point, there are two feature vectors; one for the main feature (denoted by $\{\boldsymbol{x}_i\}_{i=1}^n$) and one for the privileged information (denoted by $\{z_i\}_{i=1}^n$). There exists a metric learning method for using privileged information [17], where the distances between similar and dissimilar points are minimized and maximized, respectively, for the main features. Simultaneously, the distances of privileged features are pushed close to the distances of the main features. Having these two simultaneous goals, the following empirical risk is minimized [17]:

$$\begin{aligned} \underset{\boldsymbol{W}_1, \boldsymbol{W}_2}{\text{minimize}} \quad & \sum_{i=1}^{n} \log\Big(1 + \exp\big(y_{ij}\,(\|\boldsymbol{x}_i - \boldsymbol{x}_j\|_{\boldsymbol{W}_1}^2 - \|z_i - z_j\|_{\boldsymbol{W}_2}^2)\big)\Big) \\ \text{subject to} \quad & \boldsymbol{W}_1 \succeq \boldsymbol{0}, \quad \boldsymbol{W}_2 \succeq \boldsymbol{0}. \end{aligned} \tag{13.58}$$

13.7 Chapter Summary

Metric learning is an important family of methods for dimensionality reduction. One of the categories of metric learning is probabilistic metric learning, which is introduced in this chapter. Probabilistic metric learning models the probability that every point takes other points as its neighbours. The same idea will also be used in the neighbourhood embedding methods, introduced in Chap. 16. Spectral and deep metric learning will be introduced in Chaps. 11 and 19, respectively.

References

1. Aharon Bar-Hillel et al. "Learning a Mahalanobis metric from equivalence constraints". In: *Journal of machine learning research* 6.6 (2005).
2. Aharon Bar-Hillel et al. "Learning distance functions using equivalence relations". In: *Proceedings of the 20th international conference on machine learning (ICML-03)*. 2003, pp. 11–18.
3. Jason V Davis et al. "Information-theoretic metric learning". In: *Proceedings of the 24th international conference on Machine learning*. 2007, pp. 209–216.
4. JVDI Dhillon. "Differential entropic clustering of multivariate Gaussians". In: *Advances in Neural Information Processing Systems* 19 (2007), p. 337.
5. Lin Feng et al. "Learning a distance metric by balancing KL-divergence for imbalanced datasets". In: *IEEE Transactions on Systems, Man, and Cybernetics: Systems* 49.12 (2018), pp. 2384–2395.
6. Amir Globerson and Sam Roweis. "Metric learning by collapsing classes". In: *Advances in neural information processing systems* 18 (2005), pp. 451–458.
7. Jacob Goldberger et al. "Neighbourhood components analysis". In: *Advances in neural information processing systems*. 2005, pp. 513–520.
8. Matthieu Guillaumin, Jakob Verbeek, and Cordelia Schmid. "Is that you? Metric learning approaches for face identification". In: *2009 IEEE 12th international conference on computer vision*. IEEE. 2009, pp. 498–505.
9. Alexis Mignon and Frédéric Jurie. "PCCA: A new approach for distance learning from sparse pairwise constraints". In: *2012 IEEE conference on computer vision and pattern recognition*. IEEE. 2012, pp. 2666–2672.
10. Kevin P Murphy. Machine learning: *a probabilistic perspective*. MIT press, 2012.
11. Noam Shental et al. "Adjustment learning and relevant component analysis". In: *European conference on computer vision*. Springer. 2002, pp. 776–790.
12. Dong Wang and Xiaoyang Tan. "Bayesian neighborhood component analysis". In: *IEEE transactions on neural networks and learning systems* 29.7 (2017), pp. 3140–3151.
13. Shijun Wang and Rong Jin. "An information geometry approach for distance metric learning". In: *Artificial intelligence and statistics*. PMLR. 2009, pp. 591–598.
14. Liu Yang, Rong Jin, and Rahul Sukthankar. "Bayesian active distance metric learning". In: *Conference on Uncertainty in Artificial Intelligence (UAI)*. 2007.
15. Liu Yang et al. "An efficient algorithm for local distance metric learning". In: *Proceedings of the AAAI Conference on Artificial Intelligence*. Vol. 2. 2006, pp. 543–548.
16. Wei Yang, Kuanquan Wang, and Wangmeng Zuo. "Fast neighborhood component analysis". In: *Neurocomputing* 83 (2012), pp. 31–37.
17. Xun Yang et al. "Empirical risk minimization for metric learning using privileged information". In: *IJCAI International Joint Conference on Artificial Intelligence*. 2016.
18. Zhirong Yang and Jorma Laaksonen. "Regularized neighborhood component analysis". In: *Scandinavian Conference on Image Analysis*. Springer. 2007, pp. 253–262.

Chapter 14
Random Projection

14.1 Introduction

Linear dimensionality reduction methods project data onto the low-dimensional column space of a projection matrix. Many of these methods, such as Principal Component Analysis (PCA) (see Chap. 5) and Fisher Discriminant Analysis (FDA) (see Chap. 6), learn a projection matrix for either better representation of data or discrimination between the classes in the subspace. For example, PCA learns the projection matrix to maximize the variance of the projected data. Whereas, FDA learns the projections matrix to discriminate classes in the subspace. However, it has been demonstrated that, rather than learning the projection matrix, it is possible to sample the elements of the projection matrix randomly and have it still work. In fact, random projection preserves the distances between points with a small error after projection. To justify why random projection works well, the Johnson-Lindenstrauss (JL) lemma was proposed [43], which bounds the error of random projection. Random projection is a probabilistic dimensionality reduction method. Theories for various random projection methods take a similar approach and have similar bounds. In terms of dealing with probability bounds, it can be slightly related to the Probably Approximately Correct (PAC) learning [69].

Linear random projection theory has been developed since the 1980s. However, the theory of nonlinear random projection still requires further development since it is more complicated to analyze than linear random projection. Nonlinear random projection can be modelled as a linear random projection followed by a nonlinear function. Random Fourier Features (RFF) [61] and Random Kitchen Sinks (RKS) [62, 63] are two initial works on nonlinear random projection. Additional methods that have been proposed for nonlinear random projection include Extreme Learning Machine (ELM) [39, 40, 72], randomly weighted neural network [42], and ensemble random projections [16, 44, 67]. These methods will be introduced in this chapter.

B. Ghojogh et al., *Elements of Dimensionality Reduction and Manifold Learning*,
https://doi.org/10.1007/978-3-031-10602-6_14

14.2 Linear Random Projection

Linear random projection (as introduced in Chap. 2) is the projection of data points onto the column space of a projection matrix, where the elements of the projection matrix are independent and identically distributed random variables sampled from a distribution with zero mean and (possibly scaled) unit variance. In other words, the random projection is a function $f : \mathbb{R}^d \to \mathbb{R}^p$, $f : \boldsymbol{x} \mapsto \boldsymbol{U}^\top \boldsymbol{x}$:

$$\mathbb{R}^p \ni f(\boldsymbol{x}) := \boldsymbol{U}^\top \boldsymbol{x} = \sum_{t=1}^{p} \boldsymbol{u}_t^\top \boldsymbol{x} = \sum_{j=1}^{d} \sum_{t=1}^{p} u_{jt}\, x_j, \tag{14.1}$$

where $\boldsymbol{U} = [\boldsymbol{u}_1, \ldots, \boldsymbol{u}_p] \in \mathbb{R}^{d \times p}$ is the random projection matrix, u_{jt} is the (j, t)-th element of $\boldsymbol{U}$, and x_j is the j-th element of $\boldsymbol{x} \in \mathbb{R}^d$. The elements of $\boldsymbol{U}$ are sampled from a distribution with a mean of zero and (possibly scaled) unit variance. For example, it is possible to use the Gaussian distribution, $u_{j,t} \sim \mathcal{N}(0, 1), \forall j, t$. Some example distributions to use for random projection (to sample elements of $\boldsymbol{U}$ from) are Gaussian [28] and Cauchy [52, 64] distributions.

It is noteworthy that in some research works, the projection is normalized:

$$f : \boldsymbol{x} \mapsto \frac{1}{\sqrt{p}} \boldsymbol{U}^\top \boldsymbol{x}. \tag{14.2}$$

Consider a dataset with d-dimensional points and the sample size n; it is possible to stack the points in $\boldsymbol{X} := [\boldsymbol{x}_1, \ldots, \boldsymbol{x}_n] \in \mathbb{R}^{d \times n}$. Equation (14.1) or (14.2) is stated as:

$$\mathbb{R}^{p \times n} \ni f(\boldsymbol{X}) := \boldsymbol{U}^\top \boldsymbol{X} \quad \text{or} \quad \frac{1}{\sqrt{p}} \boldsymbol{U}^\top \boldsymbol{X}. \tag{14.3}$$

In contrast to other linear dimensionality reduction methods, which learn the projection matrix using a training dataset for better data representation of class discrimination, random projection does not learn the projection matrix but randomly samples it independent of the data. Surprisingly, projection of the data onto the column space of this random matrix works well, although the projection matrix is completely random and independent of the data. The following JL lemma [43] and its proof justify why random projection works.

14.3 The Johnson-Lindenstrauss Lemma

Theorem 14.1 (Johnson-Lindenstrauss Lemma [43]) *For any set $\mathcal{X} := \{\boldsymbol{x}_i \in \mathbb{R}^d\}_{i=1}^n$, any integer n as the sample size, and any $0 < \epsilon < 1$ as error tolerance, let p be a positive integer satisfying:*

$$p \geq \Omega(\frac{\ln(n)}{\epsilon^2 - \epsilon^3}), \tag{14.4}$$

where $\ln(.)$ *is the natural logarithm and* $\Omega(.)$ *is the lower bound complexity [note that some works state Eq. (14.4) as:*

$$p \geq \Omega(\epsilon^{-2} \ln(n)). \tag{14.5}$$

by ignoring ϵ^3 *against* ϵ^2 *in the denominator because* $\epsilon \in (0, 1)$*]. There exists a linear map* $f : \mathbb{R}^d \to \mathbb{R}^p$, $f : \boldsymbol{x} \mapsto \boldsymbol{U}^\top \boldsymbol{x}$, *with a projection matrix* $\boldsymbol{U} = [\boldsymbol{u}_1, \ldots, \boldsymbol{u}_p] \in \mathbb{R}^{d \times p}$, *such that:*

$$(1 - \epsilon)\|\boldsymbol{x}_i - \boldsymbol{x}_j\|_2^2 \leq \|f(\boldsymbol{x}_i) - f(\boldsymbol{x}_j)\|_2^2 \leq (1 + \epsilon)\|\boldsymbol{x}_i - \boldsymbol{x}_j\|_2^2, \tag{14.6}$$

for all $\boldsymbol{x}_i, \boldsymbol{x}_j \in \mathcal{X}$, *with the probability of success stated as:*

$$\mathbb{P}\Big((1 - \epsilon)\|\boldsymbol{x}_i - \boldsymbol{x}_j\|_2^2 \leq \|f(\boldsymbol{x}_i) - f(\boldsymbol{x}_j)\|_2^2 \leq (1 + \epsilon)\|\boldsymbol{x}_i - \boldsymbol{x}_j\|_2^2\Big) \geq 1 - \delta, \tag{14.7}$$

where $\delta := 2e^{-(\epsilon^2 - \epsilon^3)(p/4)}$ *and the elements of the projection matrix are independent and identically distributed (i.i.d.) random variables with a mean of zero and a (scaled) unit variance. For example, it is possible to use* $u_{ij} \sim \mathcal{N}(0, 1/p) = (1/\sqrt{p})\mathcal{N}(0, 1)$ *where* u_{ij} *denotes the* (i, j)*-th element of* $\boldsymbol{U}$.

Proof The following JL lemma proof is based on [1, 2, 20, 21, 41, 45, 56]. Moreover, additional proofs exist for the JL lemma, such as [6, 24]. The proof can be divided into three steps:

Step 1 of Proof This step demonstrates that the random projection preserves the local distances of the points, in expectation. Let $\boldsymbol{x}_i = [x_{i1}, \ldots, x_{id}]^\top \in \mathbb{R}^d$. Therefore:

$$\mathbb{E}[\|f(\boldsymbol{x}_i) - f(\boldsymbol{x}_j)\|_2^2] = \mathbb{E}[\|\boldsymbol{U}^\top \boldsymbol{x}_i - \boldsymbol{U}^\top \boldsymbol{x}_j\|_2^2] \overset{(a)}{=} \mathbb{E}\Big[\sum_{t=1}^{p} (\boldsymbol{u}_t^\top \boldsymbol{x}_i - \boldsymbol{u}_t^\top \boldsymbol{x}_j)^2\Big]$$

$$\overset{(b)}{=} \sum_{t=1}^{p} \mathbb{E}[(\boldsymbol{u}_t^\top \boldsymbol{x}_i - \boldsymbol{u}_t^\top \boldsymbol{x}_j)^2] = \sum_{t=1}^{p} \Big(\mathbb{E}[(\boldsymbol{u}_t^\top \boldsymbol{x}_i)^2] + \mathbb{E}[(\boldsymbol{u}_t^\top \boldsymbol{x}_j)^2] - 2\mathbb{E}[(\boldsymbol{u}_t^\top \boldsymbol{x}_i \boldsymbol{u}_t^\top \boldsymbol{x}_j)^2]\Big), \tag{14.8}$$

where (a) is because of the definition of the ℓ_2 norm and (b) is because the expectation is a linear operator. u_{ij} has a zero mean and unit variance. If its unit variance is normalized as $1/p$, then the covariance of the elements of the projection matrix is:

$$\mathbb{E}[u_{kt}u_{lt}] = \delta_{kl} := \begin{cases} 0 & \text{if } i \neq j \\ 1/p & \text{if } i = j, \end{cases} \tag{14.9}$$

which is a normalized Kronecker delta. The first term in this expression is:

$$\mathbb{E}\big[(\boldsymbol{u}_t^\top \boldsymbol{x}_i)^2\big] = \mathbb{E}\big[\big(\sum_{k=1}^{d} u_{kt}x_{ik}\big)^2\big] = \mathbb{E}\Big[\sum_{k=1}^{d}\sum_{l=1}^{d} u_{kt}u_{lt}x_{ik}x_{il}\Big] \overset{(a)}{=} \sum_{k=1}^{d}\sum_{l=1}^{d} \mathbb{E}[u_{kt}u_{lt}x_{ik}x_{il}]$$

$$\overset{(b)}{=} \sum_{k=1}^{d}\sum_{l=1}^{d} x_{ik}x_{il}\mathbb{E}[u_{kt}u_{lt}] \overset{(14.9)}{=} \sum_{k=1}^{d}\sum_{l=1}^{d} x_{ik}x_{il}\delta_{kl} = \frac{1}{p}\sum_{k=1}^{d} x_{ik}^2 = \frac{\|\boldsymbol{x}_i\|_2^2}{p},$$

where (a) is because of the linearity of the expectation and (b) is because the data are deterministic. Likewise, the second term of Eq. (14.8) is $\mathbb{E}\big[(\boldsymbol{u}_t^\top \boldsymbol{x}_j)^2\big] = \|\boldsymbol{x}_j\|_2^2/p$. The third term is:

$$\mathbb{E}\big[(\boldsymbol{u}_t^\top \boldsymbol{x}_i \boldsymbol{u}_t^\top \boldsymbol{x}_j)^2\big] = \mathbb{E}\big[\big(\sum_{k=1}^{d} u_{kt}x_{ik}\big)\big(\sum_{l=1}^{d} u_{lt}x_{jl}\big)\big]$$

$$= \mathbb{E}\Big[\sum_{k=1}^{d}\sum_{l=1}^{d} u_{kt}u_{lt}x_{ik}x_{jl}\Big] = \sum_{k=1}^{d}\sum_{l=1}^{d} \mathbb{E}[u_{kt}u_{lt}x_{ik}x_{jl}]$$

$$= \sum_{k=1}^{d}\sum_{l=1}^{d} x_{ik}x_{jl}\mathbb{E}[u_{kt}u_{lt}] \overset{(14.9)}{=} \sum_{k=1}^{d}\sum_{l=1}^{d} x_{ik}x_{jl}\delta_{kl} = \frac{1}{p}\sum_{k=1}^{d} x_{ik}x_{jk} = \frac{\boldsymbol{x}_i^\top \boldsymbol{x}_j}{p}.$$

Therefore, Eq. (14.8) is simplified to:

$$\mathbb{E}\big[\|f(\boldsymbol{x}_i) - f(\boldsymbol{x}_j)\|_2^2\big] = \sum_{t=1}^{p}\Big(\frac{\|\boldsymbol{x}_i\|_2^2}{p} + \frac{\|\boldsymbol{x}_j\|_2^2}{p} - 2\frac{\boldsymbol{x}_i^\top \boldsymbol{x}_j}{p}\Big)$$

$$= \sum_{t=1}^{p}\Big(\frac{\|\boldsymbol{x}_i - \boldsymbol{x}_j\|_2^2}{p}\Big) = \|\boldsymbol{x}_i - \boldsymbol{x}_j\|_2^2 \sum_{t=1}^{p}\frac{1}{p} = \|\boldsymbol{x}_i - \boldsymbol{x}_j\|_2^2, \quad \forall i, j \in \{1, \dots, n\}. \tag{14.10}$$

This shows that this projection by a random projection matrix preserves the local distances of the data points, in expectation.

Step 2 of Proof This step demonstrates that the variance of local distance preservation is bounded and small for one pair of points. The probability of error in the local distance preservation is:

$$\mathbb{P}\Big(\|f(\boldsymbol{x}_i) - f(\boldsymbol{x}_j)\|_2^2 > (1+\epsilon)\|\boldsymbol{x}_i - \boldsymbol{x}_j\|_2^2\Big) = \mathbb{P}\Big(\|\boldsymbol{U}^\top \boldsymbol{x}_i - \boldsymbol{U}^\top \boldsymbol{x}_j\|_2^2 > (1+\epsilon)\|\boldsymbol{x}_i - \boldsymbol{x}_j\|_2^2\Big)$$

$$= \mathbb{P}\Big(\|\boldsymbol{U}^\top (\boldsymbol{x}_i - \boldsymbol{x}_j)\|_2^2 > (1+\epsilon)\|\boldsymbol{x}_i - \boldsymbol{x}_j\|_2^2\Big).$$

Let $\boldsymbol{x}_D := \boldsymbol{x}_i - \boldsymbol{x}_j$, so:

$$\mathbb{P}\Big(\|\boldsymbol{U}^\top \boldsymbol{x}_D\|_2^2 > (1+\epsilon)\|\boldsymbol{x}_d\|_2^2\Big).$$

Therefore, $u_{ij} \sim (1/\sqrt{p})\mathcal{N}(0,1)$. This means that the projected data are $\mathbb{R}^p \ni \boldsymbol{y}_D = \boldsymbol{U}^\top \boldsymbol{x}_D$ and $\boldsymbol{y}_D \sim (1/\sqrt{p})\mathcal{N}(0, \|\boldsymbol{x}_D\|_2^2) = \mathcal{N}(0, (1/p)\|\boldsymbol{x}_D\|_2^2)$ holds because of the quadratic characteristic of the variance.

The following matrix is defined as $\boldsymbol{A} = [a_{ij}] \in \mathbb{R}^{d\times p}$, where $a_{ij} \sim \mathcal{N}(0,1)$. Assume $\boldsymbol{z}_D = [z_1, \dots, z_p]^\top := (\boldsymbol{A}^\top \boldsymbol{x}_D)/\|\boldsymbol{x}_D\|_2$. Therefore:

$$\begin{aligned}
\mathbb{P}\Big(\|\frac{1}{\sqrt{p}}\boldsymbol{A}^\top \boldsymbol{x}_D\|_2^2 > (1+\epsilon)\|\boldsymbol{x}_D\|_2^2\Big) &= \mathbb{P}\Big(\frac{1}{p}\|\boldsymbol{A}^\top \boldsymbol{x}_D\|_2^2 > (1+\epsilon)\|\boldsymbol{x}_D\|_2^2\Big) \\
&\overset{(a)}{=} \mathbb{P}\Big(\frac{1}{\|\boldsymbol{x}_D\|_2^2}\|\boldsymbol{A}^\top \boldsymbol{x}_D\|_2^2 > (1+\epsilon)p\Big) = \mathbb{P}\Big(\|\frac{\boldsymbol{A}^\top \boldsymbol{x}_D}{\|\boldsymbol{x}_D\|_2}\|_2^2 > (1+\epsilon)p\Big) \\
&\overset{(b)}{=} \mathbb{P}\Big(\|\boldsymbol{z}_D\|_2^2 > (1+\epsilon)p\Big) = \mathbb{P}\Big(\sum_{k=1}^{p} z_k^2 > (1+\epsilon)\,p\Big) \overset{(c)}{=} \mathbb{P}\Big(\chi_p^2 > (1+\epsilon)\,p\Big),
\end{aligned}$$

where (a) is because the sides are multiplied by $p/\|\boldsymbol{x}_D\|_2^2$, (b) is because of the definition of $\boldsymbol{z}_D$, and (c) is because the summation of p squared standard normal distributions is the chi-squared distribution with p degrees of freedom, denoted by χ_p^2. Therefore:

$$\mathbb{P}\Big(\chi_p^2 > (1+\epsilon)\,p\Big) \overset{(a)}{=} \mathbb{P}\Big(e^{\lambda\chi_p^2} > e^{\lambda(1+\epsilon)\,p}\Big) \overset{(b)}{\leq} \frac{\mathbb{E}[e^{\lambda\chi_p^2}]}{e^{\lambda(1+\epsilon)\,p}} \overset{(c)}{=} \frac{(1-2\lambda)^{-p/2}}{e^{\lambda(1+\epsilon)\,p}} \tag{14.11}$$

where (a) is because of taking the power of e with the parameter $\lambda \geq 0$, (b) is because of the Markov inequality (i.e., $\mathbb{P}(x \geq a) \leq \mathbb{E}[x]/a$), and (c) is because of the moment generating function of the chi-squared distribution $x \sim \chi_p^2$, which is $M_x(t) = \mathbb{E}[e^{\lambda x}] = (1-2\lambda)^{-p/2}$. The probability of error is minimized in Eq. (14.11) with respect to the parameter λ:

$$\frac{\partial}{\partial\lambda}\Big(\frac{(1-2\lambda)^{-p/2}}{e^{\lambda(1+\epsilon)\,p}}\Big) \overset{\text{set}}{=} 0 \implies \lambda = \frac{\epsilon}{2(1+\epsilon)}.$$

Substituting this equation into Eq. (14.11) results in:

$$\mathbb{P}\Big(\chi_p^2 > (1+\epsilon)\,p\Big) = \frac{(1+\epsilon)^{p/2}}{e^{(p\epsilon)/2}} = ((1+\epsilon)e^{-\epsilon})^{p/2} \overset{(a)}{\leq} e^{-(p/4)(\epsilon^2-\epsilon^3)},$$

where (a) is because $1 + \epsilon \leq e^{(\epsilon-(\epsilon^2-\epsilon^3)/2)}$. Above, we derived the probability of $\mathbb{P}(\|f(\boldsymbol{x}_i) - f(\boldsymbol{x}_j)\|_2^2 > (1 + \epsilon)\|\boldsymbol{x}_i - \boldsymbol{x}_j\|_2^2)$. Likewise, we can derive the same amount of probability for $\mathbb{P}(\|f(\boldsymbol{x}_i) - f(\boldsymbol{x}_j)\|_2^2 < (1 - \epsilon)\|\boldsymbol{x}_i - \boldsymbol{x}_j\|_2^2)$. For the projection of pair $\boldsymbol{x}_i$ and $\boldsymbol{x}_j$, we want to have:

$$\|f(\boldsymbol{x}_i) - f(\boldsymbol{x}_j)\|_2^2 \approx \|\boldsymbol{x}_i - \boldsymbol{x}_j\|_2^2. \tag{14.12}$$

Therefore, the total probability of error for the projection of pair $\boldsymbol{x}_i$ and $\boldsymbol{x}_j$ is:

$$\mathbb{P}\Big(\Big[\|f(\boldsymbol{x}_i) - f(\boldsymbol{x}_j)\|_2^2 > (1+\epsilon)\|\boldsymbol{x}_i - \boldsymbol{x}_j\|_2^2\Big] \cup \Big[\|f(\boldsymbol{x}_i) - f(\boldsymbol{x}_j)\|_2^2 < (1-\epsilon)\|\boldsymbol{x}_i - \boldsymbol{x}_j\|_2^2\Big]\Big)$$
$$\leq 2\,e^{-(p/4)(\epsilon^2-\epsilon^3)}. \tag{14.13}$$

This shows that the probability of error for every pair $\boldsymbol{x}_i$ and $\boldsymbol{x}_j$ is bounded. This also proves Eq. (14.7).

Step 3 of Proof This step bounds the probability of error for all pairs of data points. Consider the Bonferroni's inequality or the so-called union bound [10]:

$$\mathbb{P}\Big(\bigcup_i E_i\Big) \leq \sum_i \mathbb{P}(E_i), \tag{14.14}$$

where E_i is the i-th probability event. There are $\binom{n}{2} = n(n-1)/2$ pairs of points; therefore, the total probability of error for all pairs of points is:

$$\mathbb{P}(\text{error}) \leq \frac{n(n-1)}{2} \times 2\,e^{-(p/4)(\epsilon^2-\epsilon^3)} = n(n-1)e^{-(p/4)(\epsilon^2-\epsilon^3)} \leq \delta', \tag{14.15}$$

where the upper bound $0 \leq \delta' \ll 1$ can be selected according to n, p, and ϵ. Therefore, the probability of error is bounded. This shows that Eq. (14.6) holds with probability $1 - \delta'$. □

It is noteworthy that Eq. (14.7) can be restated as:

$$\mathbb{P}\Big(\Big|\|f(\boldsymbol{x}_i) - f(\boldsymbol{x}_j)\|_2^2 - \|\boldsymbol{x}_i - \boldsymbol{x}_j\|_2^2\Big| \geq \epsilon\Big) \leq \delta. \tag{14.16}$$

This demonstrates that there is an upper bound on the probability of error in random projection. Therefore, after linear random projection, the distance is preserved after projection for every pair of points.

The following lemma proves the required lower bound on the dimensionality of the subspace for random projection to work well.

Lemma 14.1 (Lower Bound on the Dimensionality of Subspace [41]) *The dimensionality of subspace p satisfies Eq. (14.4).*

Proof The probability of error, which is shown in Eq. (14.15), should be smaller than a constant δ:

$$\mathbb{P}(\text{error}) = n(n-1)e^{-(p/4)(\epsilon^2-\epsilon^3)} \leq \delta' \implies -(p/4)(\epsilon^2-\epsilon^3) \leq \ln(\frac{\delta'}{n(n-1)})$$

$$\implies (p/4)(\epsilon^2-\epsilon^3) \geq \ln(\frac{n(n-1)}{\delta'}) \implies (p/4) \geq \frac{1}{\epsilon^2-\epsilon^3}\ln(\frac{n(n-1)}{\delta'}) \approx \frac{\ln(\frac{n^2}{\delta'})}{\epsilon^2-\epsilon^3}$$

$$\implies p \geq \frac{8\ln(n) - 4\ln(\delta')}{\epsilon^2-\epsilon^3} = \Omega(\frac{\ln(n)}{\epsilon^2-\epsilon^3}).$$

□

Definition 14.1 (Concentration of Measure [71]) The following inequality is inspired by Eq. (14.7) [60, Lemma A.1]:

$$\mathbb{P}\Big((1-\epsilon)\|\boldsymbol{x}\|_2^2 \leq \|\boldsymbol{U}^\top\boldsymbol{x}\|_2^2 \leq (1+\epsilon)\|\boldsymbol{x}\|_2^2\Big) \geq 1 - 2e^{-c\epsilon^2 p}, \tag{14.17}$$

where $c > 0$ is a constant, ϵ is the error tolerance, and p is the dimensionality of the projected data by the projection matrix $\boldsymbol{U} \in \mathbb{R}^{d\times p}$. This is the concentration of measure for a Gaussian random matrix. This inequality holds for random projection.

Comparing Eqs. (14.7) and (14.17) demonstrates that random projection is equivalent to the concentration of the measure.

The following lemma will be used in the proof of Lemma 14.3. Lemma 14.3 provides the order of probability of success for random projection and will be used to explain why random projection is correct (i.e., works well) with high probability.

Lemma 14.2 ([20, 21]) *Let $p < d$ and $L := \|z_D\|_2^2$ where z_D is the random projection of $\boldsymbol{x}_D := \boldsymbol{x}_i - \boldsymbol{x}_j$ onto a p-dimensional subspace. If $\beta < 1$, then:*

$$\mathbb{P}(L \leq \beta p/d) \leq \exp\big(\frac{p}{2}(1-\beta+\ln\beta)\big). \tag{14.18}$$

If $\beta > 1$, then:

$$\mathbb{P}(L \geq \beta p/d) \leq \exp\big(\frac{p}{2}(1-\beta+\ln\beta)\big). \tag{14.19}$$

Proof Refer to [20, 21]. □

Lemma 14.3 ([20, 21]) *If p satisfies Eq. (14.4) as:*

$$p \geq 4(\frac{\epsilon^2}{2} - \frac{\epsilon^3}{3})^{-1}\ln(n) = \Omega(\frac{\ln(n)}{\epsilon^2 - \epsilon^3}), \tag{14.20}$$

the local distances between points are not distorted by random projection, no more than $(1 \pm \epsilon)$*, with probability* $\mathcal{O}(1/n^2)$*. In other words, the probability of error in random projection is of the order* $\mathcal{O}(1/n^2)$*.*

Proof The following proof is based on Dasugpta and Gupta's work [20].

Step 1 Consider one side of error, which is $\mathbb{P}(\|f(\boldsymbol{x}_i) - f(\boldsymbol{x}_j)\|_2^2 \leq (1 - \epsilon)(p/d)\|\boldsymbol{x}_i - \boldsymbol{x}_j\|_2^2)$. Take $L = \|f(\boldsymbol{x}_i) - f(\boldsymbol{x}_j)\|_2^2$, $\beta = (1 - \epsilon)$, and $\mu = (p/d)\|\boldsymbol{x}_i - \boldsymbol{x}_j\|_2^2$. $0 < \epsilon < 1$ results in $\beta < 1$. Using Lemma 14.2 results in:

$$\mathbb{P}(\|f(\boldsymbol{x}_i) - f(\boldsymbol{x}_j)\|_2^2 \leq (1-\epsilon)(p/d)\|\boldsymbol{x}_i - \boldsymbol{x}_j\|_2^2) \overset{(14.18)}{\leq} \exp\big(\frac{p}{2}(1 - (1-\epsilon) + \ln(1-\epsilon))\big)$$
$$= \exp\big(\frac{p}{2}(\epsilon + \ln(1-\epsilon))\big) \overset{(a)}{\leq} \exp\Big(\frac{p}{2}(\epsilon - (\epsilon + \frac{\epsilon^2}{2}))\Big) = \exp(-\frac{p\epsilon^2}{4}) \overset{(b)}{\leq} \exp(-2\ln(n)) = \frac{1}{n^2}, \tag{14.21}$$

where (a) is because:

$$\ln(1-x) \leq -(x + \frac{x^2}{2}), \quad \forall x \geq 0,$$

and (b) is because:

$$p \overset{(14.20)}{\geq} 4(\frac{\epsilon^2}{2} - \frac{\epsilon^3}{3})^{-1}\ln(n)$$
$$\implies \exp(-\frac{p\epsilon^2}{4}) \leq \exp(\frac{-6\epsilon^2}{3\epsilon^2 - 2\epsilon^3}\ln(n)) \overset{(c)}{\leq} \exp(-2\ln(n)),$$

where (c) is because:

$$3\epsilon^2 \geq 3\epsilon^2 - 2\epsilon^3 \implies \frac{-6\epsilon^2}{3\epsilon^2 - 2\epsilon^3} \leq -2.$$

Step 2 Consider another side of error which is $\mathbb{P}(\|f(\boldsymbol{x}_i) - f(\boldsymbol{x}_j)\|_2^2 \geq (1 + \epsilon)(p/d)\|\boldsymbol{x}_i - \boldsymbol{x}_j\|_2^2)$. Take $L = \|f(\boldsymbol{x}_i) - f(\boldsymbol{x}_j)\|_2^2$, $\beta = (1 + \epsilon)$, and $\mu = (p/d)\|\boldsymbol{x}_i - \boldsymbol{x}_j\|_2^2$. $0 < \epsilon < 1$ results in $\beta > 1$. Using Lemma 14.2:

$$\mathbb{P}(\|f(\boldsymbol{x}_i) - f(\boldsymbol{x}_j)\|_2^2 \geq (1+\epsilon)(p/d)\|\boldsymbol{x}_i - \boldsymbol{x}_j\|_2^2)$$
$$\overset{(14.19)}{\leq} \exp\big(\frac{p}{2}(1 - (1+\epsilon) + \ln(1+\epsilon))\big) = \exp\big(\frac{p}{2}(-\epsilon + \ln(1+\epsilon))\big)$$

$$\overset{(a)}{\leq} \exp\Big(\frac{p}{2}\big(-\epsilon+\epsilon-\frac{\epsilon^2}{2}+\frac{\epsilon^3}{3}\big)\Big) = \exp\Big(\frac{-p}{2}\big(\frac{\epsilon^2}{2}-\frac{\epsilon^3}{3}\big)\Big) \overset{(b)}{\leq} \exp(-2\ln(n)) = \frac{1}{n^2}, \tag{14.22}$$

where (a) is because:

$$\ln(1+x) \leq x - \frac{x^2}{2} + \frac{x^3}{3}, \quad \forall x > 0,$$

and (b) is because:

$$p \overset{(14.20)}{\geq} 4\big(\frac{\epsilon^2}{2}-\frac{\epsilon^3}{3}\big)^{-1}\ln(n)$$
$$\implies -\frac{p}{2} \leq -2\big(\frac{\epsilon^2}{2}-\frac{\epsilon^3}{3}\big)^{-1}\ln(n) \implies -\frac{p}{2}\big(\frac{\epsilon^2}{2}-\frac{\epsilon^3}{3}\big) \leq -2\ln(n)$$
$$\implies \exp\Big(-\frac{p}{2}\big(\frac{\epsilon^2}{2}-\frac{\epsilon^3}{3}\big)\Big) \leq \exp(-2\ln(n)).$$

Therefore, it was shown in both steps 1 and 2 that the probability of every side of error is bounded by $1/n^2$. Therefore, the probability of error in random projection is of the order $\mathcal{O}(1/n^2)$. □

Definition 14.2 (With High Probability) When the probability of success in an algorithm goes to one by increasing the input information or a parameter or parameters of the algorithm to infinity, the algorithm is correct with high probability, denoted by w.h.p.

Lemma 14.3 demonstrates that the probability of success for random projection is on the order of $\mathcal{O}(1/n^2)$. The JL Lemma, whose probability of success is of the order $\mathcal{O}(1/n^2)$, is correct with high probability because $\mathbb{P}(\text{error}) \to 0$ if $n \to \infty$ by increasing the number of data points. In other words, the more data points there are, the more accurate the random projection is in preserving the local distances after projection onto the random subspace. Equation (14.6) demonstrates that random projection does not distort the local distances by more than a factor of $(1 \pm \epsilon)$. Therefore, with a good probability:

$$\|\boldsymbol{U}^\top(\boldsymbol{x}_i - \boldsymbol{x}_j)\|_2^2 \approx \|\boldsymbol{x}_i - \boldsymbol{x}_j\|_2^2, \; \forall i, j \in \{1, \ldots, n\}. \tag{14.23}$$

Larsen and Nelson [50] demonstrate that for any function $f(.)$, which satisfies Eq. (14.6), Eq. (14.5) holds for the lower bound on the dimensionality of the subspace and does not require $f(.)$ to be necessarily a random projection. This demonstrates that the JL lemma is optimal. Moreover, it is shown in [8] that any finite subset of the Euclidean space can be embedded in a subspace with dimensionality $\mathcal{O}(\epsilon^{-2}\ln(n))$, while the distances between points are preserved, with at most $(1+\epsilon)$ distortion.

14.4 Sparse Linear Random Projection

The JL lemma, i.e., Theorem 14.1, dealt with the ℓ_2 norm. There is some research that formulates random projection and its proofs with ℓ_1 norm to have sparse random projection. The following section introduces the lemmas and theorems that justify sparse random projection. The proofs of these theories are omitted for brevity, and they can be found in the cited papers.

The following lemma demonstrates that if the dataset is k-sparse, meaning that every vector has at most k nonzero values, random projection preserves the distances well enough. As the dataset is sparse, random projection is sparse.

Lemma 14.4 ([7]) *Consider the subset of k-sparse signals, defined as:*

$$\mathcal{S}_k := \{\boldsymbol{x} \in \mathbb{R}^d : \#\{i : |x_i| > 0\} \leq k\}, \tag{14.24}$$

where $\#\{.\}$ *denotes the cardinality of the set. If* $p = \mathcal{O}(\epsilon^{-2}k\ln(d/k))$ *and* $f : \mathbb{R}^d \to \mathbb{R}^p$, $f : \boldsymbol{x} \mapsto \boldsymbol{U}^\top \boldsymbol{x}$, *then with high probability, then:*

$$(1-\epsilon)\|\boldsymbol{x}_i - \boldsymbol{x}_j\|_2^2 \leq \|f(\boldsymbol{x}_i) - f(\boldsymbol{x}_j)\|_2^2 \leq (1+\epsilon)\|\boldsymbol{x}_i - \boldsymbol{x}_j\|_2^2, \tag{14.25}$$

for all $\boldsymbol{x}_i, \boldsymbol{x}_j \in \mathcal{S}_k$. *The probability of error from Eq. (14.25) is* $2(12/\epsilon)^k e^{-(p/2)(\epsilon^2/8-\epsilon^3/24)}$.

The following definition defines the Restricted Isometry Property (RIP) and demonstrates that a random projection matrix, whose elements are sampled from a Gaussian distribution, has this property.

Definition 14.3 (Restricted Isometry Property [15]) The mapping $f : \mathbb{R}^d \to \mathbb{R}^p$, $f : \boldsymbol{x} \mapsto \boldsymbol{U}^\top \boldsymbol{x}$ has the Restricted Isometry Property (RIP) of order k and level $\epsilon \in (0, 1)$ if:

$$(1-\epsilon)\|\boldsymbol{x}\|_2^2 \leq \|\boldsymbol{U}^\top \boldsymbol{x}\|_2^2 \leq (1+\epsilon)\|\boldsymbol{x}\|_2^2, \tag{14.26}$$

for all k-sparse $\boldsymbol{x} \in \mathbb{R}^d$.

Lemma 14.5 ([15]) *A Gaussian random matrix* $\boldsymbol{U} \in \mathbb{R}^{d\times p}$ *has RIP if* $p \geq \epsilon^{-2}k\ln(d/k)$.

Theorem 14.2 ([47]) *Consider the linear mapping* $f : \mathbb{R}^d \to \mathbb{R}^p$, $f : \boldsymbol{x} \mapsto \boldsymbol{U}^\top \boldsymbol{x}$ *satisfying:*

$$(1-\epsilon)\|\boldsymbol{x}\|_2^2 \le \|\boldsymbol{U}^\top \boldsymbol{x}\|_2^2 \le (1+\epsilon)\|\boldsymbol{x}\|_2^2, \tag{14.27}$$

for all k-sparse $\boldsymbol{x} \in \mathbb{R}^d$ *and suppose* $\boldsymbol{D}_\xi \in \mathbb{R}^{d\times d}$ *is a diagonal matrix, with* $\xi = \pm 1$ *on its diagonal. Then:*

$$\mathbb{P}\Big((1-\epsilon)\|\boldsymbol{x}\|_2^2 \le \|\boldsymbol{U}^\top \boldsymbol{D}_\xi \boldsymbol{x}\|_2^2 \le (1+\epsilon)\|\boldsymbol{x}\|_2^2\Big) \ge 1 - 2e^{-c\epsilon^2 p/\ln(d)}. \tag{14.28}$$

Corollary 14.1 *Comparing Theorem 14.2 with Definition 14.1 demonstrates that a random matrix with RIP and random column sign flips implies the JL lemma up to an* $\ln(d)$ *factor.*

The following lemma is used in the proof of Corollary 14.2, which demonstrates that the ℓ_1 norm makes more sense as a distance than the Euclidean distance for high-dimensional data.

Lemma 14.6 ([30, Theorem 2]) *Suppose there are n independent and identically distributed random points in* $\mathbb{R}^d$*. Let* $dist_{min,r}$ *and* $dist_{max,r}$ *denote the minimum and maximum distances of points from the origin, respectively, with respect to a metric with the norm* ℓ_r*. Therefore:*

$$\lim_{d\to\infty} \mathbb{E}[dist_{max,r} - dist_{min,r}] = d^{1/r-1/2}. \tag{14.29}$$

Corollary 14.2 ([30]) *For high-dimensional data, the* ℓ_1 *norm (such as the Manhattan distance) is more meaningful for nearest neighbour comparisons than the* ℓ_2 *norm (such as the Euclidean distance).*

Proof Lemma 14.6 demonstrates that for a metric with ℓ_2 norm (i.e. Euclidean distance), there is $\lim_{d\to\infty} \mathbb{E}[\text{dist}_{\text{max},r} - \text{dist}_{\text{min},r}] = 1$; therefore, all points are equidistant from the origin up to a constant. For a metric by ℓ_r norm with $r > 2$, there is $1/r - 1/2 < 0$, meaning the limit becomes very small for large d values (when $d \to \infty$). Therefore, all points are completely equidistant from the origin. For the ℓ_1 norm, there is $\lim_{d\to\infty} \mathbb{E}[\text{dist}_{\text{max},r} - \text{dist}_{\text{min},r}] = d^{1/2}$; therefore, the relative difference of the nearest and farthest points from the origin increases with dimension. □

The above corollary demonstrates that the ℓ_1 norm is more meaningful for nearest neighbour comparisons than the ℓ_2 norm. The theorem below extends the JL lemma to any ℓ_r norm with $1 \le r \le \infty$.

Theorem 14.3 (Extension of the JL Lemma to Any Norm [51, Lemma 3.1], [74]) *For embedding into a subspace equipped with* ℓ_r *norm with* $1 \le r \le \infty$*, there exists a linear mapping* $f : \mathbb{R}^d \to \mathbb{R}^p$, $f : \boldsymbol{x} \mapsto \boldsymbol{U}^\top \boldsymbol{x}$*, with* $p \le d$*, which satisfies:*

$$\Big(\frac{p}{d}\Big)^{|1/r-1/2|}\|\boldsymbol{x}_i-\boldsymbol{x}_j\|_r \leq \|\boldsymbol{U}^\top(\boldsymbol{x}_i-\boldsymbol{x}_j)\|_r \leq \Big(\frac{d}{p}\Big)^{|1/r-1/2|}\|\boldsymbol{x}_i-\boldsymbol{x}_j\|_r, \tag{14.30}$$

for all $\boldsymbol{x}_i, \boldsymbol{x}_j \in \mathbb{R}^d$.

Why is the ℓ_2 norm, and not the ℓ_1 norm, often used for random projection? This is because the following corollary demonstrates that the suitable norm for random projection is the ℓ_2 norm, and it is very difficult to use other norms for random projection [12, 17].

Corollary 14.3 ([51, Section 3]) *The suitable norm for linear random projection is the ℓ_2 norm. There is a difficulty or impossibility of dimensionality reduction with a linear map in ℓ_r with $r \neq 2$.*

Proof According to Theorem 14.3, the local distances between points after projection are completely sandwiched between their local distances in the input space before projection; therefore, $\|\boldsymbol{U}^\top(\boldsymbol{x}_i-\boldsymbol{x}_j)\|_2 \approx \|\boldsymbol{x}_i-\boldsymbol{x}_j\|_2$. This demonstrates that embedding into the ℓ_2 norm works well, with constant distortion, in terms of local distance preservation. Consider ℓ_1 as an example for ℓ_r with $r \neq 2$. In this case, $(p/d)^{1/2}\|\boldsymbol{x}_i-\boldsymbol{x}_j\|_1 \leq \|\boldsymbol{U}^\top(\boldsymbol{x}_i-\boldsymbol{x}_j)\|_1 \leq (p/d)^{1/2}\|\boldsymbol{x}_i-\boldsymbol{x}_j\|_1$, which is not a constant distortion. □

Definition 14.4 (RIP-1 [9, Definition 8]) For $p \geq c\epsilon^{-2}k\ln(d)$, there exists a linear mapping $f:\mathbb{R}^d \to \mathbb{R}^p$, $f:\boldsymbol{x} \mapsto \boldsymbol{U}^\top\boldsymbol{x}$, with $p \leq d$, that satisfies:

$$(1-2\epsilon)\|\boldsymbol{x}\|_1 \leq \|\boldsymbol{U}^\top\boldsymbol{x}\|_1 \leq \|\boldsymbol{x}\|_1, \tag{14.31}$$

for all k-sparse $\boldsymbol{x} \in \mathbb{R}^d$, where ϵ is the error tolerance and c is a constant. If this holds for a matrix $\boldsymbol{U}$, it has the Restricted Isometry Property-1 (RIP-1).

It has been demonstrated that with the ℓ_1 norm, random projection is difficult to perform well and with ℓ_2 norm, it is not sparse. Therefore, a norm is needed between ℓ_1 and ℓ_2 norms to create a sparse random projection, which performs well enough and is not difficult to perform. The following defines an interpolation norm.

Definition 14.5 (Interpolation Norm [46, Appendix]) Suppose the elements of data $\boldsymbol{x} \in \mathbb{R}^d$ are rearranged from largest to smallest values. The support of $\boldsymbol{x}$ is divided into s disjoint subsets $\{\mathcal{S}_i\}_{i=1}^s$, where $\mathcal{S}_1$ corresponds to the largest values of $\boldsymbol{x}$. Let $\boldsymbol{x}_{\mathcal{S}_i}$ denote the values of $\boldsymbol{x}$ that fall in $\mathcal{S}_i$ represented as a vector. The interpolation norm, denoted by $\ell_{1,2,s}$, is defined as:

$$\|\boldsymbol{x}\|_{1,2,s} := \sqrt{\sum_{i=1}^{\lceil n/s\rceil}\|\boldsymbol{x}_{\mathcal{S}_i}\|_1^2}. \tag{14.32}$$

Note that if $s = 1$ and $s = d$, then $\|\boldsymbol{x}\|_{1,2,1} = \|\boldsymbol{x}\|_2$ and $\|\boldsymbol{x}\|_{1,2,d} = \|\boldsymbol{x}\|_1$, respectively. Additionally, for any s, if $\boldsymbol{x}$ is k-sparse, then $\|\boldsymbol{x}\|_{1,2,s} = \|\boldsymbol{x}\|_1$.

The following theorem is a version of the JL lemma, which makes use of the interpolation norm.

Theorem 14.4 ([46, Theorem 2]) *There exists the mapping* $f : \mathbb{R}^d \to \mathbb{R}^p$, $f : \boldsymbol{x} \mapsto \boldsymbol{U}^\top \boldsymbol{x}$ *such that with probability* $(1 - 2de^{-\epsilon^2 p/s})$, *there is:*

$$(0.63 - \epsilon)\|\boldsymbol{x}\|_{1,2,s} \leq \|\boldsymbol{U}^\top \boldsymbol{x}\|_1 \leq (1.63 + \epsilon)\|\boldsymbol{x}\|_{1,2,s}. \tag{14.33}$$

For $s = 1$ (ℓ_2 norm) and $s = d$ (ℓ_1 norm), Theorem 14.4 is reduced to Definition 14.1 (random projection) and Theorem 14.3 with ℓ_1 norm, respectively. For s-sparse $\boldsymbol{x}$, Theorem 14.4 is reduced to embedding into a subspace equipped with ℓ_1 norm. According to [46], in the $s = d$ case (ℓ_1 norm) in Theorem 14.4, $p \geq cd\ln(d)$, where c is a constant. This contradicts $p \leq d$, which coincides with Corollary 14.3. However, for the s-sparse data in Theorem 14.4, $p = \mathcal{O}(s\ln(d)) < \mathcal{O}(d)$ is needed [46]. This demonstrates that it is possible to embed data into a subspace equipped with ℓ_1 norm, using linear projection, if the data are sparse enough.

Baraniuk et al. [7] demonstrate that the sparse random projection and JL lemma are related to compressed sensing [22]. Some additional works on sparse random projection include sparse Cauchy random projections [52, 64], Bernoulli random projection [2], and very sparse random projection [53]. The sparse random projection can also be related to random projection on hypercubes [73, Chapter 7.2]. This will be explained in Sect. 14.5.2.1.

14.5 Applications of Linear Random Projection

The following section will explore two of the most well-known random projection applications—low-rank matrix approximation and approximate nearest neighbour search.

14.5.1 Low-Rank Matrix Approximation Using Random Projection

An application of random projection is a low-rank approximation of a matrix. According to the Eckart-Young-Mirsky theorem [23], Singular Value Decomposition (SVD) can be used for the low-rank approximation of a matrix. Consider the matrix $\boldsymbol{X} \in \mathbb{R}^{d \times n}$ where $d \geq n$. The time complexity for SVD of this matrix $\mathcal{O}(dn^2)$ grows if the matrix is large. The time complexity can be improved to $\mathcal{O}(dn\ln(n))$ using random projections [58]. In this technique, the low-rank approximation is not completely optimal but sufficient. This method requires two steps. First, it finds a smaller matrix $\boldsymbol{Y}$ by random projection:

$$\mathbb{R}^{p\times n} \ni \boldsymbol{Y} := \frac{1}{\sqrt{p}}\boldsymbol{U}^\top \boldsymbol{X}, \tag{14.34}$$

where $\boldsymbol{U} \in \mathbb{R}^{d\times p}$ is the random projection matrix, whose elements are independently drawn from a standard normal distribution. Note that p should satisfy Eq. (14.4); for example, $p \geq c\ln(n)/\epsilon^2$, where c is a positive constant. Since $p \ll d$, the matrix $\boldsymbol{Y}$ is much smaller than the matrix $\boldsymbol{X}$. The SVD of $\boldsymbol{Y}$ can be calculated as:

$$\boldsymbol{Y} = \boldsymbol{A}\boldsymbol{\Lambda}\boldsymbol{B}^\top = \sum_{i=1}^{p} \lambda_i \boldsymbol{a}_i \boldsymbol{b}_i^\top, \tag{14.35}$$

where $\boldsymbol{A} = [\boldsymbol{a}_1, \dots, \boldsymbol{a}_p] \in \mathbb{R}^{p\times p}$ and $\boldsymbol{B} = [\boldsymbol{b}_1, \dots, \boldsymbol{b}_p] \in \mathbb{R}^{n\times p}$ are the matrices of singular vectors and $\boldsymbol{\Lambda} = \textbf{diag}([\lambda_1, \dots, \lambda_p]^\top) \in \mathbb{R}^{p\times p}$ contains the singular values. Note that the SVD of $\boldsymbol{Y}$ is much faster than the SVD of $\boldsymbol{X}$ because of the smaller size of the matrix. The matrix $\boldsymbol{X}$ can be approximated as its projection by:

$$\mathbb{R}^{d\times n} \ni \widetilde{\boldsymbol{X}}_p \approx \boldsymbol{X}\Big(\sum_{i=1}^{p} \boldsymbol{b}_i \boldsymbol{b}_i^\top\Big). \tag{14.36}$$

Note that the right singular values $\boldsymbol{b}_i$'s, which are used here, are equal to the eigenvectors of $\boldsymbol{Y}^\top\boldsymbol{Y}$ (see Chap. 2). The rank of Eq. (14.36) is p because $p \ll d, n$; therefore, it is a low-rank approximation of $\boldsymbol{X}$. The following Lemma demonstrates why this is a valid low-rank approximation.

Lemma 14.7 ([58, Lemma 3]) *Let the SVD of $\boldsymbol{X}$ be $\boldsymbol{X} = \boldsymbol{C}\boldsymbol{\Sigma}\boldsymbol{E}^\top = \sum_{i=1}^{d} \sigma_i \boldsymbol{c}_i \boldsymbol{e}_i^\top$, where $\boldsymbol{C} = [\boldsymbol{c}_1, \dots, \boldsymbol{c}_d] \in \mathbb{R}^{d\times d}$, $\boldsymbol{E} = [\boldsymbol{e}_1, \dots, \boldsymbol{e}_d] \in \mathbb{R}^{n\times d}$, and $\boldsymbol{\Sigma} = \boldsymbol{diag}([\sigma_1, \dots, \sigma_d]^\top) \in \mathbb{R}^{d\times d}$ are the left singular vectors, right singular vectors, and singular values, respectively. Additionally, let the SVD of $\boldsymbol{X}$, with top p singular values, be $\boldsymbol{X}_p := \sum_{i=1}^{p} \sigma_i \boldsymbol{c}_i \boldsymbol{e}_i^\top$. If $p \geq c\ln(n)/\epsilon^2$, the singular values of $\boldsymbol{Y}$ are not much smaller than the singular values of $\boldsymbol{X}$, i.e.:*

$$\sum_{i=1}^{p} \lambda_i^2 \geq (1-\epsilon)\sum_{i=1}^{p} \sigma_i^2 = (1-\epsilon)\|\boldsymbol{X}_p\|_F^2, \tag{14.37}$$

where $\|.\|_F$ is the Frobenius norm.

Proof Refer to [58, Appendix]. □

Theorem 14.5 ([58, Theorem 5]) *The low-rank approximation $\widetilde{\boldsymbol{X}}$ can approximate $\boldsymbol{X}$ well enough:*

$$\|\boldsymbol{X} - \widetilde{\boldsymbol{X}}_p\|_F^2 \leq \|\boldsymbol{X} - \boldsymbol{X}_p\|_F^2 + 2\epsilon \|\boldsymbol{X}_p\|_F^2. \tag{14.38}$$

Proof Refer to [27]. □

Lemma 14.8 ([58]) *The time complexity of the low-rank approximation (with rank p) of the matrix $\boldsymbol{X}$ using random projection is $\mathcal{O}(dn \ln(n))$, which is better than the complexity of SVD on $\boldsymbol{X}$, which is $\mathcal{O}(dn^2)$.*

Proof Refer to [27]. □

Additional work, such as [25, 55], has been done to improve the time complexity of using random projection for low-rank applications. For more information on using random projection for low-rank approximation, refer to [73, Chapter 8].

14.5.2 Approximate Nearest Neighbour Search

14.5.2.1 Random Projection onto a Hypercube

Let $\mathbb{Z}_w^d$ denote the set of d-dimensional integer vectors with w possible values; for example, $\mathbb{Z}_2^d := \{0, 1\}^d$ is the set of binary values. Also let $\|\boldsymbol{x}_i - \boldsymbol{x}_j\|_H$ denote the Hamming distance[1] between two binary vectors $\boldsymbol{x}_i$ and $\boldsymbol{x}_j$. It is possible to have random projections of a binary dataset onto a hypercube $\mathbb{Z}_w^d$, where ℓ_1 norm or Hamming distances are used [48, 49].

Theorem 14.6 (Random Projection onto a Hypercube [48, 49]) *Let mod denote the modulo operation. Consider a binary vector $\boldsymbol{x} \in \mathbb{Z}_2^d$, which is projected as $f(\boldsymbol{x}) = \boldsymbol{U}^\top \boldsymbol{x} \bmod 2$ with a random binary projection matrix $\mathbb{R} \in \{0, 1\}^{d \times p}$, where $p = O(\ln(n)/\epsilon^2)$ (usually $p \ll d$). The elements of $\boldsymbol{U}$ are independent and identically distributed with the Bernoulli distribution, having probability $\xi = (\epsilon^2/\ell)$ to be one and probability $(1 - \xi)$ to be zero. Therefore:*

$$\text{if } \|\boldsymbol{x}_i - \boldsymbol{x}_j\|_H < \frac{\ell}{4} \implies \|f(\boldsymbol{x}_i) - f(\boldsymbol{x}_j)\|_H < (1+\epsilon) p \xi \frac{\ell}{4}, \tag{14.39}$$

$$\text{if } \frac{\ell}{4} \leq \|\boldsymbol{x}_i - \boldsymbol{x}_j\|_H \leq \frac{\ell}{2\epsilon} \implies (1-\epsilon) p \xi \leq \frac{\|f(\boldsymbol{x}_i) - f(\boldsymbol{x}_j)\|_H}{\|\boldsymbol{x}_i - \boldsymbol{x}_j\|_H} < (1+\epsilon) p \xi, \tag{14.40}$$

$$\text{if } \|\boldsymbol{x}_i - \boldsymbol{x}_j\|_H > \frac{\ell}{2\epsilon} \implies \|f(\boldsymbol{x}_i) - f(\boldsymbol{x}_j)\|_H > (1-\epsilon) p \xi \frac{\ell}{2\epsilon}, \tag{14.41}$$

[1] The Hamming distance between two discrete-value vectors is the number of elements where the two vectors are different.

for all $\boldsymbol{x}_i, \boldsymbol{x}_j \in \mathbb{Z}_2^d$, *with the probability at least* $(1 - e^{-c\epsilon^4 p})$, *where c is a positive constant.*

Proof The proof is available in [54, 73]. □

14.5.2.2 Approximate Nearest Neighbour Search by Random Projection

Consider a dataset $\mathcal{X} := \{\boldsymbol{x}_i \in \mathbb{R}^d\}_{i=1}^n$. The nearest neighbour search problem refers to finding the closest point of the dataset $\boldsymbol{x}^* \in \mathcal{X}$ to a query point $\boldsymbol{q} \in \mathbb{R}^d$. One solution to this problem is to calculate the distances of the query point from all n points of the dataset and return the point with the smallest distance. However, its time and space complexities are both $\mathcal{O}(nd)$, which is not good. There is an algorithm for nearest neighbour search [57] with time complexity $\mathcal{O}(\text{poly}(d, \ln(n)))$, where ploy() is a polynomial combination of its inputs. However, the space complexity of this algorithm is $\mathcal{O}(n^d)$.

To have better time and space complexities, an approximate nearest neighbour search is used. The approximate nearest neighbour search returns a point $\boldsymbol{x}^* \in \mathcal{X}$, which satisfies:

$$\|\boldsymbol{q} - \boldsymbol{x}^*\|_2 \leq (1 + \epsilon)\,(\min_{x \in \mathcal{X}} \|\boldsymbol{q} - \boldsymbol{x}\|_2), \tag{14.42}$$

where $\epsilon > 0$ is the error tolerance. This problem is named the ϵ-approximate nearest neighbour search problem [5]. This definition can be relaxed if the acceptable distance r is taken from the user:

$$\|\boldsymbol{q} - \boldsymbol{x}^*\|_2 \leq (1 + \epsilon)\, r. \tag{14.43}$$

If no such point is found in $\mathcal{X}$, null is returned. This relaxation is valid because the smallest r can be found using a binary search, whose number of iterations is constant with respect to d and n.

The introduced random projection onto a hypercube can be used for an ϵ-approximate nearest neighbour search [48, 49]. This algorithm is explained in the following. Assume the dataset is binary, i.e., $\mathcal{X} := \{\boldsymbol{x}_i \in \mathbb{Z}_2^d\}_{i=1}^n$. If not, the values of vector elements are quantized to binary strings and the values of a vector are reshaped to become a binary vector. Let the dimensionality of binary (or quantized and reshaped) vectors be d. Recall the binary search required for finding the smallest distance r in Eq. (14.43). In the binary search for finding the smallest distance r, several distances are tried. For every distance, k independent random projections are performed onto k hypercubes and then one of these random projections is randomly selected. In every random projection, the points $\mathcal{X}$ are projected onto a random p-dimensional hypercube to have $\{f(\boldsymbol{x}_i) \in \mathbb{Z}_2^p\}_{i=1}^n$. In the low-dimensional projected subspace, the comparison of points is very fast because (I) the subspace dimensionality p is much less than the original dimensionality d, and (II) the calculation of the Hamming distance is faster and easier than the Euclidean distance. Therefore, random projections onto hypercubes are very useful for approximate nearest neighbour search.

Lemma 14.9 *The above algorithm for approximate nearest neighbour search is correct with probability* $(1-\delta)$, *where* $0 < \delta \ll 1$. *The time complexity of the above algorithm is* $\mathcal{O}(\frac{d}{\epsilon^4}\ln(\frac{n}{\delta})\ln(d))$ *and its space complexity is* $\mathcal{O}(d^2(c_1 n \ln(d))^{c_2/\epsilon^4})$ *where* c_1 *and* c_2 *are constants.*

Proof The proof is available in [48, 49, 54, 73]. □

Random projection for approximate nearest neighbour search is also related to hashing; for example, see locality sensitive hashing [70]. There are other works on random projection for approximate nearest neighbour search [3, 41]. For more information on using random projection for approximate nearest neighbour search, refer to [73, Chapter 14:7].

14.6 Random Fourier Features and Random Kitchen Sinks for Nonlinear Random Projection

So far, this chapter has explained linear random projection in which a linear projection is used. It is possible to have nonlinear random projection, which is a much harder task to analyze theoretically. A nonlinear random projection can be modelled as a linear random projection followed by a nonlinear function. Two fundamental works on nonlinear random projection include RFF [61] and RKS [63].

14.6.1 Random Fourier Features for Learning with Approximate Kernels

When the pattern of data is nonlinear, either a nonlinear algorithm should be used or the nonlinear data should be transformed using kernels to be able to use the linear methods for nonlinear patterns. The computation of kernels is a time-consuming task because the points are pulled to the potentially high dimensional space, and then the inner products of the pulled points to the Reproducing Kernel Hilbert Space (RKHS) are calculated (see Chap. 3). Random Fourier Features (RFF) are used for accelerating kernel methods [61]. For this, RFF transforms data to a low-dimensional space in contrast to the kernels, which transform data to a potentially high-dimensional space. RFF approximates the kernel, which is the inner product of pulled data to RKHS, by the inner product of the low-dimensional feature maps of data. This feature map is a random feature map and is $z : \mathbb{R}^d \rightarrow \mathbb{R}^{2p}$ where $p \ll d$; therefore, its computation of inner products is much faster than computing the inner product of the data in RKHS. It satisfies the following approximation:

$$k(\boldsymbol{x}, \boldsymbol{y}) = \boldsymbol{\phi}(\boldsymbol{x})^\top \boldsymbol{\phi}(\boldsymbol{y}) \approx z(\boldsymbol{x})^\top z(\boldsymbol{y}), \tag{14.44}$$

where $k(.,.)$ denotes the kernel function and $\boldsymbol{\phi}(.)$ is the pulling function to RKHS. Note that z is a nonlinear mapping, which can be seen as a nonlinear random projection. It is a linear random projection $f : \mathbb{R}^d \to \mathbb{R}^p$ followed by nonlinear sine and cosine functions. The formulation of z and f functions will be discussed later.

RFF works with positive definite kernels, which are shift-invariant (also called stationary kernels), i.e., $k(\boldsymbol{x}, \boldsymbol{y}) = k(\boldsymbol{x} - \boldsymbol{y})$. Consider the inverse Fourier transform of the kernel function k:

$$k(\boldsymbol{x} - \boldsymbol{y}) = \int_{\mathbb{R}^d} \widehat{k}(\boldsymbol{u}) e^{j\boldsymbol{u}^\top(\boldsymbol{x}-\boldsymbol{y})} d\boldsymbol{u} \overset{(a)}{=} \mathbb{E}_{\boldsymbol{u}}[\zeta_{\boldsymbol{u}}(\boldsymbol{x})\zeta_{\boldsymbol{u}}(\boldsymbol{y})^*], \tag{14.45}$$

where $\boldsymbol{u} \in \mathbb{R}^d$ is the frequency, j is the imaginary unit, the superscript $*$ denotes conjugate transpose, $\mathbb{E}[.]$ is the expectation operator, (a) is because $\mathbb{R} \ni \zeta_{\boldsymbol{u}}(\boldsymbol{x}) := e^{j\boldsymbol{u}^\top \boldsymbol{x}}$ is defined, and $\widehat{k}(\boldsymbol{u})$ is the Fourier transform of kernel function:

$$\widehat{k}(\boldsymbol{u}) = \frac{1}{2\pi} \int_{\mathbb{R}^d} e^{-j\boldsymbol{u}^\top \boldsymbol{x}}\, k(\boldsymbol{x})\, d\boldsymbol{x}. \tag{14.46}$$

In Eq. (14.45), the kernel function k and the transformed kernel $\widehat{k}$ are real valued, so the sine part of $e^{j\boldsymbol{u}^\top(\boldsymbol{x}-\boldsymbol{y})}$ can be ignored (see Euler's equation) and replaced with $\cos(\boldsymbol{u}^\top(\boldsymbol{x} - \boldsymbol{y}))$; therefore, $\zeta_{\boldsymbol{u}}(\boldsymbol{x}) = \cos(\boldsymbol{u}^\top \boldsymbol{x})$. Let:

$$\mathbb{R}^2 \ni z_u(\boldsymbol{x}) := [\cos(\boldsymbol{u}^\top \boldsymbol{x}), \sin(\boldsymbol{u}^\top \boldsymbol{x})]^\top. \tag{14.47}$$

In Eq. (14.45), $\widehat{k}(\boldsymbol{u})$ can be seen as a d-dimensional probability density function. p independent and identically distributed random projection vectors, $\{\boldsymbol{u}_t \in \mathbb{R}^d\}_{t=1}^p$, are drawn from $\widehat{k}(\boldsymbol{u})$. A normalized vector version of Eq. (14.47) is defined as:

$$\mathbb{R}^{2p} \ni z(\boldsymbol{x}) := \frac{1}{\sqrt{p}}[\cos(\boldsymbol{u}_1^\top \boldsymbol{x}), \dots, \cos(\boldsymbol{u}_p^\top \boldsymbol{x}), \sin(\boldsymbol{u}_1^\top \boldsymbol{x}), \dots, \sin(\boldsymbol{u}_p^\top \boldsymbol{x})]^\top. \tag{14.48}$$

If the linear random projection is taken as:

$$\mathbb{R}^p \ni f(\boldsymbol{x}) := \boldsymbol{U}^\top \boldsymbol{x} = [\boldsymbol{u}_1^\top \boldsymbol{x}, \dots, \boldsymbol{u}_p^\top \boldsymbol{x}]^\top, \tag{14.49}$$

with the projection matrix $\boldsymbol{U} := [\boldsymbol{u}_1, \dots, \boldsymbol{u}_p] \in \mathbb{R}^{d \times p}$, then the function $z(.)$ is applying sine and cosine functions to a linear random projection $f(.)$ of the data $\boldsymbol{x}$.

According to $\cos(a - b) = \cos(a)\cos(b) + \sin(a)\sin(b)$:

$$z(\boldsymbol{x})^\top z(\boldsymbol{y}) = \cos(\boldsymbol{u}^\top(\boldsymbol{x} - \boldsymbol{y})). \tag{14.50}$$

Considering $\zeta_{\boldsymbol{u}}(\boldsymbol{x}) = \cos(\boldsymbol{u}^\top \boldsymbol{x})$:

$$k(\boldsymbol{x}, \boldsymbol{y}) = k(\boldsymbol{x} - \boldsymbol{y}) \overset{(14.45)}{=} \mathbb{E}_{\boldsymbol{u}}[\zeta_{\boldsymbol{u}}(\boldsymbol{x})\zeta_{\boldsymbol{u}}(\boldsymbol{y})^*] = \mathbb{E}_{\boldsymbol{u}}[z(\boldsymbol{x})^\top z(\boldsymbol{y})]. \tag{14.51}$$

Thus, Eq. (14.44) holds in expectation. The following demonstrates that the variance of Eq. (14.51) is small enough to have Eq. (14.44) as a valid approximation. According to Eq. (14.50), there is $-1 \leq z(\boldsymbol{x})^\top z(\boldsymbol{y}) \leq 1$; therefore, Hoeffding's inequality can be used [31]:

$$\mathbb{P}\big(|z(\boldsymbol{x})^\top z(\boldsymbol{y}) - k(\boldsymbol{x}, \boldsymbol{y})| \geq \epsilon\big) \leq 2\exp(-\frac{p\,\epsilon^2}{2}), \tag{14.52}$$

which is a bound on the probability of error in Eq. (14.44) with error tolerance $\epsilon \geq 0$. Therefore, the approximation in Eq. (14.44) holds and we can approximate kernel computation with the inner product in a random lower dimensional subspace. This improves the speed of kernel machine learning methods because p is much less than both the dimensionality of the input space, i.e., d, and the dimensionality of RKHS. Although Eq. (14.52) proves the correctness of the approximation in RFF, the following theorem demonstrates more strongly that the approximation in RFF is valid.

Theorem 14.7 ([61, Claim 1]) *Let $\mathcal{M}$ be a compact subset of $\mathbb{R}^d$ with diameter diam($\mathcal{M}$). We have:*

$$\mathbb{P}\Big(\sup_{\boldsymbol{x},\boldsymbol{y}\in\mathcal{M}} |z(\boldsymbol{x})^\top z(\boldsymbol{y}) - k(\boldsymbol{x}, \boldsymbol{y})| \geq \epsilon\Big) \leq 2^8\big(\frac{\sigma\, diam(\mathcal{M})}{\epsilon}\big)^2 \exp\big(-\frac{p\,\epsilon^2}{4(d+2)}\big), \tag{14.53}$$

where $\sigma^2 := \mathbb{E}_{\boldsymbol{u}}[\boldsymbol{u}^\top \boldsymbol{u}]$ with probability density function $\widehat{k}(\boldsymbol{u})$ used in expectation. The required lower bound on the dimensionality of the random subspace is:

$$p \geq \Omega\Big(\frac{d}{\epsilon^2} \ln\big(\frac{\sigma\, \mathrm{diam}(\mathcal{M})}{\epsilon}\big)\Big). \tag{14.54}$$

Proof The proof is available in [61, Appendix A]. □

In summary, the RFF algorithm is as follows. Given a positive definite stationary kernel function k, it is possible to calculate its Fourier transform using Eq. (14.46). Then, this Fourier transform is treated as a probability density function and p independent and identically distributed random projection vectors $\{\boldsymbol{u}_t \in \mathbb{R}^d\}_{t=1}^p$ are drawn from that. For every point $\boldsymbol{x}$, the mapping $z(\boldsymbol{x})$ is calculated using Eq. (14.48). Then, for every two points $\boldsymbol{x}$ and $\boldsymbol{y}$, their kernel is approximated by Eq. (14.44). The approximated kernel can be used in kernel machine learning algorithms. As explained, RFF can be interpreted as a nonlinear random projection because it applies nonlinear sine and cosine functions to linear random projections.[2]

14.6.2 Random Kitchen Sinks for Nonlinear Random Projection

Random Kitchen Sinks (RKS) [63] was proposed after the development of RFF [61]. RKS is a nonlinear random projection, where a nonlinear function is applied to a linear random projection. RKS models this nonlinear random projection as a random layer of neural network connecting d neurons to p neurons. A random network layer is actually a linear random projection followed by a nonlinear activation function:

$$g(\boldsymbol{x}) := \phi\Big(\sum_{j=1}^{d}\sum_{t=1}^{p} u_{jt}\, x_j\Big) \overset{(14.1)}{=} \phi(\boldsymbol{U}^\top \boldsymbol{x}) = \phi\big(f(\boldsymbol{x})\big), \tag{14.55}$$

where $g(.)$ is the function representing the whole layer, $\phi(.)$ is the possibly nonlinear activation function, $f(.)$ is the linear random projection, $\boldsymbol{U} \in \mathbb{R}^{d\times p}$ is the matrix of random weights of the network layer, u_{jt} is the (j, t)-th element of $\boldsymbol{U}$, and x_j is the j-th element of $\boldsymbol{x} \in \mathbb{R}^d$. Therefore, a random network layer can be seen as a nonlinear random projection. According to the universal approximation theorem [34, 37], Eq. (14.55) can fit any possible decision function, mapping continuous inputs to a finite set of classes to any desired level of accuracy. According to the representer theorem [4] (see Chap. 3), the function $g(.)$ in RKHS can be stated as:

$$g(\boldsymbol{x}) = \sum_{i=1}^{\infty} \alpha_i\, \phi(\boldsymbol{x}; w_i), \tag{14.56}$$

where $\{w_i\} \in \Omega$ are the parameters (Ω is the space of parameters), $\phi(.)$ is the pulling function to RKHS, $\{\alpha_i\}$ are the weights. Comparing Eqs. (14.55) and (14.56) demonstrates that there is a connection between RKS and kernels [62], if we consider the pulling function $\phi(.)$ as the activation function, the parameters $\{w_i\}$ as the parameters of the activation function, and the weights $\{\alpha_i\}$ as the weights of the network layer.

Consider a classification task, where l_i is the label of $\boldsymbol{x}_i$. The empirical risk between the labels and the output of the network layer is:

[2] There exists another method for kernel approximation, named random binning features (see [61] for more information). Similar to RFF, random binning features can also be interpreted as a nonlinear random projection.

$$R_e(g) := \frac{1}{n}\sum_{i=1}^{n} \ell\big(g(\boldsymbol{x}_i), l_i\big), \tag{14.57}$$

where ℓ is a loss function. The true risk is:

$$R_t(g) := \mathbb{E}_{\{\boldsymbol{x}_i, l_i\}}\Big[\ell\big(g(\boldsymbol{x}_i), l_i\big)\Big], \tag{14.58}$$

where $\mathbb{E}[.]$ is the expectation operator. The aim is to minimize the empirical risk; according to Eq. (14.56):

$$\underset{\{w_i, \alpha_i\}_{i=1}^p}{\text{minimize}} \quad R_e\Big(\sum_{i=1}^{p} \alpha_i\, \phi(\boldsymbol{x};\, w_i)\Big), \tag{14.59}$$

which is a joint minimization over $\{w_i\}_{i=1}^p$ and $\{\alpha_i\}_{i=1}^p$. In RKS, the parameters of the activation function, $\{w_i\}_{i=1}^p$, are sampled randomly from a distribution $\mathbb{P}(w)$ in space Ω. Then, the empirical risk is minimized over only the network weights $\{\alpha_i\}_{i=1}^p$. In other words, the optimization variables $\{w_i\}_{i=1}^p$ are eliminated by their random selection and minimized over only $\{\alpha_i\}_{i=1}^p$. In practice, the distribution $\mathbb{P}(w)$ can be any distribution in the space of parameters of activation functions. Let $\boldsymbol{\alpha} := [\alpha_1, \dots, \alpha_p]^\top \in \mathbb{R}^p$. If $\{w_i\}_{i=1}^p$ are randomly sampled for all $i \in \{1, \dots, n\}$, suppose $\boldsymbol{g}_i := [\phi(\boldsymbol{x}_i, \boldsymbol{w}_1), \dots, \phi(\boldsymbol{x}_i, \boldsymbol{w}_p)]^\top \in \mathbb{R}^p, \forall i$. According to Eqs. (14.56), (14.57), and (14.59), the optimization in RKS is:

$$\begin{aligned} &\underset{\boldsymbol{\alpha} \in \mathbb{R}^p}{\text{minimize}} \quad \frac{1}{n}\sum_{i=1}^{n} \ell\big(\boldsymbol{\alpha}^\top \boldsymbol{g}_i, l_i\big) \\ &\text{subject to} \quad \|\boldsymbol{\alpha}\|_\infty \leq \frac{c}{p}, \end{aligned} \tag{14.60}$$

where $\|.\|_\infty$ is the maximum norm and c is a positive constant. In practice, RKS relaxes the constraint of this optimization to a quadratic regularizer:

$$\underset{\boldsymbol{\alpha} \in \mathbb{R}^p}{\text{minimize}} \quad \frac{1}{n}\sum_{i=1}^{n} \ell\big(\boldsymbol{\alpha}^\top \boldsymbol{g}_i, l_i\big) + \lambda \|\boldsymbol{\alpha}\|_2^2, \tag{14.61}$$

where $\lambda > 0$ is the regularization parameter. Henceforth, let the solution of Eq. (14.60) or (14.61) be $\{\alpha_i\}_{i=1}^p$ and the randomly sampled parameters be $\{w_i\}_{i=1}^p$. They are put into Eq. (14.56), but with only p components of summation. This results in the solution of RKS, denoted by $\widehat{g'}$:

$$\widehat{g'}(\boldsymbol{x}) = \sum_{i=1}^{p} \alpha_i\, \phi(\boldsymbol{x};\, w_i). \tag{14.62}$$

The following demonstrates that Eq. (14.62) minimizes the empirical risk well even when the parameters $\{w_i\}_{i=1}^p$ are randomly selected. Therefore, RKS is a nonlinear random projection, that estimates the labels $\{l_i\}_{i=1}^n$ with a good approximation.

Consider the set of a continuous version of functions in Eq. (14.56) as:

$$\mathcal{G} := \Big\{g(\boldsymbol{x}) = \int_{\Omega} \alpha(w)\, \phi(\boldsymbol{x}; w)\, dw \,\Big|\, |\alpha(w)| \leq c\, \mathbb{P}(w)\Big\}. \tag{14.63}$$

Let g^* be a function in $\mathcal{G}$ and $\{w_i\}_{i=1}^p$ be sampled, independent and identically distributed, from $\mathbb{P}(w)$. The output of RKS, which is Eq. (14.62), lies in the random set:

$$\widehat{\mathcal{G}} := \Big\{\widehat{g} = \sum_{i=1}^{p} \alpha_i\, \phi(\boldsymbol{x}; w_i) \,\Big|\, |\alpha_i| \leq \frac{c}{p}, \forall i\Big\}. \tag{14.64}$$

Lemma 14.10 (Bound on the Approximation Error [63, Lemma 2]) *Suppose the loss $\ell(l, l')$ is L-Lipschitz. Let $g^* \in \mathcal{G}$ be the minimizer of the true risk over $\mathcal{G}$. For $\delta > 0$, with a probability of at least $(1 - \delta)$, there exists $\widehat{g} \in \widehat{G}$ satisfying:*

$$R_t(\widehat{g}) - R_t(g^*) \leq \frac{Lc}{\sqrt{p}}\Big(1 + \sqrt{2 \ln(\frac{1}{\delta})}\Big). \tag{14.65}$$

The term $R_t(\widehat{g}) - R_t(g^*)$ is the approximation error and the above equation is an upper bound on it.

Lemma 14.11 (Bound on the Estimation Error [63, Lemma 3]) *Suppose the loss can be stated as $\ell(l, l') = \ell(ll')$ and it is L-Lipschitz. For $\delta > 0$ and for all $\widehat{g} \in \widehat{\mathcal{G}}$, the following holds with the probability at least $(1 - \delta)$:*

$$|R_t(\widehat{g}) - R_e(\widehat{g})| \leq \frac{1}{\sqrt{n}}\Big(4Lc + 2|\ell(0)| + Lc\sqrt{\frac{1}{2} \ln(\frac{1}{\delta})}\Big). \tag{14.66}$$

The term $|R_t(\widehat{g}) - R_e(\widehat{g})|$ is the estimation error and the above equation is an upper bound on it.

Theorem 14.8 (Bound on Error of RKS [63, Theorem 1]) *Let the activation function $\phi(.)$ be bounded, namely, $\sup_{\boldsymbol{x},w} |\phi(\boldsymbol{x}; w)| \leq 1$. Suppose the loss can be stated as $\ell(l, l') = \ell(ll')$ and it is L-Lipschitz. The output of RKS, which is Eq. (14.62), is $\widehat{g}' \in \widehat{\mathcal{G}}$ satisfying:*

$$R_t(\widehat{g}') - \min_{g \in \mathcal{G}} R_t(g) \leq \mathcal{O}\Big(\Big(\frac{1}{\sqrt{n}} + \frac{1}{\sqrt{p}}\Big)Lc\sqrt{\ln(\frac{1}{\delta})}\Big), \tag{14.67}$$

with a probability of at least $(1 - 2\delta)$.

Proof Refer to [27]. □

14.7 Other Methods for Nonlinear Random Projection

In addition to RFF and RKS, there are other, albeit similar, approaches for nonlinear random projection, such as ELM, random weights in neural networks, and ensembles of random projections. The following section introduces these approaches.

14.7.1 *Extreme Learning Machine*

Extreme Learning Machine (ELM) was initially proposed for regression as a feed-forward neural network with one hidden layer [39]. It was then improved for a multilayer perceptron network [72] and multiclass classification [40]. ELM is a feed-forward neural network whose all layers except the last layer are random. Let the sample size of the training data be n, their dimensionality be d-dimensional, the one-to-last layer have d' neurons, and the last layer has p neurons, where p is the dimensionality of labels. Note that in the classification task, labels are one-hot encoded, so p is the number of classes. Every layer except the last layer has a possibly nonlinear activation function and behaves like an RKS; although, it is learned by backpropagation (a training algorithm for neural networks) and not Eq. (14.61). Let $\boldsymbol{\beta} \in \mathbb{R}^{d' \times p}$ be the matrix of weights for the last layer. If the n outputs of the one-to-last layer are stacked in the matrix $\boldsymbol{H} \in \mathbb{R}^{d' \times n}$ and their target desired labels are $\boldsymbol{T} \in \mathbb{R}^{p \times n}$, the aim is to have $\boldsymbol{\beta}^\top \boldsymbol{H} = \boldsymbol{T}$. After randomly sampling the weights of all layers except $\boldsymbol{\beta}$, ELM learns the weights $\boldsymbol{\beta}$ by solving a least squares problem:

$$\underset{\boldsymbol{\beta} \in \mathbb{R}^{d' \times p}}{\text{minimize}} \quad \|\boldsymbol{\beta}^\top \boldsymbol{H} - \boldsymbol{T}\|_F^2 + \lambda \|\boldsymbol{\beta}\|_F^2, \tag{14.68}$$

where $\|.\|_F$ is the Frobenius norm and $\lambda > 0$ is the regularization parameter. In other words, the last layer of ELM behaves like a (Ridge) linear regression [32].

Huang et al. [37] demonstrated that the random weights in a neural network with nonlinear activation functions are universal approximators [34]. Therefore, ELM works well enough for any classification and regression task. This demonstrates the connection between ELM and RKS because both work on random weights on a network but with slightly different approaches. This connection can be seen more if ELM is interpreted using kernels [36]. There are several surveys on ELM, such as [35, 38].

14.7.2 *Randomly Weighted Neural Networks*

A feed-forward or convolutional neural network whose layers are random also works well up to some level, for classification or prediction [42]. In other words, a stack of several RKS models works well because a random network can be seen as a stack of several nonlinear random projections. As each of these nonlinear random layers has an upper bound on their probability of error (see Theorem 14.8), the total probability of error in the network is also bounded. Note that it does not matter whether the network is feed-forward or convolutional because in both architectures, the output of the activation functions are projected by the weights of layers. Saxe et al. [65] demonstrated that convolutional pooling architectures are frequency selective and translation invariant even if their weights are random; therefore, randomly weighed convolutional networks work well enough. This also explains why random initialization of neural networks before backpropagation is a good and acceptable initialization. The randomly weighed neural network has been used for object recognition [18, 42] and facial recognition [19]. Some research has also made almost everything in the network, including weights and hyperparameters such as architecture, learning rate, and number of neurons, random[3] [59].

14.7.2.1 Distance Preservation by Deterministic Layers

Consider a layer of neural network with p neurons to d neurons. Let $\boldsymbol{U} \in \mathbb{R}^{d \times p}$ be the weight matrix of layer and $g(.)$ be the Rectified Linear Unit (ReLU) activation function [29]. The layer can be modelled as a linear projection followed by the ReLU activation function, i.e., $g(\boldsymbol{U}^\top \boldsymbol{x}_j)$. The following lemma demonstrates that for deterministic weights $\boldsymbol{U}$, the distances are preserved.

Lemma 14.12 (Preservation of Euclidean Distance by a Layer [13, 14]) *Consider a network layer $\boldsymbol{U} \in \mathbb{R}^{d \times p}$. If the outputs of the previous layer for two points are $\boldsymbol{x}_i, \boldsymbol{x}_j \in \mathbb{R}^d$, then:*

$$L_1 \|\boldsymbol{x}_i - \boldsymbol{x}_j\|_2 \leq \|g(\boldsymbol{U}^\top \boldsymbol{x}_i) - g(\boldsymbol{U}^\top \boldsymbol{x}_j)\|_2 \leq L_2 \|\boldsymbol{x}_i - \boldsymbol{x}_j\|_2, \tag{14.69}$$

where $0 < L_1 \leq L_2$ are the Lipschitz constants (see Chap. 4).

Proof The proof is available in [14]. □

14.7.2.2 Distance Preservation by Random Layers

This chapter previously demonstrated that for deterministic weights, the distances are preserved. This section demonstrates that random weights also preserve the distances, as well as the angles between points. Consider a network layer $\boldsymbol{U} \in \mathbb{R}^{d \times p}$, whose elements are independent and identically distributed random values sampled from the Gaussian distribution. The activation function is ReLU and denoted by

[3] A survey on randomness in neural networks can be found in [66].

$g(.)$. Suppose the outputs of the previous layer for two points are $\boldsymbol{x}_i, \boldsymbol{x}_j \in \mathbb{R}^d$. Assume the input data to the layer lie on a manifold $\mathcal{M}$ with the Gaussian mean width:

$$w_{\mathcal{M}} := \mathbb{E}[\sup_{\boldsymbol{x}_i, \boldsymbol{x}_j \in \mathcal{M}} \boldsymbol{q}^\top (\boldsymbol{x}_i - \boldsymbol{x}_j)],$$

where $\boldsymbol{q} \in \mathbb{R}^d$ is a random vector whose elements are independent and identically distributed, sampled from the Gaussian distribution. Assume that the manifold is normalized, meaning the input data to the layer lie on a hypersphere. Let $\mathbb{B}_r^d \subset \mathbb{R}^d$ be the ball with radius r in d-dimensional Euclidean space. Therefore, $\mathcal{M} \subset \mathbb{B}_r^d$.

Theorem 14.9 (Preservation of Euclidean Distance by a Random Layer [28, Theorem 3]) *Consider a random network layer $\boldsymbol{U} \in \mathbb{R}^{d \times p}$, where the outputs of the previous layer for two points are $\boldsymbol{x}_i, \boldsymbol{x}_j \in \mathbb{R}^d$. Suppose $\mathcal{M} \subset \mathbb{B}_r^d$. Assume the angle between $\boldsymbol{x}_i$ and $\boldsymbol{x}_j$ is denoted by:*

$$\theta_{i,j} := \cos^{-1}\Big(\frac{\boldsymbol{x}_i^\top \boldsymbol{x}_j}{\|\boldsymbol{x}_i\|_2 \|\boldsymbol{x}_j\|_2}\Big),$$

and satisfies $0 \leq \theta_{i,j} \leq \pi$. If $p \geq c\,\delta^{-4} w_{\mathcal{M}}$ with c as a constant, with high probability, then:

$$\Big| \|g(\boldsymbol{U}^\top \boldsymbol{x}_i) - g(\boldsymbol{U}^\top \boldsymbol{x}_j)\|_2^2 - \Big(\frac{1}{2}\|\boldsymbol{x}_i - \boldsymbol{x}_j\|_2^2 + \|\boldsymbol{x}_i\|_2 \|\boldsymbol{x}_j\|_2 \psi(\boldsymbol{x}_i, \boldsymbol{x}_j)\Big) \Big| \leq \delta, \tag{14.70}$$

where $\psi(\boldsymbol{x}_i, \boldsymbol{x}_j) \in [0, 1]$ is defined as:

$$\psi(\boldsymbol{x}_i, \boldsymbol{x}_j) := \frac{1}{\pi}\big(\sin(\theta_{i,j}) - \theta_{i,j}\cos(\theta_{i,j})\big). \tag{14.71}$$

Proof The proof is available in [28, Appendix A]. □

Theorem 14.10 (Preservation of Angles by a Random Layer [28, Theorem 4]) *Suppose the same assumptions as in Theorem 14.9 hold and let $\mathcal{M} \subset \mathbb{B}_1^d \backslash \mathbb{B}_\beta^d$ where $\delta \ll \beta^2 < 1$. Assume the angle between $g(\boldsymbol{U}^\top \boldsymbol{x}_i)$ and $g(\boldsymbol{U}^\top \boldsymbol{x}_j)$ is denoted by:*

$$\theta'_{i,j} := \cos^{-1}\Big(\frac{(g(\boldsymbol{U}^\top \boldsymbol{x}_i))^\top g(\boldsymbol{U}^\top \boldsymbol{x}_j)}{\|g(\boldsymbol{U}^\top \boldsymbol{x}_i)\|_2 \|g(\boldsymbol{U}^\top \boldsymbol{x}_j)\|_2}\Big),$$

and satisfies $0 \leq \theta'_{i,j} \leq \pi$. With high probability, the following holds:

$$\Big|\cos(\theta'_{i,j}) - \big(\cos(\theta_{i,j}) + \psi(\boldsymbol{x}_i, \boldsymbol{x}_j)\big)\Big| \leq \frac{15\delta}{\beta^2 - 2\delta}. \tag{14.72}$$

Proof Proof is available in [28, Appendix B]. □

Corollary 14.4 (Preservation of Distances and Angles by a Random Layer [28, Corollary 5]) *By a random layer with weights* $\boldsymbol{U}$ *and ReLU activation function* $g(.)$*, for every two points* $\boldsymbol{x}_i$ *and* $\boldsymbol{x}_j$ *as inputs to the layer, the following holds with high probability:*

$$\frac{1}{2}\|\boldsymbol{x}_i - \boldsymbol{x}_j\|_2^2 - \delta \leq \|g(\boldsymbol{U}^\top \boldsymbol{x}_i) - g(\boldsymbol{U}^\top \boldsymbol{x}_j)\|_2^2 \leq \|\boldsymbol{x}_i - \boldsymbol{x}_j\|_2^2 + \delta. \tag{14.73}$$

A similar expression can be stated for preserving the angle by the layer.

Proof The relation of $\psi(\boldsymbol{x}_i, \boldsymbol{x}_j)$ and $\cos(\boldsymbol{x}_i, \boldsymbol{x}_j)$ is depicted in Fig. 14.1 for $\theta_{i,j} \in [0, \pi]$. As this figure demonstrates:

$$\theta_{i,j} = 0, \ \cos(\boldsymbol{x}_i, \boldsymbol{x}_j) = 1 \implies \psi(\boldsymbol{x}_i, \boldsymbol{x}_j) = 0,$$
$$\theta_{i,j} = \pi, \ \cos(\boldsymbol{x}_i, \boldsymbol{x}_j) = 0 \implies \psi(\boldsymbol{x}_i, \boldsymbol{x}_j) = 1.$$

Therefore, if two points are similar, i.e., their angle is zero, then $\psi(\boldsymbol{x}_i, \boldsymbol{x}_j) = 0$. Having $\psi(\boldsymbol{x}_i, \boldsymbol{x}_j) = 0$ in Eqs. (14.70) and (14.72) demonstrates that when the two input points $\boldsymbol{x}_i$ and $\boldsymbol{x}_j$ to a layer are similar, then the followings almost hold:

$$\|g(\boldsymbol{U}^\top \boldsymbol{x}_i) - g(\boldsymbol{U}^\top \boldsymbol{x}_j)\|_2^2 \approx \|\boldsymbol{x}_i - \boldsymbol{x}_j\|_2^2,$$

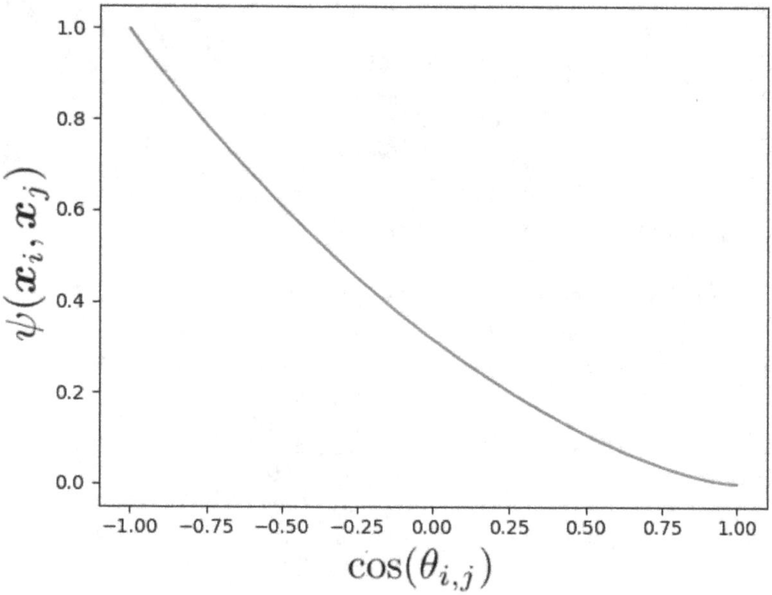

Fig. 14.1 Relation of $\psi(\boldsymbol{x}_i, \boldsymbol{x}_j)$ with $\cos(\theta_{i,j})$ for $\theta_{i,j} \in [0, \pi]$

$$\cos(\theta'_{i,j}) \approx \cos(\theta_{i,j}).$$

This proves that a network layer with random weights and ReLU activation functions preserves both the Euclidean distance and the angle of the points. □

14.7.3 Ensemble of Random Projections

Several methods exist that use an ensemble of random projections. Ensemble methods benefit from model averaging [33] and bootstrap aggregating (bagging) [11]. Bagging reduces the estimation variance, especially if the models are not correlated [26]. As the random projection models are random, they are uncorrelated; therefore, an ensemble of them can improve performance.

Giryes et al. [28] stated that a similar analysis to the analysis of the ReLU activation function can be done for max pooling [68]. Consider an ensemble of m random projection matrices $\{\boldsymbol{U}_j \in \mathbb{R}^{d\times p}\}_{j=1}^m$. It is possible to transform data as a max pooling over this ensemble [44]:

$$f(\boldsymbol{x}) := \max\{f_1(\boldsymbol{x}), \dots, f_m(\boldsymbol{x})\} = \max\{\boldsymbol{U}_1^\top \boldsymbol{x}, \dots, \boldsymbol{U}_m^\top \boldsymbol{x}\}. \quad (14.74)$$

This max pooling is performed elementwise, similar to the elementwise operation of the ReLU activation function, introduced before.

It is also possible to have an ensemble of random projections for the classification task. One of the algorithms for this is [67]. The projection matrices are sampled randomly, where their elements are independent and identically distributed. Additionally, the columns of projection matrices are normalized to have unit lengths. The ensemble of projection matrices $\{\boldsymbol{U}_j \in \mathbb{R}^{d\times p}\}_{j=1}^m$ are applied to the n training data points $\boldsymbol{X} \in \mathbb{R}^{d\times n}$ to have $\{f_1(\boldsymbol{X}), \dots, f_m(\boldsymbol{X})\} := \{\boldsymbol{U}_1^\top \boldsymbol{X}, \dots, \boldsymbol{U}_m^\top \boldsymbol{X}\}$. Let the labels of the training data be denoted by $\{l_i\}_{i=1}^n$. An arbitrary classifier model is trained using these projected data and their labels. In other words, the model is trained by $\{(\{\boldsymbol{U}_1^\top \boldsymbol{x}_i\}_{i=1}^n, \{l_i\}_{i=1}^n), \dots, (\{\boldsymbol{U}_m^\top \boldsymbol{x}_i\}_{i=1}^n, \{l_i\}_{i=1}^n)\}$. In the test phase, the test point is projected onto the column spaces of the m projection matrices. These projections are fed to the classifier and the final prediction of the label is found by majority voting. Note that there is another more sophisticated ensemble method for random projections [16], which divides the dataset into disjoint sets. In each set, an ensemble of the random projections is performed. In each set, the validation error of random projections is calculated, and the random projection with the smallest validation error is selected. After training, majority voting is performed to predict the test label.

14.8 Chapter Summary

Random projection demonstrates that if the projection/transformation matrix of the data is not learned but is randomly selected, the performance will not be worse than some bound. This chapter introduced linear and nonlinear random projections and their variants. This chapter explained why random initialization of neural networks is acceptable for deep metric learning, which is introduced in Chap. 19. Note that a survey on linear random projection was conducted by Xie et al. [75] and an informative book by Vempala [73] was written on the topic. Additional information on linear random projection can be found in [54, 74].

References

1. Dimitris Achlioptas. "Database-friendly random projections". In: *Proceedings of the twentieth ACM SIGMOD-SIGACT-SIGART symposium on Principles of database systems*. 2001, pp. 274–281.
2. Dimitris Achlioptas. "Database-friendly random projections: Johnson-Lindenstrauss with binary coins". In: *Journal of computer and System Sciences* 66.4 (2003), pp. 671–687.
3. Nir Ailon and Bernard Chazelle. "Approximate nearest neighbors and the fast Johnson-Lindenstrauss transform". In: *Proceedings of the thirty-eighth annual ACM symposium on Theory of computing*. 2006, pp. 557–563.
4. Mark A Aizerman. "Theoretical foundations of the potential function method in pattern recognition learning". In: *Automation and remote control* 25 (1964), pp. 821–837.
5. Alexandr Andoni and Piotr Indyk. "Near-optimal hashing algorithms for approximate nearest neighbor in high dimensions". In: *2006 47th annual IEEE symposium on foundations of computer science (FOCS'06)*. IEEE. 2006, pp. 459–468.
6. Rosa I Arriaga and Santosh Vempala. "Algorithmic theories of learning". In: *Foundations of Computer Science*. Vol. 5. 1999.
7. Richard Baraniuk et al. *The Johnson-Lindenstrauss lemma meets compressed sensing*. Tech. rep. Rice University, 2006.
8. Yair Bartal, Ben Recht, and Leonard J Schulman. "Dimensionality reduction: beyond the Johnson-Lindenstrauss bound". In: *Proceedings of the twenty-second annual ACM-SIAM symposium on Discrete Algorithms*. SIAM. 2011, pp. 868–887.
9. Radu Berinde et al. "Combining geometry and combinatorics: A unified approach to sparse signal recovery". In: *2008 46th Annual Allerton Conference on Communication, Control, and Computing*. IEEE. 2008, pp. 798–805.
10. Carlo Bonferroni. "Teoria statistica delle classi e calcolo delle probabilita". In: *Pubblicazioni del R Istituto Superiore di Scienze Economiche e Commericiali di Firenze* 8 (1936), pp. 3–62.
11. Leo Breiman. "Bagging predictors". In: *Machine learning* 24.2 (1996), pp. 123–140.
12. Bo Brinkman and Moses Charikar. "On the impossibility of dimension reduction in .1". In: *Journal of the ACM (JACM)* 52.5 (2005), pp. 766–788.
13. Joan Bruna, Arthur Szlam, and Yann LeCun. "Learning stable group invariant representations with convolutional networks". In: *ICLR Workshop*. 2013.
14. Joan Bruna, Arthur Szlam, and Yann LeCun. "Signal recovery from pooling representations". In: *International conference on machine learning*. PMLR. 2014, pp. 307–315.
15. Emmanuel J Candes, Justin K Romberg, and Terence Tao. "Stable signal recovery from incomplete and inaccurate measurements". In: *Communications on Pure and Applied Mathematics: A Journal Issued by the Courant Institute of Mathematical Sciences* 59.8 (2006), pp. 1207–1223.

16. Timothy I Cannings and Richard J Samworth. "Random-projection ensemble classification". In: *arXiv preprint arXiv:1504.04595* (2015).
17. Moses Charikar and Amit Sahai. "Dimension reduction in the ℓ_1 norm". In: *The 43rd Annual IEEE Symposium on Foundations of Computer Science, 2002. Proceedings*. IEEE. 2002, pp. 551–560.
18. Audrey G Chung, Mohammad Javad Shafiee, and Alexander Wong. "Random feature maps via a Layered Random Projection (LARP) framework for object classification". In: *2016 IEEE International Conference on Image Processing (ICIP)*. IEEE. 2016, pp. 246–250.
19. David Cox and Nicolas Pinto. "Beyond simple features: A large-scale feature search approach to unconstrained face recognition". In: *2011 IEEE International Conference on Automatic Face & Gesture Recognition (FG)*. IEEE. 2011, pp. 8–15.
20. Sanjoy Dasgupta and Anupam Gupta. "An elementary proof of a theorem of Johnson and Lindenstrauss". In: *Random Structures & Algorithms* 22.1 (2003), pp. 60–65.
21. Sanjoy Dasgupta and Anupam Gupta. "An elementary proof of the Johnson-Lindenstrauss lemma". In: *International Computer Science Institute, Technical Report* 22.1 (1999), pp. 1–5.
22. David L Donoho. "Compressed sensing". In: *IEEE Transactions on information theory* 52.4 (2006), pp. 1289–1306.
23. Carl Eckart and Gale Young. "The approximation of one matrix by another of lower rank". In: *Psychometrika* 1.3 (1936), pp. 211–218.
24. Peter Frankl and Hiroshi Maehara. "The Johnson-Lindenstrauss lemma and the sphericity of some graphs". In: *Journal of Combinatorial Theory, Series B* 44.3 (1988), pp. 355–362.
25. Alan Frieze, Ravi Kannan, and Santosh Vempala. "Fast Monte-Carlo algorithms for finding low-rank approximations". In: *Journal of the ACM (JACM)* 51.6 (2004), pp. 1025–1041.
26. Benyamin Ghojogh and Mark Crowley. "The theory behind overfitting, cross validation, regularization, bagging, and boosting: tutorial". In: *arXiv preprint arXiv:1905.12787* (2019).
27. Benyamin Ghojogh et al. "Johnson-Lindenstrauss lemma, linear and nonlinear random projections, random Fourier features, and random kitchen sinks: Tutorial and survey". In: *arXiv preprint arXiv:2108.04172* (2021).
28. Raja Giryes, Guillermo Sapiro, and Alex M Bronstein. "Deep neural networks with random Gaussian weights: A universal classification strategy?" In: *IEEE Transactions on Signal Processing* 64.13 (2016), pp. 3444–3457.
29. Xavier Glorot, Antoine Bordes, and Yoshua Bengio. "Deep sparse rectifier neural networks". In: *Proceedings of the fourteenth international conference on artificial intelligence and statistics*. JMLR Workshop and Conference Proceedings. 2011, pp. 315–323.
30. Alexander Hinneburg, Charu C Aggarwal, and Daniel A Keim. "What is the nearest neighbor in high dimensional spaces?" In: *26th Internat. Conference on Very Large Databases*. 2000, pp. 506–515.
31. Wassily Hoeffding. "Probability inequalities for sums of bounded random variables". In: *Journal of the American Statistical Association* 58.301 (1963), pp. 13–30.
32. Arthur E Hoerl and Robert W Kennard. "Ridge regression: Biased estimation for nonorthogonal problems". In: *Technometrics* 12.1 (1970), pp. 55–67.
33. Jennifer A Hoeting et al. "Bayesian model averaging: a tutorial". In: *Statistical science* 14.4 (1999), pp. 382–417.
34. Kurt Hornik, Maxwell Stinchcombe, and Halbert White. "Multilayer feedforward networks are universal approximators". In: *Neural networks* 2.5 (1989), pp. 359–366.
35. Gao Huang et al. "Trends in extreme learning machines: A review". In: *Neural Networks* 61 (2015), pp. 32–48.
36. Guang-Bin Huang. "An insight into extreme learning machines: random neurons, random features and kernels". In: *Cognitive Computation* 6.3 (2014), pp. 376–390.
37. Guang-Bin Huang, Lei Chen, Chee Kheong Siew, et al. "Universal approximation using incremental constructive feedforward networks with random hidden nodes". In: *IEEE Trans. Neural Networks* 17.4 (2006), pp. 879–892.

38. Guang-Bin Huang, Dian Hui Wang, and Yuan Lan. "Extreme learning machines: a survey". In: *International journal of machine learning and cybernetics* 2.2 (2011), pp. 107–122.
39. Guang-Bin Huang, Qin-Yu Zhu, and Chee-Kheong Siew. "Extreme learning machine: theory and applications". In: *Neurocomputing* 70.1-3 (2006), pp. 489–501.
40. Guang-Bin Huang et al. "Extreme learning machine for regression and multiclass classification". In: *IEEE Transactions on Systems, Man, and Cybernetics, Part B (Cybernetics)* 42.2 (2011), pp. 513–529.
41. Piotr Indyk and Rajeev Motwani. "Approximate nearest neighbors: towards removing the curse of dimensionality". In: *Proceedings of the thirtieth annual ACM symposium on Theory of computing. 1998*, pp. 604–613.
42. Kevin Jarrett et al. "What is the best multi-stage architecture for object recognition?" In: *2009 IEEE 12th international conference on computer vision*. IEEE. 2009, pp. 2146–2153.
43. William B Johnson and Joram Lindenstrauss. "Extensions of Lipschitz mappings into a Hilbert space". In: *Contemporary mathematics* 26 (1984).
44. Amir Hossein Karimi et al. "Ensembles of random projections for nonlinear dimensionality reduction". In: *Journal of Computational Vision and Imaging Systems* 3.1 (2017).
45. Amir-Hossein Karimi. "Exploring new forms of random projections for prediction and dimensionality reduction in big-data regimes". MA thesis. University of Waterloo, 2018.
46. Felix Krahmer and Rachel Ward. "A unified framework for linear dimensionality reduction in L1". In: *Results in Mathematics* 70.1 (2016), pp. 209–231.
47. Felix Krahmer and Rachel Ward. "New and improved Johnson–Lindenstrauss embeddings via the restricted isometry property". In: *SIAM Journal on Mathematical Analysis* 43.3 (2011), pp. 1269–1281.
48. Eyal Kushilevitz, Rafail Ostrovsky, and Yuval Rabani. "An algorithm for approximate closest-point queries". In: *Proc. 30th ACM Symp. on Theory of Computing*. 1998, pp. 614–623.
49. Eyal Kushilevitz, Rafail Ostrovsky, and Yuval Rabani. "Efficient search for approximate nearest neighbor in high dimensional spaces". In: *SIAM Journal on Computing* 30.2 (2000), pp. 457–474.
50. Kasper Green Larsen and Jelani Nelson. "Optimality of the Johnson-Lindenstrauss lemma". In: *2017 IEEE 58th Annual Symposium on Foundations of Computer Science (FOCS)*. IEEE. 2017, pp. 633–638.
51. James R Lee, Manor Mendel, and Assaf Naor. "Metric structures in L1: dimension, snowflakes, and average distortion". In: *European Journal of Combinatorics* 26.8 (2005), pp. 1180–1190.
52. Ping Li, Trevor J Hastie, and Kenneth W Church. "Nonlinear estimators and tail bounds for dimension reduction in ℓ_1 using Cauchy random projections". In: *Journal of Machine Learning Research* 8.Oct (2007), pp. 2497–2532.
53. Ping Li, Trevor J Hastie, and Kenneth W Church. "Very sparse random projections". In: *Proceedings of the 12th ACM SIGKDD international conference on Knowledge discovery and data mining*. 2006, pp. 287–296.
54. Edo Liberty. *The Random Projection Method; chosen chapters from DIMACS vol. 65 by Santosh S. Vempala*. Tech. rep. Yale University, 2006.
55. Per-Gunnar Martinsson and Vladimir Rokhlin. "A fast direct solver for scattering problems involving elongated structures". In: *Journal of Computational Physics* 221.1 (2007), pp. 288–302.
56. Jiři. Matoušek. "On variants of the Johnson–Lindenstrauss lemma". In: *Random Structures & Algorithms* 33.2 (2008), pp. 142–156.
57. Stefan Meiser. "Point location in arrangements of hyperplanes". In: *Information and Computation* 106.2 (1993), pp. 286–303.
58. Christos H Papadimitriou et al. "Latent semantic indexing: A probabilistic analysis". In: *Journal of Computer and System Sciences* 61.2 (2000), pp. 217–235.
59. Nicolas Pinto et al. "A high-throughput screening approach to discovering good forms of biologically inspired visual representation". In: *PLoS computational biology* 5.11 (2009), e1000579.

60. Yaniv Plan and Roman Vershynin. "One-bit compressed sensing by linear programming". In: *Communications on Pure and Applied Mathematics* 66.8 (2013), pp. 1275–1297.
61. Ali Rahimi and Benjamin Recht. "Random Features for Large-Scale Kernel Machines". In: *Advances in neural information processing systems*. Vol. 20. 2007.
62. Ali Rahimi and Benjamin Recht. "Uniform approximation of functions with random bases". In: *2008 46th Annual Allerton Conference on Communication, Control, and Computing*. IEEE. 2008, pp. 555–561.
63. Ali Rahimi and Benjamin Recht. "Weighted sums of random kitchen sinks: replacing minimization with randomization in learning". In: *Advances in neural information processing systems*. 2008, pp. 1313–1320.
64. Ana B Ramirez et al. "Reconstruction of Sparse Signals From ℓ_1 Dimensionality-Reduced Cauchy Random Projections". In: *IEEE Transactions on Signal Processing* 60.11 (2012), pp. 5725–5737.
65. Andrew M Saxe et al. "On random weights and unsupervised feature learning". In: *International Conference on Machine Learning*. 2011.
66. Simone Scardapane and Dianhui Wang. "Randomness in neural networks: an overview". In: *Wiley Interdisciplinary Reviews: Data Mining and Knowledge Discovery* 7.2 (2017), e1200.
67. Alon Schclar and Lior Rokach. "Random projection ensemble classifiers". In: *International Conference on Enterprise Information Systems*. Springer. 2009, pp. 309–316.
68. Dominik Scherer, Andreas Müller, and Sven Behnke. "Evaluation of pooling operations in convolutional architectures for object recognition". In: *International conference on artificial neural networks*. Springer. 2010, pp. 92–101.
69. Shai Shalev-Shwartz and Shai Ben-David. *Understanding machine learning: From theory to algorithms*. Cambridge university press, 2014.
70. Malcolm Slaney and Michael Casey. "Locality-sensitive hashing for finding nearest neighbors [lecture notes]". In: *IEEE Signal processing magazine* 25.2 (2008), pp. 128–131.
71. Michel Talagrand. "A new look at independence". In: *The Annals of probability* (1996), pp. 1–34.
72. Jiexiong Tang, Chenwei Deng, and Guang-Bin Huang. "Extreme learning machine for multilayer perceptron". In: *IEEE transactions on neural networks and learning systems* 27.4 (2015), pp. 809–821.
73. Santosh S Vempala. *The Random Projection Method*. Vol. 65 (Series in Discrete Mathematics and Theoretical Computer Science). American Mathematical Society, 2005.
74. Rachel Ward. *Dimension reduction via random projections*. Tech. rep. University of Texas at Austin, 2014.
75. Haozhe Xie, Jie Li, and Hanqing Xue. "A survey of dimensionality reduction techniques based on random projection". In: *arXiv preprint arXiv:1706.04371* (2017).

Chapter 15
Sufficient Dimension Reduction and Kernel Dimension Reduction

15.1 Introduction

Suppose there is a dataset that has labels, either for regression or classification. Sufficient Dimension Reduction (SDR), first proposed by Li [56], is a family of methods that find a transformation of the data to a lower dimensional space, which does not change the conditional of labels given the data [2]. Therefore, the prediction of labels or separation of classes does not change after projection onto the subspace by SDR. In other words, the subspace is sufficient for predicting the labels from the projected data onto the subspace. This sufficient subspace is called the central subspace. The SDR can be divided into three main categories—inverse regression methods, forward regression methods, and Kernel Dimension Reduction (KDR).

It is noteworthy that the terminology "dimension reduction" is often used in the literature of statistical regression, while the terminology "dimensionality reduction" is mostly used in the machine learning literature. Therefore, the methods sufficient dimension reduction and kernel dimension reduction are sometimes called sufficient dimensionality reduction and kernel dimensionality reduction, respectively. This chapter introduces SDR, KDR, and their variants. Sects. 15.3 and 15.4 demonstrate a statistical approach to regression, while Sects. 15.4.4 and 15.5 demonstrate a machine learning approach for dimensionality reduction. In this chapter, X denotes the random variable of the data, and Y is the random variable of the labels of the data. The labels can be discrete finite for classification or continuous for regression.

15.2 Preliminaries and Notations

Definition 15.1 (Regression Problem) Consider a dataset $\{(\boldsymbol{x}_i, y_i)\}_{i=1}^n$, where $\{\boldsymbol{x}_i\}_{i=1}^n$ are called the covariates, features, explanatory, or predictor variables and $\{y_i\}_{i=1}^n$ are called the labels, targets, or responses. The dimensionality of covariates

B. Ghojogh et al., *Elements of Dimensionality Reduction and Manifold Learning*,
https://doi.org/10.1007/978-3-031-10602-6_15

and labels are $\boldsymbol{x}_i \in \mathbb{R}^d$ and $y_i \in \mathbb{R}$. The labels can be either discrete finite or continuous. In the former case, the problem is a classification problem, where every label states which class the point belongs to. In the latter case, a regression problem relates every covariate $\boldsymbol{x}$ to its corresponding label y as [56]:

$$y = f(\boldsymbol{U}^\top \boldsymbol{x}, \boldsymbol{\varepsilon}) = f(\boldsymbol{u}_1^\top \boldsymbol{x}, \boldsymbol{u}_2^\top \boldsymbol{x}, \dots, \boldsymbol{u}_p^\top \boldsymbol{x}, \boldsymbol{\varepsilon}), \tag{15.1}$$

where $\boldsymbol{U} = [\boldsymbol{u}_1, \dots, \boldsymbol{u}_p] \in \mathbb{R}^{d\times p}$ is the projection matrix that projects data onto its p-dimensional column space (with $p \leq d$), $f : \mathbb{R}^{p+1} \to \mathbb{R}$ is an arbitrary unknown function, and $\boldsymbol{\varepsilon}$ is some scalar noise independent of $\boldsymbol{x}$. The matrix form of Eq. (15.1) is:

$$\boldsymbol{y} = f(\boldsymbol{U}^\top \boldsymbol{X}, \boldsymbol{\varepsilon}), \tag{15.2}$$

where $\boldsymbol{y} = [y_1, \dots, y_n]^\top \in \mathbb{R}^n$ and $\boldsymbol{X} = [\boldsymbol{x}_1, \dots, \boldsymbol{x}_n] \in \mathbb{R}^{d\times n}$.

An additional way to state a regression problem is:

$$y = f(\boldsymbol{U}^\top \boldsymbol{x}) + \varepsilon = f(\boldsymbol{u}_1^\top \boldsymbol{x}, \boldsymbol{u}_2^\top \boldsymbol{x}, \dots, \boldsymbol{u}_p^\top \boldsymbol{x}) + \varepsilon, \tag{15.3}$$

where $f : \mathbb{R}^p \to \mathbb{R}$ is an arbitrary unknown function. Note that the projection $\boldsymbol{U}^\top \boldsymbol{x}$ is a linear projection [31].

Remark 15.1 It is possible in a regression problem to take the $\boldsymbol{X}$ axis rowwise to have $\boldsymbol{X} = [\boldsymbol{x}_1, \dots, \boldsymbol{x}_n]^\top \in \mathbb{R}^{n\times d}$. In this case, the regression problem is:

$$\boldsymbol{y} = f(\boldsymbol{X}\boldsymbol{U}, \boldsymbol{\varepsilon}), \quad \text{or} \quad \boldsymbol{y} = f(\boldsymbol{X}\boldsymbol{U}) + \varepsilon. \tag{15.4}$$

This chapter sets the default for $\boldsymbol{X}$ to be columnwise unless otherwise specified.

The random variables associated with the covariates $\boldsymbol{X}$ and labels $\boldsymbol{y}$ are denoted by X and Y, respectively.

Remark 15.2 (Relation to Projection Pursuit Regression) The model of Eq. (15.3) is also used in Projection Pursuit Regression (PPR) [39]. Further explanation of Projection Pursuit (PP) can be found in [40].

Definition 15.2 (Effective Dimension Reduction (e.d.r.) Space [56, Section 1]) Equation (15.1) or (15.2) demonstrates that the d-dimensional covariates are projected onto a p-dimensional subspace, polluted with some noise, and then fed to some unknown function to produce the labels. This subspace is called the Effective Dimension Reduction (e.d.r.) space. As $\boldsymbol{U}$ is the projection matrix, its columns are bases for the e.d.r. subspace. Therefore, in some documents on regression, this matrix is signified by $\boldsymbol{B}$ to denote bases. The vectors $\{\boldsymbol{u}_i\}_{i=1}^p$ (also denoted by $\{\boldsymbol{\beta}_i\}_{i=1}^p$ in the literature) are the bases for the e.d.r. subspace.

Definition 15.3 (Dimension Reduction Subspace and the Central Subspace [18, Section 3]) Consider a reduction function $P_{\mathcal{S}} : \mathbb{R}^d \to \mathbb{R}^p$, where $p \leq d$. Let the projection of $\boldsymbol{X}$ onto a subspace $\mathcal{S}$ be denoted by $P_{\mathcal{S}}\boldsymbol{X}$ (e.g., it is $P_{\mathcal{S}}\boldsymbol{X} = \boldsymbol{U}^\top \boldsymbol{X}$ in Eq. (15.1)). The subspace is called the dimension reduction subspace if [54, 56]:

$$X \perp\!\!\!\perp Y \mid P_{\mathcal{S}}X, \tag{15.5}$$

where $\perp\!\!\!\perp$ denotes the independence of the random variables [32] and $\mid$ denotes conditioning. Considering $P_{\mathcal{S}}X = \boldsymbol{U}^\top X$, Eq. (15.5) can be restated as [14, Section 4]:

$$X \perp\!\!\!\perp Y \mid \boldsymbol{U}^\top X. \tag{15.6}$$

The aim is to find the smallest dimension reduction subspace with minimal dimension, called the minimal subspace. This smallest subspace may not be unique [17, pp. 104–105]. The central dimension reduction subspace, also called the central subspace, is denoted by $\mathcal{D}_{Y|X}$ and is defined to be the intersection of all dimension reduction subspaces for regression [13, 19]. Some weak conditions are required for the central subspace to exist (cf. [13, 17]).

Definition 15.4 (Central Mean Subspace [24]) The subspace spanned by $\mathbb{E}[\boldsymbol{x} \mid y]$ is called the central mean subspace, denoted by $\mathcal{S}_{\mathbb{E}[\boldsymbol{x} \mid y]}$. The central mean subspace is always contained in the central subspace [53].

Definition 15.5 (Sufficient Reduction [12], [2, Definition 1.1]) A reduction or projection onto the subspace $P_{\mathcal{S}} : \mathbb{R}^d \to \mathbb{R}^p$ (where $p \leq d$) is sufficient if it satisfies at least one of the following:

- inverse reduction: $X \mid (Y, P_{\mathcal{S}}X)$ is identically distributed as $X \mid P_{\mathcal{S}}X$.
- forward reduction: $Y \mid X$ is identically distributed as $Y \mid P_{\mathcal{S}}X$.
- joint reduction: $X \perp\!\!\!\perp Y \mid P_{\mathcal{S}}X$, which is Eq. (15.5).

Definition 15.6 (Effective Subspace for Regression [43]) Let $P_{\mathcal{S}}\boldsymbol{X} = \boldsymbol{U}^\top \boldsymbol{X}$ be the projection onto a p-dimensional subspace. Let $\mathbb{R}^{d \times d} \ni \boldsymbol{Q} = [\boldsymbol{U}|\boldsymbol{V}]$ be an orthogonal matrix, where $\boldsymbol{U} \in \mathbb{R}^{d \times p}$ is the truncated projection matrix onto the p-dimensional subspace and $\boldsymbol{V} \in \mathbb{R}^{d \times (d-p)}$ is the rest of matrix $\boldsymbol{Q}$. If the subspace $\mathcal{S}$ is an effective subspace for regression, there are the following equations for the conditional probabilities:

$$\mathbb{P}_{Y|P_{\mathcal{S}}X}(\boldsymbol{y} \mid P_{\mathcal{S}}\boldsymbol{X}) = \mathbb{P}_{Y|X}(\boldsymbol{y} \mid \boldsymbol{X}), \tag{15.7}$$

$$\mathbb{P}_{Y|\boldsymbol{U}^\top X, \boldsymbol{V}^\top X}(\boldsymbol{y} \mid \boldsymbol{U}^\top \boldsymbol{X}, \boldsymbol{V}^\top \boldsymbol{X}) = \mathbb{P}_{Y|\boldsymbol{U}^\top X}(\boldsymbol{y} \mid \boldsymbol{U}^\top \boldsymbol{X}), \tag{15.8}$$

which are equivalent equations. Both equations mean that all the required information for predicting the labels $\boldsymbol{Y}$ are contained in the subspace $\mathcal{S}$, which is the column space of U; therefore, this subspace is sufficient for X to be projected onto. Additionally:

$$
\begin{aligned}
I(Y, X) = I(Y, \boldsymbol{U}^\top X) \\
+ \mathbb{E}_{\boldsymbol{U}^\top X}[I(Y|\boldsymbol{U}^\top X, \boldsymbol{V}^\top X|\boldsymbol{U}^\top X)],
\end{aligned}
$$

where $I(\cdot|\cdot)$ is the mutual information between random variables.

The e.d.r. subspace (Definition 15.2), the central subspace (Definition 15.3), and the effective subspace for regression (Definition 15.6) have equivalent meanings. The e.d.r. subspace, central subspace, and effective subspace are often used in Prof. Ker-Chau Li's papers, Prof. R. Dennis Cook's papers, and the literature of kernel dimension reduction, respectively.

Definition 15.7 (Exhaustive Dimension Reduction [52, Section 3]) A dimension reduction method estimates a subspace $\mathcal{S}$ of $\mathcal{S}_{Y|X}$, i.e., $\mathcal{S} \subseteq \mathcal{S}_{Y|X}$. If $\mathcal{S} = \mathcal{S}_{Y|X}$, then the method is called exhaustive.

Definition 15.8 (Linear Regression or OLS Problem) A linear regression problem, also called the Ordinary Least Squares (OLS) problem, is a special case of Eq. (15.4), which usually considers $\boldsymbol{X}$ rowwise. It takes the function f to be an identity function, resulting in:

$$
\boldsymbol{Y} = \boldsymbol{X}\boldsymbol{U} + \boldsymbol{\varepsilon},
$$

where $\boldsymbol{X} \in \mathbb{R}^{n\times d}$, $\boldsymbol{U} \in \mathbb{R}^{d\times p}$ and the labels can be multidimensional, i.e., $\boldsymbol{Y} \in \mathbb{R}^{n\times p}$. The projection matrix, or so-called coefficients, $\boldsymbol{U}$ can be calculated using a least squares problem:

$$
\begin{aligned}
&\underset{\boldsymbol{U}}{\text{min.}}\ \|\boldsymbol{Y} - \boldsymbol{X}\boldsymbol{U}\|_F^2 \\
&\implies \frac{\partial}{\partial \boldsymbol{U}} \|\boldsymbol{Y} - \boldsymbol{X}\boldsymbol{U}\|_F^2 = -2\boldsymbol{X}^\top(\boldsymbol{Y} - \boldsymbol{X}\boldsymbol{U}) \overset{\text{set}}{=} \boldsymbol{0} \\
&\implies \boldsymbol{X}^\top \boldsymbol{Y} = \boldsymbol{X}^\top \boldsymbol{X}\boldsymbol{U} \implies \boldsymbol{U} = (\boldsymbol{X}^\top \boldsymbol{X})^{-1}\boldsymbol{X}^\top \boldsymbol{Y},
\end{aligned}
\tag{15.9}
$$

where $\|.\|_F$ denotes the Frobenius norm. There also exist some other weighted methods for linear regression, such as Iteratively Reweighted Least Squares (IRLS) [9].

Let the mean and covariance of the covariates be denoted by $\mathbb{E}[\boldsymbol{x}]$ (or $\boldsymbol{\mu}_x$) and $\boldsymbol{\Sigma}_{xx}$, respectively. The covariates can be standardized to:

$$
\boldsymbol{z} := \boldsymbol{\Sigma}_{xx}^{-1/2}(\boldsymbol{x} - \mathbb{E}[\boldsymbol{x}]).
\tag{15.10}
$$

The random variable associated with z is denoted by Z. Equation (15.1) can be restated as:

$$y = f(\boldsymbol{W}^\top z, \boldsymbol{\varepsilon}) = f(\boldsymbol{w}_1^\top z, \boldsymbol{w}_2^\top z, \ldots, \boldsymbol{w}_p^\top z, \boldsymbol{\varepsilon}), \tag{15.11}$$

where $\boldsymbol{W} = [\boldsymbol{w}_1, \ldots, \boldsymbol{w}_p] = \boldsymbol{\Sigma}_{xx}^{1/2} \boldsymbol{U} = [\boldsymbol{\Sigma}_{xx}^{1/2} \boldsymbol{u}_1, \ldots, \boldsymbol{\Sigma}_{xx}^{1/2} \boldsymbol{u}_p] \in \mathbb{R}^{d \times p}$ is the projection matrix for the standardized data. The $\boldsymbol{w}_1, \ldots, \boldsymbol{w}_p$ are called the standardized projection directions.

Corollary 15.1 (Relation of Subspace Bases for Covariates and Standardized Covariates [56, Section 2]) *According to Eqs. (15.2), (15.10), and (15.11), the relation between the projection directions (i.e., bases for covariates) and the standardized projection directions (i.e., bases for standardized covariates) are:*

$$\boldsymbol{w}_i = \boldsymbol{\Sigma}_{xx}^{1/2} \boldsymbol{u}_i \implies \boldsymbol{u}_i = \boldsymbol{\Sigma}_{xx}^{-1/2} \boldsymbol{w}_i, \quad \forall i \in \{1, \ldots, p\}. \tag{15.12}$$

Corollary 15.2 (Relation of Central Subspaces for Covariates and Standardized Covariates [18, Section 4]) *According to Eq. (15.12), the relation of central subspaces for covariates and standardized covariates is:*

$$\mathcal{S}_{Y|Z} = \boldsymbol{\Sigma}_{xx}^{1/2} \mathcal{S}_{Y|X}. \tag{15.13}$$

The mean and covariance of the covariates can be estimated with the sample mean and sample covariance matrix, respectively:

$$\mathbb{E}[\boldsymbol{x}] \approx \widehat{\boldsymbol{\mu}}_x := \frac{1}{n} \sum_{i=1}^{n} \boldsymbol{x}_i, \tag{15.14}$$

$$\boldsymbol{\Sigma}_{xx} \approx \widehat{\boldsymbol{\Sigma}}_{xx} := \frac{1}{n} \sum_{i=1}^{n} (\boldsymbol{x}_i - \bar{\boldsymbol{x}})(\boldsymbol{x}_i - \bar{\boldsymbol{x}})^\top. \tag{15.15}$$

These estimates can be used in Eq. (15.10) to estimate the standardized covariates:

$$\mathbb{R}^d \ni \widehat{z}_i := \widehat{\boldsymbol{\Sigma}}_{xx}^{-1/2} (\boldsymbol{x}_i - \widehat{\boldsymbol{\mu}}_x). \tag{15.16}$$

Additionally, according to Eq. (15.12):

$$\boldsymbol{u}_i = \widehat{\boldsymbol{\Sigma}}_{xx}^{-1/2} \boldsymbol{w}_i, \quad \forall i \in \{1, \ldots, p\}. \tag{15.17}$$

Let X', Y', and Z' be independent copies of the random variables X, Y, and Z, respectively.

Definition 15.9 (Stiefel and Grassmannian Manifolds [1]) The Stiefel manifold is defined as the set of orthogonal matrices:

$$\mathcal{S}t(p,d) := \{\boldsymbol{U} \in \mathbb{R}^{d\times p} \,|\, \boldsymbol{U}^\top \boldsymbol{U} = \boldsymbol{I}\}, \tag{15.18}$$

where $p \leq d$. The Grassmannian manifold $\mathcal{G}(p,d)$ is defined to be all p-dimensional subspaces of $\mathbb{R}^d$. The Grassmannian manifold can be seen as the quotient space[1] of the Stiefel manifold $\mathcal{S}t(p,d)$ denoted as:

$$\mathcal{G}(p,d) := \mathcal{S}t(p,d)/\mathcal{S}t(p,p). \tag{15.19}$$

The Stiefel and Grassmannian manifolds are examples of the Riemannian manifold, which is a smooth manifold endowed with a metric.

The projection matrix $\boldsymbol{U}$ onto the central subspace is usually an orthogonal matrix because the bases of the subspace, which are the columns of the projection matrix, are orthonormal vectors. Therefore, according to Definition 15.9, the projection matrix of a central subspace belongs to the Stiefel manifold. Moreover, as the central space is a subspace of the space of covariates, it belongs to Grassmannian manifold, according to Definition 15.9. Many SDR methods, such as LAD [20] and CVE [36], use Riemannian optimization [1] to find the central subspace in these manifolds [31].

15.3 Inverse Regression Methods

Definition 15.10 (Inverse Regression Problem) A forward regression problem tries to estimate the labels from the covariates by calculating the bases of the central subspace. A reverse regression problem exchanges the roles of the covariates and labels and tries to estimate the covariates from the labels by calculating the bases of the central subspace.

Inverse regression methods are based on inverse regression. They typically have strict assumptions on the distributions and conditional distributions. Some important inverse regression methods include Sliced Inverse Regression (SIR) [56], Sliced Average Variance Estimation (SAVE) [18, 29], Parametric Inverse Regression (PIR) [6], Contour Regression (CR) [53], Directional Regression (DR) [52], Principal Fitted Components (PFC) [12, 21], Likelihood Acquired Direction (LAD) [20], and graphical regression [16, 17]. These methods are introduced in the following sections. Additional methods include SDR by inverse of the k-th moment [66], SDR by minimum discrepancy [26], SDR by Intraslice covariances [27], SDR for nonelliptically distributed covariates [51], envelope models [25], and Principal Support Vector Machines (PSVM) [50], which are not covered for brevity.

[1] Additional information on the quotient spaces can be found in [1].

Definition 15.11 (The Inverse Regression Curve [56, Section 3]) The d-dimensional inverse regression curve is defined as $\mathbb{E}[\boldsymbol{x}|y]$ versus y. The centered inverse regression curve is $\mathbb{E}[\boldsymbol{x}|y] - \mathbb{E}[\boldsymbol{x}]$ versus y. Therefore, the standardized inverse regression curve is $\mathbb{E}[\boldsymbol{z}|y]$ versus y.

Lemma 15.1 (Linearity Condition [34]) *The covariates have elliptical distribution if and only if for any $\boldsymbol{b} \in \mathbb{R}^d$ the conditional expectation is linear, i.e., $\mathbb{E}[\boldsymbol{b}^\top \boldsymbol{x} \mid \boldsymbol{u}_1^\top \boldsymbol{x}, \dots, \boldsymbol{u}_p^\top \boldsymbol{x}] = c_0 + c_1 \boldsymbol{u}_1^\top \boldsymbol{x} + \dots + c_p \boldsymbol{u}_p^\top \boldsymbol{x}$. This is called the linearity condition.*

The linearity condition is assumed for almost all inverse regression methods.

Remark 15.3 (On Linearity Condition) The linearity condition holds most often [56] because the projection of high-dimensional covariates onto a low-dimensional subspace has a distribution close to a normal distribution [33, 45].

15.3.1 Sliced Inverse Regression (SIR)

Sliced Inverse Regression (SIR) was first proposed by Li [56]. SIR is explained in the following theorem.

Theorem 15.1 ([56, Condition 3.1, Theorem 3.1, and Corollary 3.1]) *Assume the linearity condition in Lemma 15.1 holds. The centered inverse regression curve $\mathbb{E}[\boldsymbol{x}|y]-\mathbb{E}[\boldsymbol{x}]$ is contained in the linear subspace spanned by $\{\boldsymbol{\Sigma}_{xx}\boldsymbol{u}_1, \dots, \boldsymbol{\Sigma}_{xx}\boldsymbol{u}_p\}$. Moreover, in this case, the standardized inverse regression curve $\mathbb{E}[\boldsymbol{z}|y]$ is contained in the linear subspace spanned by $\{\boldsymbol{w}_1, \dots, \boldsymbol{w}_p\}$.*

Definition 15.12 (Slicing Labels in Regression [56]) Let the range of labels be denoted by set $\mathcal{D}_y$. The range of labels can be divided into h slices, $\{\mathcal{D}_1, \dots, \mathcal{D}_h\}$, where:

$$\bigcup_{s=1}^{h} \mathcal{D}_s = \mathcal{D}_y,$$

$$\mathcal{D}_{s_1} \cap \mathcal{D}_{s_2} = \varnothing, \quad \forall s_1, s_2 \in \{1, \dots, h\},\ s_1 \neq s_2.$$

Let the proportion of the labels that fall in slice $\mathcal{D}_s$ be denoted by ρ_s. Therefore, $\rho_1 + \dots + \rho_h = 1$. The proportions can be used because the slices may not be equally populated.

If the regression problem is actually a classification problem whose labels are discrete finite, slicing is not needed, as the labels are already sliced into discrete finite values. In that case, the number of slices, h, is equal to the number of classes.

Lemma 15.2 (Central Subspace for Binary Labels [18, Section 4]) *Let the labels be converted to binary labels as follows:*

$$\widetilde{y} := \begin{cases} 1 \text{ if } y > c \\ 0 \text{ if } y \leq c, \end{cases}$$

for some constant c and let its associated random variable be $\widetilde{Y}$. Therefore:

$$\mathcal{S}_{\widetilde{Y}|X} \subseteq \mathcal{S}_{Y|X}. \tag{15.20}$$

Corollary 15.3 (Relation of Central Subspaces for Labels and Sliced Labels [18, Section 4]) *An extension of Lemma 15.2 is as follows. Let $\widetilde{Y}$ denote the random variable for the sliced labels into h slices, where $\widetilde{y}_i = k$ if $y_i \in \mathcal{D}_k$. Then, Eq. (15.20) holds for the sliced labels.*

The sample mean of standardized covariates whose labels fall in every slice can be determined by:

$$\mathbb{R}^d \ni \widehat{\boldsymbol{\mu}}_{z,s} := \frac{1}{n\rho_s} \sum_{y_i \in \mathcal{D}_s} \widehat{z}_i. \tag{15.21}$$

Then, the Principal Component Analysis (PCA) of the points $\{\widehat{\boldsymbol{\mu}}_{z,s}\}_{s=1}^h$ are calculated. The PCA of these points can be done by eigenvalue decomposition of their weighted covariance matrix (see Chap. 5). The weighted covariance matrix is [56]:

$$\mathbb{R}^{d\times d} \ni \widehat{V} := \sum_{s=1}^{h} \rho_s\, \widehat{\boldsymbol{\mu}}_{z,s}\, \widehat{\boldsymbol{\mu}}_{z,s}^\top.$$

Consider the top p eigenvectors of $\widehat{V}$ with the largest eigenvalues (see Chap. 2). According to Theorem 15.1, these eigenvectors are $\{\boldsymbol{w}_j\}_{j=1}^p$, which are used in Eq. (15.11). Lastly, Eq. (15.17) is used to calculate $\{\boldsymbol{u}_j\}_{j=1}^p$ from $\{\boldsymbol{w}_j\}_{j=1}^p$. The calculated $\{\boldsymbol{u}_j\}_{j=1}^p$ are the bases for the central subspace and are used in Eq. (15.1). According to Eq. (15.20), SIR is not an exhaustive method (see Definition 15.7). Note that SIR has also been extended to work for categorical (i.e., discrete finite) covariates [11].

15.3.2 Sliced Average Variance Estimation (SAVE)

Sliced Average Variance Estimation (SAVE) was first proposed by Cook [18], while it was developed further in his other works [22, 29]. Two assumptions are required for SAVE [18]:

1. Linearity condition: As in Lemma 15.1, assume that $\mathbb{E}[Z|\boldsymbol{U}^\top Z]$ is linear in $\boldsymbol{U}^\top Z$, resulting in $\mathbb{E}[Z|\boldsymbol{U}^\top Z] = P_{\mathcal{S}_{Y|Z}} Z$, which is equivalent to Z having an elliptical distribution [34].
2. Constant covariance condition: Assume that $\mathbb{V}\mathrm{ar}[Z|\boldsymbol{U}^\top Z]$ is constant, which is equivalent to Z having a normal distribution [18]. The labels can be sliced into h slices.

Theorem 15.2 ([18, Section 5]) *Let span$\{\cdots\}$ signify a space spanned by the set of bases. Then:*

$$span\Big\{\mathbb{E}\big[\boldsymbol{I} - \mathbb{V}ar[Z|\widetilde{Y}]\big]^2\Big\} \subseteq \mathcal{S}_{\widetilde{Y}|X} \overset{(15.20)}{\subseteq} \mathcal{S}_{Y|X},$$

where $\widetilde{Y}$ is the random variable for the sliced label, defined in Corollary 15.3.

The sample covariance for every slice s is:

$$\mathbb{R}^{d\times d} \ni \widehat{\boldsymbol{V}}_s := \frac{1}{n\rho_s} \sum_{y_i \in \mathcal{D}_s} (\widehat{z}_i - \widehat{\boldsymbol{\mu}}_{z,s})(\widehat{z}_i - \widehat{\boldsymbol{\mu}}_{z,s})^\top,$$

where $\widehat{\boldsymbol{\mu}}_{z,s}$ is defined in Eq. (15.21). This $\widehat{\boldsymbol{V}}_s$ is an estimation for $\mathbb{V}\mathrm{ar}[Z|\widetilde{Y}]$. Consider the following matrix:

$$\mathbb{R}^{d\times d} \ni \boldsymbol{M} := \sum_{s=1}^{h} \rho_s (\boldsymbol{I} - \widehat{\boldsymbol{V}}_s)^2,$$

where $\boldsymbol{I}$ is the identity matrix. Consider the top p eigenvectors of $\boldsymbol{M}$ with the largest eigenvalues (see Chap. 2). According to Theorem 15.2, these eigenvectors are $\{\boldsymbol{w}_j\}_{j=1}^p$, which is used in Eq. (15.11). Equation (15.17) can be used to calculate $\{\boldsymbol{u}_j\}_{j=1}^p$ from $\{\boldsymbol{w}_j\}_{j=1}^p$. The calculated $\{\boldsymbol{u}_j\}_{j=1}^p$ are the bases for the central subspace and are used in Eq. (15.1). SAVE is an exhaustive method (see Definition 15.7), according to [52].

15.3.3 Parametric Inverse Regression (PIR)

Parametric Inverse Regression (PIR) was first proposed by Bura and Cook [6]. PIR considers the rowwise regression as in Remark 15.1. Equation (15.4) is a forward regression problem. Consider the inverse regression problem (see Definition 15.10) for the standardized covariates. PIR allows the labels to be multidimensional (ℓ-dimensional), i.e., $\boldsymbol{Y} \in \mathbb{R}^{n\times\ell}$.

$$\boldsymbol{Z} = f(\boldsymbol{Y})\,\boldsymbol{B} + \boldsymbol{\varepsilon},$$

where $\boldsymbol{Z} \in \mathbb{R}^{n \times d}$ are the standardized covariates, $f : \mathbb{R}^{n \times \ell} \to \mathbb{R}^{n \times \ell}$ is some function, $\boldsymbol{B} \in \mathbb{R}^{\ell \times p}$ is the projection matrix onto a p-dimensional subspace, and ϵ denotes independent noise. According to Eq. (15.9), it is possible to calculate $\boldsymbol{B}$ as:

$$\boldsymbol{B} = \big(f(\boldsymbol{Y})^\top f(\boldsymbol{Y})\big)^{-1} f(\boldsymbol{Y})^\top \boldsymbol{Z},$$

which is the projection matrix onto a p-dimensional central subspace.

15.3.4 Contour Regression (CR)

Contour Regression (CR) was proposed in [53] with two versions—the simple and general contour regression. This chapter introduces the simple contour regression. CR claims that all the directional information of covariates exists in the set $\{\boldsymbol{x}_i - \boldsymbol{x}_j \,|\, 1 \leq i < j \leq n\}$.

Assume the covariates have an elliptical distribution (see Lemma 15.1). Additionally, assume that [53, Assumption 2.1]:

$$\mathbb{V}\text{ar}\big[\boldsymbol{w}^\top (Z' - X) \,\big|\, |Y' - Y| \leq c\big] > \mathbb{V}\text{ar}\big[\boldsymbol{v}^\top (Z' - X) \,\big|\, |Y' - Y| \leq c\big],$$

where $\boldsymbol{v} \in \mathcal{S}_{Y|Z}$, $\boldsymbol{w} \in (\mathcal{S}_{Y|Z})^\perp$, $\|\boldsymbol{v}\|_2 = \|\boldsymbol{w}\|_2 = 1$, and $c > 0$. Note that $\mathcal{S}^\perp$ denotes the orthogonal space to $\mathcal{S}$.

Theorem 15.3 ([53, Theorem 2.1 and Corollary 2.1]) *Let the above assumptions hold. Consider the matrix:*

$$\boldsymbol{M} := \mathbb{E}\big[(X' - X)(X' - X)^\top \,\big|\, |Y' - Y| \leq c\big], \tag{15.22}$$

with $c > 0$. Consider the p tailing eigenvectors of $\boldsymbol{\Sigma}_{xx}^{-1/2} \boldsymbol{M} \boldsymbol{\Sigma}_{xx}^{-1/2}$ with the smallest eigenvalues (see Chap. 2), denoted by $\gamma_{d-p+1}, \ldots, \gamma_d$. $\{\boldsymbol{u}_j\}_{j=1}^p$, used in Eq. (15.1), are the vectors $\boldsymbol{\Sigma}_{xx}^{-1/2} \gamma_{d-p+1}, \ldots, \boldsymbol{\Sigma}_{xx}^{-1/2} \gamma_d$.

It is possible to estimate Theorem 15.3 in practice. Equation (15.22) is estimated as [53]:

$$\widehat{\boldsymbol{M}} = \frac{1}{\binom{n}{c}} \sum_{i=1}^n \sum_{j=1}^n (\boldsymbol{x}_j - \boldsymbol{x}_i)(\boldsymbol{x}_j - \boldsymbol{x}_i)^\top \mathbb{I}(|\boldsymbol{y}_j - \boldsymbol{y}_i| \leq c),$$

where $\mathbb{I}(.)$ is the indicator function, which is one if its condition is satisfied and is zero otherwise. It is possible to compute the p tailing eigenvectors of $\widehat{\boldsymbol{\Sigma}}_{xx}^{-1/2} \widehat{\boldsymbol{M}} \widehat{\boldsymbol{\Sigma}}_{xx}^{-1/2}$ where the sample covariance is defined in Eq. (15.15). CR is an exhaustive method (see Definition 15.7) under some mild conditions [52].

15.3.5 Directional Regression (DR)

The CR method has demonstrated that all the directional information of covariates exists in the set $\{\boldsymbol{x}_i - \boldsymbol{x}_j \mid 1 \leq i < j \leq n\}$. Directional Regression (DR) [52] uses this fact. DR is defined as:

$$A(Y, Y') := \mathbb{E}[(Z - Z')(Z - Z')^\top \mid Y, Y'].$$

The directions $(Z - Z')$, which are aligned with $\mathcal{S}_{Y|Z}$, are affected by Y, while the directions $(Z - Z')$ aligned with $\mathcal{S}_{Y|Z}^\perp$ are not affected [52]. In DR, the labels are sliced into h slices.

Theorem 15.4 ([52, Theorem 1 and Section 4]) *With the two conditions mentioned in Sect. 15.3.2, the column space of $2\boldsymbol{I} - A(Y, Y')$ is contained in $\mathcal{S}_{Y|Z}$, where $\boldsymbol{I}$ is the identity matrix. Therefore, an estimate of $\mathcal{S}_{Y|Z}$ is:*

$$\mathbb{E}[2\boldsymbol{I} - A(Y, Y')]^2 \approx \frac{1}{\binom{h}{2}} \sum_{k<\ell} (2\boldsymbol{I} - \widehat{A}(\mathcal{D}_k, \mathcal{D}_\ell))^2,$$

where:

$$\widehat{A}(\mathcal{D}_k, \mathcal{D}_\ell) := \frac{\sum_{i<j} (\widehat{z}_i - \widehat{z}_j)(\widehat{z}_i - \widehat{z}_j)^\top \mathbb{I}(y_i \in \mathcal{D}_k, y_j \in \mathcal{D}_\ell)}{\sum_{i<j} \mathbb{I}(y_i \in \mathcal{D}_k, y_j \in \mathcal{D}_\ell)},$$

is an empirical estimate for $A(Y, Y')$. Note that $\mathbb{I}(.)$ is the indicator function and $\mathcal{D}_k$ is the k-th slice of labels.

The following is defined as [52]:

$$\mathbb{R}^{d\times d} \ni \widehat{\boldsymbol{F}} := 2\sum_{s=1}^{h} \rho_s E_n^2(\widehat{Z}\widehat{Z}^\top - \boldsymbol{I} \mid Y \in \mathcal{D}_s) + 2\Big(\sum_{s=1}^{h} \rho_s E_n(\widehat{Z} \mid Y \in \mathcal{D}_s) E_n(\widehat{Z}^\top \mid Y \in \mathcal{D}_s)\Big)^2$$
$$+ 2\Big(\sum_{s=1}^{h} \rho_s E_n(\widehat{Z}^\top \mid Y \in \mathcal{D}_s) E_n(\widehat{Z} \mid Y \in \mathcal{D}_s) \times \sum_{s=1}^{h} \rho_s E_n(\widehat{Z} \mid Y \in \mathcal{D}_s) E_n(\widehat{Z}^\top \mid Y \in \mathcal{D}_s)\Big),$$

where:

$$E_n(\widehat{Z} \mid Y \in \mathcal{D}_s) := \frac{\sum_{i=1}^{n} \widehat{z}_i \, \mathbb{I}(y_i \in \mathcal{D}_s)}{\sum_{i=1}^{n} \mathbb{I}(y_i \in \mathcal{D}_s)}.$$

Consider the top p eigenvectors of $\widehat{\boldsymbol{F}}$ with the largest eigenvalues (see Chap. 2). According to Theorem 15.4, these eigenvectors are $\{\boldsymbol{w}_j\}_{j=1}^p$ used in Eq. (15.11). Equation (15.17) is used to calculate $\{\boldsymbol{u}_j\}_{j=1}^p$ from $\{\boldsymbol{w}_j\}_{j=1}^p$. The calculated $\{\boldsymbol{u}_j\}_{j=1}^p$ are the bases for the central subspace and are used in Eq. (15.1). Under the conditions specified in Theorem 3 [52, Theorem 3], DR is an exhaustive method (see Definition 15.7).

15.3.6 Likelihood-Based Methods

There are likelihood-based methods in the family of inverse regression methods. They use Maximum Likelihood Estimation (MLE) for estimating the projection matrix onto the central subspace. Two fundamental likelihood-based methods are Principal Fitted Components (PFC) [12, 21] and Likelihood Acquired Direction (LAD) [20]. Additional likelihood-based methods, such as the envelope model [25, 68, 69], can be found in [7, 8], which are not covered here for brevity.

15.3.6.1 Principal Fitted Components (PFC)

The Principal Fitted Components (PFC) method was proposed by Cook and Forzani [21] and further developed by Cook [12]. It considers the inverse regression problem in Definition 15.10. Assume that the covariates have a normal distribution, following Lemma 15.1. Then [12, 21]:

$$\boldsymbol{X}|Y = \widehat{\boldsymbol{\mu}}_x + \boldsymbol{U}\boldsymbol{v}_y + \boldsymbol{\varepsilon}, \tag{15.23}$$

where $\boldsymbol{X} \in \mathbb{R}^{d \times n}$ are the covariates, $\boldsymbol{U} \in \mathbb{R}^{d \times p}$ is the orthogonal projection matrix onto p-dimensional central subspace, $\boldsymbol{v}_y := \boldsymbol{U}^\top(\mathbb{E}[X|Y] - \widehat{\boldsymbol{\mu}}_x) \in \mathbb{R}^p$, $\widehat{\boldsymbol{\mu}}_x$ is defined in Eq. (15.14), and $\boldsymbol{\varepsilon}$ is the independent noise.

Lemma 15.3 ([12, 21]) *In the model of Eq. (15.23), the following dimension reduction is sufficient (see Definition 15.5):*

$$P_{\mathcal{S}}\boldsymbol{X} = \boldsymbol{U}^\top \boldsymbol{\Sigma}_{\varepsilon\varepsilon}^{-1} \boldsymbol{X}, \tag{15.24}$$

where $\boldsymbol{\Sigma}_{\varepsilon\varepsilon}$ denotes the covariance of noise.

PFC estimates the matrices $\boldsymbol{U}$ and $\boldsymbol{\Sigma}_{\varepsilon\varepsilon}$ by MLE to calculate the projection onto the central subspace using Eq. (15.24). Equation (15.23) can be restated as [21]:

$$\boldsymbol{X}|Y = \boldsymbol{\mu} + \boldsymbol{U}\boldsymbol{\beta}\boldsymbol{f}(y) + \boldsymbol{\varepsilon}, \tag{15.25}$$

where $\boldsymbol{f}(y)$ is a known vector-valued function of labels. Assume that the covariates have a normal distribution, following Lemma 15.1. Therefore, the log-likelihood is [21]:

$$L(\boldsymbol{\mu}, \boldsymbol{U}, \boldsymbol{\beta}, \boldsymbol{\Sigma}_{\varepsilon\varepsilon}) = -\frac{nd}{2}\ln(2\pi) - \frac{n}{2}\ln(|\boldsymbol{\Sigma}_{\varepsilon\varepsilon}|)$$

$$- \frac{1}{2} \sum_{i=1}^{n} (\boldsymbol{x}_i - \boldsymbol{\mu} - \boldsymbol{U}\boldsymbol{\beta}(\boldsymbol{f}(y_i) - \bar{\boldsymbol{f}})^\top) \boldsymbol{\Sigma}_{\varepsilon\varepsilon}^{-1} (\boldsymbol{x}_i - \boldsymbol{\mu} - \boldsymbol{U}\boldsymbol{\beta}(\boldsymbol{f}(y_i) - \bar{\boldsymbol{f}})),$$

where $\bar{\boldsymbol{f}}$ is the mean of $\boldsymbol{f}(y_i)$'s. Optimizing this log-likelihood iteratively can provide estimations for $\boldsymbol{U}$ and $\boldsymbol{\Sigma}_{\varepsilon\varepsilon}$. Using these estimations in Eq. (15.24) results in the projection onto the central subspace.

15.3.6.2 Likelihood Acquired Direction (LAD)

Likelihood Acquired Direction (LAD) [20] is another likelihood-based method. In LAD, the labels are sliced into h slices and MLE is used to estimate the central subspace.

Theorem 15.5 ([20, Theorem 2]) *The MLE of $\mathcal{S}_{Y|X}$ maximizes the log-likelihood:*

$$\begin{aligned} L(\mathcal{S}) = &- \frac{nd}{2}(1 + \ln(2\pi)) + \frac{n}{2} \ln(|\boldsymbol{U}^\top \widehat{\boldsymbol{\Sigma}}_{xx} \boldsymbol{U}|_0) \\ &- \frac{n}{2} \ln(|\widehat{\boldsymbol{\Sigma}}_{\varepsilon\varepsilon}|) - \frac{1}{2} \sum_{s=1}^{h} \rho_s n \ln(|\boldsymbol{U}^\top \widehat{\boldsymbol{\Sigma}}_{\varepsilon\varepsilon} \boldsymbol{U}|_0), \end{aligned}$$

over the Grassmannian manifold $\mathcal{G}(p, d)$, where $|.|_0$ denotes the product of the nonzero eigenvalues of the matrix. Other notations have been defined before.

In LAD, iterative second-order Riemannian optimization [1] is used on Grassmannian manifold $\mathcal{G}(p, d)$ to find the projection matrix $\boldsymbol{U} \in \mathbb{R}^{d \times p}$ onto the subspace $\mathcal{S}$. LAD is an exhaustive dimension reduction method (see Definition 15.7) which finds the subspacc $\mathcal{S}_{Y|X}$ [20].

15.3.7 Graphical Regression

Graphical regression [16, 17] uses plots to estimate the central subspace visually.

Definition 15.13 (Sufficient Summary Plot [17]) Consider the projection of covariates onto the subspace by $P_{\mathcal{S}} X = \boldsymbol{U}^\top \boldsymbol{X}$, where Eq. (15.5) holds. A plot of $\boldsymbol{y}$ versus $\boldsymbol{U}^\top \boldsymbol{x}$ is called the (minimal) sufficient summary plot.

In this approach, the sufficient summary plot is plotted for various p values and the smallest p, which shows an acceptable projection in the plot, is chosen as the dimensionality of the central subspace [17]. This method is explained in more detail as follows [16]. Let x^j be denoted by the j-th dimension and let $\boldsymbol{x}^{\backslash j_1, j_2}$ denote all the dimensions of $\boldsymbol{x}$, except x^{j_1} and x^{j_2}. It is initiated by projecting the first two dimensions onto a subspace by c_1 and c_2 and then it is checked visually whether the following holds [16]:

$$X \perp\!\!\!\perp Y \mid (c_1 x^1 + c_2 x^2, \boldsymbol{x}^{\backslash 1,2}).$$

If this holds, then the two dimensions x^1 and x^2 are combined into one new dimension, denoted by $x^{1,2}$. This reduces the dimension of the covariate by one by projecting it onto a $(d-1)$-dimensional subspace. This is repeated for other dimensions until the conditional independence no longer holds visually. Further explanation of graphical regressions can be found in [14, 28]. Additionally, see [13] for graphical regression on binary labels.

15.4 Forward Regression Methods

Forward regression methods are not based on inverse regression. This section introduces important forward regression methods, including Principal Hessian Directions (pHd) [54], Minimum Average Variance Estimation (MAVE) [64], Conditional Variance Estimation (CVE) [36], and deep SDR [3, 47].

15.4.1 Principal Hessian Directions (pHd)

The Principal Hessian Directions (pHd) method, proposed in [54], uses the Hessian matrix (see Chap. 4) and Stein's lemma [59] to estimate the central subspace. Consider the Hessian matrix of the covariates denoted by $\boldsymbol{H}_x$. Let the mean of Hessian matrix be $\bar{\boldsymbol{H}}_x$.

Lemma 15.4 ([54, Section 2.1]) *The top p eigenvectors of $\bar{\boldsymbol{H}}_x \boldsymbol{\Sigma}_{xx}$, with the largest eigenvalues, are the bases of the p-dimensional central subspace with the projection matrix $\boldsymbol{U} \in \mathbb{R}^{d \times p}$.*

Lemma 15.5 (Stein's Lemma [59, Lemma 4]) *Let $g(.)$ be a function and a random variable X have a mean $\boldsymbol{\mu}$ and variance of one. Therefore:*

$$\mathbb{E}[X - \boldsymbol{\mu}]\, g(X) = \mathbb{E}[\frac{\partial g(X)}{\partial X}]$$

$$\mathbb{E}[X - \boldsymbol{\mu}]^2\, g(X) = \mathbb{E}[g(X)] + \mathbb{E}[\frac{\partial^2 g(X)}{\partial X^2}].$$

Theorem 15.6 ([54, Lemma 3.1, Corollary 3.1, and Theorem 3.1]) *Let:*

$$\mathbb{R}^{d\times d} \ni \boldsymbol{\Sigma}_{yxx} = \mathbb{E}[(Y-\mu_y)(X-\boldsymbol{\mu}_x)(X-\boldsymbol{\mu}_x)^\top],$$

where $\boldsymbol{\mu}_x$ and μ_y are the mean of the covariates and labels, respectively. Based on Stien's lemma:

$$\bar{\boldsymbol{H}}_x = \boldsymbol{\Sigma}_{xx}^{-1}\boldsymbol{\Sigma}_{yxx}\boldsymbol{\Sigma}_{xx}^{-1}.$$

If the covariates have a normal distribution, the bases of the central subspace are the top p eigenvectors of the following generalized eigenvalue problem:

$$\boldsymbol{\Sigma}_{yxx}\boldsymbol{u}_j = \lambda_j \boldsymbol{\Sigma}_{xx}\boldsymbol{u}_j, \quad \forall j \in \{1, \dots, p\},$$

with the largest eigenvalues, where $\boldsymbol{u}_j$'s and λ_j's are the eigenvectors and eigenvalues, respectively (see Chap. 2).

The covariance matrix $\boldsymbol{\Sigma}_{yxx}$ can be estimated by:

$$\mathbb{R}^{d\times d} \ni \widehat{\boldsymbol{\Sigma}}_{yxx} = \frac{1}{n}\sum_{i=1}^{n}(y_i-\widehat{\mu}_y)(\boldsymbol{x}_i-\widehat{\boldsymbol{\mu}}_x)(\boldsymbol{x}_i-\widehat{\boldsymbol{\mu}}_x)^\top,$$

where $\widehat{\boldsymbol{\mu}}_x$ and $\widehat{\mu}_y$ are the sample mean of the covariates and labels, respectively. Therefore, the eigenvectors of the generalized eigenvalue problem $\widehat{\boldsymbol{\Sigma}}_{yxx}\boldsymbol{u}_j = \lambda_j\widehat{\boldsymbol{\Sigma}}_{xx}\boldsymbol{u}_j, \forall j$ can be found to have the projection matrix $\boldsymbol{U} = [\boldsymbol{u}_1, \dots, \boldsymbol{u}_p]$ onto the central subspace. According to [52], pHd is not an exhaustive method (see Definition 15.7). The pHd method has been further improved upon by [15, 23, 24]. Further technical comments of pHd are available in [55].

15.4.2 Minimum Average Variance Estimation (MAVE)

The Minimum Average Variance Estimation (MAVE) was first proposed in [64].

Theorem 15.7 ([64, Section 2]) *The orthogonal projection matrix onto the central subspace is the solution to the following optimization problem:*

$$\begin{aligned} \underset{\boldsymbol{U}}{\text{minimize}} \quad & \mathbb{E}\big[Y - \mathbb{E}[Y \mid \boldsymbol{U}^\top X]\big]^2 \\ \text{subject to} \quad & \boldsymbol{U}^\top\boldsymbol{U} = \boldsymbol{I}. \end{aligned}$$

The following is defined as $\sigma_U^2(\boldsymbol{U}^\top X) := \mathbb{E}\big[(Y - \mathbb{E}[Y \mid \boldsymbol{U}^\top X])^2 | \boldsymbol{U}^\top X\big]$. Therefore:

$$\mathbb{E}\big[Y - \mathbb{E}[Y \mid \boldsymbol{U}^\top X]\big]^2 = \mathbb{E}[\sigma_U^2(\boldsymbol{U}^\top X)].$$

Therefore:

$$
\begin{aligned}
&\underset{U}{\text{minimize}} \quad \mathbb{E}[\sigma_U^2(U^\top X)] \\
&\text{subject to} \quad U^\top U = I.
\end{aligned}
\tag{15.26}
$$

It is possible to estimate $\sigma_U^2(U^\top X)$ by [64, Section 2]:

$$
\widehat{\sigma}_U^2(U^\top x_j) = \min_{a\in\mathbb{R}, b\in\mathbb{R}^p} \sum_{i=1}^{n} \Big(y_i - \big(a + b^\top U^\top (x_i - x_j)\big)\Big)^2 w_{ij},
$$

where:

$$
w_{ij} := \frac{k_h(U^\top (x_i - x_j))}{\sum_{\ell=1}^n k_h(U^\top (x_\ell - x_j))},
$$

in which $k_h(.)$ denotes a kernel function with bandwidth h (see Chap. 3). The estimated $\widehat{\sigma}_U^2(U^\top x_j)$ can be used in Eq. (15.26) to calculate the projection matrix onto the central subspace. The optimization variables of this problem are a, b, and U and any optimization method can be used in an alternative optimization approach for solving this problem (see Chap. 4). The MAVE method was improved upon in [63]. Additional methods based on MAVE include central subspace MAVE (csMAVE) [60] and ensemble MAVE [67]. An R programming language package for MAVE exists [62].

15.4.3 Conditional Variance Estimation (CVE)

Conditional Variance Estimation (CVE) was first proposed in [35, 36].

Theorem 15.8 ([36, Section 2, Section 2.1, Corollary 2]) *Consider the following definitions:*

$$
\begin{aligned}
&\widetilde{L}(V, s_0) := \mathbb{V}ar(Y|X \in s_0 + \mathbb{C}ol(V)), \\
&L(V) := \mathbb{E}[\widetilde{L}(V, X)],
\end{aligned}
$$

where $s_0 \in \mathbb{R}^d$ is a shifting point and $\mathbb{C}ol(V)$ is the column space of V. Consider the following optimization problem:

$$
\begin{aligned}
&\underset{V}{\text{minimize}} \quad L(V) \\
&\text{subject to} \quad V \in \mathcal{S}t(p, d),
\end{aligned}
\tag{15.27}
$$

whose solution is denoted by V_p. The orthogonal space to $\mathbb{C}ol(V_p)$ is equal to the columns space of the projection matrix U onto the central space:

$$\mathbb{C}ol(U) = \mathbb{C}ol(V_p)^{\perp}. \tag{15.28}$$

It is possible to estimate $L(V)$ as follows [36]:

$$d_i(V, s_0) := \|U^\top (x_i - s_0)\|_2^2,$$
$$w_i(V, s_0) := \frac{k_h(d_i(V, s_0))}{\sum_{\ell=1}^{n} k_h(d_\ell(V, s_0))},$$

where w_i is defined similarly to w_i in the MAVE method. The estimate of $\widetilde{L}(V, s_0)$ is [36]:

$$\widetilde{L}_n(V, s_0) = \sum_{i=1}^{n} w_i(V, s_0) y_i^2 - \Big(\sum_{i=1}^{n} w_i(V, s_0) y_i\Big)^2.$$

Finally, the estimate of $L(V)$ is:

$$L_n(V) = \frac{1}{n} \sum_{j=1}^{n} \widetilde{L}_n(V, x_j).$$

In Eq. (15.27), $L_n(V)$ can be used instead of $L(V)$ to find the matrix V using Riemannian optimization on the Stiefel manifold [1]. Then, using Eq. (15.28), the projection matrix U onto the central subspace can be found. It is noteworthy that there also exists an ensemble version of CVE [37].

15.4.4 Deep Sufficient Dimension Reduction

15.4.4.1 Deep Variational Sufficient Dimension Reduction (DVSDR)

Deep Variational Sufficient Dimension Reduction (DVSDR), proposed in [3], is one of the approaches for deep SDR. It formulates the network as a variational autoencoder [48] (see Chap. 20) and interprets it as SDR. Let $\mathbb{R}^{p \times n} \ni \widetilde{X} = U^\top X$ be the projected covariates onto the central subspace. According to Eq. (15.6), there is $X \perp\!\!\!\perp Y \mid \widetilde{X}$, where $\widetilde{X}$ is the random variable associated with $\widetilde{X}$. Therefore, $\widetilde{X}$ is seen as a latent factor on which both covariates X and labels Y depend, but conditioning on that, they are independent. Recall that in variational inference, there is a latent factor (see Chaps. 12 and 20). As in the variational autoencoder, there is a deep autoencoder, whose encoder and decoder model the conditional probabilities $q(\widetilde{X}|X)$ and $p(X|\widetilde{X})$, respectively. Let the weights of the encoder and decoder be denoted by ϕ and θ, respectively. The encoder layers map data from dimension d to p, and the decoder layers map data from dimension p to d. The Evidence Lower Bound (ELBO) of variational inference can be used as follows (see Chaps. 12 and 20):

$$\mathcal{L}^u_{\phi,\theta}(\boldsymbol{x}_i) := \mathbb{E}_{q_\phi(\widetilde{X}|X)}\big[\ln(p_\theta(\boldsymbol{x}_i|\widetilde{\boldsymbol{x}}_i))\big] - \text{KL}\big(q_\phi(\widetilde{\boldsymbol{x}}_i|\boldsymbol{x}_i) \,\|\, p(\widetilde{\boldsymbol{x}}_i)\big).$$

It is also possible to add additional layers in the decoder part to generate labels for classification or regression. Let the weights of these layers be denoted by ψ. These layers model $p(Y|\widetilde{X})$. Including these layers into ELBO results in:

$$\begin{aligned}\mathcal{L}^\ell_{\phi,\theta,\psi}(\boldsymbol{x}_i) &:= \mathbb{E}_{q_\phi(\widetilde{X}|X)}\big[\ln(p_\theta(\boldsymbol{x}_i|\widetilde{\boldsymbol{x}}_i))\big] \\ &+ \mathbb{E}_{q_\phi(\widetilde{X}|X)}\big[\ln(p_\psi(y_i|\widetilde{\boldsymbol{x}}_i))\big] - \text{KL}\big(q_\phi(\widetilde{\boldsymbol{x}}_i|\boldsymbol{x}_i) \,\|\, p(\widetilde{\boldsymbol{x}}_i)\big).\end{aligned}$$

A regularized loss function can be created using both of these ELBO loss functions:

$$\max_{\phi,\theta,\psi} \sum_{i=1}^{n} \big(\mathcal{L}^\ell_{\phi,\theta,\psi}(\boldsymbol{x}_i) + \mathcal{L}^u_{\phi,\theta}(\boldsymbol{x}_i)\big).$$

The weights ϕ, θ, and ψ are tuned by backpropagation. Refer to Chap. 20 for further information on how to train a variational autoencoder. If the dataset is partially labelled, $\mathcal{L}^\ell_{\phi,\theta,\psi}$ and $\mathcal{L}^u_{\phi,\theta}$ can also be used for labelled and unlabelled parts, respectively, to have semisupervised learning.

15.4.4.2 Meta-Learning for Sufficient Dimension Reduction

An additional approach for deep SDR is proposed in [47], which takes a meta learning approach [38]. Consider the regression model in Eq. (15.3). This can be modelled by a neural network, whose first layer projects data $\boldsymbol{X}$ from d dimensions to p dimensions. The weights of this layer are the matrix $\boldsymbol{U} \in \mathbb{R}^{d\times p}$. The last layer has one neuron for predicting the label y. Let the weights of the network be denoted by θ. The first layer models $f(\boldsymbol{U}^\top \boldsymbol{X})$ in (15.3). First, the covariates $\boldsymbol{X}$ are fed into the network with weights θ, predicted labels are obtained, and the weights are trained using backpropagation with least squares loss between the labels and predicted labels. Let θ' be the updated weights by this training. Then, in the metatraining phase, the covariates are fed to the network with the weights θ', predicted labels are obtained, and the weights are trained using backpropagation with least squares loss between the labels and predicted labels. This training and metatraining procedure is repeated until convergence.

15.5 Kernel Dimension Reduction

15.5.1 Supervised Kernel Dimension Reduction

Kernel Dimension Reduction (KDR) was originally proposed in [43] and further refined in [41, 42]. It is one of the methods for SDR and it makes use of Eqs. (15.7) and (15.8) for finding the sufficient subspace. KDR has a machine learning approach for dimensionality reduction [41, 43]; although, it can also be considered a statistical method for high-dimensional regression [42]. KDR uses kernels in the Reproducing Kernel Hilbert Space (RKHS) (see Chap. 3) to calculate the central subspace. It is noteworthy that another SDR method that uses RKHS is kernel Principal Support Vector Machines (PSVM) [50].

Definition 15.14 (Conditional Covariance Operator [43]) The conditional covariance operator in RKHS is defined as:

$$\boldsymbol{\Sigma}_{YY|\widetilde{X}} := \boldsymbol{\Sigma}_{YY} - \boldsymbol{\Sigma}_{Y\widetilde{X}}\boldsymbol{\Sigma}_{\widetilde{X}\widetilde{X}}^{-1}\boldsymbol{\Sigma}_{\widetilde{X}Y} \tag{15.29}$$

where $\boldsymbol{\Sigma}_{AB}$ is the covariance of the random variables A and B and $\widetilde{\boldsymbol{X}} := \boldsymbol{U}^\top \boldsymbol{X}$ is the projected covariates ($\widetilde{X}$ is its associated random variable).

Theorem 15.9 ([43, Theorem 5]) *Let $\mathbb{R}^{d\times d} \ni \boldsymbol{Q} = [\boldsymbol{U}|\boldsymbol{V}]$ be an orthogonal matrix, where $\boldsymbol{U} \in \mathbb{R}^{d\times p}$ is the truncated projection matrix onto the p-dimensional subspace and $\boldsymbol{V} \in \mathbb{R}^{d\times(d-p)}$ is the rest of the matrix $\boldsymbol{Q}$. Then:*

$$\boldsymbol{\Sigma}_{YY|\widetilde{X}} \geq \boldsymbol{\Sigma}_{YY|X},$$
$$\boldsymbol{\Sigma}_{YY|\widetilde{X}} = \boldsymbol{\Sigma}_{YY|X} \iff Y \perp\!\!\!\perp (\boldsymbol{V}^\top X) \,|\, \widetilde{X},$$

where $\widetilde{\boldsymbol{X}} := \boldsymbol{U}^\top \boldsymbol{X}$ is the projected covariates onto the p-dimensional central subspace. The central subspace can be found by minimizing the covariance of the labels conditioned on the projected covariates onto the subspace:

$$\underset{\boldsymbol{U}}{\text{minimize}} \quad \boldsymbol{\Sigma}_{YY|\widetilde{X}}. \tag{15.30}$$

KDR allows the labels to be multidimensional (ℓ-dimensional), i.e., $\boldsymbol{Y} \in \mathbb{R}^{\ell\times n}$. Consider the double-centered kernel matrices for labels and the projected covariates:

$$\mathbb{R}^{n\times n} \ni \widehat{\boldsymbol{K}}_Y = \boldsymbol{H}\boldsymbol{Y}^\top\boldsymbol{Y}\boldsymbol{H}, \tag{15.31}$$

$$\mathbb{R}^{n\times n} \ni \widehat{\boldsymbol{K}}_{\widetilde{X}} = \boldsymbol{H}\widetilde{\boldsymbol{X}}^\top\widetilde{\boldsymbol{X}}\boldsymbol{H} = \boldsymbol{H}\boldsymbol{X}^\top\boldsymbol{U}\boldsymbol{U}^\top\boldsymbol{X}\boldsymbol{H}, \tag{15.32}$$

where $\boldsymbol{H} := \boldsymbol{I} - (1/n)\boldsymbol{1}\boldsymbol{1}^\top \in \mathbb{R}^{n \times n}$ is the centering matrix (see Chap. 2). The empirical estimates for the covariance matrices are [43]:

$$\widehat{\boldsymbol{\Sigma}}_{YY} = (\widehat{\boldsymbol{K}}_Y + \epsilon \boldsymbol{I})^2, \quad \widehat{\boldsymbol{\Sigma}}_{Y\widetilde{X}} = \widehat{\boldsymbol{K}}_Y \widehat{\boldsymbol{K}}_{\widetilde{X}},$$
$$\widehat{\boldsymbol{\Sigma}}_{UU} = (\widehat{\boldsymbol{K}}_U + \epsilon \boldsymbol{I})^2, \quad \widehat{\boldsymbol{\Sigma}}_{\widetilde{X}Y} = \widehat{\boldsymbol{K}}_{\widetilde{X}} \widehat{\boldsymbol{K}}_Y,$$

where ϵ is a small positive number and adding $\epsilon \boldsymbol{I}$ makes the kernel matrices full rank and invertible. The $\boldsymbol{\Sigma}_{YY|\widetilde{X}} = \boldsymbol{\Sigma}_{YY|\boldsymbol{U}^\top X}$ can be estimated empirically as [43]:

$$\begin{aligned}\mathbb{R}^{n \times n} \ni \widehat{\boldsymbol{\Sigma}}_{YY|\widetilde{X}} &= \widehat{\boldsymbol{\Sigma}}_{YY} - \widehat{\boldsymbol{\Sigma}}_{Y\widetilde{X}} \widehat{\boldsymbol{\Sigma}}_{\widetilde{X}\widetilde{X}}^{-1} \widehat{\boldsymbol{\Sigma}}_{\widetilde{X}Y} \\ &= (\widehat{\boldsymbol{K}}_Y + \epsilon \boldsymbol{I})^2 - \widehat{\boldsymbol{K}}_Y \widehat{\boldsymbol{K}}_{\widetilde{X}} (\widehat{\boldsymbol{K}}_{\widetilde{X}} + \epsilon \boldsymbol{I})^{-2} \widehat{\boldsymbol{K}}_{\widetilde{X}} \widehat{\boldsymbol{K}}_Y.\end{aligned} \tag{15.33}$$

15.5.1.1 Supervised KDR by Projected Gradient Descent

In practice, the determinant of the estimated covariance matrix, i.e., $\det(\widehat{\boldsymbol{\Sigma}}_{YY|\widetilde{X}})$ can be used in Eq. (15.30) for optimization. According to Schur's complement [43]:

$$\det(\widehat{\boldsymbol{\Sigma}}_{YY|\widetilde{X}}) = \frac{\det(\widehat{\boldsymbol{\Sigma}}_{(Y\widetilde{X})(Y\widetilde{X})})}{\det(\widehat{\boldsymbol{\Sigma}}_{\widetilde{X}\widetilde{X}})}, \tag{15.34}$$

where:

$$\widehat{\boldsymbol{\Sigma}}_{(Y\widetilde{X})(Y\widetilde{X})} = \begin{bmatrix} \widehat{\boldsymbol{\Sigma}}_{YY} & \widehat{\boldsymbol{\Sigma}}_{Y\widetilde{X}} \\ \widehat{\boldsymbol{\Sigma}}_{\widetilde{X}Y} & \widehat{\boldsymbol{\Sigma}}_{\widetilde{X}\widetilde{X}} \end{bmatrix}.$$

Equation (15.34) is symmetrized by dividing it by the constant $\det(\widehat{\boldsymbol{\Sigma}}_{YY})$. Finally, in practice, Eq. (15.30) is stated as:

$$\underset{\boldsymbol{U}}{\text{minimize}} \quad \frac{\det(\widehat{\boldsymbol{\Sigma}}_{(Y\widetilde{X})(Y\widetilde{X})})}{\det(\widehat{\boldsymbol{\Sigma}}_{YY}) \det(\widehat{\boldsymbol{\Sigma}}_{\widetilde{X}\widetilde{X}})}. \tag{15.35}$$

This problem can be solved iteratively by gradient descent (see Chap. 4), where the gradient is [41]:

$$\begin{aligned}\frac{\partial \ln(\det(\widehat{\boldsymbol{\Sigma}}_{YY|\widetilde{X}}))}{\partial \boldsymbol{U}} &= \textbf{tr}(\widehat{\boldsymbol{\Sigma}}_{YY|\widetilde{X}} \frac{\partial \widehat{\boldsymbol{\Sigma}}_{YY|\widetilde{X}}}{\partial \boldsymbol{U}}) \\ &= 2\epsilon \, \textbf{tr}\Big(\widehat{\boldsymbol{\Sigma}}_{YY|\widetilde{X}}^{-1} \widehat{\boldsymbol{K}}_Y (\widehat{\boldsymbol{K}}_{\widetilde{X}} + \epsilon \boldsymbol{I})^{-1} \frac{\partial \widehat{\boldsymbol{K}}_{\widetilde{X}}}{\partial \boldsymbol{U}} (\widehat{\boldsymbol{K}}_{\widetilde{X}} + \epsilon \boldsymbol{I})^{-2} \widehat{\boldsymbol{K}}_{\widetilde{X}} \widehat{\boldsymbol{K}}_Y\Big).\end{aligned}$$

15.5.1.2 Supervised KDR by Riemannian Optimization

Another approach for KDR optimization [42] is to use Eq. (15.33) into Eq. (15.30), to strengthen the diagonal of $\widehat{\boldsymbol{\Sigma}}_{\widetilde{X}\widetilde{X}}$ to make it invertible. The optimization can be made constrained by putting a constraint on the projection matrix to be orthogonal, i.e., $\boldsymbol{U}^\top \boldsymbol{U} = \boldsymbol{I}$. According to Definition 15.9, this constraint means that the projection matrix belongs to the Stiefel manifold. Therefore, the optimization problem is:

$$\begin{aligned} \underset{\boldsymbol{U}}{\text{minimize}} \quad & \widehat{\boldsymbol{\Sigma}}_{YY} - \widehat{\boldsymbol{\Sigma}}_{Y\widetilde{X}}(\widehat{\boldsymbol{\Sigma}}_{\widetilde{X}\widetilde{X}} + \epsilon \boldsymbol{I})^{-1}\widehat{\boldsymbol{\Sigma}}_{\widetilde{X}Y} \\ \text{subject to} \quad & \boldsymbol{U} \in \mathcal{S}t(p, d), \end{aligned} \tag{15.36}$$

which can be solved iteratively by Riemannian optimization [1].

Equation (15.36) can be slightly altered and restated to [58]:

$$\begin{aligned} \underset{\boldsymbol{U}}{\text{minimize}} \quad & \textbf{tr}\big(\widehat{\boldsymbol{K}}_Y(\widehat{\boldsymbol{K}}_{\widetilde{X}} + n\epsilon \boldsymbol{I})^{-1}\big) \\ \text{subject to} \quad & \boldsymbol{U}^\top \boldsymbol{U} = \boldsymbol{I}. \end{aligned} \tag{15.37}$$

Equation (15.37) can also be used as the optimization problem for KDR.

15.5.1.3 Formulation of Supervised KDR by HSIC

Suppose we want to measure the dependence of two random variables. Measuring the correlation between them is easier because correlation is just "linear" dependence. Two random variables are independent if and only if any bounded continuous functions of them are uncorrelated [46]. Therefore, if the two random variables $\boldsymbol{x}_1$ and $\boldsymbol{x}_1$ are mapped to two different ("separable") RKHSs and have $\boldsymbol{\phi}(\boldsymbol{x}_1)$ and $\boldsymbol{\phi}(\boldsymbol{x}_2)$, it is possible to measure the correlation of $\boldsymbol{\phi}(\boldsymbol{x}_1)$ and $\boldsymbol{\phi}(\boldsymbol{x}_2)$ in the Hilbert space to estimate the dependence of $\boldsymbol{x}$ and $\boldsymbol{y}$ in the input space. The correlation of $\boldsymbol{\phi}(\boldsymbol{x}_1)$ and $\boldsymbol{\phi}(\boldsymbol{x}_2)$ can be computed by the Hilbert-Schmidt norm of their cross-covariance. This is the intuition behind the Hilbert-Schmidt Independence Criterion (HSIC); see Chap. 3. An empirical nonnormalized estimation of the HSIC is [44]:

$$\text{HSIC}(\boldsymbol{x}_1, \boldsymbol{x}_2) := \textbf{tr}(\boldsymbol{H}\boldsymbol{K}_{x_1}\boldsymbol{H}\boldsymbol{K}_{x_2}), \tag{15.38}$$

where $\boldsymbol{K}_{x_1}$ and $\boldsymbol{K}_{x_2}$ are the kernels over $\boldsymbol{x}_1$ and $\boldsymbol{x}_2$, respectively.

Theorem 15.10 ([61, Proposition 1]) *Let c_0 be a positive constant and $\epsilon_n^2 \to 0$ when $n \to \infty$. Then:*

$$\textbf{tr}\big(\widehat{\boldsymbol{K}}_Y(\widehat{\boldsymbol{K}}_{\widetilde{X}} + n\epsilon \boldsymbol{I})^{-1}\big) \approx -c_0 n^2 \epsilon_n^2 \textbf{tr}(\boldsymbol{H}\boldsymbol{K}_{\widetilde{X}}\boldsymbol{H}\boldsymbol{K}_Y) = -c_0 n^2 \epsilon_n^2 \textbf{tr}(\widehat{\boldsymbol{K}}_{\widetilde{X}}\boldsymbol{K}_Y), \tag{15.39}$$

where $\boldsymbol{K}_Y := \boldsymbol{Y}^\top \boldsymbol{Y}$.

Comparing Eqs. (15.37), (15.38), and (15.39) demonstrates that the optimization of KDR can also be stated as the following maximization problem:

$$\begin{aligned} \underset{U}{\text{maximize}} \quad & \text{HSIC}(\widetilde{X}, Y) = \mathbf{tr}(\widehat{K}_{\widetilde{X}} K_Y) \\ \text{subject to} \quad & U^\top U = I. \end{aligned} \tag{15.40}$$

Corollary 15.4 (Equivalency of Supervised KDR and Supervised PCA) *The optimization problem of supervised Principal Component Analysis (PCA) [4] is exactly the same as Eq. (15.40) (note that in HSIC, any of the two kernels can be double-centered and it is not important which kernel is centered, but one of them must be double-centered). This equation is the optimization of supervised PCA because it can be shown that PCA is a special case of this problem, where the information of labels is not used (see Chap. 5). Therefore, supervised KDR and supervised PCA are equivalent!*

15.5.2 Supervised KDR for Nonlinear Regression

Manifold KDR (mKDR) [58] performs KDR on manifolds for nonlinear regression. mKDR combines the ideas of KDR and Laplacian eigenmap [5] (see Chap. 9). The Laplacian eigenmap brings the nonlinear information of manifold of data into the formulation. Let $\{r_j \in \mathbb{R}^n\}_{j=1}^m$ be the m tailing eigenvectors of the Laplacian matrix of the graph of the covariates (note that the eigenvector with an eigenvalue of zero is ignored). $\{t_i \in \mathbb{R}^m\}_{i=1}^n$ are defined to be $T := [t_1, \ldots, t_n] = [r_1, \ldots, r_m]^\top \in \mathbb{R}^{m \times n}$. In mKDR, kernelization by representation theory is used to kernelize the optimization problem in Eq. (15.37). According to representation theory, if the centered projected covariates XH are pulled to RKHS, they must lie in the span of all pulled training points (see Chap. 3):

$$\widetilde{X}H = \Phi(X)T \implies \widehat{K}_{\widetilde{X}} \overset{(15.32)}{=} T^\top \Phi(X)^\top \Phi(X) T = T^\top K_x T,$$

where $\Phi(X)$ is the pull of X to the RKHS and the kernel is defined as:

$$K_x := \Phi(X)^\top \Phi(X), \tag{15.41}$$

which must be positive semidefinite according to the properties of the kernel. To make the kernel bounded, $\mathbf{tr}(K_x) = 1$ is set. Substituting these into Eq. (15.37) results in:

$$\begin{aligned} \underset{K_x}{\text{minimize}} \quad & \mathbf{tr}\big(\widehat{K}_Y (T^\top K_x T + n\epsilon I)^{-1}\big) \\ \text{subject to} \quad & K_x \succeq \mathbf{0}, \\ & \mathbf{tr}(K_x) = 1. \end{aligned} \tag{15.42}$$

Note that, as explained above, $\boldsymbol{T}$ is obtained from the Laplacian eigenmap and then used in Eq. (15.42). Equation (15.42) is solved using the projected gradient method, where every step is projected onto the positive semidefinite cone (see Chap. 4). After solving the optimization problem, the solution $\boldsymbol{K}_x$ is decomposed using eigenvalue decomposition (which can be done because it is positive semidefinite, meaning its eigenvalues are not negative):

$$\boldsymbol{K}_x = \boldsymbol{A}\boldsymbol{\Delta}\boldsymbol{A}^\top = \boldsymbol{A}\boldsymbol{\Delta}^{1/2}\boldsymbol{\Delta}^{1/2}\boldsymbol{A}^\top \overset{(15.41)}{=} \boldsymbol{\Phi}(\boldsymbol{X})^\top\boldsymbol{\Phi}(\boldsymbol{X}),$$

where $\boldsymbol{A}$ is a matrix whose columns are the eigenvectors and $\boldsymbol{\Delta}$ is the diagonal matrix of the eigenvalues. Therefore, from the above expression, $\boldsymbol{\Phi}(\boldsymbol{X}) = \boldsymbol{\Delta}^{1/2}\boldsymbol{A}^\top$. If this matrix $\boldsymbol{\Phi}(\boldsymbol{X})$ is truncated to have the p top eigenvectors with the largest eigenvalues, it is the p-dimensional embedding of the covariates into the subspace.

15.5.3 Unsupervised Kernel Dimension Reduction

Unsupervised KDR [61] does not use labels and can be used when labels are not available. Let X' be a copy of the random variable X. In unsupervised KDR, rather than Eq. (15.6), the following is considered:

$$X \perp\!\!\!\perp X' \mid \boldsymbol{U}^\top X, \tag{15.43}$$

meaning that projection onto the central subspace is sufficient for the covariates to be independent of each other. Equation (15.37) is used, but with X instead of Y:

$$\begin{aligned} \underset{\boldsymbol{U}}{\text{minimize}} \quad & \mathbf{tr}\big(\widehat{\boldsymbol{K}}_X(\widehat{\boldsymbol{K}}_{\widetilde{X}} + n\epsilon\boldsymbol{I})^{-1}\big) \\ \text{subject to} \quad & \boldsymbol{U}^\top\boldsymbol{U} = \boldsymbol{I}. \end{aligned} \tag{15.44}$$

Any numerical optimization method can be used to solve this problem to find the projection matrix $\boldsymbol{U}$ onto the central subspace.

15.6 Chapter Summary

This chapter introduced SDR as a family of methods that determine a transformation of data to a lower dimensional space, which does not change the conditional of labels given data. Therefore, the subspace of SDR is sufficient for predicting the labels from the projected data. The first part of this chapter focused on SDR variants, which were related to regression problems in high-dimensional statistics. The second part of the chapter introduced KDR with a machine learning approach. Therefore, this chapter provided the necessary background and information for readers with either statistical or machine learning goals. Additional works on SDR can be found in [2, 10, 30, 49, 57, 65].

References

1. P-A Absil, Robert Mahony, and Rodolphe Sepulchre. *Optimization algorithms on matrix manifolds*. Princeton University Press, 2009.
2. Kofi P Adragni and R Dennis Cook. "Sufficient dimension reduction and prediction in regression". In: *Philosophical Transactions of the Royal Society A: Mathematical, Physical and Engineering Sciences* 367.1906 (2009), pp. 4385–4405.
3. Ershad Banijamali, Amir-Hossein Karimi, and Ali Ghodsi. "Deep variational sufficient dimensionality reduction". In: *Third workshop on Bayesian Deep Learning (NeurIPS 2018)*. 2018.
4. Elnaz Barshan et al. "Supervised principal component analysis: Visualization, classification and regression on subspaces and submanifolds". In: *Pattern Recognition* 44.7 (2011), pp. 1357–1371.
5. Mikhail Belkin and Partha Niyogi. "Laplacian eigenmaps and spectral techniques for embedding and clustering". In: *Nips*. Vol. 14. 14. 2001, pp. 585–591.
6. Efstathia Bura and R Dennis Cook. "Estimating the structural dimension of regressions via parametric inverse regression". In: *Journal of the Royal Statistical Society: Series B (Statistical Methodology)* 63.2 (2001), pp. 393–410.
7. Efstathia Bura, Sabrina Duarte, and Liliana Forzani. "Sufficient reductions in regressions with exponential family inverse predictors". In: *Journal of the American Statistical Association* 111.515 (2016), pp. 1313–1329.
8. Efstathia Bura and Liliana Forzani. "Sufficient reductions in regressions with elliptically contoured inverse predictors". In: *Journal of the American Statistical Association* 110.509 (2015), pp. 420–434.
9. Rick Chartrand and Wotao Yin. "Iteratively reweighted algorithms for compressive sensing". In: *2008 IEEE international conference on acoustics, speech and signal processing*. IEEE. 2008, pp. 3869–3872.
10. Francesca Chiaromonte and R Dennis Cook. "Sufficient dimension reduction and graphics in regression". In: *Annals of the Institute of Statistical Mathematics* 54.4 (2002), pp. 768–795.
11. Francesca Chiaromonte, R Dennis Cook, and Bing Li. "Sufficient dimension reduction in regressions with categorical predictors". In: *Annals of Statistics* (2002), pp. 475–497.
12. R Dennis Cook. "Fisher lecture: Dimension reduction in regression". In: *Statistical Science* 22.1 (2007), pp. 1–26.
13. R Dennis Cook. "Graphics for regressions with a binary response". In: *Journal of the American Statistical Association* 91.435 (1996), pp. 983–992.
14. R Dennis Cook. "On the interpretation of regression plots". In: *Journal of the American Statistical Association* 89.425 (1994), pp. 177–189.
15. R Dennis Cook. "Principal Hessian directions revisited". In: *Journal of the American Statistical Association* 93.441 (1998), pp. 84–94.
16. R Dennis Cook. "Regression Graphics". In: *Proceedings of the 30th Interface (the 30th symposium on the Interface between Statistics and Computer Science)*. 1998.
17. R Dennis Cook. *Regression graphics: Ideas for studying regressions through graphics*. John Wiley & Sons, 1998.
18. R Dennis Cook. "SAVE: a method for dimension reduction and graphics in regression". In: *Communications in statistics-Theory and methods* 29.9-10 (2000), pp. 2109–2121.

19. R Dennis Cook. "Using dimension-reduction subspaces to identify important inputs in models of physical systems". In: *Proceedings of the section on Physical and Engineering Sciences*. 1994, pp. 18–25.
20. R Dennis Cook and Liliana Forzani. "Likelihood-based sufficient dimension reduction". In: *Journal of the American Statistical Association* 104.485 (2009), pp. 197–208.
21. R Dennis Cook and Liliana Forzani. "Principal fitted components for dimension reduction in regression". In: *Statistical Science* 23.4 (2008), pp. 485–501.
22. R Dennis Cook and Hakbae Lee. "Dimension reduction in binary response regression". In: *Journal of the American Statistical Association* 94.448 (1999), pp. 1187–1200.
23. R Dennis Cook and Bing Li. "Determining the dimension of iterative Hessian transformation". In: *The Annals of Statistics* 32.6 (2004), pp. 2501–2531.
24. R Dennis Cook and Bing Li. "Dimension reduction for conditional mean in regression". In: *The Annals of Statistics* 30.2 (2002), pp. 455–474.
25. R Dennis Cook, Bing Li, and Francesca Chiaromonte. "Envelope models for parsimonious and efficient multivariate linear regression". In: *Statistica Sinica* (2010), pp. 927–960.
26. R Dennis Cook and Liqiang Ni. "Sufficient dimension reduction via inverse regression: A minimum discrepancy approach". In: *Journal of the American Statistical Association* 100.470 (2005), pp. 410–428.
27. R Dennis Cook and Liqiang Ni. "Using intraslice covariances for improved estimation of the central subspace in regression". In: *Biometrika* 93.1 (2006), pp. 65–74.
28. R Dennis Cook and Sanford Weisberg. *An introduction to regression graphics*. Vol. 405. John Wiley & Sons, 2009.
29. R Dennis Cook and Sanford Weisberg. "Sliced inverse regression for dimension reduction: Comment". In: *Journal of the American Statistical Association* 86.414 (1991), pp. 328–332.
30. R Dennis Cook and Xiangrong Yin. "Theory & methods: special invited paper: dimension reduction and visualization in discriminant analysis (with discussion)". In: *Australian & New Zealand Journal of Statistics* 43.2 (2001), pp. 147–199.
31. John P Cunningham and Zoubin Ghahramani. "Linear dimensionality reduction: Survey, insights, and generalizations". In: *The Journal of Machine Learning Research* 16.1 (2015), pp. 2859–2900.
32. A Philip Dawid. "Conditional independence in statistical theory". In: *Journal of the Royal Statistical Society: Series B (Methodological)* 41.1 (1979), pp. 1–15.
33. Persi Diaconis and David Freedman. "Asymptotics of graphical projection pursuit". In: *The annals of statistics* (1984), pp. 793–815.
34. Morris L Eaton. "A characterization of spherical distributions". In: *Journal of Multivariate Analysis* 20.2 (1986), pp. 272–276.
35. Lukas Fertl. "Sufficient Dimension Reduction using Conditional Variance Estimation and related concepts". PhD thesis. Technischen Universität Wien, 2021.
36. Lukas Fertl and Efstathia Bura. "Conditional Variance Estimator for Sufficient Dimension Reduction". In: *arXiv preprint arXiv:2102.08782* (2021).
37. Lukas Fertl and Efstathia Bura. "Ensemble Conditional Variance Estimator for Sufficient Dimension Reduction". In: *arXiv preprint arXiv:2102.13435* (2021).
38. Chelsea Finn, Pieter Abbeel, and Sergey Levine. "Model-agnostic meta-learning for fast adaptation of deep networks". In: *International Conference on Machine Learning*. 2017, pp. 1126–1135.
39. Jerome H Friedman and Werner Stuetzle. "Projection pursuit regression". In: *Journal of the American statistical Association* 76.376 (1981), pp. 817–823.
40. Jerome H Friedman and JohnWTukey. "A projection pursuit algorithm for exploratory data analysis". In: *IEEE Transactions on computers* 100.9 (1974), pp. 881–890.
41. Kenji Fukumizu, Francis R Bach, and Michael I Jordan. "Dimensionality reduction for supervised learning with reproducing kernel Hilbert spaces". In: *Journal of Machine Learning Research* 5.Jan (2004), pp. 73–99.
42. Kenji Fukumizu, Francis R Bach, and Michael I Jordan. "Kernel dimension reduction in regression". In: *The Annals of Statistics* 37.4 (2009), pp. 1871–1905.

43. Kenji Fukumizu, Francis R Bach, and Michael I Jordan. "Kernel dimensionality reduction for supervised learning". In: *Advances in neural information processing systems*. Vol. 16. 2003.
44. Arthur Gretton et al. "Measuring statistical dependence with Hilbert-Schmidt norms". In: *International conference on algorithmic learning theory*. Springer. 2005, pp. 63–77.
45. Peter Hall and Ker-Chau Li. "On almost linearity of low dimensional projections from high dimensional data". In: *The Annals of Statistics* (1993), pp. 867–889.
46. Matthias Hein and Olivier Bousquet. "Kernels, associated structures and generalizations". In: *Max-Planck-Institut fuer biologische Kybernetik, Technical Report* (2004).
47. Daniel Kapla, Lukas Fertl, and Efstathia Bura. "Fusing Sufficient Dimension Reduction with Neural Networks". In: *arXiv preprint arXiv:2104.10009* (2021).
48. Diederik P Kingma and Max Welling. "Auto-encoding variational Bayes". In: *International Conference on Learning Representations*. 2014.
49. Bing Li. *Sufficient dimension reduction: Methods and applications with R*. CRC Press, 2018.
50. Bing Li, Andreas Artemiou, and Lexin Li. "Principal support vector machines for linear and nonlinear sufficient dimension reduction". In: *The Annals of Statistics* 39.6 (2011), pp. 3182–3210.
51. Bing Li and Yuexiao Dong. "Dimension reduction for nonelliptically distributed predictors". In: *The Annals of Statistics* 37.3 (2009), pp. 1272–1298.
52. Bing Li and Shaoli Wang. "On directional regression for dimension reduction". In: *Journal of the American Statistical Association* 102.479 (2007), pp. 997–1008.
53. Bing Li, Hongyuan Zha, and Francesca Chiaromonte. "Contour regression: a general approach to dimension reduction". In: *The Annals of Statistics* 33.4 (2005), pp. 1580–1616.
54. Ker-Chau Li. "On principal Hessian directions for data visualization and dimension reduction: Another application of Stein's lemma". In: *Journal of the American Statistical Association* 87.420 (1992), pp. 1025–1039.
55. Ker-Chau Li. "Principal Hessian Directions Revisited: Comment". In: *Journal of the American Statistical Association* 93.441 (1998), pp. 94–97.
56. Ker-Chau Li. "Sliced inverse regression for dimension reduction". In: *Journal of the American Statistical Association* 86.414 (1991), pp. 316–327.
57. Yanyuan Ma and Liping Zhu. "A review on dimension reduction". In: *International Statistical Review* 81.1 (2013), pp. 134–150.
58. Jens Nilsson, Fei Sha, and Michael I Jordan. "Regression on manifolds using kernel dimension reduction". In: *Proceedings of the 24th international conference on Machine learning*. 2007, pp. 697–704.
59. Charles M Stein. "Estimation of the mean of a multivariate normal distribution". In: *The annals of Statistics* (1981), pp. 1135–1151.
60. Hansheng Wang and Yingcun Xia. "Sliced regression for dimension reduction". In: *Journal of the American Statistical Association* 103.482 (2008), pp. 811–821.
61. Meihong Wang, Fei Sha, and Michael Jordan. "Unsupervised kernel dimension reduction". In: *Advances in neural information processing systems* 23 (2010), pp. 2379–2387.
62. Hang Weiqiang and Xia Yingcun. "MAVE: Methods for Dimension Reduction". In: *R package version* 1.10 (2019).
63. Yingcun Xia. "A constructive approach to the estimation of dimension reduction directions". In: *The Annals of Statistics* 35.6 (2007), pp. 2654–2690.
64. Yingcun Xia et al. "An adaptive estimation of dimension reduction space (with discussion)". In: *Journal of the Royal Statistical Society. Series B. Statistical Methodology* 64 (2002), pp. 363–410.
65. Xiangrong Yin. "Sufficient dimension reduction in regression". In: *High-dimensional Data Analysis*. World Scientific, 2011, pp. 257–273.
66. Xiangrong Yin and R Dennis Cook. "Estimating central subspaces via inverse third moments". In: *Biometrika* 90.1 (2003), pp. 113–125.
67. Xiangrong Yin and Bing Li. "Sufficient dimension reduction based on an ensemble of minimum average variance estimators". In: *The Annals of Statistics* (2011), pp. 3392–3416.

68. Jia Zhang and Xin Chen. "Principal envelope model". In: *Journal of Statistical Planning and Inference* 206 (2020), pp. 249–262.
69. Xin Zhang, Chong Wang, and Yichao Wu. "Functional envelope for model-free sufficient dimension reduction". In: *Journal of Multivariate Analysis* 163 (2018), pp. 37–50.

Chapter 16
Stochastic Neighbour Embedding

16.1 Introduction

Stochastic Neighbour Embedding (SNE) [13] is a manifold learning and dimensionality reduction method that can be used for feature extraction and data visualization. It takes a probabilistic approach, to fit the data in the embedding space locally hoping to preserve the global structure of data [37]. The idea of SNE is to consider every point as neighbours of other points with some probability where the closer points are neighbours with higher probability. Therefore, rather than considering k nearest neighbours in a binary manner (whether or not they are neighbours), it considers neighbours in a stochastic way (for how probable it is to be neighbours). It tries to preserve the probability of neighbourhoods in the low-dimensional embedding space. It is noteworthy that there are other similar probabilistic dimensionality reduction methods that make use of a Gaussian distribution for neighbourhood embedding. Some examples are Neighbourhood Component Analysis (NCA) [9], deep NCA [24], and Proxy-NCA [31].

SNE uses the Gaussian distribution for neighbours in both the input and embedding spaces. Student-t distributed SNE, also called t-SNE [27], considers the Student-t and Gaussian distributions in the input and embedding spaces, respectively. The reason for using the Student-t distribution in t-SNE is because of its heavier tails, meaning it can include more information from the high-dimensional data. t-SNE is one of the state-of-the-art methods for data visualization; for example, it has been used for DNA and single-cell data visualization [21]. This chapter explains SNE, symmetric SNE, t-SNE (or Cauchy-SNE), t-SNE with general degrees of freedom, their out-of-sample extensions, and their accelerations.

The goal of SNE is to embed the high-dimensional data $\{\boldsymbol{x}_i\}_{i=1}^n$ into the lower dimensional data $\{\boldsymbol{y}_i\}_{i=1}^n$, where n is the number of data points. The dimensions of high- and low-dimensional spaces are denoted by d and p, respectively, i.e., $\boldsymbol{x}_i \in \mathbb{R}^d$ and $\boldsymbol{y}_i \in \mathbb{R}^p$. Usually $p \ll d$ holds and $p \in \{2, 3\}$ is used for data visualization.

B. Ghojogh et al., *Elements of Dimensionality Reduction and Manifold Learning*,
https://doi.org/10.1007/978-3-031-10602-6_16

16.2 Stochastic Neighbour Embedding (SNE)

In SNE [13], a Gaussian probability is considered around every point $\boldsymbol{x}_i$, where point $\boldsymbol{x}_i$ is on the mean of the distribution. The Gaussian distribution is for the probability of accepting any other point as the neighbour of $\boldsymbol{x}_i$ where the farther points are neighbors with less probability. Therefore, the variable is distance, denoted by $d \in \mathbb{R}$, and the Gaussian probability is:

$$f(d) = \frac{1}{\sqrt{2\pi\sigma^2}} \exp(-\frac{d^2}{2\sigma^2}), \tag{16.1}$$

where the mean of the distribution is assumed to be zero. The fixed multiplier $\frac{1}{\sqrt{2\pi\sigma^2}}$ can be dropped; however, $\exp(-d^2/2\sigma^2)$ does not add (integrate) to one and thus it is not a probability density function. To tackle this problem, there is a trick to divide $\exp(-d^2/2\sigma^2)$ by the summation of all possible values of $\exp(-d^2/2\sigma^2)$, which results in a *softmax* function. Therefore, the probability that the point $\boldsymbol{x}_i \in \mathbb{R}^d$ takes $\boldsymbol{x}_j \in \mathbb{R}^d$ as its neighbour is:

$$\mathbb{R} \ni p_{ij} := \frac{\exp(-d_{ij}^2)}{\sum_{k \neq i} \exp(-d_{ik}^2)}, \tag{16.2}$$

where:

$$\mathbb{R} \ni d_{ij}^2 := \frac{||\boldsymbol{x}_i - \boldsymbol{x}_j||_2^2}{2\sigma_i^2}. \tag{16.3}$$

Note that this trick is also used for q_{ij} in SNE and for p_{ij} and q_{ij} in t-SNE (and its variants), as discussed later in this chapter.

It is noteworthy that the mentioned trick is also used in other methods, such as the Continuous Bag-of-Word (CBOW) model of Word2Vec [30, 35], Euclidean Embedding [7], and Parametric Embedding [16]. In this trick, the summation in the denominator can become excessively time-consuming, especially when the dataset (or corpus for Word2Vec) is large. This, plus the slow pace of gradient descent [3], is why SNE, t-SNE, and Word2Vec are very slow and even infeasible for large datasets. Word2Vec approached the problem of the slow pace by introducing the Negative Sampling Skip-Gram model [8, 29], which uses a logistic function similar to the approach of logistic regression [20]. In the logistic function, the inner product (similarity) of data points is used, rather than the distance of data points, and there is no summation in the denominator. The Negative Sampling Skip-Gram model also uses Newton's method, which is much faster than gradient descent [3], similar to logistic regression.

σ_i^2 is the variance that is considered for the Gaussian distribution used for $\boldsymbol{x}_i$. It can be set to a fixed number or determined by a binary search to make the entropy of distribution a specific value [13]. Note that according to the distribution of data in the input space, the best value for the variance of the Gaussian distributions can be found.

In the low-dimensional embedding space, a Gaussian probability distribution can be considered for the point $\boldsymbol{y}_i \in \mathbb{R}^p$ to take $\boldsymbol{y}_j \in \mathbb{R}^p$ as its neighbour:

$$\mathbb{R} \ni q_{ij} := \frac{\exp(-z_{ij}^2)}{\sum_{k \neq i} \exp(-z_{ik}^2)}, \tag{16.4}$$

where:

$$\mathbb{R} \ni z_{ij}^2 := ||\boldsymbol{y}_i - \boldsymbol{y}_j||_2^2. \tag{16.5}$$

Note that the variance of distribution is not used (or is set to $\sigma_i^2 = 0.5$ to cancel 2 in the denominator) because the variance of distribution in the embedding space is the choice of algorithm.

The aim is to have the probability distributions in both the input and embedded spaces be as similar as possible; therefore, the cost function to be minimized can be the summation of the Kullback-Leibler (KL) divergences over the n points:

$$\mathbb{R} \ni c_1 := \sum_{i=1}^{n} \text{KL}(P_i||Q_i) = \sum_{i=1}^{n} \sum_{j=1, j\neq i}^{n} p_{ij} \log(\frac{p_{ij}}{q_{ij}}), \tag{16.6}$$

where p_{ij} and q_{ij} are Eqs. (16.2) and (16.4). Note that divergences other than the KL divergence can be used for SNE optimization; e.g., see [15].

Proposition 16.1 *The gradient of c_1 with respect to $\boldsymbol{y}_i$ is:*

$$\mathbb{R}^p \ni \frac{\partial c_1}{\partial \boldsymbol{y}_i} = 2 \sum_{j=1}^{n} (p_{ij} - q_{ij} + p_{ji} - q_{ji})(\boldsymbol{y}_i - \boldsymbol{y}_j), \tag{16.7}$$

where p_{ij} and q_{ij} are Eqs. (16.2) and (16.4), and $p_{ii} = q_{ii} = 0$.

Proof This proof is inspired by [27]. Let:

$$\mathbb{R} \ni r_{ij} := z_{ij}^2 = ||\boldsymbol{y}_i - \boldsymbol{y}_j||_2^2. \tag{16.8}$$

By changing $\boldsymbol{y}_i$, there is only a change impact in z_{ij} and z_{ji} (or r_{ij} and r_{ji}) for all j's. Then, according to chain rule:

$$\mathbb{R}^p \ni \frac{\partial c_1}{\partial \boldsymbol{y}_i} = \sum_j \Big(\frac{\partial c_1}{\partial r_{ij}} \frac{\partial r_{ij}}{\partial \boldsymbol{y}_i} + \frac{\partial c_1}{\partial r_{ji}} \frac{\partial r_{ji}}{\partial \boldsymbol{y}_i}\Big).$$

According to Eq. (16.8):

$$r_{ij} = ||\boldsymbol{y}_i - \boldsymbol{y}_j||_2^2 \implies \frac{\partial r_{ij}}{\partial \boldsymbol{y}_i} = 2(\boldsymbol{y}_i - \boldsymbol{y}_j),$$

$$r_{ji} = ||\boldsymbol{y}_j - \boldsymbol{y}_i||_2^2 = ||\boldsymbol{y}_i - \boldsymbol{y}_j||_2^2 \implies \frac{\partial r_{ji}}{\partial \boldsymbol{y}_i} = 2(\boldsymbol{y}_i - \boldsymbol{y}_j).$$

Therefore:

$$\therefore \quad \frac{\partial c_1}{\partial \boldsymbol{y}_i} = 2 \sum_j \big(\frac{\partial c_1}{\partial r_{ij}} + \frac{\partial c_1}{\partial r_{ji}}\big)(\boldsymbol{y}_i - \boldsymbol{y}_j). \tag{16.9}$$

The cost function can be rewritten as:

$$c_1 = \sum_k \sum_{l \neq k} p_{kl} \log(\frac{p_{kl}}{q_{kl}}) = \sum_{k \neq l} p_{kl} \log(\frac{p_{kl}}{q_{kl}}) = \sum_{k \neq l} \big(p_{kl} \log(p_{kl}) - p_{kl} \log(q_{kl})\big),$$

whose first term is a constant with respect to q_{kl} and thus to r_{kl}. The derivative is:

$$\mathbb{R} \ni \frac{\partial c_1}{\partial r_{ij}} = -\sum_{k \neq l} p_{kl} \frac{\partial(\log(q_{kl}))}{\partial r_{ij}}.$$

According to Eq. (16.4), q_{kl} is:

$$q_{kl} := \frac{\exp(-z_{kl}^2)}{\sum_{k \neq f} \exp(-z_{kf}^2)} = \frac{\exp(-r_{kl})}{\sum_{k \neq f} \exp(-r_{kf})}.$$

The denominator of q_{kl} is taken as:

$$\beta := \sum_{k \neq f} \exp(-z_{kf}^2) = \sum_{k \neq f} \exp(-r_{kf}). \tag{16.10}$$

It is possible to restate $\log(q_{kl}) = \log(q_{kl}) + \log\beta - \log\beta = \log(q_{kl}\beta) - \log\beta$. Therefore:

$$\therefore \quad \frac{\partial c_1}{\partial r_{ij}} = -\sum_{k \neq l} p_{kl} \frac{\partial\big(\log(q_{kl}\beta) - \log\beta\big)}{\partial r_{ij}}$$

$$= -\sum_{k \neq l} p_{kl} \Big[\frac{\partial\big(\log(q_{kl}\beta)\big)}{\partial r_{ij}} - \frac{\partial\big(\log\beta\big)}{\partial r_{ij}}\Big] = -\sum_{k \neq l} p_{kl} \Big[\frac{1}{q_{kl}\beta} \frac{\partial\big(q_{kl}\beta\big)}{\partial r_{ij}} - \frac{1}{\beta}\frac{\partial\beta}{\partial r_{ij}}\Big].$$

$q_{kl}\beta$ is:

$$q_{kl}\beta = \frac{\exp(-r_{kl})}{\sum_{f\neq k}\exp(-r_{kf})} \times \sum_{k\neq f}\exp(-r_{kf}) = \exp(-r_{kl}).$$

Therefore:

$$\therefore \quad \frac{\partial c_1}{\partial r_{ij}} = -\sum_{k\neq l} p_{kl}\Big[\frac{1}{q_{kl}\beta}\frac{\partial\big(\exp(-r_{kl})\big)}{\partial r_{ij}} - \frac{1}{\beta}\frac{\partial \beta}{\partial r_{ij}}\Big].$$

$\partial\big(\exp(-r_{kl})\big)/\partial r_{ij}$ is nonzero for only $k = i$ and $l = j$; therefore:

$$\frac{\partial\big(\exp(-r_{ij})\big)}{\partial r_{ij}} = -\exp(-r_{ij}),$$

$$\frac{\partial \beta}{\partial r_{ij}} = \frac{\partial \sum_{k\neq f}\exp(-r_{kf})}{\partial r_{ij}} = \frac{\partial \exp(-r_{ij})}{\partial r_{ij}} = -\exp(-r_{ij}).$$

Therefore:

$$\therefore \quad \frac{\partial c_1}{\partial r_{ij}} = -\Big(p_{ij}\Big[\frac{-1}{q_{ij}\beta}\exp(-r_{ij})\Big] + 0 + \cdots + 0\Big) - \sum_{k\neq l} p_{kl}\Big[\frac{1}{\beta}\exp(-r_{ij})\Big].$$

$\sum_{k\neq l} p_{kl} = 1$ holds because the summation of all possible probabilities is one. Thus:

$$\begin{aligned}\frac{\partial c_1}{\partial r_{ij}} &= -p_{ij}\Big[\frac{-1}{q_{ij}\beta}\exp(-r_{ij})\Big] - \Big[\frac{1}{\beta}\exp(-r_{ij})\Big] \\ &= \underbrace{\frac{\exp(-r_{ij})}{\beta}}_{=q_{ij}}\Big[\frac{p_{ij}}{q_{ij}} - 1\Big] = p_{ij} - q_{ij}. \end{aligned} \tag{16.11}$$

Similarly:

$$\frac{\partial c_1}{\partial r_{ji}} = p_{ji} - q_{ji}. \tag{16.12}$$

Substituting the obtained derivatives in Eq. (16.9) results in:

$$\frac{\partial c_1}{\partial \boldsymbol{y}_i} = 2\sum_j (p_{ij} - q_{ij} + p_{ji} - q_{ji})(\boldsymbol{y}_i - \boldsymbol{y}_j),$$

which is the gradient mentioned in the proposition. □

The update of the embedded point $\boldsymbol{y}_i$ is performed by gradient descent. Every iteration is:

$$\begin{aligned} \Delta \boldsymbol{y}_i^{(t)} &:= -\eta \, \frac{\partial c_1}{\partial \boldsymbol{y}_i} + \alpha(t)\, \Delta \boldsymbol{y}_i^{(t-1)}, \\ \boldsymbol{y}_i^{(t)} &:= \boldsymbol{y}_i^{(t-1)} + \Delta \boldsymbol{y}_i^{(t)}, \end{aligned} \tag{16.13}$$

where momentum, or $\alpha(t)$, is used for better convergence [33]. It can be smaller for initial iterations and larger for additional iterations. For example [27]:

$$\alpha(t) := \begin{cases} 0.5 & t < 250, \\ 0.8 & t \geq 250. \end{cases} \tag{16.14}$$

In Hinton and Roweis's original paper on SNE [13], the momentum term is not mentioned, but it is suggested in [27]. η is the learning rate, which can be a small positive constant (e.g., $\eta = 0.1$) or can be updated adaptively [17]. Moreover, in both [13] and [27], it is mentioned that in SNE, the Gaussian noise (random jitter) should be added to the solution of the first iterations before going to the next iterations. Adding noise will help avoid the local optimum solutions.

16.3 Symmetric Stochastic Neighbour Embedding

In symmetric SNE [27], a Gaussian probability is considered for every point $\boldsymbol{x}_i$. The probability that point $\boldsymbol{x}_i \in \mathbb{R}^d$ takes $\boldsymbol{x}_j \in \mathbb{R}^d$ as its neighbour is:

$$\mathbb{R} \ni p_{ij} := \frac{\exp(-d_{ij}^2)}{\sum_{k \neq l} \exp(-d_{kl}^2)}, \tag{16.15}$$

where:

$$\mathbb{R} \ni d_{ij}^2 := \frac{||\boldsymbol{x}_i - \boldsymbol{x}_j||_2^2}{2\sigma_i^2}. \tag{16.16}$$

Note that the denominator of Eq. (16.15) for all points is fixed; thus, it is symmetric for i and j. Compare this with Eq. (16.2) which is not symmetric. σ_i^2 is the variance that is considered for the Gaussian distribution used for $\boldsymbol{x}_i$. It can be set to a fixed number or determined by a binary search to make the entropy of the distribution a specific value [13].

Equation (16.15) has a problem with outliers. If point $\boldsymbol{x}_i$ is an outlier, its p_{ij} will be extremely small because the denominator is fixed for every point and the numerator will be small for the outlier. However, if Eq. (16.2) is used for p_{ij}, the denominator will not be the same for all the points. Therefore, the denominator for an outlier will also be small, waving out the problem of a small numerator. For this problem, instead of Eq. (16.15), the following is used:

$$\mathbb{R} \ni p_{ij} := \frac{p_{i|j} + p_{j|i}}{2n}, \tag{16.17}$$

where:

$$\mathbb{R} \ni p_{j|i} := \frac{\exp(-d_{ij}^2)}{\sum_{k \neq i} \exp(-d_{ik}^2)}, \tag{16.18}$$

is the probability that $\boldsymbol{x}_i \in \mathbb{R}^d$ takes $\boldsymbol{x}_j \in \mathbb{R}^d$ as its neighbour.

In the low-dimensional embedding space, a Gaussian probability distribution is considered for the point $\boldsymbol{y}_i \in \mathbb{R}^p$ to take $\boldsymbol{y}_j \in \mathbb{R}^p$ as its neighbour. This is then made symmetric, with a fixed denominator for all points:

$$\mathbb{R} \ni q_{ij} := \frac{\exp(-z_{ij}^2)}{\sum_{k \neq l} \exp(-z_{kl}^2)}, \tag{16.19}$$

where:

$$\mathbb{R} \ni z_{ij}^2 := ||\boldsymbol{y}_i - \boldsymbol{y}_j||_2^2. \tag{16.20}$$

Note that Eq. (16.19) does not have a problem with outliers, as seen with Eq. (16.15) because even for an outlier, the embedded points are initialized close together.

The goal is to have the probability distributions in both the input and embedded spaces to be as similar as possible; therefore, the cost function to be minimized can be the summation of the Kullback-Leibler (KL) divergences over the n points:

$$\mathbb{R} \ni c_2 := \sum_{i=1}^{n} \text{KL}(P_i||Q_i) = \sum_{i=1}^{n} \sum_{j=1, j \neq i}^{n} p_{ij} \log(\frac{p_{ij}}{q_{ij}}), \tag{16.21}$$

where p_{ij} and q_{ij} are Eqs. (16.17) and (16.19).

Proposition 16.2 *The gradient of c_2 with respect to $\boldsymbol{y}_i$ is:*

$$\mathbb{R}^p \ni \frac{\partial c_2}{\partial \boldsymbol{y}_i} = 4 \sum_{j=1}^{n} (p_{ij} - q_{ij})(\boldsymbol{y}_i - \boldsymbol{y}_j), \tag{16.22}$$

where p_{ij} and q_{ij} are Eqs. (16.17) and (16.19), and $p_{ii} = q_{ii} = 0$.

Proof This proof is inspired by [27]. Similar to Eq. (16.9):

$$\frac{\partial c_2}{\partial \boldsymbol{y}_i} = 2\sum_j \Big(\frac{\partial c_2}{\partial r_{ij}} + \frac{\partial c_2}{\partial r_{ji}}\Big)(\boldsymbol{y}_i - \boldsymbol{y}_j). \tag{16.23}$$

Similar to the derivation of Eqs. (16.11) and (16.12), the following can be derived:

$$\frac{\partial c_2}{\partial r_{ij}} = p_{ij} - q_{ij}, \text{ and} \tag{16.24}$$

$$\frac{\partial c_2}{\partial r_{ji}} = p_{ji} - q_{ji},$$

respectively. In the symmetric SNE, there is:

$$\frac{\partial c_2}{\partial r_{ji}} = p_{ji} - q_{ji} \overset{(a)}{=} p_{ij} - q_{ij}, \tag{16.25}$$

where (a) is because in symmetric SNE, p_{ij} and q_{ij} are symmetric for i and j according to Eqs. (16.17) and (16.19). Substituting Eqs. (16.24) and (16.25) in Eq. (16.23) results in:

$$\frac{\partial c_2}{\partial \boldsymbol{y}_i} = 4\sum_j (p_{ij} - q_{ij})(\boldsymbol{y}_i - \boldsymbol{y}_j),$$

which is the gradient mentioned in the proposition. □

The update of the embedded point $\boldsymbol{y}_i$ is done by gradient descent, where every iteration is done by Eq. (16.13), where c_1 is replaced by c_2. Note that the momentum term can be omitted in the symmetric SNE. In symmetric SNE, as in SNE, Gaussian noise (random jitter) should be added to the solution of the first iterations before going to the next iterations to avoid the local optimum solutions.

16.4 t-Distributed Stochastic Neighbour Embedding (t-SNE)

16.4.1 The Crowding Problem

In SNE [13], Gaussian distributions are considered for both the input and embedded spaces. This works for the input space because it already has a high dimensionality. However, when the high-dimensional data are embedded into a low-dimensional space, it is very difficult to fit the information of all the points in the same neighbourhood area. To illustrate this concept, suppose the dimensionality is similar to the size of a room, as depicted in Fig. 16.1. In high dimensionality, there is a large hall including a huge crowd of people. Now, if you have to fit all of these same

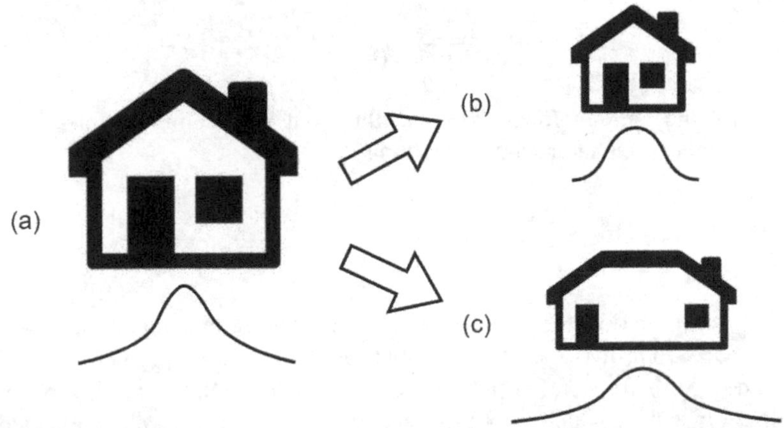

Fig. 16.1 Illustration of the crowding problem: (**a**) The high dimensional data in the high dimensional space (large room). (**b**) The information does not fit in the low dimensional embedding space (small room). (**c**) Therefore, an embedding distribution with heavier tails (a larger room) can be used in the embedding space to better fit the information

people into a small room it will be impossible. This problem is referred to as the *crowding problem.*

The main idea of t-SNE [27] is to address the crowding problem, which exists in SNE [13]. To solve the problem of fitting everyone into a small room, t-SNE enlarges the room to solve the crowding problem (see Fig. 16.1). Therefore, in the formulation of t-SNE, the Student-t distribution [10] is used, rather than the Gaussian distribution, for the low-dimensional embedded space. This is because the Student-t distribution has heavier tails than the Gaussian distribution, which is similar to a larger room, and can fit the information of the high-dimensional data in the low dimensional embedding space.

The q_{ij} in t-SNE is:

$$q_{ij} = \frac{(1 + z_{ij}^2)^{-1}}{\sum_{k \neq l}(1 + z_{kl}^2)^{-1}},$$

which is based on the standard Cauchy distribution:

$$f(z) = \frac{1}{\pi(1 + z^2)}, \tag{16.26}$$

where π is cancelled from the numerator and the normalizing denominator in q_{ij} (see the explanations of this trick in Sect. 16.2). If the Student-t distribution [10] with the general degrees of freedom δ is used, then:

$$f(z) = \frac{\Gamma(\frac{\delta+1}{2})}{\sqrt{\delta \times \pi}\ \Gamma(\frac{\delta}{2})}(1 + \frac{z^2}{\delta})^{-\frac{\delta+1}{2}}, \tag{16.27}$$

where Γ is the gamma function. Cancelling out the scaling factors from the numerator and denominator would result in [26]:

$$q_{ij} = \frac{(1 + z_{ij}^2/\delta)^{-(\delta+1)/2}}{\sum_{k \neq l}(1 + z_{kl}^2/\delta)^{-(\delta+1)/2}}. \tag{16.28}$$

However, as the first degree of freedom has the heaviest tails among the different degrees of freedom, it is the most suitable for the crowding problem. Therefore, the first degree of freedom, which is the Cauchy distribution, is used. Note that the t-SNE algorithm, which uses the Cauchy distribution, may also be called the Cauchy-SNE. Later, t-SNE with general degrees of freedom was proposed [26], which is explained in Sect. 16.5.

16.4.2 t-SNE Formulation

In t-SNE [27], a Gaussian probability around every point $\boldsymbol{x}_i$ is considered in the input space because the crowding problem does not exist in the high dimensional data. The probability that point $\boldsymbol{x}_i \in \mathbb{R}^d$ takes $\boldsymbol{x}_j \in \mathbb{R}^d$ as its neighbour is:

$$\mathbb{R} \ni p_{j|i} := \frac{\exp(-d_{ij}^2)}{\sum_{k \neq i}\exp(-d_{ik}^2)}, \tag{16.29}$$

where:

$$\mathbb{R} \ni d_{ij}^2 := \frac{||\boldsymbol{x}_i - \boldsymbol{x}_j||_2^2}{2\sigma_i^2}. \tag{16.30}$$

Note that Eq. (16.29) is not symmetric for i and j because of the denominator. The symmetric p_{ij} is taken as the scaled average of $p_{i|j}$ and $p_{j|i}$:

$$\mathbb{R} \ni p_{ij} := \frac{p_{i|j} + p_{j|i}}{2n}. \tag{16.31}$$

In the low-dimensional embedding space, a Student's t-distribution with one degree of freedom (Cauchy distribution) is considered for the point $\boldsymbol{y}_i \in \mathbb{R}^p$ to take $\boldsymbol{y}_j \in \mathbb{R}^p$ as its neighbour:

$$\mathbb{R} \ni q_{ij} := \frac{(1 + z_{ij}^2)^{-1}}{\sum_{k \neq l}(1 + z_{kl}^2)^{-1}}, \tag{16.32}$$

where:

$$\mathbb{R} \ni z_{ij}^2 := ||\boldsymbol{y}_i - \boldsymbol{y}_j||_2^2. \tag{16.33}$$

The aim is to have the probability distributions in both the input and embedded spaces be as similar as possible. Therefore, the cost function to be minimized can be the summation of the Kullback-Leibler (KL) divergences over the n points:

$$\mathbb{R} \ni c_3 := \sum_{i=1}^{n} \text{KL}(P_i||Q_i) = \sum_{i=1}^{n} \sum_{j=1, j\neq i}^{n} p_{ij} \log(\frac{p_{ij}}{q_{ij}}), \tag{16.34}$$

where p_{ij} and q_{ij} are Eqs. (16.31) and (16.32).

Proposition 16.3 *The gradient of c_3, with respect to $\boldsymbol{y}_i$, is:*

$$\frac{\partial c_3}{\partial \boldsymbol{y}_i} = 4 \sum_{j=1}^{n} (p_{ij} - q_{ij})(1 + ||\boldsymbol{y}_i - \boldsymbol{y}_j||_2^2)^{-1}(\boldsymbol{y}_i - \boldsymbol{y}_j), \tag{16.35}$$

where p_{ij} and q_{ij} are Eqs. (16.31) and (16.32), and $p_{ii} = q_{ii} = 0$.

Proof The following proof is according to [27]. Let:

$$\mathbb{R} \ni r_{ij} := z_{ij}^2 = ||\boldsymbol{y}_i - \boldsymbol{y}_j||_2^2. \tag{16.36}$$

By changing $\boldsymbol{y}_i$, only z_{ij} and z_{ji} are impacted for all j's. According to chain rule:

$$\mathbb{R}^p \ni \frac{\partial c_3}{\partial \boldsymbol{y}_i} = \sum_j \big(\frac{\partial c_3}{\partial r_{ij}} \frac{\partial r_{ij}}{\partial \boldsymbol{y}_i} + \frac{\partial c_3}{\partial r_{ji}} \frac{\partial r_{ji}}{\partial \boldsymbol{y}_i}\big).$$

According to Eq. (16.36):

$$r_{ij} = ||\boldsymbol{y}_i - \boldsymbol{y}_j||_2^2 \implies \frac{\partial r_{ij}}{\partial \boldsymbol{y}_i} = 2(\boldsymbol{y}_i - \boldsymbol{y}_j),$$

$$r_{ji} = ||\boldsymbol{y}_j - \boldsymbol{y}_i||_2^2 = ||\boldsymbol{y}_i - \boldsymbol{y}_j||_2^2 \implies \frac{\partial r_{ji}}{\partial \boldsymbol{y}_i} = 2(\boldsymbol{y}_i - \boldsymbol{y}_j).$$

Therefore:

$$\therefore \quad \frac{\partial c_3}{\partial \boldsymbol{y}_i} = 2 \sum_j \big(\frac{\partial c_3}{\partial r_{ij}} + \frac{\partial c_3}{\partial r_{ji}}\big)(\boldsymbol{y}_i - \boldsymbol{y}_j). \tag{16.37}$$

The cost function can be rewritten as:

$$c_3 = \sum_k \sum_{l \neq k} p_{kl} \log(\frac{p_{kl}}{q_{kl}}) = \sum_{k \neq l} p_{kl} \log(\frac{p_{kl}}{q_{kl}}) = \sum_{k \neq l} \big(p_{kl} \log(p_{kl}) - p_{kl} \log(q_{kl})\big),$$

whose first term is a constant with respect to q_{kl} and to r_{kl}. Therefore:

$$\mathbb{R} \ni \frac{\partial c_3}{\partial r_{ij}} = -\sum_{k \neq l} p_{kl} \frac{\partial(\log(q_{kl}))}{\partial r_{ij}}.$$

According to Eq. (16.32), q_{kl} is:

$$q_{kl} := \frac{(1 + z_{kl}^2)^{-1}}{\sum_{m \neq f}(1 + z_{mf}^2)^{-1}}, \overset{\cdot}{=} \frac{(1 + r_{kl})^{-1}}{\sum_{m \neq f}(1 + r_{mf})^{-1}},$$

The denominator of q_{kl} is taken as:

$$\beta := \sum_{m \neq f} (1 + z_{mf}^2)^{-1} = \sum_{m \neq f} (1 + r_{mf})^{-1}. \tag{16.38}$$

It is possible to restate $\log(q_{kl}) = \log(q_{kl}) + \log \beta - \log \beta = \log(q_{kl}\beta) - \log \beta$. Therefore:

$$\begin{aligned}
\therefore \quad \frac{\partial c_3}{\partial r_{ij}} &= -\sum_{k \neq l} p_{kl} \frac{\partial\big(\log(q_{kl}\beta) - \log \beta\big)}{\partial r_{ij}} \\
&= -\sum_{k \neq l} p_{kl} \Big[\frac{\partial\big(\log(q_{kl}\beta)\big)}{\partial r_{ij}} - \frac{\partial\big(\log \beta\big)}{\partial r_{ij}}\Big] = -\sum_{k \neq l} p_{kl} \Big[\frac{1}{q_{kl}\beta} \frac{\partial(q_{kl}\beta)}{\partial r_{ij}} - \frac{1}{\beta} \frac{\partial \beta}{\partial r_{ij}}\Big].
\end{aligned}$$

$q_{kl}\beta$ is:

$$q_{kl}\beta = \frac{(1 + r_{kl})^{-1}}{\sum_{m \neq f}(1 + r_{mf})^{-1}} \times \sum_{m \neq f} (1 + r_{mf})^{-1} = (1 + r_{kl})^{-1}.$$

Therefore:

$$\therefore \quad \frac{\partial c_3}{\partial r_{ij}} = -\sum_{k \neq l} p_{kl} \Big[\frac{1}{q_{kl}\beta} \frac{\partial\big((1 + r_{kl})^{-1}\big)}{\partial r_{ij}} - \frac{1}{\beta} \frac{\partial \beta}{\partial r_{ij}}\Big].$$

The $\partial\big((1 + r_{kl})^{-1}\big)/\partial r_{ij}$ is non-zero for only $k = i$ and $l = j$; therefore:

$$\frac{\partial\big((1 + r_{ij})^{-1}\big)}{\partial r_{ij}} = -(1 + r_{ij})^{-2},$$

$$\frac{\partial \beta}{\partial r_{ij}} = \frac{\partial \sum_{m \neq f}(1 + r_{mf})^{-1}}{\partial r_{ij}} = \frac{\partial (1 + r_{ij})^{-1}}{\partial r_{ij}} = -(1 + r_{ij})^{-2}.$$

Therefore:

$$\therefore \quad \frac{\partial c_3}{\partial r_{ij}} = -\Big(p_{ij}\Big[\frac{-1}{q_{ij}\beta}(1 + r_{ij})^{-2}\Big] + 0 + \dots + 0\Big) - \sum_{k \neq l} p_{kl}\Big[\frac{1}{\beta}(1 + r_{ij})^{-2}\Big].$$

$\sum_{k \neq l} p_{kl} = 1$ because the summation of all possible probabilities is one. Thus:

$$\begin{aligned}
\frac{\partial c_3}{\partial r_{ij}} &= -p_{ij}\Big[\frac{-1}{q_{ij}\beta}(1 + r_{ij})^{-2}\Big] - \Big[\frac{1}{\beta}(1 + r_{ij})^{-2}\Big] \\
&= (1 + r_{ij})^{-1} \underbrace{\frac{(1 + r_{ij})^{-1}}{\beta}}_{=q_{ij}} \Big[\frac{p_{ij}}{q_{ij}} - 1\Big] = (1 + r_{ij})^{-1}(p_{ij} - q_{ij}).
\end{aligned}$$

Similarly:

$$\frac{\partial c_3}{\partial r_{ji}} = (1 + r_{ji})^{-1}(p_{ji} - q_{ji}) \overset{(a)}{=} (1 + r_{ij})^{-1}(p_{ij} - q_{ij}),$$

where (a) is because in t-SNE, p_{ij}, q_{ij}, and r_{ij} are symmetric for i and j according to Eqs. (16.31), (16.32), and (16.36). Substituting the obtained derivatives in Eq. (16.37) results in:

$$\frac{\partial c_3}{\partial \boldsymbol{y}_i} = 4 \sum_j (p_{ij} - q_{ij})(1 + r_{ij})^{-1}(\boldsymbol{y}_i - \boldsymbol{y}_j),$$

which is the gradient mentioned in the proposition.[1] □

The update of the embedded point $\boldsymbol{y}_i$ is done by gradient descent, whose iterations are similar to Eq. (16.13), but c_1 is replaced by c_3. For t-SNE, there is no need to add noise (random jitter) to the solution of the initial iterations [27] because it is more robust than SNE. $\alpha(t)$ is the momentum, which can be updated according to Eq. (16.14). η is the learning rate, which can be a small positive constant (e.g., $\eta = 0.1$) or can be updated according to [17] (in [27], the initial η is 100).

[1] Note that in [27], the proof uses z_{ij} rather than $r_{ij} = z_{ij}^2$ in Eq. (16.36) and the rest of the proof. In the opinions of the authors, it is better to use z_{ij}^2 rather than z_{ij} for the proof.

16.4.3 Early Exaggeration

In t-SNE, it is better to multiply all p_{ij}'s by a constant (e.g., 4) in the initial iterations:

$$p_{ij} := p_{ij} \times 4, \tag{16.39}$$

which is called *early exaggeration*. This heuristic helps the optimization focus on the large p_{ij}'s (close neighbours) more in the early iterations. This is because large p_{ij} values are affected more by multiplying by 4 than small p_{ij} values. After the neighbours are embedded close to one another, this multiplication is no longer needed, and the far-away points can be embedded using the probabilities without multiplication. Note that the early exaggeration is optional and not mandatory.

16.5 General Degrees of Freedom in t-SNE

It is possible to have general degrees of freedom for Student-t distribution in t-SNE [26]. As seen in Eqs. (16.27) and (16.28), it is possible to have any degrees of freedom for q_{ij} (note that α is a positive integer). Recall Eq. (16.28):

$$q_{ij} = \frac{(1 + z_{ij}^2/\delta)^{-(\delta+1)/2}}{\sum_{k\neq l}(1 + z_{kl}^2/\delta)^{-(\delta+1)/2}}. \tag{16.40}$$

If $\delta \to \infty$, the Student-t distribution formulated in Eq. (16.27) tends toward the Gaussian distribution used in SNE [13]. SNE and t-SNE use degrees $\delta \to \infty$ and $\delta = 1$ in Eq. (16.40), respectively. Note that the kernel q_{ij} in the low-dimensional space has no need to be a probability distribution, but it is enough for it to be a decaying function. Kobak et al. [22] demonstrated that the degree $\delta < 1$ works well for embedding.

There are three ways to determine δ [26]:

1. δ can be fixed. For example, $\delta = 1$ is used in the original t-SNE [27], which uses the Cauchy distribution in Eq. (16.32).
2. The main problem with taking the first approach is that it does not consider the relationship between the crowding problem and the dimensionality of the embedded space. Recall the crowding problem discussed in Sect. 16.4.1. On the one hand, as Eq. (16.27) demonstrates, the degree of freedom is in the power, so the tail thickness of the Student-t distribution decreases exponentially with δ. On the other hand, the volume of a hypersphere grows exponentially with the dimension. For example, in two and three dimensions, the volume is πr^2 and $(4/3)\pi r^3$, respectively, where r is the radius. The crowding volume in the embedded space, which stores the embedded data points, is $\propto \pi r^h$ and

grows exponentially with h. Therefore, the relation of δ and h (dimensionality of embedded space) is linear, i.e., $h \propto \delta$. To be consistent with the original t-SNE [27], $\delta = h - 1$ is considered, resulting in $\delta = 1$ for $h = 2$ [26].

3. The main problem of taking the second approach is that δ might not "only" depend on h. In this approach, the best α is found to minimize the cost c_3, i.e. Eq. (16.34), [26] where p_{ij} is obtained using Eqs. (16.29) and (16.31) and q_{ij} is Eq. (16.28). Gradient descent [3] is used for the optimization of both δ and $\{\boldsymbol{y}_i\}_{i=1}^n$, where the abovementioned gradients are used. The parametric t-SNE [26] uses the restricted Boltzmann machine [12, 14] (see Chap. 18) to learn the optimal δ and $\{\boldsymbol{y}_i\}_{i=1}^n$ by a neural network. It is possible to use an alternating optimization approach [18] (see Chap. 4) to solve for both δ and $\{\boldsymbol{y}_i\}_{i=1}^n$ simultaneously. In this approach, $\{\boldsymbol{y}_i\}_{i=1}^n$ are updated with gradient descent using Eq. (16.47). Then, the degree δ is updated with gradient descent using Eq. (16.46), and this procedure is repeated until convergence.

   Note that the degree δ is an integer greater than or equal to one. However, the gradient in Eq. (16.46) is a float number. To update the degree using gradient descent in the alternating optimization approach, it is possible to update the degree using the sign of the gradient:

$$\delta := \delta - \text{sign}(\frac{\partial c_3}{\partial \delta}), \tag{16.41}$$

because the direction of updating is opposite to the gradient direction.

Recall p_{ij}, q_{ij}, and c_3:

$$\mathbb{R} \ni p_{j|i} := \frac{\exp(-d_{ij}^2)}{\sum_{k \neq i} \exp(-d_{ik}^2)}, \tag{16.42}$$

$$\mathbb{R} \ni p_{ij} := \frac{p_{i|j} + p_{j|i}}{2n}, \tag{16.43}$$

$$\mathbb{R} \ni q_{ij} = \frac{(1 + z_{ij}^2/\delta)^{-(\delta+1)/2}}{\sum_{k \neq l}(1 + z_{kl}^2/\delta)^{-(\delta+1)/2}}, \tag{16.44}$$

$$\mathbb{R} \ni c_3 := \sum_i \text{KL}(P_i||Q_i) = \sum_i \sum_{j \neq i} p_{ij} \log(\frac{p_{ij}}{q_{ij}}). \tag{16.45}$$

Proposition 16.4 *The gradient of c_3 with respect to δ is:*

$$\frac{\partial c_3}{\partial \delta} = \sum_{i \neq j} \Big(\frac{-(1+\delta)z_{ij}^2}{2\delta^2(1 + \frac{z_{ij}^2}{\delta})} + \frac{1}{2}\log(1 + \frac{z_{ij}^2}{\delta})\Big)(p_{ij} - q_{ij}), \tag{16.46}$$

where p_{ij} and q_{ij} are Eqs. (16.43) and (16.44), respectively, and $z_{ij}^2 := ||\boldsymbol{y}_i - \boldsymbol{y}_j||_2^2$.

Regardless of how δ is found, the cost function c_3 (Eq. (16.45)) needs to be optimized using gradient descent.

Proposition 16.5 *The gradient of c_3, with respect to $\boldsymbol{y}_i$, is:*

$$\frac{\partial c_3}{\partial \boldsymbol{y}_i} = \frac{2\delta + 2}{\delta} \times \sum_j (p_{ij} - q_{ij})(1 + \frac{||\boldsymbol{y}_i - \boldsymbol{y}_j||_2^2}{\delta})^{-1}(\boldsymbol{y}_i - \boldsymbol{y}_j), \quad (16.47)$$

where p_{ij} and q_{ij} are Eqs. (16.43) and (16.44), respectively.

Proof Let:

$$\mathbb{R} \ni r_{ij} := z_{ij}^2 = ||\boldsymbol{y}_i - \boldsymbol{y}_j||_2^2. \quad (16.48)$$

By changing $\boldsymbol{y}_i$, there is only change impact in z_{ij} and z_{ji} for all j's. Considering Eq. (16.48) and according to chain rule:

$$\mathbb{R}^p \ni \frac{\partial c_3}{\partial \boldsymbol{y}_i} = \sum_j \Big(\frac{\partial c_3}{\partial r_{ij}} \frac{\partial r_{ij}}{\partial \boldsymbol{y}_i} + \frac{\partial c_3}{\partial r_{ji}} \frac{\partial r_{ji}}{\partial \boldsymbol{y}_i}\Big).$$

According to Eq. (16.48):

$$r_{ij} = ||\boldsymbol{y}_i - \boldsymbol{y}_j||_2^2 \implies \frac{\partial r_{ij}}{\partial \boldsymbol{y}_i} = 2(\boldsymbol{y}_i - \boldsymbol{y}_j),$$

$$r_{ji} = ||\boldsymbol{y}_j - \boldsymbol{y}_i||_2^2 = ||\boldsymbol{y}_i - \boldsymbol{y}_j||_2^2 \implies \frac{\partial r_{ji}}{\partial \boldsymbol{y}_i} = 2(\boldsymbol{y}_i - \boldsymbol{y}_j).$$

Therefore:

$$\therefore \quad \frac{\partial c_3}{\partial \boldsymbol{y}_i} = 2 \sum_j \Big(\frac{\partial c_3}{\partial r_{ij}} + \frac{\partial c_3}{\partial r_{ji}}\Big)(\boldsymbol{y}_i - \boldsymbol{y}_j). \quad (16.49)$$

The cost function can be rewritten as:

$$c_3 = \sum_k \sum_{l \neq k} p_{kl} \log(\frac{p_{kl}}{q_{kl}}) = \sum_{k \neq l} p_{kl} \log(\frac{p_{kl}}{q_{kl}}) = \sum_{k \neq l} \big(p_{kl} \log(p_{kl}) - p_{kl} \log(q_{kl})\big),$$

whose first term is a constant with respect to q_{kl} and thus to r_{kl}. The derivative is:

$$\mathbb{R} \ni \frac{\partial c_3}{\partial r_{ij}} = -\sum_{k \neq l} p_{kl} \frac{\partial(\log(q_{kl}))}{\partial r_{ij}}.$$

According to Eq. (16.44), q_{kl} is:

$$q_{kl} = \frac{(1 + r_{kl}/\delta)^{-(\delta+1)/2}}{\sum_{m \neq f}(1 + r_{mf}/\delta)^{-(\delta+1)/2}},$$

The denominator of q_{kl} is denoted by:

$$\beta := \sum_{m \neq f} (1 + r_{mf}/\delta)^{-(\delta+1)/2}. \tag{16.50}$$

It is possible to restate $\log(q_{kl}) = \log(q_{kl}) + \log\beta - \log\beta = \log(q_{kl}\beta) - \log\beta$. Therefore:

$$\therefore \quad \frac{\partial c_3}{\partial r_{ij}} = -\sum_{k \neq l} p_{kl} \frac{\partial\big(\log(q_{kl}\beta) - \log\beta\big)}{\partial r_{ij}}$$

$$= -\sum_{k \neq l} p_{kl} \Big[\frac{\partial\big(\log(q_{kl}\beta)\big)}{\partial r_{ij}} - \frac{\partial\big(\log\beta\big)}{\partial r_{ij}}\Big] = -\sum_{k \neq l} p_{kl} \Big[\frac{1}{q_{kl}\beta}\frac{\partial(q_{kl}\beta)}{\partial r_{ij}} - \frac{1}{\beta}\frac{\partial\beta}{\partial r_{ij}}\Big].$$

$q_{kl}\beta$ is:

$$q_{kl}\beta = \frac{(1 + r_{kl}/\delta)^{-(\delta+1)/2}}{\sum_{m \neq f}(1 + r_{mf}/\delta)^{-(\delta+1)/2}} \times \sum_{m \neq f} (1 + r_{mf}/\delta)^{-(\delta+1)/2}$$

$$= (1 + r_{kl}/\delta)^{-(\delta+1)/2}.$$

$\partial\big((1 + r_{kl}/\delta)^{-(\delta+1)/2}\big)/\partial r_{ij}$ is non-zero for only $k = i$ and $l = j$; therefore:

$$\frac{\partial(q_{kl}\beta)}{\partial r_{ij}} = \frac{\partial\big((1 + r_{kl}/\delta)^{-(\delta+1)/2}\big)}{\partial r_{ij}} = -\frac{\delta+1}{2\delta}(1 + \frac{r_{ij}}{\delta})^{-\frac{\delta+3}{2}},$$

$$\frac{\partial\beta}{\partial r_{ij}} = \frac{\partial \sum_{m \neq f}(1 + r_{mf}/\delta)^{-(\delta+1)/2}}{\partial r_{ij}}$$

$$= \frac{\partial\big((1 + r_{kl}/\delta)^{-(\delta+1)/2}\big)}{\partial r_{ij}} = -\frac{\delta+1}{2\delta}(1 + \frac{r_{ij}}{\delta})^{-\frac{\delta+3}{2}}.$$

Therefore:

$$\frac{\partial c_3}{\partial r_{ij}} = -\Bigg(p_{ij}\Big[\frac{-1}{q_{ij}\beta}\frac{\delta+1}{2\delta}(1 + \frac{r_{ij}}{\delta})^{-\frac{\delta+3}{2}}\Big] + 0$$

$$+ \cdots + 0\Bigg) - \sum_{k \neq l} p_{kl}\Big[\frac{1}{\beta}\frac{\delta+1}{2\delta}(1 + \frac{r_{ij}}{\delta})^{-\frac{\delta+3}{2}}\Big].$$

There is $\sum_{k \neq l} p_{kl} = 1$ because the summation of all possible probabilities is one. Thus:

$$
\begin{aligned}
\frac{\partial c_3}{\partial r_{ij}} &= -p_{ij}\Big[\frac{-1}{q_{ij}\beta}\frac{\delta+1}{2\delta}(1+\frac{r_{ij}}{\delta})^{-\frac{\delta+3}{2}}\Big] - \Big[\frac{1}{\beta}\frac{\delta+1}{2\delta}(1+\frac{r_{ij}}{\delta})^{-\frac{\delta+3}{2}}\Big] \\
&= \frac{\delta+1}{2\delta}(1+\frac{r_{ij}}{\delta})^{-\frac{\delta+3}{2}}\frac{1}{\beta}(\frac{p_{ij}}{q_{ij}} - 1) \\
&= \frac{\delta+1}{2\delta}\underbrace{\frac{(1+\frac{r_{ij}}{\delta})^{-\frac{\delta+1}{2}}}{\beta}}_{=q_{ij}}(1+\frac{r_{ij}}{\delta})^{-1}(\frac{p_{ij}}{q_{ij}} - 1) \\
&= \frac{\delta+1}{2\delta}(1+\frac{r_{ij}}{\delta})^{-1}(p_{ij} - q_{ij}).
\end{aligned} \tag{16.51}
$$

Similarly:

$$
\frac{\partial c_3}{\partial r_{ji}} = \frac{\delta+1}{2\delta}(1+\frac{r_{ji}}{\delta})^{-1}(p_{ji} - q_{ji}) \overset{(a)}{=} \frac{\delta+1}{2\delta}(1+\frac{r_{ij}}{\delta})^{-1}(p_{ij} - q_{ij}),
$$

where (a) is because in t-SNE with general degrees of freedom, p_{ij}, q_{ij}, and r_{ij} are symmetric for i and j according to Eqs. (16.43), (16.44), and (16.48). Substituting the obtained derivatives in Eq. (16.49) results in:

$$
\frac{\partial c_3}{\partial \boldsymbol{y}_i} = 2\sum_j \frac{\delta+1}{\delta}(1+\frac{r_{ij}}{\delta})^{-1}(p_{ij} - q_{ij})(\boldsymbol{y}_i - \boldsymbol{y}_j),
$$

which is the gradient mentioned in the proposition. □

Comparing Eqs. (16.35) and (16.47) demonstrates that the original t-SNE [27] is a special case with $\delta = 1$.

16.6 Out-of-Sample Embedding

Recall that there are n high-dimensional data points $\{\boldsymbol{x}_i\}_{i=1}^n$ and the aim is to embed them into the lower dimensional data $\{\boldsymbol{y}_i\}_{i=1}^n$, where $\boldsymbol{x}_i \in \mathbb{R}^d$ and $\boldsymbol{y}_i \in \mathbb{R}^p$. Assume there are n_t out-of-sample data points $\{\boldsymbol{x}_i^{(t)}\}_{i=1}^{n_t}$ and the aim is to embed them into the lower dimensional data $\{\boldsymbol{y}_i^t\}_{i=1}^{n_t}$, where $\boldsymbol{x}_i^{(t)} \in \mathbb{R}^d$ and $\boldsymbol{y}_i^{(t)} \in \mathbb{R}^p$. There are several different methods for out-of-sample extension of SNE and t-SNE methods; this section will explain *kernel mapping* [5, 6].

A map that maps any data point as $\boldsymbol{x} \mapsto \boldsymbol{y}(\boldsymbol{x})$ is defined as:

$$\mathbb{R}^p \ni \boldsymbol{y}(\boldsymbol{x}) := \sum_{j=1}^{n} \boldsymbol{\alpha}_j \frac{k(\boldsymbol{x}, \boldsymbol{x}_j)}{\sum_{\ell=1}^{n} k(\boldsymbol{x}, \boldsymbol{x}_\ell)}, \tag{16.52}$$

and $\boldsymbol{\alpha}_j \in \mathbb{R}^p$, and $\boldsymbol{x}_j$ and $\boldsymbol{x}_\ell$ denote the j-th and ℓ-th training data points, respectively. The $k(\boldsymbol{x}, \boldsymbol{x}_j)$ is a kernel, such as the Gaussian kernel:

$$k(\boldsymbol{x}, \boldsymbol{x}_j) = \exp(\frac{-||\boldsymbol{x} - \boldsymbol{x}_j||_2^2}{2\sigma_j^2}), \tag{16.53}$$

where σ_j is calculated as [5]:

$$\sigma_j := \gamma \times \min_i(||\boldsymbol{x}_j - \boldsymbol{x}_i||_2), \tag{16.54}$$

where γ is a small positive number.

Assume the training data points have already been embedded using SNE or t-SNE; therefore, the set $\{\boldsymbol{y}_i\}_{i=1}^n$ is available. While mapping the training data points, the aim is to minimize the following least squares cost function to obtain $\boldsymbol{y}(\boldsymbol{x}_i)$ close to $\boldsymbol{y}_i$ for the i-th training point:

$$\underset{\boldsymbol{\alpha}_j\text{'s}}{\text{minimize}} \quad \sum_{i=1}^{n} ||\boldsymbol{y}_i - \boldsymbol{y}(\boldsymbol{x}_i)||_2^2, \tag{16.55}$$

where the summation is over the training data points. This cost function can be written in matrix form as:

$$\underset{\boldsymbol{A}}{\text{minimize}} \quad ||\boldsymbol{Y} - \boldsymbol{K}\boldsymbol{A}||_F^2, \tag{16.56}$$

where $\mathbb{R}^{n\times p} \ni \boldsymbol{Y} := [\boldsymbol{y}_1, \dots, \boldsymbol{y}_n]^\top$ and $\mathbb{R}^{n\times p} \ni \boldsymbol{A} := [\boldsymbol{\alpha}_1, \dots, \boldsymbol{\alpha}_n]^\top$. $\boldsymbol{K} \in \mathbb{R}^{n\times n}$ is the kernel matrix, whose (i, j)-th element is:

$$\boldsymbol{K}(i, j) := \frac{k(\boldsymbol{x}_i, \boldsymbol{x}_j)}{\sum_{\ell=1}^{n} k(\boldsymbol{x}_i, \boldsymbol{x}_\ell)}. \tag{16.57}$$

Equation (16.56) is always nonnegative; thus, its smallest value is zero. Therefore, the solution to this equation is:

$$\boldsymbol{Y} - \boldsymbol{K}\boldsymbol{A} = \boldsymbol{0} \implies \boldsymbol{Y} = \boldsymbol{K}\boldsymbol{A} \overset{(a)}{\implies} \boldsymbol{A} = \boldsymbol{K}^\dagger \boldsymbol{Y}, \tag{16.58}$$

where $\boldsymbol{K}^\dagger$ is the pseudoinverse of $\boldsymbol{K}$:

$$\boldsymbol{K}^\dagger = (\boldsymbol{K}^\top \boldsymbol{K})^{-1} \boldsymbol{K}^\top, \tag{16.59}$$

and (a) is because $\boldsymbol{K}^\dagger \boldsymbol{K} = \boldsymbol{I}$.

Lastly, the mapping of Eq. (16.52) for the n_t out-of-sample data points is:

$$\boldsymbol{Y}^{(t)} = \boldsymbol{K}^{(t)} \boldsymbol{A}, \tag{16.60}$$

where $\mathbb{R}^{n_t \times p} \ni \boldsymbol{Y}^{(t)} := [\boldsymbol{y}_1^{(t)}, \dots, \boldsymbol{y}_{n_t}^{(t)}]^\top$ and the (i, j)-th element of the out-of-sample kernel matrix $\boldsymbol{K}^{(t)} \in \mathbb{R}^{n_t \times n}$ is:

$$\boldsymbol{K}^{(t)}(i, j) := \frac{k(\boldsymbol{x}_i^{(t)}, \boldsymbol{x}_j)}{\sum_{\ell=1}^{n} k(\boldsymbol{x}_i^{(t)}, \boldsymbol{x}_\ell)}, \tag{16.61}$$

where $\boldsymbol{x}_i^{(t)}$ is the i-th out-of-sample data point, and $\boldsymbol{x}_j$ and $\boldsymbol{x}_\ell$ are the j-th and ℓ-th training data points, respectively. In Eq. (16.58), if, $\boldsymbol{Y}$ is the embedding of training data using SNE/t-SNE, then the out-of-sample embedding of SNE/t-SNE are obtained.

16.7 Accelerating SNE and t-SNE

The SNE and t-SNE methods are slow due to numerical iterative optimization. Different methods have been proposed for accelerating these methods [23], which are based on landmarks. In these methods, a random sample is taken from the dataset to obtain a subset of data. The sampled data points are called *landmarks* (see Chap. 3). In the following, three methods for accelerating t-SNE and/or SNE that use landmarks are mentioned.[2]

16.7.1 Acceleration Using Out-of-Sample Embedding

Kernel mapping [5], which was introduced in Sect. 16.6, can speed up SNE and t-SNE. Landmarks are considered to be the training data points that are used to train SNE or t-SNE. Thereafter, the nonlandmark data points are treated as out-of-sample points. Kernel SNE or kernel t-SNE (or Fisher kernel t-SNE for supervised cases) is used to embed the out-of-sample data points.

Another method is to consider the landmarks as training points and then embed the training points using SNE or t-SNE. Then, the nonlandmarks, which are out-of-sample points, are embedded using optimization [4], as mentioned in Sect. 16.6.

[2] Some of the acceleration methods for SNE and t-SNE are *tree-based* algorithms [25, 34], also referred to as *Barnes-Hut t-SNE* [25, 40]. These methods are not covered in this book for the sake of brevity.

16.7.2 Acceleration Using Random Walk

Another way of accelerating t-SNE is by using a random walk [27]. First, a k-Nearest neighbour (kNN) graph is constructed using all points, including landmarks and nonlandmarks. This method has an acceptable robustness to the choice of k; for example, $k = 20$ can be used [27]. Additionally, note that the calculation of kNN is time-consuming for large datasets; however, it is only done once. Then, multiple random walks are performed in this kNN graph [38]. For every random walk, a random landmark is selected as the starting point and then edges are selected randomly to walk through. This continues until another landmark is reached, where the random walk is terminated. After performing all the random walks, the fraction of the random walks that pass through point $\boldsymbol{x}_i$ (either landmark or nonlandmark) and then reach point $\boldsymbol{x}_j$ (either landmark or nonlandmark) is a good approximation for $p_{j|i}$. In t-SNE, this approximation is used in place of Eq. (16.29), which is then used in Eq. (16.31). The rest of t-SNE is similar to the original t-SNE. Therefore, for p_{ij} in Eq. (16.35), the approximation is used rather than Eq. (16.31), which makes the t-SNE faster.

16.8 Recent Improvements of t-SNE

This section will identify a number of recent improvements to t-SNE, but will not explain them in detail for brevity. Recall that the variance σ_i^2 is determined for every point $\boldsymbol{x}_i$ using a binary search. This cancels the local density information of the points because a point in denser regions will have a smaller σ_i^2. A dense t-SNE [32] resolves this problem by using a density radius to include the density information. LargeVis [39] and UMAP [28] (see Chap. 17) are methods that closely related to t-SNE. Parametric t-SNE [26] and parametric kernel t-SNE [5] implement the t-SNE formulation in a neural network structure. The optimization of t-SNE can be seen as optimizing attractive and repulsive forces between points. Some discussions on the attractive and repulsive forces in t-SNE can be found in [23, 36]. Many algorithms, such as t-SNE, which are based on these forces, can be unified into a family of neighbourhood embedding methods [1, 2]. Lastly, note that a combination of a variational autoencoder [19] (see Chap. 20) and SNE exists [11].

16.9 Chapter Summary

SNE is a fundamental method and the first to propose using probability distributions for every point to take other points as its neighbours. t-SNE improved SNE by addressing its crowding problem. Currently, t-SNE and UMAP (introduced in Chap. 17) are state-of-the-art methods for high-dimensional data visualization. They are especially used in the visualization of DNA data, as well as embedding output of deep neural networks. SNE, t-SNE, and UMAP belong to a family of methods, named neighbourhood embedding methods. This chapter introduced SNE and t-SNE as two neighbourhood embedding methods. UMAP will be introduced in Chap. 17, and its relation with t-SNE will be discussed. The idea and formulation of SNE is similar to Neighbourhood Component Analysis (NCA), which will be introduced in Chap. 19.

References

1. Jan Niklas Böhm. "Dimensionality Reduction with Neighborhood Embeddings". PhD thesis. University of Tübingen, 2020.
2. Jan Niklas Böhm, Philipp Berens, and Dmitry Kobak. "A unifying perspective on neighbor embeddings along the attraction-repulsion spectrum". In: *arXiv preprint arXiv:2007.08902* (2020).
3. Stephen Boyd and Lieven Vandenberghe. *Convex optimization*. Cambridge university press, 2004.
4. Kerstin Bunte, Michael Biehl, and Barbara Hammer. "A general framework for dimensionality-reducing data visualization mapping". In: *Neural Computation* 24.3 (2012), pp. 771–804.
5. Andrej Gisbrecht, Alexander Schulz, and Barbara Hammer. "Parametric nonlinear dimensionality reduction using kernel t-SNE". In: *Neurocomputing* 147 (2015), pp. 71–82.
6. Andrej Gisbrecht et al. "Out-of-sample kernel extensions for nonparametric dimensionality reduction." In: *European Symposium on Artificial Neural Networks, Computational Intelligence and Machine Learning*. Vol. 2012. 2012, pp. 531–536.
7. Amir Globerson et al. "Euclidean embedding of co-occurrence data". In: *Journal of Machine Learning Research* 8.Oct (2007), pp. 2265–2295.
8. Yoav Goldberg and Omer Levy. "word2vec Explained: deriving Mikolov et al.'s negative-sampling word-embedding method". In: *arXiv preprint arXiv:1402.3722* (2014).
9. Jacob Goldberger et al. "Neighbourhood components analysis". In: *Advances in neural information processing systems*. 2005, pp. 513–520.
10. William Sealy Gosset (Student). "The probable error of a mean". In: *Biometrika* (1908), pp. 1–25.
11. Jacob M Graving and Iain D Couzin. "VAE-SNE: a deep generative model for simultaneous dimensionality reduction and clustering". In: *BioRxiv* (2020).
12. Geoffrey E Hinton. "A practical guide to training restricted Boltzmann machines". In: *Neural networks: Tricks of the trade*. Springer, 2012, pp. 599–619.
13. Geoffrey E Hinton and Sam T Roweis. "Stochastic neighbor embedding". In: *Advances in neural information processing systems*. 2003, pp. 857–864.
14. Geoffrey E Hinton and Ruslan R Salakhutdinov. "Reducing the dimensionality of data with neural networks". In: *science* 313.5786 (2006), pp. 504–507.
15. Daniel Jiwoong Im, Nakul Verma, and Kristin Branson. "Stochastic Neighbor Embedding under f-divergences". In: *arXiv preprint arXiv:1811.01247* (2018).
16. Tomoharu Iwata et al. "Parametric embedding for class visualization". In: *Advances in neural information processing systems*. 2005, pp. 617–624.
17. Robert A Jacobs. "Increased rates of convergence through learning rate adaptation". In: *Neural networks* 1.4 (1988), pp. 295–307.
18. Prateek Jain and Purushottam Kar. "Non-convex optimization for machine learning". In: *arXiv preprint arXiv:1712.07897* (2017).

19. Diederik P Kingma and Max Welling. “Auto-encoding variational Bayes”. In: *International Conference on Learning Representations*. 2014.
20. David G Kleinbaum et al. *Logistic regression*. Springer, 2002.
21. Dmitry Kobak and Philipp Berens. “The art of using t-SNE for single-cell transcriptomics”. In: *Nature communications* 10.1 (2019), pp. 1–14.
22. Dmitry Kobak et al. “Heavy-tailed kernels reveal a finer cluster structure in t-SNE visualisations”. In: *Joint European Conference on Machine Learning and Knowledge Discovery in Databases*. Springer. 2019, pp. 124–139.
23. George C Linderman et al. “Efficient algorithms for t-distributed stochastic neighborhood embedding”. In: *arXiv preprint arXiv:1712.09005* (2017).
24. Xueliang Liu et al. “Deep Neighborhood Component Analysis for Visual Similarity Modeling”. In: *ACM Transactions on Intelligent Systems and Technology (TIST)* 11.3 (2020), pp. 1–15.
25. Laurens van der Maaten. “Accelerating t-SNE using tree-based algorithms”. In: *The Journal of Machine Learning Research* 15.1 (2014), pp. 3221–3245.
26. Laurens van der Maaten. “Learning a parametric embedding by preserving local structure”. In: *Artificial Intelligence and Statistics*. 2009, pp. 384–391.
27. Laurens van der Maaten and Geoffrey Hinton. “Visualizing data using t-SNE”. In: *Journal of machine learning research* 9.Nov (2008), pp. 2579–2605.
28. Leland McInnes, John Healy, and James Melville. “UMAP: Uniform manifold approximation and projection for dimension reduction”. In: *arXiv preprint arXiv:1802.03426* (2018).
29. Tomas Mikolov et al. “Distributed representations of words and phrases and their compositionality”. In: *Advances in neural information processing systems*. 2013, pp. 3111–3119.
30. Tomas Mikolov et al. “Efficient estimation of word representations in vector space”. In: *arXiv preprint arXiv:1301.3781* (2013).
31. Yair Movshovitz-Attias et al. “No fuss distance metric learning using proxies”. In: *Proceedings of the IEEE International Conference on Computer Vision*. 2017, pp. 360–368.
32. Ashwin Narayan, Bonnie Berger, and Hyunghoon Cho. “Assessing single-cell transcriptomic variability through density-preserving data visualization”. In: *Nature Biotechnology* 39.6 (2021), pp. 765–774.
33. Ning Qian. “On the momentum term in gradient descent learning algorithms”. In: *Neural networks* 12.1 (1999), pp. 145–151.
34. Isaac Robinson and Emma Pierce-Hoffman. “Tree-SNE: Hierarchical Clustering and Visualization Using t-SNE”. In: *arXiv preprint arXiv:2002.05687* (2020).
35. Xin Rong. “word2vec parameter learning explained”. In: *arXiv preprint arXiv:1411.2738* (2014).
36. Tim Sainburg, Leland McInnes, and Timothy Q Gentner. “Parametric UMAP: learning embeddings with deep neural networks for representation and semi-supervised learning”. In: (2020).
37. Lawrence K Saul and Sam T Roweis. “Think globally, fit locally: unsupervised learning of low dimensional manifolds”. In: *Journal of machine learning research* 4.Jun (2003), pp. 119–155.
38. Frank Spitzer. *Principles of random walk*. Vol. 34. Springer Science & Business Media, 2013.
39. Jian Tang et al. “Visualizing large-scale and high-dimensional data”. In: *Proceedings of the 25th international conference on world wide web*. 2016, pp. 287–297.
40. Laurens Van Der Maaten. “Barnes-Hut-SNE”. In: *arXiv preprint arXiv:1301.3342* (2013).

Chapter 17
Uniform Manifold Approximation and Projection (UMAP)

17.1 Introduction

Probabilistic methods are a category of dimensionality reduction. Among the probabilistic methods, there are neighbour embedding algorithms where the probabilities of neighbourhoods are used. In these algorithms, attractive and repulsive forces are utilized for neighbour and non-neighbour points, respectively. Well-known neighbour embedding methods include Student's t-distributed Stochastic Neighbour Embedding (t-SNE) [14] (see Chap. 16), LargeVis [25], and Uniform Manifold Approximation and Projection (UMAP) [17]. Interestingly, both t-SNE and UMAP are state-of-the-art methods for data visualization.

UMAP assumes and approximates that the data points are uniformly distributed on an underlying manifold; it projects, or embeds, the data onto a subspace for dimensionality reduction. The theory behind UMAP is based on algebraic topology and category theory. UMAP constructs fuzzy topological representations for both the high-dimensional data and the low-dimensional embedding of the data and changes the embedding so that its fuzzy topological representation becomes similar to that of the high-dimensional data. UMAP has been widely used for DNA and single-cell data visualization and feature extraction [1, 7]. Additional applications of UMAP include visualizing deep features [4], art [27], and visualizing BERT features in natural language processing [5, 13]. This chapter introduces UMAP and its variants.

B. Ghojogh et al., *Elements of Dimensionality Reduction and Manifold Learning*,
https://doi.org/10.1007/978-3-031-10602-6_17

17.2 UMAP

17.2.1 Data Graph in the Input Space

Consider a training dataset $\boldsymbol{X} = [\boldsymbol{x}_1, \ldots, \boldsymbol{x}_n] \in \mathbb{R}^{d \times n}$, where n is the sample size and d is the dimensionality. A k-Nearest Neighbours (kNN) graph is constructed for this dataset. It has been empirically observed that UMAP requires fewer neighbours than t-SNE [23]. Its default value is $k = 15$. The j-th neighbour of $\boldsymbol{x}_i$ is denoted by $\boldsymbol{x}_{i,j}$. Let $\mathcal{N}_i$ denote the set of neighbour points for point $\boldsymbol{x}_i$, i.e., $\mathcal{N}_i := \{\boldsymbol{x}_{i,1}, \ldots, \boldsymbol{x}_{i,k}\}$. The neighbourhood relationship between points is treated stochastically. Inspired by SNE [9] and t-SNE [14] (see Chap. 16), the Gaussian or Radial Basis Function (RBF) kernel is used to measure the similarity between points in the input space. The probability that a point $\boldsymbol{x}_i$ has the point $\boldsymbol{x}_j$ as its neighbour can be computed by the similarity of these points:

$$p_{j|i} := \begin{cases} \exp\big(-\frac{\|\boldsymbol{x}_i - \boldsymbol{x}_j\|_2 - \rho_i}{\sigma_i}\big) & \text{if } \boldsymbol{x}_j \in \mathcal{N}_i \\ 0 & \text{Otherwise,} \end{cases} \tag{17.1}$$

where $\|.\|_2$ denotes the ℓ_2 norm. ρ_i is the distance from $\boldsymbol{x}_i$ to its nearest neighbour:

$$\rho_i := \min\{\|\boldsymbol{x}_i - \boldsymbol{x}_{i,j}\|_2 \,|\, 1 \leq j \leq k\}. \tag{17.2}$$

σ_i is the scale parameter, which is calculated such that the total similarity of point $\boldsymbol{x}_i$ to its k nearest neighbours is normalized. By binary search, σ_i is found to satisfy:

$$\sum_{j=1}^{k} \exp\Big(-\frac{\|\boldsymbol{x}_i - \boldsymbol{x}_{i,j}\|_2 - \rho_i}{\sigma_i}\Big) = \log_2(k). \tag{17.3}$$

Note that t-SNE [14] has a similar search for its scale using entropy as perplexity. These searches make the neighbourhoods of various points behave similarly because the scale for a point in a dense region of the dataset becomes small, while the scale of a point in a sparse region of the data becomes large. In other words, UMAP and t-SNE both assume (or approximate) that points are uniformly distributed on an underlying low-dimensional manifold. Equation (17.1) is a directional similarity measure. To have a symmetric measure with respect to i and j, it is symmetrized as:

$$\mathbb{R} \ni p_{ij} := p_{j|i} + p_{i|j} - p_{j|i}\, p_{i|j}. \tag{17.4}$$

This is a symmetric measure of similarity between points $\boldsymbol{x}_i$ and $\boldsymbol{x}_j$ in the input space.

17.2.2 Data Graph in the Embedding Space

Let the embeddings of the points be $\boldsymbol{Y} = [\boldsymbol{y}_1, \ldots, \boldsymbol{y}_n] \in \mathbb{R}^{p \times n}$, where p is the dimensionality of the embedding space, which is smaller than the input dimensionality, i.e., $p \ll d$. Note that $\boldsymbol{y}_i$ is the embedding corresponding to $\boldsymbol{x}_i$. In the embedding space, the probability that a point $\boldsymbol{y}_i$ has the point $\boldsymbol{y}_j$ as its neighbour can be computed by the similarity of these points:

$$\mathbb{R} \ni q_{ij} := (1 + a \, \|\boldsymbol{y}_i - \boldsymbol{y}_j\|_2^{2b})^{-1}, \tag{17.5}$$

which is symmetric with respect to i and j. The variables $a > 0$ and $b > 0$ are hyperparameters determined by the user. By default, $a \approx 1.929$ and $b \approx 0.7915$ [17] are used, although it has been empirically proven that setting $a = b = 1$ does not qualitatively impact the results [3].

17.2.3 Optimization Cost Function

UMAP aims to make the data graph in the low-dimensional embedding space similar to the data graph in the high-dimensional embedding space. In other words, Eqs. (17.4) and (17.5) are treated as probability distributions and minimize the difference of these distributions to make similarities of points in the embedding space as the similarities of points in the input space. A measure for the difference of these similarities of the graphs is the fuzzy cross-entropy, which is defined as:

$$c_1 := \sum_{i=1}^{n} \sum_{j=1, j \neq i}^{n} \Big(p_{ij} \ln(\frac{p_{ij}}{q_{ij}}) + (1 - p_{ij}) \ln(\frac{1 - p_{ij}}{1 - q_{ij}})\Big), \tag{17.6}$$

where ln(.) is the natural logarithm. The definition of this cross-entropy can be found in Sect. 17.3 (see Eq. (17.17)).

The first term in Eq. (17.6) is the *attractive force*, which attracts the embeddings of the neighbour points toward each other. This term should only appear when $p_{ij} \neq 0$, which means either $\boldsymbol{x}_j$ is a neighbour of $\boldsymbol{x}_i$, or $\boldsymbol{x}_i$ is a neighbour of $\boldsymbol{x}_j$, or both (see Eq. (17.4)). The second term in Eq. (17.6) is the *repulsive force* that repulses the embeddings of the nonneighbour points away from each other. As the number of all permutations of nonneighbour points is exceedingly large, computation of the second term is nontractable in big data. Inspired by Word2Vec [18] and LargeVis [25], UMAP uses *negative sampling*, where for every point $\boldsymbol{x}_i$, m points are sampled randomly from the training dataset and are treated as nonnegative (negative) points for $\boldsymbol{x}_i$. As the dataset is usually large, i.e. $m \ll n$, the sampled points will be actual negative points with high probability. The summation over the second term in Eq. (17.6) is computed only for these negative samples, rather than *all* negative points.

UMAP changes the data graph in the embedding space to make it similar to the data graph in the input space. Equation (17.6) is the cost function that is minimized in UMAP, where the optimization variables are $\{y_i\}_{i=1}^n$:

$$
\begin{aligned}
\min_{\{\mathbf{y}_i\}_{i=1}^n} c_1 := \min_{\{\mathbf{y}_i\}_{i=1}^n} \sum_{i=1}^{n} \sum_{j=1, j\neq i}^{n} \Big(& p_{ij} \ln(p_{ij}) - p_{ij} \ln(q_{ij}) \\
& + (1 - p_{ij}) \ln(1 - p_{ij}) - (1 - p_{ij}) \ln(1 - q_{ij}) \Big) \\
= \min_{\{\mathbf{y}_i\}_{i=1}^n} - \sum_{i=1}^{n} \sum_{j=1, j\neq i}^{n} \Big(& p_{ij} \ln(q_{ij}) + (1 - p_{ij}) \ln(1 - q_{ij}) \Big)
\end{aligned}
$$

According to Eqs. (17.1), (17.4), and (17.5), in contrast to q_{ij}, p_{ij} is independent of the optimization variables $\{y_i\}_{i=1}^n$. Therefore, the constant terms can be dropped to revise the cost function:

$$
c_2 := - \sum_{i=1}^{n} \sum_{j=1, j\neq i}^{n} \Big(p_{ij} \ln(q_{ij}) + (1 - p_{ij}) \ln(1 - q_{ij}) \Big), \tag{17.7}
$$

which should be minimized. Two important terms in this cost function include the following:

$$
c^a_{i,j} := - \ln(q_{ij}), \tag{17.8}
$$

$$
c^r_{i,j} := - \ln(1 - q_{ij}), \tag{17.9}
$$

and the cost can be stated as:

$$
c_2 := \sum_{i=1}^{n} \sum_{j=1, j\neq i}^{n} \Big(p_{ij}\, c^a_{i,j} + (1 - p_{ij})\, c^r_{i,j} \Big) \tag{17.10}
$$

$$
\overset{(a)}{=} 2 \sum_{i=1}^{n} \sum_{j=i+1}^{n} \Big(p_{ij}\, c^a_{i,j} + (1 - p_{ij})\, c^r_{i,j} \Big), \tag{17.11}
$$

where (a) is because $p_{ij} = p_{ji}$, $c^a_{i,j} = c^a_{j,i}$, and $c^r_{i,j} = c^r_{j,i}$ are symmetric.

Equations (17.8) and (17.9) are the attractive and repulsive forces in Eq. (17.7), respectively. The attractive force attracts the neighbour points towards each other in the embedding space, while the repulsive force pushes the nonneighbour points (i.e., points with low probability of being neighbours) away from each other in the embedding space. According to Eq. (17.10), $c^a_{i,j}$ and $c^r_{i,j}$ occur with probability p_{ij} and $(1 - p_{ij})$, respectively. Every point, indexed by i, is considered the *anchor* point, for which several *positive* points with large p_{ij} and several *negative* points with small p_{ij} are used.

```
 1  Input: Training data {x_i}_{i=1}^n
 2  Construct kNN graph
 3  Initialize {y_i}_{i=1}^n by Laplacian eigenmap
 4  Calculate p_{ij} and q_{ij} for ∀i, j ∈ {1, …, n} by Eqs. (17.4) and (17.5)
 5  η ← 1, ν ← 0
 6  while not converged do
 7  │   ν ← ν + 1    // epoch index
 8  │   for i from 1 to n do
 9  │   │   for j from 1 to n do
10  │   │   │   u ~ U(0, 1)
11  │   │   │   if u ≤ p_{ij} then
12  │   │   │   │   y_i ← y_i − η (∂c^a_{i,j} / ∂y_i)
13  │   │   │   │   y_j ← y_j − η (∂c^a_{i,j} / ∂y_j)
14  │   │   │   │   for m iterations do
15  │   │   │   │   │   l ~ U{1, …, n}
16  │   │   │   │   │   y_i ← y_i − η (∂c^r_{i,l} / ∂y_i)
17  │   │   │   │   │   // The next line does not exist in original UMAP:
18  │   │   │   │   │   y_l ← y_l − η (∂c^r_{i,l} / ∂y_l)
19  │   η ← 1 − ν/ν_max
20  Return {y_i}_{i=1}^n
```

Algorithm 17.1: UMAP algorithm

17.2.4 The Training Algorithm of UMAP

Optimization in UMAP is explained in Algorithm 17.1, which demonstrates that a kNN graph is constructed from the training data $\{x_i\}_{i=1}^n$. UMAP uses the Laplacian eigenmap [2] (see Chap. 9), also called spectral embedding, for initializing the embeddings of the points denoted by $\{y_i\}_{i=1}^n$. Using Eqs. (17.4) and (17.5), p_{ij} and q_{ij} are calculated for all points. Stochastic Gradient Descent (SGD) is used for optimization, where optimization is performed iteratively (see Chap. 4). In every epoch, iteration occurs over points twice with indices i and j, where the i-th point is called the *anchor*. For every pair of points x_i and x_j, their embeddings x_i and x_j are updated with the probability p_{ij} (recall Eq. (17.7)). If p_{ij} is large, it means that the points x_i and x_j are most likely neighbours (in this case, the j-th point is called the *positive* point) and their embeddings are highly likely to be updated so that they can become closer in the embedding space based on the attractive force. For its implementation, a uniform value can be sampled from the continuous uniform distribution $U(0, 1)$ and if that is less than p_{ij}, the embeddings are updated. The embeddings y_i and y_j are updated by the gradients $\partial c^a_{i,j}/\partial y_i$ and $\partial c^a_{i,j}/\partial y_j$, respectively, where η is the learning rate.

As previously explained in Sect. 17.2.3, negative sampling is used for repulsive forces. If m denotes the size of negative sample, m indices are sampled from the discrete uniform distribution $U\{1, \dots, n\}$. These are the indices of the points that are considered as *negative* samples $\{\boldsymbol{y}_l\}$, where $|\{\boldsymbol{y}_l\}| = m$. As the size of the dataset is usually large enough to satisfy $n \gg m$, these negative points are probably valid because many of the points are nonneighbours of the considered anchor. In negative sampling, the original UMAP [17] updates only the embedding of the anchor $\boldsymbol{y}_i$ by the gradient of the repulsive force $\partial c^a_{i,j}/\partial \boldsymbol{y}_i$. Additionally, the embedding of negative point $\boldsymbol{y}_l$ can be updated by the gradient of the repulsive force $\partial c^a_{i,j}/\partial \boldsymbol{y}_l$ [6]. The mentioned gradients are computed in the following lemmas.

Lemma 17.1 ([17]) *The gradients of the attractive and repulsive cost functions in UMAP are:*

$$\frac{\partial c^a_{i,j}}{\partial \boldsymbol{y}_i} = \frac{2ab\|\boldsymbol{y}_i - \boldsymbol{y}_j\|_2^{2(b-1)}}{(1 + a\,\|\boldsymbol{y}_i - \boldsymbol{y}_j\|_2^{2b})}(\boldsymbol{y}_i - \boldsymbol{y}_j), \tag{17.12}$$

$$\frac{\partial c^r_{i,j}}{\partial \boldsymbol{y}_i} = \frac{-2b}{(\varepsilon + \|\boldsymbol{y}_i - \boldsymbol{y}_j\|_2^2)(1 + a\,\|\boldsymbol{y}_i - \boldsymbol{y}_j\|_2^{2b})}(\boldsymbol{y}_i - \boldsymbol{y}_j), \tag{17.13}$$

where ε is a small positive number, e.g. $\varepsilon = 0.001$, for stability to prevent division by zero when $\boldsymbol{y}_i \approx \boldsymbol{y}_j$. Likewise:

$$\frac{\partial c^a_{i,j}}{\partial \boldsymbol{y}_j} = \frac{2ab\|\boldsymbol{y}_i - \boldsymbol{y}_j\|_2^{2(b-1)}}{(1 + a\,\|\boldsymbol{y}_i - \boldsymbol{y}_j\|_2^{2b})}(\boldsymbol{y}_j - \boldsymbol{y}_i),$$

$$\frac{\partial c^r_{i,j}}{\partial \boldsymbol{y}_j} = \frac{-2b}{(\varepsilon + \|\boldsymbol{y}_i - \boldsymbol{y}_j\|_2^2)(1 + a\,\|\boldsymbol{y}_i - \boldsymbol{y}_j\|_2^{2b})}(\boldsymbol{y}_j - \boldsymbol{y}_i).$$

Proof For the first equation, there is:

$$\frac{\partial c^a_{i,j}}{\partial \boldsymbol{y}_i} = \frac{\partial c^a_{i,j}}{\partial q_{ij}} \times \frac{\partial q_{ij}}{\partial \boldsymbol{y}_i} = \frac{-1}{q_{ij}} \times \Big(\frac{-1}{(1 + a\,\|\boldsymbol{y}_i - \boldsymbol{y}_j\|_2^{2b})^2} \times 2ab(\boldsymbol{y}_i - \boldsymbol{y}_j) \times \|\boldsymbol{y}_i - \boldsymbol{y}_j\|_2^{2(b-1)}\Big)$$

$$\overset{(17.5)}{=} \frac{2ab\|\boldsymbol{y}_i - \boldsymbol{y}_j\|_2^{2(b-1)}}{(1 + a\,\|\boldsymbol{y}_i - \boldsymbol{y}_j\|_2^{2b})}(\boldsymbol{y}_i - \boldsymbol{y}_j). \tag{17.14}$$

For the second equation, there is:

$$\frac{\partial c^r_{i,j}}{\partial \boldsymbol{y}_i} = \frac{\partial c^r_{i,j}}{\partial q_{ij}} \times \frac{\partial q_{ij}}{\partial \boldsymbol{y}_i}$$

$$= \frac{1}{1 - q_{ij}} \times \Big(\frac{-1}{(1 + a\,\|\boldsymbol{y}_i - \boldsymbol{y}_j\|_2^{2b})^2} \times 2ab(\boldsymbol{y}_i - \boldsymbol{y}_j) \times \|\boldsymbol{y}_i - \boldsymbol{y}_j\|_2^{2(b-1)}\Big)$$

$$= \frac{-2ab\|\boldsymbol{y}_i - \boldsymbol{y}_j\|_2^{2(b-1)}}{(1 - q_{ij})(1 + a\,\|\boldsymbol{y}_i - \boldsymbol{y}_j\|_2^{2b})^2}(\boldsymbol{y}_i - \boldsymbol{y}_j). \tag{17.15}$$

The term in the numerator can be simplified as:

$$-2ab\|\boldsymbol{y}_i - \boldsymbol{y}_j\|_2^{2(b-1)} = -2b(a\|\boldsymbol{y}_i - \boldsymbol{y}_j\|_2^{2b})\|\boldsymbol{y}_i - \boldsymbol{y}_j\|_2^{-2} \overset{(17.5)}{=} -2b\,(q_{ij}^{-1} - 1)\|\boldsymbol{y}_i - \boldsymbol{y}_j\|_2^{-2}.$$

The term in the denominator can be simplified as:

$$(1 - q_{ij})(1 + a\,\|\boldsymbol{y}_i - \boldsymbol{y}_j\|_2^{2b})^2 \overset{(17.5)}{=} (1 - q_{ij})q_{ij}^{-2} = q_{ij}^{-2} - q_{ij}^{-1} = q_{ij}^{-1}(q_{ij}^{-1} - 1).$$

Therefore, Eq. (17.15) can be simplified as:

$$\frac{\partial c_{i,j}^r}{\partial \boldsymbol{y}_i} = \frac{-2b\,(q_{ij}^{-1} - 1)\|\boldsymbol{y}_i - \boldsymbol{y}_j\|_2^{-2}}{q_{ij}^{-1}(q_{ij}^{-1} - 1)}(\boldsymbol{y}_i - \boldsymbol{y}_j)$$

$$= \frac{-2b}{\|\boldsymbol{y}_i - \boldsymbol{y}_j\|_2^2\, q_{ij}^{-1}}(\boldsymbol{y}_i - \boldsymbol{y}_j) \overset{(17.5)}{=} \frac{-2b}{\|\boldsymbol{y}_i - \boldsymbol{y}_j\|_2^2\,(1 + a\,\|\boldsymbol{y}_i - \boldsymbol{y}_j\|_2^{2b})}(\boldsymbol{y}_i - \boldsymbol{y}_j).$$

If ε is added for stability to the squared distance in the denominator, the equation is obtained. □

17.2.5 *Supervised and Semisupervised Embedding*

The UMAP algorithm, explained in Sect. 17.2.4, is unsupervised. There can be both supervised and semisupervised embeddings by UMAP [23]. For the supervised UMAP, the UMAP cost function, Eq. (17.7), can be used, regularized by a classification cost function such as cross-entropy or triplet loss. In semisupervised scenarios, a part of the dataset has labels and other parts do not. It is possible to iteratively alternate between UMAP's cost function, Eq. (17.7), and a classification cost function. In this way, the embeddings are updated by UMAP and fine-tuned by the class labels; this procedure is repeated iteratively until convergence of the embedding.

17.3 Justifying UMAP's Cost Function by Algebraic Topology and Category Theory

UMAP is a neighbour embedding method, where the probability of neighbours for every point is used for optimization of the embedding. However, its cost function,

Eq. (17.6), can be justified by algebraic topology [8, 16] and category theory [15, 22]. Specifically, its theory heavily uses fuzzy category theory [24]. This section introduces the theory behind UMAP.

The following concepts are used in the theory of UMAP:

- A *simplex* is a generalization of the triangle to arbitrary dimensions.
- A *simplicial complex* is a set of points, line segments, triangles, and their d-dimensional counterparts [16].
- A *fuzzy set* is a mapping $\mu : \mathcal{A} \to [0, 1]$ from carrier set $\mathcal{A}$, where the mapping is called the membership function [28]. A fuzzy set can be signified by $(\mathcal{A}, \mu)$.
- In category theory, a *category* is a collection of objects that are linked by arrows. For example, objects can be sets and the arrows can be functions between the sets [15]. A *morphism* is a mapping from a mathematical structure to another structure without changing its type (e.g., morphisms are functions in set theory). A *functor* is defined as a mapping between categories. *Adjunction* is the relationship between two functors. The two functors having an adjunction are *adjoint* functors, one of which is the *left adjoint* and the other of which is the *right adjoint*.
- A *topology* is a geometrical object that is preserved by continuous deformations, such as stretching and twisting, but without tearing and making or closing holes. A *topological space* is a set of topologies whose operations are continuous deformations.
- The category $\mathbf{\Delta}$, whose objects are finite-order sets, is defined as $[n] = \{1, \dots, n\}$, with order-preserving maps as its morphisms [17, Definition 1]. A simplicial set is a functor from $\mathbf{\Delta}$ to the category of sets [17, Definition 2].
- Consider a category of fuzzy sets, denoted by **Fuzz** [17, Definition 4]. The category of fuzzy simplicial sets, denoted by **sFuzz**, is the category of objects with functors from $\mathbf{\Delta}$ to **Fuzz** and natural transformations as its morphisms [17, Definition 5].
- An extended-pseudometric space is a set $\mathcal{X}$ and a mapping $d : \mathcal{X} \times \mathcal{X} \to \mathbb{R}_{\geq 0} \cup \{\infty\}$, where for $x, y \in \mathcal{X}$, there are $d(x, y) \geq 0$ and $x = y \implies d(x, y) = 0$ and $d(x, y) = d(y, x)$ and $d(x, z) \leq d(x, y) + d(y, z)$ [17, Definition 6]. The mapping d can be seen as a distance metric or pseudometric.
- Let **Fin-sFuzz** be the subcategory of the bounded fuzzy simplicial sets. Let **FinEPMet** be the subcategory of finite extended-pseudometric spaces.

Theorem 17.1 ([17, Theorem 1]) *The functors* ***FinReal****:* ***Fin-sFuzz*** *→* ***FinEPMet*** *and* ***FinSing****:* ***FinEPMet*** *→* ***Fin-sFuzz*** *form an adjunction with* ***FinReal*** *and* ***FinSing*** *as the left and right adjoints, respectively.*

Proof Proof is available in [17, Appendix B]. □

The above theorem demonstrates that it is possible to convert an extended-pseudometric space to a fuzzy simplicial set and vice versa. In other words, there is a fuzzy simplicial representation of the data space. Therefore, the following corollary holds.

Corollary 17.1 (Fuzzy Topological Representation [17, Definition 9]) *Consider a dataset $\mathcal{X} := \{\boldsymbol{x}_i \in \mathbb{R}^d\}_{i=1}^n$ lying on an underlying manifold $\mathcal{M}$. Let $\{(\mathcal{X}, d_i)\}_{i=1}^n$ be a family of extended-pseudometric spaces with common carrier set $\mathcal{X}$ such that:*

$$d_i(\boldsymbol{x}_j, \boldsymbol{x}_l) := \begin{cases} d_{\mathcal{M}}(\boldsymbol{x}_j, \boldsymbol{x}_l) - \rho_i & \text{if } i = j \text{ or } i = l, \\ \infty & \text{Otherwise}, \end{cases} \tag{17.16}$$

where $d_{\mathcal{M}}(.,.)$ is the geodesic (shortest) distance on the manifold, and ρ_i is the distance to the nearest neighbour of $\boldsymbol{x}_i$ (see Eq. (17.2)). The fuzzy topological representation of dataset $\mathcal{X}$ is:

$$\bigcup_{i=1}^{n} \mathbf{FinSing}((\mathcal{X}, d_i)),$$

where $\bigcup$ is the fuzzy set union.

UMAP creates a fuzzy topological representation for the high-dimensional dataset. Then, it initializes a low-dimensional embedding of the dataset and creates a fuzzy topological representation for the low-dimensional embedding of the dataset. Then, it tries to modify the low-dimensional embedding of the dataset in such a way that the fuzzy topological representation of the embedding becomes similar to the fuzzy topological representation of the high-dimensional data. A measure of the difference between two fuzzy topological representations $(\mathcal{A}, \mu_1)$ and $(\mathcal{A}, \mu_2)$ is their cross-entropy defined as [17]:

$$c\big((\mathcal{A}, \mu_1), (\mathcal{A}, \mu_2)\big) := \sum_{a \in \mathcal{A}} \Big(\mu_1(a) \ln\Big(\frac{\mu_1(a)}{\mu_2(a)}\Big) + (1 - \mu_1(a)) \ln\Big(\frac{1 - \mu_1(a)}{1 - \mu_2(a)}\Big)\Big). \tag{17.17}$$

UMAP minimizes this cross-entropy by iteratively changing the embedding. Therefore, it uses cross-entropy as its cost function, i.e., Eq. (17.6).

17.4 Neighbour Embedding: Comparison with t-SNE and LargeVis

UMAP has a connection with t-SNE[1] [14] (see Chap. 16) and LargeVis [25] (see [17, Appendix C] for more information). UMAP, t-SNE, and LargeVis are neighbour embedding methods, in which attractive and repulsive forces are used [3]. As explained in Sect. 17.2, for every point considered as an anchor, attractive forces

[1] An empirical comparison of UMAP and t-SNE is available in [21].

are used for pushing the neighbour (also called positive) points to the anchor points and repulsive forces are used for pulling nonneighbour (also called negative) points away from the anchor points.

Comparison of Probabilities In t-SNE, the probabilities in the input and embedding spaces are (see Chap. 16):

$$p_{j|i} := \frac{\exp\big(-\frac{\|\boldsymbol{x}_i - \boldsymbol{x}_j\|_2}{\sigma_i}\big)}{\sum_{k=1,k\neq i}^{n} \exp\big(-\frac{\|\boldsymbol{x}_i - \boldsymbol{x}_k\|_2}{\sigma_i}\big)}, \tag{17.18}$$

$$p_{ij} := \frac{p_{j|i} + p_{i|j}}{2n}, \tag{17.19}$$

$$q_{ij} := \frac{(1 + \|\boldsymbol{y}_i - \boldsymbol{y}_j\|_2^2)^{-1}}{\sum_{k=1}^{n} \sum_{l=1,l\neq k}^{n} (1 + \|\boldsymbol{y}_k - \boldsymbol{y}_l\|_2^2)^{-1}}, \tag{17.20}$$

where $p_{i|i} = 0, \forall i$. The $p_{j|i}$ probabilities can be computed for the kNN graph, where $p_{j|i}$ is set to zero for nonneighbour points in the kNN graph. LargeVis uses the same p_{ij} probabilities as t-SNE but approximates the kNN to decrease the computation time and increase efficiency. In LargeVis, the probability in the embedding space is:

$$q_{ij} := (1 + \|\boldsymbol{y}_i - \boldsymbol{y}_j\|_2^2)^{-1}. \tag{17.21}$$

Comparing Eqs. (17.1) and (17.18) demonstrates that UMAP, t-SNE, and LargeVis all use Gaussian or RBF kernel for computing probabilities in the input space. Comparing Eqs. (17.4) and (17.19) demonstrates that UMAP and t-SNE/LargeVis use different approaches for symmetrizing the probabilities in the input space. Comparing Eqs. (17.5), (17.20), and (17.21) demonstrates that, in contrast to t-SNE, UMAP and LargeVis do not normalize the probabilities in the embedding space by all pairs of points. This advantage makes UMAP faster than t-SNE and makes it more suitable for minibatch optimization in deep learning (see Sect. 17.7). Comparing Eqs. (17.5), (17.20), and (17.21) also demonstrates that UMAP, t-SNE, and LargeVis all use a Cauchy distribution for computing probabilities in the embedding space. If $a = b = 1$ is used in Eq. (17.5), it is exactly the same as Eq. (17.20) up to the scale of normalization.

Comparison of Cost Functions The cost function in t-SNE, to be minimized, is the KL-divergence [12] between the probabilities in the input and embedding spaces:

$$c_4 := \sum_{i=1}^{n} \sum_{j=1,j\neq i}^{n} p_{ij} \ln(\frac{p_{ij}}{q_{ij}}) = \sum_{i=1}^{n} \sum_{j=1,j\neq i}^{n} \big(p_{ij} \ln(p_{ij}) - p_{ij} \ln(q_{ij})\big), \tag{17.22}$$

where Eqs. (17.19) and (17.20) are used. The cost function of LargeVis, to be minimized, is the negative likelihood function stated below:

$$c_5 := -\sum_{i=1}^{n} \sum_{j=1, j \neq i}^{n} \big(p_{ij} \ln(q_{ij}) + \lambda \ln(1 - q_{ij})\big), \tag{17.23}$$

where λ is the regularization parameter and Eqs. (17.19) and (17.21) are used. Comparing Eqs. (17.7), (17.22), and (17.23) demonstrates that UMAP, t-SNE, and LargeVis have similar, but not equal, cost functions. The first term in all these cost functions is responsible for the attractive forces, and the second term is responsible for the repulsive forces; therefore, they can all be considered as neighbour embedding methods [3].

17.5 Discussion on Repulsive Forces and Negative Sampling in the UMAP's Cost Function

17.5.1 UMAP's Emphasis on Repulsive Forces

The gradient of UMAP's cost function, Eq. (17.10), in epoch ν is denoted by $(\partial c_2/\partial \boldsymbol{y}_i)|_\nu$ and can be stated as [6]:

$$\frac{\partial c_2}{\partial \boldsymbol{y}_i}\bigg|_\nu = \sum_{j=1}^{n} \Big(\mathbb{I}_{ij}^\nu \frac{\partial c_{i,j}^a}{\partial \boldsymbol{y}_i} + \mathbb{I}_{ji}^\nu \frac{\partial c_{j,i}^a}{\partial \boldsymbol{y}_i} + \mathbb{I}_{ij}^\nu \sum_{l=1}^{n} \mathbb{I}_{ijl}^\nu \frac{\partial c_{i,l}^r}{\partial \boldsymbol{y}_i}\Big), \tag{17.24}$$

where $c_{i,j}^a$ and $c_{i,j}^r$ are defined in Eqs. (17.8) and (17.8), respectively, and $\mathbb{I}_{ij}^\nu$ is a binary random variable equal to one if points $\boldsymbol{y}_i$ and $\boldsymbol{y}_j$ are randomly selected in epoch ν; otherwise, it is equal to zero. Additionally, $\mathbb{I}_{ijl}^\nu$ is a binary random variable equal to one if point $\boldsymbol{y}_l$ is one of the negative samples that is randomly sampled for the pair $\boldsymbol{y}_i$ and $\boldsymbol{y}_j$ in epoch ν; otherwise it is equal to zero. Recall that the original UMAP does not update the embedding of the negative sample itself, so there is no available term for that update in this gradient.

Equation (17.24) is only for one epoch. The expectation of Eq. (17.24) over all epochs is [6]:

$$\mathbb{E}\Big[\frac{\partial c_2}{\partial \boldsymbol{y}_i}\Big|_\nu\Big] = \mathbb{E}\Big[\sum_{j=1}^{n} \Big(\mathbb{I}_{ij}^\nu \frac{\partial c_{i,j}^a}{\partial \boldsymbol{y}_i} + \mathbb{I}_{ji}^\nu \frac{\partial c_{j,i}^a}{\partial \boldsymbol{y}_i} + \mathbb{I}_{ij}^\nu \sum_{l=1}^{n} \mathbb{I}_{ijl}^\nu \frac{\partial c_{i,l}^r}{\partial \boldsymbol{y}_i}\Big)\Big]$$

$$\overset{(a)}{=} \sum_{j=1}^{n} \Big(\mathbb{E}[\mathbb{I}_{ij}^\nu] \frac{\partial c_{i,j}^a}{\partial \boldsymbol{y}_i} + \mathbb{E}[\mathbb{I}_{ji}^\nu] \frac{\partial c_{j,i}^a}{\partial \boldsymbol{y}_i}\Big) + \sum_{j=1}^{n} \sum_{l=1}^{n} \mathbb{E}[\mathbb{I}_{ij}^\nu\, \mathbb{I}_{ijl}^\nu] \frac{\partial c_{i,l}^r}{\partial \boldsymbol{y}_i}$$

$$\overset{(b)}{=} \sum_{j=1}^{n} \Big(p_{ij} \frac{\partial c^a_{i,j}}{\partial \boldsymbol{y}_i} + p_{ji} \frac{\partial c^a_{j,i}}{\partial \boldsymbol{y}_i} \Big) + \sum_{j=1}^{n} \sum_{l=1}^{n} p_{ij} \frac{m}{n} \frac{\partial c^r_{i,l}}{\partial \boldsymbol{y}_i}$$

$$\overset{(c)}{=} \sum_{j=1}^{n} \Big(p_{ij} \frac{\partial c^a_{i,j}}{\partial \boldsymbol{y}_i} + p_{ji} \frac{\partial c^a_{j,i}}{\partial \boldsymbol{y}_i} \Big) + \frac{m}{n} \underbrace{\sum_{j=1}^{n} p_{ij}}_{=d_i} \sum_{l=1}^{n} \frac{\partial c^r_{i,l}}{\partial \boldsymbol{y}_i}$$

$$\overset{(d)}{=} \sum_{j=1}^{n} \Big(p_{ij} \frac{\partial c^a_{i,j}}{\partial \boldsymbol{y}_i} + p_{ji} \frac{\partial c^a_{j,i}}{\partial \boldsymbol{y}_i} \Big) + \frac{d_i m}{n} \sum_{j=1}^{n} \frac{\partial c^r_{i,j}}{\partial \boldsymbol{y}_i} \overset{(e)}{=} 2 \sum_{j=1}^{n} \Big(p_{ij} \frac{\partial c^a_{i,j}}{\partial \boldsymbol{y}_i} + \frac{d_i m}{2n} \frac{\partial c^r_{i,j}}{\partial \boldsymbol{y}_i} \Big), \tag{17.25}$$

where (a) is because the expectation is a linear operator, (b) is because $\mathbb{E}[\mathbb{I}^{\nu}_{ij}] = p_{ij}$ and $\mathbb{E}[\mathbb{I}^{\nu}_{ij}\, \mathbb{I}^{\nu}_{ijl}] = \mathbb{E}[\mathbb{I}^{\nu}_{ijl} \,|\, \mathbb{I}^{\nu}_{ij}] \times \mathbb{E}[\mathbb{I}^{\nu}_{ij}] = \frac{m}{n} \times p_{ij}$, (c) is because the degree of the i-th point (node) in the kNN graph is defined as $d_i := \sum_{j=1}^{n} p_{ij}$, (d) is because the dummy variable l is changed to j in the last summation, and (e) is because $p_{ij} = p_{ji}$ and $c^a_{i,j} = c^a_{j,i}$ are symmetric.

On the other hand, according to Eq. (17.11), the gradient of UMAP's cost function, Eq. (17.7), can be stated as [6]:

$$\frac{\partial c_2}{\partial \boldsymbol{y}_i} = 2 \sum_{j=1}^{n} \Big(p_{ij} \frac{\partial c^a_{i,j}}{\partial \boldsymbol{y}_i} + (1 - p_{ij}) \frac{\partial c^r_{i,j}}{\partial \boldsymbol{y}_i} \Big). \tag{17.26}$$

Equation (17.26) is the gradient of the original UMAP's loss function, while Eq. (17.25) is the expected gradient of UMAP's loss function. Comparing Eqs. (17.25) and (17.26) demonstrates that the original UMAP places more emphasis on negative samples (or repulsive forces) compared to the expected UMAP's loss function. This is because, for a negative sample, there is $1 - p_{ij} \approx 1$, while $d_i m/n \approx 0$ is because $m \ll n$. Therefore, UMAP is mistakenly putting more emphasis on negative sampling (or repulsive forces) than required [6]. This has been empirically investigated in [3], which demonstrated that negative sampling (or repulsive forces) in UMAP is weighed more than needed.

17.5.2 UMAP's Effective Cost Function

The original UMAP does not update the embedding of negative samples themselves. If they are updated, i.e., line 18 is performed in Algorithm 17.1, then the gradient of UMAP's cost function, Eq. (17.10), in epoch ν can be stated as [6]:

$$\frac{\partial c_2}{\partial \boldsymbol{y}_i}\Big|_\nu = \sum_{j=1}^{n} \Big(\mathbb{I}_{ij}^\nu \frac{\partial c_{i,j}^a}{\partial \boldsymbol{y}_i} + \mathbb{I}_{ji}^\nu \frac{\partial c_{j,i}^a}{\partial \boldsymbol{y}_i} + \mathbb{I}_{ij}^\nu \sum_{l=1}^{n} \mathbb{I}_{ijl}^\nu \frac{\partial c_{i,l}^r}{\partial \boldsymbol{y}_i} + \sum_{k=1}^{n} \mathbb{I}_{jk}^\nu \mathbb{I}_{jki}^\nu \frac{\partial c_{j,i}^r}{\partial \boldsymbol{y}_i} \Big). \tag{17.27}$$

If reverse engineering is used to find the cost function from its gradient, the cost function at epoch ν becomes:

$$c_2\big|_\nu := \sum_{i=1}^{n} \sum_{j=1, j\neq i}^{n} \Big(\mathbb{I}_{ij}^\nu c_{i,j}^a + \sum_{l=1}^{n} \mathbb{I}_{ij}^\nu \mathbb{I}_{ijl}^\nu c_{i,l}^r \Big). \tag{17.28}$$

This is the cost at one epoch. The expectation of this cost over all epochs is [6]:

$$\begin{aligned}
c_2 = \mathbb{E}[c_2\big|_\nu] &= \sum_{i=1}^{n} \sum_{j=1, j\neq i}^{n} \Big(\mathbb{E}[\mathbb{I}_{ij}^\nu]\, c_{i,j}^a + \sum_{l=1}^{n} \mathbb{E}[\mathbb{I}_{ij}^\nu \mathbb{I}_{ijl}^\nu]\, c_{i,l}^r \Big) \\
&= \sum_{i=1}^{n} \sum_{j=1, j\neq i}^{n} \Big(\mathbb{E}[\mathbb{I}_{ij}^\nu]\, c_{i,j}^a \Big) + \sum_{i=1}^{n} \sum_{j=1, j\neq i}^{n} \sum_{l=1}^{n} \Big(\mathbb{E}[\mathbb{I}_{ij}^\nu \mathbb{I}_{ijl}^\nu]\, c_{i,l}^r \Big) \\
&\overset{(a)}{=} \sum_{i=1}^{n} \sum_{j=1, j\neq i}^{n} \Big(p_{ij}\, c_{i,j}^a \Big) + \sum_{i=1}^{n} \sum_{j=1, j\neq i}^{n} \sum_{l=1}^{n} \Big(\frac{m}{n} p_{ij}\, c_{i,l}^r \Big) \\
&= \sum_{i=1}^{n} \sum_{j=1, j\neq i}^{n} \Big(p_{ij}\, c_{i,j}^a \Big) + \frac{m}{n} \Big(\underbrace{\sum_{i=1}^{n} \sum_{j=1}^{n} p_{ij}}_{=d_i} + \underbrace{\sum_{j=1}^{n} \sum_{i=1}^{n} p_{ji}}_{=d_j} \Big) \sum_{l=1}^{n} c_{i,l}^r \\
&\overset{(b)}{=} \sum_{i=1}^{n} \sum_{j=1, j\neq i}^{n} \Big(p_{ij}\, c_{i,j}^a \Big) + \sum_{i=1}^{n} \sum_{j=i+1}^{n} \frac{(d_i + d_j)m}{n} c_{i,j}^r \\
&\overset{(c)}{=} 2 \sum_{i=1}^{n} \sum_{j=i+1}^{n} \Big(p_{ij}\, c_{i,j}^a + \frac{(d_i + d_j)m}{2n} c_{i,j}^r \Big),
\end{aligned} \tag{17.29}$$

where (a) is because $\mathbb{E}[\mathbb{I}_{ij}^\nu] = p_{ij}$ and $\mathbb{E}[\mathbb{I}_{ij}^\nu \mathbb{I}_{ijl}^\nu] = \mathbb{E}[\mathbb{I}_{ijl}^\nu \,|\, \mathbb{I}_{ij}^\nu] \times \mathbb{E}[\mathbb{I}_{ij}^\nu] = \frac{m}{n} \times p_{ij}$, (b) is because the dummy variable l is changed to j in the last summation, and (c) is because of symmetry of terms in the first summations with respect to i and j. Equation (17.29) can be considered as UMAP's effective cost function [6] because it also updates the embedding of the negative samples by line 18 in Algorithm 17.1.

Comparing UMAP's cost, Eq. (17.11), with UMAP's effective cost, Eq. (17.29), shows that the weight of negative sample (or repulsive forces) should be $\frac{(d_i+d_j)m}{2n}$ rather than $(1 - p_{ij})$ if we also update the embeddings of negative samples. As we have $m \ll n$ and p_{ij} is small for negative samples, this weight is much less than the weight in the original UMAP.

17.6 DensMAP for Density-Preserving Embedding

As explained in Sect. 17.2, UMAP uses a binary search for the scale of each point, σ_i, to satisfy Eq. (17.3). A similar binary search is used in t-SNE [14], which searches for its scale using entropy as perplexity. The binary search makes the neighbourhoods of the various points behave similarly, so UMAP and t-SNE both assume that the points are uniformly distributed on an underlying low-dimensional manifold. Therefore, UMAP ignores the density of the data around every point by cancelling the effect of the density with the binary search for the scales of points. DensMAP [20] regularizes the cost function of UMAP to take into account and add back the density information around each point. It has been empirically demonstrated that considering density information results in better embedding [20], although calculation of the regularization term introduces more computation.

If the neighbours of a point are very close to that point, that region is considered to be dense for that point. Therefore, a measure of local density can be the local radius defined as the expected (average) distances from its neighbours. The local densities in the input and embedding spaces can be denoted by:

$$R_p(\boldsymbol{x}_i) := \mathbb{E}_{j\sim p}\big[\|\boldsymbol{x}_i - \boldsymbol{x}_j\|_2^2\big] = \frac{\sum_{j=1}^n p_{ij}\|\boldsymbol{x}_i - \boldsymbol{x}_j\|_2^2}{\sum_{j=1}^n \|\boldsymbol{x}_i - \boldsymbol{x}_j\|_2^2}, \tag{17.30}$$

$$R_q(\boldsymbol{y}_i) := \mathbb{E}_{j\sim q}\big[\|\boldsymbol{y}_i - \boldsymbol{y}_j\|_2^2\big] = \frac{\sum_{j=1}^n q_{ij}\|\boldsymbol{y}_i - \boldsymbol{y}_j\|_2^2}{\sum_{j=1}^n \|\boldsymbol{y}_i - \boldsymbol{y}_j\|_2^2}. \tag{17.31}$$

As the volume of points is proportional to the powers of the radius (e.g., notice that the volume of the three dimensional sphere is proportional to the radius to the power three), the relation of local densities in the input and embedding spaces can be:

$$R_q(\boldsymbol{y}_i) = \alpha\,\big(R_p(\boldsymbol{x}_i)\big)^\beta \implies r_q^i = \beta\, r_p^i + \gamma, \tag{17.32}$$

where $r_q^i := \ln(R_q(\boldsymbol{y}_i))$, $r_p^i := \ln(R_p(\boldsymbol{y}_i))$, and $\gamma := \ln(\alpha)$. Therefore, the relation of the logarithms of the local densities should be affine dependence. A measure of linear (or affine) dependence is a correlation, so the correlation of the logarithms of the local densities is utilized:

$$\mathrm{Corr}(r_q, r_p) := \frac{\mathrm{Cov}(r_q, r_p)}{\sqrt{\mathrm{Var}(r_q)\mathrm{Var}(r_p)}}, \tag{17.33}$$

where the covariance and variance of densities are:

$$\mathrm{Cov}(r_q, r_p) := \frac{1}{n-1}\sum_{i=1}^n \Big[(r_q^i - \mu_q)(r_p^i - \mu_p)\Big]$$

$$\mathrm{Var}(r_q) := \frac{1}{n-1} \sum_{i=1}^{n} (r_q^i - \mu_q)^2,$$

where $\mu_q := (1/n) \sum_{j=1}^{n} r_q^j$, $\mu_p := (1/n) \sum_{j=1}^{n} r_p^j$, and $\mathrm{Var}(r_p)$ is defined similarly. The cost function of densMAP, to be minimized, is the UMAP's cost function regularized by the maximization of the correlation of local densities [20]:

$$c_6 := c_2 - \lambda \, \mathrm{Corr}(r_q, r_p), \tag{17.34}$$

where λ is the regularization parameter, which weights the correlation compared to UMAP's original cost. The gradient of c_2 is the gradient of the original UMAP discussed before (see Eq. (17.10) and Lemma 17.1). The gradient of the correlation term is [20]:

$$\frac{\partial c_6}{\partial \boldsymbol{y}_i} = \sum_{i=1}^{n} \sum_{j=1, j \neq i}^{n} \frac{\partial \mathrm{Corr}(r_q, r_p)}{\partial d_{ij}^2} (y_i - y_j), \tag{17.35}$$

where $d_{ij}^2 := \|\boldsymbol{y}_i - \boldsymbol{y}_j\|_2^2$ and:

$$\frac{\partial \mathrm{Corr}(r_q, r_p)}{\partial d_{ij}^2} = \frac{1}{(n-1)\mathrm{Var}(r_q)^{(3/2)}} \times$$

$$\Big[\mathrm{Var}(r_q)\Big(r_p^i \frac{\partial r_q^i}{\partial d_{ij}^2} + r_p^j \frac{\partial r_q^j}{\partial d_{ij}^2}\Big) - \mathrm{Cov}(r_q, r_p)\Big((r_q^i - \mu_q)\frac{\partial r_q^i}{\partial d_{ij}^2} + (r_p^j - \mu_q)\frac{\partial r_q^j}{\partial d_{ij}^2}\Big)\Big],$$

and:

$$\frac{\partial r_q^i}{\partial d_{ij}^2} = (1 + a d_{ij}^{2b})^{-2}\Big[ab d_{ij}^{2(b-1)} + e^{-r_q^i}(1 + a(1-b)d_{ij}^2)\Big].$$

Proofs of these derivatives are available in [20, Supplementary Note 2]. As in UMAP, DensMAP uses stochastic gradient descent for optimization. Except for the cost function, the algorithm for DensMAP is the same as UMAP.

17.7 Parametric UMAP for Embedding by Deep Learning

Dimensionality reduction algorithms can have parametric versions, where the cost function of the algorithm is used as the loss function of a neural network and the parameters (i.e., weights) of the network are trained by backpropagating the error of the loss function. Inspired by parametric t-SNE [26], it is possible to have a parametric UMAP [23]. Parametric UMAP uses UMAP's cost function, Eq. (17.6),

Input: Training data $\{\boldsymbol{x}_i\}_{i=1}^n$, learning rate η
Construct kNN graph
Initialize weights θ of network randomly
Initialize $\{\boldsymbol{y}_i\}_{i=1}^n \leftarrow \{f_\theta(\boldsymbol{x}_i)\}_{i=1}^n$
Calculate p_{ij} and q_{ij} for $\forall i, j \in \{1, \dots, n\}$ by Eqs. (17.4) and (17.5)
// make batches:
for *s from* 1 *to* $\lfloor n/b \rfloor$ **do**
    $\mathcal{B}_s \leftarrow \{\boldsymbol{x}_{(s-1)b+1}, \dots, \boldsymbol{x}_{sb}\}$
    // We denote $\mathcal{B}_s = \{\boldsymbol{x}_1^{(s)}, \dots, \boldsymbol{x}_b^{(s)}\}$
$\nu \leftarrow 0$
while *not converged* **do**
    $\nu \leftarrow \nu + 1$ // epoch index
    // optimize over every minibatch:
    for *s from* 1 *to* $\lfloor n/b \rfloor$ **do**
        $\{\boldsymbol{y}_i^{(s)}\}_{i=1}^b \leftarrow \{f_\theta(\boldsymbol{x}_i^{(s)})\}_{i=1}^b$
        $c_s^a \leftarrow 0, c_s^r \leftarrow 0$
        for *i from* 1 *to* *b* **do**
            for *j from* 1 *to* *b* **do**
                $u \sim U(0, 1)$
                if $u \leq p_{ij}$ **then**
                    $q_{ij} = (1 + a\|\boldsymbol{y}_i^{(s)} - \boldsymbol{y}_j^{(s)}\|_2^{2b})^{-1}$
                    $c_s^a = c_s^a + (-\ln(q_{ij}))$
                    for *m iterations* **do**
                        $l \sim U\{1, \dots, b\}$
                        $q_{il} = (1 + a\|\boldsymbol{y}_i^{(s)} - \boldsymbol{y}_l^{(s)}\|_2^{2b})^{-1}$
                        $c_s^r = c_s^r + (-\ln(1 - q_{il}))$
        $\theta \leftarrow$ backpropagate with loss $(c_s^a + c_s^r)$
Return $\{\boldsymbol{y}_i\}_{i=1}^n \leftarrow \{f_\theta(\boldsymbol{x}_i)\}_{i=1}^n$

Algorithm 17.2: Parametric UMAP algorithm

as the loss function of a neural network for deep learning. It optimizes this cost in minibatches rather than on the whole dataset; therefore, it can be used for embedding large datasets. Additionally, because of the nonlinearity of neural networks, the parametric UMAP can handle highly nonlinear data better than UMAP. Note that the learnable parameters in UMAP are the embedding of points, but the learnable parameters in the parametric UMAP are the weights of the neural network, so the embedding is obtained by the network. An advantage of parametric UMAP over parametric t-SNE is that the probability of UMAP in the embedding space, Eq. (17.5), does not have a normalization factor; therefore, there is no need to normalize over the whole dataset. This makes UMAP easy to use in deep learning.

The algorithm for parametric UMAP is demonstrated in Algorithm 17.2. As this algorithm shows, minibatches of the data points are made, where $\mathcal{B}_s$ denotes the s-th batch. Every epoch iterates over minibatches and for every minibatch, it iterates twice over the points of the batch to have the anchor points and the positive points. The cost function of UMAP can be optimized over the batch and not the entire data. The attractive and repulsive cost functions are denoted by c_s^a and c_s^r, respectively. According to Eq. (17.10), the loss function of the neural network in a parametric UMAP should be $(c_s^a + c_s^r)$. Backpropagating this loss function trains the parameters θ of the neural network denoted by $f_\theta(.)$. After training, the embeddings are obtained as $\{f_\theta(\boldsymbol{x}_i)\}_{i=1}^n$. Note that as discussed in Sect. 17.2.5, it is possible to combine UMAP's cost function with cross-entropy loss or triplet loss to produce a supervised or semisupervised embedding. Moreover, combining UMAP's cost function with the reconstruction loss can train an autoencoder to embed data in its middle code layer.

17.8 Progressive UMAP for Streaming and Out-of-Sample Data

Although the original UMAP cannot support out-of-sample (test) data embedding, Progressive UMAP [11] can. It can also be used for embedding streaming (online) data. Although UMAP is generally faster than t-SNE, it requires substantive time to embed big data. Progressive UMAP can be used to embed portions of the data and then complete the embedding by embedding the rest of the data as streaming data.

The algorithm for progressive UMAP (see Algorithm 17.3) uses PANENE [10] to construct a streaming kNN graph. PANENE uses randomized kd-trees [19] for approximating and updating the kNN graph. When a new batch of data, $\boldsymbol{X}_{\text{new}}$, arrives, some data points are affected in the kNN graph for one of the following reasons: (1) their neighbours are changed, (2) they become new neighbours of some points, or (3) they are no longer neighbours of some points. The affected points are denoted by $\boldsymbol{X}_{\text{updated}}$. If $\boldsymbol{X}_{\text{new}}$ is the very first batch of data, a Laplacian eigenmap is used to embed the data, as also performed in the original UMAP. If the incoming batch is not the first batch, the nearest neighbour of every new point among the previously accumulated data is found. Then, the initial embedding of every new point is set as the embedding of its nearest neighbour point, with the addition of Gaussian noise. As in the original UMAP, for every new or updated point, ρ_i and σ_i are calculated and updated by Eqs. (17.2) and (17.3). For every new point $\boldsymbol{x}_i$, p_{ij} and q_{ij} are calculated by Eqs. (17.4) and (17.5), where j indexes all existing and new points. Then, for every new or updated point, the embedding of the point is updated by gradient descent, where m negative samples are used.

17.9 Chapter Summary

UMAP and t-SNE belong to a family of methods, named the neighbourhood embedding methods. Currently, UMAP and t-SNE (introduced in Chap. 16) are

---

1 Input: New batch of training data $X_{\text{new}} = \{x_i\}_{i=1}^n$
2 X_{new}, X_{updated} ← Update kNN graph by PANENE method
3 if *it is initial batch* **then**
4 | Initialize $\{y_i\}_{i=1}^n$ by Laplacian eigenmap
5 else
6 | **for** *each new point x_i in X_{new}* **do**
7 | | Find the nearest neighbour to the previously accumulated data
8 | | Initialize y_i to the embedding of the nearest neighbour point plus Gaussian noise

9 for *each new or updated point x_i in X_{new} or $X_{updated}$* **do**
10 | Calculate/update ρ_i and σ_i by Eqs. (17.2) and (17.3)
11 | Calculate p_{ij} and q_{ij} for $\forall j$ by Eqs. (17.4) and (17.5)

12 while *not converged* **do**
13 | **for** *each new or updated point x_i in X_{new} or $X_{updated}$* **do**
14 | | **for** *j from 1 to n* **do**
15 | | | $u \sim U(0, 1)$
16 | | | **if** $u \leq p_{ij}$ **then**
17 | | | | $y_i \leftarrow y_i - \eta \frac{\partial c_{i,j}^a}{\partial y_i}$
18 | | | | **for** *m iterations* **do**
19 | | | | | $l \sim U\{1, \ldots, n\}$
20 | | | | | $y_i \leftarrow y_i - \eta \frac{\partial c_{i,l}^r}{\partial y_i}$

21 Return embedding $\{y_i\}$ for the new and updated points

Algorithm 17.3: Progressive UMAP algorithm

state-of-the-art methods for high-dimensional data visualization. They are especially used in the visualization of DNA data, as well as the visualization of output embedding in deep neural networks. This chapter introduced UMAP as a neighbourhood embedding method. The justification of UMAP, lying in the algebraic topology and category theory from pure mathematics, was also explained briefly.

References

1. Etienne Becht et al. "Dimensionality reduction for visualizing single-cell data using UMAP". In: *Nature biotechnology* 37.1 (2019), pp. 38–44.
2. Mikhail Belkin and Partha Niyogi. "Laplacian eigenmaps and spectral techniques for embedding and clustering". In: *Advances in neural information processing systems*. Vol. 14. 14. 2001, pp. 585–591.
3. Jan Niklas Böhm, Philipp Berens, and Dmitry Kobak. "A unifying perspective on neighbor embeddings along the attraction-repulsion spectrum". In: *arXiv preprint arXiv:2007.08902* (2020).
4. Shan Carter et al. "Activation atlas". In: *Distill* 4.3 (2019), e15.

5. Andy Coenen et al. "Visualizing and measuring the geometry of BERT". In: *arXiv preprint arXiv:1906.02715* (2019).
6. Sebastian Damrich and Fred A Hamprecht. "On UMAP's true loss function". In: *arXiv preprint arXiv:2103.14608* (2021).
7. Michael W Dorrity et al. "Dimensionality reduction by UMAP to visualize physical and genetic interactions". In: *Nature communications* 11.1 (2020), pp. 1–6.
8. Greg Friedman. "Survey article: an elementary illustrated introduction to simplicial sets". In: *The Rocky Mountain Journal of Mathematics* (2012), pp. 353–423.
9. Geoffrey E Hinton and Sam T Roweis. "Stochastic neighbor embedding". In: *Advances in neural information processing systems*. 2003, pp. 857–864.
10. Jaemin Jo, Jinwook Seo, and Jean-Daniel Fekete. "PANENE: A progressive algorithm for indexing and querying approximate k-nearest neighbors". In: *IEEE transactions on visualization and computer graphics* 26.2 (2018), pp. 1347–1360.
11. Hyung-Kwon Ko, Jaemin Jo, and Jinwook Seo. "Progressive Uniform Manifold Approximation and Projection." In: *EuroVis (Short Papers)*. 2020, pp. 133–137.
12. Solomon Kullback and Richard A Leibler. "On information and sufficiency". In: *The annals of mathematical statistics* 22.1 (1951), pp. 79–86.
13. Yoav Levine et al. "Sensebert: Driving some sense into BERT". In: *arXiv preprint arXiv:1908.05646* (2019).
14. Laurens van der Maaten and Geoffrey Hinton. "Visualizing data using t-SNE". In: *Journal of machine learning research* 9.Nov (2008), pp. 2579–2605.
15. Saunders Mac Lane. *Categories for the working mathematician*. Vol. 5. Springer Science & Business Media, 2013.
16. J Peter May. *Simplicial objects in algebraic topology*. Vol. 11. University of Chicago Press, 1992.
17. Leland McInnes, John Healy, and James Melville. "UMAP: Uniform manifold approximation and projection for dimension reduction". In: *arXiv preprint arXiv:1802.03426* (2018).
18. Tomas Mikolov et al. "Distributed representations of words and phrases and their compositionality". In: *Advances in neural information processing systems*. 2013, pp. 3111–3119.
19. Marius Muja and David G Lowe. "Fast approximate nearest neighbors with automatic algorithm configuration". In: *VISAPP (1)* 2.331–340 (2009), p. 2.
20. Ashwin Narayan, Bonnie Berger, and Hyunghoon Cho. "Density-preserving data visualization unveils dynamic patterns of single-cell transcriptomic variability". In: *Nature Biotechnology* 39 (2021), pp. 765–774.
21. Tim Repke and Ralf Krestel. "Robust Visualisation of Dynamic Text Collections: Measuring and Comparing Dimensionality Reduction Algorithms". In: *Proceedings of the 2021 Conference on Human Information Interaction and Retrieval*. ACM, 2021, pp. 255–259.
22. Emily Riehl. *Category theory in context*. Courier Dover Publications, 2017.
23. Tim Sainburg, Leland McInnes, and Timothy Q Gentner. "Parametric UMAP: learning embeddings with deep neural networks for representation and semi-supervised learning". In: *arXiv preprint arXiv:2009.12981* (2020).
24. David I Spivak. *Metric realization of fuzzy simplicial sets*. Tech. rep. Self published notes, 2012.
25. Jian Tang et al. "Visualizing large-scale and high-dimensional data". In: *Proceedings of the 25th international conference on world wide web*. 2016, pp. 287–297.
26. Laurens Van Der Maaten. "Learning a parametric embedding by preserving local structure". In: *Artificial Intelligence and Statistics*. PMLR. 2009, pp. 384–391.
27. Marc Vermeulen et al. "Application of Uniform Manifold Approximation and Projection (UMAP) in spectral imaging of artworks". In: *Spectrochimica Acta Part A: Molecular and Biomolecular Spectroscopy* 252 (2021), p. 119547.
28. Lofti A Zadeh. "Fuzzy sets". In: *Information and Control* 8.3 (1965), pp. 338–353.

Part IV
Neural Network-Based Dimensionality Reduction

Chapter 18
Restricted Boltzmann Machine and Deep Belief Network

18.1 Introduction

Centuries ago, the Boltzmann distribution [6], also called the Gibbs distribution [16], was proposed. This energy-based distribution was found to be useful for statistically modelling physical systems [30]. One of these systems was the Ising model, which modelled interacting particles with binary spins [31, 41]. Later, it was discovered that the Ising model could be a neural network [42]. Therefore, the Hopfield network was proposed, which implemented an Ising model in a network for modelling memory [28]. Inspired by the Hopfield network [28, 42], which was itself inspired by the physical Ising model [31, 41], Hinton et al. proposed the Boltzmann Machine (BM) and Restricted Boltzmann Machine (RBM) [1, 27]. These models are energy-based models [40]. Their names come from the Boltzmann distribution [6, 16] used in these models. A BM has weighted links between two layers of neurons, as well as links between neurons of every layer. RBM restricts these links to eliminate links between neurons of a layer. BM and RBM take one of the layers as the layer of data and the other layer as a representation or embedding of data. BM and RBM are special cases of the Ising model, whose weights (coupling parameters) are learned. BM and RBM are also special cases of the Hopfield network, whose weights are learned by maximum likelihood estimation rather than the Hebbian learning method [19], which is used in the Hopfield network.

The Hebbian learning method, used in the Hopfield network, was weak in generalizing to unseen data. Therefore, backpropagation [53] was proposed for training neural networks. Backpropagation was gradient descent plus the chain rule technique. However, researchers found that neural networks cannot become deep in their number of layers. This is because, in deep networks, gradients become exceedingly small after applying chain rules from the last layer to the first layer. This problem was called vanishing gradients. Vanishing gradients, plus the glory of theory in kernel support vector machines [7], resulted in the decline of interest

B. Ghojogh et al., *Elements of Dimensionality Reduction and Manifold Learning*,
https://doi.org/10.1007/978-3-031-10602-6_18

in neural networks until 2006. This decline is referred to as the winter of neural networks.

During the winter of neural networks, Hinton tried to save neural networks from being forgotten in the history of machine learning. Therefore, he and his colleagues, including Max Welling, returned to his previously proposed RBM and posited a learning method for RBM [23, 66]. They proposed training the weights of BM and RBM using maximum likelihood estimation. BM and RBM can be seen as generative models, where new values for neurons can be generated using Gibbs sampling [13]. Hinton noticed RBM because he knew that the set of weights between every two layers of a neural network is an RBM. In 2006, he thought it was possible to train a network in a greedy way[1] [4], where the weights of every layer of the network are trained using RBM training [24, 25]. This stack of RBM models, with greedy training, was named Deep Belief Network (DBN) [22, 24]. DBN allowed the networks to become deep by preparing a good initialization of weights (using RBM training) for backpropagation. This starting point for backpropagation optimization was no longer plagued by the problem of vanishing gradients. Since this breakthrough in 2006, the winter of neural networks started to end gradually because the networks could delve deep to become more nonlinear and were able to handle more nonlinear data.

DBN has been used in varying applications, including speech recognition [44–46] and action recognition [64]. Hinton was very excited about the success of RBM and was thinking that the future of neural networks belongs to DBN. However, two important techniques were proposed—the ReLU activation function [17] and the dropout technique [62]. These two regularization methods prevented overfitting [14] and resolved vanishing gradients, even without RBM pretraining. Therefore, backpropagation could be used alone if the new regularization methods were utilized. Neural networks have become a success due to their various applications [39], for example their use in image recognition [32].

18.2 Background

18.2.1 Probabilistic Graphical Model and Markov Random Field

A Probabilistic Graphical Model (PGM) is a graph-based representation of a complex distribution in a possibly high-dimensional space [34]. In other words, PGM is a combination of graph theory and probability theory. In a PGM, the random variables are represented by nodes or vertices. Edges exist between two

[1] A greedy algorithm makes every decision based on the most benefit at the current step and does not consider the final outcome at the final step. This greedy approach hopes that the final step will obtain a good result by small best steps based on their current benefits.

variables that interact with one another in terms of probability. Different conditional probabilities can be represented by a PGM. There are two types of PGM—the Markov network (also called the Markov random field) and the Bayesian network [34]. In the Markov network and Bayesian network, the edges of the graph are undirected and directed, respectively. BM and RBM are Markov networks (Markov random field) because their links are undirected [21].

18.2.2 Gibbs Sampling

Gibbs sampling, first proposed in [13], draws samples from a d-dimensional multivariate distribution $\mathbb{P}(X)$ using d conditional distributions [5]. This sampling algorithm assumes that the conditional distributions of every dimension of data conditioned on the rest of the coordinates are easy to draw from.

In Gibbs sampling, the aim is to sample from a multivariate distribution $\mathbb{P}(X)$ where $X \in \mathbb{R}^d$. Consider the notation $\mathbb{R}^d \ni \boldsymbol{x} := [x_1, x_2, \dots, x_d]^\top$. A random d-dimensional vector in the range of data is selected to be the starting point. Then, the first dimension of the first sample is drawn from the distribution of the first dimension conditioned on the other dimensions. This process is followed for all dimensions, where the j-th dimension is sampled as [15]:

$$x_j \sim \mathbb{P}(x_j \,|\, x_1, \dots, x_{j-1}, x_{j+1}, \dots, x_d). \tag{18.1}$$

This is done for all dimensions until all of the dimensions of the first sample are drawn. Then, starting from the first sample, this procedure is repeated for the dimensions of the second sample. This is iteratively performed for all samples; however, some initial samples are not yet valid because the algorithm has started from a not-necessarily valid vector. All samples are accepted after several iterations, called the burn-in iterations.

It is noteworthy that Gibbs sampling can be seen as a special case of the Metropolis-Hastings algorithm, which accepts the proposed samples with the probability one (see [15] for proof). Gibbs sampling is used in BM and RBM to generate visible and hidden samples.

18.2.3 Statistical Physics and Ising Model

18.2.3.1 Boltzmann (Gibbs) Distribution

Assume there are several particles $\{x_i\}_{i=1}^d$ in statistical physics. These particles can be seen as random variables that can randomly have some notion of state. For example, if the particles are electrons, they can have states $+1$ and -1 for counterclockwise and clockwise spins, respectively. The *Boltzmann distribution* [6], also called the *Gibbs distribution* [16], can show the probability that a physical system can have a specific state. i.e., every particle has a specific state. The probability mass function of this distribution is [30]:

$$\mathbb{P}(x) = \frac{e^{-\beta E(x)}}{Z}, \tag{18.2}$$

where $E(x)$ is the energy of variable x and Z is the normalization constant so that the probabilities sum to one. This normalization constant is called the *partition function*, which is hard to compute as it sums over all possible configurations of the states (values) that the particles can have. If $\mathbb{R}^d \ni \boldsymbol{x} := [x_1, \dots, x_d]^\top$, then:

$$Z := \sum_{\boldsymbol{x} \in \mathbb{R}^d} e^{-\beta E(x)}. \tag{18.3}$$

The coefficient $\beta \geq 0$ is defined as:

$$\beta := \frac{1}{k_\beta T} \propto \frac{1}{T}, \tag{18.4}$$

where k_β is the Boltzmann constant and $T \geq 0$ is the absolute thermodynamic temperature in Kelvin. If the temperature tends to absolute zero, $T \to 0$, then $\beta \to \infty$ and $\mathbb{P}(x) \to 0$, meaning that the absolute zero temperature occurs extremely rarely in the universe.

The *free energy* is defined as:

$$F(\beta) := \frac{-1}{\beta} \ln(Z), \tag{18.5}$$

where ln(.) is the natural logarithm. The *internal energy* is defined as:

$$U(\beta) := \frac{\partial}{\partial \beta}\big(\beta F(\beta)\big). \tag{18.6}$$

Therefore:

$$U(\beta) = \frac{\partial}{\partial \beta}(-\ln(Z)) = \frac{-1}{Z}\frac{\partial Z}{\partial \beta} \overset{(18.3)}{=} \sum_{\boldsymbol{x} \in \mathbb{R}^d} E(x)\, \frac{e^{-\beta E(x)}}{Z} \overset{(18.2)}{=} \sum_{\boldsymbol{x} \in \mathbb{R}^d} \mathbb{P}(x) E(x). \tag{18.7}$$

The *entropy* is defined as:

$$H(\beta) := -\sum_{\boldsymbol{x} \in \mathbb{R}^d} \mathbb{P}(x) \ln\big(\mathbb{P}(x)\big) \overset{(18.2)}{=} -\sum_{\boldsymbol{x} \in \mathbb{R}^d} \mathbb{P}(x)\big(-\beta E(x) - \ln(Z)\big)$$

$$= \beta \sum_{x \in \mathbb{R}^d} \mathbb{P}(x) E(x) + \ln(Z) \underbrace{\sum_{x \in \mathbb{R}^d} \mathbb{P}(x)}_{=1} \overset{(a)}{=} -\beta F(\beta) + \beta U(\beta), \quad (18.8)$$

where (a) is because of Eqs. (18.7) and (18.3).

Lemma 18.1 *A physical system prefers to be in low energy; therefore, the system continually loses energy.*

Proof On the one hand, according to the second law of thermodynamics, the entropy of a physical system always increases over time [9]. Entropy is a measure of randomness and disorder in the system. On the other hand, when a system loses energy to its surroundings, it becomes less ordered. Therefore, by passing time, the energy of the system decreases, which results in it having more entropy. □

Corollary 18.1 *According to Eq. (18.2) and Lemma 18.1, the probability* $\mathbb{P}(x)$ *of states in a system tends to increase over time.*

This corollary makes sense because systems tend to become more probable. This idea is also used in simulated annealing[2] [33], where the temperature of the system is cooled gradually. Simulated annealing and temperature-based learning have been used in BM models [2, 3, 51].

18.2.3.2 Ising Model

The Ising model [31, 41], also known as the Lenz-Ising model, is a model in which the particles can have -1 or $+1$ spins [8]. Therefore, $x_i \in \{-1, +1\}, \forall i \in \{1, \dots, d\}$. It uses the Boltzmann distribution, Eq. (18.2), where the energy function is defined as:

$$E(x) := \mathcal{H}(x) = -\sum_{(i,j)} J_{ij}\, x_i\, x_j, \quad (18.9)$$

where $\mathcal{H}(x)$ is called the Hamiltonian, $J_{ij} \in \mathbb{R}$ is the coupling parameter, and the summation is over the particles that interact with each other. Note that as energy is proportional to the reciprocal of the squared distance, nearby particles are only assumed to be interacting. Therefore, usually the interaction graph of the particles is a chain (one-dimensional grid), mesh grid (lattice), closed chain (loop), or torus (multidimensional loop).

[2] Simulated annealing is a metaheuristic optimization algorithm in which a temperature parameter controls the amount of global search versus local search. This gradually reduces the temperature to decrease the exploration and increase the exploitation of the search space.

Based on the characteristics of the model, the coupling parameter has different values. If for all interacting i and j, $J_{ij} \geq 0$ or $J_{ij} < 0$, the model is named *ferromagnetic* and *anti-ferromagnetic*, respectively. If J_{ij} can be both positive and negative, the model is called *spin glass*. If the coupling parameters are all constant, the model is *homogeneous*. According to Lemma 18.1, the energy decreases over time. According to Eq. (18.9), in the ferromagnetic model ($J_{ij} \geq 0$), the energy of an Ising model decreases if the interacting x_i and x_j have the same state (spin) because of the negative sign before summation. Likewise, in anti-ferromagnetic models, the nearby particles tend to have different spins over time.

According to Eq. (18.9), in ferromagnetic models, the energy is zero if $J_{ij} = 0$. This results in $\mathbb{P}(x) = 1$ according to Eq. (18.2). With similar analysis and according to the previous discussion, in ferromagnetic models, $J_{ij} \to \infty$ yields the same spins for all particles. This means that there will eventually be all $+1$ spins with probability of half or all -1 spins with a probability of half.

Ising models can be modelled as normal factor graphs. For more information on this, refer to [47, 48]. The BM and RBM are Ising models whose coupling parameters are considered weights, and these weights are learned using maximum likelihood estimation [21]. Therefore, it is possible to say that BM and RBM are energy-based learning methods [40].

18.2.4 Hopfield Network

It was proposed in [42] to use the Ising model in a neural network structure. Hopfield extended this idea to model memory by a neural network. The resulting network was the Hopfield network [28]. This network has units or neurons denoted by $\{x_i\}_{i=1}^d$. The states or outputs of these units are all binary $x_i \in \{-1, +1\}, \forall i$. Let w_{ij} denote the weight of the link connecting unit i to unit j. The weights of the Hopfield network are learned using Hebbian learning (Hebb's law of association) [19]:

$$w_{ij} := \begin{cases} x_i \times x_j & \text{if } i \neq j, \\ 0 & \text{otherwise.} \end{cases} \tag{18.10}$$

After training, the outputs of the units can be determined for an input if the weighted summation of the inputs to unit passes a threshold θ:

$$x_i := \begin{cases} +1 \text{ if } \sum_{j=1}^d w_{ij} x_j \geq \theta, \\ -1 \text{ otherwise.} \end{cases} \tag{18.11}$$

In the original proposal of the Hopfield network [28], the binary states are $x_i \in \{0, 1\}, \forall i$ so the Hebbian learning is $w_{ij} := (2x_i - 1) \times (2x_j - 1), \forall i \neq j$. The Hopfield network is an Ising model, meaning it uses Eq. (18.9) as its energy. This energy is also used in the Boltzmann distribution, which is Eq. (18.2).

It is noteworthy that there are also Hopfield networks with continuous states [29]. Modern Hopfield networks, such as [52], are often based on dense associative memories [38]. Other recent works on associative memories are [36, 37]. The BM and RBM models are Hopfield networks whose weights are learned using maximum likelihood estimation and not Hebbian learning.

18.3 Restricted Boltzmann Machine

18.3.1 Structure of Restricted Boltzmann Machine

Boltzmann Machine (BM) is a generative model and a Probabilistic Graphical Model (PGM) [5] which is a building block of many probabilistic models. Its name is because of the Boltzmann distribution [6, 16] used in this model. It was first introduced to be used in machine learning in [1, 27] and then in [23, 66]. A BM consists of a visible (or observation) layer $\boldsymbol{v} = [v_1, \dots, v_d] \in \mathbb{R}^d$ and a hidden layer $\boldsymbol{h} = [h_1, \dots, h_p] \in \mathbb{R}^p$. The visible layer is the layer that can be seen; for example, it can be the layer of data. The hidden layer is the layer of latent variables that represent meaningful features or embeddings for the visible data. In other words, there is a meaningful connection between the hidden and visible layers, although their dimensionality might differ, i.e., $d \neq p$. In the PGM of BM, there are connection links between elements of $\boldsymbol{v}$ and elements of $\boldsymbol{h}$. Each of the elements of $\boldsymbol{v}$ and $\boldsymbol{h}$ have a bias. There are also links between the elements of $\boldsymbol{v}$ and between the elements of $\boldsymbol{h}$ [56]. Consider the following notations: w_{ij} (the link between v_i and h_j), l_{ij} (the link between v_i and v_j), j_{ij} (the link between h_i and h_j), b_i (the bias link for v_i), and c_i (the bias link for h_i). The dimensionality of these links are $\boldsymbol{W} = [w_{ij}] \in \mathbb{R}^{d \times p}$, $\boldsymbol{L} = [l_{ij}] \in \mathbb{R}^{d \times d}$, $\boldsymbol{J} = [j_{ij}] \in \mathbb{R}^{p \times p}$, $\boldsymbol{b} = [b_1, \dots, b_d] \in \mathbb{R}^d$, and $\boldsymbol{c} = [c_1, \dots, c_p] \in \mathbb{R}^p$. Note that $\boldsymbol{W}$ is a symmetric matrix, i.e., $w_{ij} = w_{ji}$. As there is no link from a node to itself, the diagonal elements of $\boldsymbol{L}$ and $\boldsymbol{J}$ are zero, i.e., $l_{ii} = j_{ii} = 0, \forall i$. A Restricted Boltzmann Machine (RBM) is a BM that does not have links within a layer, i.e., there are no links between the elements of $\boldsymbol{v}$ and no links between the elements of $\boldsymbol{h}$. In other words, the links are restricted in RBM to $\boldsymbol{L} = \boldsymbol{J} = \boldsymbol{0}$. Figure 18.1 depicts both BM and RBM with their layers and links. This section focuses on RBM.

Recall that RBM is an Ising model. As seen in Eq. (18.9), the energy of an Ising model can be modelled as [1, 27]:

$$\mathbb{R} \ni E(\boldsymbol{v}, \boldsymbol{h}) := -\boldsymbol{b}^\top \boldsymbol{v} - \boldsymbol{c}^\top \boldsymbol{h} - \boldsymbol{v}^\top \boldsymbol{W} \boldsymbol{h}, \tag{18.12}$$

which is based on the interactions between linked units. As introduced in Eq. (18.2), the visible and hidden variables make a joint Boltzmann distribution [20]:

$$\mathbb{P}(\boldsymbol{v}, \boldsymbol{h}) = \frac{1}{Z} \exp(-E(\boldsymbol{v}, \boldsymbol{h})) \overset{(18.12)}{=} \frac{1}{Z} \exp(\boldsymbol{b}^\top \boldsymbol{v} + \boldsymbol{c}^\top \boldsymbol{h} + \boldsymbol{v}^\top \boldsymbol{W} \boldsymbol{h}), \tag{18.13}$$

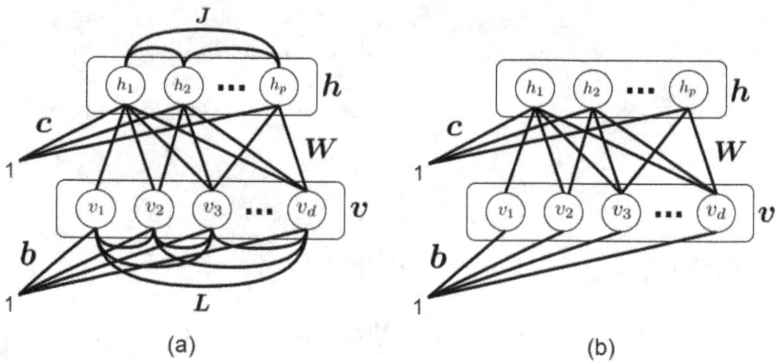

Fig. 18.1 The structures of (**a**) a Boltzmann machine and (**b**) a restricted Boltzmann machine

where **Z** is the partition function:

$$Z := \sum_{\boldsymbol{v}\in\mathbb{R}^d} \sum_{\boldsymbol{h}\in\mathbb{R}^p} \exp(-E(\boldsymbol{v}, \boldsymbol{h})). \tag{18.14}$$

According to Lemma 18.1, the BM and RBM try to reduce the energy of a model. Training the BM or RBM reduces its energy [1, 27].

18.3.2 Conditional Distributions

Proposition 18.1 (Conditional Independence of Variables) *In RBM, given the visible variables, the hidden variables are conditionally independent. Likewise, given the hidden variables, the visible variables are conditionally independent. This does not hold in BM because of the links within each layer.*

Proof According to Bayes' rule:

$$\begin{aligned}
\mathbb{P}(\boldsymbol{h}|\boldsymbol{v}) &= \frac{\mathbb{P}(\boldsymbol{h}, \boldsymbol{v})}{\mathbb{P}(\boldsymbol{v})} = \frac{\mathbb{P}(\boldsymbol{v}, \boldsymbol{h})}{\sum_{\boldsymbol{h}\in\mathbb{R}^p} \mathbb{P}(\boldsymbol{v}, \boldsymbol{h})} \overset{(18.13)}{=} \frac{\frac{1}{Z}\exp(\boldsymbol{b}^\top \boldsymbol{v} + \boldsymbol{c}^\top \boldsymbol{h} + \boldsymbol{v}^\top \boldsymbol{W}\boldsymbol{h})}{\sum_{\boldsymbol{h}\in\mathbb{R}^p} \frac{1}{Z}\exp(\boldsymbol{b}^\top \boldsymbol{v} + \boldsymbol{c}^\top \boldsymbol{h} + \boldsymbol{v}^\top \boldsymbol{W}\boldsymbol{h})} \\
&= \frac{\frac{1}{Z}\exp(\boldsymbol{b}^\top \boldsymbol{v})\exp(\boldsymbol{c}^\top \boldsymbol{h})\exp(\boldsymbol{v}^\top \boldsymbol{W}\boldsymbol{h})}{\frac{1}{Z}\sum_{\boldsymbol{h}\in\mathbb{R}^p} \exp(\boldsymbol{b}^\top \boldsymbol{v})\exp(\boldsymbol{c}^\top \boldsymbol{h})\exp(\boldsymbol{v}^\top \boldsymbol{W}\boldsymbol{h})} \\
&\overset{(a)}{=} \frac{\exp(\boldsymbol{b}^\top \boldsymbol{v})\exp(\boldsymbol{c}^\top \boldsymbol{h})\exp(\boldsymbol{v}^\top \boldsymbol{W}\boldsymbol{h})}{\exp(\boldsymbol{b}^\top \boldsymbol{v})\sum_{\boldsymbol{h}\in\mathbb{R}^p} \exp(\boldsymbol{c}^\top \boldsymbol{h})\exp(\boldsymbol{v}^\top \boldsymbol{W}\boldsymbol{h})} = \frac{\exp(\boldsymbol{c}^\top \boldsymbol{h})\exp(\boldsymbol{v}^\top \boldsymbol{W}\boldsymbol{h})}{\sum_{\boldsymbol{h}\in\mathbb{R}^p} \exp(\boldsymbol{c}^\top \boldsymbol{h})\exp(\boldsymbol{v}^\top \boldsymbol{W}\boldsymbol{h})},
\end{aligned}$$

where (a) is because the term $\exp(\boldsymbol{b}^\top \boldsymbol{v})$ does not have $\boldsymbol{h}$ in it. Note that $\sum_{\boldsymbol{h}\in\mathbb{R}^p}$ denotes summation over all possible p-dimensional hidden variables for the sake of marginalization. Let $Z' := \sum_{\boldsymbol{h}\in\mathbb{R}^p} \exp(\boldsymbol{c}^\top \boldsymbol{h})\exp(\boldsymbol{v}^\top \boldsymbol{W}\boldsymbol{h})$. Therefore:

$$\mathbb{P}(\boldsymbol{h}|\boldsymbol{v}) = \frac{1}{Z'}\exp(\boldsymbol{c}^\top \boldsymbol{h} + \boldsymbol{v}^\top \boldsymbol{W}\boldsymbol{h}) = \frac{1}{Z'}\exp\Big(\sum_{j=1}^{p} c_j h_j + \sum_{j=1}^{p} \boldsymbol{v}^\top \boldsymbol{W}_{:j} h_j\Big)$$

$$= \frac{1}{Z'}\prod_{j=1}^{p}\exp(c_j h_j + \boldsymbol{v}^\top \boldsymbol{W}_{:j} h_j), \tag{18.15}$$

where $\boldsymbol{W}_{:j} \in \mathbb{R}^d$ denotes the j-th column of matrix $\boldsymbol{W}$. Equation (18.15) demonstrates that given the visible variables, the hidden variables are conditionally independent because their joint distribution is the product of every distribution. Similar expressions can be written for the probability $\mathbb{P}(\boldsymbol{v}|\boldsymbol{h})$:

$$\mathbb{P}(\boldsymbol{v}|\boldsymbol{h}) = \frac{1}{Z''}\exp(\boldsymbol{b}^\top \boldsymbol{v} + \boldsymbol{v}^\top \boldsymbol{W}\boldsymbol{h}) = \frac{1}{Z''}\exp\Big(\sum_{i=1}^{d} b_i v_i + \sum_{i=1}^{d} v_i \boldsymbol{W}_{i:}\boldsymbol{h}\Big)$$

$$= \frac{1}{Z''}\prod_{i=1}^{d}\exp(b_i v_i + v_i \boldsymbol{W}_{i:}\boldsymbol{h}), \tag{18.16}$$

where $\boldsymbol{W}_{i:} \in \mathbb{R}^p$ denotes the i-th row of the matrix $\boldsymbol{W}$ and $Z'' := \sum_{\boldsymbol{v}\in\mathbb{R}^d}\exp(\boldsymbol{b}^\top \boldsymbol{v})\exp(\boldsymbol{v}^\top \boldsymbol{W}\boldsymbol{h})$. This equation demonstrates that given the hidden variables, the visible variables are conditionally independent. $\square$

According to Eq. (18.15) and considering the rule $\mathbb{P}(\boldsymbol{h}|\boldsymbol{v}) = \mathbb{P}(\boldsymbol{h}, \boldsymbol{v})/\mathbb{P}(\boldsymbol{v})$:

$$\mathbb{P}(\boldsymbol{h}|\boldsymbol{v}) = \frac{1}{Z'}\prod_{j=1}^{p}\exp(c_j h_j + \boldsymbol{v}^\top \boldsymbol{W}_{:j} h_j) = \frac{1}{Z'}\prod_{j=1}^{p}\mathbb{P}(h_j, \boldsymbol{v})$$

$$\implies \mathbb{P}(h_j, \boldsymbol{v}) = \exp(c_j h_j + \boldsymbol{v}^\top \boldsymbol{W}_{:j} h_j) = \exp(c_j h_j + \sum_{i=1}^{d} v_i w_{ij} h_j). \tag{18.17}$$

Similarly, according to Eq. (18.16) and considering the rule $\mathbb{P}(\boldsymbol{v}|\boldsymbol{h}) = \mathbb{P}(\boldsymbol{h}, \boldsymbol{v})/\mathbb{P}(\boldsymbol{h})$:

$$\mathbb{P}(\boldsymbol{h}|\boldsymbol{v}) = \frac{1}{Z''}\prod_{i=1}^{d}\exp(b_i v_i + v_i \boldsymbol{W}_{i:}\,\boldsymbol{h}) = \frac{1}{Z''}\prod_{i=1}^{d}\mathbb{P}(\boldsymbol{h}, v_i)$$

$$\implies \mathbb{P}(\boldsymbol{h}, v_i) = \exp(b_i v_i + v_i \boldsymbol{W}_{i:}\,\boldsymbol{h}) = \exp(b_i v_i + \sum_{j=1}^{p} v_i w_{ij} h_j). \tag{18.18}$$

18.3.3 Sampling Hidden and Visible Variables

18.3.3.1 Gibbs Sampling

Gibbs sampling can be used for sampling and generating the hidden and visible units. If ν denotes the iteration index of Gibbs sampling, then the followings are iteratively sampled:

$$\boldsymbol{h}^{(\nu)} \sim \mathbb{P}(\boldsymbol{h}|\boldsymbol{v}^{(\nu)}), \tag{18.19}$$

$$\boldsymbol{v}^{(\nu+1)} \sim \mathbb{P}(\boldsymbol{v}|\boldsymbol{h}^{(\nu)}), \tag{18.20}$$

until the burn-in convergence. As explained in Sect. 18.2.2, only several iterations of Gibbs sampling are usually sufficient. After the burn-in iterations, the samples are approximate samples from the joint distribution $\mathbb{P}(\boldsymbol{v}, \boldsymbol{h})$. As the variables are conditionally independent, this Gibbs sampling can be implemented as demonstrated in Algorithm 18.1. In this algorithm, $h_j^{(\nu)} \sim \mathbb{P}(h_j|\boldsymbol{v}^{(\nu)})$ can be implemented as drawing a sample from uniform distribution $u \sim U[0, 1]$ and comparing it to the value of Probability Density Function (PDF), $\mathbb{P}(h_j|\boldsymbol{v}^{(\nu)})$. If u is less than or equal to this value, then $h_j = 1$; otherwise, $h_j = 0$. The implementation of sampling v_i has a similar procedure. Alternatively, the inverse of the cumulative distribution function of these distributions can be used for drawing samples (see [15] for more details about sampling).

18.3.3.2 Generations and Evaluations by Gibbs Sampling

Gibbs sampling for generating both observation and hidden units is used for both the training and evaluation phases of RBM. The use of Gibbs sampling in training RBM will be explained in Sects. 18.3.4 and 18.3.5. After the RBM model is trained, any number of p-dimensional hidden variables can be generated as a meaningful representation of the d-dimensional observation using Gibbs sampling. Moreover, using Gibbs sampling, other d-dimensional observations can be generated in addi-

Input: visible dataset $\boldsymbol{v}$, (initialization: optional)
Get initialization or do random initialization of $\boldsymbol{v}$
while *until burn-in* **do**
  for j *from* 1 *to* p **do**
    $h_j^{(\nu)} \sim \mathbb{P}(h_j|\boldsymbol{v}^{(\nu)})$
  for i *from* 1 *to* d **do**
    $v_i^{(\nu+1)} \sim \mathbb{P}(v_i|\boldsymbol{h}^{(\nu)})$

Algorithm 18.1: Gibbs sampling in RBM

tion to the original dataset. These newly generated observations are d-dimensional representations for the p-dimensional hidden variables. This demonstrates that BM and RBM are generative models.

18.3.4 Training Restricted Boltzmann Machine by Maximum Likelihood Estimation

The weights of the links, $\boldsymbol{W}$, $\boldsymbol{b}$, and $\boldsymbol{c}$, should be learned so that they can be used for sampling/generating the hidden and visible units. Consider a dataset of n visible vectors $\{\boldsymbol{v}_i \in \mathbb{R}^d\}_{i=1}^n$. Note that $\boldsymbol{v}_i$ should not be confused with v_i, where the former is the i-th visible data instance and the latter is the i-th visible unit. The j-th dimension of $\boldsymbol{v}_i$ is signified by $\boldsymbol{v}_{i,j}$; in other words, $\boldsymbol{v}_i = [\boldsymbol{v}_{i,1}, \ldots, \boldsymbol{v}_{i,d}]^\top$. The log-likelihood of the visible data is:

$$\begin{aligned}
\ell(\boldsymbol{W}, \boldsymbol{b}, \boldsymbol{c}) &= \sum_{i=1}^{n} \log(\mathbb{P}(\boldsymbol{v}_i)) = \sum_{i=1}^{n} \log\Big(\sum_{\boldsymbol{h} \in \mathbb{R}^p} \mathbb{P}(\boldsymbol{v}_i, \boldsymbol{h})\Big) \\
&\overset{(18.13)}{=} \sum_{i=1}^{n} \log\Big(\sum_{\boldsymbol{h} \in \mathbb{R}^p} \frac{1}{Z} \exp(-E(\boldsymbol{v}_i, \boldsymbol{h}))\Big) = \sum_{i=1}^{n} \log\Big(\frac{1}{Z} \sum_{\boldsymbol{h} \in \mathbb{R}^p} \exp(-E(\boldsymbol{v}_i, \boldsymbol{h}))\Big) \\
&= \sum_{i=1}^{n} \Big[\log\Big(\sum_{\boldsymbol{h} \in \mathbb{R}^p} \exp(-E(\boldsymbol{v}_i, \boldsymbol{h}))\Big) - \log Z \Big] \\
&= \sum_{i=1}^{n} \log\Big(\sum_{\boldsymbol{h} \in \mathbb{R}^p} \exp(-E(\boldsymbol{v}_i, \boldsymbol{h}))\Big) - n \log Z \\
&\overset{(18.14)}{=} \sum_{i=1}^{n} \log\Big(\sum_{\boldsymbol{h} \in \mathbb{R}^p} \exp(-E(\boldsymbol{v}_i, \boldsymbol{h}))\Big) - n \log \sum_{\boldsymbol{v} \in \mathbb{R}^d} \sum_{\boldsymbol{h} \in \mathbb{R}^p} \exp(-E(\boldsymbol{v}, \boldsymbol{h})).
\end{aligned} \tag{18.21}$$

Maximum Likelihood Estimation (MLE) is used to find the parameters $\theta := \{\boldsymbol{W}, \boldsymbol{b}, \boldsymbol{c}\}$. The derivative of the log-likelihood, with respect to parameter θ, is:

$$\nabla_\theta \ell(\theta) = \nabla_\theta \sum_{i=1}^{n} \log\Big(\sum_{\boldsymbol{h} \in \mathbb{R}^p} \exp(-E(\boldsymbol{v}_i, \boldsymbol{h}))\Big) - n \nabla_\theta \log \sum_{\boldsymbol{v} \in \mathbb{R}^d} \sum_{\boldsymbol{h} \in \mathbb{R}^p} \exp(-E(\boldsymbol{v}, \boldsymbol{h})). \tag{18.22}$$

The first term of this derivative is:

$$\nabla_\theta \sum_{i=1}^{n} \log\Big(\sum_{\boldsymbol{h} \in \mathbb{R}^p} \exp(-E(\boldsymbol{v}_i, \boldsymbol{h}))\Big) = \sum_{i=1}^{n} \nabla_\theta \log\Big(\sum_{\boldsymbol{h} \in \mathbb{R}^p} \exp(-E(\boldsymbol{v}_i, \boldsymbol{h}))\Big)$$

$$= \sum_{i=1}^{n} \frac{\nabla_\theta \sum_{\boldsymbol{h}\in\mathbb{R}^p} \exp(-E(\boldsymbol{v}_i, \boldsymbol{h}))}{\sum_{\boldsymbol{h}\in\mathbb{R}^p} \exp(-E(\boldsymbol{v}_i, \boldsymbol{h}))}$$

$$= \sum_{i=1}^{n} \frac{\sum_{\boldsymbol{h}\in\mathbb{R}^p} \exp(-E(\boldsymbol{v}_i, \boldsymbol{h}))\nabla_\theta(-E(\boldsymbol{v}_i, \boldsymbol{h}))}{\sum_{\boldsymbol{h}\in\mathbb{R}^p} \exp(-E(\boldsymbol{v}_i, \boldsymbol{h}))} \overset{(a)}{=} \sum_{i=1}^{n} \mathbb{E}_{\sim\mathbb{P}(\boldsymbol{h}|\boldsymbol{v}_i)}[\nabla_\theta(-E(\boldsymbol{v}_i, \boldsymbol{h}))], \tag{18.23}$$

where (a) is because the definition of expectation is $\mathbb{E}_{\sim\mathbb{P}}[\boldsymbol{x}] := \sum_{i=1} \mathbb{P}(\boldsymbol{x}_i)\,\boldsymbol{x}_i$. However, if $\mathbb{P}$ is not an actual distribution and does not sum to one, it should be normalized to behave like a distribution in the expectation: $\mathbb{E}_{\sim\mathbb{P}}[\boldsymbol{x}] := (\sum_{i=1} \mathbb{P}(\boldsymbol{x}_i)\,\boldsymbol{x}_i)/(\sum_{i=1} \mathbb{P}(\boldsymbol{x}_i))$. The second term of the derivative of log-likelihood is:

$$-\,n\nabla_\theta \log \sum_{\boldsymbol{v}\in\mathbb{R}^d} \sum_{\boldsymbol{h}\in\mathbb{R}^p} \exp(-E(\boldsymbol{v}, \boldsymbol{h})) = -n\frac{\nabla_\theta \sum_{\boldsymbol{v}\in\mathbb{R}^d} \sum_{\boldsymbol{h}\in\mathbb{R}^p} \exp(-E(\boldsymbol{v}, \boldsymbol{h}))}{\sum_{\boldsymbol{v}\in\mathbb{R}^d} \sum_{\boldsymbol{h}\in\mathbb{R}^p} \exp(-E(\boldsymbol{v}, \boldsymbol{h}))}$$

$$= -n\frac{\sum_{\boldsymbol{v}\in\mathbb{R}^d} \sum_{\boldsymbol{h}\in\mathbb{R}^p} \nabla_\theta \exp(-E(\boldsymbol{v}, \boldsymbol{h}))}{\sum_{\boldsymbol{v}\in\mathbb{R}^d} \sum_{\boldsymbol{h}\in\mathbb{R}^p} \exp(-E(\boldsymbol{v}, \boldsymbol{h}))}$$

$$= -n\frac{\sum_{\boldsymbol{v}\in\mathbb{R}^d} \sum_{\boldsymbol{h}\in\mathbb{R}^p} \exp(-E(\boldsymbol{v}, \boldsymbol{h}))\nabla_\theta(-E(\boldsymbol{v}, \boldsymbol{h}))}{\sum_{\boldsymbol{v}\in\mathbb{R}^d} \sum_{\boldsymbol{h}\in\mathbb{R}^p} \exp(-E(\boldsymbol{v}, \boldsymbol{h}))} \overset{(a)}{=} -n\,\mathbb{E}_{\sim\mathbb{P}(\boldsymbol{h},\boldsymbol{v})}[\nabla_\theta(-E(\boldsymbol{v}, \boldsymbol{h}))], \tag{18.24}$$

where (a) is for the definition of expectation, which was already explained above. In summary, the derivative of the log-likelihood is:

$$\nabla_\theta \ell(\theta) = \sum_{i=1}^{n} \mathbb{E}_{\sim\mathbb{P}(\boldsymbol{h}|\boldsymbol{v}_i)}[\nabla_\theta(-E(\boldsymbol{v}_i, \boldsymbol{h}))] - n\,\mathbb{E}_{\sim\mathbb{P}(\boldsymbol{h},\boldsymbol{v})}[\nabla_\theta(-E(\boldsymbol{v}, \boldsymbol{h}))]. \tag{18.25}$$

Setting this derivative to zero does not result in a closed-form solution. Therefore, the parameters should be learned iteratively using the gradient ascent for MLE.

Now, consider each of the parameters $\theta = \{\boldsymbol{W}, \boldsymbol{b}, \boldsymbol{c}\}$. The derivative with respect to these parameters in Eq. (18.25) are:

$$\nabla_{\boldsymbol{W}}(-E(\boldsymbol{v}, \boldsymbol{h})) \overset{(18.12)}{=} \frac{\partial}{\partial \boldsymbol{W}}(\boldsymbol{b}^\top \boldsymbol{v} + \boldsymbol{c}^\top \boldsymbol{h} + \boldsymbol{v}^\top \boldsymbol{W}\boldsymbol{h}) = \boldsymbol{v}\boldsymbol{h}^\top,$$

$$\nabla_{\boldsymbol{b}}(-E(\boldsymbol{v}, \boldsymbol{h})) \overset{(18.12)}{=} \frac{\partial}{\partial \boldsymbol{b}}(\boldsymbol{b}^\top \boldsymbol{v} + \boldsymbol{c}^\top \boldsymbol{h} + \boldsymbol{v}^\top \boldsymbol{W}\boldsymbol{h}) = \boldsymbol{v},$$

$$\nabla_{\boldsymbol{c}}(-E(\boldsymbol{v}, \boldsymbol{h})) \overset{(18.12)}{=} \frac{\partial}{\partial \boldsymbol{c}}(\boldsymbol{b}^\top \boldsymbol{v} + \boldsymbol{c}^\top \boldsymbol{h} + \boldsymbol{v}^\top \boldsymbol{W}\boldsymbol{h}) = \boldsymbol{h}.$$

Therefore, Eq. (18.25) for these parameters becomes:

$$\nabla_{\boldsymbol{W}} \ell(\theta) = \sum_{i=1}^{n} \mathbb{E}_{\sim \mathbb{P}(\boldsymbol{h}|\boldsymbol{v}_i)}[\boldsymbol{v}\boldsymbol{h}_i^\top] - n\, \mathbb{E}_{\sim \mathbb{P}(\boldsymbol{h},\boldsymbol{v})}[\boldsymbol{v}\boldsymbol{h}^\top]$$

$$= \sum_{i=1}^{n} \boldsymbol{v}_i\, \mathbb{E}_{\sim \mathbb{P}(\boldsymbol{h}|\boldsymbol{v}_i)}[\boldsymbol{h}^\top] - n\, \mathbb{E}_{\sim \mathbb{P}(\boldsymbol{h},\boldsymbol{v})}[\boldsymbol{v}\boldsymbol{h}^\top],$$

$$\nabla_{\boldsymbol{b}} \ell(\theta) = \sum_{i=1}^{n} \mathbb{E}_{\sim \mathbb{P}(\boldsymbol{h}|\boldsymbol{v}_i)}[\boldsymbol{v}_i] - n\, \mathbb{E}_{\sim \mathbb{P}(\boldsymbol{h},\boldsymbol{v})}[\boldsymbol{v}] = \sum_{i=1}^{n} \boldsymbol{v}_i - n\, \mathbb{E}_{\sim \mathbb{P}(\boldsymbol{h},\boldsymbol{v})}[\boldsymbol{v}],$$

$$\nabla_{\boldsymbol{c}} \ell(\theta) = \sum_{i=1}^{n} \mathbb{E}_{\sim \mathbb{P}(\boldsymbol{h}|\boldsymbol{v}_i)}[\boldsymbol{h}] - n\, \mathbb{E}_{\sim \mathbb{P}(\boldsymbol{h},\boldsymbol{v})}[\boldsymbol{h}].$$

If:

$$\widehat{\boldsymbol{h}}_i := \mathbb{E}_{\sim \mathbb{P}(\boldsymbol{h}|\boldsymbol{v}_i)}[\boldsymbol{h}], \tag{18.26}$$

these derivatives can be summarized as follows:

$$\mathbb{R}^{d \times p} \ni \nabla_{\boldsymbol{W}} \ell(\theta) = \sum_{i=1}^{n} \boldsymbol{v}_i \widehat{\boldsymbol{h}}_i^\top - n\, \mathbb{E}_{\sim \mathbb{P}(\boldsymbol{h},\boldsymbol{v})}[\boldsymbol{v}\boldsymbol{h}^\top], \tag{18.27}$$

$$\mathbb{R}^{d} \ni \nabla_{\boldsymbol{b}} \ell(\theta) = \sum_{i=1}^{n} \boldsymbol{v}_i - n\, \mathbb{E}_{\sim \mathbb{P}(\boldsymbol{h},\boldsymbol{v})}[\boldsymbol{v}], \tag{18.28}$$

$$\mathbb{R}^{p} \ni \nabla_{\boldsymbol{c}} \ell(\theta) = \sum_{i=1}^{n} \widehat{\boldsymbol{h}}_i - n\, \mathbb{E}_{\sim \mathbb{P}(\boldsymbol{h},\boldsymbol{v})}[\boldsymbol{h}]. \tag{18.29}$$

Setting these derivatives to zero does not result in a closed form solution. Therefore, the solution needs to be found iteratively using gradient descent, where the above gradients are used. In the derivatives of log-likelihood, there are two types of expectation. The conditional expectation $\mathbb{E}_{\sim \mathbb{P}(\boldsymbol{h}|\boldsymbol{v}_i)}[.]$ is based on the observation or data, which is $\boldsymbol{v}_i$. The joint expectation $\mathbb{E}_{\sim \mathbb{P}(\boldsymbol{h},\boldsymbol{v})}[.]$, however, has nothing to do with the observation and is merely about the RBM model.

18.3.5 Contrastive Divergence

According to Eq. (18.23), the conditional expectation used in Eq. (18.26) includes one summation. Moreover, according to Eq. (18.24), the joint expectations used in Eqs. (18.27), (18.28), and (18.29) contain two summations. This double summation makes the computation of the joint expectation intractable because it sums over all possible values for both the hidden and visible units. Therefore, exact computation of MLE is difficult and should be approximated. One way to approximate the computation of the joint expectations in MLE is *contrastive divergence* [23]. Contrastive divergence improves the efficiency and reduces the variance of the estimation in RBM [23, 66].

The idea of contrastive divergence is as follows. First, a point $\widetilde{\boldsymbol{v}}$ is obtained using Gibbs sampling, starting from observation $\boldsymbol{v}_i$ (see Sect. 18.3.3 for Gibbs sampling in RBM). Then, the expectation is computed using only point $\widetilde{\boldsymbol{v}}$. The intuitive reason for why contrastive divergence works is explained in the following. The gradients need to be minimized to find the solution of MLE. In the joint expectations in Eqs. (18.27), (18.28), and (18.29), rather than considering all possible values of observations, contrastive divergence considers only one of the data points (observations). If this observation is a wrong belief, which is undesirable in the generation of observations by RBM, contrastive divergence is performing a task called *negative sampling* [20]. In negative sampling, rather than training the model to avoid generating all incorrect observations, the model is trained iteratively but less ambitiously in every iteration. Each iteration tries to teach the model to not generate only one of the wrong outputs. Gradually, the model learns to generate correct observations by avoiding generating these negative samples.

Let $\widetilde{\boldsymbol{h}} = [\widetilde{h}_1, \dots, \widetilde{h}_m]^\top$ be the corresponding sampled $\boldsymbol{h}$ to $\widetilde{\boldsymbol{v}} = [\widetilde{v}_1, \dots, \widetilde{v}_m]^\top$ in Gibbs sampling. According to the above explanations, contrastive divergence approximates the joint expectation in the derivative of log-likelihood, Eq. (18.25), by Monte-Carlo approximation [15] evaluated at $\widetilde{\boldsymbol{v}}_i$ and $\widetilde{\boldsymbol{h}}_i$ for the i-th observation and hidden units, where $\widetilde{\boldsymbol{v}}_i$ and $\widetilde{\boldsymbol{h}}_i$ are found by Gibbs sampling. Therefore:

$$\mathbb{E}_{\sim\mathbb{P}(\boldsymbol{h},\boldsymbol{v})}[\nabla_\theta(-E(\boldsymbol{v},\boldsymbol{h}))] \approx \frac{1}{n}\sum_{i=1}^{n} \nabla_\theta(-E(\boldsymbol{v}_i,\boldsymbol{h}_i))\Big|_{\boldsymbol{v}_i=\widetilde{\boldsymbol{v}}_i,\boldsymbol{h}_i=\widetilde{\boldsymbol{h}}_i}. \tag{18.30}$$

Experiments have shown that a small number of iterations in Gibbs sampling suffice for contrastive divergence. Hinton has even used one iteration of Gibbs sampling for this task [23]. This small number of required iterations has the support of the literature because Gibbs sampling is a special case of Metropolis-Hastings algorithms [15], which are fast to converge [11].

By using the approximation in Eq. (18.30), Eqs. (18.27), (18.28), and (18.29) become:

$$\nabla_{\boldsymbol{W}}\ell(\theta) = \sum_{i=1}^{n} \boldsymbol{v}_i\widehat{\boldsymbol{h}}_i^\top - \sum_{i=1}^{n} \widetilde{\boldsymbol{v}}_i\widetilde{\boldsymbol{h}}_i^\top, \tag{18.31}$$

$$\nabla_{\boldsymbol{b}}\ell(\theta) = \sum_{i=1}^{n} \boldsymbol{v}_i - \sum_{i=1}^{n} \widetilde{\boldsymbol{v}}_i, \tag{18.32}$$

$$\nabla_{\boldsymbol{c}}\ell(\theta) = \sum_{i=1}^{n} \widehat{\boldsymbol{h}}_i - \sum_{i=1}^{n} \widetilde{\boldsymbol{h}}_i. \tag{18.33}$$

Input: training data $\{\boldsymbol{x}_i\}_{i=1}^n$
Randomly initialize $\boldsymbol{W}, \boldsymbol{b}, \boldsymbol{c}$
while *not converged* **do**
  Sample a minibatch $\{\boldsymbol{v}_1, \ldots, \boldsymbol{v}_m\}$ from training dataset $\{\boldsymbol{x}_i\}_{i=1}^n$ (note that it is possible to set $m = n$)
  // Gibbs sampling for each data point:
  Initialize $\widehat{\boldsymbol{v}}_i^{(0)} \leftarrow \boldsymbol{v}_i$ for all $i \in \{1, \ldots, m\}$
  for i *from* 1 *to* m **do**
    Algorithm 18.1 $\leftarrow \widehat{\boldsymbol{v}}_i^{(0)}$
    $\{h_i\}_{i=1}^p, \{v_i\}_{i=1}^d \leftarrow$ Last iteration of Algorithm 18.1
    $\widetilde{\boldsymbol{h}}_i \leftarrow [h_1, \ldots, h_p]^\top$
    $\widetilde{\boldsymbol{v}}_i \leftarrow [v_1, \ldots, v_d]^\top$
    $\widehat{\boldsymbol{h}}_i \leftarrow \mathbb{E}_{\sim \mathbb{P}(\boldsymbol{h}|\boldsymbol{v}_i)}[\boldsymbol{h}]$
  // gradients:
  $\nabla_{\boldsymbol{W}} \ell(\theta) \leftarrow \sum_{i=1}^m \boldsymbol{v}_i \widehat{\boldsymbol{h}}_i^\top - \sum_{i=1}^m \widetilde{\boldsymbol{h}}_i \widetilde{\boldsymbol{v}}_i^\top$
  $\nabla_{\boldsymbol{b}} \ell(\theta) \leftarrow \sum_{i=1}^m \boldsymbol{v}_i - \sum_{i=1}^m \widetilde{\boldsymbol{v}}_i$
  $\nabla_{\boldsymbol{c}} \ell(\theta) \leftarrow \sum_{i=1}^m \widehat{\boldsymbol{h}}_i - \sum_{i=1}^m \widetilde{\boldsymbol{h}}_i$
  // gradient descent for updating solution:
  $\boldsymbol{W} \leftarrow \boldsymbol{W} - \eta \nabla_{\boldsymbol{W}} \ell(\theta)$
  $\boldsymbol{b} \leftarrow \boldsymbol{b} - \eta \nabla_{\boldsymbol{b}} \ell(\theta)$
  $\boldsymbol{c} \leftarrow \boldsymbol{c} - \eta \nabla_{\boldsymbol{c}} \ell(\theta)$
Return $\boldsymbol{W}, \boldsymbol{b}, \boldsymbol{c}$

Algorithm 18.2: Training RBM using contrastive divergence

These equations make sense because when the observation variable and hidden variable given the observation variable become equal to the approximations by Gibbs sampling, the gradient should be zero and the training should stop. Note that some works in the literature restate Eqs. (18.31), (18.32), and (18.33) as [20, 23, 64]:

$$\forall i, j: \nabla_{w_{ij}} \ell(\theta) = \langle v_i h_j \rangle_{\text{data}} - \langle v_i h_j \rangle_{\text{recon.}}, \tag{18.34}$$

$$\forall i: \quad \nabla_{b_i} \ell(\theta) = \langle v_i \rangle_{\text{data}} - \langle v_i \rangle_{\text{recon.}}, \tag{18.35}$$

$$\forall j: \quad \nabla_{c_j} \ell(\theta) = \langle h_j \rangle_{\text{data}} - \langle h_j \rangle_{\text{recon.}}, \tag{18.36}$$

where $\langle . \rangle_{\text{data}}$ and $\langle . \rangle_{\text{recon.}}$ denote the expectation over the data and reconstruction of the data, respectively.

The training algorithm of RBM, using contrastive divergence, can be found in Algorithm 18.2. In this algorithm, minibatch gradient descent is used with the batch size m. If the training dataset is not large, $m = n$ can be set to have gradient descent. This algorithm is iterative until convergence, where in every iteration, a minibatch is sampled, where there are an observation $\boldsymbol{v}_i \in \mathbb{R}^d$ and a hidden variable $\boldsymbol{h}_i \in \mathbb{R}^p$ for every i-th training data point. For every data point, Gibbs sampling is applied as demonstrated in Algorithm 18.1. After Gibbs sampling, gradients are calculated using Eqs. (18.31), (18.32), and (18.33) and then the variables are updated using a gradient descent step.

18.3.6 Boltzmann Machine

This section introduces training of BM which has more links compared to RBM [56]. As explained in Sect. 18.3.1, BM has additional links $\boldsymbol{L} = [l_{ij}] \in \mathbb{R}^{d \times d}$ and $\boldsymbol{J} = [j_{ij}] \in \mathbb{R}^{p \times p}$ (see Fig. 18.1 for its structure). The weights $\boldsymbol{W} \in \mathbb{R}^{d \times p}$ and biases $\boldsymbol{b} \in \mathbb{R}^d$ and $\boldsymbol{c} \in \mathbb{R}^p$ are trained by gradient descent using the gradients in Eqs. (18.31), (18.32), and (18.33). The additional weights $\boldsymbol{L}$ and $\boldsymbol{J}$ are updated similarly using the following gradients [56]:

$$\nabla_{\boldsymbol{L}} \ell(\theta) = \sum_{i=1}^{n} \boldsymbol{v}_i \boldsymbol{v}_i^\top - \sum_{i=1}^{n} \widetilde{\boldsymbol{v}}_i \widetilde{\boldsymbol{v}}_i^\top, \tag{18.37}$$

$$\nabla_{\boldsymbol{J}} \ell(\theta) = \sum_{i=1}^{n} \mathbb{E}_{\sim \mathbb{P}(\boldsymbol{h}|\boldsymbol{v}_i)}[\boldsymbol{h}\boldsymbol{h}^\top] - \sum_{i=1}^{n} \widetilde{\boldsymbol{h}}_i \widetilde{\boldsymbol{h}}_i^\top. \tag{18.38}$$

These equations can be restated as:

$$\forall i, j: \nabla_{l_{ij}} \ell(\theta) = \langle v_i v_j \rangle_{\text{data}} - \langle v_i v_j \rangle_{\text{recon.}}, \tag{18.39}$$

$$\forall i, j: \nabla_{j_{ij}} \ell(\theta) = \langle h_i h_j \rangle_{\text{data}} - \langle h_i h_j \rangle_{\text{recon.}}, \tag{18.40}$$

where $\langle . \rangle_{\text{data}}$ and $\langle . \rangle_{\text{recon.}}$ denote expectation over the data and reconstruction of data, respectively.

18.4 Distributions of Visible and Hidden Variables

18.4.1 Modelling with Exponential Family Distributions

According to Proposition 18.1, the units $\boldsymbol{v} \in \mathbb{R}^d$ and $\boldsymbol{h} \in \mathbb{R}^p$ have conditional independence, meaning their distribution is the product of each conditional distribution. It is possible to choose distributions from the exponential family of distributions for the visible and hidden variables [66]:

$$\mathbb{P}(\boldsymbol{v}) = \prod_{i=1}^{d} r_i(v_i) \exp\Big(\sum_{a} \theta_{ia} f_{ia}(v_i) - A_i(\{\theta_{ia}\})\Big), \tag{18.41}$$

$$\mathbb{P}(\boldsymbol{h}) = \prod_{j=1}^{p} s_j(h_j) \exp\Big(\sum_b \lambda_{jb}\, g_{jb}(h_j) - B_j(\{\lambda_{jb}\})\Big), \tag{18.42}$$

where $\{f_{ia}(v_i), g_{jb}(h_j)\}$ are the sufficient statistics, $\{\theta_i, \lambda_j\}$ are the canonical parameters of the models, $\{A_i, B_j\}$ are the log-normalization factors, and $\{r_i(v_i), s_j(h_j)\}$ are the normalization factors, which are additional features multiplied by constants. The log-normalization factors can be ignored because they are difficult to compute.

For the joint distribution of visible and hidden variables, a quadratic term should be introduced for their cross-interaction [66]:

$$\mathbb{P}(\boldsymbol{v}, \boldsymbol{h}) \propto \exp\Big(\sum_{i=1}^{d}\sum_a \theta_{ia}\, f_{ia}(v_i) + \sum_{j=1}^{p}\sum_b \lambda_{jb}\, g_{jb}(h_j)$$
$$+ \sum_{i=1}^{d}\sum_{j=1}^{p}\sum_a\sum_b \boldsymbol{W}_{ia}^{jb}\, f_{ia}(v_i)\, g_{jb}(h_j)\Big). \tag{18.43}$$

According to Proposition 18.1, the visible and hidden units have conditional independence. Therefore, the conditional distributions can be written as the multiplication of exponential family distributions [66]:

$$\mathbb{P}(\boldsymbol{v}|\boldsymbol{h}) = \prod_{i=1}^{d} \exp\Big(\sum_a \widehat{\theta}_{ia}\, f_{ia}(v_i) - A_i(\{\widehat{\theta}_{ia}\})\Big), \tag{18.44}$$

$$\mathbb{P}(\boldsymbol{h}|\boldsymbol{v}) = \prod_{j=1}^{p} \exp\Big(\sum_b \widehat{\lambda}_{jb}\, g_{jb}(h_j) - B_j(\{\widehat{\lambda}_{jb}\})\Big), \tag{18.45}$$

where:

$$\widehat{\theta}_{ia} := \theta_{ia} + \sum_{j=1}^{p}\sum_b \boldsymbol{W}_{ia}^{jb}\, g_{jb}(h_j), \tag{18.46}$$

$$\widehat{\lambda}_{jb} := \lambda_{jb} + \sum_{i=1}^{d}\sum_a \boldsymbol{W}_{ia}^{jb}\, f_{ia}(v_i). \tag{18.47}$$

Therefore, it is possible to choose one of the distributions in the exponential family for the conditional distributions of the visible and hidden variables. Section 18.4.2 introduces different cases where the units can have either discrete or continuous values. In all cases, the distributions are from exponential families.

18.4.2 Binary States

The hidden and visible variables can have a discrete number of values, also called states. Most often, inspired by the Hopfield network, BM and RBM have binary states for the hidden and visible units, i.e., $v_i, h_j \in \{0, 1\}, \forall i, j$. Therefore, it is possible to say:

$$\mathbb{P}(h_j = 1|\boldsymbol{v}) = \frac{\mathbb{P}(h_j = 1, \boldsymbol{v})}{\mathbb{P}(h_j = 0, \boldsymbol{v}) + \mathbb{P}(h_j = 1, \boldsymbol{v})}. \tag{18.48}$$

In binary states, the joint probability in Eq. (18.17) is simplified to:

$$\mathbb{P}(h_j = 0, \boldsymbol{v}) = \exp(c_j \times 0 + \boldsymbol{v}^\top \boldsymbol{W}_{:j} \times 0) = \exp(0) = 1,$$
$$\mathbb{P}(h_j = 1, \boldsymbol{v}) = \exp(c_j \times 1 + \boldsymbol{v}^\top \boldsymbol{W}_{:j} \times 1) = \exp(c_j + \boldsymbol{v}^\top \boldsymbol{W}_{:j}).$$

Therefore, Eq. (18.48) becomes:

$$\begin{aligned} \mathbb{P}(h_j = 1|\boldsymbol{v}) &= \frac{\exp(c_j + \boldsymbol{v}^\top \boldsymbol{W}_{:j})}{1 + \exp(c_j + \boldsymbol{v}^\top \boldsymbol{W}_{:j})} \\ &= \frac{1}{1 + \exp\big(-(c_j + \boldsymbol{v}^\top \boldsymbol{W}_{:j})\big)} = \sigma(c_j + \boldsymbol{v}^\top \boldsymbol{W}_{:j}), \end{aligned} \tag{18.49}$$

where:

$$\sigma(x) := \frac{\exp(x)}{1 + \exp(x)} = \frac{1}{1 + \exp(-x)},$$

is the sigmoid (or logistic) function and $\boldsymbol{W}_{:j} \in \mathbb{R}^d$ denotes the j-th column of the matrix $\boldsymbol{W}$. If the visible units also have binary states, then similarly there will be:

$$\mathbb{P}(v_i = 1|\boldsymbol{h}) = \sigma(b_i + \boldsymbol{W}_{i:}\, \boldsymbol{h}), \tag{18.50}$$

where $\boldsymbol{W}_{i:} \in \mathbb{R}^p$ denotes the i-th row of the matrix $\boldsymbol{W}$. As there are only two states $\{0, 1\}$, from Eqs. (18.49) and (18.50), the following holds:

$$\mathbb{P}(h_j|\boldsymbol{v}) = \sigma(c_j + \boldsymbol{v}^\top \boldsymbol{W}_{:j}) = \sigma(c_j + \sum_{i=1}^{d} v_i\, w_{ij}), \tag{18.51}$$

$$\mathbb{P}(v_i|\boldsymbol{h}) = \sigma(b_i + \boldsymbol{W}_{i:}\, \boldsymbol{h}) = \sigma(b_i + \sum_{j=1}^{p} w_{ij}\, h_j). \tag{18.52}$$

According to Proposition 18.1, the units have conditional independence, meaning their distribution is the product of each conditional distribution:

$$\mathbb{P}(\boldsymbol{h}|\boldsymbol{v}) = \prod_{j=1}^{p} \mathbb{P}(h_j|\boldsymbol{v}) = \prod_{j=1}^{p} \sigma(c_j + \boldsymbol{v}^\top \boldsymbol{W}_{:j}), \tag{18.53}$$

$$\mathbb{P}(\boldsymbol{v}|\boldsymbol{h}) = \prod_{i=1}^{d} \mathbb{P}(v_i|\boldsymbol{h}) = \prod_{i=1}^{d} \sigma(b_i + \boldsymbol{W}_{i:}\boldsymbol{h}). \tag{18.54}$$

Therefore, in the Gibbs sampling of Algorithm 18.1, samples are drawn from the distributions of Eqs. (18.51) and (18.52). Note that the sigmoid function is between zero and one, meaning the uniform distribution $u \sim U[0, 1]$ can be used for sampling from it, as explained in Sect. 18.3.3. Moreover, for binary states, $\mathbb{E}_{\sim\mathbb{P}(h_j|\boldsymbol{v})}[h_j] = \sigma(c_j + \boldsymbol{v}^\top \boldsymbol{W}_{:j})$. Therefore, if the sigmoid function is applied elementwise on the elements of $\widehat{\boldsymbol{h}}_i \in \mathbb{R}^p$, this can be used for Eq. (18.26) in training binary-state RBM:

$$\widehat{\boldsymbol{h}}_i = \mathbb{E}_{\sim\mathbb{P}(\boldsymbol{h}|\boldsymbol{v}_i)}[\boldsymbol{h}] = \sigma(\boldsymbol{c} + \boldsymbol{v}_i^\top \boldsymbol{W}). \tag{18.55}$$

This equation is also used in Algorithm 18.2.

18.4.3 Continuous Values

In some cases, the hidden units can be set to have continuous values as continuous representations for the visible unit. According to the definition of conditional probability:

$$\mathbb{P}(v_i = 1|\boldsymbol{h}) = \frac{\mathbb{P}(v_i = 1, \boldsymbol{h})}{\sum_{v_i \in \mathbb{R}^d} \mathbb{P}(\boldsymbol{h}, v_i)}. \tag{18.56}$$

According to Eq. (18.18):

$$\mathbb{P}(\boldsymbol{h}, v_i = 1) = \exp(b_i \times 1 + 1 \times \boldsymbol{W}_{i:}\boldsymbol{h}) = \exp(b_i + \boldsymbol{W}_{i:}\boldsymbol{h}). \tag{18.57}$$

Therefore, Eq. (18.56) becomes:

$$\mathbb{P}(v_i = 1|\boldsymbol{h}) = \frac{\exp(b_i + \boldsymbol{W}_{i:}\boldsymbol{h})}{\sum_{v_i \in \mathbb{R}^d} \mathbb{P}(\boldsymbol{h}, v_i)}. \tag{18.58}$$

This is a softmax function that can approximate a Gaussian (normal) distribution and it sums to one. Therefore, it can be written as the normal distribution with a variance of one:

$$\mathbb{P}(v_i|\boldsymbol{h}) = \mathcal{N}(b_i + \boldsymbol{W}_{i:}\boldsymbol{h}, 1) = \mathcal{N}(b_i + \sum_{j=1}^{p} w_{ij}h_j, 1). \tag{18.59}$$

According to Proposition 18.1, the units have conditional independence, meaning that their distribution is the product of each conditional distribution:

$$\mathbb{P}(\boldsymbol{v}|\boldsymbol{h}) = \prod_{i=1}^{d} \mathbb{P}(v_i|\boldsymbol{h}) = \prod_{i=1}^{d} \mathcal{N}(b_i + \boldsymbol{W}_{i:}\boldsymbol{h}, 1). \tag{18.60}$$

Usually, when the hidden units have continuous values, the visible units have binary states [46, 66]. In this case, the conditional distribution $\mathbb{P}(\boldsymbol{h}|\boldsymbol{v})$ is obtained by Eq. (18.53). If the visible units have continuous values, their distribution can be similarly calculated as:

$$\mathbb{P}(\boldsymbol{h}|\boldsymbol{v}) = \prod_{j=1}^{p} \mathbb{P}(h_j|\boldsymbol{v}) = \prod_{j=1}^{p} \mathcal{N}(c_j + \boldsymbol{v}^\top \boldsymbol{W}_{:j}, 1). \tag{18.61}$$

These normal distributions can be used for sampling in Gibbs sampling of Algorithm 18.1. In this case, Eq. (18.26), used in Algorithm 18.2 for training RBM, is:

$$\widehat{\boldsymbol{h}}_i = \mathbb{E}_{\sim \mathbb{P}(\boldsymbol{h}|\boldsymbol{v}_i)}[\boldsymbol{h}] = \mathcal{N}(\boldsymbol{c} + \boldsymbol{v}_i^\top \boldsymbol{W}, \boldsymbol{I}_{p\times p}), \tag{18.62}$$

where $\boldsymbol{I}$ denotes the identity matrix.

18.4.4 Discrete Poisson States

In some cases, the units have discrete states with more than two values, as discussed in Sect. 18.4.2. In this case, the well-known Poisson distribution can be used for discrete random variables:

$$\mathrm{Ps}(t, \lambda) = \frac{e^{-\lambda}\lambda^t}{t!}.$$

Assume every visible unit can have a value $t \in \{0, 1, 2, 3, \dots\}$. If the conditional Poisson distribution is considered for the visible units, then [57]:

$$\mathbb{P}(v_i = t|\boldsymbol{h}) = \mathrm{Ps}(t, \frac{\exp(b_i + \boldsymbol{W}_{i:}\boldsymbol{h})}{\sum_{k=1}^{d} \exp(b_k + \boldsymbol{W}_{k:}\boldsymbol{h})}). \tag{18.63}$$

Similarly, if there are discrete states for the hidden units, then:

$$\mathbb{P}(h_j = t|\boldsymbol{v}) = \text{Ps}(t, \frac{\exp(c_j + \boldsymbol{v}^\top \boldsymbol{W}_{:j})}{\sum_{k=1}^{p} \exp(c_k + \boldsymbol{v}^\top \boldsymbol{W}_{:k})}), \tag{18.64}$$

These Poisson distributions can be used for sampling in Gibbs sampling of Algorithm 18.1. In this case, Eq. (18.26), used in Algorithm 18.2 for training RBM, can be calculated using a multivariate Poisson distribution [12]. It is noteworthy that RBM has been used for *semantic hashing*,[3] where the hidden variables are used as a hashing representation of data [57]. Semantic hashing uses Poisson distribution and sigmoid function for the conditional visible and hidden variables, respectively.

18.5 Conditional Restricted Boltzmann Machine

RBM is suitable for static data because it does not include temporal (time) information. The Conditional RBM (CRBM), proposed in [64], incorporates the temporal information into the configuration of RBM. It considers visible variables of previous time steps as conditional variables. CRBM adds two sets of directed links to the RBM. The first set of directed links is the autoregressive links from the past $\mathcal{T}_1$ visible units to the visible units of the current time step. The second set of directed links is the links from the past τ_2 visible units to the hidden units of the current time step. In general, $\mathcal{T}_1$ is not necessarily equal to $\mathcal{T}_2$ but, for simplicity, they are usually set to $\mathcal{T}_1 = \mathcal{T}_2 = \mathcal{T}$ [64]. The links from the visible units at time $t - \tau$ to the visible units at the current time are denoted by $\boldsymbol{G}^{(t-\tau)} = [g_{ij}] \in \mathbb{R}^{d \times d}$. Additionally, the links from the visible units at time $t - \tau$ to the hidden units at the current time are denoted by $\boldsymbol{Q}^{(t-\tau)} = [q_{ij}] \in \mathbb{R}^{d \times p}$. The structure of CRBM is shown in Fig. 18.2. Note that each arrow in this figure is a set of links representing a matrix or vector of weights.

The updating rule for the weights $\boldsymbol{W}$ and biases $\boldsymbol{b}$ and $\boldsymbol{c}$ are the same as in Eqs. (18.31), (18.32), and (18.33). The directed links from the previous visible units to current visible units and current hidden units are considered dynamically changing biases. Recall that for updating the biases $\boldsymbol{b}$ and $\boldsymbol{c}$, Eqs. (18.32) and (18.33) were used. Similarly, for updating the added links from τ previous time steps, the following equations are used:

$$\forall i : \mathbb{R}^d \ni \nabla_{\boldsymbol{G}_{i:}^{(t-\tau)}} \ell(\theta) = v_i^{(t-\tau)} \Big(\sum_{k=1}^{n} \boldsymbol{v}_k^{(t)} - \sum_{k=1}^{n} \widetilde{\boldsymbol{v}}_k^{(t)} \Big), \tag{18.65}$$

$$\forall i : \mathbb{R}^p \ni \nabla_{\boldsymbol{Q}_{i:}^{(t-\tau)}} \ell(\theta) = v_i^{(t-\tau)} \Big(\sum_{k=1}^{n} \widehat{\boldsymbol{h}}_k - \sum_{k=1}^{n} \widetilde{\boldsymbol{h}}_k \Big), \tag{18.66}$$

[3] In hashing, a hash function is used to map data of arbitrary size to fixed-size values.

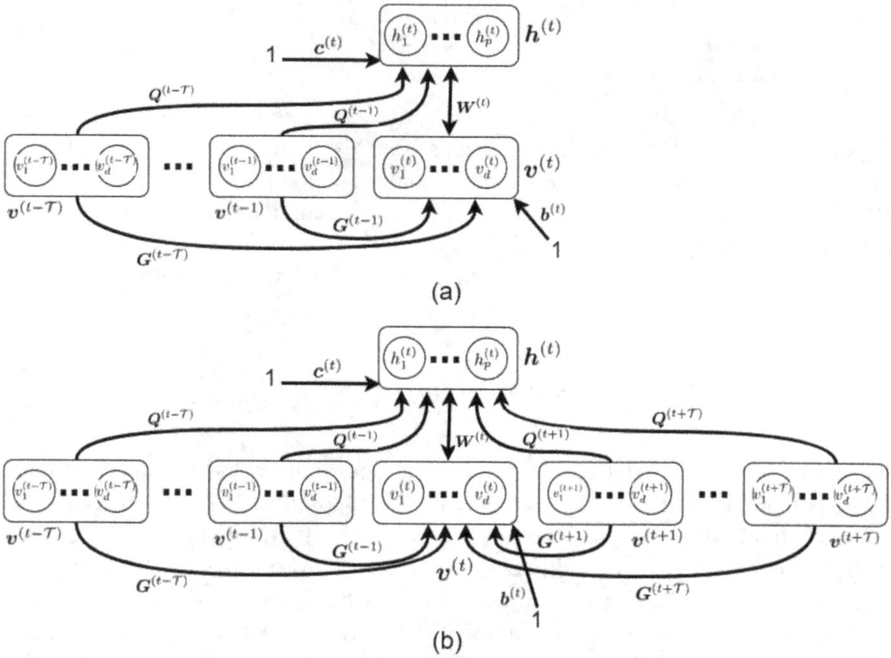

Fig. 18.2 The structures of (a) a conditional Boltzmann machine and (b) an interpolating conditional restricted Boltzmann machine. Note that each arrow in this figure represents a set of links

where $i \in \{1, \dots, d\}$, $\tau \in \{1, 2, \dots, \mathcal{T}\}$, $\boldsymbol{G}_{i:}^{(t-\tau)}$ denotes the i-th row of $\boldsymbol{G}^{(t-\tau)}$, and $\boldsymbol{Q}_{i:}^{(t-\tau)}$ denotes the i-th row of $\boldsymbol{Q}^{(t-\tau)}$. These equations are similar to Eqs. (18.32) and (18.33), but they are multiplied by the visible values in the previous time steps. This is because RBM multiplies biases by one (see Fig. 18.1), while these newly introduced biases have previous visible values instead of one (see Fig. 18.2). These equations can be restated as [64]:

$$\forall i, j: \nabla_{g_{ij}^{(t-\tau)}} \ell(\theta) = v_i^{(t-\tau)} \Big(\langle v_i^{(t)} \rangle_{\text{data}} - \langle v_i^{(t)} \rangle_{\text{recon.}} \Big), \tag{18.67}$$

$$\forall i, j: \nabla_{q_{ij}^{(t-\tau)}} \ell(\theta) = v_i^{(t-\tau)} \Big(\langle h_j^{(t)} \rangle_{\text{data}} - \langle h_j^{(t)} \rangle_{\text{recon.}} \Big), \tag{18.68}$$

where $\langle . \rangle_{\text{data}}$ and $\langle . \rangle_{\text{recon.}}$ denote expectation over the data and reconstruction of the data, respectively. Further to the weights and biases of RBM, the additional links are learned by gradient descent using the above gradients. Algorithm 18.2 can be used for training CRBM if learning the added links is also included in the algorithm.

Interpolating CRBM (ICRBM) [46] is an improvement over the CRBM where some links have been added from visible variables in the future. Figure 18.2 depicts the structure of ICRBM. Its training and formulation are similar, but its theory is not covered here for the sake of brevity. Note that CRBM has been used in various time-series applications, such as action recognition [64] and acoustics [46].

18.6 Deep Belief Network

18.6.1 *Stacking RBM Models*

A neural network can be trained using RBM training [24, 25]. Training a neural network using RBM training can result in a good initialization of weights for training the network using backpropagation. Before the development of ReLU [17] and dropout [62], multilayer perceptron networks could not become deep for the problem of vanishing gradients. This was because random initial weights were not sufficiently suitable for starting optimization in backpropagation, especially in deep networks. Therefore, a method was proposed for pretraining neural networks that initializes the network to a suitable set of weights and then the pretrained weights are fine-tuned using backpropagation [24, 25].

A neural network consists of several layers. Let ℓ denote the number of layers, where the first layer receives the input data, and let p_ℓ be the number of neurons in the ℓ-th layer. By convention, $p_1 = d$. Every two successive layers can be considered one RBM. This is demonstrated in Fig. 18.3. Starting from the first pair of layers as an RBM, the training dataset $\{\boldsymbol{x}_i \in \mathbb{R}^d\}_{i=1}^n$ is introduced as the visible variable $\{\boldsymbol{v}_i\}_{i=1}^n$ of the first pair of layers. The weights and biases of

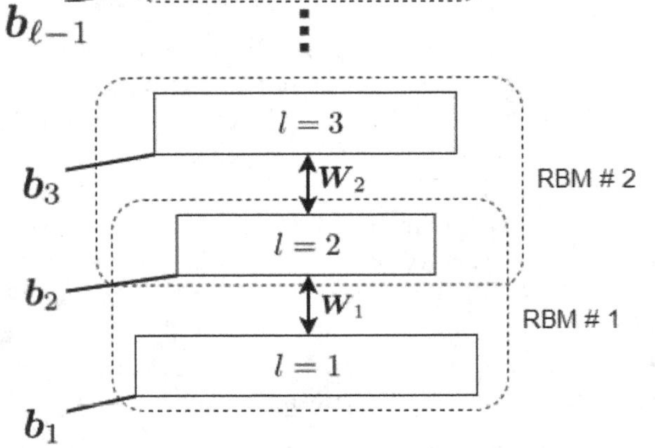

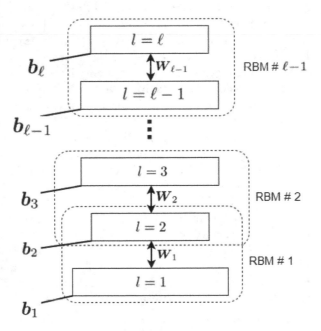

Fig. 18.3 Pretraining a deep belief network by considering every pair of layers as an RBM

Input: training data $\{\boldsymbol{x}_i\}_{i=1}^n$
// pre-training:
for *l from* 1 *to* $\ell - 1$ **do**
    if $l = 1$ **then**
        $\{\boldsymbol{v}_i\}_{i=1}^n \leftarrow \{\boldsymbol{x}_i\}_{i=1}^n$
    else
        // generate n hidden variables of previous RBM:
        $\{\boldsymbol{h}_i\}_{i=1}^n \leftarrow$ Algorithm 18.1 for $(l-1)$-th RBM $\leftarrow \{\boldsymbol{v}_i\}_{i=1}^n$
        $\{\boldsymbol{v}_i\}_{i=1}^n \leftarrow \{\boldsymbol{h}_i\}_{i=1}^n$
    $\boldsymbol{W}_l, \boldsymbol{b}_l, \boldsymbol{b}_{l+1} \leftarrow$ Algorithm 18.2 for l-th RBM $\leftarrow \{\boldsymbol{v}_i\}_{i=1}^n$
// fine-tuning using backpropagation:
Initialize network with weights $\{\boldsymbol{W}_l\}_{l=1}^{\ell-1}$ and biases $\{\boldsymbol{b}_l\}_{l=2}^{\ell}$.
$\{\boldsymbol{W}_l\}_{l=1}^{\ell-1}, \{\boldsymbol{b}_l\}_{l=1}^{\ell} \leftarrow$ Backpropagate the error of loss for several epochs.

Algorithm 18.3: Training a deep belief network

this first layer are trained as an RBM using Algorithm 18.2. After training this RBM, n p_2-dimensional hidden variables are generated using Gibbs sampling in Algorithm 18.1. Now, the hidden variables of the first RBM are considered visible variables for the second RBM (the second pair of layers). This RBM is trained by Algorithm 18.2, and then, the hidden variables are generated using Gibbs sampling in Algorithm 18.1. This procedure is repeated until all pairs of layers are trained using RBM training. This layerwise training of the neural network has a greedy approach [4]. This greedy training of layers prepares good initialized weights and biases for the whole neural network. After this initialization, the weights and biases can be fine-tuned using backpropagation [53].

The explained training algorithm was first proposed in [24, 25] and was used for dimensionality reduction. By increasing ℓ to any large number, the network becomes large and deep. As layers are trained one by one as RBM models, it is possible to make the network as deep as is desired without concern for vanishing gradients because the weights are initialized well for backpropagation. As this network can become deep and is pretrained by belief propagation (RBM training), it is referred to as the *Deep Belief Network* (DBN) [22, 24]. DBN can be seen as a stack of RBM models. The pretraining of a DBN using RBM training is depicted in Fig. 18.3. This algorithm is summarized in Algorithm 18.3. In this algorithm, $\boldsymbol{W}_l \in \mathbb{R}^{p_l \times p_{l+1}}$ denotes the weights connecting layer l to layer $(l+1)$ and $\boldsymbol{b}_l \in \mathbb{R}^{p_l}$ denotes the biases of layer l. Note that as weights are between every two layers, the sets of weights are $\{\boldsymbol{W}_l\}_{l=1}^{\ell-1}$.

Note that pretraining of DBN is an unsupervised task because RBM training is unsupervised. Fine-tuning of a DBN can be either unsupervised or supervised depending on the loss function for backpropagation. If the DBN is an autoencoder with a low-dimensional middle layer in the network, both its pretraining and fine-tuning stages are unsupervised because the loss function of backpropagation is also a mean squared error. This DBN autoencoder can learn a low-dimensional embedding

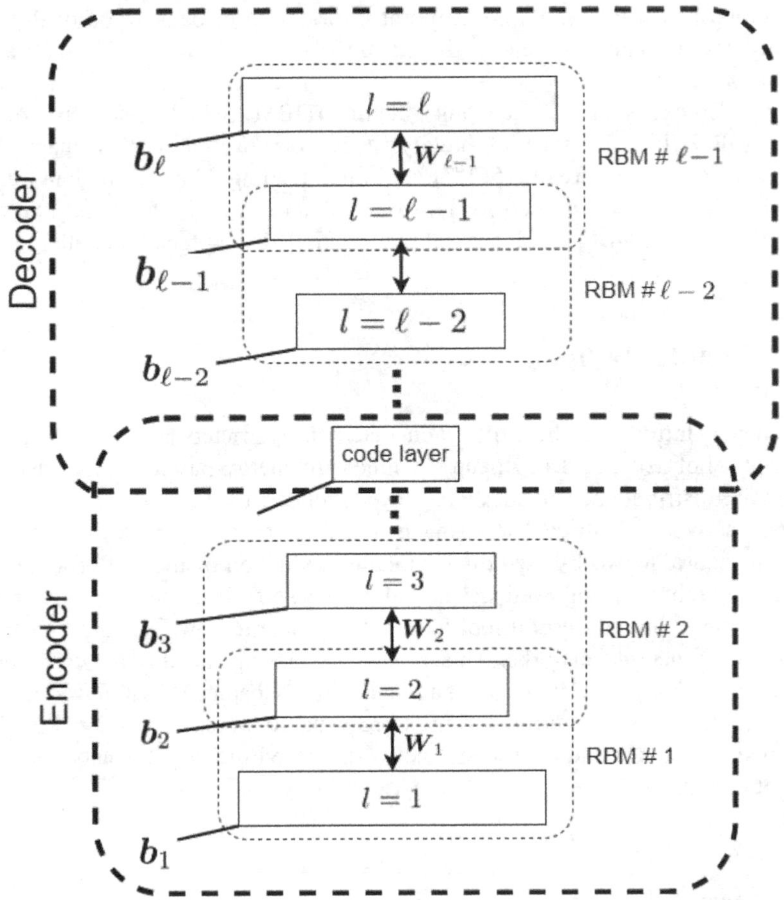

Fig. 18.4 A DBN autoencoder, where the numbers of neurons in the corresponding layers of the encoder and decoder are usually set equal. The coder layer is a low-dimensional embedding for the representation of data

or representation of the data and can be used for dimensionality reduction [25]. The DBN autoencoder has also been used for hashing [57]. The structure of this network is depicted in Fig. 18.4.

18.6.2 Other Improvements over RBM and DBN

There have been various improvements over DBN, such as the convolutional DBN [35] and the use of DBN for hashing [57]. Greedy training of DBN, using RBM training, has been used in training the t-SNE network for general degrees of freedom [65] (see Chap. 16 for t-SNE). In addition to CRBM [64], recurrent RBM [63] has been proposed to handle the temporal information of data. Additionally, note that there exist some other energy-based models in addition to BM, such as the Helmholtz machine [10].

There also exists Deep Boltzmann Machine (DBM), which is slightly different from a DBN. For the sake of brevity, it is not covered in this chapter; the interested reader is referred to [56] for additional reading. Various efficient training algorithms have been proposed for DBM [18, 26, 43, 49, 54, 55, 58–60]. Some of the applications of DBM are document processing [61] and facial modelling [50].

18.7 Chapter Summary

This chapter introduced the Boltzmann machine, restricted Boltzmann machine, and deep belief network. Boltzmann machines are energy-based methods using the Boltzmann distribution, and they are closely related to the Ising model and the Hopfield network. Restricted Boltzmann machines and deep belief networks are important neural network-based dimensionality reduction methods. Before dropout and ReLU techniques allowed the neural networks to become deep, Boltzmann machines were the only useful tool to have deep neural networks, and they ended the winter of neural networks. This is because the layers of a network can be considered separate restricted Boltzmann machines. Chapter 19 will introduce deep metric learning, where neural networks are used for metric learning. Restricted Boltzmann machines and deep belief networks paved the way for networks to be used for dimensionality reduction.

References

1. David H Ackley, Geoffrey E Hinton, and Terrence J Sejnowski. "A learning algorithm for Boltzmann machines". In: *Cognitive science* 9.1 (1985), pp. 147–169.
2. Diego Alberici, Pierluigi Contucci, and Emanuele Mingione. "Deep Boltzmann machines: rigorous results at arbitrary depth". In: *Annales Henri Poincaré*. Springer. 2021, pp. 1–24.
3. Diego Alberici et al. "Annealing and replica-symmetry in deep Boltzmann machines". In: *Journal of Statistical Physics* 180.1 (2020), pp. 665–677.
4. Yoshua Bengio et al. "Greedy layer-wise training of deep networks". In: *Advances in neural information processing systems*. 2007, pp. 153–160.
5. Christopher M Bishop. "Pattern recognition". In: *Machine learning* 128.9 (2006).
6. Ludwig Boltzmann. "Studien uber das Gleichgewicht der lebenden Kraft". In: *Wissenschafliche Abhandlungen* 1 (1868), pp. 49–96.
7. Bernhard E Boser, Isabelle M Guyon, and Vladimir N Vapnik. "A training algorithm for optimal margin classifiers". In: *Proceedings of the fifth annual workshop on Computational learning theory*. 1992, pp. 144–152.
8. Stephen G Brush. "History of the Lenz-Ising model". In: *Reviews of modern physics* 39.4 (1967), p. 883.
9. Sean Carroll. *From eternity to here: the quest for the ultimate theory of time*. Penguin, 2010.
10. Peter Dayan et al. "The Helmholtz machine". In: *Neural computation* 7.5 (1995), pp. 889–904.

11. Raaz Dwivedi et al. "Log-concave sampling: Metropolis-Hastings algorithms are fast!" In: *Conference on learning theory*. PMLR. 2018, pp. 793–797.
12. Carol Bates Edwards. *Multivariate and multiple Poisson distributions*. Iowa State University, 1962.
13. Stuart Geman and Donald Geman. "Stochastic relaxation, Gibbs distributions, and the Bayesian restoration of images". In: *IEEE Transactions on pattern analysis and machine intelligence*. PAMI-6.6 (1984), pp. 721–741.
14. Benyamin Ghojogh and Mark Crowley. "The theory behind overfitting, cross validation, regularization, bagging, and boosting: tutorial". In: *arXiv preprint arXiv:1905.12787* (2019).
15. Benyamin Ghojogh et al. "Sampling algorithms, from survey sampling to Monte Carlo methods: Tutorial and literature review". In: *arXiv preprint arXiv:2011.00901* (2020).
16. J Willard Gibbs. *Elementary principles in statistical mechanics*. Courier Corporation, 1902.
17. Xavier Glorot, Antoine Bordes, and Yoshua Bengio. "Deep sparse rectifier neural networks". In: *Proceedings of the fourteenth international conference on artificial intelligence and statistics*. JMLR Workshop and Conference Proceedings. 2011, pp. 315–323.
18. Ian Goodfellow et al. "Multi-prediction deep Boltzmann machines". In: *Advances in Neural Information Processing Systems* 26 (2013), pp. 548–556.
19. Donald Hebb. *The Organization of Behavior*. Wiley & Sons, New York, 1949.
20. Geoffrey E Hinton. "A practical guide to training restricted Boltzmann machines". In: *Neural networks: Tricks of the trade*. Springer, 2012, pp. 599–619.
21. Geoffrey E Hinton. "Boltzmann machine". In: *Scholarpedia* 2.5 (2007), p. 1668.
22. Geoffrey E Hinton. "Deep belief networks". In: *Scholarpedia* 4.5 (2009), p. 5947.
23. Geoffrey E Hinton. "Training products of experts by minimizing contrastive divergence". In: *Neural computation* 14.8 (2002), pp. 1771–1800.
24. Geoffrey E Hinton, Simon Osindero, and Yee-Whye Teh. "A fast learning algorithm for deep belief nets". In: *Neural computation* 18.7 (2006), pp. 1527–1554.
25. Geoffrey E Hinton and Ruslan R Salakhutdinov. "Reducing the dimensionality of data with neural networks". In: *Science* 313.5786 (2006), pp. 504–507.
26. Geoffrey E Hinton and Russ R Salakhutdinov. "A better way to pretrain deep Boltzmann machines". In: *Advances in Neural Information Processing Systems* 25 (2012), pp. 2447–2455.
27. Geoffrey E Hinton and Terrence J Sejnowski. "Optimal perceptual inference". In: *Proceedings of the IEEE conference on Computer Vision and Pattern Recognition*. Vol. 448. IEEE, 1983.
28. John J Hopfield. "Neural networks and physical systems with emergent collective computational abilities". In: *Proceedings of the national academy of sciences* 79.8 (1982), pp. 2554–2558.
29. John J Hopfield. "Neurons with graded response have collective computational properties like those of two-state neurons". In: *Proceedings of the national academy of sciences* 81.10 (1984), pp. 3088–3092.
30. Kerson Huang. *Statistical Mechanics*. John Wiley & Sons, 1987.
31. Ernst Ising. "Beitrag zur theorie des ferromagnetismus". In: *Zeitschrift für Physik* 31.1 (1925), pp. 253–258.
32. Andrej Karpathy and Li Fei-Fei. "Deep visual-semantic alignments for generating image descriptions". In: *Proceedings of the IEEE conference on computer vision and pattern recognition*. 2015, pp. 3128–3137.
33. Scott Kirkpatrick, C Daniel Gelatt, and Mario P Vecchi. "Optimization by simulated annealing". In: *science* 220.4598 (1983), pp. 671–680.
34. Daphne Koller and Nir Friedman. *Probabilistic graphical models: principles and techniques*. MIT press, 2009.
35. Alex Krizhevsky and Geoff Hinton. "Convolutional deep belief networks on CIFAR-10". In: *Unpublished manuscript* 40.7 (2010), pp. 1–9.
36. Dmitry Krotov. "Hierarchical Associative Memory". In: *arXiv preprint arXiv:2107.06446* (2021).
37. Dmitry Krotov and John Hopfield. "Large associative memory problem in neurobiology and machine learning". In: *International Conference on Learning Representations (ICLR)*. 2021.

38. Dmitry Krotov and John J Hopfield. "Dense associative memory for pattern recognition". In: *Advances in neural information processing systems* 29 (2016), pp. 1172–1180.
39. Yann LeCun, Yoshua Bengio, and Geoffrey Hinton. "Deep learning". In: *nature* 521.7553 (2015), pp. 436–444.
40. Yann LeCun et al. "A tutorial on energy-based learning". In: *Predicting structured data* 1 (2006).
41. Wilhelm Lenz. "Beitršge zum verstšndnis der magnetischen eigenschaften in festen kšrpern". In: *Physikalische Z* 21 (1920), pp. 613–615.
42. William A Little. "The existence of persistent states in the brain". In: *Mathematical biosciences* 19.1–2 (1974), pp. 101–120.
43. Jan Melchior, Asja Fischer, and Laurenz Wiskott. "How to center deep Boltzmann machines". In: *The Journal of Machine Learning Research* 17.1 (2016), pp. 3387–3447.
44. Abdel-rahman Mohamed, George Dahl, Geoffrey Hinton, et al. "Deep belief networks for phone recognition". In: *Nips workshop on deep learning for speech recognition and related applications*. Vol. 1. 9. Vancouver, Canada. 2009, p. 39.
45. Abdel-rahman Mohamed, George E Dahl, and Geoffrey Hinton. "Acoustic modeling using deep belief networks". In: *IEEE transactions on audio, speech, and language processing* 20.1 (2011), pp. 14–22.
46. Abdel-rahman Mohamed and Geoffrey Hinton. "Phone recognition using restricted Boltzmann machines". In: *2010 IEEE International Conference on Acoustics, Speech and Signal Processing*. IEEE. 2010, pp. 4354–4357.
47. Mehdi Molkaraie. "Marginal Densities, Factor Graph Duality and High-Temperature Series Expansions". In: *International Conference on Artificial Intelligence and Statistics*. 2020, pp. 256–265.
48. Mehdi Molkaraie. "The primal versus the dual Ising model". In: *2017 55th Annual Allerton Conference on Communication, Control, and Computing (Allerton)*. IEEE. 2017, pp. 53–60.
49. Grégoire Montavon and Klaus-Robert Müller. "Deep Boltzmann machines and the centering trick". In: *Neural networks: tricks of the trade*. Springer, 2012, pp. 621–637.
50. Chi Nhan Duong et al. "Beyond principal components: Deep Boltzmann machines for face modeling". In: *Proceedings of the IEEE Conference on Computer Vision and Pattern Recognition*. 2015, pp. 4786–4794.
51. Leandro Aparecido Passos and Joao Paulo Papa. "Temperature-based deep Boltzmann machines". In: *Neural Processing Letters* 48.1 (2018), pp. 95–107.
52. Hubert Ramsauer et al. "Hopfield networks is all you need". In: *arXiv preprint arXiv:2008.02217* (2020).
53. David E Rumelhart, Geoffrey E Hinton, and Ronald J Williams. "Learning representations by back-propagating errors". In: *Nature* 323.6088 (1986), pp. 533–536.
54. Ruslan Salakhutdinov. "Learning deep Boltzmann machines using adaptive MCMC". In: *Proceedings of the 27th International Conference on Machine Learning*. 2010, pp. 943–950.
55. Ruslan Salakhutdinov and Geoffrey Hinton. "An efficient learning procedure for deep Boltzmann machines". In: *Neural computation* 24.8 (2012), pp. 1967–2006.
56. Ruslan Salakhutdinov and Geoffrey Hinton. "Deep Boltzmann machines". In: *Artificial intelligence and statistics*. PMLR. 2009, pp. 448–455.
57. Ruslan Salakhutdinov and Geoffrey Hinton. "Semantic hashing". In: *International Journal of Approximate Reasoning* 50.7 (2009), pp. 969–978.
58. Ruslan Salakhutdinov and Hugo Larochelle. "Efficient learning of deep Boltzmann machines". In: *Proceedings of the thirteenth international conference on artificial intelligence and statistics*. JMLR Workshop and Conference Proceedings. 2010, pp. 693–700.
59. Nitish Srivastava and Ruslan Salakhutdinov. "Multimodal Learning with Deep Boltzmann Machines". In: *Advances in neural information processing systems*. Vol. 1. 2012, p. 2.
60. Nitish Srivastava and Ruslan Salakhutdinov. "Multimodal learning with deep Boltzmann machines". In: *Journal of Machine Learning Research* 15.1 (2014), pp. 2949–2980.
61. Nitish Srivastava, Ruslan R Salakhutdinov, and Geoffrey E Hinton. "Modeling documents with deep Boltzmann machines". In: *arXiv preprint arXiv:1309.6865* (2013).

62. Nitish Srivastava et al. “Dropout: a simple way to prevent neural networks from overfitting”. In: *The journal of machine learning research* 15.1 (2014), pp. 1929–1958.
63. Ilya Sutskever, Geoffrey E Hinton, and Graham W Taylor. “The recurrent temporal restricted Boltzmann machine”. In: *Advances in neural information processing systems*. 2009, pp. 1601–1608.
64. Graham W Taylor, Geoffrey E Hinton, and Sam T Roweis. “Modeling human motion using binary latent variables”. In: *Advances in neural information processing systems*. 2007, pp. 1345–1352.
65. Laurens Van Der Maaten. “Learning a parametric embedding by preserving local structure”. In: *Artificial Intelligence and Statistics*. 2009, pp. 384–391.
66. Max Welling, Michal Rosen-Zvi, and Geoffrey E Hinton. “Exponential Family Harmoniums with an Application to Information Retrieval.” In: *Advances in neural information processing systems*. Vol. 4. 2004, pp. 1481–1488.

Chapter 19
Deep Metric Learning

19.1 Introduction

It was mentioned in Chap. 11 that metric learning can be divided into spectral, probabilistic, and deep metric learning. Chapters 11 and 13 explained that both spectral and probabilistic metric learning methods use the generalized Mahalanobis distance, i.e., Eq. (11.53) in Chap. 11, and learn the weight matrix in the metric. Deep metric learning, however, takes a different approach. Deep metric learning methods usually do not use a generalized Mahalanobis distance; instead, they learn an embedding space using a neural network. This chapter introduces and reviews deep metric learning methods.

In deep metric learning, the neural network learns a p-dimensional embedding space for discriminating between classes or dissimilar points while decreasing the distance between similar points. The network embeds the data in the embedding space (or subspace) of the metric. Then, any distance metric $d(.,.) : \mathbb{R}^p \times \mathbb{R}^p \rightarrow \mathbb{R}$ can be used in this embedding space. In the loss functions of the network, the distance function $d(.,.)$ can be used in the embedding space. For example, an option for the distance function is the squared ℓ_2 norm or squared Euclidean distance:

$$d\big(\mathbf{f}(\boldsymbol{x}_i^1), \mathbf{f}(\boldsymbol{x}_i^2)\big) := \|\mathbf{f}(\boldsymbol{x}_i^1) - \mathbf{f}(\boldsymbol{x}_i^2)\|_2^2, \tag{19.1}$$

where $\mathbf{f}(\boldsymbol{x}_i) \in \mathbb{R}^p$ denotes the output of the network for the input $\boldsymbol{x}_i$ as its p-dimensional embedding. The network is trained using minibatch methods, such as the minibatch stochastic gradient descent, where the minibatch size is denoted by b. The weights of the network are denoted by the learnable parameter θ.

B. Ghojogh et al., *Elements of Dimensionality Reduction and Manifold Learning*,
https://doi.org/10.1007/978-3-031-10602-6_19

19.2 Reconstruction Autoencoders

19.2.1 Types of Autoencoders

An autoencoder is a model consisting of an encoder $E(.)$ and a decoder $D(.)$. There are several types of autoencoders. All types of autoencoders learn a code layer in the middle of the encoder and decoder. Inferential autoencoders, such as a variational autoencoder (see Chap. 20) and adversarial autoencoder (see Chap. 21), learn a stochastic latent space in the code layer between the encoder and decoder. Another type of autoencoder is the reconstruction autoencoder consisting of an encoder, transforming data to a code, and a decoder, transforming the code back into data. Therefore, the decoder reconstructs the input data to the encoder. The code is a representation of the data. Each of the encoder and decoder can be multiple layers of neural network with activation functions.

19.2.2 Reconstruction Loss

The input data point to the encoder is signified by $\boldsymbol{x} \in \mathbb{R}^d$, where d is the dimensionality of the data. The reconstructed data point is the output of the decoder and is denoted by $\widehat{\boldsymbol{x}} \in \mathbb{R}^d$. The representation code, which is the output of the encoder and the input of the decoder, is denoted by $\mathbf{f}(\boldsymbol{x}) := E(\boldsymbol{x}) \in \mathbb{R}^p$. It can be said that $\widehat{\boldsymbol{x}} = D(E(\boldsymbol{x})) = D(\mathbf{f}(\boldsymbol{x}))$. If the dimensionality of the code is greater than the dimensionality of the input data, i.e., $p > d$, then the autoencoder is called an overcomplete autoencoder [14]. Otherwise, if $p < d$, the autoencoder is an undercomplete autoencoder [14]. The loss function of the reconstruction autoencoder attempts to make the reconstructed data close to the input data:

$$\underset{\theta}{\text{minimize}} \sum_{i=1}^{b} \Big(d\big(\boldsymbol{x}_i, \widehat{\boldsymbol{x}}_i\big) + \lambda \Omega(\theta)\Big), \tag{19.2}$$

where $\lambda \geq 0$ is the regularization parameter and $\Omega(\theta)$ is the penalty or regularization on the weights. Here, the distance function $d(., .)$ is defined on $\mathbb{R}^d \times \mathbb{R}^d$. Note that the penalty term can be a regularization on the code $\mathbf{f}(\boldsymbol{x}_i)$. If the distance metric used is the squared Euclidean distance, this loss is named the regularized Mean Squared Error (MSE) loss.

19.2.3 Denoising Autoencoder

A problem with an overcomplete autoencoder is that its training only copies each feature of the data input to one of the neurons in the code layer and then copies it back to the corresponding feature of the output layer. This is because the number of neurons in the code layer is greater than the number of neurons in the input and output layers. In other words, the network just memorizes or gets overfitted. This copying happens by making some of the weights equal to one (or a scale of one depending on the activation functions) and the rest of the weights equal to zero. To avoid this problem in overcomplete autoencoders, noise can be added to the input data, and then the data without noise are reconstructed. For this, Eq. (19.2) is used, while the input to the network is the minibatch plus some noise. This forces the overcomplete autoencoder to not just copy data to the code layer. The overcomplete autoencoder can be used for denoising, as it reconstructs the data without noise for a noisy input. This network is called the Denoising Autoencoder (DAE) [14].

19.2.4 Metric Learning by Reconstruction Autoencoder

The undercomplete reconstruction autoencoder can be used for metric learning and dimensionality reduction, especially when $p \ll d$. Equation (19.2) is the loss function for learning a low-dimensional representation code and reconstructing the data by the autoencoder. The code layer between the encoder and decoder is the embedding space of the metric.

Note that if the activation functions of all of the layers are linear, the undercomplete autoencoder is reduced to principal component analysis (see Chap. 5). Let $\boldsymbol{U}_l$ denote the weight matrix of the l-th layer of the network, ℓ_e be the number of layers of the encoder, and ℓ_d be the number of layers of the decoder. With the linear activation function, the encoder and decoder are:

$$\text{encoder:} \quad \mathbb{R}^p \ni \mathbf{f}(\boldsymbol{x}_i) = \underbrace{\boldsymbol{U}_{\ell_e}^\top \boldsymbol{U}_{\ell_e-1}^\top \dots \boldsymbol{U}_1^\top}_{\boldsymbol{U}_e^\top} \boldsymbol{x}_i,$$

$$\text{decoder:} \quad \mathbb{R}^d \ni \widehat{\boldsymbol{x}}_i = \underbrace{\boldsymbol{U}_1 \dots \boldsymbol{U}_{\ell_d-1} \boldsymbol{U}_{\ell_d}}_{\boldsymbol{U}_d} \mathbf{f}(\boldsymbol{x}_i),$$

where the ℓ concatenated linear projections can be replaced by a linear projection with projection matrices $\boldsymbol{U}_e$ and $\boldsymbol{U}_d$ in the encoder and decoder, respectively.

To learn complicated data patterns, nonlinear activation functions can be used between the encoder and decoder layers, which results in nonlinear metric learning and dimensionality reduction. It is noteworthy that a nonlinear neural network can be seen as an ensemble or concatenation of dimensionality reduction (or feature extraction) and kernel methods. The justification of this claim is as follows. Let the dimensionality for a layer of network be $\boldsymbol{U} \in \mathbb{R}^{d_1 \times d_2}$, so it connects d_1 neurons to d_2 neurons. Two cases can happen:

- If $d_1 \geq d_2$, this layer acts as a dimensionality reduction or feature extraction because it has reduced the dimensionality of its input data. If this layer has a nonlinear activation function, the dimensionality reduction is nonlinear; otherwise, it is linear.

- If $d_1 < d_2$, this layer acts as a kernel method that maps its input data to the high-dimensional feature space in a Reproducing Kernel Hilbert Space (RKHS). This kernelization can help with the nonlinear separation of some classes that are not separable linearly (see Chap. 3). An example of the use of kernelization in machine learning is the kernel support vector machine [50].

Therefore, a neural network is a complicated feature extraction method as a concatenation of dimensionality reduction and kernel methods, where each layer of the network learns its own features from the data.

19.3 Supervised Metric Learning by Supervised Loss Functions

Various loss functions exist for supervised metric learning by neural networks. Supervised loss functions can teach the network to separate classes in the embedding space [44]. For this, a network whose last layer is for classification of data points, can be used. The features of the one-to-last layer can be used for feature embedding. The last layer after the embedding features is named the classification layer. The structure of this network is visualized in Fig. 19.1. Let the i-th point in the minibatch be denoted by $\boldsymbol{x}_i \in \mathbb{R}^d$ and its label be denoted by $y_i \in \mathbb{R}$. Suppose the network has one output neuron and its output for the input $\boldsymbol{x}_i$ is denoted by $\mathbf{f}_o(\boldsymbol{x}_i) \in \mathbb{R}$. This output is the estimated class label by the network. The output of the one-to-last layer is denoted by $\mathbf{f}(\boldsymbol{x}_i) \in \mathbb{R}^p$, where p is the number of neurons in that layer which is equivalent to the dimensionality of the embedding space. The last layer of the network, connecting the p neurons to the output neuron, is a fully-connected layer. The network until the one-to-last layer can be any feed-forward or convolutional network depending on the type of data. If the network is convolutional, it should be flattened at the one-to-last layer. The network learns to classify the classes, by the supervised loss functions, so the features of the one-to-last layer will be discriminating features and suitable for embedding.

19.3.1 *Mean Squared Error and Mean Absolute Value Losses*

One of the supervised losses is the Mean Squared Error (MSE), which uses the squared ℓ_2 norm to ensure that the estimated labels closely resemble the true labels:

$$\underset{\theta}{\text{minimize}} \quad \sum_{i=1}^{b} (\mathbf{f}_o(\boldsymbol{x}_i) - y_i)^2. \tag{19.3}$$

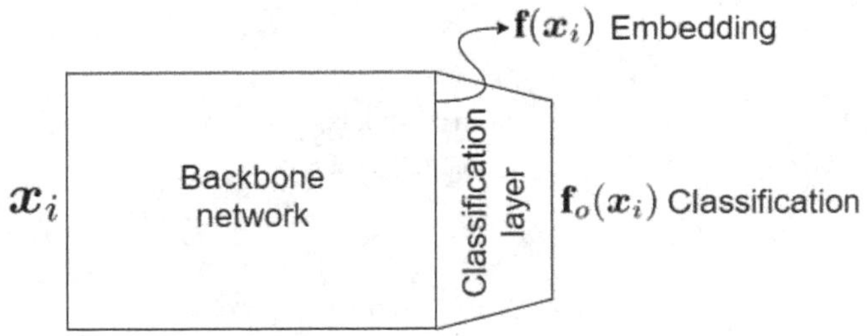

Fig. 19.1 Network structure for metric learning with a supervised loss function

One problem with this loss function is that it is known to exaggerate the outliers because of being squared, but its advantage is its differentiability. Another loss function is the Mean Absolute Error (MAE), which ensures that the estimated labels closely resemble the true labels using the ℓ_1 norm or the absolute value:

$$\underset{\theta}{\text{minimize}} \quad \sum_{i=1}^{b} |\mathbf{f}_o(\boldsymbol{x}_i) - y_i|. \tag{19.4}$$

The distance used in this loss function is also named the Manhattan distance. This loss function does not have the problem of MSE and can be used for imposing sparsity in the embedding. It is not differentiable at point $\mathbf{f}(\boldsymbol{x}_i) = y_i$, but as the derivatives are calculated numerically by the neural network, this is not a big issue nowadays.

19.3.2 Huber and KL-Divergence Loss

Another loss function is the Huber loss, which is a combination of the advantages of MSE and MAE:

$$\underset{\theta}{\text{minimize}} \quad \sum_{i=1}^{b} \begin{cases} 0.5(\mathbf{f}_o(\boldsymbol{x}_i) - y_i)^2 & \text{if } |\mathbf{f}_o(\boldsymbol{x}_i) - y_i| \leq \delta \\ \delta(|\mathbf{f}_o(\boldsymbol{x}_i) - y_i| - 0.5\delta) & \text{otherwise.} \end{cases} \tag{19.5}$$

The KL-divergence loss function makes the distribution of the estimated labels close to the distribution of the true labels:

$$\underset{\theta}{\text{minimize}} \quad \text{KL}(\mathbb{P}(\mathbf{f}(\boldsymbol{x}))\|\mathbb{P}(y)) = \sum_{i=1}^{b} \mathbf{f}(\boldsymbol{x}_i) \log(\frac{\mathbf{f}(\boldsymbol{x}_i)}{y_i}). \tag{19.6}$$

19.3.3 Hinge Loss

If there are two classes, i.e. $c = 2$, the true labels can be $y_i \in \{-1, 1\}$. In this case, a possible loss function is the Hinge loss, defined as:

$$\underset{\theta}{\text{minimize}} \quad \sum_{i=1}^{b} \big[m - y_i\, \mathbf{f}_o(\boldsymbol{x}_i)\big]_+, \tag{19.7}$$

where $[\cdot]_+ := \max(\cdot, 0)$ and $m > 0$ is the margin. If the signs of the estimated and true labels are different, the loss is positive (i.e., $\big[m - y_i\, \mathbf{f}_o(\boldsymbol{x}_i)\big]_+ = m - y_i\, \mathbf{f}_o(\boldsymbol{x}_i) > 0$) which should be minimized. If the signs are the same and $|\mathbf{f}_o(\boldsymbol{x}_i)| \geq m$, then the loss function is zero (because $m - y_i\, \mathbf{f}_o(\boldsymbol{x}_i) < 0$). If the signs are the same but $|\mathbf{f}_o(\boldsymbol{x}_i)| < m$, the loss is positive and should be minimized because the estimation is correct, but not with enough margin from the incorrect estimation.

19.3.4 Cross-Entropy Loss

For any number of classes, denoted by c, it is possible to have a cross-entropy loss. For this loss, there are c neurons, rather than one neuron, at the last layer. In contrast to the MSE, MAE, Huber, and KL-divergence losses, which use the linear activation function at the last layer, cross-entropy requires the softmax or sigmoid activation function at the last layer so that the output values are between zero and one. For this loss, there are c outputs, i.e., $\mathbf{f}_o(\boldsymbol{x}_i) \in \mathbb{R}^c$ (continuous values between zero and one), and the true labels are one-hot encoded, i.e., $\boldsymbol{y}_i \in \{0, 1\}^c$. This loss is defined as:

$$\underset{\theta}{\text{minimize}} \quad -\sum_{i=1}^{b} \sum_{l=1}^{c} (\boldsymbol{y}_i)_l \log\big(\mathbf{f}_o(\boldsymbol{x}_i)_l\big), \tag{19.8}$$

where $(\boldsymbol{y}_i)_l$ and $\mathbf{f}_o(\boldsymbol{x}_i)_l$ denote the l-th element of $\boldsymbol{y}_i$ and $\mathbf{f}_o(\boldsymbol{x}_i)$, respectively. Minimizing this loss separates the classes for classification, that results in a discriminating embedding (an embedding which discriminates the classes) in the one-to-last layer [1, 44].

The reason why cross-entropy can be suitable for metric learning is theoretically justified in [1] and explained in the following. Consider the mutual information between the true labels Y and the estimated labels $\mathbf{f}_o(X)$:

$$\begin{aligned} I(\mathbf{f}_o(X); Y) &= H(\mathbf{f}_o(X)) - H(\mathbf{f}_o(X)|Y) && (19.9)\\ &= H(Y) - H(Y|\mathbf{f}_o(X)), && (19.10) \end{aligned}$$

where $H(.)$ denotes entropy. On the one hand, Eq. (19.9) has a generative view that exists in the metric learning loss functions generating embedding features. Equation (19.10), on the other hand, has a discriminative view used in the cross-entropy loss function. Therefore, the metric learning losses and the cross-entropy loss are related. It is shown in [1, Proposition 1] that the cross-entropy is an upper-bound on the metric learning losses, so its minimization for classification also provides embedding features.

It is noteworthy that another supervised loss function is the triplet loss, introduced in the next section. Triplet loss can be used for both hard labels (for classification) and soft labels (for the similarity and dissimilarity of points). The triplet loss also does not need a last classification layer; therefore, the embedding layer can be the last layer for this loss.

19.4 Metric Learning by Siamese Networks

19.4.1 Siamese and Triplet Networks

One of the important deep metric learning methods is the Siamese network, which is widely used for feature extraction. The Siamese network, originally proposed in [2], is a network consisting of several equivalent subnetworks sharing their weights. The number of subnetworks in a Siamese network can be any number, but it usually is two or three. A Siamese network with three subnetworks is also called a triplet network [22]. The weights of the subnetworks in a Siamese network are trained such that the intraclass and interclass variances are decreased and increased, respectively. In other words, similar points are pushed toward each other, while dissimilar points are pulled away from one another. Siamese networks have been used in various applications such as computer vision [41] and natural language processing [62].

19.4.2 Pairs and Triplets of Data Points

Depending on the number of subnetworks in the Siamese network, there are loss functions for training. The loss functions of Siamese networks usually require pairs or triplets of data points. Siamese networks do not use the data points one by one but instead use pairs or triplets of points out of the dataset for training the network. To make the pairs or triplets, every data point is considered as the anchor point, denoted by $\boldsymbol{x}_i^a$. Then, one of the similar points to the anchor point is taken as the positive (or neighbour) point, denoted by $\boldsymbol{x}_i^p$. One of the dissimilar points to the anchor point is also taken as the negative (or distant) point, denoted by $\boldsymbol{x}_i^n$. If the class labels are available, they can be used to find the positive and negative points. The positive point is found from one of the points in the same class as the anchor point. The negative point is found from one of the points in a different class from the anchor point's class. Another approach is to augment the anchor point using one of the augmentation methods (such as illumination change or rotation for image data) to obtain positive points for the anchor point [3, 26].

For Siamese networks with two subnetworks, pairs of anchor-positive points $\{(\boldsymbol{x}_i^a, \boldsymbol{x}_i^p)\}_{i=1}^{n_t}$ and anchor-negative points $\{(\boldsymbol{x}_i^a, \boldsymbol{x}_i^n)\}_{i=1}^{n_t}$, can be made, where n_t is the number of pairs. For Siamese networks with three subnetworks, triplets of the anchor-positive-negative points $\{(\boldsymbol{x}_i^a, \boldsymbol{x}_i^p, \boldsymbol{x}_i^n)\}_{i=1}^{n_t}$ can be made, where n_t is the number of triplets. If every point of the dataset is considered an anchor, the number of pairs/triplets is the same as the number of data points, i.e., $n_t = n$.

Various loss functions of Siamese networks use pairs or triplets of data points to push the positive point towards the anchor point and pull the negative point away from it. Doing this iteratively for all pairs or triplets will make the intraclass variances smaller and the interclass variances larger for better discrimination between classes or clusters. In this section, loss functions for training a Siamese network will be introduced.

19.4.3 Implementation of Siamese Networks

A Siamese network with two and three subnetworks is depicted in Fig. 19.2. The output of the Siamese network for input $\boldsymbol{x} \in \mathbb{R}^d$ is denoted by $\mathbf{f}(\boldsymbol{x}) \in \mathbb{R}^p$, where p is the dimensionality of embedding (or the number of neurons at the last layer of the network), which is usually much less than the dimensionality of the data, i.e., $p \ll d$. Note that the subnetworks of a Siamese network can be any fully-connected or convolutional network depending on the type of data. The network structure used for the subnetworks is usually called the backbone network.

The weights of subnetworks are shared in the sense that the values of their weights are equal. Implementation of a Siamese network can be performed in two ways:

1. It is possible to implement several subnetworks in the memory. In the training phase, every data point is fed in pairs or triplets to one of the subnetworks, and the outputs of the subnetworks are taken to be $\mathbf{f}(\boldsymbol{x}_i^a)$, $\mathbf{f}(\boldsymbol{x}_i^p)$, and $\mathbf{f}(\boldsymbol{x}_i^n)$. These are used in the loss function and to update the weights of only one of the subnetworks through backpropagation. Then, the updated weights are copied to the other subnetworks. This process is repeated for all minibatches and epochs until convergence. In the test phase, the test point $\boldsymbol{x}$ is fed to only one of the subnetworks and the output $\mathbf{f}(\boldsymbol{x})$ is taken as its embedding.
2. It is possible to implement only one subnetwork in the memory. In the training phase, the data points are fed in pairs or triplets to the subnetwork one by one, and the outputs of the subnetwork are taken to be $\mathbf{f}(\boldsymbol{x}_i^a)$, $\mathbf{f}(\boldsymbol{x}_i^p)$, and $\mathbf{f}(\boldsymbol{x}_i^n)$. These are used in the loss function to update the weights of the subnetwork by backpropagation. This is repeated for all minibatches and epochs until

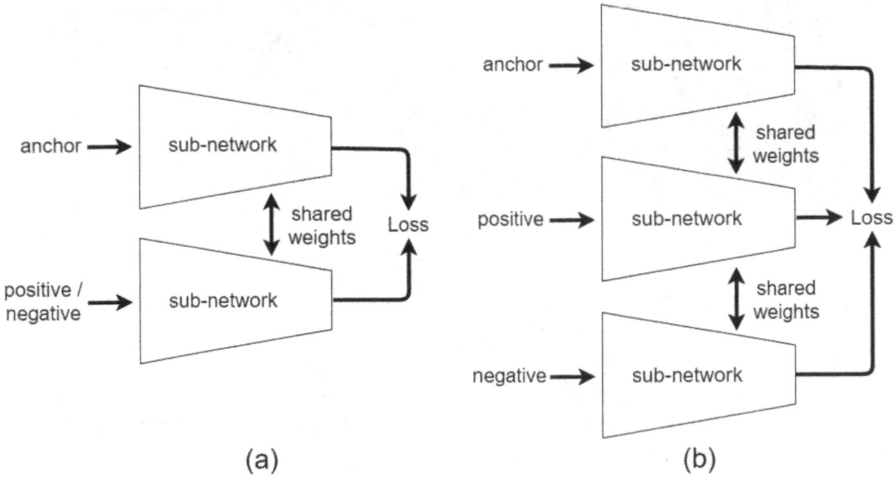

Fig. 19.2 The structure of the Siamese network with (**a**) two and (**b**) three subnetworks

convergence. In the test phase, the test point $\boldsymbol{x}$ is fed to the subnetwork and the output $\mathbf{f}(\boldsymbol{x})$ is taken as its embedding.

The advantage of the first approach is that all of the subnetworks are ready and there is no need to feed the points of pairs or triplets one by one. However, its main disadvantage is that it uses more memory. As the number of points in the pairs or triplets is small (i.e., only two or three), the second approach is recommended because it is memory-efficient.

19.4.4 Contrastive Loss

One loss function for Siamese networks is the contrastive loss that uses the anchor-positive and anchor-negative pairs of points. Suppose that in each minibatch, there are b pairs of points $\{(\boldsymbol{x}_i^1, \boldsymbol{x}_i^2)\}_{i=1}^b$, some of which are anchor-positive, and some are anchor-negative pairs. The points in an anchor-positive pair are similar, i.e., $(\boldsymbol{x}_i^1, \boldsymbol{x}_i^2) \in \mathcal{S}$, and the points in an anchor-negative pair are dissimilar, i.e., $(\boldsymbol{x}_i^1, \boldsymbol{x}_i^2) \in \mathcal{D}$, where $\mathcal{S}$ and $\mathcal{D}$ denote the similar and dissimilar sets.

19.4.4.1 Contrastive Loss

The following is defined as:

$$y_i := \begin{cases} 0 \text{ if } (\boldsymbol{x}_i^1, \boldsymbol{x}_i^2) \in \mathcal{S} \\ 1 \text{ if } (\boldsymbol{x}_i^1, \boldsymbol{x}_i^2) \in \mathcal{D}. \end{cases} \quad \forall i \in \{1, \dots, n_t\}. \tag{19.11}$$

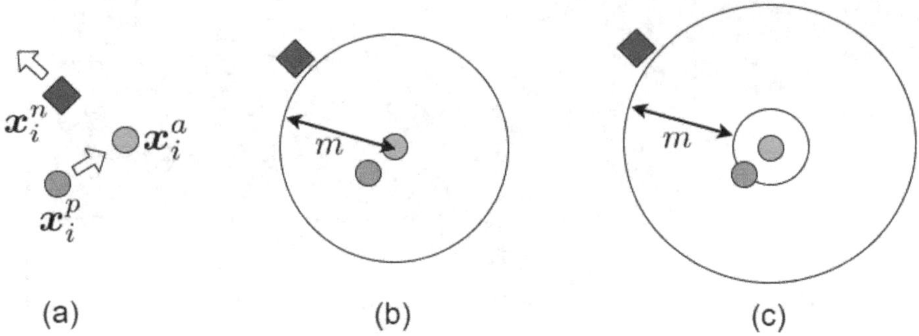

Fig. 19.3 Visualization of what contrastive and triplet losses do: (**a**) a triplet of anchor (green circle), positive (blue circle), and negative (red diamond) points, (**b**) the effect of contrastive loss making a margin between the anchor and negative point, and (**c**) the effect of triplet loss making a margin between the positive and negative points

The main contrastive loss was proposed in [15] and is:

$$\underset{\theta}{\text{minimize}} \quad \sum_{i=1}^{b} \Big((1 - y_i) d\big(\mathbf{f}(\boldsymbol{x}_i^1), \mathbf{f}(\boldsymbol{x}_i^2)\big) + y_i \big[- d\big(\mathbf{f}(\boldsymbol{x}_i^1), \mathbf{f}(\boldsymbol{x}_i^2)\big) + m \big]_+ \Big), \tag{19.12}$$

where $m > 0$ is the margin and $[.]_+ := \max(., 0)$ is the standard Hinge loss. The first term of the loss minimizes the embedding distances between similar points and the second term maximizes the embedding distances between dissimilar points. As shown in Fig. 19.3b, it tries to make the distances between similar points as small as possible and the distances between dissimilar points at least greater than a margin m (because the term inside the Hinge loss should become close to zero).

19.4.4.2 Generalized Contrastive Loss

The y_i, defined in Eq. (19.11), is used in the contrastive loss, i.e., Eq. (19.12). This variable is binary and a hard measure of similarity and dissimilarity. Rather than this hard measure, it is possible to have a soft measure of similarity and dissimilarity, denoted by ψ_i, which states how similar $\boldsymbol{x}_i^1$ and $\boldsymbol{x}_i^2$ are. This measure is between zero (completely similar) and one (completely dissimilar). It can be either given by the dataset as a hand-set measure or can be computed using any similarity measure, such as the cosine function:

$$[0, 1] \ni \psi_i := \frac{1}{2}\big(- \cos(\boldsymbol{x}_i^1, \boldsymbol{x}_i^2) + 1 \big). \tag{19.13}$$

In this case, the pairs $\{(\boldsymbol{x}_i^1, \boldsymbol{x}_i^2)\}_{i=1}^b$ need not be completely similar or dissimilar points, but they can be any two random points from the dataset with some level of similarity or dissimilarity. The generalized contrastive loss generalizes the contrastive loss using this soft measure of similarity [30]:

$$\underset{\theta}{\text{minimize}} \sum_{i=1}^{b} \Big((1 - \psi_i) d\big(\mathbf{f}(\boldsymbol{x}_i^1), \mathbf{f}(\boldsymbol{x}_i^2)\big) + \psi_i \big[- d\big(\mathbf{f}(\boldsymbol{x}_i^1), \mathbf{f}(\boldsymbol{x}_i^2)\big) + m \big]_+ \Big). \tag{19.14}$$

19.4.5 Triplet Loss

One of the losses for Siamese networks with three subnetworks is the triplet loss [41], which uses the triplets in minibatches, denoted by $\{(\boldsymbol{x}_i^a, \boldsymbol{x}_i^p, \boldsymbol{x}_i^n)\}_{i=1}^b$, and is defined as:

$$\underset{\theta}{\text{minimize}} \sum_{i=1}^{b} \Big[d\big(\mathbf{f}(\boldsymbol{x}_i^a), \mathbf{f}(\boldsymbol{x}_i^p)\big) - d\big(\mathbf{f}(\boldsymbol{x}_i^a), \mathbf{f}(\boldsymbol{x}_i^n)\big) + m \Big]_+, \tag{19.15}$$

where $m > 0$ is the margin and $[.]_+ := \max(., 0)$ is the standard Hinge loss. As demonstrated in Fig. 19.3c, because of the Hinge loss used, this loss makes the distances between dissimilar points greater than the distances between similar points by at least a margin m. In other words, there will be a distance of at least margin m between the positive and negative points. This loss desires to eventually have:

$$d\big(\mathbf{f}(\boldsymbol{x}_i^a), \mathbf{f}(\boldsymbol{x}_i^p)\big) + m \leq d\big(\mathbf{f}(\boldsymbol{x}_i^a), \mathbf{f}(\boldsymbol{x}_i^n)\big), \tag{19.16}$$

for all triplets. The triplet loss is closely related to the cost function for spectral large margin metric learning [56, 57] (see Chap. 11). It is also noteworthy that using the triplet loss as a regularization technique for cross-entropy loss has been shown to increase the robustness of the network to some adversarial attacks [34].

19.4.6 Tuplet Loss

In triplet loss, i.e. Eq. (19.15), one positive point and one negative point are used per anchor point. The tuplet loss [45] uses several negative points per anchor point. If k denotes the number of negative points per anchor point and $\boldsymbol{x}_i^{n,j}$ denotes the j-th negative point for $\boldsymbol{x}_i$, the tuplet loss is [45]:

$$\underset{\theta}{\text{minimize}} \sum_{i=1}^{b} \sum_{j=1}^{k} \Big[d\big(\mathbf{f}(\boldsymbol{x}_i^a), \mathbf{f}(\boldsymbol{x}_i^p)\big) - d\big(\mathbf{f}(\boldsymbol{x}_i^a), \mathbf{f}(\boldsymbol{x}_i^{n,j})\big) + m \Big]_+. \tag{19.17}$$

This loss function pushes multiple negative points away from the anchor point simultaneously.

19.4.7 Neighbourhood Component Analysis Loss

Neighbourhood Component Analysis (NCA) [13] was originally proposed as a spectral metric learning method (see Chap. 11). After the success of deep learning, it was used as the loss function of Siamese networks, where the negative log-likelihood is minimized using a Gaussian distribution or the softmax form within the minibatch. Assume there are c classes in every minibatch. The class index of $\boldsymbol{x}_i$ is denoted by $c(\boldsymbol{x}_i)$ and the data points of the j-th class in the minibatch are denoted by $\mathcal{X}_j$. The NCA loss is:

$$\underset{\theta}{\text{minimize}} \; -\sum_{i=1}^{b} \log\Big(\frac{\exp\big(-d\big(\mathbf{f}(\boldsymbol{x}_i^a), \mathbf{f}(\boldsymbol{x}_i^p)\big)\big)}{\sum_{j=1, j\neq c(\boldsymbol{x}_i)}^{c} \sum_{\boldsymbol{x}_j^n \in \mathcal{X}_j} \exp\big(-d\big(\mathbf{f}(\boldsymbol{x}_i^a) - \mathbf{f}(\boldsymbol{x}_j^n)\big)\big)}\Big). \tag{19.18}$$

The numerator minimizes the distances between similar points and the denominator maximizes the distances between dissimilar points.

19.4.8 Proxy Neighbourhood Component Analysis Loss

The computation of terms, especially the normalization factor in the denominator, is time- and memory-consuming in the NCA loss function (see Eq. (19.18)). Proxy-NCA loss functions define some proxy points in the embedding space of the network and use them in the NCA loss to accelerate computation and make the process more memory-efficient [35]. The proxies are representatives of the classes in the embedding space and can be defined in various ways. The simplest way is to define the proxy of every class as the mean of the embedded points of that class. Of course, new minibatches are used during training. It is possible to accumulate the embedded points of the minibatches and update the proxies after training the network by every minibatch. Another approach for defining proxies is to cluster the embedded points into c clusters (e.g., by K-means) and use the centroid of clusters.

Let the set of proxies be denoted by $\mathcal{P}$, whose cardinality is the number of classes, i.e., c. Every embedded point is assigned to one of the proxies by Movshovitz-Attias et al. [35]:

$$\Pi(\mathbf{f}(\boldsymbol{x}_i)) := \arg\min_{\boldsymbol{\pi}\in\mathcal{P}} \|\mathbf{f}(\boldsymbol{x}_i) - \boldsymbol{\pi}\|_2^2, \tag{19.19}$$

or every point can be assigned to the proxy of its own class. Let π_j denote the proxy associated with the j-th class. The Proxy-NCA loss is the NCA loss, Eq. (19.18), but by using proxies [35]:

$$\underset{\theta}{\text{minimize}} \; -\sum_{i=1}^{b} \log\Big(\frac{\exp\big(-d\big(\mathbf{f}(\boldsymbol{x}_i^a), \Pi(\mathbf{f}(\boldsymbol{x}_i^p))\big)\big)}{\sum_{j=1, j\neq c(\boldsymbol{x}_i)}^{c} \exp\big(-d\big(\mathbf{f}(\boldsymbol{x}_i^a) - \boldsymbol{\pi}_j\big)\big)}\Big). \tag{19.20}$$

It is demonstrated in [35] that the Proxy-NCA loss, i.e., Eq. (19.20), is an upper bound on the NCA loss, i.e. Eq. (19.18). Therefore, its minimization also achieves the goal of NCA. Comparing Eqs. (19.18) and (19.20) demonstrates that Proxy-NCA is faster and more efficient than NCA because it uses only the proxies of the negative classes, rather than using all of the negative points in the minibatch. Proxy-NCA has also been used in feature extraction from medical images [47]. It is noteworthy that it is possible to incorporate temperature scaling [20] in the Proxy-NCA loss. The obtained loss is named Proxy-NCA++ [46] and is defined as:

$$\underset{\theta}{\text{minimize}} \; -\sum_{i=1}^{b} \log\Big(\frac{\exp\big(-d\big(\mathbf{f}(\boldsymbol{x}_i^a), \Pi(\mathbf{f}(\boldsymbol{x}_i^p))\big) \times \frac{1}{\tau}\big)}{\sum_{j=1, j\neq c(\boldsymbol{x}_i)}^{c} \exp\big(-d\big(\mathbf{f}(\boldsymbol{x}_i^a) - \boldsymbol{\pi}_j\big) \times \frac{1}{\tau}\big)}\Big), \tag{19.21}$$

where $\tau > 0$ is the temperature which is a hyper-parameter.

19.4.9 Softmax Triplet Loss

Consider a minibatch containing points from c classes, where $c(\boldsymbol{x}_i)$ is the class index of $\boldsymbol{x}_i$ and $\mathcal{X}_j$ denotes the points of the j-th class in the minibatch. The softmax function or the Gaussian distribution can be used for the probability that point $\boldsymbol{x}_i$ takes $\boldsymbol{x}_j$ as its neighbour. Similar to Eq. (19.18), it is possible to have the softmax function used in NCA [13]:

$$p_{ij} := \frac{\exp\big(-d\big(\mathbf{f}(\boldsymbol{x}_i), \mathbf{f}(\boldsymbol{x}_j)\big)\big)}{\sum_{k\neq i, k=1}^{b} \exp\big(-d\big(\mathbf{f}(\boldsymbol{x}_i), \mathbf{f}(\boldsymbol{x}_k)\big)\big)}, \quad j \neq i. \tag{19.22}$$

Another approach for the softmax form is to use the inner product in the exponent [63]:

$$p_{ij} := \frac{\exp\big(\mathbf{f}(\boldsymbol{x}_i)^\top \mathbf{f}(\boldsymbol{x}_j)\big)}{\sum_{k=1, k\neq i}^{b} \exp\big(\mathbf{f}(\boldsymbol{x}_i)^\top \mathbf{f}(\boldsymbol{x}_k)\big)}, \quad j \neq i. \tag{19.23}$$

The loss function for training the network can be the negative log-likelihood, which can be called the softmax triplet loss [63]:

$$\underset{\theta}{\text{minimize}} \quad -\sum_{i=1}^{b}\Big(\sum_{\boldsymbol{x}_j \in \mathcal{X}_{c(\boldsymbol{x}_i)}} \log(p_{ij}) - \sum_{\boldsymbol{x}_j \notin \mathcal{X}_{c(\boldsymbol{x}_i)}} \log(1 - p_{ij})\Big). \tag{19.24}$$

This decreases and increases the distances between similar points and dissimilar points, respectively.

19.4.10 Triplet Global Loss

The triplet global loss [29] uses the mean and variance of the anchor-positive pairs and anchor-negative pairs. It is defined as follows:

$$\underset{\theta}{\text{minimize}} \quad (\sigma_p^2 + \sigma_n^2) + \lambda\,[\mu_p - \mu_n + m]_+, \tag{19.25}$$

where $\lambda > 0$ is the regularization parameter, $m > 0$ is the margin, and the means of the pairs are:

$$\mu_p := \frac{1}{b}\sum_{i=1}^{b} d\big(\mathbf{f}(\boldsymbol{x}_i^a), \mathbf{f}(\boldsymbol{x}_i^p)\big),$$

$$\mu_n := \frac{1}{b}\sum_{i=1}^{b} d\big(\mathbf{f}(\boldsymbol{x}_i^a), \mathbf{f}(\boldsymbol{x}_i^n)\big),$$

and the variances of pairs are:

$$\sigma_p^2 := \frac{1}{b}\sum_{i=1}^{b} \Big(d\big(\mathbf{f}(\boldsymbol{x}_i^a), \mathbf{f}(\boldsymbol{x}_i^p)\big) - \mu_p\Big)^2,$$

$$\sigma_n^2 := \frac{1}{b}\sum_{i=1}^{b} \Big(d\big(\mathbf{f}(\boldsymbol{x}_i^a), \mathbf{f}(\boldsymbol{x}_i^n)\big) - \mu_n\Big)^2.$$

The first term of this loss minimizes the variances of the anchor-positive and anchor-negative pairs. The second term, however, discriminates anchor-positive pairs from anchor-negative pairs. Therefore, the negative points are separated from the positive points.

19.4.11 Angular Loss

For a triplet $(\boldsymbol{x}_i^a, \boldsymbol{x}_i^p, \boldsymbol{x}_i^n)$, consider a triangle whose vertices are the anchor, positive, and negative points. To satisfy Eq. (19.16) in the triplet loss, the angle at vertex $\boldsymbol{x}_i^n$ should be small so that edge $d\big(\mathbf{f}(\boldsymbol{x}_i^a), \mathbf{f}(\boldsymbol{x}_i^n)\big)$ becomes larger than the edge $d\big(\mathbf{f}(\boldsymbol{x}_i^a), \mathbf{f}(\boldsymbol{x}_i^p)\big)$. Therefore, an upper bound $\alpha > 0$ is needed on the angle at the vertex $\boldsymbol{x}_i^n$. If $\boldsymbol{x}_i^c := (\boldsymbol{x}_i^a + \boldsymbol{x}_i^p)/2$, the angular loss is defined as [51]:

$$\underset{\theta}{\text{minimize}} \quad \sum_{i=1}^{b} \Big[d\big(\mathbf{f}(\boldsymbol{x}_i^a), \mathbf{f}(\boldsymbol{x}_i^p)\big) - 4\tan^2\big(\alpha\, d\big(\mathbf{f}(\boldsymbol{x}_i^a), \mathbf{f}(\boldsymbol{x}_i^c)\big)\big) \Big]_+. \tag{19.26}$$

This loss reduces the distance between the anchor and the positive points and increases the distance between the anchor and $\boldsymbol{x}_i^c$ and the upper bound α. This increases the distance of the anchor and the negative points for discriminating between dissimilar points.

19.4.12 SoftTriple Loss

If the points are normalized to have unit lengths, Eq. (19.16) can be restated by using the inner products:

$$\mathbf{f}(\boldsymbol{x}_i^a)^\top \mathbf{f}(\boldsymbol{x}_i^n) + m \leq \mathbf{f}(\boldsymbol{x}_i^a)^\top \mathbf{f}(\boldsymbol{x}_i^p), \tag{19.27}$$

whose margin is not exactly equal to the margin in Eq. (19.16). Consider a Siamese network whose last layer's weights are $\{\boldsymbol{w}_l \in \mathbb{R}^p\}_{l=1}^c$, where p is the dimensionality of the one-to-last layer and c is the number of classes and the number of output neurons. Consider k centers for the embedding of every class; therefore, $\boldsymbol{w}_l^j \in \mathbb{R}^p$ is defined as $\boldsymbol{w}_l$ for its j-th center. It is shown in [38] that softmax loss results in Eq. (19.27). Therefore, the SoftTriple loss can be used for training a Siamese network [38]:

$$\underset{\theta}{\text{minimize}} \quad -\sum_{i=1}^{b} \log\Big(\frac{\exp(\lambda(s_{i,y_i} - \delta))}{\exp(\lambda(s_{i,y_i} - \delta)) + \sum_{l=1, l\neq y_i}^{c} \exp(\lambda s_{i,l})}\Big), \tag{19.28}$$

where $\lambda > 0$ and $\delta > 0$ are hyperparameters, y_i is the label of $\boldsymbol{x}_i$, and:

$$s_{i,l} := \sum_{j=1}^{k} \frac{\exp\big(\mathbf{f}(\boldsymbol{x}_i)^\top \boldsymbol{w}_l^j\big)}{\sum_{t=1}^{k} \exp\big(\mathbf{f}(\boldsymbol{x}_i)^\top \boldsymbol{w}_l^t\big)} \mathbf{f}(\boldsymbol{x}_i)^\top \boldsymbol{w}_l^k.$$

This loss increases and decreases the intraclass and interclass distances, respectively.

19.4.13 Fisher Siamese Losses

Fisher Discriminant Analysis (FDA) (see Chap. 6) decreases the intraclass variance and increases the interclass variance by maximizing the Fisher criterion. This idea is very similar to the idea of loss functions for Siamese networks. Therefore, the methods of FDA and Siamese loss functions can be combined.

Consider a Siamese network whose last layer is denoted by the projection matrix $\boldsymbol{U}$. Consider the features of the one-to-last layer in the minibatch. The covariance matrices of similar points and dissimilar points (one-to-last layer features) in the minibatch are denoted by $\boldsymbol{S}_W$ and $\boldsymbol{S}_B$. These covariances become $\boldsymbol{U}^\top \boldsymbol{S}_W \boldsymbol{U}$ and $\boldsymbol{U}^\top \boldsymbol{S}_B \boldsymbol{U}$, respectively, after the later layer's projection because of the quadratic characteristic of covariance. As in FDA, the Fisher criterion is maximized or, equivalently, the negative Fisher criterion is minimized:

$$\underset{\boldsymbol{U}}{\text{minimize}} \quad \mathbf{tr}(\boldsymbol{U}^\top \boldsymbol{S}_W \boldsymbol{U}) - \mathbf{tr}(\boldsymbol{U}^\top \boldsymbol{S}_B \boldsymbol{U}).$$

This problem is ill-posed because it increases the total covariance of the embedded data to increase the term $\mathbf{tr}(\boldsymbol{U}^\top \boldsymbol{S}_B \boldsymbol{U})$. Therefore, minimization of the total covariance is added as the regularization term:

$$\underset{\boldsymbol{U}}{\text{minimize}} \quad \mathbf{tr}(\boldsymbol{U}^\top \boldsymbol{S}_W \boldsymbol{U}) - \mathbf{tr}(\boldsymbol{U}^\top \boldsymbol{S}_B \boldsymbol{U}) + \epsilon \mathbf{tr}(\boldsymbol{U}^\top \boldsymbol{S}_T \boldsymbol{U}),$$

where $\epsilon \in (0, 1)$ is the regularization parameter and $\boldsymbol{S}_T$ is the covariance of all points in the minibatch at the one-to-last layer. The total scatter can be written as the summation of $\boldsymbol{S}_W$ and $\boldsymbol{S}_B$; therefore:

$$\begin{aligned}
&\mathbf{tr}(\boldsymbol{U}^\top \boldsymbol{S}_W \boldsymbol{U}) - \mathbf{tr}(\boldsymbol{U}^\top \boldsymbol{S}_B \boldsymbol{U}) + \epsilon \mathbf{tr}(\boldsymbol{U}^\top \boldsymbol{S}_T \boldsymbol{U}) \\
&= \mathbf{tr}\big(\boldsymbol{U}^\top (\boldsymbol{S}_W - \boldsymbol{S}_W + \epsilon \boldsymbol{S}_W + \epsilon \boldsymbol{S}_B) \boldsymbol{U}\big) = (2 - \lambda)\mathbf{tr}(\boldsymbol{U}^\top \boldsymbol{S}_W \boldsymbol{U}) - \lambda \mathbf{tr}(\boldsymbol{U}^\top \boldsymbol{S}_B \boldsymbol{U}),
\end{aligned}$$

where $\lambda := 1 - \epsilon$. Inspired by Eq. (19.15), it is possible to have the following loss, named the Fisher discriminant triplet loss [12]:

$$\underset{\theta}{\text{minimize}} \quad \Big[(2 - \lambda)\mathbf{tr}(\boldsymbol{U}^\top \boldsymbol{S}_W \boldsymbol{U}) - \lambda \mathbf{tr}(\boldsymbol{U}^\top \boldsymbol{S}_B \boldsymbol{U}) + m\Big]_+, \tag{19.29}$$

where $m > 0$ is the margin. Backpropagating the error of this loss can update both $\boldsymbol{U}$ and other layers of the network. Note that the summation over the minibatch is integrated into the computation of covariance matrices $\boldsymbol{S}_W$ and $\boldsymbol{S}_B$. Inspired by Eq. (19.12), it is possible to obtain the Fisher discriminant contrastive loss [12]:

$$\underset{\theta}{\text{minimize}} \quad (2 - \lambda)\mathbf{tr}(\boldsymbol{U}^\top \boldsymbol{S}_W \boldsymbol{U}) + \big[- \lambda \mathbf{tr}(\boldsymbol{U}^\top \boldsymbol{S}_B \boldsymbol{U}) + m\big]_+. \tag{19.30}$$

Note that the variable y_i used in the contrastive loss (see Eq. (19.11)) is already used in the computation of the covariances $\boldsymbol{S}_W$ and $\boldsymbol{S}_B$. There are additional loss functions inspired by Fisher discriminant analysis, but they are not used for Siamese networks. These methods are introduced in Sect. 19.5.

19.4.14 Deep Adversarial Metric Learning

In deep adversarial metric learning [7], negative points are generated in an adversarial learning approach (see Chap. 21 for adversarial learning). In this method, there is a generator $G(.)$, which tries to generate negative points to fool the metric learning. Using triplet inputs $\{(\boldsymbol{x}_i^a, \boldsymbol{x}_i^p, \boldsymbol{x}_i^n)\}_{i=1}^b$, the loss function of the generator is [7]:

$$\begin{aligned}\mathcal{L}_G := \sum_{i=1}^{b} \Big(&\|G(\boldsymbol{x}_i^a, \boldsymbol{x}_i^p, \boldsymbol{x}_i^n) - \boldsymbol{x}_i^a\|_2^2 + \lambda_1 \|G(\boldsymbol{x}_i^a, \boldsymbol{x}_i^p, \boldsymbol{x}_i^n) - \boldsymbol{x}_i^n\|_2^2 \\ &+ \lambda_2 \big[d(\mathbf{f}(\boldsymbol{x}_i^a), \mathbf{f}(G(\boldsymbol{x}_i^a, \boldsymbol{x}_i^p, \boldsymbol{x}_i^n))) - d(\mathbf{f}(\boldsymbol{x}_i^a), \mathbf{f}(\boldsymbol{x}_i^p)) + m\big]_+ \Big),\end{aligned} \tag{19.31}$$

where $\lambda_1 > 0$ and $\lambda_2 > 0$ are the regularization parameters. This loss makes the generated negative point close to the real negative point (to be a negative sample) and the anchor point (for fooling metric learning adversarially). The Hinge loss makes the embedding of the generated negative point different from the embeddings of the anchor and positive points, so it also acts like a real negative. If $\mathcal{L}_M$ denotes any loss function for the Siamese network, such as the triplet loss, the total loss function in deep adversarial metric learning minimizes $\mathcal{L}_G + \lambda_3 \mathcal{L}_M$, where $\lambda_3 > 0$ is the regularization parameter [7]. It is noteworthy that another adversarial metric learning method exists that is not for Siamese networks but for cross-modal data [59].

19.4.15 Triplet Mining

In every minibatch containing data points from c classes, triplets of data points can be selected and used in different ways. For example, all similar and dissimilar points can be used for every anchor point as positive and negative points, respectively. Another approach is to only use some of the similar and dissimilar points within the minibatch. These approaches for selecting and using triplets are called triplet mining [43]. Triplet mining methods can be used for the triplet loss, i.e., Eq. (19.15). Suppose b is the minibatch size, $c(\boldsymbol{x}_i)$ is the class index of $\boldsymbol{x}_i$, $\mathcal{X}_j$ denotes the points of the j-th class in the minibatch, and $\mathcal{X}$ denotes the data points in the minibatch. The following reviews some of the most important triplet mining methods.

19.4.15.1 Batch-All

Batch-all triplet mining [4] considers every point in the minibatch as an anchor point. All points in the minibatch that are in the same class as the anchor point are used as positive points. All points in the minibatch that are in a different class from the class of anchor points are used as negative points:

$$\underset{\theta}{\text{minimize}} \sum_{i=1}^{b} \sum_{\boldsymbol{x}_j \in \mathcal{X}_{c(\boldsymbol{x}_i)}} \sum_{\boldsymbol{x}_k \in \mathcal{X} \setminus \mathcal{X}_{c(\boldsymbol{x}_i)}} \Big[d\big(\mathbf{f}(\boldsymbol{x}_i), \mathbf{f}(\boldsymbol{x}_j)\big) - d\big(\mathbf{f}(\boldsymbol{x}_i), \mathbf{f}(\boldsymbol{x}_k)\big) + m\Big]_+. \tag{19.32}$$

Batch-all mining makes use of all data points in the minibatch to utilize all available information.

19.4.15.2 Batch-Hard

Batch-hard triplet mining [19] considers every point in the minibatch as an anchor point. The hardest positive, which is the farthest point from the anchor point in the same class, is used as the positive point. The hardest negative, which is the closest point to the anchor point from another class, is used as the negative point:

$$\underset{\theta}{\text{minimize}} \sum_{i=1}^{b} \Big[\max_{\boldsymbol{x}_j \in \mathcal{X}_{c(\boldsymbol{x}_i)}} d\big(\mathbf{f}(\boldsymbol{x}_i), \mathbf{f}(\boldsymbol{x}_j)\big) - \min_{\boldsymbol{x}_k \in \mathcal{X} \setminus \mathcal{X}_{c(\boldsymbol{x}_i)}} d\big(\mathbf{f}(\boldsymbol{x}_i), \mathbf{f}(\boldsymbol{x}_k)\big) + m\Big]_+. \tag{19.33}$$

Bath-hard mining uses the hardest points so that the network learns the hardest cases. By learning the hardest cases, other cases are expected to be learned easily and properly. Learning the hardest cases can also be justified by opposition-based learning [48]. Batch-hard mining has been used in many applications, such as person reidentification [53].

19.4.15.3 Batch-Semi-Hard

Batch-semihard triplet mining [41] considers every point in the minibatch as an anchor point. All points in the minibatch that are in the same class as the anchor point are used as positive points. The hardest negative (closest to the anchor point from another class), which is farther than the positive point, is used as the negative point:

$$\underset{\theta}{\text{minimize}} \sum_{i=1}^{b} \sum_{\boldsymbol{x}_j \in \mathcal{X}_{c(\boldsymbol{x}_i)}} \Big[d\big(\mathbf{f}(\boldsymbol{x}_i), \mathbf{f}(\boldsymbol{x}_j)\big)$$

$$- \min_{\boldsymbol{x}_k \in \mathcal{X} \setminus \mathcal{X}_{c(\boldsymbol{x}_i)}} \big\{ d\big(\mathbf{f}(\boldsymbol{x}_i), \mathbf{f}(\boldsymbol{x}_k)\big) \,|\, d\big(\mathbf{f}(\boldsymbol{x}_i), \mathbf{f}(\boldsymbol{x}_k)\big) > d\big(\mathbf{f}(\boldsymbol{x}_i), \mathbf{f}(\boldsymbol{x}_j)\big) \big\} + m \Big]_+ . \tag{19.34}$$

19.4.15.4 Easy-Positive

Easy-positive triplet mining [61] considers every point in the minibatch as an anchor point. The easiest positive (closest to the anchor point from the same class) is used as the positive point. All points in the minibatch that are in a different class from the class of anchor points are used as negative points:

$$\underset{\theta}{\text{minimize}} \sum_{i=1}^{b} \sum_{\boldsymbol{x}_k \in \mathcal{X} \setminus \mathcal{X}_{c(\boldsymbol{x}_i)}} \Big[\min_{\boldsymbol{x}_j \in \mathcal{X}_{c(\boldsymbol{x}_i)}} d\big(\mathbf{f}(\boldsymbol{x}_i), \mathbf{f}(\boldsymbol{x}_j)\big) - d\big(\mathbf{f}(\boldsymbol{x}_i), \mathbf{f}(\boldsymbol{x}_k)\big) + m \Big]_+ . \tag{19.35}$$

This triplet mining approach can be used in the NCA loss function, such as in Eq. (19.23). For example [61]:

$$\underset{\theta}{\text{minimize}} \sum_{i=1}^{b} \Big(\frac{\min_{\boldsymbol{x}_j \in \mathcal{X}_{c(\boldsymbol{x}_i)}} \exp\big(\mathbf{f}(\boldsymbol{x}_i)^\top \mathbf{f}(\boldsymbol{x}_j)\big)}{\min_{\boldsymbol{x}_j \in \mathcal{X}_{c(\boldsymbol{x}_i)}} \exp\big(\mathbf{f}(\boldsymbol{x}_i)^\top \mathbf{f}(\boldsymbol{x}_j)\big) + \sum_{\boldsymbol{x}_k \in \mathcal{X} \setminus \mathcal{X}_{c(\boldsymbol{x}_i)}} \exp\big(\mathbf{f}(\boldsymbol{x}_i)^\top \mathbf{f}(\boldsymbol{x}_k)\big)} \Big), \tag{19.36}$$

where the embeddings for all points of the minibatch are normalized to have a length of one.

19.4.15.5 Lifted Embedding Loss

The lifted embedding loss [37] is related to the anchor-positive distance and the smallest (hardest) anchor-negative distance:

$$\underset{\theta}{\text{minimize}} \sum_{i=1}^{b} \sum_{\boldsymbol{x}_j \in \mathcal{X}_{c(\boldsymbol{x}_i)}} \Big(\Big[d(\mathbf{f}(\boldsymbol{x}_i), \mathbf{f}(\boldsymbol{x}_j))$$

$$+ \max\Big(\max_{\boldsymbol{x}_k \in \mathcal{X} \setminus \mathcal{X}_{c(\boldsymbol{x}_i)}} \{m - d(\mathbf{f}(\boldsymbol{x}_i), \mathbf{f}(\boldsymbol{x}_k))\}, \max_{\boldsymbol{x}_l \in \mathcal{X} \setminus \mathcal{X}_{c(\boldsymbol{x}_j)}} \{m - d(\mathbf{f}(\boldsymbol{x}_j), \mathbf{f}(\boldsymbol{x}_l))\} \Big) \Big]_+ \Big)^2, \tag{19.37}$$

This loss uses triplet mining because of the extreme distances. Alternatively, another version of this loss function uses logarithm and exponential operators [37]:

$$\underset{\theta}{\text{minimize}} \quad \sum_{i=1}^{b} \sum_{\boldsymbol{x}_j \in \mathcal{X}_{c(\boldsymbol{x}_i)}} \Big(\Big[d(\mathbf{f}(\boldsymbol{x}_i), \mathbf{f}(\boldsymbol{x}_j))$$

$$+ \log\Big(\sum_{\boldsymbol{x}_k \in \mathcal{X} \setminus \mathcal{X}_{c(\boldsymbol{x}_i)}} \exp\big(m - d(\mathbf{f}(\boldsymbol{x}_i), \mathbf{f}(\boldsymbol{x}_k))\big) + \sum_{\boldsymbol{x}_l \in \mathcal{X} \setminus \mathcal{X}_{c(\boldsymbol{x}_j)}} \exp\big(m - d(\mathbf{f}(\boldsymbol{x}_j), \mathbf{f}(\boldsymbol{x}_l))\big)\Big)\Big]_+\Big)^2. \tag{19.38}$$

19.4.15.6 Hard Mining Center-Triplet Loss

Let the minibatch contain data points from c classes. Hard mining center–triplet loss [33] considers the mean of every class as an anchor point. The hardest (farthest) positive point and the hardest (closest) negative point are used in this loss as [33]:

$$\underset{\theta}{\text{minimize}} \sum_{l=1}^{c} \Big[\max_{\boldsymbol{x}_j \in \mathcal{X}_{c(\bar{\boldsymbol{x}}^l)}} d\big(\mathbf{f}(\bar{\boldsymbol{x}}^l), \mathbf{f}(\boldsymbol{x}_j)\big) - \min_{\boldsymbol{x}_k \in \mathcal{X} \setminus \mathcal{X}_{c(\bar{\boldsymbol{x}}^l)}} d\big(\mathbf{f}(\bar{\boldsymbol{x}}^l), \mathbf{f}(\boldsymbol{x}_k)\big) + m \Big]_+. \tag{19.39}$$

where $\bar{\boldsymbol{x}}^l$ denotes the mean of the l-th class.

19.4.15.7 Triplet Loss with Cross-Batch Memory

A version of triplet loss can be [54]:

$$\underset{\theta}{\text{minimize}} \sum_{i=1}^{b} \Big(- \sum_{\boldsymbol{x}_j \in \mathcal{X}_{c(\boldsymbol{x}_i)}} \mathbf{f}(\boldsymbol{x}_i)^\top \mathbf{f}(\boldsymbol{x}_j) + \sum_{\boldsymbol{x}_k \in \mathcal{X} \setminus \mathcal{X}_{c(\boldsymbol{x}_i)}} \mathbf{f}(\boldsymbol{x}_i)^\top \mathbf{f}(\boldsymbol{x}_k) \Big). \tag{19.40}$$

This triplet loss can use a cross-batch memory, where a few of the latest minibatches are accumulated. Every incoming minibatch updates the memory. Let the capacity of the memory be w points and the minibatch size be b. Let $\widetilde{\boldsymbol{x}}_i$ denote the i-th data point in the memory. The triplet loss with cross-batch memory is defined as [54]:

$$\underset{\theta}{\text{minimize}} \sum_{i=1}^{b} \Big(- \sum_{\widetilde{\boldsymbol{x}}_j \in \mathcal{X}_{c(\boldsymbol{x}_i)}} \mathbf{f}(\boldsymbol{x}_i)^\top \mathbf{f}(\widetilde{\boldsymbol{x}}_j) + \sum_{\widetilde{\boldsymbol{x}}_k \in \mathcal{X} \setminus \mathcal{X}_{c(\boldsymbol{x}_i)}} \mathbf{f}(\boldsymbol{x}_i)^\top \mathbf{f}(\widetilde{\boldsymbol{x}}_k) \Big), \tag{19.41}$$

which takes the positive and negative points from the memory, rather than from the incoming minibatch.

19.4.16 Triplet Sampling

Rather than using the extreme (hardest or easiest) positive and negative points [43], positive and negative points can be sampled from the points in the minibatch or from some distributions. There are several sampling approaches for the positive and negative points [9]:

- Sampled by extreme distances of points (this approach was introduced in Sect. 19.4.15),
- Sampled randomly from classes,
- Sampled by distribution but from existing points,
- Sampled stochastically from distributions of classes.

These approaches are used for triplet sampling. The first, second, and third approaches sample the positive and negative points from the set of points in the minibatch. This type of sampling is called survey sampling [11]. The third and fourth approaches stochastically sample points from distributions. These triplet sampling approaches are used in the following methods.

19.4.16.1 Distance Weighted Sampling

Distance weighted sampling [58] samples by distribution from existing points. The distribution of the pairwise distances is proportional to [58]:

$$\mathbb{P}\big(d(\mathbf{f}(\boldsymbol{x}_i), \mathbf{f}(\boldsymbol{x}_j))\big) \sim \big(d(\mathbf{f}(\boldsymbol{x}_i), \mathbf{f}(\boldsymbol{x}_j))\big)^{p-2} \times \Big(1 - 0.25\big(d(\mathbf{f}(\boldsymbol{x}_i), \mathbf{f}(\boldsymbol{x}_j))\big)^2\Big)^{(b-3)/2},$$

where b is the number of points in the minibatch and p is the dimensionality of the embedding space (i.e., the number of neurons in the last layer of the Siamese network). In every minibatch, every point is considered once as an anchor point. For an anchor point, all points of the minibatch, which are in a different class, are considered as candidates for the negative point. A negative point, denoted by $\boldsymbol{x}_*^n$, is sampled from these candidates [58]:

$$\boldsymbol{x}_*^n \sim \min\Big(\lambda, \mathbb{P}^{-1}\big(d(\mathbf{f}(\boldsymbol{x}_i), \mathbf{f}(\boldsymbol{x}_j))\big)\Big), \quad \forall j \neq i,$$

where $\lambda > 0$ is a hyperparameter to ensure that all candidates have a chance to be chosen. This sampling is performed for every minibatch. The loss function in distance weighted sampling is [58]:

$$\underset{\theta}{\text{minimize}} \sum_{i=1}^{b} \sum_{\boldsymbol{x}_j \in \mathcal{X}_{c(\boldsymbol{x}_i)}} \Big[d\big(\mathbf{f}(\boldsymbol{x}_i), \mathbf{f}(\boldsymbol{x}_j)\big) - d\big(\mathbf{f}(\boldsymbol{x}_i), \mathbf{f}(\boldsymbol{x}_*^n)\big) + m\Big]_+. \tag{19.42}$$

19.4.16.2 Sampling by Bayesian Updating Theorem

Triplets can be sampled from the distributions of classes, which is the fourth approach of sampling mentioned above. One method for this sampling is using the Bayesian updating theorem [42], which updates the posterior by Bayes' rule using new data. In this method, p-dimensional Gaussian distributions are assumed for every class in the embedding space, where p is the dimensionality of the embedding space. The embedded points are accumulated for every class when the new minibatches are introduced to the network. The distributions of classes are updated based on both the existing points available thus far and the new data points. It can be demonstrated that the posterior of the mean and covariance of a Gaussian distribution is a normal inverse Wishart distribution [36]. The mean and covariance of a Gaussian distribution have a generalized Student-t distribution and inverse Wishart distribution, respectively [36]. Let the available data have a sample size of n_0, a mean of $\boldsymbol{\mu}^0$, and a covariance of $\boldsymbol{\Sigma}^0$. Additionally, let the incoming data have a sample size of n', a mean of $\boldsymbol{\mu}'$, and a covariance of $\boldsymbol{\Sigma}'$. The mean and covariance are updated by the expected values of these distributions [42]:

$$\boldsymbol{\mu}^0 \leftarrow \mathbb{E}(\boldsymbol{\mu} \,|\, \boldsymbol{x}^0) = \frac{n'\boldsymbol{\mu}' + n_0\boldsymbol{\mu}^0}{n' + n_0},$$

$$\boldsymbol{\Sigma}^0 \leftarrow \mathbb{E}(\boldsymbol{\Sigma} \,|\, \boldsymbol{x}^0) = \frac{\boldsymbol{\Upsilon}^{-1}}{n'+n_0-p-1}, \quad \forall n'+n_0 > p+1,$$

where:

$$\mathbb{R}^{d\times d} \ni \boldsymbol{\Upsilon} := n'\boldsymbol{\Sigma}' + n_0\boldsymbol{\Sigma}^0 + \frac{n_1' n_0}{n_1' + n_0}(\boldsymbol{\mu}^0 - \boldsymbol{\mu}')(\boldsymbol{\mu}^0 - \boldsymbol{\mu}')^\top.$$

The updated mean and covariance are used for Gaussian distributions of the classes. Then, triplets are sampled from the distributions of the classes rather than from the points in the minibatch. Every point of the new minibatch is considered an anchor point and a positive point is sampled from the distribution of the same class. $c-1$ negative points are sampled from the distributions of $c-1$ other classes. If this triplet sampling procedure is used with triplet and contrastive loss functions, the approach is named Bayesian Updating with Triplet loss (BUT) and Bayesian Updating with NCA loss (BUNCA) [42].

19.4.16.3 Hard Negative Sampling

Let the anchor, positive, and negative points be denoted by $\boldsymbol{x}^a$, $\boldsymbol{x}^p$, and $\boldsymbol{x}^n$, respectively. Consider the following distributions for the negative and positive points [39]:

$$\begin{aligned}
&\mathbb{P}(\boldsymbol{x}^n) \propto \alpha \mathbb{P}_n(\boldsymbol{x}^n) + (1-\alpha)\mathbb{P}_p(\boldsymbol{x}^n), \\
&\mathbb{P}_n(\boldsymbol{x}) \propto \exp\big(\beta \mathbf{f}(\boldsymbol{x}^a)^\top \mathbf{f}(\boldsymbol{x})\big)\, \mathbb{P}(\boldsymbol{x} | c(\boldsymbol{x}) \neq c(\boldsymbol{x}^a)), \\
&\mathbb{P}_p(\boldsymbol{x}) \propto \exp\big(\beta \mathbf{f}(\boldsymbol{x}^a)^\top \mathbf{f}(\boldsymbol{x})\big)\, \mathbb{P}(\boldsymbol{x} | c(\boldsymbol{x}) = c(\boldsymbol{x}^a)),
\end{aligned} \tag{19.43}$$

where $\alpha \in (0, 1)$ is a hyperparameter. The loss function with hard negative sampling is [39]:

$$\begin{aligned}
\underset{\theta}{\text{minimize}} \quad & -\sum_{i=1}^{b} \mathbb{E}_{\boldsymbol{x}^p \sim \mathbb{P}_p(\boldsymbol{x})} \log \Bigg(\exp\big(\mathbf{f}(\boldsymbol{x}_i^a)^\top \mathbf{f}(\boldsymbol{x}^p)\big) \\
& \Big(\exp\big(\mathbf{f}(\boldsymbol{x}_i^a)^\top \mathbf{f}(\boldsymbol{x}^p)\big) + \mathbb{E}_{\boldsymbol{x}^n \sim \mathbb{P}(\boldsymbol{x}^n)} \big[\exp\big(\mathbf{f}(\boldsymbol{x}_i^a)^\top \mathbf{f}(\boldsymbol{x}^n)\big) \big] \Big)^{-1} \Bigg),
\end{aligned} \tag{19.44}$$

where the positive and negative points are sampled from positive and negative distributions, defined in Eq. (19.43). The expectations can be estimated using the Monte Carlo approximation [11]. This triplet sampling samples stochastically from the distributions of classes.

19.5 Deep Discriminant Analysis Metric Learning

Deep discriminant analysis metric learning methods use the idea of Fisher discriminant analysis (see Chap. 6) in deep learning to learn an embedding space that separates classes. Some of these methods are deep probabilistic discriminant analysis [31], discriminant analysis with virtual samples [27], Fisher Siamese losses [12], and deep Fisher discriminant analysis [5, 6]. The Fisher Siamese losses were already introduced in Sect. 19.4.13.

19.5.1 Deep Probabilistic Discriminant Analysis

Deep probabilistic discriminant analysis [31] minimizes the inverse Fisher criterion:

$$\begin{aligned}
\underset{\theta}{\text{minimize}} \quad & \frac{\mathbb{E}[\mathbf{tr}(\text{cov}(\mathbf{f}(\boldsymbol{x})|y))]}{\mathbf{tr}(\text{cov}(\mathbb{E}[\mathbf{f}(\boldsymbol{x})|y]))} = \frac{\sum_{i=1}^{b} \mathbb{E}[\text{var}(\mathbf{f}(\boldsymbol{x}_i)|y_i)]}{\sum_{i=1}^{b} \text{var}(\mathbb{E}[\mathbf{f}(\boldsymbol{x}_i)|y_i])} \\
& \overset{(a)}{=} \frac{\sum_{i=1}^{b} \mathbb{E}[\text{var}(\mathbf{f}(\boldsymbol{x}_i)|y_i)]}{\sum_{i=1}^{b} \big(\text{var}(\mathbf{f}(\boldsymbol{x}_i)) - \mathbb{E}[\text{var}(\mathbf{f}(\boldsymbol{x}_i)|y_i)]\big)} \\
& \overset{(b)}{=} \frac{\sum_{i=1}^{b} \sum_{l=1}^{c} \mathbb{P}(y=l)\text{var}(\mathbf{f}(\boldsymbol{x}_i)|y_i=l)}{\sum_{i=1}^{b} \big(\text{var}(\mathbf{f}(\boldsymbol{x}_i)) - \sum_{l=1}^{c} \mathbb{P}(y=l)\text{var}(\mathbf{f}(\boldsymbol{x}_i)|y_i=l)\big)},
\end{aligned} \tag{19.45}$$

where b is the minibatch size, c is the number of classes, y_i is the class label of $\boldsymbol{x}_i$, cov(.) denotes the covariance, var(.) denotes the variance, $\mathbb{P}(y = l)$ is the prior of the l-th class (estimated by the ratio of the class population to the total number of points in the minibatch), (a) is due to the law of total variance, and (b) is because of the definition of expectation. The numerator and denominator represent the intraclass and interclass variances, respectively.

19.5.2 Discriminant Analysis with Virtual Samples

In discriminant analysis metric learning with virtual samples [27], any backbone network is considered until the one-to-last layer of the neural network and a last layer with the linear activation function is used. Let the outputs of the one-to-last layer be denoted by $\{\mathbf{f}'(\boldsymbol{x}_i)\}_{i=1}^b$ and the weights of the last layer be $\boldsymbol{U}$. The intraclass scatter $\boldsymbol{S}_W$ and the interclass scatter $\boldsymbol{S}_B$ are computed for the one-to-last layer's features $\{\mathbf{f}'(\boldsymbol{x}_i)\}_{i=1}^b$. If the last layer is seen as a Fisher discriminant analysis model with the projection matrix $\boldsymbol{U}$, the solution is the eigenvalue problem (see Chap. 2) for $\boldsymbol{S}_W^{-1}\boldsymbol{S}_B$. Let λ_j denote the j-th eigenvalue of this problem.

Assume $\mathcal{S}_b$ and $\mathcal{D}_b$ denote the similar and dissimilar points in the minibatch, where $|\mathcal{S}_b| = |\mathcal{D}_b| = q$. The following is defined as [27]:

$$\boldsymbol{g}_p := [\exp(-\mathbf{f}'(\boldsymbol{x}_i)^\top \mathbf{f}'(\boldsymbol{x}_j)) \mid (\boldsymbol{x}_i, \boldsymbol{x}_j) \in \mathcal{S}_b]^\top \in \mathbb{R}^q,$$

$$\boldsymbol{g}_n := [\exp(-\mathbf{f}'(\boldsymbol{x}_i)^\top \mathbf{f}'(\boldsymbol{x}_j)) \mid (\boldsymbol{x}_i, \boldsymbol{x}_j) \in \mathcal{D}_b]^\top \in \mathbb{R}^q,$$

$$s_{ctr} := \frac{1}{2q} \sum_{i=1}^{q} \big(\boldsymbol{g}_p(i) + \boldsymbol{g}_n(i)\big),$$

where $\boldsymbol{g}(i)$ is the i-th element of $\boldsymbol{g}$. q numbers are sampled, namely virtual samples, from the uniform distribution $U(s_{ctr} - \epsilon\bar{\lambda}, s_{ctr} + \epsilon\bar{\lambda})$, where ϵ is a small positive number and $\bar{\lambda}$ is the mean of the eigenvalues λ_j's. The q virtual samples are put in a vector $\boldsymbol{r} \in \mathbb{R}^q$.

The loss function for discriminant analysis with virtual samples is [27]:

$$\underset{\theta, \boldsymbol{U}}{\text{minimize}} \quad \frac{1}{q} \sum_{i=1}^{q} \Big[\frac{1}{q}\, \boldsymbol{g}_p(i)\, \|\boldsymbol{r}\|_1 - \frac{1}{q}\, \boldsymbol{g}_n(i)\, \|\boldsymbol{r}\|_1 + m\Big]_+ - 10^{-5} \frac{\mathbf{tr}(\boldsymbol{U}^\top \boldsymbol{S}_B \boldsymbol{U})}{\mathbf{tr}(\boldsymbol{U}^\top \boldsymbol{S}_W \boldsymbol{U})}, \tag{19.46}$$

where $\|.\|_1$ is the ℓ_1 norm, $[.]_+ := \max(., 0)$, $m > 0$ is the margin, and the second term is the maximization of the Fisher criterion.

19.5.3 Deep Fisher Discriminant Analysis

It is demonstrated in [17] that the solution to the following least squares problem is equivalent to the solution of Fisher discriminant analysis:

$$\underset{\boldsymbol{w}_0 \in \mathbb{R}^c, \boldsymbol{W} \in \mathbb{R}^{d \times c}}{\text{minimize}} \quad \frac{1}{2}\|\boldsymbol{Y} - \mathbf{1}_{n \times 1}\boldsymbol{w}_0^\top - \boldsymbol{X}\boldsymbol{W}\|_F^2, \tag{19.47}$$

where $\|.\|_F$ is the Frobenius norm, $\boldsymbol{X} \in \mathbb{R}^{n \times d}$ is the rowwise stack of the data points, $\boldsymbol{Y} := \boldsymbol{H}\boldsymbol{E}\boldsymbol{\Pi}^{-(1/2)} \in \mathbb{R}^{n \times c}$ where $\boldsymbol{H} := \boldsymbol{I} - (1/n)\mathbf{1}\mathbf{1}^\top \in \mathbb{R}^{n \times n}$ is the centering matrix, $\boldsymbol{E} \in \{0, 1\}^{n \times c}$ is the one-hot-encoded labels stacked rowwise, and $\boldsymbol{\Pi} \in \mathbb{R}^{c \times c}$ is the diagonal matrix whose (l, l)-th element is the cardinality of the l-th class.

Deep Fisher discriminant analysis [5, 6] implements Eq. (19.47) by a nonlinear neural network with the loss function:

$$\underset{\theta}{\text{minimize}} \quad \frac{1}{2}\|\boldsymbol{Y} - \mathbf{f}(\boldsymbol{X}; \theta)\|_F^2, \tag{19.48}$$

where θ is the weights of the network, $\boldsymbol{X} \in \mathbb{R}^{n \times d}$ denotes the rowwise stack of points in the minibatch of size b, $\boldsymbol{Y} := \boldsymbol{H}\boldsymbol{E}\boldsymbol{\Pi}^{-(1/2)} \in \mathbb{R}^{b \times c}$ is computed in every minibatch, and $\mathbf{f}(.) \in \mathbb{R}^{b \times c}$ is the rowwise stack of output embeddings of the network. After training, the output $\mathbf{f}(\boldsymbol{x})$ is the embedding for the input point $\boldsymbol{x}$.

19.6 Multimodal Deep Metric Learning

Data may have several modals where a separate set of features is available for every modality of data. In other words, there can be several features for every data point. Note that the dimensionality of the features may differ. Multimodal deep metric learning [40] learns a metric for multimodal data. Let m denote the number of modalities. Consider m stacked autoencoders, each of which is for one of the modalities. The l-th autoencoder obtains the l-th modality of the i-th data point, denoted by $\boldsymbol{x}_i^l$, and reconstructs it as an output, denoted by $\widehat{\boldsymbol{x}}_i^l$. The embedding layer, or the layer between the encoder and decoder, is shared between all m autoencoders. The output of this shared embedding layer is denoted by $\mathbf{f}(\boldsymbol{x}_i)$. The loss function for training the m stacked autoencoders with the shared embedding layer can be [40]:

$$\begin{aligned}
\underset{\theta}{\text{minimize}} \quad & \sum_{i=1}^{b}\sum_{l=1}^{m} \|\boldsymbol{x}_i^l - \widehat{\boldsymbol{x}}_i^l\|_2^2 + \lambda_1 \sum_{i=1}^{b} \sum_{\boldsymbol{x}_j \in \mathcal{X}_{c(\boldsymbol{x}_i)}} \big[d(\mathbf{f}(\boldsymbol{x}_i), \mathbf{f}(\boldsymbol{x}_j)) - m_1\big]_+ \\
& + \lambda_2 \sum_{i=1}^{b} \sum_{\boldsymbol{x}_j \in \mathcal{X} \setminus \mathcal{X}_{c(\boldsymbol{x}_i)}} \big[-d(\mathbf{f}(\boldsymbol{x}_i), \mathbf{f}(\boldsymbol{x}_j)) + m_2\big]_+,
\end{aligned} \tag{19.49}$$

where $\lambda_1 > 0$ and $\lambda_2 > 0$ are the regularization parameters and $m_1 > 0$ and $m_2 > 0$ are the margins. The first term is the reconstruction loss, and the second and third terms are for metric learning, which collapses each class to a margin m_1 and discriminates classes by a margin m_2. This loss function is optimized in a stacked autoencoder setup [21, 52]. Then, it is fine-tuned by backpropagation (see Chap. 18). After training, the embedding layer can be used for embedding the data points.[1]

19.7 Geometric Metric Learning by Neural Network

It is possible to use neural networks for Riemannian metric learning with a geometric approach. There are some works, such as [23], [18], and [16], that have implemented neural networks on the Riemannian manifolds. Layered geometric learning [16] implements Geometric Mean Metric Learning (GMML) [64] (see Chap. 11) in a neural network framework. In this method, every layer of the network is a metric layer, that projects the output of its previous layer onto the subspace of its own metric (see Proposition 11.2 in Chap. 11).

For the l-th layer of the network, the weight matrix (i.e., the projection matrix of the metric) and the output of the layer for the i-th data point are denoted by $\boldsymbol{U}_l$ and $\boldsymbol{x}_{i,l}$, respectively. Therefore, the metric in the l-th layer models $\|\boldsymbol{x}_{i,l} - \boldsymbol{x}_{j,l}\|_{\boldsymbol{U}_l\boldsymbol{U}_l^\top}$. Consider a dataset of n points $\boldsymbol{X} \in \mathbb{R}^{d\times n}$. The output of the l-th layer is denoted by $\boldsymbol{X}_l \in \mathbb{R}^{d\times n}$. The projection of a layer onto its metric subspace is $\boldsymbol{X}_l = \boldsymbol{U}_l^\top \boldsymbol{X}_{l-1}$.

Every layer solves the optimization problem of GMML [64], i.e., Eq. (11.60) in Chap. 11:

$$
\begin{aligned}
&\underset{\boldsymbol{W}}{\text{minimize}} \quad \sum_{(\boldsymbol{x}_i,\boldsymbol{x}_j)\in\mathcal{S}} \|\boldsymbol{x}_i - \boldsymbol{x}_j\|_{\boldsymbol{W}}^2 + \sum_{(\boldsymbol{x}_i,\boldsymbol{x}_j)\in\mathcal{D}} \|\boldsymbol{x}_i - \boldsymbol{x}_j\|_{\boldsymbol{W}^{-1}}^2 \\
&\text{subject to} \quad \boldsymbol{W} \succeq \boldsymbol{0}.
\end{aligned}
\tag{19.50}
$$

For this, the process starts from the first layer and proceeds to the last layer by feed-propagation. The l-th layer computes $\boldsymbol{\Sigma}_{\mathcal{S}}$ and $\boldsymbol{\Sigma}_{\mathcal{D}}$ for $\boldsymbol{X}_{l-1}$, by Eq. (11.14) in Chap. 11:

$$
\begin{aligned}
\boldsymbol{\Sigma}_{\mathcal{S}} &:= \sum_{(\boldsymbol{x}_i,\boldsymbol{x}_j)\in\mathcal{S}} (\boldsymbol{x}_i - \boldsymbol{x}_j)(\boldsymbol{x}_i - \boldsymbol{x}_j)^\top, \\
\boldsymbol{\Sigma}_{\mathcal{D}} &:= \sum_{(\boldsymbol{x}_i,\boldsymbol{x}_j)\in\mathcal{D}} (\boldsymbol{x}_i - \boldsymbol{x}_j)(\boldsymbol{x}_i - \boldsymbol{x}_j)^\top.
\end{aligned}
\tag{19.51}
$$

[1] There is also another multimodal deep metric learning method [59] for the interested readers.

Then, the optimization solution (19.50) is computed, which is Eq. (11.63) in Chap. 11. That solution is $\boldsymbol{W}_l = \boldsymbol{\Sigma}_{\mathcal{S}}^{-1} \sharp_{(1/2)} \boldsymbol{\Sigma}_{\mathcal{D}} = \boldsymbol{\Sigma}_{\mathcal{S}}^{(-1/2)} (\boldsymbol{\Sigma}_{\mathcal{S}}^{(1/2)} \boldsymbol{\Sigma}_{\mathcal{D}} \boldsymbol{\Sigma}_{\mathcal{S}}^{(1/2)})^{(1/2)}$ $\boldsymbol{\Sigma}_{\mathcal{S}}^{(-1/2)}$. Then, using Eq. (11.9) in Chap. 11 which is $\boldsymbol{W}_l = \boldsymbol{U}_l \boldsymbol{U}_l^\top$, the obtained $\boldsymbol{W}_l$ is decomposed to find $\boldsymbol{U}_l$. Then, the data points are projected onto the metric subspace as $\boldsymbol{X}_l = \boldsymbol{U}_l^\top \boldsymbol{X}_{l-1}$.

If the aim is to have the output of the layers lie on the positive semidefinite manifold, the activation function of every layer can be a projection onto the positive semidefinite cone (see Chap. 4):

$$\boldsymbol{X}_l := \boldsymbol{V}\, \mathbf{diag}(\max(\lambda_1, 0), \ldots, \max(\lambda_d, 0))\, \boldsymbol{V}^\top,$$

where $\boldsymbol{V}$ and $\{\lambda_1, \ldots, \lambda_d\}$ are the eigenvectors and eigenvalues of $\boldsymbol{X}_l$, respectively. This activation function is called the eigenvalue rectification layer in [23]. Lastly, it is noteworthy that backprojection [10] takes a similar approach, but in the Euclidean and Hilbert spaces and not in the Riemannian space.

19.8 Few-Shot Metric Learning

Few-shot learning refers to learning from a few data points, rather than from a large dataset. It is used for domain generalization to be able to use for unseen data in the test phase [55]. The training phase of few-shot learning is episodic, where in every iteration or so-called episode of training, there is a support set and a query set. In other words, the training dataset is divided into minibatches, where every minibatch contains a support set and a query set [49]. Consider a training dataset with c_{tr} classes and a test dataset with c_{te} classes. The test and training datasets are usually disjointed in few-shot learning so it is useful for domain generalization. In every episode, also called the task or the minibatch, training is performed using a set of training classes by randomly sampling from the classes.

The support set is $\mathcal{S}_s := \{(\boldsymbol{x}_{s,i}, y_{s,i})\}_{i=1}^{|\mathcal{S}_s|}$, where $\boldsymbol{x}$ and y denote the data point and its label, respectively. The query set is $\mathcal{S}_q := \{(\boldsymbol{x}_{q,i}, y_{q,i})\}_{i=1}^{|\mathcal{S}_q|}$. The training data of every episode (minibatch) is the union of the support and query sets. At every episode, c_s classes are randomly selected out of the total c_{tr} classes of the training dataset, where $c_s \ll c_{\text{tr}}$ usually holds. Then, k_s training data points are sampled from these c_s selected classes. These $c_s \times k_s = |\mathcal{S}_s|$ data points form the support set. This few-shot setup is called c_s-way, k_s-shot in which the support set contains c_s classes and k_s points in every class. In the query set, the number of classes and the number of points from every class may or may not be the same as in the support set.

In every episode of the training phase of few-shot learning, the network weights are updated by backpropagating error using the support set. Then, the query set is fed into the network with the updated weights, and the error is backpropagated using the query set. This second backpropagation with the query set updates the weights of the network at the end of episode. In other words, the query set is used to evaluate how effective the update by the support set is. The learning procedure for few-shot learning is called meta-learning [8].

There are several families of methods used for few-shot learning, including deep metric learning methods. Various metric learning methods have been proposed for learning from few-shot data. For example, the Siamese network, introduced in Sect. 19.4, has been used for few-shot learning [28, 32]. The following section introduces two metric learning methods for few-shot learning.

19.8.1 Multiscale Metric Learning

Multiscale metric learning [24] obtains the embedding space by learning multiple scales of middle features in the training process. There are multiple steps in multiscale metric learning:

1. Produce several different scales of features for both the support and query sets using a pretrained network with multiple output layers.
2. Take the average of the k_s features in every class within every scale of the support set. This results in c_s features for every scale in the support set. This and the features of the query set are fed to step three.
3. Feed every scale to a subnetwork, where larger scales are fed to subnetworks with more layers as they contain more information to process. These subnetworks are concatenated to provide a scalar output for every data point with multiple scales of features. Therefore, a scalar score is obtained for every data point in the support and query sets.

Finally, a combination of a classification loss function, such as the cross-entropy loss (see Eq. (19.8)), and triplet loss (see Eq. 19.15) is used in the support-query setup explained before.

19.8.2 Metric Learning with Continuous Similarity Scores

Another few-shot metric learning method is [60], which takes pairs of data points as the input support and query sets. For the pair $(\boldsymbol{x}_i, \boldsymbol{x}_j)$, consider the binary similarity score, y_{ij}, defined as:

$$y_{ij} := \begin{cases} 1 \text{ if } (\boldsymbol{x}_i, \boldsymbol{x}_j) \in \mathcal{S} \\ 0 \text{ if } (\boldsymbol{x}_i, \boldsymbol{x}_j) \in \mathcal{D}. \end{cases} \tag{19.52}$$

where $\mathcal{S}$ and $\mathcal{D}$ denote the sets of similar and dissimilar points, respectively. The continuous similarity score, y'_{ij}, can be defined as [60]:

$$y'_{ij} := \begin{cases} (\beta - 1)d(\boldsymbol{x}_i, \boldsymbol{x}_j) + 1 & \text{if } (\boldsymbol{x}_i, \boldsymbol{x}_j) \in \mathcal{S} \\ -\alpha d(\boldsymbol{x}_i, \boldsymbol{x}_j) + \alpha & \text{if } (\boldsymbol{x}_i, \boldsymbol{x}_j) \in \mathcal{D}, \end{cases} \tag{19.53}$$

where $0 < \alpha < \beta < 1$ and $d(\boldsymbol{x}_i, \boldsymbol{x}_j)$ is the normalized squared Euclidean distance (distances are normalized within every minibatch). The ranges of these continuous similarities are:

$$y'_{ij} \in \begin{cases} [\beta, 1] \text{ if } (\boldsymbol{x}_i, \boldsymbol{x}_j) \in \mathcal{S} \\ [0, \alpha] \text{ if } (\boldsymbol{x}_i, \boldsymbol{x}_j) \in \mathcal{D}. \end{cases}$$

In every episode (minibatch), the pairs are fed into a network with several feature vector outputs. For every pair $(\boldsymbol{x}_i, \boldsymbol{x}_j)$, these feature vectors are fed into another network, which outputs a scalar similarity score s_{ij}. The loss function of metric learning in this method is [60]:

$$\begin{aligned} \underset{\theta}{\text{maximize}} \quad & \sum_{(\boldsymbol{x}_i, \boldsymbol{x}_j) \in \mathcal{X}} (1 + \lambda)(s_{ij} - y'_{ij})^2, \\ \text{subject to} \quad & \beta \leq s_{ij}, y'_{ij} \leq 1 \quad \text{if} \quad y_{ij} = 1, \\ & 0 \leq s_{ij}, y'_{ij} \leq \alpha \quad \text{if} \quad y_{ij} = 0, \end{aligned} \tag{19.54}$$

where $\lambda > 0$ is the regularization parameter and $\mathcal{X}$ is the minibatch of the support or query set depending on whether it is the phase of support or query.

19.9 Chapter Summary

Deep metric learning is used for dimensionality reduction and feature extraction from various types of data. For example, it has been widely used in digital pathology analysis, medical image analysis, object recognition, geographical image analysis, etc. Its application is endless because it can extract informative features and learn an appropriate embedding space for any type of data. This chapter introduced deep metric learning. It especially focused on reconstruction autoencoders, supervised loss functions, Siamese networks, and several other variants. Note that autoencoders are categorized into reconstruction autoencoders and inferential autoencoders. The former category was introduced in this chapter. Variational autoencoder and adversarial autoencoder, as two inferential autoencoders, will be introduced in Chaps. 20 and 21, respectively. An additional survey on deep metric learning is [25].

References

1. Malik Boudiaf et al. "A unifying mutual information view of metric learning: cross-entropy vs. pairwise losses". In: *European Conference on Computer Vision*. Springer. 2020, pp. 548–564.
2. Jane Bromley et al. "Signature verification using a "Siamese" time delay neural network". In: *International Journal of Pattern Recognition and Artificial Intelligence* 7.04 (1993), pp. 669–688.
3. Ting Chen et al. "A simple framework for contrastive learning of visual representations". In: *International conference on machine learning*. 2020, pp. 1597–1607.
4. Shengyong Ding et al. "Deep feature learning with relative distance comparison for person re-identification". In: *Pattern Recognition* 48.10 (2015), pp. 2993–3003.
5. David Dıaz-Vico and José R Dorronsoro. "Deep least squares Fisher discriminant analysis". In: *IEEE transactions on neural networks and learning systems* 31.8 (2019), pp. 2752–2763.
6. David Dıaz -Vico et al. "Deep Fisher discriminant analysis". In: *International Work-Conference on Artificial Neural Networks*. Springer. 2017, pp. 501–512.
7. Yueqi Duan et al. "Deep adversarial metric learning". In: *Proceedings of the IEEE Conference on Computer Vision and Pattern Recognition*. 2018, pp. 2780–2789.
8. Chelsea Finn, Pieter Abbeel, and Sergey Levine. "Model-agnostic meta-learning for fast adaptation of deep networks". In: *International Conference on Machine Learning*. 2017, pp. 1126–1135.
9. Benyamin Ghojogh. "Data Reduction Algorithms in Machine Learning and Data Science". PhD thesis. University of Waterloo, 2021.
10. Benyamin Ghojogh, Fakhri Karray, and Mark Crowley. "Backprojection for training feedforward neural networks in the input and feature spaces". In: *International Conference on Image Analysis and Recognition*. Springer. 2020, pp. 16–24.
11. Benyamin Ghojogh et al. "Fisher discriminant triplet and contrastive losses for training Siamese networks". In: *2020 international joint conference on neural networks (IJCNN)*. IEEE. 2020, pp. 1–7.
12. Benyamin Ghojogh et al. "Sampling algorithms, from survey sampling to Monte Carlo methods: Tutorial and literature review". In: *arXiv preprint arXiv:2011.00901* (2020).
13. Jacob Goldberger et al. "Neighbourhood components analysis". In: *Advances in neural information processing systems*. 2005, pp. 513–520.
14. Ian Goodfellow, Yoshua Bengio, and Aaron Courville. *Deep learning*. MIT press, 2016.
15. Raia Hadsell, Sumit Chopra, and Yann LeCun. "Dimensionality reduction by learning an invariant mapping". In: *2006 IEEE Computer Society Conference on Computer Vision and Pattern Recognition (CVPR'06)* Vol. 2. IEEE. 2006, pp. 1735–1742.
16. Hamideh Hajiabadi et al. "Layered Geometric Learning". In: *International Conference on Artificial Intelligence and Soft Computing*. Springer. 2019, pp. 571–582.
17. Peter E Hart, David G Stork, and Richard O Duda. *Pattern classification*. Wiley Hoboken, 2000.
18. Michael B Hauser. "Principles of Riemannian geometry in neural networks". In: *Advances in neural information processing systems*. 2017, pp. 2807–2816.
19. Alexander Hermans, Lucas Beyer, and Bastian Leibe. "In defense of the triplet loss for person reidentification". In: *arXiv preprint arXiv:1703.07737* (2017).
20. Geoffrey Hinton, Oriol Vinyals, and Jeff Dean. "Distilling the knowledge in a neural network". In: *NIPS 2014 Deep Learning Workshop*. 2014.
21. Geoffrey E Hinton and Ruslan R Salakhutdinov. "Reducing the dimensionality of data with neural networks". In: *Science* 313.5786 (2006), pp. 504–507.
22. Elad Hoffer and Nir Ailon. "Deep metric learning using triplet network". In: *International workshop on similarity-based pattern recognition*. Springer. 2015, pp. 84–92.
23. Zhiwu Huang and Luc Van Gool. "A Riemannian network for SPD matrix learning". In: *Thirty-First AAAI Conference on Artificial Intelligence*. 2017.

24. Wen Jiang et al. "Multi-scale metric learning for few-shot learning". In: *IEEE Transactions on Circuits and Systems for Video Technology* 31.3 (2020), pp. 1091–1102.
25. Mahmut Kaya and Hasan Şakir Bilge. "Deep metric learning: A survey". In: *Symmetry* 11.9 (2019), p. 1066.
26. Siavash Khodadadeh, Ladislau Bölöni, and Mubarak Shah. "Unsupervised meta-learning for few-shot image classification". In: *Advances in neural information processing systems*. 2019.
27. Dae Ha Kim and Byung Cheol Song. "Virtual sample-based deep metric learning using discriminant analysis". In: *Pattern Recognition*. 110 (2021), p. 107643.
28. Gregory Koch, Richard Zemel, Ruslan Salakhutdinov, et al. "Siamese neural networks for one-shot image recognition". In: *ICML deep learning workshop*. Vol. 2. Lille. 2015.
29. Vijay Kumar BG, Gustavo Carneiro, and Ian Reid. "Learning local image descriptors with deep Siamese and triplet convolutional networks by minimising global loss functions". In: *Proceedings of the IEEE conference on computer vision and pattern recognition*. 2016, pp. 5385–5394.
30. Marıa Leyva-Vallina, Nicola Strisciuglio, and Nicolai Petkov. "Generalized Contrastive Optimization of Siamese Networks for Place Recognition". In: *arXiv preprint arXiv:2103.06638* (2021).
31. Li Li, Miloš Doroslovački, and Murray H Loew. "Discriminant analysis deep neural networks". In: *2019 53rd annual conference on information sciences and systems (CISS)*. IEEE. 2019, pp. 1–6.
32. Xiaomeng Li et al. "Revisiting metric learning for few-shot image classification". In: *Neurocomputing*. 406 (2020), pp. 49–58.
33. Xinbi Lv, Cairong Zhao, and Wei Chen. "A novel hard mining center-triplet loss for person re-identification". In: *Chinese Conference on Pattern Recognition and Computer Vision (PRCV)*. Springer. 2019, pp. 199–210.
34. Chengzhi Mao et al. "Metric learning for adversarial robustness". In: *Advances in neural information processing systems*. (2019).
35. Yair Movshovitz-Attias et al. "No fuss distance metric learning using proxies". In: *Proceedings of the IEEE International Conference on Computer Vision*. 2017, pp. 360–368.
36. Kevin P Murphy. *Conjugate Bayesian analysis of the Gaussian distribution*. Tech. rep. University of British Colombia, 2007.
37. Hyun Oh Song et al. "Deep metric learning via lifted structured feature embedding". In: *Proceedings of the IEEE conference on computer vision and pattern recognition*. 2016, pp. 4004–4012.
38. Qi Qian et al. "SoftTriple loss: Deep metric learning without triplet sampling". In: *Proceedings of the IEEE/CVF International Conference on Computer Vision*. 2019, pp. 6450–6458.
39. Joshua Robinson et al. "Contrastive learning with hard negative samples". In: *International Conference on Learning Representations*. 2021.
40. Seyed Mahdi Roostaiyan, Ehsan Imani, and Mahdieh Soleymani Baghshah. "Multi-modal deep distance metric learning". In: *Intelligent Data Analysis* 21.6 (2017), pp. 1351–1369.
41. Florian Schroff, Dmitry Kalenichenko, and James Philbin. "FaceNet: A unified embedding for face recognition and clustering". In: *Proceedings of the IEEE conference on computer vision and pattern recognition*. 2015, pp. 815–823.
42. Milad Sikaroudi et al. "Batch-incremental triplet sampling for training triplet networks using Bayesian updating theorem". In: *2020 25th International Conference on Pattern Recognition (ICPR)*. IEEE. 2021, pp. 7080–7086.
43. Milad Sikaroudi et al. "Offline versus online triplet mining based on extreme distances of histopathology patches". In: *International Symposium on Visual Computing*. Springer. 2020, pp. 333–345.
44. Milad Sikaroudi et al. "Supervision and source domain impact on representation learning: A histopathology case study". In: *2020 42nd Annual International Conference of the IEEE Engineering in Medicine & Biology Society (EMBC)*. IEEE. 2020, pp. 1400–1403.
45. Kihyuk Sohn. "Improved deep metric learning with multi-class n-pair loss objective". In: *Advances in neural information processing systems*. 2016, pp. 1857–1865.

46. Eu Wern Teh, Terrance DeVries, and Graham W Taylor. "ProxyNCA++: Revisiting and revitalizing proxy neighborhood component analysis". In: *European Conference on Computer Vision (ECCV)*. Springer. 2020, pp. 448–464.
47. Eu Wern Teh and Graham W Taylor. "Learning with less data via weakly labeled patch classification in digital pathology". In: *2020 IEEE 17th International Symposium on Biomedical Imaging (ISBI)*. IEEE. 2020, pp. 471–475.
48. Hamid R Tizhoosh. "Opposition-based learning: a new scheme for machine intelligence". In: *International conference on computational intelligence for modelling, control and automation and international conference on intelligent agents, web technologies and internet commerce (CIMCA-IAWTIC'06)*. Vol. 1. IEEE. 2005, pp. 695–701.
49. Eleni Triantafillou et al. "Meta-dataset: A dataset of datasets for learning to learn from few examples". In: *International Conference on Learning Representations*. 2020.
50. Vladimir Vapnik. *The nature of statistical learning theory*. Springer science & business media, 1995.
51. Jian Wang et al. "Deep metric learning with angular loss". In: *Proceedings of the IEEE International Conference on Computer Vision*. 2017, pp. 2593–2601.
52. Wei Wang et al. "Effective multi-modal retrieval based on stacked auto-encoders". In: *Proceedings of the VLDB Endowment* 7.8 (2014), pp. 649–660.
53. Xiao Wang et al. "Improved Hard Example Mining by Discovering Attribute-based Hard Person Identity". In: *arXiv preprint arXiv:1905.02102* (2019).
54. Xun Wang et al. "Cross-batch memory for embedding learning". In: *Proceedings of the IEEE/CVF Conference on Computer Vision and Pattern Recognition*. 2020, pp. 6388–6397.
55. Yaqing Wang et al. "Generalizing from a few examples: A survey on few-shot learning". In: *ACM Computing Surveys (CSUR)* 53.3 (2020), pp. 1–34.
56. Kilian Q Weinberger, John Blitzer, and Lawrence K Saul. "Distance metric learning for large margin nearest neighbor classification". In: *Advances in neural information processing systems*. 2006, pp. 1473–1480.
57. Kilian Q Weinberger and Lawrence K Saul. "Distance metric learning for large margin nearest neighbor classification". In: *Journal of machine learning research* 10.2 (2009).
58. Chao-Yuan Wu et al. "Sampling matters in deep embedding learning". In: *Proceedings of the IEEE International Conference on Computer Vision*. 2017, pp. 2840–2848.
59. Xing Xu et al. "Deep adversarial metric learning for cross-modal retrieval". In: *World Wide Web* 22.2 (2019), pp. 657–672.
60. Xinyi Xu et al. "Zero-shot Metric Learning". In: *International Joint Conference on Artificial Intelligence*. 2019, pp. 3996–4002.
61. Hong Xuan, Abby Stylianou, and Robert Pless. "Improved embeddings with easy positive triplet mining". In: *Proceedings of the IEEE/CVF Winter Conference on Applications of Computer Vision*. 2020, pp. 2474–2482.
62. Liu Yang et al. "Beyond 512 tokens: Siamese multi-depth transformer-based hierarchical encoder for long-form document matching". In: *Proceedings of the 29th ACM International Conference on Information & Knowledge Management*. 2020, pp. 1725–1734.
63. Mang Ye et al. "Unsupervised embedding learning via invariant and spreading instance feature". In: *Proceedings of the IEEE/CVF Conference on Computer Vision and Pattern Recognition*. 2019, pp. 6210–6219.
64. Pourya Zadeh, Reshad Hosseini, and Suvrit Sra. "Geometric mean metric learning". In: *International conference on machine learning*. 2016, pp. 2464–2471.

Chapter 20
Variational Autoencoders

20.1 Introduction

Chapter 12 explained that learning models can be divided into discriminative and generative models [3, 18]. The Variational Autoencoder (VAE), introduced in this chapter, is a generative model. Variational inference is a technique that finds a lower bound on the log-likelihood of the data and maximizes the lower bound rather than the log-likelihood in the Maximum Likelihood Estimation (MLE) (see Chap. 12). This lower bound is usually referred to as the Evidence Lower Bound (ELBO). Learning the parameters of latent space can be done using Expectation Maximization (EM) [2], as done in factor analysis [8] (see Chap. 12). The Variational Autoencoder (VAE) [15] implements variational inference in an autoencoder neural network setup, where the encoder and decoder model the E-step (expectation step) and M-step (maximization step) of EM, respectively. However, VAE is usually trained using backpropagation in practice [13, 21]. Variational inference and VAE are found in many Bayesian analysis applications. For example, variational inference has been used in 3D human motion analysis [23] and VAE has been used in forecasting [27].

Variational inference was introduced in Chap. 12 so it is recommended to study that chapter before reading this chapter. Recall the following equations from Chap. 12.

ELBO is (see Eq. (12.4) in Chap. 12):

$$\mathcal{L}(q, \boldsymbol{\theta}) := -\text{KL}\big(q(\boldsymbol{z}_i) \,\|\, \mathbb{P}(\boldsymbol{x}_i, \boldsymbol{z}_i \,|\, \boldsymbol{\theta})\big), \tag{20.1}$$

where $\boldsymbol{z}_i$ and $\boldsymbol{x}_i$ are the latent and observed variables, respectively, $\boldsymbol{\theta}$ is the learnable parameter, and $q(\boldsymbol{z}_i)$ is the distribution of the latent variable.

Expectation Maximization (EM) in variational inference is (see Eqs. (12.8) and (12.9) in Chap. 12):

B. Ghojogh et al., *Elements of Dimensionality Reduction and Manifold Learning*,
https://doi.org/10.1007/978-3-031-10602-6_20

$$\text{E-step:} \quad q^{(t)} := \arg\max_{q} \quad \mathcal{L}(q, \boldsymbol{\theta}^{(t-1)}), \tag{20.2}$$

$$\text{M-step:} \quad \boldsymbol{\theta}^{(t)} := \arg\max_{\boldsymbol{\theta}} \quad \mathcal{L}(q^{(t)}, \boldsymbol{\theta}), \tag{20.3}$$

where t denotes the iteration index. These steps were simplified to (see Eqs. (12.12) and (12.13) in Chap. 12):

$$q^{(t)}(z_i) \leftarrow \mathbb{P}(z_i \mid \boldsymbol{x}_i, \boldsymbol{\theta}^{(t-1)}), \tag{20.4}$$

$$\boldsymbol{\theta}^{(t)} \leftarrow \arg\max_{\boldsymbol{\theta}} \mathbb{E}_{\sim q^{(t)}(z_i)}\big[\log \mathbb{P}(\boldsymbol{x}_i, z_i \mid \boldsymbol{\theta})\big]. \tag{20.5}$$

Additionally, recall that the prior distribution on the latent variable in factor analysis was assumed to be the standard Gaussian (see Eq. (12.28) in Chap. 12):

$$\mathbb{P}(z_i) = \mathcal{N}(\boldsymbol{0}, \boldsymbol{I}). \tag{20.6}$$

20.2 Parts of the Variational Autoencoder

The Variational Autoencoder (VAE) [15] applies variational inference, i.e., maximizes the ELBO, but in an autoencoder setup and makes it differentiable for backpropagation training [22]. Figure 20.1 demonstrates that VAE includes an encoder and a decoder, each of which can have several network layers. A latent space is learned between the encoder and decoder. The latent variable z_i is sampled from the latent space. The following section explains the encoder and decoder parts. The input of the encoder in VAE is the data point $\boldsymbol{x}_i$ and the output of the decoder in VAE is its reconstruction $\boldsymbol{x}_i$.

20.2.1 Encoder of Variational Autoencoder

The encoder of the VAE models the distribution $q(z_i) = \mathbb{P}(z_i \mid \boldsymbol{x}_i, \boldsymbol{\theta}_e)$, where the parameters of distribution $\boldsymbol{\theta}_e$ are the weights of the encoder layers in VAE. The input and output of the encoder are $\boldsymbol{x}_i \in \mathbb{R}^d$ and $z_i \in \mathbb{R}^p$, respectively. Figure 20.1 depicts that the output neurons of the encoder are supposed to determine the parameters of the conditional distribution $\mathbb{P}(z_i \mid \boldsymbol{x}_i, \boldsymbol{\theta}_e)$. If this conditional distribution has m number of parameters, then there are m sets of output neurons from the encoder, denoted by $\{\boldsymbol{e}_j\}_{j=1}^m$. The dimensionality of these sets may differ depending on the size of the parameters.

For example, let the latent space be p-dimensional, i.e., $z_i \in \mathbb{R}^p$. If the distribution $\mathbb{P}(z_i \mid \boldsymbol{x}_i, \boldsymbol{\theta}_e)$ is a multivariate Gaussian distribution, then there are two sets of output neurons for the encoder, where one set has p neurons for the mean of this distribution $\boldsymbol{\mu}_{z|x} = \boldsymbol{e}_1 \in \mathbb{R}^d$ and the other set has $(p \times p)$ neurons for the covariance of this distribution $\boldsymbol{\Sigma}_{z|x} =$ matrix form of $\boldsymbol{e}_2 \in \mathbb{R}^{p \times p}$. If the covariance matrix is diagonal, the second set has p neurons rather than $(p \times p)$ neurons. In this case, $\boldsymbol{\Sigma}_{z|x} = \mathbf{diag}(\boldsymbol{e}_2) \in \mathbb{R}^{d \times d}$. Any distribution with any number of parameters can be chosen for $\mathbb{P}(z_i \mid \boldsymbol{x}_i, \boldsymbol{\theta}_e)$, but the multivariate Gaussian with a diagonal covariance is well-used:

$$q(z_i) = \mathbb{P}(z_i \mid \boldsymbol{x}_i, \boldsymbol{\theta}_e) = \mathcal{N}(z_i \mid \boldsymbol{\mu}_{z|x}, \boldsymbol{\Sigma}_{z|x}). \tag{20.7}$$

Let the network weights for the output sets of the encoder, $\{\boldsymbol{e}_j\}_{j=1}^m$, be denoted by $\{\boldsymbol{\theta}_{e,j}\}_{j=1}^m$. As the input of the encoder is $\boldsymbol{x}_i$, the j-th output set of the encoder can be written as $\boldsymbol{e}_j(\boldsymbol{x}_i, \boldsymbol{\theta}_{e,j})$. In the case of a multivariate Gaussian distribution for the latent space, the parameters are $\boldsymbol{\mu}_{z|x} = \boldsymbol{e}_1(\boldsymbol{x}_i, \boldsymbol{\theta}_{e,1})$ and $\boldsymbol{\Sigma}_{z|x} = \mathbf{diag}(\boldsymbol{e}_2(\boldsymbol{x}_i, \boldsymbol{\theta}_{e,2}))$.

20.2.2 Sampling the Latent Variable

When the data point $\boldsymbol{x}_i$ is fed as an input to the encoder, the parameters of the conditional distribution $q(z_i)$ are obtained. Therefore, the distribution of latent space, which is $q(z_i)$, is determined corresponding to the data point $\boldsymbol{x}_i$. Now, in the latent space, the corresponding latent variable is sampled from the distribution of latent space:

$$z_i \sim q(z_i) = \mathbb{P}(z_i \mid \boldsymbol{x}_i, \boldsymbol{\theta}_e). \tag{20.8}$$

This latent variable is fed as input to the decoder, which is explained in the following.

20.2.3 Decoder of Variational Autoencoder

As Fig. 20.1 demonstrates, the decoder of the VAE models the conditional distribution $\mathbb{P}(\boldsymbol{x}_i \mid z_i, \boldsymbol{\theta}_d)$, where $\boldsymbol{\theta}_d$ are the weights of the decoder layers in the VAE. The input and output of the decoder are $z_i \in \mathbb{R}^p$ and $\boldsymbol{x}_i \in \mathbb{R}^d$, respectively. The output neurons of the decoder are supposed to either generate the reconstructed data point or determine the parameters of the conditional distribution $\mathbb{P}(\boldsymbol{x}_i \mid z_i, \boldsymbol{\theta}_d)$; the former is more common. In the latter case, if this conditional distribution has l parameters, then there are l sets of output neurons from the decoder, denoted by $\{\boldsymbol{d}_j\}_{j=1}^l$. The dimensionality of these sets may differ depending the size of every parameter. The example of a multivariate Gaussian distribution can also be mentioned for the decoder. Let the network weights for the output sets of the decoder, $\{\boldsymbol{d}_j\}_{j=1}^l$, be

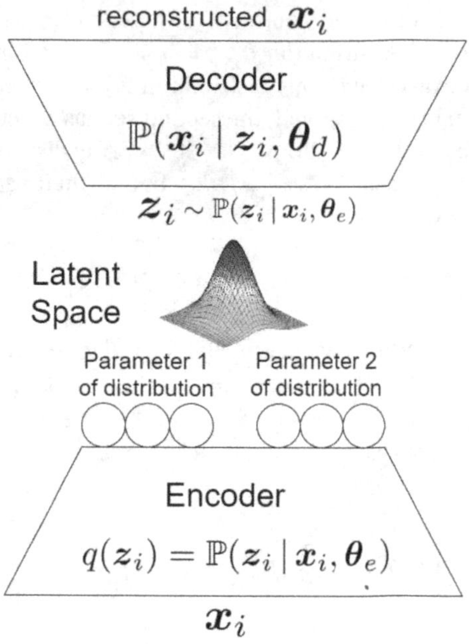

Fig. 20.1 The structure of a variational autoencoder

denoted by $\{\boldsymbol{\theta}_{d,j}\}_{j=1}^{l}$. As the input of the decoder is z_i, the j-th output set of the decoder can be written as $\boldsymbol{d}_j(z_i, \boldsymbol{\theta}_{d,j})$.

20.3 Training Variational Autoencoder with Expectation Maximization

EM is used for training the VAE. Recall Eqs. (20.2) and (20.3) for EM in variational inference. Inspired by that, VAE uses EM for training, where the ELBO is a function of encoder weights $\boldsymbol{\theta}_e$, decoder weights $\boldsymbol{\theta}_d$, and data point $\boldsymbol{x}_i$:

$$\text{E-step:} \quad \boldsymbol{\theta}_e^{(t)} := \arg\max_{q} \quad \mathcal{L}(\boldsymbol{\theta}_e, \boldsymbol{\theta}_d^{(t-1)}, \boldsymbol{x}_i), \tag{20.9}$$

$$\text{M-step:} \quad \boldsymbol{\theta}_d^{(t)} := \arg\max_{q} \quad \mathcal{L}(\boldsymbol{\theta}_e^{(t)}, \boldsymbol{\theta}_d, \boldsymbol{x}_i). \tag{20.10}$$

This iterative optimization algorithm can be simplified by alternating optimization [14] (see Chap. 4), where a gradient ascent optimization step is completed every iteration. Consider minibatch stochastic gradient ascent in which the training data are taken in batches, where b denotes the minibatch size. Therefore, the optimization is:

$$\text{E-step:} \quad \boldsymbol{\theta}_e^{(t)} := \boldsymbol{\theta}_e^{(t-1)} + \eta_e \frac{\partial \sum_{i=1}^{b} \mathcal{L}(\boldsymbol{\theta}_e, \boldsymbol{\theta}_d^{(t-1)}, \boldsymbol{x}_i)}{\partial \boldsymbol{\theta}_e}, \tag{20.11}$$

$$\text{M-step:} \quad \boldsymbol{\theta}_d^{(t)} := \boldsymbol{\theta}_d^{(t-1)} + \eta_d \frac{\partial \sum_{i=1}^{b} \mathcal{L}(\boldsymbol{\theta}_e^{(t)}, \boldsymbol{\theta}_d, \boldsymbol{x}_i)}{\partial \boldsymbol{\theta}_d}, \tag{20.12}$$

where η_e and η_d are the learning rates for $\boldsymbol{\theta}_e$ and $\boldsymbol{\theta}_d$, respectively.

The ELBO is simplified as:

$$\begin{aligned}\sum_{i=1}^{b} \mathcal{L}(q, \boldsymbol{\theta}) &\overset{(20.1)}{=} -\sum_{i=1}^{b} \text{KL}\big(q(z_i) \,\|\, \mathbb{P}(\boldsymbol{x}_i, z_i \,|\, \boldsymbol{\theta}_d)\big) \\ &\overset{(20.4)}{=} -\sum_{i=1}^{b} \text{KL}\big(\mathbb{P}(z_i \,|\, \boldsymbol{x}_i, \boldsymbol{\theta}_e) \,\|\, \mathbb{P}(\boldsymbol{x}_i, z_i \,|\, \boldsymbol{\theta}_d)\big).\end{aligned} \tag{20.13}$$

Note that the parameter of $\mathbb{P}(\boldsymbol{x}_i, z_i \,|\, \boldsymbol{\theta}_d)$ is $\boldsymbol{\theta}_d$ because z_i is generated after the encoder and before the decoder.

There are different ways to approximate the KL divergence in Eq. (20.13) [6, 12]. The ELBO can be simplified in at least two different ways, which are explained in the following.

20.3.1 Simplification Type 1

The first type of ELBO simplification is:

$$\begin{aligned}\sum_{i=1}^{b} \mathcal{L}(q, \boldsymbol{\theta}) &= -\sum_{i=1}^{b} \text{KL}\big(\mathbb{P}(z_i \,|\, \boldsymbol{x}_i, \boldsymbol{\theta}_e) \,\|\, \mathbb{P}(\boldsymbol{x}_i, z_i \,|\, \boldsymbol{\theta}_d)\big) \\ &= -\sum_{i=1}^{b} \mathbb{E}_{\sim q^{(t-1)}(z_i)}\Big[\log\Big(\frac{\mathbb{P}(z_i \,|\, \boldsymbol{x}_i, \boldsymbol{\theta}_e)}{\mathbb{P}(\boldsymbol{x}_i, z_i \,|\, \boldsymbol{\theta}_d)}\Big)\Big] \\ &= -\sum_{i=1}^{b} \mathbb{E}_{\sim \mathbb{P}(z_i \,|\, \boldsymbol{x}_i, \boldsymbol{\theta}_e)}\Big[\log\Big(\frac{\mathbb{P}(z_i \,|\, \boldsymbol{x}_i, \boldsymbol{\theta}_e)}{\mathbb{P}(\boldsymbol{x}_i, z_i \,|\, \boldsymbol{\theta}_d)}\Big)\Big].\end{aligned} \tag{20.14}$$

This expectation can be approximated using the Monte Carlo approximation [9], where ℓ samples $\{z_{i,j}\}_{j=1}^{\ell}$, corresponding to the i-th data point, are drawn from the conditional distribution as:

$$z_{i,j} \sim \mathbb{P}(z_i \,|\, \boldsymbol{x}_i, \boldsymbol{\theta}_e), \quad \forall j \in \{1, \dots, \ell\}. \tag{20.15}$$

The Monte Carlo approximation [9], in general, approximates the expectation as:

$$\mathbb{E}_{\sim\mathbb{P}(z_i \,|\, \boldsymbol{x}_i, \boldsymbol{\theta}_e)}\big[f(z_i)\big] \approx \frac{1}{\ell}\sum_{j=1}^{\ell} f(z_{i,j}), \tag{20.16}$$

where $f(z_i)$ is a function of z_i. Here, the approximation is:

$$\begin{aligned}
\sum_{i=1}^{b} \mathcal{L}(q, \boldsymbol{\theta}) &\approx \sum_{i=1}^{b} \widetilde{\mathcal{L}}(q, \boldsymbol{\theta}) \\
&= -\sum_{i=1}^{b} \frac{1}{\ell} \sum_{j=1}^{\ell} \log\Big(\frac{\mathbb{P}(z_{i,j} \,|\, \boldsymbol{x}_i, \boldsymbol{\theta}_e)}{\mathbb{P}(\boldsymbol{x}_i, z_{i,j} \,|\, \boldsymbol{\theta}_d)}\Big) \\
&= \sum_{i=1}^{b} \frac{1}{\ell} \sum_{j=1}^{\ell} \Big[\log\big(\mathbb{P}(\boldsymbol{x}_i, z_{i,j} \,|\, \boldsymbol{\theta}_d)\big) - \log\big(\mathbb{P}(z_{i,j} \,|\, \boldsymbol{x}_i, \boldsymbol{\theta}_e)\big)\Big].
\end{aligned} \tag{20.17}$$

20.3.2 Simplification Type 2

The second type of ELBO simplification is:

$$\begin{aligned}
\sum_{i=1}^{b} \mathcal{L}(q, \boldsymbol{\theta}) &= -\sum_{i=1}^{b} \mathrm{KL}\big(\mathbb{P}(z_i \,|\, \boldsymbol{x}_i, \boldsymbol{\theta}_e) \,\|\, \mathbb{P}(\boldsymbol{x}_i, z_i \,|\, \boldsymbol{\theta}_d)\big) \\
&= -\sum_{i=1}^{b} \int \mathbb{P}(z_i \,|\, \boldsymbol{x}_i, \boldsymbol{\theta}_e) \log\Big(\frac{\mathbb{P}(z_i \,|\, \boldsymbol{x}_i, \boldsymbol{\theta}_e)}{\mathbb{P}(\boldsymbol{x}_i, z_i \,|\, \boldsymbol{\theta}_d)}\Big)\, dz_i \\
&= -\sum_{i=1}^{b} \int \mathbb{P}(z_i \,|\, \boldsymbol{x}_i, \boldsymbol{\theta}_e) \log\Big(\frac{\mathbb{P}(z_i \,|\, \boldsymbol{x}_i, \boldsymbol{\theta}_e)}{\mathbb{P}(\boldsymbol{x}_i \,|\, z_i, \boldsymbol{\theta}_d)\, \mathbb{P}(z_i)}\Big)\, dz_i \\
&= -\sum_{i=1}^{b} \int \mathbb{P}(z_i \,|\, \boldsymbol{x}_i, \boldsymbol{\theta}_e) \log\Big(\frac{\mathbb{P}(z_i \,|\, \boldsymbol{x}_i, \boldsymbol{\theta}_e)}{\mathbb{P}(z_i)}\Big)\, dz_i \\
&\quad + \sum_{i=1}^{b} \int \mathbb{P}(z_i \,|\, \boldsymbol{x}_i, \boldsymbol{\theta}_e) \log\big(\mathbb{P}(\boldsymbol{x}_i \,|\, z_i, \boldsymbol{\theta}_d)\big)\, dz_i \\
&= -\sum_{i=1}^{b} \mathrm{KL}\big(\mathbb{P}(z_i \,|\, \boldsymbol{x}_i, \boldsymbol{\theta}_e) \,\|\, \mathbb{P}(z_i)\big)
\end{aligned}$$

$$+ \sum_{i=1}^{b} \mathbb{E}_{\sim \mathbb{P}(z_i \mid \boldsymbol{x}_i, \boldsymbol{\theta}_e)} \Big[\log \big(\mathbb{P}(\boldsymbol{x}_i \mid z_i, \boldsymbol{\theta}_d)\big) \Big]. \tag{20.18}$$

The second term in the above equation can be estimated using Monte Carlo approximation [9], where ℓ samples $\{z_{i,j}\}_{j=1}^{\ell}$ are drawn from $\mathbb{P}(z_i \mid \boldsymbol{x}_i, \boldsymbol{\theta}_e)$:

$$\begin{aligned}\sum_{i=1}^{b} \mathcal{L}(q, \boldsymbol{\theta}) &\approx \sum_{i=1}^{b} \widetilde{\mathcal{L}}(q, \boldsymbol{\theta}) \\ &= - \sum_{i=1}^{b} \mathrm{KL}\big(\mathbb{P}(z_i \mid \boldsymbol{x}_i, \boldsymbol{\theta}_e) \,\|\, \mathbb{P}(z_i)\big) + \sum_{i=1}^{b} \frac{1}{\ell} \sum_{j=1}^{\ell} \log \big(\mathbb{P}(\boldsymbol{x}_i \mid z_{i,j}, \boldsymbol{\theta}_d)\big).\end{aligned} \tag{20.19}$$

The first term in the above equation can be converted to the expectation and then computed using the Monte Carlo approximation [9] again, where ℓ samples $\{z_{i,j}\}_{j=1}^{\ell}$ are drawn from $\mathbb{P}(z_i \mid \boldsymbol{x}_i, \boldsymbol{\theta}_e)$:

$$\begin{aligned}\sum_{i=1}^{b} \mathcal{L}(q, \boldsymbol{\theta}) &\approx \sum_{i=1}^{b} \widetilde{\mathcal{L}}(q, \boldsymbol{\theta}) \\ &= - \sum_{i=1}^{b} \mathbb{E}_{\sim \mathbb{P}(z_i \mid \boldsymbol{x}_i, \boldsymbol{\theta}_e)} \Big[\log \big(\frac{\mathbb{P}(z_i \mid \boldsymbol{x}_i, \boldsymbol{\theta}_e)}{\mathbb{P}(z_i)}\big) \Big] + \sum_{i=1}^{b} \frac{1}{\ell} \sum_{j=1}^{\ell} \log \big(\mathbb{P}(\boldsymbol{x}_i \mid z_{i,j}, \boldsymbol{\theta}_d)\big) \\ &\approx - \sum_{i=1}^{b} \frac{1}{\ell} \sum_{j=1}^{\ell} \log \big(\mathbb{P}(z_{i,j} \mid \boldsymbol{x}_i, \boldsymbol{\theta}_e)\big) - \log \big(\mathbb{P}(z_{i,j})\big) + \sum_{i=1}^{b} \frac{1}{\ell} \sum_{j=1}^{\ell} \log \big(\mathbb{P}(\boldsymbol{x}_i \mid z_{i,j}, \boldsymbol{\theta}_d)\big).\end{aligned} \tag{20.20}$$

If there are families of distributions, such as Gaussian distributions, for $\mathbb{P}(z_{i,j} \mid \boldsymbol{x}_i, \boldsymbol{\theta}_e)$ and $\mathbb{P}(z_{i,j})$, the first term in Eq. (20.19) can be computed analytically. In the following, Eq. (20.19) is simplified further for Gaussian distributions.

20.3.3 Simplification Type 2 for Special Case of Gaussian Distributions

The KL divergence can be computed in the first term of Eq. (20.19) analytically for univariate or multivariate Gaussian distributions. For this, the following lemmas are needed.

Lemma 20.1 *The KL divergence between two univariate Gaussian distributions $p_1 \sim \mathcal{N}(\mu_1, \sigma_1^2)$ and $p_2 \sim \mathcal{N}(\mu_2, \sigma_2^2)$ is:*

$$KL(p_1 \| p_2) = \log(\frac{\sigma_2}{\sigma_1}) + \frac{\sigma_1^2 + (\mu_1 - \mu_2)^2}{2\sigma_2^2} - \frac{1}{2}. \tag{20.21}$$

Proof See Appendix 20.1 for proof. □

Lemma 20.2 *The KL divergence between two multivariate Gaussian distributions $p_1 \sim \mathcal{N}(\boldsymbol{\mu}_1, \boldsymbol{\Sigma}_1)$ and $p_2 \sim \mathcal{N}(\boldsymbol{\mu}_2, \boldsymbol{\Sigma}_2)$ with dimensionality p is:*

$$KL(p_1 \| p_2) = \frac{1}{2}\Big(\log(\frac{|\boldsymbol{\Sigma}_2|}{|\boldsymbol{\Sigma}_1|}) - p + \mathbf{tr}(\boldsymbol{\Sigma}_2^{-1}\boldsymbol{\Sigma}_1) + (\boldsymbol{\mu}_2 - \boldsymbol{\mu}_1)^\top \boldsymbol{\Sigma}_2^{-1}(\boldsymbol{\mu}_2 - \boldsymbol{\mu}_1)\Big). \tag{20.22}$$

Proof See [6, Section 9] for proof. □

Consider the case in which there are:

$$\mathbb{P}(\boldsymbol{z}_i \,|\, \boldsymbol{x}_i, \boldsymbol{\theta}_e) \sim \mathcal{N}(\boldsymbol{\mu}_{z|x}, \boldsymbol{\Sigma}_{z|x}), \tag{20.23}$$

$$\mathbb{P}(\boldsymbol{z}_i) \sim \mathcal{N}(\boldsymbol{\mu}_z, \boldsymbol{\Sigma}_z), \tag{20.24}$$

where $\boldsymbol{z}_i \in \mathbb{R}^p$. Note that the parameters $\boldsymbol{\mu}_{z|x}$ and $\boldsymbol{\Sigma}_{z|x}$ are trained in the neural network, while the parameters $\mathbb{P}(\boldsymbol{z}_{i,j})$ can be set to $\boldsymbol{\mu}_z = \boldsymbol{0}$ and $\boldsymbol{\Sigma}_z = \boldsymbol{I}$, which are inspired by Eq. (20.6) in factor analysis. According to Lemma 20.2, the approximation of the ELBO, i.e., Eq. (20.19), can be simplified to:

$$\begin{aligned}
\sum_{i=1}^{b} \mathcal{L}(q, \boldsymbol{\theta}) &\approx \sum_{i=1}^{b} \widetilde{\mathcal{L}}(q, \boldsymbol{\theta}) \\
&= -\sum_{i=1}^{b} \frac{1}{2}\Big(\log(\frac{|\boldsymbol{\Sigma}_z|}{|\boldsymbol{\Sigma}_{z|x}|}) - p + \mathbf{tr}(\boldsymbol{\Sigma}_z^{-1}\boldsymbol{\Sigma}_{z|x}) + (\boldsymbol{\mu}_z - \boldsymbol{\mu}_{z|x})^\top \boldsymbol{\Sigma}_z^{-1}(\boldsymbol{\mu}_z - \boldsymbol{\mu}_{z|x})\Big) \\
&\quad + \sum_{i=1}^{b} \frac{1}{\ell} \sum_{j=1}^{\ell} \log\big(\mathbb{P}(\boldsymbol{x}_i \,|\, \boldsymbol{z}_{i,j}, \boldsymbol{\theta}_d)\big).
\end{aligned} \tag{20.25}$$

20.3.4 Training Variational Autoencoder with Approximations

VAE can be trained with EM, where the Monte Carlo approximations are applied to the ELBO. Equations (20.11) and (20.12) are replaced by the following equations:

$$\text{E-step:} \quad \boldsymbol{\theta}_e^{(t)} := \boldsymbol{\theta}_e^{(t-1)} + \eta_e \frac{\partial \sum_{i=1}^{b} \widetilde{\mathcal{L}}(\boldsymbol{\theta}_e, \boldsymbol{\theta}_d^{(t-1)}, \boldsymbol{x}_i)}{\partial \boldsymbol{\theta}_e}, \tag{20.26}$$

$$\text{M-step:} \quad \boldsymbol{\theta}_d^{(t)} := \boldsymbol{\theta}_d^{(t-1)} + \eta_d \frac{\partial \sum_{i=1}^{b} \widetilde{\mathcal{L}}(\boldsymbol{\theta}_e^{(t)}, \boldsymbol{\theta}_d, \boldsymbol{x}_i)}{\partial \boldsymbol{\theta}_d}, \tag{20.27}$$

where the approximated ELBO was introduced in the previous sections.

20.3.5 Prior Regularization

Some works regularize the ELBO, Eq. (20.13), with a penalty on the prior distribution $\mathbb{P}(z_i)$. Using this, the learned distribution of latent space $\mathbb{P}(z_i \,|\, \boldsymbol{x}_i, \boldsymbol{\theta}_e)$ is guided to have a specific prior distribution $\mathbb{P}(z_i)$. Some examples for prior regularization in VAE include geodesic priors [11] and optimal priors [24]. Note that this regularization can inject domain knowledge into the latent space. It can also be useful for making the latent space more interpretable.

20.4 The Reparameterization Trick

Sampling the ℓ samples for the latent variables, i.e., Eq. (20.8), blocks the gradient flow because computing the derivatives through $\mathbb{P}(z_i \,|\, \boldsymbol{x}_i, \boldsymbol{\theta}_e)$ by chain rule results in a high variance estimate of the gradient. To overcome this problem, the reparameterization technique [15, 21, 25] is used. In this technique, instead of sampling $z_i \sim \mathbb{P}(z_i \,|\, \boldsymbol{x}_i, \boldsymbol{\theta}_e)$, z_i is assumed to be a random variable but is a deterministic function of another random variable $\boldsymbol{\epsilon}_i$ as follows:

$$z_i = g(\boldsymbol{\epsilon}_i, \boldsymbol{x}_i, \boldsymbol{\theta}_e), \tag{20.28}$$

where $\boldsymbol{\epsilon}_i$ is a stochastic variable sampled from a distribution as:

$$\boldsymbol{\epsilon}_i \sim \mathbb{P}(\boldsymbol{\epsilon}). \tag{20.29}$$

Equations (20.14) and (20.18) both contain an expectation of a function $f(z_i)$. Using this technique, this expectation is replaced:

$$\mathbb{E}_{\sim\mathbb{P}(z_i \,|\, \boldsymbol{x}_i, \boldsymbol{\theta}_e)}[f(z_i)] \rightarrow \mathbb{E}_{\sim\mathbb{P}(z_i \,|\, \boldsymbol{x}_i, \boldsymbol{\theta}_e)}[f(g(\boldsymbol{\epsilon}_i, \boldsymbol{x}_i, \boldsymbol{\theta}_e))]. \tag{20.30}$$

Using the reparameterization technique, the encoder, which implements $\mathbb{P}(z_i \,|\, \boldsymbol{x}_i, \boldsymbol{\theta}_e)$, is replaced by $g(\boldsymbol{\epsilon}_i, \boldsymbol{x}_i, \boldsymbol{\theta}_e)$, where in the latent space between the encoder and decoder, $\boldsymbol{\epsilon}_i \sim \mathbb{P}(\boldsymbol{\epsilon})$ and $z_i = g(\boldsymbol{\epsilon}_i, \boldsymbol{x}_i, \boldsymbol{\theta}_e)$.

A simple example for the reparameterization technique is when z_i and ϵ_i are univariate Gaussian variables:

$$
\begin{aligned}
z_i &\sim \mathcal{N}(\mu, \sigma^2), \\
\epsilon_i &\sim \mathcal{N}(0, 1), \\
z_i &= g(\epsilon_i) = \mu + \sigma \epsilon_i .
\end{aligned}
$$

For more advanced reparameterization techniques, refer to [7].

20.5 Training Variational Autoencoder with Backpropagation

In practice, VAE is trained by backpropagation [21], where the backpropagation algorithm [22] is used for training the weights of the network. Recall that in training VAE with EM, the encoder and decoder are trained separately using the E-step and the M-step of EM, respectively. However, in training VAE with backpropagation, the whole network is trained together and not in separate steps. Suppose the whole weights if VAE are denoted by $\boldsymbol{\theta} := \{\boldsymbol{\theta}_e, \boldsymbol{\theta}_d\}$. Backpropagation trains VAE using the minibatch stochastic gradient descent with the negative ELBO, $\sum_{i=1}^{b} -\widetilde{\mathcal{L}}(\boldsymbol{\theta}, \boldsymbol{x}_i)$, as the loss function:

$$
\boldsymbol{\theta}^{(t)} := \boldsymbol{\theta}^{(t-1)} - \eta \frac{\partial \sum_{i=1}^{b} -\widetilde{\mathcal{L}}(\boldsymbol{\theta}, \boldsymbol{x}_i)}{\partial \boldsymbol{\theta}}, \tag{20.31}
$$

where η is the learning rate. Note that minimization occurs here because neural networks usually minimize the loss function.

20.6 The Test Phase in the Variational Autoencoder

In the test phase, the test data point $\boldsymbol{x}_i$ is fed to the encoder to determine the parameters of the conditional distribution of the latent space, i.e., $\mathbb{P}(z_i \mid \boldsymbol{x}_i, \boldsymbol{\theta}_e)$. Then, from this distribution, the latent variable z_i is sampled from the latent space, and the corresponding reconstructed data point $\boldsymbol{x}_i$ is generated by the decoder. Therefore, VAE is a generative model that generates data points [18].

20.7 Other Notes and Other Variants of the Variational Autoencoder

There have been multiple improvements to VAE. This section briefly reviews some of these improvements. VAE has been known to generate blurry images when the data points represent images. This blurry artifact may be because of the following reasons:

- sampling for the Monte Carlo approximations
- lower bound approximation by ELBO
- restrictions on the family of distributions, where usually simple Gaussian distributions are used.

There are also other interpretations of why VAE generates blurry artifacts; for example, see [29]. This work also proposed a generalized ELBO. Generative adversarial networks [10] (see Chap. 21) usually generate clearer images; therefore, it is possible to combine variational and adversarial inferences [17] to take advantage of both models.

Variational discriminant analysis [28] has also been proposed for classification and discrimination between classes. Two other tutorials on VAE include [5] and [19]. Some of the recent papers on VAE are nearly optimal VAE [1], deep VAE [13], Hamiltonian VAE [4], and Nouveau VAE [26], which is a hierarchical VAE. For image data and image and caption modelling, a fusion of VAE and convolutional neural network is also proposed [20]. The influential factors in VAE are also analyzed in [16].

20.8 Chapter Summary

Variational autoencoders are inferential autoencoders that focus on inference from data. They are an autoencoder implementation of variational inference and Bayesian analysis, introduced in Chap. 12. The expectation maximization algorithm of variational inference is performed using backpropagation in variational autoencoders. This chapter introduced these autoencoders and explained how they can be trained and tested. Another inferential autoencoder is the adversarial autoencoder, which will be introduced in Chap. 21.

Appendix 20.1: Proof for Lemma 20.1

$$\begin{aligned}
\mathrm{KL}(p_1 \| p_2) &= \int p_1(x) \log\big(\frac{p_1(x)}{p_2(x)}\big) dx \\
&= \int p_1(x) \log(p_1(x))\, dx - \int p_1(x) \log(p_2(x))\, dx.
\end{aligned}$$

According to integration by parts:

$$\int p_1(x)\log(p_1(x))\,dx = -\frac{1}{2}(1+\log(2\pi\sigma_1^2)).$$

There is also:

$$\begin{aligned}
-\int p_1(x)\log(p_2(x))\,dx &= -\int p_1(x)\log\Big(\frac{1}{\sqrt{2\pi\sigma_2^2}}e^{-\frac{(x-\mu_2)^2}{2\sigma_2^2}}\Big)\,dx \\
&= \frac{1}{2}\log(2\pi\sigma_2^2)\underbrace{\int p_1(x)dx}_{=1} - \int p_1(x)\log\big(e^{-\frac{(x-\mu_2)^2}{2\sigma_2^2}}\big)\,dx \\
&= \frac{1}{2}\log(2\pi\sigma_2^2) + \int p_1(x)\frac{(x-\mu_2)^2}{2\sigma_2^2}\,dx \\
&= \frac{1}{2}\log(2\pi\sigma_2^2) + \frac{1}{2\sigma_2^2}\Big(\int p_1(x)x^2dx - \int p_1(x)2x\mu_2dx + \int p_1(x)\mu_2^2dx\Big) \\
&= \frac{1}{2}\log(2\pi\sigma_2^2) + \frac{1}{2\sigma_2^2}\Big(\mathbb{E}_{\sim p_1(x)}[x^2] - 2\mu_2\mathbb{E}_{\sim p_1(x)}[x] + \mu_2^2\Big).
\end{aligned}$$

It is known that:

$$\mathbb{V}\mathrm{ar}[x] = \mathbb{E}[x^2] - \mathbb{E}[x]^2 \implies \mathbb{E}[x^2] = \sigma_1^2 + \mu_1^2$$

Therefore:

$$\begin{aligned}
-\int p_1(x)\log(p_2(x))\,dx &= \frac{1}{2}\log(2\pi\sigma_2^2) + \frac{1}{2\sigma_2^2}\Big(\sigma_1^2 + \mu_1^2 - 2\mu_2\mu_1 + \mu_2^2\Big) \\
&= \frac{1}{2}\log(2\pi\sigma_2^2) + \frac{1}{2\sigma_2^2}(\sigma_1^2 + (\mu_1 - \mu_2)).
\end{aligned}$$

Lastly:

$$\begin{aligned}
\mathrm{KL}(p_1\|p_2) &= -\frac{1}{2}(1+\log(2\pi\sigma_1^2)) + \frac{1}{2}\log(2\pi\sigma_2^2) + \frac{1}{2\sigma_2^2}(\sigma_1^2 + (\mu_1 - \mu_2)) \\
&= \log(\frac{\sigma_2}{\sigma_1}) + \frac{\sigma_1^2 + (\mu_1-\mu_2)^2}{2\sigma_2^2} - \frac{1}{2}.
\end{aligned}$$

□

References

1. Jincheng Bai, Qifan Song, and Guang Cheng. "Nearly Optimal Variational Inference for High Dimensional Regression with Shrinkage Priors". In: *arXiv preprint arXiv:2010.12887* (2020).
2. Christopher M Bishop. *Pattern recognition and machine learning*. Springer, 2006.
3. Guillaume Bouchard and Bill Triggs. "The tradeoff between generative and discriminative classifiers". In: *16th IASC International Symposium on Computational Statistics*. 2004.
4. Anthony L Caterini, Arnaud Doucet, and Dino Sejdinovic. "Hamiltonian variational auto-encoder". In: *Advances in Neural Information Processing Systems* 31 (2018), pp. 8167–8177.
5. Carl Doersch. "Tutorial on variational autoencoders". In: *arXiv preprint arXiv:1606.05908* (2016).
6. John Duchi. *Derivations for linear algebra and optimization*. Tech. rep. Berkeley, California, 2007.
7. Mikhail Figurnov, Shakir Mohamed, and Andriy Mnih. "Implicit reparameterization gradients". In: *Advances in Neural Information Processing Systems* 31 (2018), pp. 441–452.
8. Benjamin Fruchter. *Introduction to factor analysis*. Van Nostrand, 1954.
9. Benyamin Ghojogh et al. "Sampling Algorithms, from Survey Sampling to Monte Carlo Methods: Tutorial and Literature Review". In: *arXiv preprint arXiv:2011.00901* (2020).
10. Ian Goodfellow et al. "Generative adversarial nets". In: *Advances in neural information processing systems*, 2014, pp. 2672–2680.
11. Gaëtan Hadjeres, Frank Nielsen, and François Pachet. "GLSR-VAE: Geodesic latent space regularization for variational autoencoder architectures". In: *2017 IEEE Symposium Series on Computational Intelligence*. IEEE. 2017, pp. 1–7.
12. John R Hershey and Peder A Olsen "Approximating the Kullback Leibler divergence between Gaussian mixture models". In: *2007 IEEE International Conference on Acoustics, Speech and Signal Processing*. Vol. 4. IEEE. 2007, pp. IV–317.
13. Xianxu Hou et al. "Deep feature consistent variational autoencoder". In: *2017 IEEE Winter Conference on Applications of Computer Vision*. IEEE. 2017, pp. 1133–1141.
14. Prateek Jain and Purushottam Kar. "Non-convex optimization for machine learning". In: *Foundations and Trends®in Machine Learning* 10.3–4 (2017), pp. 142–336.
15. Diederik P Kingma and Max Welling. "Auto-encoding variational Bayes". In: *International Conference on Learning Representations*. 2014.
16. Shiqi Liu et al. "Discovering influential factors in variational autoencoders". In: *Pattern Recognition* 100 (2020), p. 107166.
17. Lars Mescheder, Sebastian Nowozin, and Andreas Geiger. "Adversarial variational bayes: Unifying variational autoencoders and generative adversarial networks". In: *International Conference on Machine Learning*. 2017.
18. Andrew Y Ng and Michael I Jordan. "On discriminative vs. generative classifiers: A comparison of logistic regression and naive Bayes". In: *Advances in neural information processing systems*. 2002, pp. 841–848.
19. Stephen Odaibo. "Tutorial: Deriving the Standard Variational Autoencoder (VAE) Loss Function". In: *arXiv preprint arXiv:1907.08956* (2019).
20. Yunchen Pu et al. "Variational autoencoder for deep learning of images, labels and captions". In: *Advances in neural information processing systems*. 2016, pp. 2352–2360.
21. Danilo Jimenez Rezende, Shakir Mohamed, and Daan Wierstra. "Stochastic backpropagation and approximate inference in deep generative models". In: *International Conference on Machine Learning*. 2014.
22. David E Rumelhart, Geoffrey E Hinton, and Ronald J Williams. "Learning representations by backpropagating errors". In: *Nature* 323.6088 (1986), pp. 533–536.
23. Cristian Sminchisescu and Allan Jepson. "Generative modeling for continuous non-linearly embedded visual inference". In: *Proceedings of the twenty-first international conference on Machine learning*. 2004, p. 96.

24. Hiroshi Takahashi et al. "Variational autoencoder with implicit optimal priors". In: *Proceedings of the AAAI Conference on Artificial Intelligence*. Vol. 33. 2019, pp. 5066–5073.
25. Michalis Titsias and Miguel Làzaro-Gredilla. "Doubly stochastic variational Bayes for non-conjugate inference". In: *International conference on machine learning*. 2014, pp. 1971–1979.
26. Arash Vahdat and Jan Kautz. "NVAE: A deep hierarchical variational autoencoder". In: *Advances in Neural Information Processing Systems* 33 (2020).
27. Jacob Walker et al. "An uncertain future: Forecasting from static images using variational autoencoders". In: *European Conference on Computer Vision*. Springer. 2016, pp. 835–851.
28. Weichang Yu, Lamiae Azizi, and John T Ormerod. "Variational nonparametric discriminant analysis". In: *Computational Statistics & Data Analysis* 142 (2020), p. 106817.
29. Shengjia Zhao, Jiaming Song, and Stefano Ermon. "Towards deeper understanding of variational autoencoding models". In: *arXiv preprint arXiv:1702.08658* (2017).

Chapter 21
Adversarial Autoencoders

21.1 Introduction

Suppose there is a generative model that takes random noise as input and generates a data point. The aim is to have the generated data point to be of good quality; therefore, there is a need to judge its quality. One way to judge it is to observe the generated sample and assess its quality visually. In this case, the judge is a human. However, it is not possible to take the derivative of a human's judgment for optimization. Generative Adversarial Network (GAN), proposed in [11], has the same idea, but it can take the derivative of the judgment using a classifier as the judge rather than a human. Therefore, there is a generator generating a sample and a binary classifier (or discriminator) that classifies the generated sample to be a real or generated (fake) sample. This classifier can be a pretrained network that is already trained by both real and generated (fake) data points. However, GAN puts a step ahead and lets the classifier be trained simultaneously with training the generator. This is the core idea of adversarial learning where the classifier, also called the discriminator, and the generator compete with one another; hence, they make each other gradually stronger by this competition [12].

It is noteworthy that the term "adversarial" is used in two main streams of research in machine learning and they should not be confused. These two research areas include the following:

- Adversarial attack, which is also called learning with adversarial examples or adversarial machine learning: this line of research inspects some examples that can be changed slightly, but wisely to fool a trained learning model. For example, perturbation of specific pixels in the input image may change the decision of the learning model. The reason for this can be analyzed theoretically. Examples of this research area include [17, 20–22, 30].

B. Ghojogh et al., *Elements of Dimensionality Reduction and Manifold Learning*,
https://doi.org/10.1007/978-3-031-10602-6_21

- Adversarial learning for generation: this line of research is categorized as generative models [33] and/or methods based on that. GAN is in this line of research. This chapter focuses on this research area.

A good tutorial on GAN is [10], but it does not cover the most recent methods in adversarial learning. Additionally, an honorary introduction of GAN, by several main contributors of GAN, is [12]. Some existing surveys on GAN include: [2, 9, 16, 37, 42]. More information on GAN models can be found in [7].

21.2 Generative Adversarial Network (GAN)

21.2.1 *Adversarial Learning: The Adversarial Game*

The original GAN, also called the vanilla GAN, was proposed in [11]. Consider a d-dimensional dataset with n data points, i.e., $\{\boldsymbol{x}_i \in \mathbb{R}^d\}_{i=1}^n$. In GAN, there is a generator G, which takes a p-dimensional random noise $\boldsymbol{z} \in \mathbb{R}^p$ as the input and outputs a d-dimensional generated point $\boldsymbol{x} \in \mathbb{R}^d$. Therefore, it is the mapping $G : \boldsymbol{z} \to \boldsymbol{x}$, where:

$$G(\boldsymbol{z}) = \boldsymbol{x}. \tag{21.1}$$

The random noise can be seen as a latent factor on which the generated data point is conditioned. The probabilistic graphical model of the generator is a variable $\boldsymbol{x}$ conditioned on a latent variable $\boldsymbol{z}$ (see [10, Fig. 13] for its visualization).

Let the distribution of the random noise be denoted by $\boldsymbol{z} \sim p_z(\boldsymbol{z})$. The aim is to have the generated $\widehat{\boldsymbol{x}}$ be similar to the original (or real) data point $\boldsymbol{x}$ in the dataset. A module is needed to judge the quality of the generated point to see how similar it is to the real point. This module can be a human, but it is not possible to take derivative of human's judgment for optimization! A classifier, also called the discriminator or critic, can fill the role of the judging module. The discriminator, denoted by $D : \boldsymbol{x} \to [0, 1]$, is a binary classifier that classifies the generated point as a real or generated point:

$$D(\boldsymbol{x}) := \begin{cases} 1 & \text{if } \boldsymbol{x} \text{ is real,} \\ 0 & \text{if } \boldsymbol{x} \text{ is generated (fake).} \end{cases} \tag{21.2}$$

The perfect discriminator outputs one for real points and zero for generated points. The discriminator's output is in the range [0, 1], where the output for the real data is closer to one and the output for the fake data is closer to zero. If the generated point closely resembles a real data point, the classifier may make a mistake and output a value close to one. Therefore, if the classifier makes a mistake for the generated point, the generator has accurately generated the data point.

The discriminator can be a pretrained classifier, but the problem can be made to be more sophisticated. Consider training the discriminator simultaneously while still training the generator. This makes the discriminator D and the generator G gradually stronger, while they compete against each other (as in a game of tug-of-war). On the one hand, the generator tries to generate realistic points to fool the discriminator; while on the other hand, the discriminator tries to discriminate the fake (i.e., generated) point from a real point. When one of them improves in its training, the other one tries to become stronger to be able to compete. Therefore, there is an adversarial game[1] between the generator and the discriminator. This game is zero-sum because whatever one of them loses, the other wins.

21.2.2 Optimization and Loss Function

The probability distributions of the dataset and noise are expressed by $p_{\text{data}}(\boldsymbol{x})$ and $p_z(z)$, respectively. The structure of GAN is depicted in Fig. 21.1, which demonstrates that the discriminator is trained by real points from the dataset as well as generated points from the generator. The discriminator and generator are trained simultaneously. The optimization loss function for both the discriminator and generator is:

$$\min_G \max_D \; V(D, G) := \mathbb{E}_{\boldsymbol{x}\sim p_{\text{data}}(\boldsymbol{x})}\Big[\log\big(D(\boldsymbol{x})\big)\Big] + \mathbb{E}_{z\sim p_z(z)}\Big[\log\Big(1 - D\big(G(z)\big)\Big)\Big], \tag{21.3}$$

where $\mathbb{E}[.]$ denotes the expectation operator and the loss function $V(D, G)$ is also called the value function of the game. In practice, the Monte Carlo approximation

Fig. 21.1 The structure of GAN

[1] Game theory is a field of mathematics, where two or more agents compete or cooperate with each other in a game to maximize their payoffs. In an adversarial game, the agents compete with each other adversarially, where their payoffs are against each other.

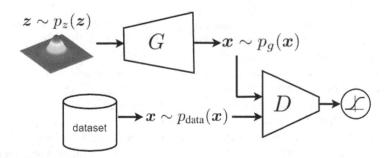

[8] of the expectation can be used, where the expectations are replaced with averages over the minibatch. This loss function is in the form of a cross-entropy loss.

The first term in Eq. (21.3) is the expectation over the real data. This term is only used for the discriminator, while it is a constant for the generator. According to Eq. (21.2), $D(\boldsymbol{x})$ outputs one (the larger label) for the real data. Therefore, the discriminator maximizes this term because it assigns the larger label to the real data.

The second term in Eq. (21.3) is the expectation over noise. It inputs the noise $\boldsymbol{z}$ to the generator to obtain $G(\boldsymbol{z})$. The output of the generator, which is the generated point, is fed as an input into the discriminator (see Fig. 21.1) to obtain $D\big(G(\boldsymbol{z})\big)$. The discriminator wants to minimize $D\big(G(\boldsymbol{z})\big)$ because the smaller label is assigned to the generated data, according to Eq. (21.2). In other words, the discriminator wants to maximize $1-D\big(G(\boldsymbol{z})\big)$. As logarithm is a monotonic function, it is possible to say that the discriminator wants to maximize $\mathbb{E}_{\boldsymbol{z}\sim p_z(\boldsymbol{z})}[\log(1-D(G(\boldsymbol{z})))]$, which is the second term in Eq. (21.3). As opposed to the discriminator, the generator minimizes $\mathbb{E}_{\boldsymbol{z}\sim p_z(\boldsymbol{z})}[\log(1-D(G(\boldsymbol{z})))]$, which is the second term in Eq. (21.3). This is because the generator wants to fool the discriminator to label the generated data as real data.

Equation (21.3) is a minimax optimization problem [3] and can be solved using alternating optimization (see Chap. 4), where optimization occurs over D and G iteratively until convergence (i.e., Nash equilibrium). The original GAN [11] uses a stochastic gradient descent step (see Chap. 4) to update each variable in the alternating optimization. If the loss function in Eq. (21.3) is signified by $V(D, G)$, the alternating optimization is as follows:

$$D^{(k+1)} := D^{(k)} + \eta^{(k)} \frac{\partial}{\partial D}\Big(V(D, G^{(k)})\Big), \tag{21.4}$$

$$G^{(k+1)} := G^{(k)} - \eta^{(k)} \frac{\partial}{\partial G}\Big(V(D^{(k+1)}, G)\Big), \tag{21.5}$$

where k is the index of the iteration and $\eta^{(k)}$ is the learning rate at iteration k. Throughout this chapter, derivatives with respect to D and G indicate the derivatives with respect to the parameters (weights) of the D and G networks, respectively. Equations (21.4) and (21.5) are one step of gradient ascent and gradient descent, respectively. Note that the gradients here are the average of the gradients in the minibatch. Every minibatch includes both real and generated data. The paper [11] suggests that Eq. (21.4) can be performed several times before performing Eq. (21.5). However, the experiments of that paper perform Eq. (21.4) only once before performing Eq. (21.5). Another way to solve the optimization problem in GANs is simultaneous optimization [28] in which Eqs. (21.4) and (21.5) are performed simultaneously and not one after the other.

Remark 21.4 (Minimax Versus Maximin in GAN [10, Section 5]) Equation (21.3) demonstrates that the optimization of GAN is a minimax problem:

$$\min_{G} \max_{D} \quad V(D, G). \tag{21.6}$$

By changing the order of optimization, GAN can be seen as a maximin problem [10]:

$$\max_{D} \min_{G} \quad V(D, G). \tag{21.7}$$

In fact, under some conditions, Eqs. (21.6) and (21.7) are equivalent [3].

21.2.3 Network Structure of GAN

In practice, the discriminator and generator are two (deep) neural networks. The structure of GAN is depicted in Fig. 21.1. The first layer of the discriminator network is d-dimensional, and its last layer is one dimensional with a scalar output. In the original GAN, the maxout activation function [13] is used for all layers except the last layer, which has the sigmoid activation function to output a probability for modelling Eq. (21.2). The closer the output of D is to one, the more likely its input is to be real.

The generator network has a p-dimensional input layer for noise and a d-dimensional output layer for generating data. In the generator, a combination of ReLU [32] and sigmoid activation functions are used. The space of noise as the input to the generator is called the latent space or the latent factor. Equations (21.4) and (21.5) are performed using backpropagation in the neural networks.

21.2.4 Optimal Solution of GAN

Theorem 21.1 ([11, Proposition 1]) *For a fixed generator G, the optimal discriminator is:*

$$D^*(\boldsymbol{x}) = \frac{p_{data}(\boldsymbol{x})}{p_{data}(\boldsymbol{x}) + p_g(\boldsymbol{x})}, \tag{21.8}$$

where $p_{data}(\boldsymbol{x})$ is the probability distribution of the real dataset evaluated at point $\boldsymbol{x}$ and $p_g(\boldsymbol{x})$ is the probability distribution of the output of the generator evaluated at point $\boldsymbol{x}$.

Proof According to the definition of expectation, the loss function in Eq. (21.3) can be stated as:

$$V(D, G) = \int_{x} p_{\text{data}}(x) \log(D(x))dx + \int_{z} p_z(z) \log(1 - D(G(z)))dz.$$

According to Eq. (21.1):

$$G(z) = x \implies z = G^{-1}(x) \implies dz = (G^{-1})'(x)dx,$$

where $(G^{-1})'(x)$ is the derivative of $(G^{-1})(x)$ with respect to x. Therefore:

$$V(D, G) = \int_{x} p_{\text{data}}(x) \log(D(x))dx + \int_{x} p_z(G^{-1}(x)) \log(1 - D(x))(G^{-1})'(x)dx.$$

The relation between the distributions of the input and output of the generator is:

$$p_g(x) = p_z(z) \times G^{-1}(x) = p_z(G^{-1}(x))\, G^{-1}(x), \tag{21.9}$$

where $G^{-1}(x)$ is the Jacobian of the distribution at point x. Therefore:

$$\begin{aligned} V(D, G) &= \int_{x} p_{\text{data}}(x) \log(D(x))dx + \int_{x} p_g(x) \log(1 - D(x))dx \\ &= \int_{x} \Big(p_{\text{data}}(x) \log(D(x)) + p_g(x) \log(1 - D(x))\Big)dx. \end{aligned} \tag{21.10}$$

For optimization in Eq. (21.3), taking the derivative with respect to $D(x)$ results in:

$$\begin{aligned} &\frac{\partial V(D, G)}{\partial D(x)} \overset{(a)}{=} \frac{\partial}{\partial D(x)}\Big(p_{\text{data}}(x) \log(D(x)) + p_g(x) \log(1 - D(x))\Big) \\ &= \frac{p_{\text{data}}(x)}{D(x)} - \frac{p_g(x)}{1 - D(x)} = \frac{p_{\text{data}}(x)(1 - D(x)) - p_g(x)D(x)}{D(x)(1 - D(x))} \overset{\text{set}}{=} 0 \\ &\implies p_{\text{data}}(x) - p_{\text{data}}(x)D(x) - p_g(x)D(x) = 0 \\ &\implies D(x) = \frac{p_{\text{data}}(x)}{p_{\text{data}}(x) + p_g(x)}, \end{aligned}$$

where (a) is because taking the derivative with respect to $D(x)$ considers a specific x and therefore removes the integral (summation). □

Theorem 21.2 ([11, Theorem 1]) *The optimal solution of GAN is when the distribution of the generated data becomes equal to the distribution of the data:*

$$p_{g^*}(x) = p_{data}(x). \tag{21.11}$$

Proof Putting the optimum $D^*(x)$, i.e., Eq. (21.8), in Eq. (21.10) results in:

$$
\begin{aligned}
V(D^*, G) &= \int_{x} \Big(p_{\text{data}}(\boldsymbol{x}) \log(D^*(\boldsymbol{x})) + p_g(\boldsymbol{x}) \log(1 - D^*(\boldsymbol{x})) \Big) d\boldsymbol{x} \\
&\overset{(21.8)}{=} \int_{x} \Big[p_{\text{data}}(\boldsymbol{x}) \log\Big(\frac{p_{\text{data}}(\boldsymbol{x})}{p_{\text{data}}(\boldsymbol{x}) + p_g(\boldsymbol{x})} \Big) + p_g(\boldsymbol{x}) \log\Big(\frac{p_g(\boldsymbol{x})}{p_{\text{data}}(\boldsymbol{x}) + p_g(\boldsymbol{x})} \Big) \Big] d\boldsymbol{x} \\
&= \int_{x} \Big[p_{\text{data}}(\boldsymbol{x}) \log\Big(\frac{p_{\text{data}}(\boldsymbol{x})}{2 \times \frac{p_{\text{data}}(\boldsymbol{x}) + p_g(\boldsymbol{x})}{2}} \Big) + p_g(\boldsymbol{x}) \log\Big(\frac{p_g(\boldsymbol{x})}{2 \times \frac{p_{\text{data}}(\boldsymbol{x}) + p_g(\boldsymbol{x})}{2}} \Big) \Big] d\boldsymbol{x} \\
&= \int_{x} \Big[p_{\text{data}}(\boldsymbol{x}) \log\Big(\frac{p_{\text{data}}(\boldsymbol{x})}{\frac{p_{\text{data}}(\boldsymbol{x}) + p_g(\boldsymbol{x})}{2}} \Big) \\
&\qquad + p_g(\boldsymbol{x}) \log\Big(\frac{p_g(\boldsymbol{x})}{\frac{p_{\text{data}}(\boldsymbol{x}) + p_g(\boldsymbol{x})}{2}} \Big) \Big] d\boldsymbol{x} + \log(\frac{1}{2}) + \log(\frac{1}{2}) \\
&= \int_{x} \Big[p_{\text{data}}(\boldsymbol{x}) \log\Big(\frac{p_{\text{data}}(\boldsymbol{x})}{\frac{p_{\text{data}}(\boldsymbol{x}) + p_g(\boldsymbol{x})}{2}} \Big) + p_g(\boldsymbol{x}) \log\Big(\frac{p_g(\boldsymbol{x})}{\frac{p_{\text{data}}(\boldsymbol{x}) + p_g(\boldsymbol{x})}{2}} \Big) \Big] d\boldsymbol{x} - \log(4) \\
&\overset{(a)}{=} \text{KL}\Big(p_{\text{data}}(\boldsymbol{x}) \Big\| \frac{p_{\text{data}}(\boldsymbol{x}) + p_g(\boldsymbol{x})}{2} \Big) + \text{KL}\Big(p_g(\boldsymbol{x}) \Big\| \frac{p_{\text{data}}(\boldsymbol{x}) + p_g(\boldsymbol{x})}{2} \Big) - \log(4),
\end{aligned}
\tag{21.12}
$$

where (a) is because of the definition of KL divergence. The Jensen-Shannon Divergence (JSD) is defined as [34]:

$$
\text{JSD}(P \| Q) := \frac{1}{2}\text{KL}(P \| \frac{1}{2}(P + Q)) + \frac{1}{2}\text{KL}(Q \| \frac{1}{2}(P + Q)), \tag{21.13}
$$

where P and Q denote the probability densities. In contrast to KL divergence, the JSD is symmetric. The obtained $V(D^*, G)$ can be restated as:

$$
V(D^*, G) = 2\,\text{JSD}\big(p_{\text{data}}(\boldsymbol{x}) \parallel p_g(\boldsymbol{x})\big) - \log(4), \tag{21.14}
$$

According to Eq. (21.3), the generator minimizes $V(D^*, G)$. As the JSD is nonnegative, the above loss function is minimized if:

$$
\text{JSD}\big(p_{\text{data}}(\boldsymbol{x}) \parallel p_{g^*}(\boldsymbol{x})\big) = 0 \implies p_{\text{data}}(\boldsymbol{x}) = p_{g^*}(\boldsymbol{x}).
$$

□

Corollary 21.1 ([11, Theorem 1]) *From Eqs. (21.11) and (21.14), the optimal loss function in GAN can be concluded to be:*

$$
V(D^*, G^*) = -\log(4). \tag{21.15}
$$

It is noteworthy that Eq. (21.13) in GAN can be generalized to [18]:

$$
\text{JSD}_{\pi}(P \| Q) := \pi\,\text{KL}\big(P \| \pi P + (1 - \pi) Q\big) + (1 - \pi)\,\text{KL}\big(Q \| \pi P + (1 - \pi) Q\big), \tag{21.16}
$$

with $\pi \in (0, 1)$. Its special case is Eq. (21.13), with $\pi = 0.5$.

Corollary 21.2 *From Eqs. (21.8) and (21.11), it is possible to conclude that at convergence (i.e., Nash equilibrium), the discriminator cannot distinguish between generated and real data:*

$$\begin{aligned} D^*(\boldsymbol{x}) &= 0.5, \quad \forall \boldsymbol{x} \sim p_{data}(\boldsymbol{x}), \\ D^*(\boldsymbol{x}) &= 0.5, \quad \boldsymbol{x} = G^*(\boldsymbol{z}), \forall \boldsymbol{z} \sim p_z(\boldsymbol{z}). \end{aligned} \tag{21.17}$$

Lemma 21.1 (Label Smoothing in GAN [38, Section 3.4]) *It is demonstrated that replacing labels* 0 *and* 1*, respectively, with smoother values of* 0.1 *and* 0.9 *[39] can improve the neural network against adversarial attacks [14]. If the labels of discriminator D are smoothed for the real and generated data to be α and β, respectively, the optimal discriminator becomes [38]:*

$$D^*(\boldsymbol{x}) = \frac{\alpha p_{data}(\boldsymbol{x}) + \beta p_g(\boldsymbol{x})}{p_{data}(\boldsymbol{x}) + p_g(\boldsymbol{x})}, \tag{21.18}$$

which generalizes Eq. (21.8). The presence of $p_g(\boldsymbol{x})$ causes a problem because, for an $\boldsymbol{x}$ with a small $p_{data}(\boldsymbol{x})$ and large $p_g(\boldsymbol{x})$, the point does not change the generator well enough to get close to the real data. Therefore, it is recommended to set $\beta = 0$ to have one-sided label smoothing. In this case, the optimal discriminator is:

$$D^*(\boldsymbol{x}) = \frac{\alpha p_{data}(\boldsymbol{x})}{p_{data}(\boldsymbol{x}) + p_g(\boldsymbol{x})}. \tag{21.19}$$

Proof (Sketch) Using α and β in the proof of Theorem 21.1 results in Eq. (21.18). □

21.2.5 Convergence and Equilibrium Analysis of GAN

Theorem 21.3 ([11, Proposition 2]) *If the discriminator and generator have enough capacity and, at every iteration of the alternating optimization, the followings occur:*

- *the discriminator is allowed to reach its optimum value as in Eq. (21.8), and*
- *$p_g(\boldsymbol{x})$ is updated to minimize $V(D^*, G)$, as stated in Eq. (21.14),*

$p_g(\boldsymbol{x})$ converges to $p_{data}(\boldsymbol{x})$, as stated in Eq. (21.11).

Proof The KL divergences in Eq. (21.12) are convex functions with respect to $p_g(\boldsymbol{x})$. Therefore, with sufficiently small updates of $p_g(\boldsymbol{x})$, it converges to $p_{\text{data}}(\boldsymbol{x})$. Note that Eq. (21.12), which is used here, holds if Eq. (21.8) holds, i.e., the discriminator is allowed to reach its optimum value. □

The GAN loss, i.e., Eq. (21.3), can be restated as [31]:

$$\min_{G} \max_{D} \quad V(D, G) := \mathbb{E}_{\boldsymbol{x} \sim p_{\text{data}}(\boldsymbol{x})}\Big[f\big(D(\boldsymbol{x})\big)\Big] + \mathbb{E}_{z \sim p_z(z)}\Big[f\Big(-D\big(G(z)\big)\Big)\Big], \tag{21.20}$$

where f is the negative logistic function, i.e., $f(x) := -\log(1 + \exp(-x))$. In fact, the function $f(.)$ can be any concave function. This formulation is slightly different from the original GAN in the sense that, here, the discriminator D outputs a real-valued scalar (without any activation function), while the discriminator of Eq. (21.3) outputs values in the range (0, 1) after a sigmoid activation function. If D outputs 0.5 and 0, it means that it is completely confused in Eqs. (21.3) and (21.20), respectively. Equation (21.20) is a concave-concave loss function in most of the domain of the discriminator [31, Proposition 3.1].

Theorem 21.4 ([31, Theorem 3.1]) *After satisfying several reasonable assumptions (see [31] for details), a GAN with the loss function of Eq. (21.20) is locally exponentially stable.*

Lemma 21.2 (Nash Equilibrium in GAN [4]) *The Nash equilibrium is the state of the game where no player can improve its gain by choosing a different strategy. At the Nash equilibrium of GAN:*

$$V(D, G^*) \leq V(D^*, G^*) \leq V(D^*, G), \tag{21.21}$$

which is obvious because $V(G, D)$ is being minimized and maximized by the generator and discriminator, respectively, in Eq. (21.3).

Empirical experiments have demonstrated that GAN may not reach its Nash equilibrium in practice [4]. Regularization can help the convergence of GAN to the Nash equilibrium [27]. It is demonstrated in [27] that an effective regularization for GAN is through noise injection [6], in which independent Gaussian noise is added to the training data points.

Definition 21.1 (Proximal Equilibrium [4]) The proximal operator (see Chap. 4) can be used in the loss function of GAN:

$$\min_{G} \max_{D} \Big(V_{\text{prox}}(D, G) := \max_{\widetilde{D}} (V(\widetilde{D}, G) - \lambda \|\widetilde{D} - D\|_2^2) \Big),$$

where $V(D, G)$ is defined in Eq. (21.3) and $\lambda > 0$ is the regularization parameter. The equilibrium of the game having this loss function is called the proximal equilibrium.

Theorem 21.5 (Convergence of GAN Based on the Jacobian [28]) *Let the updated solution of GAN optimization at every iteration be obtained by an operator $F(D, G)$, such as a step of gradient descent. The convergence of GAN can be explained based on the Jacobian of $F(D, G)$, with respect to D and G. If the* *absolute values of some eigenvalues of the Jacobian are larger than one, GAN will not converge to the Nash equilibrium. If all eigenvalues have absolute values less than one, GAN will converge to the Nash equilibrium with a linear rate of* $\mathcal{O}(|\lambda_{\max}|^k)$*, where* $\lambda_{\max}$ *is the eigenvalue with the largest absolute eigenvalue and* k *is the iteration index. If all eigenvalues have a unit absolute value, GAN may or may not converge.*

Refer to [5] for duality in GAN, which is not explained here for brevity. Moreover, some papers have specifically combined GAN with game theory (see [1, 35, 40, 41]).

21.3 Mode Collapse Problem in GAN

It is expected that GAN will learn a meaningful latent space of z, so that every specific value of z maps to a specific generated data point $\boldsymbol{x}$. Additionally, nearby z values in the latent space should be mapped to similar but slightly different generations. The mode collapse problem [29], also known as the Helvetica scenario [10], refers to when the generator cannot learn a perfectly meaningful latent space; rather, it learns to map several different z values to the same generated data point. Mode collapse usually occurs in GAN when the distribution of the training data, $p_{\text{data}}(\boldsymbol{x})$, has multiple modes. This problem is a common problem in GAN models.

Figure 21.2 illustrates an example of the mode collapse, which demonstrates the training steps of a GAN model when the training data are a mixture of Gaussians [29]. In different training steps, GAN learns to map all z values to one of the modes of the mixture. When the discriminator learns to reject the generation of a mode, the generator learns to map all z values to another mode. However, it never learns to generate all modes of the mixture. GAN is expected to map some part, and not all parts, of the latent space to one of the modes, so that all modes are covered by the whole latent space.

Another statement of the mode collapse is as follows [43, Fig. 1]. Assume $p_{\text{data}}(\boldsymbol{x})$ is multimodal, while the latent space $p_z(z)$ has only one mode. Consider two points $\boldsymbol{x}_1$ and $\boldsymbol{x}_2$ from two modes of data, whose corresponding latent noises are z_1 and z_2, respectively. According to the mean value theorem, there is a latent noise with the absolute gradient value $\|\boldsymbol{x}_2 - \boldsymbol{x}_1\|/\|z_2 - z_1\|$, where $\|.\|$ is a norm. As this gradient is Lipschitz continuous, there is a problem when the two modes are far apart resulting in a large $\|\boldsymbol{x}_2 - \boldsymbol{x}_1\|$. In this case, the latent noises between z_1

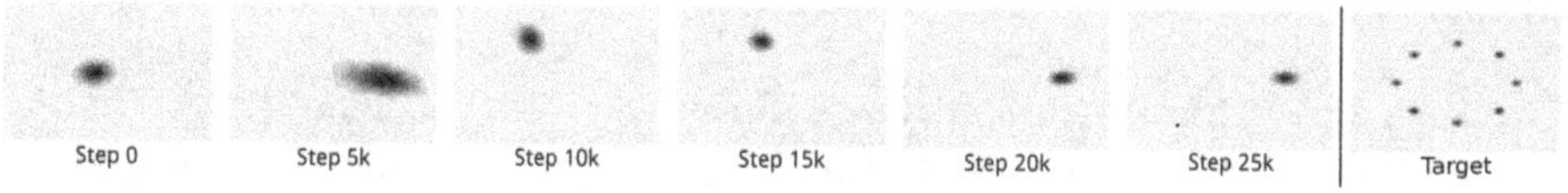

Fig. 21.2 An example of mode collapse in a GAN. The image is from [29]

and z_2 generate data points between $\boldsymbol{x}_1$ and $\boldsymbol{x}_2$ which are not in the modes of data and are, therefore, not valid.

Various methods exist that resolve the mode collapse problem in GANs and adversarial learning. Some of them make the latent space a mixture distribution to imitate the generation of the multimodal training data. One of the first methods for the mode collapse problem was minibatch discrimination [38, Section 3.2]. Additional methods for resolving the mode collapse problem are listed and detailed in the survey [7].

21.4 Autoencoders Based on Adversarial Learning

Previously, variational Bayes was used in an autoencoder setting to obtain a variational autoencoder [19] (see Chap. 20). Likewise, adversarial learning can be used in an autoencoder setting [24]. Several adversarial-based autoencoders exist which are introduced in the following section.

21.4.1 Adversarial Autoencoder (AAE)

21.4.1.1 Unsupervised AAE

The Adversarial Autoencoder (AAE) was proposed in [26]. In contrast to the variational autoencoder [19] (see Chap. 20), which uses KL divergence and evidence lower bound, AAE uses adversarial learning to impose a specific distribution on the latent variable in its coding layer. The structure of AAE is depicted in Fig. 21.3. Each of the blocks B_1, B_2, and B_3 in this figure has several network layers with nonlinear activation functions. AAE has an encoder (i.e., block B_1) and a decoder (i.e., block B_2). The input of the encoder is a real data point $\boldsymbol{x} \in \mathbb{R}^d$ and the output of the decoder is the reconstructed data point $\widehat{\boldsymbol{x}} \in \mathbb{R}^d$. One of the low-dimensional middle layers is the latent (or code) layer, denoted by $\boldsymbol{z} \in \mathbb{R}^p$, where $p \ll d$. The encoder and decoder model conditional distributions $p(\boldsymbol{z}|\boldsymbol{x})$ and $p(\boldsymbol{x}|\boldsymbol{z})$, respectively. Let the distribution of the latent variable in the autoencoder be denoted by $q(\boldsymbol{z})$. This is the posterior distribution of the latent variable. The blocks B_1 and B_3 are the generator G and discriminator D of an adversarial network, respectively. There is also a prior distribution, denoted by $p(\boldsymbol{z})$, on the latent variable, which is chosen by the user. This prior distribution can be a p-dimensional normal distribution $\mathcal{N}(\boldsymbol{0}, \boldsymbol{I})$, for example. The encoder of the autoencoder (i.e., block B_1) is the generator G, which generates the latent variable from the posterior distribution:

$$G(\boldsymbol{x}) = \boldsymbol{z} \sim q(\boldsymbol{z}). \tag{21.22}$$

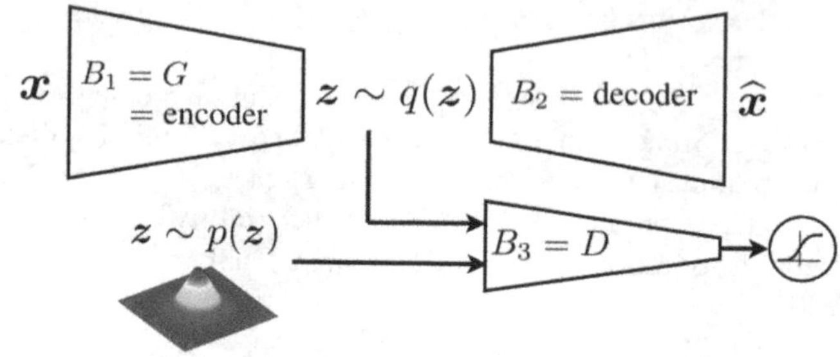

Fig. 21.3 The structure of the unsupervised AAE

The discriminator D (i.e., block B_3) has a single output neuron with a sigmoid activation function. It classifies the latent variable z to be a real latent variable from the prior distribution $p(z)$ or a generated latent variable by the encoder of the autoencoder:

$$D(z) := \begin{cases} 1 \text{ if } z \text{ is real, i.e., } z \sim p(z) \\ 0 \text{ if } z \text{ is generated, i.e., } z \sim q(z). \end{cases} \tag{21.23}$$

As explained above, block B_1 is shared between the autoencoder and the adversarial network. This adversarial learning makes both the autoencoder and adversarial network gradually stronger because the autoencoder tries to generate the latent variable, which is similar to the real latent variable from the prior distribution. In this way, it tries to fool the discriminator. The discriminator, on the other hand, tries to improve so as not to be fooled by the encoder of the autoencoder.

In AAE, alternating optimization is used (see Chap. 4), where the reconstruction and regularization phases are repeated iteratively. In the reconstruction phase, the mean squared error is minimized between the data $\boldsymbol{x}$ and the reconstructed data $\widehat{\boldsymbol{x}}$. In the regularization phase, the discriminator and generator are updated using the GAN approach. For each of these updates, stochastic gradient descent is used with backpropagation (see Chap. 4). Overall, the two phases are performed as follows:

$$B_1', B_2^{(k+1)} := \arg\min_{B_1, B_2} \|\widehat{\boldsymbol{x}} - \boldsymbol{x}\|_2^2, \tag{21.24}$$

$$\begin{cases} B_3^{(k+1)} := B_3^{(k)} - \eta^{(k)} \frac{\partial}{\partial B_3}\Big(V(B_3, B_1')\Big), \\ B_1^{(k+1)} := B_1' - \eta^{(k)} \frac{\partial}{\partial B_1}\Big(V(B_3^{(k+1)}, B_1)\Big), \end{cases} \tag{21.25}$$

where $B_1 = G$ and $B_3 = D$ (see Fig. 21.3). Equation (21.24) is the reconstruction phase, and Eq. (21.25) is the regularization phase.

21.4.1.2 Sampling the Latent Variable

There are several approaches for sampling the latent variable z from the coding layer of the autoencoder with posterior $q(z)$. The following explains these approaches [26]:

- Deterministic approach: the latent variable is the output of the encoder directly, i.e., $z_i = B_1(\boldsymbol{x}_i)$. The stochasticity in $q(z)$ is in the distribution of the dataset, $p_{\text{data}}(\boldsymbol{x})$.
- Gaussian posterior: this approach is similar to the variational autoencoder [19] (see Chap. 20). The encoder outputs the mean $\boldsymbol{\mu}$ and covariance $\boldsymbol{\Sigma}$, and the latent variable is sampled from the Gaussian distribution, i.e., $z_i \sim \mathcal{N}(\boldsymbol{\mu}(\boldsymbol{x}_i), \boldsymbol{\Sigma}(\boldsymbol{x}_i))$. The stochasticity in $q(z)$ is in both $p_{\text{data}}(\boldsymbol{x})$ and the Gaussian distribution as the output of the encoder.
- Universal approximator posterior: the data point $\boldsymbol{x}$ and some noise $\boldsymbol{\eta}$, with a fixed distribution such as Gaussian, are concatenated as the input to the encoder. Therefore, the latent variable is $z_i = B_1(\boldsymbol{x}_i, \boldsymbol{\eta}_i)$, where $\boldsymbol{\eta}_i \sim \mathcal{N}(\boldsymbol{0}, \boldsymbol{I})$. The stochasticity in $q(z)$ is in both $p_{\text{data}}(\boldsymbol{x})$ and the noise $\boldsymbol{\eta}$.

21.4.1.3 Supervised AAE

There are two variants for supervised AAE [26] where the class labels are utilized. These two structures are illustrated in Fig. 21.4. Let c denote the number of classes. In the first variant, the one-hot encoded label $\boldsymbol{y} \in \mathbb{R}^c$ is fed to the discriminator, i.e., block B_3, by concatenating it to the latent variable z. In this way, the discriminator learns the label of point $\boldsymbol{x}$, as well as discrimination of real and generated latent variables. This makes the generator or the encoder generate the latent variables corresponding to the label of the point for fooling the discriminator.

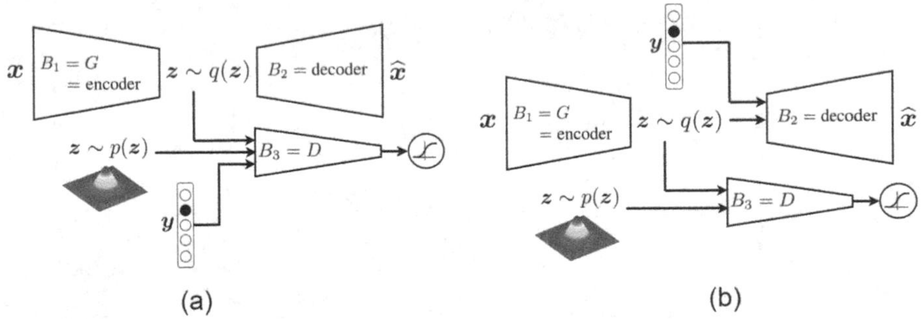

Fig. 21.4 Two structures for supervised AAE

In the second variant of supervised AAE, the one-hot encoded label $\boldsymbol{y}$ is fed to the decoder, i.e., block B_2, by concatenating it to the latent variable $\boldsymbol{z}$. In this way, the decoder learns to reconstruct the data point by using the label of the point. This also ensures that the encoder, which is also the generator, generates the latent variable $\boldsymbol{z}$ based on the label of the point. Therefore, the discriminator improves while competing with the generator, in adversarial learning. Note that the two variants can also be combined, i.e., the one-hot encoded label can be fed to both the discriminator and the decoder.

21.4.1.4 Semisupervised AAE

Consider a partially labelled dataset. The labelled part of the data has c classes. AAE can be used for semisupervised learning with a partially labelled dataset. The structure for semisupervised AAE is depicted in Fig. 21.5. This structure includes an autoencoder (blocks B_1 and B_2), adversarial learning for generating the latent variable (blocks B_1 and B_3), and adversarial learning for generating class labels (blocks B_1 and B_4). The encoder generates the label $\boldsymbol{y} \in \mathbb{R}^c$ and the latent variable $\boldsymbol{z} \in \mathbb{R}^p$. The last layer of the encoder for the label has a softmax activation function to output a c-dimensional vector, whose entries sum to one (behaving as a probability). The last layer of the encoder for the latent variable has a linear activation function.

Semisupervised AAE has three phases—reconstruction, regularization, and semisupervised classification. In the reconstruction phase, the reconstruction error is minimized. The regularization phase trains the discriminator and generator to generate the latent variable $\boldsymbol{z}$. The semisupervised classification phase generates the one-hot encoded class label $\boldsymbol{y}$ for point $\boldsymbol{x}$. If point $\boldsymbol{x}$ has a label, its label is used for training B_1 and B_4. However, if point $\boldsymbol{x}$ does not have a label, a label $\boldsymbol{y} \in \mathbb{R}^c$ is randomly sampled from a categorical distribution, i.e., $\boldsymbol{y} \sim \text{Cat}(\boldsymbol{y})$. This categorical distribution results in a one-hot encoded vector, where the prior probability of every class is estimated based on the proportion of a class's population to the total number of labelled points. An iteration of the alternating optimization for semisupervised learning is:

$$B_1', B_2^{(k+1)} := \arg\min_{B_1, B_2} \|\widehat{\boldsymbol{x}} - \boldsymbol{x}\|_2^2, \tag{21.26}$$

$$\begin{cases} B_3^{(k+1)} := B_3^{(k)} - \eta^{(k)} \frac{\partial}{\partial B_3}\Big(V_z(B_3, B_1')\Big), \\ B_1'' := B_1' - \eta^{(k)} \frac{\partial}{\partial B_1}\Big(V_z(B_3^{(k+1)}, B_1)\Big), \end{cases} \tag{21.27}$$

$$\begin{cases} B_4^{(k+1)} := B_4^{(k)} - \eta^{(k)} \frac{\partial}{\partial B_4}\Big(V_y(B_4, B_1'')\Big), \\ B_1^{(k+1)} := B_1'' - \eta^{(k)} \frac{\partial}{\partial B_1}\Big(V_y(B_4^{(k+1)}, B_1)\Big), \end{cases} \tag{21.28}$$

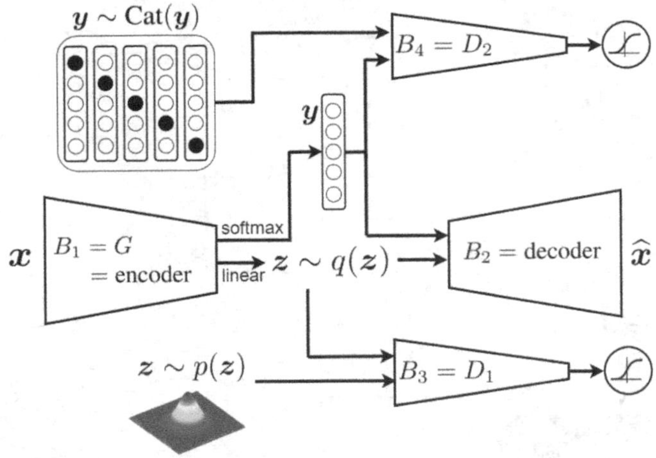

Fig. 21.5 The structure of semisupervised AAE

where $V_z(D, G)$ and $V_y(D, G)$ are the loss functions defined in Eq. (21.3), in which the generated variables are the latent variable $\boldsymbol{z}$ and the one-hot encoded label $\boldsymbol{y}$, respectively.

21.4.1.5 Unsupervised Clustering with AAE

The structure of Fig. 21.5 can be used for clustering, but rather than the classes, it is assumed that there are c clusters. All points in the dataset are unlabelled and the cluster indices are sampled randomly by the categorical distribution. The cluster labels and the latent code are both trained in the three phases, explained in Sect. 21.4.1.4.

21.4.1.6 Dimensionality Reduction with AAE

The AAE can be used for dimensionality reduction and representation learning. The structure of AAE for this purpose is depicted in Fig. 21.6. The encoder generates both the label $\boldsymbol{y} \in \mathbb{R}^c$ and latent variable $\boldsymbol{z} \in \mathbb{R}^p$, where $p \ll d$. Everything remains similar to before, except a network layer $\boldsymbol{W} \in \mathbb{R}^{c \times p}$ is added after the generated label by the encoder. The low-dimensional representation $\widetilde{\boldsymbol{x}} \in \mathbb{R}^p$ is obtained as:

$$\mathbb{R}^p \ni \widetilde{\boldsymbol{x}} = \boldsymbol{W}^\top \boldsymbol{y} + \boldsymbol{z}, \tag{21.29}$$

where $\boldsymbol{z}$ is the latent variable generated by the encoder. The three phases explained in Sect. 21.4.1.4 train the AAE for dimensionality reduction.

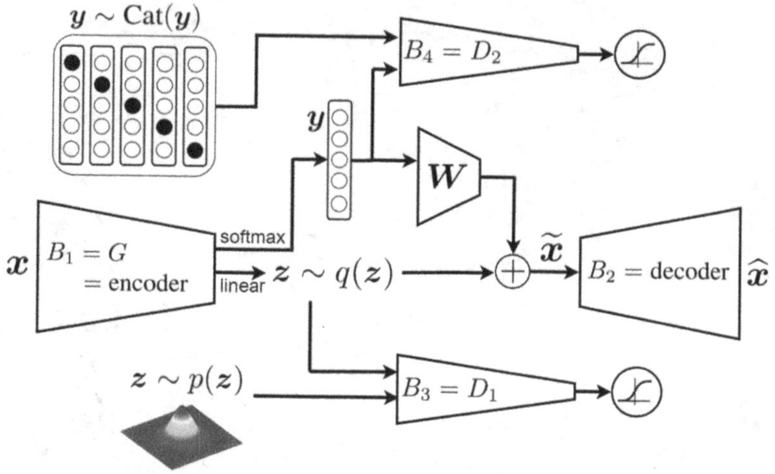

Fig. 21.6 The structure of AAE for dimensionality reduction

21.4.2 PixelGAN Autoencoder

In variational inference (see Chaps. 12 and 20), the Evidence Lower Bound (ELBO) can be restated as [15]:

$$\begin{aligned}\mathbb{E}_{x \sim p_{\text{data}}(x)}[\log(p(x))] > &- \mathbb{E}_{x \sim p_{\text{data}}(x)}[\mathbb{E}_{q(z|x)}[-\log(p(x|z))]] \\ &- \text{KL}(q(z)\|p(z)) - \mathbb{I}(z; x),\end{aligned} \tag{21.30}$$

where $\mathbb{I}(.;.)$ denotes the mutual information. The first and second terms in this lower bound are the reconstruction error and the marginal KL divergence on the latent space. The PixelGAN autoencoder [25] uses this lower bound but ignores its third term, which is the mutual information. The third term is ignored because the optimization of that term makes z independent of x. The reconstruction error is minimized in the reconstruction phase of training, and the KL divergence part is addressed by adversarial learning.

The structure of PixelGAN is illustrated in Fig. 21.7. Block B_1 is the encoder, which has the data point x added with some noise n as the input and outputs the latent code $z \sim q(z|x)$. Block B_2 is the decoder that is a PixelCNN network [36], from which PixelGAN has borrowed its name. This decoder outputs the reconstructed data $\widehat{x}$. The generated latent code z is used as the adaptive biases of the layers in the PixelCNN. Blocks B_1 and B_3 are the generator and discriminator in adversarial learning, respectively, where the aim is to make the distribution of the generated latent code z similar to a prior distribution $p(z)$. In summary, blocks B_1 and B_2 are used for the reconstruction phase and blocks B_1 and B_3 are used for the adversarial learning phase.

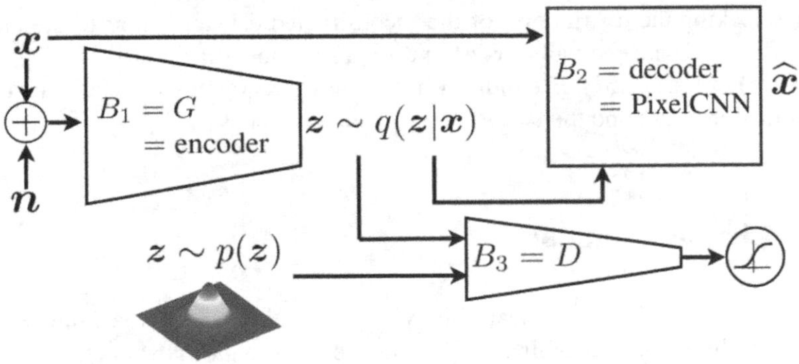

Fig. 21.7 The structure of PixelGAN

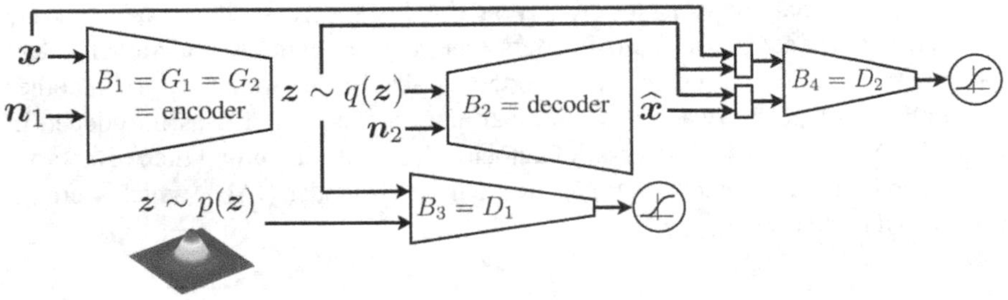

Fig. 21.8 The structure of IAE

21.4.3 Implicit Autoencoder (IAE)

In variational inference (see Chaps. 12 and 20), the Evidence Lower Bound (ELBO) can be restated as [23]:

$$\begin{aligned}\mathbb{E}_{x \sim p_{\text{data}}(x)}[\log(p(x))] \geq &- \text{KL}(q(x, z)\|q(\widehat{x}, z)) \\ &- \text{KL}(q(z)\|p(z)) - H_{\text{data}}(x),\end{aligned} \tag{21.31}$$

where $H_{\text{data}}(x)$ is the entropy of the data, $\widehat{x}$ is the reconstructed data, and z is a latent factor. The proof is straightforward and can be found in [23, Appendix A]. The first and second terms are the reconstruction and regularization terms, respectively. The Implicit Autoencoder (IAE) [23] implements the above distributions in Eq. (21.31), implicitly using networks. The structure of IAE is demonstrated in Fig. 21.8. Block B_1 is the encoder that takes data x and noise n_1 as the inputs and outputs the latent code $z \sim q(z)$. Block B_2 takes the generated latent code z, as well as some noise n_2, and outputs the reconstructed data $\widehat{x}$. Blocks B_1 and B_3 are the generator G_1 and discriminator D_1 of the first adversarial learning method used for making the distribution of the latent code z similar to a prior distribution $p(z)$. Blocks B_1 and B_4 are the generator G_2 and discriminator D_2 of the second adversarial learning method used for making the distribution of the reconstructed data $\widehat{x}$ similar to the data x. The inputs to B_4 are the pairs (x, z) and $(\widehat{x}, z)$ to model the distributions $q(x, z)$ and $q(\widehat{x}, z)$ in Eq. (21.31). In summary, three phases of training are performed—the reconstruction phase and the two adversarial learning phases.

21.5 Chapter Summary

The proposal of adversarial learning was a breakthrough in machine learning because it allowed the machine to generate realistic data that humans think are real! After the proposal of adversarial learning for generating data, it was used in dimensionality reduction by the proposal of adversarial autoencoders. This chapter introduced the concept of adversarial learning and then explained the different types of adversarial autoencoders. One category of inferential autoencoders for inference from data using autoencoders is adversarial autoencoders. Another inferential autoencoder was the variational autoencoder, which was introduced in Chap. 20. Several types of adversarial autoencoders exist, including the Adversarial Autoencoder (AAE), PixelGAN, and Implicit Autoencoder (IAE), which were all introduced in this chapter.

References

1. Sanjeev Arora et al. "Generalization and equilibrium in generative adversarial nets (GANs)". In: *International Conference on Machine Learning*. 2017, pp. 224–232.
2. Antonia Creswell et al. "Generative adversarial networks: An overview". In: *IEEE Signal Processing Magazine* 35.1 (2018), pp. 53–65.
3. Ding-Zhu Du and Panos M Pardalos. *Minimax and applications*. Vol. 4. Springer Science & Business Media, 2013.
4. Farzan Farnia and Asuman Ozdaglar. "Do GANs always have Nash equilibria?" In: *International Conference on Machine Learning*. 2020, pp. 3029–3039.
5. Farzan Farnia and David Tse. "A convex duality framework for GANs". In: *Advances in neural information processing systems*. Vol. 31. 2018.
6. Benyamin Ghojogh and Mark Crowley. "The theory behind overfitting, cross validation, regularization, bagging, and boosting: tutorial". In: *arXiv preprint arXiv:1905.12787* (2019).
7. Benyamin Ghojogh et al. "Generative Adversarial Networks and Adversarial Autoencoders: Tutorial and Survey". In: *arXiv preprint arXiv:2111.13282* (2021).
8. Benyamin Ghojogh et al. "Sampling algorithms, from survey sampling to Monte Carlo methods: Tutorial and literature review". In: *arXiv preprint arXiv:2011.00901* (2020).
9. Liang Gonog and Yimin Zhou. "A review: Generative adversarial networks". In: *2019 14th IEEE Conference on Industrial Electronics and Applications (ICIEA)*, IEEE. 2019, pp. 505–510.
10. Ian Goodfellow. "NIPS 2016 tutorial: Generative adversarial networks". In: *Advances in neural information processing systems, Tutorial rack*. 2016.
11. Ian Goodfellow et al. "Generative adversarial nets". In: *Advances in neural information processing systems*. Vol. 27. 2014.

12. Ian Goodfellow et al. "Generative adversarial networks". In: *Communications of the ACM* 63.11 (2020), pp. 139–144.
13. Ian Goodfellow et al. "Maxout networks". In: *International conference on machine learning*. 2013, pp. 1319–1327.
14. Tamir Hazan, George Papandreou, and Daniel Tarlow. "Adversarial Perturbations of Deep Neural Networks". In: (2017).
15. Matthew D Hoffman and Matthew J Johnson. "ELBO surgery: yet another way to carve up the variational evidence lower bound". In: *Workshop in Advances in Approximate Bayesian Inference, NIPS*. 2016.
16. Yongjun Hong et al. "How generative adversarial networks and their variants work: An overview". In: *ACM Computing Surveys (CSUR)* 52.1 (2019), pp. 1–43.
17. Ling Huang et al. "Adversarial machine learning". In: *Proceedings of the 4th ACM workshop on Security and artificial intelligence*. 2011, pp. 43–58.
18. Ferenc Huszár. "How (not) to train your generative model: Scheduled sampling, likelihood adversary?" In: *arXiv preprint arXiv:1511.05101* (2015).
19. Diederik P Kingma and Max Welling. "Auto-encoding variational Bayes". In: *International Conference on Learning Representations*. 2014.
20. Alexey Kurakin, Ian Goodfellow, Samy Bengio, et al "Adversarial examples in the physical world". In: *International Conference on Learning Representations, Workshop Track*. 2017.
21. Alexey Kurakin, Ian Goodfellow, and Samy Bengio. "Adversarial machine learning at scale". In: *International Conference on Learning Representations*. 2017.
22. Aleksander Madry et al. "Towards deep learning models resistant to adversarial attacks". In: *International Conference on Learning Representations*. 2018.
23. Alireza Makhzani. "Implicit autoencoders". In: *arXiv preprint arXiv:1805.09804* (2018).
24. Alireza Makhzani. "Unsupervised representation learning with autoencoders". PhD thesis. University of Toronto, 2018.
25. Alireza Makhzani and Brendan Frey. "PixelGAN autoencoders". In: *Advances in neural information processing systems*. 2017.
26. Alireza Makhzani et al. "Adversarial autoencoders". In: *arXiv preprint arXiv:1511.05644* (2015).
27. Lars Mescheder, Andreas Geiger, and Sebastian Nowozin. "Which training methods for GANs do actually converge?" In: *International conference on machine learning*. PMLR. 2018, pp. 3481–3490.
28. Lars Mescheder, Sebastian Nowozin, and Andreas Geiger. "The numerics of GANs". In: *Advances in neural information processing systems*. 2017.
29. Luke Metz et al. "Unrolled generative adversarial networks". In: *International Conference on Learning Representations*. 2017.
30. Seyed-Mohsen Moosavi-Dezfooli, Alhussein Fawzi, and Pascal Frossard. "DeepFool: a simple and accurate method to fool deep neural networks". In: *Proceedings of the IEEE conference on computer vision and pattern recognition*. 2016, pp. 2574–2582.
31. Vaishnavh Nagarajan and J Zico Kolter. "Gradient descent GAN optimization is locally stable". In: *Advances in neural information processing systems*. 2017.
32. Vinod Nair and Geoffrey E Hinton. "Rectified linear units improve restricted Boltzmann machines". In: *International Conference on Machine Learning*. 2010.
33. Andrew Y Ng and Michael I Jordan. "On discriminative vs. generative classifiers: A comparison of logistic regression and naive Bayes". In: *Advances in neural information processing systems*. 2002, pp. 841–848.
34. Frank Nielsen. "A family of statistical symmetric divergences based on Jensen's inequality". In: *arXiv preprint arXiv:1009.4004* (2010).
35. Frans A Oliehoek et al. "GANGs: Generative adversarial network games". In: *arXiv preprint arXiv:1712.00679* (2017).
36. Aaron van den Oord et al. "Conditional image generation with PixelCNN decoders". In: *Advances in neural information processing systems*. 2016, pp. 4790–4798.

37. Zhaoqing Pan et al. "Recent progress on generative adversarial networks (GANs): A survey". In: *IEEE Access* 7 (2019), pp. 36322–36333.
38. Tim Salimans et al. "Improved techniques for training GANs". In: *Advances in neural information processing systems*. 29 (2016), pp. 2234–2242.
39. Christian Szegedy et al. "Rethinking the inception architecture for computer vision". In: *Proceedings of the IEEE conference on computer vision and pattern recognition*. 2016, pp. 2818–2826.
40. Hamidou Tembine. "Deep learning meets game theory: Bregman-based algorithms for interactive deep generative adversarial networks". In: *IEEE transactions on cybernetics* 50.3 (2019), pp. 1132–1145.
41. Thomas Unterthiner et al. "Coulomb GANs: Provably optimal Nash equilibria via potential fields". In: *International Conference on Learning Representations*. 2018.
42. Kunfeng Wang et al. "Generative adversarial networks: introduction and outlook". In: *IEEE/CAA Journal of Automatica Sinica* 4.4 (2017), pp. 588–598.
43. Chang Xiao, Peilin Zhong, and Changxi Zheng. "BourGAN: Generative networks with metric embeddings". In: *Advances in neural information processing systems*. 2018.

Index

B. Ghojogh et al., *Elements of Dimensionality Reduction and Manifold Learning*,
https://doi.org/10.1007/978-3-031-10602-6

T

U

V

CPSIA information can be obtained
at www.ICGtesting.com
Printed in the USA
LVHW020800120423
744136LV00002B/84

9 783031 106019